Advances in Intelligent Systems and Computing

Volume 1117

The series "Advances in Intelligent Systems and Computing" contains publications on theory, applications, and design methods of Intelligent Systems and Intelligent Computing. Virtually all disciplines such as engineering, natural sciences, computer and information science, ICT, economics, business, e-commerce, environment, healthcare, life science are covered. The list of topics spans all the areas of modern intelligent systems and computing such as: computational intelligence, soft computing including neural networks, fuzzy systems, evolutionary computing and the fusion of these paradigms, social intelligence, ambient intelligence, computational neuroscience, artificial life, virtual worlds and society, cognitive science and systems, Perception and Vision, DNA and immune based systems, self-organizing and adaptive systems, e-Learning and teaching, human-centered and human-centric computing, recommender systems, intelligent control, robotics and mechatronics including human-machine teaming, knowledge-based paradigms, learning paradigms, machine ethics, intelligent data analysis, knowledge management, intelligent agents, intelligent decision making and support, intelligent network security, trust management, interactive entertainment, Web intelligence and multimedia.

The publications within "Advances in Intelligent Systems and Computing" are primarily proceedings of important conferences, symposia and congresses. They cover significant recent developments in the field, both of a foundational and applicable character. An important characteristic feature of the series is the short publication time and world-wide distribution. This permits a rapid and broad dissemination of research results.

** Indexing: The books of this series are submitted to ISI Proceedings, EI-Compendex, DBLP, SCOPUS, Google Scholar and Springerlink **

More information about this series at http://www.springer.com/series/11156

Mohammed Atiquzzaman · Neil Yen ·
Zheng Xu
Editors

Big Data Analytics for Cyber-Physical System in Smart City

BDCPS 2019, 28–29 December 2019,
Shenyang, China

Set 1

 Springer

Editors
Mohammed Atiquzzaman
School of Computer Science
University of Oklahoma
Norman, OK, USA

Neil Yen
University of Aizu
Fukushima, Japan

Zheng Xu
Shanghai University
Shanghai, China

ISSN 2194-5357   ISSN 2194-5365   (electronic)
Advances in Intelligent Systems and Computing
ISBN 978-981-15-2567-4   ISBN 978-981-15-2568-1   (eBook)
https://doi.org/10.1007/978-981-15-2568-1

This Springer imprint is published by the registered company Springer Nature Singapore Pte Ltd.
The registered company address is: 152 Beach Road, #21-01/04 Gateway East, Singapore 189721, Singapore

Foreword

With the rapid development of big data and current popular information technology, the problems include how to efficiently use systems to generate all the different kinds of new network intelligence and how to dynamically collect urban information. In this context, Internet of things and powerful computers can simulate urban operations while operating with reasonable safety regulations. However, achieving sustainable development for a new urban generation currently requires major breakthroughs to solve a series of practical problems facing cities.

A smart city involves a wide use of information technology for multidimensional aggregation. The development of smart cities is a new concept. Using Internet of things technology on the Internet, networking, and other advanced technology, all types of cities will use intelligent sensor placement to create object-linked information integration. Then, using intelligent analysis to integrate the collected information along with the Internet and other networking, the system can provide analyses that meet the demand for intelligent communications and decision support. This concept represents the way smart cities will think.

Cyber-physical system (CPS) as a multidimensional and complex system is a comprehensive calculation, network, and physical environment. Through the combination of computing technology, communication technology, and control technology, the close integration of the information world and the physical world is realized. IOT not only is closely related to people's life and social development, but also has a wide application in military affairs, including aerospace, military reconnaissance, intelligence grid system, intelligent transportation, intelligent medical, environmental monitoring, industrial control, etc. Intelligent medical system as a typical application of IOT will be used as a node of medical equipment to provide real-time, safe, and reliable medical services for people in wired or wireless way. In the intelligent transportation system, road, bridge, intersection, traffic signal, and other key information will be monitored in real time. The vast amount of information is analyzed, released, and calculated by the system, so that the road vehicles can share road information in real time. Personnel of road management can observe and monitor the real-time situation of the key sections in the system and even release the information to guide the vehicle so as to improve the existing urban traffic conditions. The Internet of things, which has been widely used in the industry, is a simple application of IOT. It can realize the function of object identification, positioning, and monitoring through the access to the network.

BDSPS 2019 which is held on December 28–29, 2019, Shenyang, China, is dedicated to address the challenges in the areas of CPS, thereby presenting a consolidated view to the interested researchers in the related fields. The conference looks for significant contributions to CPS in theoretical and practical aspects.

Each paper was reviewed by at least two independent experts. The conference would not have been a reality without the contributions of the authors. We sincerely thank all the authors for their valuable contributions. We would like to express our appreciation to all members of the Program Committee for their valuable efforts in the review process that helped us to guarantee the highest quality of the selected papers for the conference.

We would like to express our thanks to our distinguished keynote speakers, Professor Bo Fei, Shanghai University of Medicine & Health Sciences, China, and Professor Tiejun Cui, Shenyang Ligong University, China. We would also like to acknowledge the strong support of Shenyang Ligong University, as well as the general chairs, publication chairs, organizing chairs, program committee members, and all volunteers.

Our special thanks are due also to the editors of Springer book series "Advances in Intelligent Systems and Computing," Dr. Thomas Ditzinger, Dr. Ramesh Nath Premnath, and Arumugam Deivasigamani for their assistance throughout the publication process.

Organization

General Chairs

Tharam Dillon	La Trobe University, Australia
Bo Fei	Shanghai University of Medicine & Health Sciences, China

Program Committee Chairs

Mohammed Atiquzzaman	University of Oklahoma, USA
Zheng Xu	Shanghai University, China
Neil Yen	University of Aizu, Japan

Publication Chairs

Juan Du	Shanghai University, China
Ranran Liu	The University of Manchester, UK
Xinzhi Wang	Tsinghua University, China

Publicity Chairs

Junyu Xuan	University of Technology Sydney, Australia
Vijayan Sugumaran	Oakland University, USA
Yu-Wei Chan	Providence University, Taiwan, China

Local Organizing Chairs

Qingjun Wang	Shenyang Ligong University, China
Chang Liu	Shenyang Ligong University, China

Program Committee Members

William Bradley Glisson	University of South Alabama, USA
George Grispos	University of Limerick, Ireland
Abdullah Azfar	KPMG, Sydney, Australia
Aniello Castiglione	Università di Salerno, Italy
Wei Wang	The University of Texas at San Antonio, USA
Neil Yen	University of Aizu, Japan
Meng Yu	The University of Texas at San Antonio, USA
Shunxiang Zhang	Anhui University of Science and Technology, China
Guangli Zhu	Anhui University of Science and Technology, China
Tao Liao	Anhui University of Science and Technology, China
Xiaobo Yin	Anhui University of Science and Technology, China
Xiangfeng Luo	Shanghai University, China
Xiao Wei	Shanghai University, China
Huan Du	Shanghai University, China
Zhiguo Yan	Fudan University, China
Rick Church	UC Santa Barbara, USA
Tom Cova	The University of Utah, USA
Susan Cutter	University of South Carolina, USA
Zhiming Ding	Beijing University of Technology, China
Yong Ge	University of North Carolina at Charlotte, USA
T. V. Geetha	Anna University, India
Danhuai Guo	Computer Network Information Center, Chinese Academy of Sciences, China
Jianping Fang	University of North Carolina at Charlotte, USA
Jianhui Li	Computer Network Information Center, Chinese Academy of Sciences, China
Yi Liu	Tsinghua University, China
Foluso Ladeinde	SUNU Korea
Kuien Liu	Pivotal Inc., USA
Feng Lu	Institute of Geographic Sciences and Natural Resources Research, Chinese Academy of Sciences, China
Ricardo J. Soares Magalhaes	The University of Queensland, Australia
D. Manjula	Anna University, India
Alan Murray	Drexel University, USA
S. Murugan	Sathyabama Institute of Science and Technology, India
Yasuhide Okuyama	University of Kitakyushu, Japan
S. Padmavathi	Amrita University, India

Contents

Design of Home Stay Soft Outfit Display Platform Based on Android System

Ming Yang[⊠]

Jilin Engineering Normal University, Changchun 130052, Jilin, China
tougao_007@163.com

Abstract. With the continuous development of modern science and technology society and the continuous advancement of network technology, it has become a trend to switch from the traditional PC-based Internet access method to the mobile Internet. After years of development, the mobile Internet is forming a stable market service structure. According to the characteristics of mobile Internet, mobile client will be an important factor affecting the development of mobile Internet. The mobile phone client refers to a special terminal application specially developed by the mobile Internet enterprise for user convenience. The purpose of this paper is to design and implement a home-based soft (installation) platform based on Android system. The wireless client designed and implemented is based on such an example application software. The client software system is written in the Java language based on the C/S architecture and the Android platform during the development process. Mainly to achieve the data request, parsing and display of the server. The server accessed by the platform designed in this paper is a mobile wireless mobile device, so the mobile customer gives a description of the requirements, and the specification includes the protocol framework and interface description. The client accesses the server's data through these interfaces. The experimental results show that the platform of this paper can be used to browse the soft furnishings of mobile homes, which has excellent performance.

Keywords: Android system · Home furnishings · Platform design · C/S architecture

1 Introduction

With the advent of the era of tourism popularization, the tourism industry has increasingly become the main social and economic activity and leisure and entertainment lifestyle of modern human society. The tourism industry has gradually become one of the most powerful and largest industries in the world economic development. It has increasingly highlighted its important position in the national economy [1–3]. The progress of the tourism industry is based on the level of economic development of the entire country, and is subject to the level of economic development, while at the same time directly or indirectly lead the progress of the national economy [4]. Just because tourism is important to the development of the people's livelihood for the country's economy, it is necessary to establish a mobile phone client for the wireless travel and

© Springer Nature Singapore Pte Ltd. 2020
M. Atiquzzaman et al. (Eds.): BDCPS 2019, AISC 1117, pp. 1–8, 2020.
https://doi.org/10.1007/978-981-15-2568-1_1

accommodation system to understand the information of the hotel, promote the development of the tourism and lodging industry and improve the quality of tourism. Make management information. Comply with social informationization and trend [5].

With the development of technology and the reduction of mobile phone production costs, the popularity of smart phones is getting higher and higher. People have realized many life scenarios such as online shopping, mobile payment, information inquiry, mobile banking, game socialization, etc. through smart phones. People's traditional way of life [6]. The operating systems currently installed on smartphones in the market mainly include: Android developed by Google, iOS developed by Apple, and Windows Phone developed by Microsoft. The Android system has been widely used [7]. Android is a Linux-based open operating system, which is led and developed by Google and Open Handset Alliance, using the Dalvik Java virtual machine developed by Google. The open source and versatility of Android has made it more and more popular among mobile phone manufacturers [8]. Hundreds of companies such as Samsung, Huawei, ZTE, and Meizu have launched Android-based smartphones. According to well-known market data research company Kantar Worldpanel, as of the first quarter of 2018, in the domestic market, the market share of Android mobile phones has reached 86.4%. More and more developers are participating in the research and development of Android system mobile applications. The functions of smart phones are becoming more and more powerful, and can even replace the work of other devices to some extent [9, 10].

The purpose of this paper is to design and implement a home-based soft (installation) display platform based on the Android system. The wireless client designed and implemented is based on a sample application software written based on the C/S architecture and the Java language of the Android platform, mainly to implement server data request, parsing and display functions. The server accessed by the platform designed in this paper is a mobile wireless travel agency, so the demand is given by the mobile client. The specification includes a protocol framework and an interface description through which the terminal accesses the host information.

2 Proposed Method

2.1 Feasibility Analysis

The feasibility study is how to solve the problem with the least time and capital cost is not a good deal. This paper will discuss the feasibility of the design of the hotel's furnishings system from two aspects: technical feasibility and operational feasibility.

(1) Technical feasibility

The development of the mobile home-furnishing platform enables visitors to browse the B&B anytime and anywhere, which is technically feasible. This system is divided into terminal and PC management terminal. The mobile terminal is designed based on Android technology under Android Studio technology, and uses a relatively stable SQLite database to temporarily store information related to the home furnishings. The PC terminal is developed on the basis of J2EE technology using the Eclipse tool and deployed on the Apache Tomcat server. Only use the open source MySQL database to save the data on the host side, and communicate with the mobile terminal and the host through the JSON format data. The software development environment of this system is working. Due to the high-speed development of modern information technology on the hardware, the replacement speed of mobile terminal hardware is very frequent, the storage capacity is also increasing, the performance is constantly becoming reliable, and the price is continuously reduced, and the hardware quality can meet the home furnishings system. The need for research and development and use.

(2) Operational feasibility

The popularity of mobile terminals and the large-area access of mobile Inter provide very convenient conditions for the promotion and use of application systems. Visitors can make full use of the time of leisure, and can log in to the mobile homepage platform for display style everywhere. Pick. Mobile applications have become a trend in the travel industry, and more people will benefit from the mobile home furnishings platform.

2.2 Android System Framework

The Android system framework uses a hierarchical structure that divides Android into four layers and five blocks. They are Applications, Applications Framework, Libraries, System Runtime, and Linux Kernel.

Among them, the application and application framework can be designed and written in Java program. The Dalvik virtual machine in the system runtime can be realized by running a Java program. The function library is a program library written in C/C++ language, and the bottom layer is the kernel part. Includes the Linux kernel and Driver.

(1) Application

The application layer is located at the top of the Android system framework and is used to provide three aspects of the application package. Including the system's own contacts, notes and mail applications, users download microblogs, QQ and WeChat applications from third parties, as well as some applications developed by users.

(2) Application framework

The application framework layer is located in the second layer of the Android system framework, and is used to provide the application interface required for the entire system to build the APP, and is also the basis for ensuring the normal use and unified management of the mobile APP. The application framework layer consists of ten parts: Activity Manager, Package Manager, XMPP Service, and View System. Users can not only view the core program interfaces that come with the native system through this layer, but also use these interfaces to create user-specific applications, such as creating applications that record user motion.

(3) Function library

Libraries is a library of functions written in C/C++, including the Media Framework, the lightweight database engine, the Secure Sockets Layer, the underlying 2D graphics engine, and nine parts of the standard C language library that inherits from the Berkeley software distribution. Libraries provide important support for the entire system, such as support for databases, support for 2D graphics, and support for multimedia.

(4) System operation time

The Android runtime includes the Core Libraries and the Dalvik virtual machine, which provides most of the functionality of the Java programming language core library. Google has customized the Dalvik virtual machine for mobile devices to optimize the memory and CPU performance of mobile devices. The Dalvik virtual machine allows multiple instances to run side by side, so every APP in the system can run in the Instance of a virtual machine. The advantage of this mode of operation is that the Crash generated by the application running does not affect the normal operation of other virtual machines.

3   Experiments

3.1   C/S Architecture Design

At present, the overall architecture of software development is divided into C/S (client/server) architecture and B/S (browser/server) architecture. In the C/S architecture, the client and the server are usually on two computers that are far apart. The client submits the user request to the server. After receiving the data from the client, the server returns the processed result to the client. The client displays the results to the user in a specific form. The C/S architecture diagram is shown in Fig. 1.

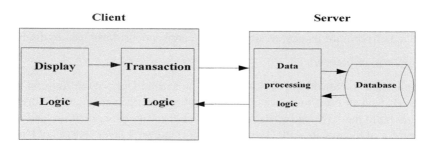

Fig. 1. C/S architecture diagram

The C/S architecture can realize the reasonable allocation of tasks to the client and the server by utilizing the advantages of the hardware at both ends, and reduce the communication overhead of the system. It is necessary to install dedicated client software. At the same time, the C/S architecture can fully utilize the processing capabilities of the client terminal devices. Many tasks can be processed on the client and submitted to the server, which can reduce the load on the server, improve the response speed, and be transparent to the data storage management function. Corresponding to the C/S architecture is the B/S architecture. In essence, the B/S architecture is also a C/S architecture, which can be seen as a development of the traditional Layer 2 mode C/S architecture. A special case of the application of the three-layer mode C/S architecture on the Web. Users can access the server through a browser without having to install any software on the system. Compared with the B/S architecture, the C/S architecture has a rich operation interface and a higher security guarantee. It is usually used in the LAN and has a faster response.

4   Discussion

4.1   System Implementation

The system interface is shown in Fig. 2. Users can freely view the homestays we provide, and set their own filters to filter out the homestays that suit his wishes. You can also sort by to find the homestay. In short, we offer a variety of options, just to let the user choose the favorite B&B, sleep in a house like home. We will also recommend some good homestays, and good landlords, generally these are good landlords or homestays with high ratings or new arrivals. And the new ones will offer some specials, in order to attract people to stay in the new landlord's home.

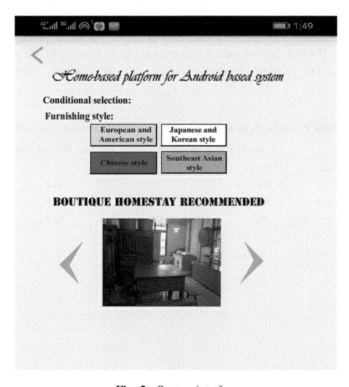

Fig. 2. System interface

4.2   System Testing

Because the software is manually developed by programmers, the development process will be due to the design is not comprehensive or careless and other reasons that lead to system errors of one kind or another. How to find these errors before the system goes online becomes an urgent problem. Therefore, software testing is an important part of system development. By testing the software before formal operation, problems in software development can be found and losses caused by program errors can be avoided after the program goes online. Software testing is the process of operating a program under certain conditions to find program errors, measure software quality, and evaluate whether it meets design requirements. The test of the function of the software is the black box test, which is the function that the known product should have. Through the test, it can test whether the function can work normally. The tester does not consider the internal structure, internal characteristics and implementation details of the program, and tests the interface of the program. During testing, the program can be viewed as a black box that cannot be opened. Common system testing methods include unit testing and performance testing.

The home stay furnishings system implemented in this paper is improved on the basis of Android native system, and LVM logical volume management is realized mainly based on Device Mapper technology and Thin Provisioning technology. This article compares the performance of Android native, external, and hidden volumes, using dd commands in Linux and bonni++. Bonni++ is a tool for testing Linux system performance and is widely used for its ease of use and concise output display style. Bonni++ will test a file of a known size or use the default 100 MB file size if the file size is not specified. So we specified three files of 200M, 300M, and 500M sizes for the test operation. The test method used is to run each file 20 times, take the average of the results, and use this as an evaluation parameter. The test results are shown in Table 1 and Fig. 3.

Table 1. Throughput tested using dd commands and bonni++ tools

	Android native system	External volume	Hidden volume
DD-read	39.4	40.2	39.3
DD-write	46.3	50.7	48.7
Bonnie++ read	630.1	548	552.6
Bonnie++ write	54.2	59.3	55.8

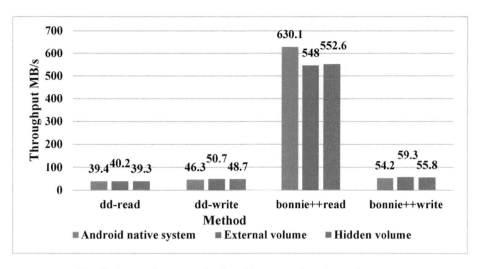

Fig. 3. Throughput tested using dd commands and bonni++ tools

5 Conclusions

Compared with the traditional mobile internet, the client designed in this paper has the following advantages: first, mobile (wireless) Internet is inconvenient to operate, the content homogenization is more serious, and the client is needed to help users simplify the operation. Mobile (wireless) Internet relies on mobile carriers, so there are always limitations in screen size, input mode and so on. Chaotic content placement and serious homogeneity have been a problem that perplexes users. And the use of the client makes users only need to operate locally, and the client provides a more friendly interface and more functions. Users do not have to bother to input the web site and complex click operations, they can get the required information. Secondly, the operation on the client can effectively save users' online costs. Because users download the client, they can directly operate the content they are concerned about, instead of entering the download home page every time, or opening unnecessary pages to reduce traffic costs. At present, in the mobile Internet era where traffic is used to calculate fees, reducing traffic is an important issue for users. Although the monthly traffic contract system reduces the burden of users, most users still care about how to reduce unnecessary waste.

References

1. Álvarez-García, J., de la Cruz del Río-Rama, M., Simonetti, B.: Quality management and customer results: the tourist accommodation sector in Spain. Int. J. Qual. Reliab. Manag. **34**(5), 701–719 (2017)
2. Shen, C.C., Liu, D.J.: Correlation between the homestay experience and brand equity: using the Yuehetang Rural Residence as a case study. J. Hospitality Tourism Technol. **6**(1), 59–72 (2015)

3. Hughes, B.O., Moshabela, M., Owen, J., et al.: The relevance and role of homestays in medical education: a scoping study. Med. Educ. Online **22**(1), 1320185 (2017)
4. Ismail, M.N.I., Hanafiah, M.H., Aminuddin, N., et al.: Community-based homestay service quality, visitor satisfaction, and behavioral intention. Procedia Soc. Behav. Sci. **22**(2), 398–405 (2016)
5. Ahmed, S., Cross, J.: A tourist growth model to predict accommodation nights spent in Australian Hotel Industry. SIRC **86**(2S), 1–6 (2015)
6. Ke, X., Li, Y., Deng, R.H.: ICCDetector: ICC-based malware detection on Android. IEEE Trans. Inf. Forensics Secur. **11**(6), 1252–1264 (2017)
7. Wang, Z., Li, C., Yuan, Z., et al.: DroidChain: a novel Android malware detection method based on behavior chains. Pervasive Mob. Comput. **32**, 3–14 (2016)
8. Hudha, M.N., Aji, S.D., Huda, C.: E-Rubric: scientific work based on Android for experimental physic. IOP Conf. Ser. Mater. Sci. Eng. **28**(8), 12100 (2018)
9. Shahram, T., Mostafa, L., Marjan, G., et al.: Designing and implementation of fuzzy case-based reasoning system on Android platform using electronic discharge summary of patients with chronic kidney diseases. Acta Informatica Medica **24**(4), 266–270 (2016)
10. Liang, Z., Wu, S., Yang, C., et al.: Portable epileptic seizure monitoring intelligent system based on Android system. J. Biomed. Eng. **33**(1), 31 (2016)

Design of Educational Administration Comprehensive Information Management System Based on Thinkphp

Hong Zhang$^{(\boxtimes)}$

Economic and Technological Development Zones, 1-005, Huayuan Road,
Changchun, China
452855684@qq.com

Abstract. In view of the low efficiency of teaching management in traditional colleges and universities and the duplication of tasks, this paper realizes the educational administration information management system based on ThinkPHP open source framework technology. The system mainly completed the management of the school's information statistics, student experimental reports and graduation papers, effectively improved the efficiency of teaching managers, and more scientific and standardized the management of students' experimental reports and graduation papers.

Keywords: Educational affairs · Information management · Thinkphp · Design

1 Introduction

Since the reform and opening up, and especially since the beginning of the new period, the educational administration of Chinese universities has gradually undergone in-depth reforms under the background of university education reform, and important achievements have been made, but there are still certain problems in practical work. It mainly shows that the educational administration system is relatively rigid, the structure of educational administration personnel needs to be optimized, the supervision system of educational administration is not perfect, and there are certain defects in teaching work [1]. The more rigid management system of university education is mainly due to the influence of the traditional teaching management model of Chinese universities for a long time. The color of planned economy still remains in the management work of university education administration. The specialized setting of colleges and universities is not scientific enough, and there is a relative lag with the needs of society and the needs of the times. There is a certain degree of disconnection between college and university students after graduation and the needs of society. In the educational administration of universities, there is insufficient understanding and implementation of the people-oriented educational concept and the people-centered scientific development concept [2]. The academic management work places too much emphasis on obedience and subordinate relations, and lacks flexibility. In addition, there is still a lack of experience of educational administration personnel and lack of scientific guidance in the construction of educational administration team. Educational

© Springer Nature Singapore Pte Ltd. 2020
M. Atiquzzaman et al. (Eds.): BDCPS 2019, AISC 1117, pp. 9–15, 2020.
https://doi.org/10.1007/978-981-15-2568-1_2

administration lacks scientific and effective supervision and management system; There are some problems in the teaching process such as insufficient investment and insufficient science in teaching evaluation system [3].

2 Total System Design

The system is mainly developed for the statistical work of the whole school's data status information, as well as student experimental reports and graduation thesis management work. It mainly includes experimental project statistics, experimental report management, graduation thesis management, class time statistics, scientific research paper statistics, book patent statistics, award-winning information statistics, and textbook statistics. Through this system, the statistics of daily teaching information are convenient and fast, the management of student experimental reports is more standardized and orderly, the topics of students' graduation papers become more flexible and efficient, and the process guidance and process supervision for completing graduation papers are more scientific and effective.

The system is developed using ThinkPHP + DWZ framework technology and is built on SAE. ThinkPHP is a fast, simple, object-oriented, lightweight open source PHP framework that uses the MVC (Model View Control Model View) model to separate the model from the view code, making the development more efficient and easier to maintain later. The framework can meet complex project requirements and portal development standards in enterprise development [5]. The DWZ Rich Client Framework is based on the Ajax RIA open source framework implemented by jQuery. It is simple, practical, and easy to expand [6]. Data storage uses a Mysql relational database, Mysql's SQL language uses the most common standardized language for accessing databases, and Mysql is small, fast, and inexpensive. Combining with PHP is the best choice for developing small and medium-sized websites [7]. The lightweight framework development technology has greatly improved the development efficiency. The open source features and architecture in the SAE cloud environment make the system easy to deploy and can save a lot of money [8].

3 Module Design

The system is mainly divided into six main functional modules: course management, class time management, experimental projects, experimental reports, graduation papers, and teacher and student information management. Each module is read from the database according to the module name under the overall framework, and the module is dynamically generated according to the user's permissions. Each module consists of a separate MVC, thus reducing the coupling between the relevant modules [9]. The functional descriptions of the modules are as follows:

3.1   Curriculum Management Module

The administrator introduces the course data of the current semester of the college into the system through Excel, and can also add the modified course data separately. The course data is obtained from the school's square educational management system.

3.2   Hours Management Module

The school committee of each class regularly reports the actual class time of the teacher of each course at this stage. The teacher verifies whether the class time is wrong. If the class time is wrong, the teacher can directly revise the class time submission, but it will not cover the class time submitted by the student. If it is correct, After all the teachers confirmed, the administrator derived the statistics of the class hours at this stage.

3.3   Pilot Project Modules

The instructor of the experimental course enters the experimental project of the course according to the course taught this semester, and accurately adds the project number of the experimental project to generate the experimental number of the experimental project in the statistical data. When adding experimental items, the requirements of the experimental project, experimental time, whether to submit experimental reports, and the deadline for submitting experimental reports can be set. After all the experimental projects were imported, the administrator derived the experimental project data for this semester to fill in the experimental statistics.

3.4   Experimental Reporting Module

According to the experimental projects recorded by the teachers, the students uploaded the experimental reports within the specified time. The students uploaded the experimental reports in the form of PDF documents. After the teachers read the experimental reports online, the students can view the results and comments of each experimental project. The final electronic version of the experimental report is archived in PDF format according to the corresponding catalogue.

3.5   Graduate Thesis Management Module

The teacher publishes the topics and requirements of the graduation thesis (design) according to the profession, and sets the maximum number of topics that can be selected. Students choose the topic within a specified time. Each topic is determined by the teacher from the number of topics selected, so as to achieve the purpose of students and teachers choosing each other. After the topic is completed, the administrator (director of the teaching and research office) controls the entire process of writing the paper. Students submit electronic manuscripts for each stage of the paper within a specified period of time. Teachers can directly review and feedback to students online.

3.6 Information Management Module for Teachers and Students

For the revision of personal information of teachers and students, teachers shall fill in statistical information such as scientific research papers, patents for works, award-winning information, and teaching materials.

4 Database Design

The system uses Mysql database. According to the functional requirements of the system, the system mainly designed the schedule, experimental list, experimental report table, class schedule, paper title chart, paper result table, scientific research work information table, student list, teacher list, administrator list, and role table. An example is given of the experimental project sheet, the structure of which is shown in Table 1.

Table 1. Experimental list

Field name	Data type	Whether primary key	Instructions
ld	Lnt(10)	Y	Sequence number
Teache_id	Lnt(10)	N	Teacher ID
Course_id	Lnt(10)	N	Course ID
Item_id	Tinyint(4)	N	Item number
Item_name	Tinyint(4)	N	Project name
Hours	Tinyint(4)	N	Project class hours
Type	Tinyint(4)	N	Type
Report	Bool	N	Whether to submit a report on the experiment
End_time	Int(10)	N	Deadline for experiments
Creat_time	Int(10)	N	Creation time
Update_time	Int(10)	N	Update Time
Status	Bool	N	Status of the project
Remark	Text	N	Remarks

5 System Implementation

The system is designed according to the modular design idea, and each module completes independent functions, which is conducive to the functional expansion of the system in the later development. Since the ThinkPHP framework is based on the MVC design mode, the front page's page display and logic are separate, and the page is mainly composed of Html + Css + JavaScript [4]. And the extension attribute implementation of dwz, the logic is implemented by ThinkPHP code [10].

5.1   Secure Access Achieved

The system users mainly include students, teachers, and administrators. They respectively create the corresponding user's project directory and use a single entry file to detect the user's login. The password used by the login user is encrypted using the more secure Sha1 encryption. At the same time, the use of verification code technology, to a certain extent, can effectively prevent malicious password cracking behavior. The administrator user uses RBAC (role permission management) technology to design the access of different characters in detail, and gives corresponding operating permissions according to the permission level.

```
public function checkLogin() {
...
$map = array();
$map['account'] = $_POST['account'];
$map["status"]= array('gt',0);
if($_SESSION['verify'] != md5($_POST['verify'])) {
$this->error;
}
import ( '@.ORG.Util.RBAC' );
$auth = RBAC::authenticate($map);
if(null === $auth) {
$this->error;
}else {
if($auth['password'] != sha1($_POST['password'])){
$this->error;
}
...
RBAC::saveAccessList();
$this->success;
}
```

5.2   Logical Business Implementation

Each Action class in the system is inherited from the Common Action class. Use the getAction Name method to obtain the name of the current Action operation, and then pass it to the Model operation to implement database basic operations such as adding, deleting, modifying, and checking. Because this system involves more modules, Only take the concrete implementation of the experimental module as an example.

The experimental module mainly involves the curriculum, the experimental list, and the experimental report table. The teachers of the experimental class enter the experimental project according to the requirements of the course. They can be added in the form of a single data or imported in the form of an excel table. If the project needs to upload the test report to be marked as an example, the data is added to the test list. According to the experimental project introduced by the teacher, the students uploaded the experimental report according to regulations, and the experimental report uploaded information to the experimental report table.

5.3   Online Access to Documents

Can not directly implement office files in the browser [7], so students upload PDF documents directly when they upload documents, and then use the Jquery Media plug-in to achieve an online preview of PDF file experimental reports and graduation papers. Since the IE kernel browser does not support this plug-in, users using IE need to install PDF reader software such as Foxit Reader and Adobe Reader. Other kernel browsers can browse PDF files directly online.

5.4   Deployment of SAE Platform Achieved

The SAE platform limits the user's use of local IO due to security considerations, which is inconvenient for systems that have files uploaded or cached for file reading and writing. In order to solve this problem, SAE officially provides TmpFS functionality. TmpFS allows developers to temporarily read and write local IO through standard IO functions, so temporarily read and write data can set the path to SAE_TMP_PATH, while persistent data storage still uses Storage or Mysql storage. At the same time, in order to cope with high traffic, the SAE platform provides Memcache services. It only needs to initialize Memcache and call Memcache_init in the program to use the Memcache service. This can reduce the number of database reads and ease database pressure.

6   Conclusion

This paper mainly introduces the design and implementation of the educational administration integrated information management system using ThinkPHP + DWZ framework technology. After more than a year of trial operation and continuous iterative development, the system is now operating in good condition, stable performance, high security, friendly interactive interface, users use fast, convenient, greatly improve the work efficiency. Structuring the system on the SAE cloud platform, close to zero cost, makes it easier to deploy and use.

References

1. Shi, W., Pan, L.: Influence of filling ratio and working fluid thermal properties on starting up and heat transferring performance of closed loop plate oscillating heat pipe with parallel channels. J. Therm. Sci. **26**(01), 73–81 (2017)
2. Nikolayev, V.S.: Effect of tube heat conduction on the single branch pulsating heat pipe start-up. Int. J. Heat Mass Transfer **95**, 477–487 (2016)
3. Anderson, H.E., Caldwell, J.H., Weir, R.F.: An automated method for the quantification of transgene expression in motor axons of the peripheral nerve. J. Neurosci. Methods **308**, 346–353 (2018)
4. Anderson, H.E., Schaller, K.L., Caldwell, J.H., Weir, R.F.: Intravascular injections of adenoassociated viral vector serotypes rh10 and PHP.B transduce murine sciatic nerve axons. Neurosci. Lett. **706**, 51–55 (2019)

5. Schaller, K.L., Caldwell, J.H.: Expression and distribution of voltage-gated sodium channels in the cerebellum. Cerebellum **2**(1), 2–9 (2003)
6. Shao, Y., Fu, Y.-X., Wang, Q.-F., Cheng, Z.-Q., Zhang, G.-Y., Hu, S.-Y.: Khubchandani's procedure combined with stapled posterior rectal wall resection for rectocele. World J. Gastroenterol. **25**(11), 1421–1431 (2019)
7. Wang, X., Jia, L.: Experimental study on heat transfer performance of pulsating heat pipe with refrigerants. J. Therm. Sci. **25**(05), 449–453 (2016)
8. Syk, E., Glimelius, B., Nilsson, P.J.: Factors influencing local failure in rectal cancer: analysis of 2315 patients from a population-based series. Dis. Colon Rectum **53**(5), 744–752 (2010)
9. Caldwell, J.H., Klevanski, M., Saar, M., Müller, U.C.: Roles of the amyloid precursor protein family in the peripheral nervous system. Mech. Dev. **130**(6–8), 433–446 (2013)
10. Rothschild, B.: What qualifies as rheumatoid arthritis? World J. Rheumatol. (01) (2013)

Discussion on Application of Informatization in MCU Course

Mingxin Qiu[✉]

Department of Electrical and Mechanical Engineering, Shandong Vocational
College of Light Industry, Zibo 255300, Shandong, China
qmx122634737@163.com

Abstract. MCU course is the core course of electromechanical major. Mastering this course will promote the employment of students. The content of the MCU course is too abstract, it is relatively difficult to students to understand. The traditional teaching mode can not meet the teaching requirements of the course. Therefore, it is necessary to integrate information technology into it, give full play to the advantages of information technology in education, and avoid the problems exposed in MCU teaching. Relying on Internet of Things technology and network technology to enhance the effectiveness of MCU teaching. Teachers constantly improve the application ability of information technology, integrate abundant teaching resources into teaching, transform boring theory teaching into interesting information teaching, and significantly improve the quality of teaching.

Keywords: Information technology · MCU teaching · Internet of Things technology · MOOC

1 Introduction

The MCU course pays great attention to practice. Many students have a strong sense of logic when they first learn this course. For those students with relatively weak theoretical basis, it is difficult to learn. Therefore, students' learning initiative usually lower. With the effective development of information technology, the application of educational information has gradually become an important trend in teaching reform. How to introduce information technology into the teaching of single-chip microcomputer has become an important topic for educators to deeply explore.

2 Overview of Informatization Teaching

2.1 Advantages of Information-Based Teaching

Under the traditional teaching mode, the relationship between teachers and students is limited to lecturer and the recipient [1, 2]. The common mode is "teacher speaking – student listening – student asking – teacher answering". The teacher acts as the leader and the role of master in the whole teaching process. The entire curriculum is based on

M. Atiquzzaman et al. (Eds.): BDCPS 2019, AISC 1117, pp. 16–21, 2020.
https://doi.org/10.1007/978-981-15-2568-1_3

teachers, knowledge and class. As for the acceptance of the content of the class and the use of knowledge, the students are not included in the scope of consideration, which is required by the new curriculum. The original intention of being a subject of teaching does not match. With the application of information technology, information technology teaching has begun to receive widespread attention and solved the disadvantages of traditional teaching effectively. First of all, information technology pays more attention to the whole process of teaching, instead of paying too much attention to class teaching as in the past. Outside the classroom, students can also use network software and information teaching tools to achieve full control. Secondly, the teaching resources are gradually transformed from textbooks to short videos, micro-courses, simulations and other related resources. The wide application of these information resources can enrich classroom teaching content, activate the classroom teaching atmosphere, stimulate students' enthusiasm for learning, and improve efficiency [4]. Finally, in information-based teaching, students are placed in the main position of teaching, learning is more autonomous, which is conducive to actively acquiring new knowledge and improving students' independent exploration and learning ability.

2.2 Informational Teaching Resources

(1) Teaching design software

In the traditional single-chip teaching process, the program of the teacher's lecture is: basic application circuit - circuit corresponding program – burning to the chip - watching the code effect. This process is too long, it does not have high controllability, and it is easy to produce a high error rate. Students do not have a deep understanding of the single-chip microcomputer, and lack of hands-on experience in programming and debugging. However, with the development of information technology, various simulation software came into being [3]. If the teacher can use these software in the teaching of single-chip microcomputer, it will inevitably significantly improve the teaching effect.

(2) Resource Communication Software

Informatization technology could achieve the whole process of teaching. Teachers could publish course-related resources and assessment content on various cloud education platforms [6, 7]. Its purpose is to let more students can fully understand this section to tell the content, curriculum so as to advance into the state, timely find problems in the course, and to bring problems to the classroom teaching, to achieve targeted learning. In the course teaching, cloud education platform is used to carry out related activities such as attendance, discussion and answering, which can activate the classroom teaching atmosphere and improve classroom teaching efficiency. After the class, the cloud education platform can be used to post homework assignments, and instant messaging technologies such as WeChat and QQ can be used to realize the communication between teachers and students, which is helpful to solve students' doubts and expand tasks.

3   The Status Quo of High-Tech Single-Chip Teaching

3.1   The Teaching Mode Is Imperfect

Before the introduction of MCU course, it is necessary to be taught digital electronic technology, analog electronic technology and computer basics. Many students have little knowledge of the pre-course, so in the MCU course, they feel that they are in the fog and do not know what to do, and slowly lose the interest to continue learning. Teachers believe that the content of repeated explanations has been fully mastered, but it is not known that many knowledge students only master the surface and lack of in-depth analysis of the nature of knowledge. The reason for this situation is that teachers do not guide students to build a complete knowledge system [9]. Considering the individual differences among students, some students have poor ability to accept and understand. Teachers should adopt differentiated teaching in the process of teaching. The traditional unified teaching mode cannot balance each student, and it is easy to cause different knowledge among different students.

3.2   The Effect of Classroom Demonstration Is not Ideal

In traditional teaching, teachers who want to demonstrate the experimental process in class usually need to prepare the corresponding hardware facilities after class, which will consume a lot of energy and time and is cumbersome. In the classroom demonstration session, the demonstration effect of the experiment is affected by many factors, and the goal of teaching is usually impossible to get. In addition, some students sit behind the teacher, they can not see the whole experiment procedures and parameters, which led to their lack of comprehensive understanding of the whole experiment process.

3.3   The Training Effect Is not High

The MCU course pays great attention to practicality, and the training part occupies a considerable share in MCU teaching. In the traditional practical training operation, students' lack of in-depth understanding of the theoretical part of class teaching is likely to cause operational errors, leading to the damage of experimental equipment. Due to the pressure of funds, higher vocational colleges usually cannot replace equipment in time, which affects the progress of training [8]. With the expansion of college enrollment, the quality of students in higher vocational colleges is continuously decreasing, and some students have a poor grasp of basic knowledge, which will adversely affect the development of practical training courses.

In addition, due to the limited experimental equipment, teachers will ask students to form different groups to conduct experiments, which will lead to the existence of a large number of students in the group, do not participate in the training, and completely mix data with other students. MCU training is mainly for verification experiments. It does not have too high openness and design, so it lacks the cultivation of students' practical ability and self-awareness. There are also some students completely in accordance with the requirements of the teacher step by step to connect the circuit.

They do not understand the knowledge content themselves and cannot achieve the fundamental goal of the experiment.

4   The Application Strategy of Information Technology in the Teaching of Single-Chip Microcomputer

4.1   Application Strategy of Internet of Things Technology in MCU Teaching

Internet of Things technology can be widely used in the construction of training platforms. Higher vocational colleges can use IoT technology to make up for the limitations of hardware facilities. The MCU training device is taken as the core, and the resources, cases and teaching materials are set up to match with it [9].

By using the Internet of Things technology, STC microcontroller can be developed and applied to the core chip teaching in the practical training device, transforming the shackles of traditional hardware resources. In the process of practical training, students can build relevant hardware circuits based on their own ideas and Internet of things technology. During extracurricular practice, students can also use the Internet of Things technology to design small devices to enhance students' understanding and awareness of the Internet of Things and improve students' understanding and cognition of the Internet of things. For example, in the study of counters, teachers use the Internet of Things technology to create a matching teaching context, explain the internal structure of the counter, analyze its working principle, and illustrate examples. Teachers should encourage students to use the after-school time to conduct independent innovation design in groups and cultivate teamwork awareness and ability [10].

Relying on the Internet of Things technology to build a practical soft platform, incorporating data, case and textbook content, so that students can optimize teaching resources in the process of learning the soft platform, and students can debug software and hardware while learning, and improve the effectiveness of teaching content. For example, when explaining the SP1 interface of the single-chip microcomputer, the teacher can use the Internet of Things technology to put the resources of the SP1 interface, the interface function, the module usage, and the case application into the compression package. The students use the Internet of Things to learn the teaching resources after class. IoT software for practical operations. In addition, teachers can not use large theoretical knowledge to teach in classroom teaching, pay attention to explain some practical content, add content in wireless communication, explain the network topology structure, and ensure that students can carry out offline simulation operations independently.

4.2   Application Strategy of MOOC in MCU Teaching

(1) Learning before class

Pre-class preparation is the core of MCU teaching reform. Teachers need to choose reasonable curriculum content and design effective teaching methods and modes according to the teaching content and objectives of MCU courses. According to the basic characteristics of online teaching, the teacher divides the course into one small knowledge point, sets the content of each knowledge point to within 5–20 min, and records related videos, and upload the relevant teaching plan, PPT courseware and test questions to the platform. Teachers should guide students' online learning before the start of the course. Teachers should design corresponding thinking questions based on practical cases, fully guide students to start from the problem, watch the video materials of the MOOC, Students in the class will build corresponding WeChat groups, to learn MOOC in groups. For example, in the learning of counters, the class hours can be set to 3, the contents related to counters can be divided into small points, and different teaching methods can be used according to the degree of difficulty and characteristics of the points (Table 1).

Table 1. Counter knowledge point design

Knowledge point	Duration/Min	Teaching methods
Counter structure	10	MOOC+Class Discussion
working principle	10	Classroom face-to-face
Working mode register	10	MOOC+Class Discussion
Control register	10	MOOC+Class Discussion
Counter initialization	10	Classroom face-to-face
Counting application	10	MOOC+Class Discussion
Timing application	10	MOOC+Class Discussion
Analysis and debugging	15	Classroom face-to-face

(2) Classroom teaching

Students have already learned enough about the content of the course through pre-class study, and put forward corresponding questions. Teacher can develop targeted teaching in the classroom teaching activities, which are common for students to preview before class the difficulty and key content can be in the form of blackboard writing, explaining, simple content can be in the form of group discussion in the teaching. For example, in the process of counter experiment and training, the teacher can ask students to complete the experiment in groups, debug the program, and ask each student to elaborate on his own harvest in the experiment and the final experimental results. Other students give comments and effective supplements to the answers of the student, and finally the teacher gives a summary and evaluation.

(3) After class summary

After class, teachers post classroom discussions and the difficult issues in the MOOC. Students review classroom teaching content by watching video materials, and communicate with teachers by using instant communication devices such as WeChat and QQ to answer questions. Students finish homework by watching video [5].

This information-based teaching model based on the classroom is effective in reversing the passive acceptance of students in the traditional teaching mode, making students more motivated and more interested in participating in class discussion activities, which can significantly enhance students' Interest and effectively enhance their ability to solve problems on their own.

5 Conclusion

In summary, the application of information technology represented by the Internet of Things and network technology in the teaching of high-end single-chip microcomputers is increasingly widespread. In the practical teaching work, teachers need to continuously improve the practical application ability of information technology, integrate rich teaching resources into teaching, transform boring theoretical teaching into interesting information-based teaching, and significantly improve teaching quality and cultivate more high-quality talents to meet the market demand.

References

1. Sokol Randi, G., Slawson David, C., Shaughnessy, A.F.: Teaching evidence-based medicine application: transformative concepts of information mastery that foster evidence-informed decision-making. BMJ Evid. Based Med. **24**(4), 149–154 (2019)
2. Chang, J.-J., Lin, W.-S., Chen, H.-R.: How attention level and cognitive style affect learning in a MOOC environment? Based on the perspective of brainwave analysis. Comput. Hum. Behav. **100**, 209–217 (2019)
3. Shuobo, X., Dishi, X., Lele, L.: Construction of regional informatization ecological environment based on the entropy weight modified AHP hierarchy model. Sustain. Comput. Inf. Syst. **22**, 26–31 (2019)
4. MacAulay, M.: Antiviral marketing: the informationalization of HIV prevention. Can. J. Commun. **44**(2), 239–261 (2019)
5. You, Z., Wu, C.: A framework for data-driven informatization of the construction company. Adv. Eng. Inform. **39**, 269–277 (2019)
6. Sahoo, K.S., Puthal, D., Tiwary, M., Rodrigues, J.J.P.C., Sahoo, B., Dash, R.: An early detection of low rate DDoS attack to SDN based data center networks using information distance metrics. Future Gener. Comput. Syst. **89**, 685–697 (2018)
7. XiangbinYan, P.: Effect of the dynamics of human behavior on the competitive spreading of information. Comput. Hum. Behav. **89**, 1–7 (2018)
8. Howell, J.A., Roberts, L.D., Mancini, V.O.: Learning analytics messages: impact of grade, sender, comparative information and message style on student affect and academic resilience. Comput. Hum. Behav. **89**, 8–15 (2018)
9. Marta, S., Shruti, C., Walter, F., Francesa, F., Jashodhara, D., Lorena, R.A.: Does information and communication technology add value to citizen-led accountability initiatives in health? Experiences from India and Guatemala. Health Hum. Rights **20**(2), 169–184 (2018)
10. Sajay, A., Vroman, K.G., Catherine, L., Joseph, G.: Multi-stakeholder perspectives on information communication technology training for older adults: implications for teaching and learning. Disabil. Rehabil. Assistive Technol. **14**(5), 453–461 (2019)

Analysis of ASTM Standards Developing Mode and Its Enlightenment

Qing Xu[✉]

China National Institute of Standardization, No. 4 Zhichun Road, Haidian District, Beijing 100191, China
xuqing@cnis.ac.cn

Abstract. The background of domestic association standards presenting mainly contains policy requirement, market requirement and overseas experience. In terms of the third background, association standards have been accepted practice for a long time. This paper studies the organizational structure of ASTM, the division of responsibilities and the standardization technical organization of ASTM. The standard development policy of ASTM is studied from the aspects of standard development procedure, standard type and appeal mechanism. In addition, it analyzes the characteristics of ASTM standard developing model and its enlightenment to Chinese association standardization work in order to provide reference and help for the association standardization work in China.

Keywords: ASTM · Association standards · Standard development procedure · Appeal mechanism

1 Introduction

<Plan for Furthering the Standardization Reforms> released in March 2015 and since then, China's standardization reform has undergone tremendous changes. There are many highlights, including the simplifying of compulsory standard, optimization of recommend standard, and the release of enterprise standard [1], among which the development of association standards is the most outstanding. The proposal of association standard breaks the original onefold supply structure, fully stimulates the vitality of market participants, timely fills the gap of standards in relevant industry in China, and satisfies the market and innovation demand. Nearly three years of time, is the most spectacular period in the history of China's standardization work, especially for China's more than 300,000 social organizations, witnessed the association standard from shaping to existing, and experienced from can do to how to do the association standardization work. Association standard is rising in China as a member of the new standard system. It has a unique background contains policy requirement, market requirement and overseas experience. In terms of overseas experience, association standards have been accepted practice for a long time [2]. ASTM was founded in 1898, and headquartered in Philadelphia with offices in Belgium, Canada, China, Peru and Washington. In 2001, ASTM was renamed as ASTM International (ASTM for short), which is one of the largest International professional standardization organizations in the world [3]. ASTM is committed to serving the needs of society around the world.

© Springer Nature Singapore Pte Ltd. 2020
M. Atiquzzaman et al. (Eds.): BDCPS 2019, AISC 1117, pp. 22–27, 2020.
https://doi.org/10.1007/978-981-15-2568-1_4

ASTM integrates consensus standards and innovative services to help the world function better [4]. ASTM is also a standard-developing organization accredited by ANSI. As of November 2017, more than 240 standards development organizations have been accredited by ANSI [5]. This paper studies the organizational structure of ASTM, the division of responsibilities and the standardization technical organization of ASTM. The standard development policy of ASTM is studied from the aspects of standard setting procedure, standard type and appeal mechanism. In addition, it analyzes the characteristics of ASTM standard developing model and its enlightenment to Chinese association standardization work.

2 Organization Structure and Assignment of Responsibility of ASTM

As a global standard-development platform, the organizational structure of ASTM includes the Board and its Board Committees and Standing Committees [6]. The board of directors is responsible for the major decisions. It is composed of 25 members, including 1 chairman, 2 vice chairmen and 18 directors. According to its functions, the board of directors consists of two board committees and four standing committees.

2.1 Board Committees Include

Executive Committee. When the board is not in session, the executive committee is responsible for exercising all the general powers of the board, except the power to fill board vacancies and modify the ASTM board procedures.

Finance and Audit Committee. The finance and audit committee oversees the financial operations of ASTM and makes recommendations to the board on financial policy matters.

2.2 Standing Committees Include

Committee on Standards. Committee on Standards is responsible for reviewing and approving all technical committee recommendations for standard action, checking that ASTM standard development procedural requirements are met, and resolving judicial disputes over standards.

Committee on Technical Committee Operations. It is responsible for developing and maintenance of *<Regulations Governing ASTM Technical Committees>* and act upon the proposed changes and be responsible for the interpretation and implementation of these regulations.

Committee on Publications. It advises the board on the development of publishing policies and is responsible for ASTM's publishing plans.

Committee on Certification Programs. The committee is responsible for advising the board on the development of the certification program and for approving or dissolving the ASTM certification program.

3 Analysis of Standardization Technical Organization of ASTM

According to ASTM Technical Committee Officer Handbook [7], the technical organization of ASTM is divided into Main committee, Subcommittee and Task group. The name and scope and work of the main technical committee shall be approved by the board of directors. Each main technical committee may draw up its own bylaws and obtain the approval from the operation committee of the main technical committee. Main technical committee may have subcommittees and task groups. Subcommittees deal with specific topics within the work of their respective technical committees and may establish sections and task groups as required. Subcommittee is the basic unit of standard developing. All standard voting items originate from the subcommittee, and it is also responsible for handling of negative vote. The subcommittee must have a chairman, a vice-chairman and a secretary. Task groups are small organizations responsible for specific tasks (such as drafting standards or conducting interlaboratory research and learning). They are generally composed of 4–6 members, and the chairman of the task group must be appointed by the chairman of the subcommittee or section to which the group belongs. Upon completion of a specific task, the task group is disbanded unless the subcommittee consider that it may set additional standards.

4 ASTM Standard Developing Model Analysis

4.1 ASTM Standard Development Procedures

ASTM standard development procedures are very flexible. After 111 years of practice, it can adapt to a variety of activities. The procedures include following steps:

Proposing. Any company, organization, industry association, professional association, university, government agency or individual may propose standard development activities to ASTM. Requests to propose standard development shall be made in writing to the chairman of the subcommittee. Requests will be made by the chairman at the next subcommittee meeting and will be discussed for adoption.

Project Approval. If the request to propose a standard development activity is approved, the chairman of the subcommittee appoints the chairman of the task group and establishes the task group. At the same time, Work Item Registration will be conducted to publicize the information of the standard, so as to collect more opinions and participants.

Draft Standard. The chairman of task group organizes members to draft standards and prepare to submit ballot to the subcommittee.

Subcommittee Ballot. After Work Item Registration, task group submits the standards to subcommittee for a vote.

Main Committee Ballot and Society Review. After passing the ballot by subcommittee, the standard is submitted to technical committee for balloting then submitted to ASTM for approval. In the balloting of technical committee, the votes which are determined as unconvincing or irrelevant negative votes must be submitted together.

Committee on Standards Review. After the standards are voted on by the technical committee, they are submitted to standards committee for review, which determines whether all procedural requirements are met.

Approval and Publication. After the standard has been approved by standards committee, it will be approved and released by ASTM.

4.2 ASTM Standards Categories

<Regulations Governing ASTM Technical Committees> divides ASTM standards into five categories [8].

Classification. A systematic arrangement or division of materials, products, systems, or services into groups based on similar characteristics such as origin, composition, properties, or use.

Guide. A compendium of information or series of options that does not recommend a specific course of action.

Practice. A set of instructions for performing one or more specific operations that does not produce a test result. Examples of practices include, but are not limited to: application, assessment, cleaning, collection, decontamination, inspection, installation, preparation, sampling, screening, and training.

Specification. An explicit set of requirements to be satisfied by a material, product, system or service.

Terminology. A document comprising definitions of terms; explanations of symbols, abbreviations, or acronyms.

Test Method. A definitive procedure that produces a test result. Examples of test methods include, but are not limited to: identification, measurement, and evaluation of one or more qualities, characteristics, or properties.

4.3 Appeal Mechanism

The ASTM standard developing has very strict voting procedures. In the subcommittee (SC) voting stage, each vote needs to receive 60% of the total number of official votes to be valid. After that, if 2/3 votes are obtained, the vote is passed. In the technical committee (TC) voting stage, it is also required that each ballot shall be valid only upon receipt of 60% of the total number of official votes cast, after which, if 90% of the votes are in favour, the vote shall be passed [9].

The written request must come to ASTM Headquarters within 30 days after the negative voter has been notified of the committee action. A two-thirds affirmative vote of the affirmative and negative votes returned by subcommittee voting members in favor of the not persuasive motion is required to confirm the action [10].

5 The Enlightenments to China's Standardization Development

ASTM standards have been established for more than one hundred years, and their standards have been widely used in the world. The advantages of ASTM standards are characterized by open participation, transparent procedures, balanced interests, consensus process and attention to negative vote. The foregoing analysis and research are of great reference significance to China's association standardization in the primary stage of development. To sum up, there are several implications as follows:

5.1 Focus on Openness

Standardization is a process of constant consensus building among various parties. Social organizations should adhere to the concept of openness in carrying out standardization, and involve multiple parties in the developing of standards, so as to achieve consensus through multiple concerns and participation.

5.2 Establish Scientific and Transparent Standards Development Procedures

Standard development procedure is the core of standard developing in standardization organization. To carry out the work of association standardization, social organizations should first formulate scientific and transparent standard development procedures based on the actual situation of social organizations. In the registration process of the national association standard information platform, standard development procedures are also required when submitting registration, and also are the focus of registration review.

5.3 Attach Importance to the Handling of Negative Votes

In some standards development process, many interests are involved. In the process of developing standards, social organizations should attach importance to the negative votes and opposing opinions during consultation, especially to properly handle the opposing opinions on substantive issues.

5.4 Establish an Effective Appeal Mechanism

The appeal mechanism is a weak point in the standardization work carried out by Chinese social organizations. At present, only a few social organizations have established the appeal mechanism in the corresponding rules and regulations. Next, in order to make association standardization work more fair, reasonable, scientific and efficient, social organizations should establish an effective appeal mechanism.

6 Conclusion

ASTM standard development procedures have experienced 111 years of practice and it can adapt to a variety of activities. ASTM ensures that it well follow the basic principles of standardization activities through scientific and reasonable standard developing mode, thus guaranteeing the quality of ASTM standards. China's association standards have experienced more than three years of development, social organizations in carrying out standardization activities should fully absorb the excellent experience of foreign standardization associations and pay greater attention to openness, scientific and transparent standards development procedures, handling of negative votes and effective appeal mechanism as to continuously improve the ability and level of association standardization work.

Acknowledgment. This research was financially supported by Standardization Administration of the People's Republic of China project <Research on Chinese and foreign standardization system architecture> (Project Number: 572018B-6575).

References

1. The Plan for Furthering the Standardization Reforms (2015)
2. Xianghua, Z., Yiyi, W., Qing, X.: Guidelines and Good Practices for Social Organization Standardization, pp. 2–4
3. https://www.astm.org/ABOUT/factsheet.html. Accessed 9 Jan 2019
4. https://www.astm.org/ABOUT/overview.html. Accessed 9 Jan 2019
5. https://www.ansi.org/standards_activities/overview/overview?menuid=3. Accessed 8 Jan 2019
6. https://www.astm.org/ABOUT/GOV/BOD_GOV.html. Accessed 9 Jan 2019
7. https://www.astm.org/TechCommitteeOfficer_Handbook.html. Accessed 9 Jan 2019
8. https://www.astm.org/Regulations.html. Accessed 9 Jan 2019
9. China National Institute of Standardization, International Standardization Development Report, pp. 225–227 (2016)
10. https://www.astm.org/TechCommitteeOfficer_Handbook.html. Accessed 5 Jan 2019

Methods of Improving the Efficiency of Computer Labs

Hui-yong Guo[1] and Qiong Zhang[2(✉)]

[1] School of Management, Wuhan Donghu University, Wuhan, China
345430543@qq.com
[2] School of Economics, Wuhan Donghu University, Wuhan, China
515275934@qq.com

Abstract. With the rapid development of modern science and technology, computer technology has penetrated into various fields and become an indispensable tool for various industries. And the wide application of computers in daily life, proficiency in computer technology has become a must for all colleges and universities. Therefore, in addition to arranging a large number of computer theory courses for students, colleges and universities should also arrange enough experimental courses for students, and how to arrange reasonable time for each student. The management of traditional laboratories consumes a lot of manpower and energy and experimental course management. Many problems such as chaos are exposed. This paper will analyze and discuss the existing computer experiment management system, and explore ways to improve the efficiency of computer lab use.

Keywords: Computer laboratory · Management system · Efficiency · Method

1 Introduction

With the rapid development of computer technology and information technology, the number of computer labs in colleges and universities is increasing. How to effectively manage and maintain computer labs has become a concern of managers, experimenters and technicians [1, 2]. At present, the scale of running schools in many schools is constantly expanding, the number of students is increasing, and the amount of information that needs to be processed is exponentially increasing. It is necessary to have a corresponding software system to improve the efficiency of laboratory management [3–5]. Therefore, designing and developing a computer lab management system to standardize the management process of computer labs, scientifically arranging various experimental equipment, and accurately querying the operation of the laboratory can not only reduce the duplication of labor in computer lab management, but also reduce The workload of management personnel, improve work efficiency and service level, and strengthen the macro control and management of the procurement, maintenance and use of experimental equipment by the competent authorities, save costs and improve the effective utilization of experimental equipment [6–9]. This paper discusses the laboratory management system through the computer network, data storage, fast data processing and other related technologies to carry out all-round management of the laboratory, so that the laboratory automation operation and information management.

M. Atiquzzaman et al. (Eds.): BDCPS 2019, AISC 1117, pp. 28–34, 2020.
https://doi.org/10.1007/978-981-15-2568-1_5

2   Analysis of Computer Laboratory Management System

2.1   Status of Domestic Laboratory Management Systems

University Computer Laboratory Management System In the 21st century, which was produced at the beginning of this century, the university computer laboratory management system is mainly integrated by a unified visual interface and a central server. With the continuous development of Internet technology, the laboratory management system of related products of the University of Computer Science is constantly increasing its strength, but its main function is data storage and distribution, which is far from the need of scientific management. However, information management has been removed. The efficiency of laboratory management, laboratory management can be part of automation, simplifying the work of workers. In the meantime, the Internet industry computer as a new technology, the cost and cost are very high, and the popularity in the country is very narrow. Due to the development of information technology, the development of laboratory management information system has been gradually recognized at the end of the 20th century, and has been applied to only a few chemical enterprises [10]. Its application scope is only data storage and distribution, which is far below the popularity of the effect, and can not make better use of the system. With the continuous development of information technology, in recent years, the development of university computer laboratory management system has been more and more mature by the commercial software of University Computer College Computer Laboratory Management system. The university laboratory management system needs to be able to transform advanced management ideas and business models into the functional framework of the management system. From the perspective of the actual teaching work of the university, the significance of the university computer laboratory management system is explained. It helps the school to establish an effective and flexible management platform, so that the system administrators can view all links and facilitate the integration of teaching, scientific research and equipment. Harmonize and optimize to realize scientific and modern management.

2.2   The Necessity of Laboratory Management System

The computer experiment management system of colleges and universities is an indispensable part of education and teaching. Its construction is very important for college education and scientific research. With the gradual advancement of the reform of university laboratory teaching management, it is urgent to meet the standardized and complicated laboratory management system to meet the laboratory teaching. In this process, many colleges and universities began to explore new ways and new channels. Some colleges and universities decomposed part of the transactional work of the experimental department into the next-level experimental teaching unit, so that data processing and analysis can be performed efficiently and quickly. In terms of practical needs, the university laboratory management system revolves around the practical teaching and training program, combining teacher and student data, intelligently generating laboratory teaching plan data and laboratory scheduling data. Through the laboratory management system to achieve the standardization and informationization of practical teaching, experimental instruments and experimental consumables management, improve the management level and service level of experimental teaching, especially open experimental teaching, for laboratory evaluation, laboratory construction and experimental teaching quality Management and other decision-making provide data support, intelligently generate data reports of the Ministry of Education, help colleges and universities to easily complete the data reporting work for each academic year to achieve real-time communication of information and reports in the daily office of the laboratory.

2.3   Problems in the Laboratory Management System

Judging from the overall appearance of China's colleges and universities, many colleges and universities still pay more attention to theoretical teaching, and lack of high recognition for practice and experimental teaching. This phenomenon will lead to the traditional management mode of computer hardware labs, the management innovation and reform, and the lack of good professional knowledge. When the management mode of computer hardware labs in colleges and universities is outdated and the professional skills of managers are insufficient, in the long run, it will not only lead to a weak management awareness, but also a low level of enthusiasm, which will seriously affect the full utilization of computer hardware laboratory equipment. Because some lab courses are more convenient, more intuitive, and faster to operate on a computer. Install the lab course software on your computer. But then another problem arises. In the classroom, using computers to do things that are not related to the course, such as inserting a USB flash drive, MP3 into a computer, running a game within a computer program, etc., not only affects the progress of the computer, but also makes the computer The use efficiency is reduced, and the effect expected by the experimental teaching cannot be achieved.

3   Optimization Design Theory of Laboratory Management System

3.1   Design of Laboratory Management System

The design principle of the computer laboratory management system in colleges and universities is based on software applications and extended objects. The system uses the object-oriented design method to establish the solution domain model, and transforms the stage requirements into cost and quality requirements. The goal of this abstracted system implementation is to increase productivity. The main tasks are data design, architecture design, and improved maintainability and quality. This kind of object-oriented design should determine the object hierarchy and structure existing in the system space problem through the identification of the object, determine the hierarchical structure that should exist in the system solution space through the hierarchical structure, and determine the external structure and data structure.

3.2 Requirements Elements for Laboratory Management System Implementation

First of all, security. Security is the basic requirement of the computer laboratory management system, and is the principle to ensure the effective operation of the system. Security is the primary consideration in the design of the computer room management system. The security of the operating system includes not only its own security and data security, but also the ability to prevent malicious attacks and system errors caused by operations; then, design the database reasonably. According to the data management model analysis, management system development mainly involves two major problems, namely network and database. The network is the platform for resource sharing and the basis for server connection, while the database is the place for all information storage. The quality of database design is directly related to the quality of system software, affecting the speed of data access. Finally, the rational use of new types technology. In the development process of laboratory management system software, the rational use of new technologies should follow the principles of innovation, economy and forward-looking. The software has certain advanced nature and will not be eliminated quickly. It must not only have functional satisfaction, but also have certain economic practicability and development. In today's society, there are many new technologies. If you don't want to be eliminated by society, you must develop software. The development of software can bring good prospects. The rational use of software is also a must. Always follow the big steps, our goal is to effectively use software to help us achieve more requirements.

3.3 Module Construction of Laboratory Management System

Laboratory management system includes four modules: user module, experiment construction module, experiment teaching module and auxiliary function module. The auxiliary function modules include: routine office module, laboratory evaluation module and system maintenance module. As shown in Fig. 1.

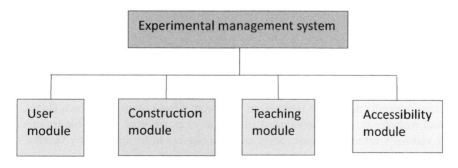

Fig. 1. Module plan

According to the needs analysis of the laboratory management system, the system integrated management platform has the following functional modules: user module, experimental building module, experimental teaching module and auxiliary function module. The auxiliary functions include regular office module, laboratory team building module, system maintenance module, laboratory evaluation module, etc. Through the establishment of the module, the module structure can be arbitrarily deleted, inquired or modified, without affecting the normal operation of the system. The overall structure of the system is interconnected and interdependent, and the overall operation of the system is maintained by the administrator. Different functional module roles and permissions are not the same, the addition, deletion and modification of user modules, the addition and deletion of experiments in experimental construction, and the laboratory management functions of experimental teaching modules.

4   Methods for Improving the Efficiency of Computer Labs

According to the National Computer Virus Emergency Response Center, as shown in Fig. 2 the national computer virus infection rate was 92.7% in 2017, 90.3% in 2018, and 84.9% in 2019. Computer viruses have become the most important factor threatening the security of computer systems.

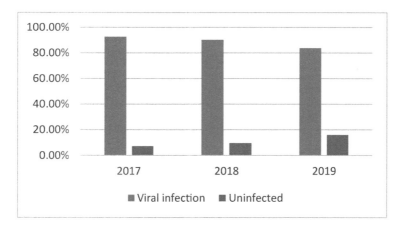

Fig. 2. Computer virus infection in the past three years

Cause analysis of computer viruses, computer laboratories in Colleges and universities generally access the Internet, and allow students to bring their own U disk, log on bad websites or use toxic external storage devices are very vulnerable to infection with various viruses, computer rooms are unavoidable to be attacked by viruses.

4.1   Network Virus Prevention and Management

At present, most of the computer labs in higher vocational colleges are linked to the external Internet through the campus network. The threat of network hacking will directly affect the security of all computers in the computer lab. Such as remote control,

Trojans, viruses, etc. Just mentioned earlier, you can use the automatic restore system to restore the hard disk. If the system is infected by a virus or a Trojan, you can automatically remove the virus Trojan by simply restarting the computer. However, there are now some new types of viruses, such as the "Dog Dog" virus, which can easily bypass the automatic restore software system and other hardware protection facilities and force the writing of virus data to the hard disk. The main means to prevent such viruses is to install a more powerful restore system and timely fill in system vulnerabilities. It must also be reminded that the main route of transmission of viruses between computer labs is through removable storage devices such as USB flash drives and mobile hard drives. Therefore, it is recommended to disable all USB computers in the computer lab to prevent the spread of virus Trojans from the source. There are many ways to disable the USB port. The most common and most effective method is to shield the USB port from the motherboard BIOS settings.

4.2   Implementation of Remote Management Mode

When the computer lab is open or in class, there are more students. Most students can shut down the computer normally when they leave. However, there are also many students who have not developed good habits. When they leave, they always forget to turn off the computer. Not shutting down the computers one by one, wasting a lot of time. If you turn off the main power directly, it may adversely affect the software and hardware of your computer, or directly damage your computer. In addition, sometimes students use the Internet to do things that do not match the content of the classroom, such as games. At this time, the teacher can notify the students to pay attention to classroom discipline by sending remote messages without affecting other students. In some cases, you can also send a remote message to inform students about the precautions and emergency information of the machine. These features depend on the implementation of remote control and management functions.

4.3   Strengthen Management Skills Training

Advanced technical means are inseparable from high-level management talents, so we must pay attention to the training of the administrator team. Strengthen the construction of the laboratory team, and regularly train the management personnel through internal and external, short-term and long-term, general and special-purpose means, while paying attention to the introduction of highly educated and practical personnel, optimizing the composition and age of the management team. Composition. A large number of high-quality computer personnel must be introduced. The management and maintenance of computer labs must rely not only on strict systems and regulations, but also on attracting high-level management talents. The training plan for such talents cannot be completed overnight. Only in accordance with the talent training strategy introduced from the inside and from the outside can we ensure the strengthening of the computer talent reserve. In addition, we must also follow the actual situation to the computer administrators for a series of problems in the computer lab management process. Training, in addition to ensuring the rationality of the talent team structure, always ensure the vitality of computer management personnel.

5   Conclusion

With the continuous development of higher education, the state and colleges will pay more attention to the informationization process of computer laboratory management, and invest more human, financial and material resources. The construction of the experimental management system can save a lot of manpower, material resources and financial resources, so that the laboratory administrator can get rid of the heavy work and realize the intelligence in the management process, which can greatly reduce unnecessary labor and reduce unnecessary The mistakes have brought great convenience to the school, teachers and classmates, thus improving the efficiency of the use of computer labs.

Acknowledgments. This research is funded by the school-level teaching research project of Wuhan Donghu University (A talent training research of real estate valuation based on the integration of industry and education) (Item Number: 34).

References

1. Lowe, D., Yeung, H., Tawfik, M., Sancristobal, E., Castro, M., Orduña, P., Richter, T.: Interoperating remote laboratory management systems (RLMSs) for more efficient sharing of laboratory resources. Comput. Stand. Interfaces **43**, 21–29 (2016)
2. Yu, H., Xu, H., Xu, Y., et al.: Construction and evaluation of PHP-based management and training system for electrical power laboratory. Comput. Appl. Eng. Educ. **24**(3), 371–381 (2016)
3. Dorwal, P., Sachdev, R., Gautam, D., Jain, D., Sharma, P., Tiwari, A.K., Raina, V.: Role of WhatsApp messenger in the laboratory management system: a boon to communication. J. Med. Syst. **40**(1), 14 (2016)
4. Bendou, H., Sizani, L., Reid, T., et al.: Baobab laboratory information management system: development of an open-source laboratory information management system for biobanking. Biopreservation Biobanking **15**(2), 116–120 (2017)
5. Liu, C.H., Chen, W.K., Ho, S.H.: NFS: an algorithm for avoiding restarts to improve the efficiency of crawling android applications. In: 2018 IEEE 42nd Annual Computer Software and Applications Conference (COMPSAC). IEEE Computer Society (2018)
6. Stankova, E.N., Barmasov, A.V., Dyachenko, N.V., et al.: The use of computer technology as a way to increase efficiency of teaching physics and other natural sciences. In: International Conference on Computational Science and Its Applications. Springer, Cham (2016)
7. Koyatsu, J., Ueda, Y.: Quality management system in pathological laboratory. Rinsho Byori Jpn. J. Clin. Pathol. **63**(7), 817 (2015)
8. Kelecha, T.: Outcome of the strengthening laboratory management towards accreditation (SLMTA) on laboratory quality management systems in selected Clinical Laboratories, Addis Ababa, Ethiopia. Am. J. Clin. Pathol. **144**(suppl 2), A194–A194 (2015)
9. Zhu, L., Wang, X.L., Zhang, P.P., et al.: The design of laboratory management system based on embedded video and campus card. Appl. Mech. Mater. **742**, 625–630 (2015)
10. Craig, T., Holland, R., D'Amore, R., et al.: Leaf LIMS: a flexible laboratory information management system with a synthetic biology focus. ACS Synth. Biol. **6**(12), 2273–2280 (2017)

Application of Computer Aided Instruction in College Physical Education

Jizhe Chen$^{(\boxtimes)}$ and Ying Lu

Southwest Petroleum University, Chengdu 6105000, China
1164512202@qq.com, 944592896@qq.com

Abstract. Starting from the characteristics and laws of sports teaching in Colleges, the paper discusses the application of computer-aided teaching methods in sports teaching in Colleges by the method of documentation. It is believed that, with the rapid development of information technology, the traditional teaching mode of sports in Colleges can hardly meet the requirements of cultivating high-quality and skilled talents in higher education. The teachers in Colleges must speed up the transformation of their ideas and realize the transformation from traditional teaching In order to keep up with the trend of the times in the tide of higher education reform, teachers of education should change to those who master new computer teaching methods.

Keywords: Computer aided instruments (CAI) · College sports · Information technology

1 Introduction

At present, mass sports in China is developing like a raging fire. The development of mass sports requires lots number of sports talents. In addition, the physical level of young people in China has been declining. For improving the physical level of young people, this situation must be changed. In 2015, Dengfeng Wang director of the Department of Sports and Health of the Ministry of Education, released a number to the world. At present, there are 300,000 PE teachers in short supply. The Blue Paper on Employment published by Max Research Institute and Social Sciences Literature Publishing House in 2018 shows that the employment problem of undergraduate PE graduates is not optimistic. From 2010 to 2014, they are all red card warning majors for undergraduate employment for five consecutive years. Since 2015, although it has turned into a yellow card specialty, the employment problem of sports specialty is still serious. And the top 10 majors with the highest unemployment rate in 2016 and 2017, announced by the Ministry of Education, all include sports. The sports talents highlights the shortcomings of the existing training mode of this specialty in Colleges. It can be said that the 21st century is a double-edged sword. Challenges and opportunities for sports major coexist. Therefore, how to change the existing talent training mode is an important issue to be solved urgently in front of us. To fundamentally solve the above problems, we must construct a different system of training mode for sports professionals in universities, in order to achieve creative training of innovative talents with profound sports professional knowledge, higher sports cultural literacy and

© Springer Nature Singapore Pte Ltd. 2020
M. Atiquzzaman et al. (Eds.): BDCPS 2019, AISC 1117, pp. 35–42, 2020.
https://doi.org/10.1007/978-981-15-2568-1_6

multiple sports development potentials. So, it puts forward the employment-oriented talent training mode of Sports Specialty in ordinary universities.

1.1 Study on Training Objectives and Specifications

Han put forward four criteria for training objectives of sports specialty: (1) to have flexible comprehensive application ability; (2) to have good ideological quality; (3) to have a rich knowledge framework; (4) to have strong physical quality [1]. Yan and Tian proposed that the primary school should be expanded on the basis of the original secondary school, and the primary school sports teachers and secondary schools are the main training objectives of this major [2]. Liu and Fang put forward that the standard of talent training should be based on the all-round development of morality, intelligence and physique, grasp the basic knowledge, skills and theory of sports, and have strong adaptability and creativity of social sports workers [3].

1.2 Study on the Ways of Culture

Li put forward nine basic elements of the educational ability structure system of college students majoring in sports through questionnaires of experts: (1) the ability to explain new technology; (2) the ability of teaching organization; (3) the ability of action demonstration; (4) the ability to analyze textbooks; (5) the ability to innovate teaching; (6) the ability to analyze, observe and correct errors; (7) the ability to compile text-books; (8) the ability to quote. The ability of guidance and propaganda; (9) the ability of amateur coaching other sports activities [4]. Zhang put forward that in the 21st century, the training mode of sports professionals in universities and Colleges in China should gradually change from orientated to open. However, due to the different geographical conditions in different parts of China, the preliminary rational choice is to focus on the training of specialized sports education departments and normal universities, supplemented by the participation of sports institutions and comprehensive universities, the mixed training mode of non-directional and directional combination. Gradually, a compound talent training mode, which emphasizes the integration of professional theory and quality improvement, and integrates all-round and specialized talents, has been established [5].

1.3 Research on Curriculum and Teachers

Shu and Wang believe that the present situation and problems of the cultivation of higher sports professionals are: the traditional Soviet theoretical teaching methods; the mode of simply teaching competitive skills; secondly, the proportion of academic subjects and disciplines, elective courses and compulsory courses is too imbalanced; the proportion of related education courses is low, and the hours of practice are less [6]. Huang and Ji pointed out that the current training mode of sports professionals in China has the defects of low proportion of public courses and professional basic courses, high proportion of professional courses, serious disconnection between teacher education theory and application practice and dual separation [7]. Peng and Wang pointed out the following problems in the talent cultivation mode of Sports Majors in China: too low requirement for "elective courses"; inappropriate handling of the relationship between theory and practice; relatively backward cultivation mode; and the policy of enlarging enrollment, which aggravated the task of the teachers in Colleges, could not complete the teaching task and affected the teaching effect [8]. Zhao et al. put forward the idea and conception of constructing innovative applied talents training mode of Sports Specialty in Colleges, in order to train versatile talents with "broad foundation, rich knowledge and high quality" of sports [9]. Fang et al. put forward that the curriculum of Sports Specialty in schools should show obvious marketability. It should have the cooperation of interdisciplinary comprehensive knowledge, realize the communication and integration between sports and related disciplines, and even different disciplines, so as to develop students' comprehensive quality [10].

1.4 Research on Students' Employment

Han et al. investigated the future of graduates majoring in Sports in recent years, and pointed out that there are differences between the personal abilities of graduates majoring in sports and the actual needs of their work units. The specific manifestations are: (1) the relationship between reality and ideal can not be well handled; (2) the lack of dedication and career spirit; (3) the comprehensive application ability needs to be further improved, such as language. The expression is not strong, the training and teaching ability is not specialized, and the scientific research innovation is not outstanding [11]. Fan et al. pointed out the following problems faced by graduates majoring in sports when facing employment: 1. inadequate understanding of career development goals; 2. too much emphasis on the role of regional factors in job hunting; 3. weak psychological adaptability and social adaptability of job seekers; 4. insufficient understanding of employment orientation and employment situation [12]; and Yang proposed No. The demand for talents in the society presents diversified and multi-level development [13].

2 Constructing Employment-Oriented Training Model for Sports Education Professionals

2.1 Guiding Ideas and Basic Principles for Constructing Employment-Oriented Talents Training Model

At the beginning of the 21st century, the guiding ideology of constructing the training mode of sports professionals in China should be guided by the spirit and requirements of the Decision of the Central Committee of the Communist Party of China and the State Council on Deepening Educational Reform and Promoting Quality Education in an All-round Way and the Decision of the State Council on the Reform and Development of Basic Education. It should further change ideas, renew concepts, open up new ideas and embody To implement the concept and guiding ideology of "health first" and implement quality education in an all-round way, we should focus on the requirements of economic development, knowledge innovation, modern information technology and social development in the 21st century, comprehensively strengthen the foundation, enhance adaptability, broaden professional calibre, cultivate innovative spirit, pay attention to individual development and improve comprehensive quality. To cultivate a wide caliber, wide adaptability, high quality, strong foundation, strong ability and innovation-oriented compound sports talents.

2.2   Optimizing the Course System and Reforming the Sports Course in Colleges and Universities

The reform of sports curriculum should combine physical exercise with the cultivation of college students' consciousness, habit, interest and ability of lifelong sports, and combine sports with sports, intelligence, mind and morality, so as to promote the overall development of College Students' comprehensive quality. Therefore, the reform of school sports curriculum needs to deal with the following aspects: (1) combining practice with theory. Theory and practice are complementary and unified. Theory provides people with a scientific way of thinking, while practice provides the most powerful proof for the practice of theory. In order to shape and guide people's values and pursuit of values, the setting of sports curriculum in Colleges must integrate theory with practice. (2) Highlighting students' principal position. Normal classroom teaching order needs teachers' teaching and students' learning to accomplish together, both of which are indispensable. Among these two factors, teachers' teaching only plays the role of side guidance and assistance, while the role of schools is to teach and educate people, cultivate virtue and serve the society. Therefore, the real subject of teaching order should be students. Therefore, the setting of sports curriculum in Colleges should be combined with the reality of students in contemporary society, remove red tape, reduce difficulty, and add some content which is conducive to mobilizing students' enthusiasm, inspiring students' wisdom and cultivating students' innovative ability. In a word, the establishment of sports curriculum in Colleges should fully reflect the main position of students. (3) Pay attention to the comprehensive cultivation of students' abilities and comprehensively improve the quality of students. The 21st century is a century of all-round development of material culture, and the current social situation can no longer meet the growing material and cultural needs of people. For the students majoring in Sports in Colleges, besides strengthening the ability of teaching, training and necessary scientific research, they should also pay attention to the ability of communicating with people, organizing, planning and managing, and the ability of integrating and processing various resources. This requires that core courses should be highlighted, basic disciplines should be strengthened, the proportion of elective courses should be enlarged and students should be guided to choose courses across majors so as to increase students' perspective of all-round view. In addition, schools should actively shoulder the responsibility of advocating students to study for double degrees and take part in postgraduate entrance exams, which is conducive to increasing students' all-round ability to meet their demands for fast-paced changes in society.

2.3   Optimizing Teachers' Team and Improving Teachers' Professional Quality and Ideological Quality

Teachers are the spirit of universities. College teachers have irreplaceable significance and role in improving innovative ability, shaping all-round talents and enhancing national cohesion. In the final analysis, the competition of universities is the competition of educational quality, and its foundation lies in the combination of theory and practice. Studies have shown that theoretical practice and practical theory are the motive force and source of promoting the reform and progress of University teachers. Therefore, the research on the status of university teachers will play a vital role in University education. Vigorously introducing high-quality talents such as Ph. D. and encouraging young teachers to further study their degrees are the most important ways to improve the structure of academic qualifications of sports major in Colleges in China. At the same time, teachers should also strive to improve their own professional competence, constantly absorb the most cutting-edge professional knowledge, from a single technology-oriented to management-oriented, organizational planning-oriented development, with its exquisite technology, profound knowledge, noble personality, for students to establish a learning model, to be worthy of the requirements of the times of education pioneers.

2.4   Optimizing Teaching Methods

Change the traditional single solidified teaching mode and transform it into an open teaching mode. It emphasizes the process of teaching rather than the conclusion, and advocates such teaching methods as "mu-lesson, micro-lesson and case-based". Take the Mu lesson style as an example:

The so-called MOOC refers to massive Open Online Course, which is a large-scale open online course. It comprehensively utilizes lens language, teaching theory and technical means, according to students' hobbies and characteristics, decomposes, integrates and rebuilds the traditional course content, arranges a learning module with rich content and organic combination, and then displays it through Internet technology. Give students the teaching mode of learning experience, interaction and so on, so as to absorb knowledge. Mu lesson is being accepted by more and more students. The reason why it can be so welcomed lies in: on the one hand, Mu lesson is refreshing with its teaching idea of flipping classroom, interactive teaching method and modular teaching content; on the other hand, it breaks through the space-time limit of teaching with its open and large-scale online teaching, allowing students to pass through online. Exchange and enjoy quality teaching resources from all over the world, and witness the elegance of famous teachers from different disciplines. This not only gives full play to the main role of students, but also allows students to digest, absorb and create in the learning process, and fully cultivate their self-study ability, thinking ability and comprehensive application ability.

2.5   Optimizing the Way of Talents Training

1. Standardize graduation design, strictly prohibit fraud and cultivate scientific research consciousness. 2. Increase the practice link, strengthen the classroom teaching practice ability, strengthen the practical operation ability, as well as a number of social sports competition practice, sports investigation practice, referee practice as one of the teaching practice system. 3. Establish a hierarchical education model that integrates sports training and competitions, after-school counseling.

2.6   Focus on Cultivating and Improving the Employment Competitiveness of Graduates

With the fierce competition in the talent market and the increasing saturation of employers, it is far from enough for the undergraduates majoring in sports to improve their employment competitiveness. As far as colleges are concerned, 1. They should make long-term career planning and career guidance for students, and run the career guidance throughout the whole process of college education for students majoring in sports; 2. They should establish a long-term cooperative mechanism with employers so as to provide students with as much employment information as possible, so that they can timely understand the changes and needs of the employment market, and correspondingly achieve the same goal. Advance with the times; 3. Colleges should help students actively change their roles to adapt to the new work life; 4. Encourage students to find jobs first, then choose careers, and at the same time encourage them to start their own businesses.

At present, as far as students themselves are concerned, 1. The social demand for sports talents is gradually developing towards the direction of comprehensive ability, such as foreign language level, computer application ability, professional theory and professional skills, and the comprehensive application ability of relevant knowledge. This is not only the goal of higher education personnel training in the future, but also the requirement for graduates to adapt to social development in the new era. 2. Graduates majoring in sports should evaluate themselves objectively and locate themselves. Excessive expectation is often an important reason that many graduates of sports major can't get employment smoothly. The lack of experience in the world and society is a true portrayal of graduates majoring in sports. In addition to the influence of society and family, most graduates have a high orientation and compare with each other, which will inevitably affect the rationality and accuracy of graduates' employment choices. Therefore, the graduates of Sports Majors in Colleges should enhance their comprehensive ability and autonomous consciousness on the basis of strengthening their professional accomplishment, make objective evaluation and positioning of themselves, and make objective, timely and accurate judgments on the living space suitable for their own development. Only in this way can they continuously improve their employment competitiveness and adapt to the new environment, and achieve long-term development.

3 Conclusion

To construct the employment-oriented talent training model for sports major in general colleges, we should implement and implement the guiding ideology and basic principles of the relevant countries, and keep pace with the times: optimizing the curriculum system, reforming the sports curriculum in Colleges; optimizing the teaching staff, improving the ideological and professional quality of teachers themselves; and optimizing teaching methods and personnel training approaches according to local conditions. Focus on training and improving the employment competitiveness of graduates. The key is to set up the modern education concept of people-oriented, to take the training of sports teachers as the main body, to pay attention to the training of students' comprehensive quality and ability as its own responsibility, and to strive to build a employment-oriented talent training model with Chinese characteristics, so as to realize the Chinese dream and make unremitting efforts.

Acknowledgments. This research was supported by Southwest Petroleum University Humanities and Social Sciences Research Foundation Project (project number: 2018RW028).

References

1. Han, Z.: Adjustment and optimization of the undergraduate training program of sports major in Chinese universities. J. Beijing Sports Univ. **39**(07), 89–94 (2016)
2. Yan, Z., Tian, X.: Research on the reform of college sports professional training from the perspective of ability standard. Educ. Theory Pract. **38**(21), 9–11 (2018)
3. Liu, S., Fang, Q.: Thoughts on the development of undergraduate sports professional education from the perspective of supply-side reform. J. Xi'an Inst. Sports **35**(03), 361–365 (2018)
4. Li, R.: On higher sports curriculum setting guided by employment and talent training. J. Guangzhou Inst. Sports **34**(04), 114–116 (2014)
5. Zhang, L.: Research on the cultivation of entrepreneurship sports talents in colleges. Educ. Theory Pract. **38**(15), 18–19 (2018)
6. Shu, Z., Wang, H.: Reshaping and optimizing the curriculum system of sports specialty for "excellent sports teachers". J. Wuhan Inst. Sports **51**(04), 75–81 (2017)
7. Huang, H., Ji, K.: Research on the reform of undergraduate course system of sports in Chinese universities. Sports Sci. **24**(3), 51–57 (2004)
8. Peng, Q., Wang, C.: Research on the reform of sports professional training in local applied universities based on lifelong sports service: taking Huaihua College as an example. J. Sports **25**(04), 105–109 (2018)
9. Zhao, Q., Li, M., Li, H., Li, W.: Characteristics, ideas and enlightenment: a study on the curriculum of sports major in Ohio State University. J. Beijing Sports Univ. **39**(12), 105–111 (2016)
10. Fang, Q., Wang, R., Xu, J., Xie, Z., Li, F.: Basic theory and discipline system construction of sports: logical approach, research progress and perspective. Sports Sci. **37**(06), 3–23 (2017)
11. Han, Z., Sun, B., Liu, M.: The United States, Russia, Germany and Japan. Sports Culture Guide (08), 169–173 (2017)

12. Fan, X., Wang, X., Yin, Z., Yi, S.: Research on the reform and development of Ireland's sports major: a case study of Limorick University. J. Beijing Univ. Sports **38**(04), 105–110 (2015)
13. Yang, H.: From athletes, coaches, teachers to multi-class and multi-level—reflections on the training of sports talents in China. Sports Sci. **38**(07), 6–8 (2018)

Development of the Model of Combining Medical Care with Old-Age Care Driven by "Internet +"

Yan Gao$^{(\boxtimes)}$ and Rui Li

Jilin Engineering Normal University, Changchun, Jilin, China
123042979@qq.com

Abstract. Population aging is a major strategic problem of China in the 21st century, Give full play to the advantages of "Internet plus" combining medical and nursing care, We will optimize the allocation of pension resources, Improve information asymmetry, achieve effective online and offline connectivity, Improve efficiency and reduce medical expenses, improve quality of life and promote mental health. This paper aims to explore how to use the advantages of Internet technology to make the combination of medical care and nursing more compact and make the model of combining medical care and nursing care play a greater role in the perspective of "Internet +". Based on the needs of the elderly under the active guidance of the government and the support of the Internet service cloud platform, public and private cooperation should be strengthened to fully tap the potential of the medical and nursing market.

Keywords: Internet + · Combination of medical and nursing care · Pension model · Community home-based care

1 Introduction

The 13th five-year plan period is a decisive stage for China to complete the building of a moderately prosperous society in all respects, and an important strategic window period for the reform and development of China's undertakings for the aged and the construction of the pension system. It is expected that by 2020, the number of people over 60 will increase to about 255 million, accounting for 17.8% of the total population [1]. The number of elderly people will increase to about 29 million, the number of elderly people living alone or living in empty nests will increase to about 118 million, and the old-age dependency ratio will increase to about 28%. With the aging of the population, higher requirements have been put forward for the development of old-age insurance, medical care, old-age care service industry, protection of the rights and interests of the elderly, and spiritual care for the elderly.

On November 20, 2015, the state council office forward nine departments such as health and family planning commission "about promoting health and pension service of combining the guidance", put forward the "opinions", "in 2020, in line with the national conditions of medical keep combination of systems and mechanisms and policies and regulations system basic establishment, health care and pension service

© Springer Nature Singapore Pte Ltd. 2020
M. Atiquzzaman et al. (Eds.): BDCPS 2019, AISC 1117, pp. 43–49, 2020.
https://doi.org/10.1007/978-981-15-2568-1_7

resources share orderly, covering urban and rural areas, appropriate scale, reasonable functions, integrated continuous medical raise with basic service network". With special emphasis on the combined medical service information in the medical system, the importance of the build process for example, by relying on the community of all kinds of "full services and information network platform", "use the old basic information file, electronic health records, electronic medical records", "the organization of medical institutions to carry out the long-distance medical service in pension institutions", "encourage around explore new model based on combining medical keep Internet service", "d have combined with" service for pension institutions to provide information and technical support. Therefore, with the development of the Internet and the arrival of the era of big data information, it is an irresistible trend to build the model of "Internet + medical care" and form an intelligent platform of medical care information sharing.

2 The Significance of the Development of the Model of Combining Medical Care with Elderly Care Driven by "Internet +"

2.1 Effectively Realize Information Communication and Resource Sharing Between Supply and Demand Parties

Driven by the "Internet +", the model of combining medical care with old-age care breaks through the limitations of time and space, effectively realizing the communication and connection of information between supply and demand. The elderly can choose the services they need at any time and anywhere through the information platform, and the outside world can also timely grasp the elderly's pension and medical needs [2]. Medical combined with the establishment of the service information platform, the meaning of the more important is can get through the elderly, community health centers, hospitals, the connection between the endowment institutions, realize the different regions, different institutions endowment and complete sharing medical resources and effective integration, at the same time also can provide the elderly with quick online registration, two-way referral channel, significantly reduce the cost of medical treatment of the elderly.

2.2 Improve the Quality and Efficiency of Medical and Nursing Services

Combining international medical have Internet, but is firstly established by grassroots community for the old people including the individual information, family situation, condition, electronic health records and other information service demand, on the basis of building the elderly health information platform of big data and real-time information uploaded to the cloud, to master the elderly in comprehensive and detailed basic information at the same time for the elderly, such as emergency rescue, life for help, home care, health examination, the leisure entertainment more intelligent, personalized and professional online endowment service. After the completion of the service, the elderly group can also provide effective feedback to the service through the platform evaluation system, so as to prompt the supplier to continuously improve the service deficiencies and improve the service quality.

3 "Internet +" Drives the Development of the Mode of Combining Medical Care with Elderly Care

3.1 The Distribution of Interests and Rights and Responsibilities

As a new thing driven by "Internet plus", the combination of medical care and nursing is short of specific institutional guarantee and policy convergence, the quality evaluation system and supervision and management mechanism are not perfect, and the internal interest coordination mechanism is not perfect, and there are problems in the distribution of interests and powers and responsibilities [3]. Due to the overlapping management of competent departments and the lack of a clear and unified management system, there are contradictions between interests and powers and responsibilities. It is difficult for pension institutions to connect with medical institutions and resources sharing is not smooth, which leads to insufficient motivation of institutions and reduces their enthusiasm [4].

3.2 There Are Obvious Barriers Between Supply and Demand in Information Communication and Sharing

The combination medical raise pension mode is the important challenges facing the supply-side and demand-side in information communication and sharing among the elderly obvious barriers, suppliers' pension institutions, hospitals, community health center is difficult to accurately grasp the elderly living habits and health information, such as the elderly health archives information between different pension and health care institutions also does not have corresponding sharing mechanism, thus in medical nursing services between supply and demand, the supplier to form the information isolated island. Together with the unblocked referral channel and the imperfect medical insurance settlement system, the elderly face the dual problems of providing for the aged and seeking medical care.

3.3 There Are Barriers to the Use of the Internet Among the Elderly, and It Is Difficult to Fully Express Their Effective Needs

There are obvious barriers for the elderly to use the Internet, and even huge deficiencies in the use of smartphone communication terminals and APP software. The reasons for the above phenomenon are various. First, most of the elderly have a low acceptance of new things. Even though the Internet technology has been developed in China for more than 20 years, they still have difficulty in accepting compared with the young people. Second, online fraud and security problems occur from time to time. The elderly have a certain degree of fear about the Internet and always worry that they will fall into the trap of online consumption. Third, many old people believe more in the offline experience of old-age care services and have no sense of online services. Only when online services and offline services are organically integrated, will the sense of identity and acceptance of "Internet +" old-age care service mode be enhanced [5, 6]. The above factors directly affect the promotion of the "Internet +" pension service model, and it is difficult for the effective needs of the elderly to be fully expressed in a timely manner. Therefore, the promotion of the new model faces insurmountable obstacles.

3.4 The Cooperation Enthusiasm of Relevant Institutions Is Not High

The combination of medical care and nursing faces huge contradiction between supply and demand of services. There is a general shortage of funds, facilities, venues and service personnel, resulting in a general situation of "small horse-drawn carts" with insufficient power, low efficiency, slow development and poor results. Due to the constraints of practical factors such as strong public welfare, insufficient resources, low profit and high risk, many professional pension institutions, medical institutions and nursing institutions are not willing to cooperate with community home-based pension service centers [7]. Medical institutions with high diagnosis and treatment level and good reputation lack the motivation to provide medical services for pension institutions due to their tight medical resources and potential risks such as medical disputes. For the sake of profit, the professional private pension institutions consider that the development of pension services is not profitable enough, thus affecting their investment enthusiasm.

4 Suggestions on the Development and Optimization of the "Combination of Medical Care and Nursing Care" Pension Model Driven by "Internet +"

4.1 Increase Government Support and Improve Top-Level Design

First of all, the government should introduce laws and regulations to clearly define the responsibilities and obligations that the government, service providers, communities and the elderly themselves should assume, so that all participants can have equal responsibilities and rights in the process of combining medical and nursing services.

Second, the government should set up special organizations, experts in the related fields of solicit, service representatives, representatives of social organizations and consumers on the basis of the opinions, according to the different service content to set up the different service standards and charge standards, design specification of the service process and evaluation mechanism, and according to the standard prior to unified bidding and training service providers, to ensure maximum d have combined with the specification and quality of service.

Third, the government should improve the policy support mechanism. In view of the particularity of investment in the pension industry, for projects with large one-time investment and long payback period, the government can offer certain preferential policies in various aspects, such as low loan interest rate and long repayment period in investment and financing, and preferential right of use of land for projects combining medical care and nursing, etc. (Fig. 1).

Fig. 1. Development of the mode of combining medical care with old-age care driven by "Internet +"

4.2   Build an Internet Service Cloud Platform

The development of big data Internet technology can provide technical solutions for complex management activities, improve management efficiency and save management costs [8]. Using "Internet + " as a tool to facilitate the combination of medical and nursing services, three technologies should be improved.

First, the establishment of the elderly personal information database, including health records, service files and so on. It can be registered through household survey or online platforms such as App, website, etc., and the id number and social security card number are the registration identification number. Wearable devices, such as smart bracelet, electronic nanny, automatic measuring instrument and other devices, GPS positioning and other real-time monitoring and uploading of basic signs and location information of the elderly, facilitate timely and effective intervention of medical care services.

Second, the establishment of the main network service platform. With the help of the Internet and the Internet of things, the supply resources of pension services from different enterprises and public welfare organizations are integrated to effectively match supply and demand. Access to meal ordering, shopping, learning and entertainment, mutual assistance, social organization participation and other networks in daily care, and develop private customized projects; Access to cooperative medical institutions and network medical service platform for medical care, providing telephone network appointment, registration, online outpatient service, disease prevention and nursing advice.

Third, cooperate to develop convenient service terminal equipment. A variety of easy-to-operate devices are provided based on the differences in the ability of the elderly to use the Internet. Smart phones and tablets are suitable for the young and educated elderly, wearable devices are suitable for the elderly with poor physical conditions, and telephone hotlines are suitable for the elderly who are not familiar with smart terminals. At the same time, strengthening the training of the elderly on the use of smart devices can also enable their children to customize the required services for the elderly through the Internet [9]. In short, through the integrated community management service platform and the Internet platform, the combination of physical community and virtual community can be realized, and various medical care and pension resources can be deeply integrated, various policies can be effectively connected, and the supply and demand of services can be coordinated to meet the diversified service needs of the elderly.

4.3 Strengthen Personnel Training

First of all, elderly medical care and pension service professionals are the human resource guarantee for the development of combined medical care and pension service mode, which is directly related to whether the elderly can get high-quality services. It is the key to strengthen the training of nursing professionals and improve the quality of pension and medical service. We should try to set up market-oriented training institutions, increase training efforts, improve the skill level of elderly care personnel and management personnel, and meet the needs of the medical care combined with elderly care services.

Secondly, the Internet in the combination of medical and nursing care should drive the large-scale network facilities construction, and a large number of professional talents proficient in network planning, design, construction, integration, management and maintenance are needed. However, at present, China still lacks professional talents who master network technology, and it is difficult to meet the needs of network development. Therefore, the government should provide material resources, training funds and grant support for network personnel training, and actively encourage colleges and universities to develop disciplines and departments of Internet technology application, medical and nursing combined with related content, and cultivate innovative and technical talents.

4.4 Increase Social Participation

The participation of the whole society, including communities, enterprises and families, is an important driving force for the development and promotion of "Internet plus" combined medical and nursing services. The advantages and prospects of the Internet plus combined medical and nursing services should be vigorously publicized in the community. In view of the elderly group's low acceptance of intelligent devices, we can start with the children of the elderly and encourage them to participate in the use of intelligent devices to help the elderly choose their own intelligent service types.

In addition, through certain incentives, such as registration, technology research and development, product certification, taxation offer certain preferential policies, encourage private enterprises, social organizations to participate in medical have combined with the project's construction and the technology development process, adopting direct subsidies, enterprise financing, social donations, etc. way to expand the scale of all kinds of public welfare and charity organization and sources of funds, promote the combination of volunteer organization and team, in order to joint the social from all walks of life force, better as the "Internet + d" to raise the pension mode service [10].

5 Summary

We will encourage Internet enterprises to enter the old-age care industry, promote the construction of an Internet service cloud platform, integrate the basic information files and electronic health files of the elderly, and realize the sharing of old-age care and medical information resources. Relying on the Internet, it gives full play to the role of big data. Through collecting data from the elderly's needs and preferences, health risk factors, prevention and treatment of chronic diseases and other aspects, it can accurately grasp the conditions of the elderly and provide accurate guidance and more convenient and quick information technology support for the development of the elderly care industry.

All in all, the medical development to achieve depth fusion, the difficulties, but as long as a reasonable guide, the straightening out mechanism, with the help of "Internet +" technology to build open, participation, interaction and integration of new public service platform, can realize the medical resources unified coordination and sharing of cohesion, to achieve the goal of social endowment, take on task of endowment service in China.

Acknowledgements. This work was supported by The Social Science Planning Project of Jilin Province in 2019:Research on the strategy of home-based care service in urban community of jilin province under the mode of combining medical care and nursing(NO.2019C45)

References

1. Shen, X.: Research and discussion on community wisdom retirement in the context of "Internet +". Comput. Knowl. Technol. **14**(19), 302–303 (2018)
2. Liao, S., Zhu, H., Tan, B.: The dilemma and countermeasure of the health management service of "the combination of health care and elderly care based on Internet +" among the elderly with chronic diseases in community. Chin. Gen. Pract. **22**(07), 770–776 (2019)
3. Zhou, H., Dong, Y.: The mechanism and realization path of "Internet +" to promote the precision of the old-age service. Acad. J. Zhongzhou **03**, 60–65 (2019)
4. W, Hu: The plan for the development of the combination of health care and elderly care with institutional reform under the background of healthy China. Adm. Reform **02**, 48 56 (2019)
5. Geng, Y., Wei, Y., Zhou, J.: Research on the development of "old-age service based on Internet +". Macroecon. Manage. **01**, 71–77 (2019)
6. Zhang, B., Han, J.: Research on the intelligent old-age service based on "Internet +". Macroecon. Manage. **12**, 40–44 (2018)
7. Chen, Q.: Research on the innovation and development mode of the old-age service based on "Internet +" - comments on stepping into the development new era of the old-age service industry. Journalism Lover **09**, 111 (2018)
8. Dong, P.: The impact of the rapid development of aging on economic development and social governance in the internet era and its countermeasures. Lanzhou Acad. J. **11**, 165–174 (2018)
9. Liao, Y., Qin, Y.: Internet + health pension: creating a new model of intelligent aged care service. World Telecom **08**, 75–77 (2015)
10. Xie, J., Zhang, L.: Intelligent wearable device and its application. China Med. Devices Inf. **21**(03), 18–23 (2015)

Design and Development of Mobile Automobile Exhaust Detection System Based on Internet of Things

Qiang Liu[1,2], Zhongchang Liu[1], Yongqiang Han[1(✉)], Jun Wang[2],
Zhou Yang[2], and Tonglin Bai[3]

[1] State Key Laboratory of Automotive Simulation and Control, Jilin University,
Changchun 130025, China
liuqiang@jlenu.edu.cn, {liuzc,hanyq}@jlu.edu.cn
[2] Research Center of Automobile Safety Technology of Jilin Engineering
Normal University, Innovative Research Team of Jilin Engineering Normal
University (IRTJLENU), Jilin Engineering Normal University, Changchun
130052, China
wangjun0620@163.com, lq200379@tom.com
[3] Dehui Secondary Vocational and Technical School, Dehui 130300, China
1329034744@qq.com

Abstract. The rapid development of automobile industry in modern society, and automobile has become an indispensable walking tool in people's Daily life, which brings a lot of convenience to people's life and work. But in the car to provide us with convenience at the same time the air quality is also deteriorating, causing a lot of problems for the environment. In view of the urgent and arduous task of controlling automobile exhaust emissions and protecting the environment, this paper proposes to design and develop a mobile automobile exhaust detection system based on the internet of things, so that the exhaust emissions generated by the operation of the automobile will be mixed into the air, which makes the air pollution more and more serious. In order to effectively control the exhaust emissions.

Keywords: Internet of things · Exhaust detection · Mobile communications

1 Introduction

So far, China's car ownership has reached more than 200 million, ranking in the world's top ten. Nevertheless, the number of cars in our country has not reached saturation, car ownership continues to grow, calculated on the average annual car ownership will increase by tens of millions of units. In the case of increasing the number of cars, the air pollution caused by automobile exhaust emissions is becoming more and more serious, the air quality is deteriorating, and it is urgent to protect the environment, so by testing the exhaust emissions of automobiles so that the vehicle exhaust emissions can be controlled within the standards set by the State to control the exhaust emissions, Achieve the goal of reducing air pollution.

© Springer Nature Singapore Pte Ltd. 2020
M. Atiquzzaman et al. (Eds.): BDCPS 2019, AISC 1117, pp. 50–57, 2020.
https://doi.org/10.1007/978-981-15-2568-1_8

Contaminants in automotive exhaust emissions include hydrocarbons (HC), carbon monoxide (CO), nitrogen oxides (NOx), carbon dioxide (CO_2), sulphur dioxide (SO_2) and other particulate matter [1]. At present, there are four kinds of detection methods for automobile exhaust gas: simple transient condition method, filter paper smoke method, diesel vehicle loading deceleration LUGDOWN method, double idle speed detection method. At present, the most commonly used methods in China are simple transient condition method and double idle speed detection method, both of which are the first to simulate the road conditions of the car, and then further test the concentration of its emission gas, but this detection method is not accurate enough, the operation is also more difficult, And it is not possible to accurately reproduce the car's usual discharge status when it is actually moving.

The content of this paper is not the same as that of exhaust gas detection in the traditional sense, the innovation point is to combine the current hot internet of things with automobile exhaust detection technology, mainly from the three-layer structure of the Internet of things including perceptual layer, network layer, application layer to develop and design the system. As long as the use of embedded processor to obtain the relevant exhaust analysis device, navigation module, OBD module, sensor and other data, and connect to the remote control platform, and then make the owner on the mobile side of the fault query results, for the car maintenance has brought convenience.

According to the subject of this paper, the following research is carried out: firstly, the exhaust emission system of automobile is studied and designed within the scope of international regulations, followed by the hardware problem of the perceptual terminal and the remote management platform, and finally the diagnosis of the fire fault of the engine.

2 The Internet of Things

The first is the research on the Internet of things, the internet of Things is a new noun, it is closely related to people's lives, in the next generation of information technology, it is a very important component. It is also an important development course in the "informationization" era that we are now stepping into. The English name is "The Internet of Things (IoT)" [2]. The system hierarchy of the Internet of things can be divided into three layers: application layer, network layer and perceptual layer, perception layer as the core component of the Internet of things, its role is mainly information acquisition, perceptual layer is responsible for receiving, processing and transmitting perceived information, and the application layer mainly relies on cloud computing platform to deal with specific information.

The function of the Internet of things is mainly through a series of information-aware devices and technologies such as radio frequency Identification technology (RFID) [3], Global positioning system, infrared sensors, laser scanners and so on, so that objects and networks are combined with each other, so that objects and objects between the exchange of information.

Secondly, we have made a study on the automobile exhaust detection technology, we can analyze the components of these substances in the exhaust gas, we can judge the operating conditions of the engine and the failure of the engine. Because when the content of a component in the exhaust gas rises or decreases, the engine is bound to fail, so through the analysis of the exhaust gas, we can judge the failure of the engine. And we have to judge the engine failure in the process of car use to test the emissions of exhaust gas.

3   Automobile Exhaust Detection System

Exhaust analysis system first to detect and analyze the emission of cars, and then dilute the exhaust gas, and the most important in the system is the exhaust analyzer, the exhaust analyzer is mainly composed of exhaust dilution system, sensor, signal circuit, control Unit, processing unit, output equipment, etc. The main use is non-spectroscopy infrared analysis and chemiluminescence method. Optical infrared analysis is actually used to measure concentrations, but also the concentration of asymmetric molecules with dipole moments, because the absorption capacity of infrared rays varies depending on the wavelength of absorption, as well as carbon monoxide (CO), hydrocarbons (HC), carbon dioxide (CO_2) in the exhaust gas. Such gases have their own different special absorption peaks in the infrared band, so the non-optical infrared absorption method is mainly used to detect these gases. The Chemiluminescence method is mainly used to detect the concentration of NOx in the exhaust gas, the main principle is to produce nitrogen dioxide and oxygen through the reaction of nitric oxide and ozone, and some of the NO_2 in the formation will be in the active condition, When the NO_2 of activation becomes the ground state, the quantitative strength and the concentration of NO are proportional to the light quantum, and then the intensity of the light quantum within the wavelength range is changed into an electrical signal by the relevant device, and then the total concentration of NOx can be measured by the corresponding circuit and computer processing.

4   System Hardware and Software Design

After studying the exhaust gas detection, we mainly design the hardware and software part of the system. As shown in Fig. 1. The main components of mobile automobile exhaust detection system based on Internet of things designed in this paper are three: perceptual terminal [4], mobile communication and remote management. The perceptual terminal is also composed of many parts, including embedded microprocessor, OBD module, navigation part, communication part, storage part, exhaust analyzer, exhaust dilution sampling device and other hardware equipment. In these compositions, the mobile communication network is provided by the appropriate vendor, and the management tasks are performed by the server.

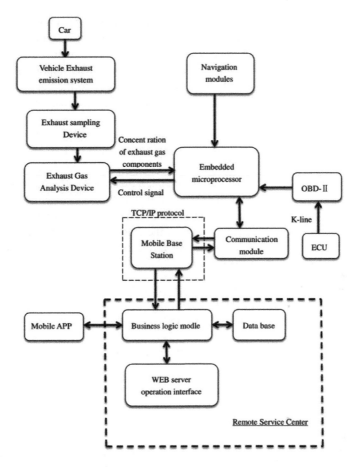

Fig. 1. System structure diagram

Embedded microprocessor applies a variety of digits, that is (2^2-2^6), for microprocessors, the more bits represent its stronger performance, we will generally choose 32-bit or 64-bit processor; The OBD system used in this paper is actually a car self-diagnosis system, And OBD-II is the abbreviation of II type car diagnosis system, its work function is to monitor the work of engine and exhaust system in real time, once found that there may be faults will be alerted, and storage of related fault information to facilitate the work of maintenance personnel. The OBD module works first through the interface to read the data in the electronic control unit, and then transfer to the processor for processing; The main function of the navigation system in the traffic is to obtain the vehicle's location information data, which is more suitable for China's road conditions is the Beidou navigation and positioning system, so at this stage the use of Beidou navigation and positioning system, In this kind of navigation system mainly rely on the car installed on the starter to receive and send information, and then send the obtained information to the remote management platform; The system to be designed in this paper should be based on data information transmission anytime and anywhere, so it is necessary to communicate on the terminal to join the mobile network. We mainly use the fourth generation mobile communication network for information exchange; The sensor module mainly contains two kinds of sensors: wet temperature sensor and pressure sensor; the Development Board of 5 V,2A DC power supply is used in this paper, but the voltage is converted when the power supply to the car is received. We mainly use the LM2596 switch voltage regulator to achieve the design of voltage conversion and conversion voltage.

In the software aspect this article mainly chooses the embedded Linux system to do the operating system, and this system has many advantages, such as this system is not required to pay the fee, any user can obtain and use, we can also according to their actual needs to modify the system and manufacture their own software products, There will be no copyright issues. And the system also has the advantage of being able to customize according to individual needs, without changing the stability of the premise can greatly reduce the cost of the system, in the start of system development can be revised with the existing program code, can save time and economic costs.

The process for designing the software is as follows:

(1) First, through a series of tools at the host end to the operation of the core and lead the code of the program generation;
(2) To download the data transfer to the processor through the corresponding interface;
(3) After completing the above steps, use the debugger to debug the Designed software.

5   Remote Management Platform

Finally, we designed the Internet of things remote management platform. Remote management platform is also called remote Management system, mainly the network communication technology as the basic management system, remote management technology is an advanced and important technology in the field of control, and the function has become more and more powerful and widely used with the growth of time. The IoT remote management platform is a proprietary management platform tailored to the user to support the development of the management work. In this platform, it is mainly through the management platform to obtain relevant information, and the information obtained analysis and processing then integrate the user terminal fault reasons, user information and communication status and other information, and then feed this information back to the relevant users.

First of all, we should choose the development mode, there are two development modes in the remote platform, namely B/S mode and C/S mode. Among them, B/S mode is more convenient and fast, we just need to install a browser (Browser) on the client side, as for the problem of processing other data on the server to complete it, but the only drawback is that its response speed is slow; the second C/S mode has a different client and the first one, It is installed on the server, it has the advantage of faster response than the first, it is very capable of processing, but the disadvantage is the need to install the corresponding dedicated client software, and this software installation is difficult, its compatibility is also relatively poor, want to develop to use the cost is also very difficult. Therefore, based on the above point of view analysis, and according to the actual needs of the system and the existing experimental conditions, we will choose b/s mode to develop and design the system.

Secondly, the function module of the platform should be designed, and in the design of the system, the function module of the system platform is divided into roughly five parts according to the type and actual demand of the parameters required by the system, which are the user Management module, vehicle query module, network Information Communication module, System Management module and historical data query module. The function of the user management module is mainly to assign and verify the maintenance of the user's personal rights and to maintain the relevant information of the user [5]. The function of the vehicle query module is mainly to query and detect the relevant data of the vehicle according to the license plate number provided, and to provide the detected results; The main function of the network communication module is to be responsible for the data of the access sensing terminal, and the main function of the System Management module is to maintain the system, while the main function of the historical data query module is to provide the historical data related to query, analysis and invocation.

Next, you need to design the remote management platform. The Internet of things remote management platform can be logged in a browser or mobile client, while internal workers can access the platform on the browser and manipulate the mobile client's APP is primarily used by the car owner to query the test results [6].

In this article, the main use is the J2EE model, which is mainly divided into three layers, namely the presentation layer (UI), the Business Logic layer (BLL) and the data Access Layer (DAL), where the three-layer structure each has its own function: the function of the presentation layer is to judge the interface control format, Mainly to transfer information to the next layer of business logic layer, the function of the business logic layer is to receive the data from the page, and transform the type of data, the business logic processing, the function of the data access layer is to provide the interface to the database that needs to be accessed, and to process the database; In this paper, the MVC design pattern is used in the development and design of WEB applications, because it can be developed by reference to existing experience. The main function of MVC mode is to separate the data layer from the representation layer, which is mainly composed of three parts: model, view, controller.

The structure flow of the system is mainly used to access the external part of the system integration interface, and then transfer the information to the Remote service center system, mainly the five modules of user management, vehicle query, System setup, historical data query and network communication, and then transfer data to J2EE Remote Center control platform, It mainly includes the business logic components, security control, log management and data storage components of each module, and then stores the obtained information data to the database. The database used in this article is SQL database, this type of database function is very comprehensive compared to other databases, and almost comprehensive support for all development tools, maneuverability is very strong, at present in all aspects are widely used.

After detecting the data, we should analyze the data and judge the engine fire fault situation, in this paper, the BP Neural network algorithm is mainly used to analyze and judge the fire fault diagnosis. BP Neural network algorithm mainly uses the reverse propagation error, forward propagation signal method to calculate, in the actual calculation time weight is in accordance with the reverse order from output to input to modify and correct, the BP neural network algorithm used in this paper needs to carry out a very complex operation, And the value of each layer of weight increase or decrease can be calculated and then the new value is obtained by operation, so that the result of the operation to the output has been met with the requirements. There are three main methods for diagnosing fire faults when there is a problem with the engine, namely waveform analysis method, instantaneous speed detection method and measurement analysis method, and the measurement analysis method can be divided into three kinds, namely oxygen sensor method, ion current method and exhaust component analysis method. In this paper, the main use of the measurement analysis method of the exhaust component analysis method. The emission status of the car's exhaust gas in our daily life can reflect the gasoline combustion state of the engine, the concentration of each exhaust component will fluctuate greatly when the engine has a fire failure during the car operation, for example, when the gas oil ratio is very low, the concentration of the O_2 will increase, and then the concentration of carbon monoxide will be reduced a lot, And when the car's fuel supply system fails, it will increase the concentration of hydrocarbons, we will be the fault of the car exhaust component concentration and their respective state of the BP neural network system to diagnose the fire fault model can be based on the established model to judge the corresponding failure of its occurrence.

6 Summary

The main center of this article is the development and design of mobile automobile exhaust detection system based on Internet of things by looking for data we analyzed various components of automobile exhaust gas, in which the exhaust gas contains carbon monoxide, carbon dioxide, nitrogen oxygen compounds, hydrocarbons and other components, And understand the various methods that can detect the exhaust gas analysis of these components, and have a comprehensive understanding of the Internet of things, to understand that the internet of Things is the use of wireless communication technology in the true sense of the realization of the exchange of information between objects and objects, saving time and reducing costs.

After the completion of the preparatory work, the design of the system, we first design an overall structural framework, and then select the hardware that the system can apply to and carry out the relevant design, and then build the perceptual terminal and use mobile communication as a communication method between the terminal and the remote platform, Using J2EE [7] technology to design the remote platform, and BP neural network system to study and diagnose the fire fault, by detecting the components of various automobile exhaust gas to judge the fire failure of the engine.

Through these technical theories, we have a deeper understanding of the design and development of mobile automobile exhaust detection system based on IoT, and can develop the system design in the subject, so that it is no longer just an idea or idea, But can really contribute to the control of automobile exhaust emissions and the prevention of pollution [8].

There are still many shortcomings in this design, in addition to the article we will also consult the data, enrich the theoretical knowledge of the system and put into practice, according to the shortcomings of the corresponding solutions, so that the system is designed to be more comprehensive, can be widely used, and reduce exhaust emissions and failures, to control pollution, Convenient for people to travel, to add convenience to people's daily life [9, 10].

Acknowledgments. This work was supported by the National Key Research and Development Program of China (No. 2017YFB0306605 and No. 2017YFB0103503). This work was also supported by the program for Research Center of Automobile Safety Technology of Jilin Engineering Normal University and Innovative Research Team of Jilin Engineering Normal University (IRTJLENU).

References

1. Bellur Nagarajaiah, S., Prakash, J.: Nutritional composition, acceptability, and shelf stability of carrot pomace-incorporated cookies with special reference to total and β-carotene retention. Cogent Food Agric. **1** (2015)
2. Nord, J.H., Koohang, A., Paliszkiewicz, J.: The internet of things: review and theoretical framework. Expert Syst. Appl
3. Chen, P.Y., Chen, W.T., Wu, C.H., Tseng, Y.C., Huang, C.F.: A group tour guide system by RFIDs and wireless sensor networks. In: Information Processing in Sensor Networks (IPSN), pp. 46–48 (2007)
4. Bereaved Family Members' Perceptions of the Distressing Symptoms of Terminal Patients With Cancer. Okamoto Yoshiaki The American Journal of Hospice & Palliative Care (2018)
5. Drühe, M.A.: Exhaust-gas measuring techniques. Bosch Professional Automotive Information, pp. 298–303 (2015)
6. Kreh, A., Hinner, B., Pelka, R.: Exhaust-gas Measuring Techniques, pp. 352–359. Springer Fachmedien, Wiesbaden (2014)
7. Liao, F.: Design and Implementation of Public Transportation Inquiry System Based on J2EE (2018)
8. Liu, Q., Liu, Z.C., Han, Y., Tian, J., Wang, J., Fang, J.: Experimental investigation of the loading strategy of an automotive diesel engine under transient operation conditions. Energies **11**, 1293 (2018)
9. Han, Y., Li, R., Liu, Z., Tian, J., Wang, X., Kang, J.: Feasibility analysis and performance characteristics investigation of spatial recuperative expander based on organic Rankine cycle for waste heat recovery. Energy Convers. Manage. **121**, 335–348 (2016)
10. Li, R., Liu, Z., Han, Y., Tan, M., Xu, Y., Tian, J., Chong, D., Chai, J., Liu, J., Li, Z.: Experimental and numerical investigation into the effect of fuel type and fuel/air molar concentration on autoignition temperature of n-heptane, methanol, ethanol, and butanol. Energy Fuels **31**, 2572–2584 (2017)

Exploration of TCM Culture Communication Based on the New Media

Xiangli Wang and Xueqin Du[✉]

Jiangxi University of Traditional Chinese Medicine, Nanchang, China
602477325@qq.com

Abstract. As one of the oldest cultural factors in the ancient Chinese civilization tradition, TCM is Chinese original medicine and an important part of traditional Chinese culture. Traditional science and technology is still playing an important role and is most likely to drive China's medical science and technology, leading the world. The cultural background of foreign audiences is different from that of traditional Chinese medicine culture, so they are confronted with certain obstacles in the process of understanding and accepting traditional Chinese medicine culture. Under the influence of such obstacles, it is difficult for foreign audiences to have a high degree of recognition of traditional Chinese medicine culture like domestic audiences. In the 21st century, the rapid development of new media brings new chances to the spread of TCM culture. Therefore, the paper aims at discussing narrative exploration on the external communication of TCM culture from the perspective of new media.

Keywords: TCM · New media · Culture communication

1 Introduction

Since 2003, President Jinping Xi has repeatedly stressed the importance of telling China's story and spreading China's voice to the international community. To successfully complete this task, we must first spread traditional Chinese culture. "going global" of traditional Chinese culture depends on the spread of traditional Chinese medicine culture. Therefore, the most direct and effective way to tell a good Chinese story is to do a good job in the communication of traditional Chinese medicine culture. As an engine to promote the revival of the cultural spirit of the contemporary Chinese nation, the dissemination of traditional Chinese medicine culture is an essential part of building the "soft power" of the country.

2 The Present Situation and Predicament of TCM Culture Communication Abroad

During recent years, Chinese medicine have been in the spreading boom and domestic scholars have shown great concern on the exploring external situation of Chinese medicine and spreading the culture of Chinese medicine. However, only superficial scientific and technological knowledge of TCM has been introduced to the west, and

M. Atiquzzaman et al. (Eds.): BDCPS 2019, AISC 1117, pp. 58–62, 2020.
https://doi.org/10.1007/978-981-15-2568-1_9

the theoretical and cultural connotations of TCM have not been effectively excavated, communicated and spread. The external communication of TCM culture is the revitalization of "aphasia" traditional Chinese medicine and the engine to promote the revival of Chinese cultural spirit. In the process of mutual influence, mutual learning, mutual benefit and common development between TCM culture and the world medical culture, TCM culture can effectively prevent the cultural infiltration of cultural entitlement, promote the transformation of TCM culture from nationality to the world, and gradually enhance the influence and discourse power of TCM in the world medical system [1].

Therefore, a detailed analysis of the difficulties in the external communication of traditional Chinese medicine culture can be divided into the following factors: 1. The difficulties in the international communication of traditional Chinese medicine culture lie in cultural differences; 2. The complexity of language makes international communication difficult. 3. TCM is treated with misunderstanding or even distorted with colored glasses due to ignorance. At the same time, in order to improve the international communication of traditional Chinese medicine culture, three strategic suggestions are made here: 1. Expand the multi-path of international communication; 2. Recognize the international communication forms of traditional Chinese medicine, and play the "combined method" of communication. 3. Actively building a system of trust [2].

3   Research on New Media Communication of the Culture of Traditional Chinese Medicine

The most common and main feature of culture may be communication, because without communication, culture cannot develop, survive, extend and succeed. In addition to the obvious characteristics such as "magnanimity", "timeliness", "multimedia", "hypertext" and "high-speed mobility" brought by communication means and contents, New media also has the characteristics such as high interactivity of communication behavior, non-linear communication mode, diversification of communication means, personalization of communication mode and diversity of communication content [3]. In the era of rapid development of multimedia technology, new media communication integrates many influencing factors into the communication process of traditional Chinese medicine culture, which not only reduces the burden of understanding brought by obscure traditional Chinese medicine to the audience, but also makes traditional Chinese medicine culture more appealing and attractive.

3.1   Domestic Research on New Media Communication of Culture of Traditional Chinese Medicine

The arrival of the new media era brings fundamental changes to the mode of production and dissemination of information, breaking the monopoly of traditional media on information dissemination. The diversification of new media communication channels, and the low-threshold communication make it possible for the audience to become information disseminators. On the one hand, they can accept the information spread by other media, and on the other hand, they can actively participate in the production and dissemination of information.

Jun [4] analyzed the social background of the current popular health preservation TV programs, which integrates traditional Chinese medicine culture communication in China. Taking YangSheng Tang (A health show on BTV) as an example, starting from the perspective of health communication, the Agenda Setting theory and Process Model theory were used to analyze and study them. His paper summarizes the shortcomings and difficulties in communication of traditional culture by TV programs, and puts forward counter measures and suggestions for improvement. Li clarified the connotation of TCM culture and TV programs, and pointed out their important practical significance and dissemination value in the process of cultural transmission [5]. However, as far as the current situation is concerned, domestic scholars only emphasize the communication strategies and opinions of the media, but seldom discuss the media and audience from the overall perspective.

3.2 International Research on New-Media of Culture of Traditional Chinese Medicine

In 2017, the General Office Of Communist Party of China Central Committee and The General Office Of State Council issued the *opinions on the implementation of the project* to carry forward and develop outstanding traditional Chinese culture, raising the inheritance and promotion of traditional culture to an unprecedented height. In the process of inheritance, external communication is an important and long-term systematic project. In the new media environment, international communication strategies will change from broad communication to specific communication, from political indoctrination to cultural soft power communication, and from passive disclosure to active reporting of emergencies. Specific strategies and methods for international communication through new media are proposed.

Zhang [6] from the Development Research Center of CCTV, discussed in *"transformation of international communication strategy of traditional culture + new media"* about what kind of traditional culture we should spread abroad. We should not only spread superficial culture to deep culture, but also attack and defend both sides. We should be good at exploring the international value of traditional culture, and provide Chinese views for global public issues such as peace, environmental protection, health and poverty from the perspective of traditional culture. At the same time, he emphasizes spreading traditional culture by means of new media, strengthening communication and cooperation with overseas mainstream new media, platforms and integrating marketing with external resources, publishing and pushing on relevant BBS, maintaining fans and emphasizing the content.

4 Narrative Exploration Based on New Media

The cross-cultural communication of TCM under the background of the network is more convenience, interaction, equality and diversity. Although the Internet may bring more options for the cross-cultural communication of TCM, it cannot exist in isolation from reality. In the network environment, it is also necessary to choose appropriate ways to spread TCM knowledge and culture based on the actual situation.

4.1 Daily Narrative to Spread TCM Culture

So far, although the external communication of traditional Chinese medicine culture has achieved certain achievements, Liu [7], a scholar from the Beijing University of Traditional Chinese Medicine, has studied 21 institutions of TCM in Beijing, none of which uses popular foreign new media such as twitter for external communication. Among the 21 institutions, only Beijing University of Traditional Chinese Medicine has a twitter account, but has never tweeted.

Therefore, the application of new communication paths of traditional Chinese medicine culture should be strengthened. The application of microblog and other new media can make the communication of traditional Chinese medicine culture more in line with the development trend of the times. The Chinese medicine industry should actively participate in the construction of new media. The application of new media has been widely promoted in related masters, colleges, management and other groups. We need to release TCM information according to social needs at the same time, strengthen communication with the public in a free and open platform, and make new media become an emerging path of TCM culture communication [8].

The TCM theory such as essence and qi, Yin and Yang, five elements, and the theory of "harmony between man and nature" and "correspondence between man and nature" are full of heterogeneous factors for the western world which increases the unacceptability of western readers to traditional Chinese medicine culture. Therefore, the daily narrative carries on the communication and dissemination of discourse in the way of telling, has its inherent linguistic characteristics and advantages of dissemination [9]. There is a great research space on how to overcome the linguistic particularity, cultural diversity and complexity in the cultural communication of traditional Chinese medicine and spread the cultural essence contained in the culture of TCM to the world through narrations.

4.2 Spread of TCM Culture in Image Narrative Mode

The creation of TCM cultural productions in the new media era needs to pay more attention to the participation and interaction of users. There are many Internet search engines and video broadcast platforms, such as Google, You Ku, IQIYI, Baidu video, YouTube and so on.

It is more intuitive and interesting to make and broadcast the relevant video on these platforms. Take the traditional Chinese medicine culture documentary "Ben Cao China" for example, each episode has a theme and presents the inheritance and development of TCM culture through events and stories. In addition, Chinese medicine, a series of programs launched by the Chinese international channel, invites influential traditional Chinese medicine practitioners to participate in the program, and conducts follow-up interviews on real cases or interviews on hot traditional Chinese medicine events, which have a wide influence both at home and abroad. These narrative characteristics are adapted to the new form of works in the new media era, and all of them have considerable reference significance.

5 Conclusion

Modern media has entered the era of new media. The emergence of new media makes the communication of traditional Chinese medicine culture more convenient, faster and more extensive, broadens the communication channels of traditional Chinese medicine culture, expands the scope of communication, improves the communication effect, and brings new opportunities for the dissemination of TCM theories [10]. But at the same time, new media also brings great challenges to the communication of TCM culture due to its own characteristics. Therefore, further integration of Internet new media platform resources and excellent academic resources to form a new media communication matrix of traditional Chinese medicine will help better promote the transmission, inheritance and the development of it.

Acknowledgements. This research was supported by 2017 Research Project of Humanities and Social Sciences in Universities in Jiangxi province, "English Translation of TCM Stories from a Narrative Perspective and Research on Their External Communication" (project number: YY17271);

2019 Jiangxi University of Traditional Chinese Medicine Special Field Innovation Fund Project for Graduate Student, "Narrative Exploration of TCM Culture Communication Based on New Media"(project number: JZYC19S37).

References

1. Xu, Y.: Realistic portrayal of the external communication of traditional Chinese medicine culture. J. Tradit. Chin. Med. Manage. **23**(03), 8–11 (2015)
2. Xiao, Y.: A study on the current predicament of international communication of traditional Chinese medicine culture and its communication path. Heilongjiang university of traditional Chinese medicine (2016)
3. Tang, X.: Research on new media communication of traditional Chinese medicine culture. Xiangtan university (2015)
4. Jun, X.: TV program research on TCM health preservation from the perspective of health communication. Chengdu university of technology (2013)
5. Li, W.: Communication research of TV programs of traditional Chinese medicine. Hunan university of traditional Chinese medicine (2013)
6. Zhang, T.: Transformation of international communication strategy of "traditional culture + new media". West. Radio Telev. (06), 1–2 + 6 (2018)
7. Liu, P.: Research on external communication of Chinese medicine in Beijing in the new media era. Mod. Distance Educ. Chin. Med. China **16**(23), 30–32 (2008)
8. Zhen, X., Wang, X.: Analysis and countermeasures on the transmission path of traditional Chinese medicine culture. J. Chengdu Univ. Tradit. Chin. Med. **35**(03), 94–96 (2012)
9. Fu, X.: From western narratology to Chinese narratology. Chin. Comp. Lit. (04), 1–24 (2014)
10. Wang, Z., Zeng, X.: Strategies and approaches of international communication of traditional Chinese medicine culture. Media (21), 76–78 (2018)

Design of a Scanning Frame Vertical Shaft

Xuexin Zhang[1(✉)], Xiaoyan Gao[2], and Jun Liu[1]

[1] Basic Department, Changchun Institute of Engineering and Technology,
Changchun, China
876728255@qq.com, 332552083@qq.com
[2] Information Application Department, Changchun Institute of Engineering
and Technology, Changchun, China
Gaoxy1995@sina.com

Abstract. In the scanning frame, the stability of the vertical axis is always an important index to determine the accuracy of the whole structure. Therefore, the design of vertical axis is very important in the whole scanning frame design. This paper designs a vertical shaft with high stability and precision.

Keywords: Near field scanning rack · Y-axis design · The error analysis

1 Introduction

In the design of the overall structure of the scanning frame, the X-axis, as the base of the whole scanning frame bears the largest load, but because it is directly in contact with the ground and the center of gravity is lower, the overall stability is very high, and relatively little interference is caused by external vibration [1–4]. In the inverted "T" structure, the Y-axis is installed perpendicular to the ground on the X-axis, with a high center of gravity, and is affected by the interference of the X-axis, so the overall stability of the Y-axis structure is much worse than that of the X-axis [5–8]. Therefore, the structure design of the Y-axis directly affects the positioning accuracy of the probe and the planarity of the scanning plane, and takes a large proportion in the error [9, 10].

2 Y Axis Design and Selection

In this project, the Y-axis adopts the structure of parallel welding of two rectangular steel tubes, and the reinforcing bars are welded between the steel tubes to ensure the parallelism and planarity of the two steel tubes as far as possible. The two ends of the steel tubes are connected with steel plates respectively. The welded rectangular steel tube is vertically installed on the skateboard of the X-axis, and the oblique support beam is installed on the back side to provide the Y-axis oblique support and enhance stability. Meanwhile, the verticality and orthogonality of the Y-axis relative to the X-axis can be fine-tuned by adjusting the tightness of the fixing bolt. Two high-precision guide rails are arranged on two rectangular square tubes on the Y-axis, and the two guide rails are parallel. The slide plate is fixed on the slide block of the guide rail for loading the z-axis and scanning probe. In order to ensure the planarity and parallelism

© Springer Nature Singapore Pte Ltd. 2020
M. Atiquzzaman et al. (Eds.): BDCPS 2019, AISC 1117, pp. 63–67, 2020.
https://doi.org/10.1007/978-981-15-2568-1_10

of the two guide rails installed, the Y-axis guide rail mounting surface is processed with high precision through a large gantry milling machine. At the same time, in order to adjust the parallelism of the guide rail in the Z direction, an adjustment block is installed on both sides of the Y axis guide rail, which can fine-tune the parallelism of the Y axis Z phase and play the role of fixing the guide rail. A Y-axis servo motor and a screw nut pair connected to the Y-axis servo motor are installed between the two rectangular tubes to control the motion of the Y-axis slider. The schematic diagram is shown below.

In order to ensure the stability of Y-axis and the scanning stroke of X-axis and Y-axis, the following design restrictions are given:

(1) The size and length of the X-axis slide plate shall not be greater than 700 mm.
(2) The center distance between two guide rails of Y-axis shall not be less than 360 mm.
(3) The effective stroke of Y-axis shall be more than 2000 mm, so the length of rectangular square pipe shall be at least 2500 mm.
(4) The Y axis guide rail is 28 mm wide, and the total width of adjusting baffle blocks on both sides is 30 mm. The operating margin should be removed, and the width of rectangular square pipe should be at least 70 mm.

On this basis, in line with the principle of improving accuracy and reducing total mass, the specification and thickness of Y-axis rectangular square pipe are modeled and analyzed, and the optimal design scheme is given.

Table 1 shows the standard rectangular steel pipe dimensions that meet the design requirements.

Table 1. Dimensions of rectangular square pipes

Number	Size (mm)	Thick 1 (mm)	Thick 2 (mm)	Thick 3 (mm)	Thick 4 (mm)
1	80 × 120	4	5	6	8
2	80 × 140	4	5	6	8
3	80 × 160	4	4.5		
4	100 × 100	4	5	6	8
5	100 × 150	4	5	6	8
6	100 × 180	4.5	4		
7	100 × 200		5	6	8

As can be seen from the figure, items 3 and 6 are not commonly used standard parts, and the thickness is too thin, so they are not suitable for Y-axis square pipe material. Among the remaining dimensions, 80 × 120 (mm), 100 × 100 (mm) and 100 × 200 (mm) were selected as the Y-axis square pipe dimensions respectively. Meanwhile, the pipe wall thickness was taken as 5 mm and 8 mm respectively for modeling, and 6 Y-axis CAD models could be obtained.

Six models were meshed with the same parameters. Materials are selected system default structural steel. Fixed constraints are applied to the base surface. Apply 100 N tension outward to two rectangular tubes on the Y-axis respectively. Under the premise of ensuring the consistency of internal parameters and external environment, force analysis was carried out on the six models, and the conclusions were shown in Table 2 (Figs. 1, 2, 3, 4, 5 and 6).

Table 2. Analysis data

Number	Size (mm)	Thick (mm)	Quality of Y (kg)	Meshing the number of nodes	Meshing the number of cells	Maximum formal variable (mm)	Maximum equivalent elastic strain (mm)
1	80 × 120	5	226.68	12510	5449	0.02469	2.299e−6
2	80 × 120	8	270.11	12115	5121	0.01951	1.772e−6
3	100 × 100	5	232.29	11802	5016	0.02425	2.0104e−6
4	100 × 100	8	275.72	11796	5014	0.01914	1.5507e−6
5	100 × 200	5	270.33	14596	6296	0.01823	1.3324e−6
6	100 × 200	8	338.73	13893	5944	0.01446	1.1514e−6

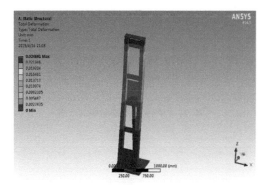

Fig. 1. Force analysis of no. 1

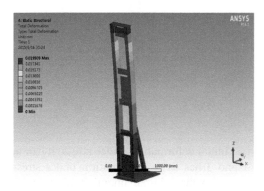

Fig. 2. Force analysis of no. 2

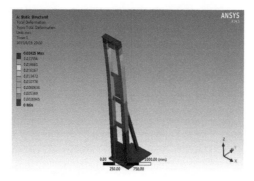

Fig. 3. Force analysis of no. 3

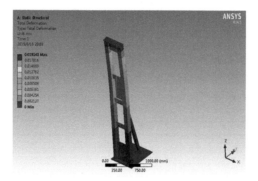

Fig. 4. Force analysis of no. 4

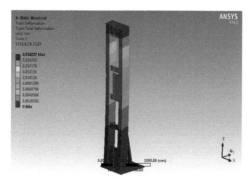

Fig. 5. Force analysis of no. 5

By analyzing the maximum deformation under stress, it can be seen that the model with serial Numbers 1 and 3 has large deformation under stress and poor structural stiffness, which is contrary to the original intention of improving the accuracy and stability of the structure as much as possible. For the rest of the 2, 4, 5, and 6 model,

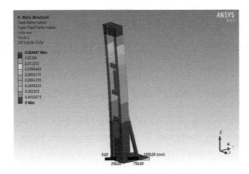

Fig. 6. Force analysis of no. 6

6 forced deformation model is much smaller than the other three groups, but the model weight much more than the other three groups, the quality of more than about 68 kg, increasing the quality of the whole scanning plane, the pressure of the X axis, increased the force deformation of the X axis, and therefore, consideration, based on the weight. For the rest of the 2, 4, 5 in the model, model quality is 270 kg, the biggest stress variables and the maximum equivalent elastic should be 5 model is slightly better than the other models, at the same time the quality of the 5 model in an acceptable range, based on the above reasons, finally Y rectangular square tube size is 100 * 200 mm, thickness is 5 mm steel as the material.

References

1. Li, K., Hu, Y., Yuan, F.: Simulator flatness measurement and compensation of spatial target based on weighted least squares. Opt. Electron. Eng. (04) (2015)
2. Zhang, F.: Error analysis and compensation ultralow sidelobe antenna planar near field measurement. Xi'an Electronic Science and Technology (doctoral thesis) (12) (1999)
3. Wang, X.: Using a laser tracker measuring straightness long way. Appl. Opt. (2013)
4. Liu, C., Li, H.: Straightness space is based on the principle of coordinate conversion error evaluation. Mod. Manufact. Eng. (3) (2013)
5. Wang, S.: Positioning accuracy analysis of ball screw feed system. Dalian university of technology (master thesis) (12) (2006)
6. Ding, H.: Research on positioning error dynamic measurement and compensation system of x-y table. Shandong university (master thesis) (05) (2008)
7. Jiao, H., Chen, W., Wang, C.: Analysis and discussion on the error analysis of scanning frame for large scale compression field test. J. Beijing Technol. Bus. Univ. (11) (2006)
8. Shi Man, B., Kong, X.: Discussion on the design of foundation drawing of machine tool to improve the accuracy of machine tool. Stand. Qual. Mach. Ind. (11) (2011)
9. Sun, X., Yan, K., Ding, G.: Analysis on positioning accuracy of machine tool by comprehensive error of ball screw. Mach. Manufact. Technol. (11) (2008)
10. Lei, T.: Automatically detects the thickness of the non-contact development system. University of Electronic Science and Technology (Master's Thesis) (06) (2013)

Development Status and Challenges of Unmanned Vehicle Driving Technology

Zhenyu Wang[1] and Mengqiao Wang[2(✉)]

[1] Zhengzhou Technical College, Zhengzhou 450000, Henan, China
[2] Beijing University of Posts and Telecommunications, Beijing 100000, China
wangmengqiao@bupt.edu.cn

Abstract. Nowadays, artificial intelligence instead of the human race driving is no longer just a dream, but gradually becomes a reality. Internet of Things and smart technologies are gradually being used in vehicles, so cars can have their own ideas. Driverless cars are one of the typical representatives. Driverless cars, that is, unmanned cars on the road, mainly through the sensor system installed on the vehicle, intelligent software and various environmental sensing devices, to sense the state and environment of the vehicle itself and its surrounding environment. At the same time, according to the software set the motion track to reach the destination. Compared to traditional cars, unmanned vehicles have the characteristics of high safety, high reliability and intelligence, which are the main development directions of future cars. This paper will introduce the unmanned car driving technology from three aspects: application value, domestic and international development status and main challenges.

Keywords: Self-driving car · Autopilot technology · Self-learning · Internet of things interaction · High precision GPS

1 Introduction

Driverless cars should have the ability to learn and explore themselves to cope with complex road conditions. Future cars will become wheeled robots with learning capabilities. The division of labor between people and such wheeled robots can be divided into four states. First, human driving, wheeled robot recording and learning. The second is wheeled robot driving, artificially supervised. The third is that wheeled robots learn to drive independently, and humans do not intervene. The fourth is that wheeled robots teach people to drive [1]. In this way, future cars can not only learn human driving skills, but also independently explore and master driving skills in complex environments. With the continuous improvement of a large number of driving data and intelligent technology, the core technology of unmanned driving will continue to improve, and the technology will gradually mature. Judging from the automobile automation level department of the National Highway Traffic Safety Administration, the current automation level of autonomous vehicles is mainly concentrated in the second and third levels, that is, the degree of automatic driving is limited (Fig. 1).

© Springer Nature Singapore Pte Ltd. 2020
M. Atiquzzaman et al. (Eds.): BDCPS 2019, AISC 1117, pp. 68–74, 2020.
https://doi.org/10.1007/978-981-15-2568-1_11

Fig. 1. RinspeedXchangE concept car

2 Application Value of Driverless Technology

2.1 Improve Driving Safety

Since the driverless car can achieve complete autonomous control, this can make humans get rid of the heavy and fatigued driving behavior, thus greatly improving driving safety. According to the World Health Organization, about 1.25 million people die every year from various traffic accidents, and more than 50 million people suffer different degrees of damage, directly or indirectly causing hundreds of billions of dollars in economic losses [2]. Compared with human drivers, the response speed, sensing range, operational error probability and traffic rules of the driverless car are better, which can greatly reduce the collision probability of the car and the surrounding environment, thus greatly improving driving safety.

2.2 Improve Travel Efficiency

The driverless car can receive traffic information of the entire road network in real time, and is equipped with an accurate self-positioning device and an intelligent network connection system between the vehicle and the vehicle. Coordinated selection of paths and control of many vehicles effectively solves the problem of vehicle congestion.

2.3 Engage in Dangerous Work

In some cases, driverless cars can be used to replace dangerous work for humans. For example, the German "scavenger" unmanned mine-sweeping vehicle can replace humans to investigate explosives. His mine-clearing efficiency is 450–600 times that of an experienced mine-clearing engineer [3]. For example, the cooling of the reactor at the Fukushima nuclear power plant in Japan and the cleaning of the nuclear leak were independently completed by unmanned fire engines.

3   Development Status of Driverless Technology

3.1   Development Status of Foreign Driverless Cars

As the first company to develop driverless technology, Google has developed a fully automated vehicle that automatically starts and stops. Google's artificial intelligence foundation lays a solid foundation for the development of autonomous driving technology and has established its own laboratory dedicated to the research of driverless cars. Google is primarily engaged in the research and development of core technologies for autonomous driving. It does not perform the overall manufacture of the car. Unmanned prototypes have been modified on other existing models. Google is far ahead in autopilot and has completed tens of thousands of kilometers of road testing. Its Waymo company plans to enter the driverless taxi industry. In addition, Tesla and Uber have also made great achievements in the field of driverless driving. Tesla developed its own autopilot system, Autopilot, and applied it to the Modle S model, which enables autopilot functionality. Compared to Google, Tes Pull is better at commercializing autonomous driving technology. In Europe, Mercedes and BMW are also exploring unmanned prototypes and plans to develop mature commercial vehicles in the future. In addition, Ford and the Massachusetts Institute of Technology jointly initiated research on motion control in automated driving systems and attempted to use drones as part of the sensor system. Lexus, Volvo, Audi, General Motors, etc. are also conducting research on related technologies, and many foreign companies are actively participating in this thriving new market [4] (Fig. 2).

Fig. 2. Honda and Google reach cooperation in the field of driverless driving

3.2   Development Status of Domestic Driverless Cars

Compared with developed countries such as Europe, the United States and Japan, China's research on driverless car technology started late, but it has made great progress. China's first driverless car was successfully developed by the National Defense University of Science and Technology in 1992 [5]. After years of development, Internet company Baidu continued its efforts to achieve fully automated driving-related testing in 2015 under mixed road conditions in cities, loops and highways. In 2018, the L4 unmanned pure electric bus "Apollon" was declared the world's first mass-produced driverless car.

At present, domestic Baidu, Changan and other enterprises and military science and technology universities, military transportation colleges and other military institutions are at the forefront of domestic research and development. For example, Changan Automobile has created a long-distance driving record for domestic driverless cars from Chongqing to Beijing. Baidu Motors experimented on the road of the first unmanned car in Beijing and achieved success. It is estimated that by 2020, unmanned vehicles will go to Chongli in Beijing for road testing (Table 1).

Table 1. Vehicle manufacturing enterprise competitiveness ranking [6]

Top	Group	Territorial
Top1	Alphabet-WayMo	Silicon Valley
Top2	Baidu	Beijing China
Top3	Tesla	Silicon Valley
Top4	General Motors	Detroit, United States
Top5	Uber	Silicon Valley/Pittsburgh

4 The Main Challenges Facing Driverless Technology

4.1 Technical Issues

At present, the exploration of domestic unmanned technology is still based on technology emerging abroad. Real-time, high-precision GPS information is the key to helping unmanned vehicles complete body positioning, path planning, and autonomous driving [7]. The development of high-precision GPS technology in China is immature and cannot be used for civilian vehicles. However, the accuracy of ordinary civilian GPS is flawed and cannot be accurately located. In addition, the ability of laser radar to scan and discriminate does not meet the required requirements. For example, in the case of inclement weather and low visibility, the ability of radar to discriminate on road conditions will be reduced, resulting in safety hazards.

4.2 Accurately Judge the Behavior of the Person

Unmanned vehicles can largely solve the dangers caused by factors such as fatigue driving. At present, unmanned vehicles can guarantee the accuracy of identification of passing vehicles, traffic lights and traffic signs. However, under city traffic conditions, the road environment and human behavior have become very complex and unpredictable. This requires artificial intelligence to accurately identify a wide variety of road information. When there is a person's behavior, such as when the traffic police signal the vehicle to pass, especially when these gesture signals collide with the traffic light or the stop sign, it will challenge the self-judgment of the driverless car. Once the artificial intelligence makes a mistake, there is a danger of an accident. Therefore, how to strengthen the identification of road information by artificial intelligence will become a difficult point for unmanned driving in the future (Fig. 3).

Fig. 3. City road conditions are complicated

4.3   Reduce Costs

At present, the cost of a driverless car is generally more than one million yuan. In addition to the manufacturing cost of the car itself, the driverless car consists of three main components: laser radar, high-precision GPS, high-precision inertial navigation system. Among them, the price of each laser radar is about 100,000 dollars. The cost of the latter two adds up to $150,000. As a result, the cost of a truly fully automated vehicle is as high as $250,000, which is 10 to 20 times the cost of a typical car [8].

4.4   Policies and Regulations

The openness and integrity of the legal and policy environment determine the direction, speed and room for growth of a new industry. Driverless cars want to truly achieve the landing, in addition to technical barriers, national laws and government regulation is also a barrier to break through. Only after the legal level has determined the status of unmanned driving and the corresponding legal responsibilities, its subsequent market is possible. If you want to travel in a driverless car, you must first resolve the policy and regulatory issues. Because in the process of actually driving a driverless car, an accident will inevitably occur. At this time, there will be problems such as how to divide the responsibility of the accident and how to make a fair ruling. So far, no country in the world has enacted comprehensive laws and regulations for the travel of driverless cars [9]. Therefore, we still have a long way to go before the full-scale launch of driverless cars.

4.5   Information Security

Due to the high level of intelligent network connectivity of driverless cars, this poses a huge challenge in terms of information security. To this end, it is necessary to accelerate breakthroughs in core technology areas, such as driverless vehicle-mounted operating systems, 5G networks and vehicle communication, control system security "write protection" mechanism and data information fault-tolerant protection mechanism, thereby enhancing the encounter of unmanned vehicles. The main information. Emergency handling during security incidents to prevent hackers from posing a security threat to driving (Fig. 4).

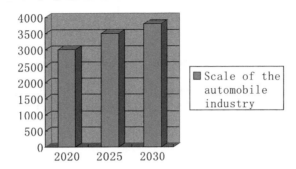

Fig. 4. Future national phase plan for smart cars (10,000 units) [10]

5   Conclusion

The car of the future is not only a means of transportation, but also the development of a new generation of data sets. The key technologies of driverless vehicles are important for their future development. Although the key technologies of driverless driving have made major breakthroughs, there are still many defects in related technologies, and there is still much room for improvement. China's related research work started late, but it has developed rapidly. It requires the government and relevant researchers to work harder to achieve a true integration of driverless cars with our daily lives.

Unmanned vehicles currently face network security, high costs, consumer trust, reliability, and legal and regulatory challenges. With the continuous advancement of computer science, the future development of unmanned vehicles is unprecedented. The continuous development of technology will certainly break the bottleneck of industrialization.

References

1. Anania, E.C., Rice, S., Walters, N.W., Pierce, M., Winter, S.R., Milner, M.N.: The effects of positive and negative information on consumers' willingness to ride in a driverless vehicle. Transp. Policy **72**, 218–224 (2018)

2. Hofacker, A.: Let's go driverless challenges of the first season. ATZextra Worldwide **22**(2), 22–27 (2017)
3. Winter, S.R., Keebler, J.R., Rice, S., Mehta, R., Baugh, B.S.: Patient perceptions on the use of driverless ambulances: an affective perspective. Transp. Res. Part F: Traffic Psychol. Behav. **58**, 431–441 (2018)
4. Gehrie, E.A.: The impact of driverless cars on the US blood supply. Transfus. Apheresis Sci. **56**(2), 233 (2017)
5. Reynolds, M.: Pedestrians signal to stop driverless cars. New Sci. **3113**, 16 (2017)
6. Rice, S., Winter, S.R.: Do gender and age affect willingness to ride in driverless vehicles: if so, then why? Technol. Soc. **58**, 101145 (2019)
7. Nagalingam, S.R.: GPUs revolutionising the driverless automotive ecosystem. Auto Tech. Rev. **6**(1), 14–15 (2017)
8. Bae, C., Kim, H., Son, Y., Lee, H., Han, S., Suh, M.: Development of a web-based RCM system for the driverless Rubber-Tired K-AGT system. J. Mech. Sci. Technol. **23**(4), 1142–1156 (2009)
9. Hofacker, A.: Let's go driverless Herausforderungen der ersten Saison. ATZextra **22**(7), 24–29 (2017)
10. Curl, A., Fitt, H.: Will driverless cars be good for us? now is the time for public health to act together with urban and transport planning. J. Glob. Health (2019)

Analysis of Bayesian LASSO Using High Dimensional Data

Xuan Huang[1(✉)] and Yinsong Ye[2]

[1] Chengdu College of University of Electronic Science and Technology
of China, Chengdu, China
christyhuang8429@163.com
[2] Chongqing University of Posts and Telecommunications, Chongqing, China

Abstract. The sparse models play an important role in the field of machine learning. It can dimensionality reduction and effectively solve the over-fitting problem in modeling. The Bayesian method forms a priori distribution by fusing different information to obtain high-quality statistical inference. This paper summarizes the representative sparse model LASSO and Bayesian theory-based model Bayesian LASSO, and discussed the relationship between the two models. Through numeric experiments, the effects of the two models on variable selection were compared, and the parameter estimation of Bayesian LASSO under different prior conditions is further analyzed. Results attained showed that the two models have good effects in the variable selection. When the number of samples is small, especially when the number of samples is much smaller than the number of features, the effect of Bayesian LASSO is more prominent. It is also possible to estimate the model parameters and calculate the Bayesian confidence interval for each regression coefficient at a certain level of confidence, which is more flexible and convenient.

Keywords: Sparsity · Feature selection · LASSO · Bayesian LASSO · Parameter estimation

1 Introduction

In recent years, the emergence of high-dimensional data has hindered the development of machine learning. In high-dimensional data, only a small part of variables are related to the output to be predicted, and the rest is noise. If the high-dimensional data is modeled directly, many variables in the model obscure the inherent causal link of the data itself, making the model difficult to interpret. In addition, high-dimensional models will inevitably increase computational complexity. Therefore, dimensionality reduction is a key step in data preprocessing and can extract core information of data. Modeling with key features simplifies the model while improving computational efficiency. It has become the mainstream research direction.

The sparse model can extract important low-dimensional information from high dimensional data and automatically perform variable selection. It not only effectively solves many problems of high-dimensional data space, such as: the interpretability, complexity and computationality of the model, but also solves the problems of data

M. Atiquzzaman et al. (Eds.): BDCPS 2019, AISC 1117, pp. 75–82, 2020.
https://doi.org/10.1007/978-981-15-2568-1_12

storage and visualization. Especially in the fields of Machine Learning, Signal Processing and Image Processing, the data is generally sparse or nearly sparse, so the sparse model has been widely used [3, 9, 11].

Tibshirani found that the ℓ_2 norm penalty in the ridge regression model was replaced by the ℓ_1 norm, it has the ability to produce sparse solution, which is called Least Absolute Shrinkage and Selection Operator model (LASSO) [10]. The most attractive aspect of LASSO is that it can fit the model parameters while selecting variables, which selects the model by estimating the predicted parameters. At the same time, Tibshirani mentioned that LASSO based on linear regression model can be represented as a maximum a posteriori estimate, where the regularization term of LASSO corresponds to the prior and the loss function corresponds to the likelihood function [10]. Based on this theory, Park et al. proposed The Bayesian LASSO [7, 8], which uses the principle of different prior distributions to perform the maximum posterior estimation of the regularized sparse model. The Bayesian LASSO can get a more sparse solution than the ordinary LAS. SO, and it can maintain better performance when the correlation between variables is strong.

A brief overview of this paper is as follows: Sect. 2 briefly describes some related works about sparse model and its extension. We conducted extensive experiments to compare and evaluate the effects of sparse model on feature selection in Sect. 3, followed by the conclusions and future works in Sect. 4.

2 Related Work

In this section, we will briefly review some approaches relate to sparse model. First, let us introduce several notations. In this paper, matrices are written as capital letters and vectors are denoted as lowercase. Support $X \in R^{m \times n}$ represents observation data, which m, n represent the number of observed samples and the number of features, respectively. The linear regression model is expressed as $y = X\beta + \varepsilon$, where $\beta \in R^n$ denotes a vector of regression coefficient, $\varepsilon \in R^m$ denotes an error vector and $\varepsilon_i (i = 1 \ldots m)$ obeys independent and homogeneously distributed, $y \in R^m$ is the vector of responses. The ℓ_1 norm is denote by $\|\beta\|_1 = \sum_{i=1}^{n} |\beta_i|$.

2.1 LASSO

Tibshirani replaces the ℓ_2-norm in the ridge regression model with the ℓ_1 norm penalty, and builds the LASSO model based on the linear regression, the model is define as:

$$f(\beta) = \arg\min_{\beta \in R^m} \frac{1}{2} \|y - X\beta\|_2^2 + \lambda \|\beta\|_1 \tag{1}$$

which minimizes the sum of squared error with the ℓ_1-norm penalty on the weight vector. In the above formula, $\lambda \geq 0$ is the penalty term, it is an adjustable parameter. The greater the lambda value, the greater the penalty, and more coefficients are shrink to 0, and the more sparse the model is, or vice versa.

The ℓ_1 norm penalty of the model is the sum of the absolute values of the regression coefficients, so that the regression coefficient with smaller absolute value is automatically set to 0, which can generate sparse solutions. Compared with the traditional feature selection method, LASSO can realize variable selection and parameter shrinkage simultaneously. In addition, the loss function and penalty function of LASSO are convex functions, so the model has an optimal solution. However, the disadvantage is that both the target variable and the noise variable are penalized to the same extent, resulting in a biased estimate of the regression coefficient of the target variable. LASSO approximates the unbiased values only under the additional of irrepresentable condition, sparse Riesz condition and restricted eigenvalue condition, and there is consistency of parameter estimation and consistency of variable selection [5]. In view of the biased estimation problem of LASSO, some scholars have proposed an approximately unbiased sparse model, such as: the Adaptive LASSO [3], the Relaxed LASSO, smoothly clipped absolute deviation penalty model, (SCAD) and so on. These algorithms all reduce the degree of compression of the regression coefficients of the target variable by adjusting the degree of penalty of the vector components of the regression coefficients.

In addition, LASSO cannot perform group variable selection when performing feature selection. That is, some highly related variables cannot be selected as a whole at the same time, or removed from the model at the same time. Some researchers have proposed to solve this problem by adding ridge penalty. Models such as the elastic net, the elastic SCAD [6], and the pairwise elastic net [1] fall into this category. Some scholars have proposed the fused lasso, applying the ℓ_1-norm penalty to the difference of the regression coefficients to solve the problem, or using pairwise ℓ_∞ norm penalty to achieve the automatic group effect [4]. It is worth noting that Group LASSO [12], Group SCAD, Sparse Group LASSO (SGL), etc. All these need to be grouped manually in advance to achieve grouping of arbitrary variables.

2.2 Bayesian LASSO

Tibshirani proposed an interpretation of Bayesian statistical theory for LASSO based on linear regression model in [10], the LASSO estimate can be viewed as the mode of the posterior distribution of the parameter vector. When he sets the prior distribution of the model variables to the Laplace prior distribution, the estimate of the posterior distribution mode is consistent with the LASSO estimation.

The essence of Bayesian LASSO is the maximum posterior distribution estimate of the parameter vector [7]. In the linear regression model $y = X\beta + \varepsilon$, in order to solve β, it can be assumed that β is a random vector of independent and identical Laplace distribution. So the prior is expressed as:

$$P(\beta|\sigma^2) = \prod_{j=1}^{n} \frac{\lambda}{2\sigma} e^{-\lambda|\beta_j|/\sigma} \tag{2}$$

which corresponds to the regularization term of the LASSO. The likelihood function:

$$P(y|X, \beta) = \prod_{i=1}^{m} N(y_i|x_i^T \beta, \sigma^2) \tag{3}$$

corresponds to the loss function of the LASSO model. The posterior probability distribution of β can be obtained by

$$P(\beta|y) = P(y|X, \beta)P(\beta|\sigma^2) \tag{4}$$

With a given confidence level, a Bayesian confidence interval for β can be calculated to guide variable selection [9]. Laplace distribution and Gaussian distribution are non-conjugated, which increases the computational complexity [13]. Chandran represents the Laplace distribution as a product integral of the normal distribution and the Gamma distribution [2]:

$$P(\beta|\sigma^2) = \prod_{j=1}^{n} \int_0^{\infty} \frac{\lambda}{\sqrt{2\pi\sigma^2\tau_j^2}} e^{-\lambda|\beta_j|^2/2\sigma^2\tau_j^2} \frac{\lambda^2}{2} e^{-\frac{\lambda^2}{2}\tau_j^2} d\tau_j^2 \tag{5}$$

$$= \prod_{j=1}^{n} \int_0^{\infty} N(\beta_j|0, \sigma^2\tau_j^2) Gamma(\tau_j^2|1, \frac{\lambda^2}{2}) d\tau_j^2 \tag{6}$$

The posterior estimate of Bayesian LASSO is expressed as:

$$y|X, \beta, \sigma^2 \sim N_m(X\beta, \sigma^2 I_m) \tag{7}$$

$$\beta|\sigma^2, \tau_1^2, \ldots, \tau_n^2 \sim N_n(0_n, \sigma^2 D_\tau) \tag{8}$$

$$D_\tau = diag(\tau_1^2, \ldots, \tau_n^2) \tag{9}$$

$$\sigma^2, \tau_1^2, \ldots, \tau_p^2 \sim \frac{1}{\sigma^2} d\sigma^2 \prod_{j=1}^{n} \frac{\lambda^2}{2} e^{-\lambda^2\tau_j^2/2} d\tau_j^2, \text{ where } \sigma^2, \tau_1^2, \ldots, \tau_n^2 > 0 \tag{10}$$

$$\pi(\sigma^2) = \frac{\gamma^a}{\Gamma(a)} (\sigma^2)^{-a-1} e^{-\gamma/\sigma^2}, (a > 0, \gamma > 0) \tag{11}$$

Through the above model, the posterior of beta is:

$$\beta|y, X, \tau_j^2 \sim N_n((X^T X + D_\tau^{-1})^{-1} X^T y, \sigma^2(X^T X + D_\tau^{-1})) \tag{12}$$

The posterior of σ^2 is:

$$\sigma^2 \sim invgamma((m-1)/2 + n/2 + a, (y - X\beta)^T(y - X\beta)/2 + \beta^T D_\tau^{-1}\beta/2 + \gamma) \tag{13}$$

2.3   Gibbs Sampling

Bayesian LASSO uses the posterior probability density function for statistical inference, but for high-dimensional data, the calculation is very tricky. The Markov chain Monte Carlo (MCMC) simulate direct draws the model parameters from the complex distribution, as more and more sample values are generated, the distribution of values is closer to the desired distribution. It is applicable to various models. Therefore, the sampling algorithms based on MCMC can be used instead of the integral calculation of the parameter posterior density estimation in the model.

The Metropolis-Hastings and the Gibbs sampling arc both sampling methods based on MCMC technology. The Metropolis-Hastings algorithm is not efficient enough when the observed variables arc large and the dimensions are high. The Gibbs sampling algorithm [3, 8] is a special case of the Metropolis-Hasting algorithm. It use an iterative method that samples only one variable at a time, which can effectively solve the sampling problem in high-dimensional space.

3   Methodology

In this section, the simulated and the real data sets were selected for several experiments for verification respectively, and their results are reported and discussed.

First, randomly generate three simulated data sets, as shown in Table 1, Where M and N represent the number of samples and the number of variables in the data set, respectively. Hcrc, the data sets are fitted with the LASSO and Bayesian LASSO (BLASSO) models respectively, and the variables with a coefficient greater than 0 are selected. Table 1 compares the results of feature selection bctween the two models. Then, we randomly selected 10 variables from each data set as related variables with set the correlation coefficient to 0.9 [5, 12]. In this paper, 50 experiments were performed on each data set, and the average feature numbers corresponding to the two sparse models were compared under the condition that the variables were correlated.

Table 1. Comparison of variable selection effect in simulated dataset

Model	Sample number M	Feature number N	Uncorrelated		Correlation variable selection
			Selected variables	Noise variables	
LASSO	20	10	5.40	1.88	4.16
BLASSO			5.40	1.86	4.22
LASSO	20	30	7.24	1.66	4.52
BLASSO			7.26	1.62	4.56
LASSO	20	100	12.40	1.18	5.08
BLASSO			11.50	1.08	5.06

As observed in Table 1, both LASSO and BLASSO can effectively select useful variables to get better regression effects, BLASSO does not produce the desired results on the simulation dataset, and its calculations are more complicated. But as the sample features increase, especially when the number of features is much larger than the number of samples, BLASSO brings further advantages. In addition, after adding the strong correlation constraint to the features s between the simulated data, the effect of the feature selection of the two models is significantly reduced, and none of them had group sparsity.

In another experiment, the data selected the Wine Quality Data Set from the University of California Irvine (UCI) Machine Learning Repository, which has 1599 observation samples, each containing 12 red wine characteristics such as fixed acidity, volatile acidity, chlorides, total sulfur dioxide, PH, density, quality, etc. And score of quality is between 0 and 10. We attempt to find out the key characteristics of the red wine to explain its attributes most directly related to the quality of red wine.

We generalize Bayesian LASSO to the generalized linear model via penalized maximum likelihood, and then use the red wine data to test the effect of Bayesian lasso on the feature selection under different a priori conditions. In the experiment, we used the packages of the sparse model from the Comprehensive R Archive Network (CRAN) to solve the sparse model, and set total number of MCMC samples to be collected to 1000, square of the initial lasso penalty parameter to 1, and specify the variance of the response variable as initial variance of the sample. The gamma distribution prior for the lamda2 parameter is set as G (1, 1.78) [13].

Table 2 shows the results of parameter estimation under Laplace a priori (S), horseshoe prior (S1) and normal Gamma prior (S2). The results show that the parameters of fixed acidity, residual sugar, free sulfur dioxide and total sulfur dioxide are almost shrunk to 0, which are almost independent of the quality of red wine. Bayesian LASSO also produces interval estimates for all of the parameters, such as estimated 95% credible interval for lambda is approximately (.092, .257). These may help to choose lambda to make the results more stable.

Table 2. Comparison of parameter estimation of BLASSO for the different priors

Variable	S	S1	S2
Fixed acidity	0.002	0.003	0.002
Volatile acidity	−1.036	−1.051	−1.038
Citric acidity	−0.032	−0.046	−0.035
Residual sugar	0.002	0.002	0.001
Chlorides	−1.918	−1.854	−1.890
Free sulfur dioxide	0.002	0.002	0.002
Total sulfur dioxide	−0.002	−2.748e − 03	−0.002
Density	−1.086	−0.981	−1.621
PH	−0.428	−0.417	−0.440
Sulphates	0.872	0.866	0.863
Alcohol	0.288	0.290	0.288
Intercept	5.345	5.194	5.923
Median value of variance	0.419	0.421	0.421
Median value of Lambda2	0.024	0.029	0.023
The number of non-zero entries of beta	7	8	8

4 Conclusions

The emergence of high-dimensional data creates obstacles for the development of machine learning. The sparse model can extract valuable low-dimensional information from high-dimensional data, and automatically select features while fitting model parameters. Compared with the traditional feature selection method, it has better interference immunity effect and can generate sparseness solution, effectively solve the "Curse of Dimensionality", so it has attracted the attention of some scholars.

This paper first compares the feature selection effects of the sparse models on the simulated data. We generalize Bayesian LASSO to the generalized linear model, and then inference for model parameters by Gibbs sampling from the Bayesian posterior distribution. The experimental results show that both LASSO and Bayesian LASSO can effectively reduce the data dimension. However, when the number of features is large, especially the number of features is much larger than the number of samples, Bayesian LASSO brings further advantages. And it can automatically generate interval estimates for all parameters.

Future works will focus on following three parts: first, consider more different prior information, and study the consistency of feature selection for sparse models; second, the influence of parameter lambda on the complexity and feature selection effect of the algorithm is further studied. Third, the selected features are used for classification, and the effects of the subset of features are applied to different classifiers.

References

1. Lorbert, A., Ramadge, P.J.: The pairwise elastic net support vector machine for automatic fMRI feature selection. In: International Conference on Acoustics, Speech, and Signal Processing, pp. 1036–1040 (2013)
2. Chandran, M.: Analysis of Bayesian group-lasso in regression models. University of Florida, Florida, USA (2011)
3. Tung, D.T., Tran, M.N., Cuong, T.M.: Bayesian adaptive lasso with variational Bayes for variable selection in high-dimensional generalized linear mixed models. Commun. Stat. Simul. Comput. **48**(2), 530–543 (2019)
4. Bondell, H.D., Reich, B.J.: Simultaneous regression shrinkage, variable selection, and supervised clustering of predictors with OSCAR. Biometrics **64**(1), 115–123 (2008)
5. Liu, J., Cui, L., Liu, Z., et al.: Survey on the regularized sparse models. Chin. J. Comput. **38**(7), 1307–1325 (2015)
6. Zeng, L., Xie, J.: Group variable selection via SCAD-L2. Statistics **48**(1), 49–66 (2014)
7. Park, T., Casella, G.: The Bayesian lasso. J. Am. Stat. Assoc. **103**(482), 681–686 (2008)
8. Alhamzawi, R., Taha Mohammad Ali, A.: A new Gibbs sampler for Bayesian lasso. Commun. Stat. Simul. Comput. 1–17 (2018)
9. Shang, H., Feng, M., Zhang, B., et al.: Variable selection and outlier detection based on Bayesian lasso method. Appl. Res. Comput. **32**(12), 3586–3589 (2015)
10. Tibshiranit, R.: Regression shrinkage and selection via the lasso. J. Roy. Stat. Soc. Ser. B **58**(1), 267–288 (1996)

11. Lu, W., Yu, Z., Gu, Z., et al.: Variable selection using the Lasso-Cox model with Bayesian regularization. In: Conference on Industrial Electronics and Applications, Wuhan, China, pp. 924–927 (2018)
12. Xu, X., Ghosh, M.: Bayesian variable selection and estimation for group lasso. Bayesian Anal. **10**(4), 909–936 (2015)
13. Botev, Z., Chen, Y., L'Ecuyer, P., et al.: Exact posterior simulation from the linear LASSO regression. In: Winter Simulation Conference, Gothenburg, Sweden, pp. 1706–1717 (2018)

Review of Neural Network Models for Air Quality Prediction

Kai Zhou[1(✉)] and Ruichao Xie[2]

[1] Army Academy of Artillery and Air Defense, Hefei, China
1763331669@qq.com
[2] Hefei 230031, Anhui, China

Abstract. With the diversified development of air quality data acquisition and processing technology, people began to gradually adopt the new technology represented by machine learning to predict the air quality data to make up for the shortcomings of traditional forecasting methods. However, many machine learning models applied to air quality prediction generally use batch learning and prediction methods, that is, after a sample study and prediction, new samples will not be learned, and air quality prediction will be increased. The error, which deviates from the track of real-time prediction, is difficult to apply effectively to actual engineering. Therefore, in view of the problems existing in the current air quality prediction field, we review the previous research on the air quality neural network prediction model.

Keywords: Big data · Neural networks · Predictive model · Air quality

1 Introduction

Air quality data is typical real-time streaming data, mainly from ground monitoring, meteorological satellites and other collection sites. Mathematical analysis of air quality data has become one of the feasible ways to predict air quality. The traditional air quality prediction is mainly divided into numerical prediction and statistical prediction. The so-called numerical prediction is a mainstream prediction method in the past few decades. It mainly uses the existing air quality data to derive a series of physics. And chemical state equations, these equations are usually high-order differential equations, and the future air quality values are obtained by importing the corresponding parameters, but this kind of prediction method requires a large amount of computational power, and the impact of consideration is quite limited, such as human activities, etc. The parameters of numerical prediction are difficult to grasp and quantify. The statistical prediction is to analyze the existing data through mathematical modeling, such as nonlinear numerical analysis, multivariate statistics, gray analysis, Chebeshev development, etc. However, statistical prediction has long cycle and complicated operation, which is difficult to be timely and accurate. Provide information about air quality data. With the passage of time and the diversified development of air quality data acquisition and processing technology, people began to gradually adopt new technologies such as machine learning to predict air quality data to make up for the shortcomings of traditional forecasting methods. However, traditional machine learning

© Springer Nature Singapore Pte Ltd. 2020
M. Atiquzzaman et al. (Eds.): BDCPS 2019, AISC 1117, pp. 83–90, 2020.
https://doi.org/10.1007/978-981-15-2568-1_13

and other forecasting methods generally use batch learning and prediction methods, that is, after a sample learning and prediction, new samples will not be learned, which increases the error of air quality prediction and deviates from Real-time predicted orbits are difficult to apply effectively to actual engineering.

It can be seen that based on the actual demand of air quality prediction, we urgently need to find a big data framework that can realize real-time processing to apply to the field of air quality prediction. Therefore, based on the previous studies, this paper compares, studies and adopts the advantages and disadvantages of different real-time computing frameworks, and proposes an air quality prediction framework based on STORM and deep learning to achieve the purpose of online learning and real-time prediction.

2 Research Status of Predictive Model Based on Neural Network

Beginning in 1997, China began to use the ANN model in the field of environmental prediction. Initially, BP neural network was used to construct the prediction model. The SO2 concentration in a city was predicted in the short-term and compared with the prediction results of the traditional fuzzy recognition algorithm. The experiment proved the nerve. The network has good prediction accuracy. The advantage of the ANN model for predicting air quality is gradually emerging. However, due to the hardware conditions at the time, the large-scale development of the neural network model was limited, and it was only after 2005 that it developed rapidly. The method based on neural network model prediction is largely determined by factors such as the type of sample data and the logical structure of the model (Fig. 1).

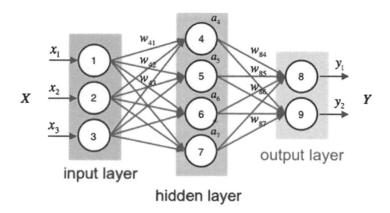

Fig. 1. BPNN

In 2007, Giuseppe et al. used an improved BP neural network model for air quality prediction, and proposed an error cost function to adjust the output error of the model. This adjustment is based on the dynamic adjustment of time series features, and the model was proved by experiments. It has a good predictive advantage in bad weather conditions and is highly stable [1].

In 2008, Kaminski et al. collated the monthly average data of SO2 in a city for 30 years, and used RBF and MLP neural network models to conduct prediction experiments. The results were also obtained. The experiment also showed that the more abundant the training samples, the final results will be The closer to the true value. In the same year, Pietro et al. considered the important influence of neural network input factors, analyzed air quality related factors such as traffic, and added training samples to establish multi-layer perceptron (MLP), radial basis function (RBF) and modular network (MNN). The neural network prediction model, through experiments, the difference between the prediction accuracy of the three models is obtained. Overall, the RBF network model is better than the other two models (Fig. 2).

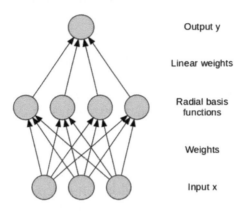

Fig. 2. Radial basis neural network

In 2014, Cheng Huayi improved the BP neural network training algorithm by using the hybrid algorithm composed of genetic algorithm and PSO algorithm, optimized its initial weight, improved its generalization ability, and effectively avoided the premature convergence of ANN. The model was used to predict the concentration of SO2, and the expected effect was achieved [1]. In the same year, Jiang Hao proposed a classification method of pollutant data based on Kohonen neural network, and self-organizing the original data, and the experimental results are ideal [2].

In 2017, Zhao proposed a multiple linear regression optimization model by optimizing the modeling method and increasing the input factor. Through experimental comparison, increasing the input factors of seasonal factors and other pollutant concentrations can accurately predict the future atmospheric pollutant concentration and apply to the short-term prediction of atmospheric pollutant concentration. A medium- and long-term prediction model based on genetic neural network is also proposed [3]. In the same year, Lin Kaichun and others based on random forest and neural network optimization, established a model for predicting air quality index. The random forest algorithm is used for feature selection, the optimal feature subset is selected, and the acquired optimal features and subsets are used to train the neural network model to predict the air quality index. The test results show that the accuracy of the air quality level prediction is increased to 80.56% [4].

In 2018, Pu and others proposed an improved air quality prediction model (KABC-BP) combined with artificial bee colony algorithm and backpropagation neural network [5], and proved that the accuracy is ideal through experiments; Tong Liu et al. Regional numerical simulation of Hong Kong air quality time series prediction based on current numerical prediction models (multi-scale air quality [CMAQ] model, extended integrated air quality model and nested air quality prediction model system) and observations from Hong Kong monitoring stations, An autoregressive comprehensive moving average (ARIMA) model and numerical prediction (ARIMAX) were used to improve the prediction of air pollutants including PM2.5, NO2 and O3 [6]. The results showed a significant improvement in multiple assessment indicators predicted daily (1–3 days) and hourly (1–72 h). The forecast for maximum O3 for 1 h and 8 h per day has also improved.

In 2019, Kang et al. proposed a deep stack self-encoding air quality prediction model, and adjusted the parameters, and compared with SVR and linear regression models [7]. He Xiaolu et al. proposed a high air quality prediction model based on fuzzy time series SVR. By fuzzy processing the historical indicators of four different seasons, the SVR regression prediction model was established to predict LOW, R and UP, and the k-fold crossover was adopted. Verification method for super-parameter tuning [8]. Miskell Georgia et al. proposed an airborne prediction model for gradient-enhanced binary classifiers that successfully used machine learning algorithms to predict reliable short-term high-concentration events, or predict peaks of fine-particle air pollution (PM2.5) one hour earlier (<60 min). The goal is to use a gradient-enhanced machine to predict the occurrence of short-term peaks with a binary classifier (1 = peak, 0 = no peak) and the result is successful [9].

3   Prediction Model Based on Dynamic Neural Network Model

In 2009, Zhang et al. proposed a pollution source data prediction model based on ELMAN neural network. Taking SO2 as an example, the correlation between SO2 concentration and other air quality factors such as air temperature, wind speed, humidity and pressure was explored. The simulation experiment proves that the model is better than the BP model and has good prediction results [10]. In 2010, Chen et al. used wavelet decomposition and reconstruction to establish a segmented BP neural network prediction model. The prediction results show that the model has high prediction accuracy and good generalization ability for SO2 concentration prediction, and it is obvious. Better than the general neural network model [11, 12].

In 2015, Liu improved the prediction model based on the modeling effects of multiple linear regression, BP neural network and support vector machine. The non-mechanical modeling method combining fuzzy time series with support vector machine is proposed, which solves the unstable prediction result caused by incomplete factors in the mechanism modeling method [13].

In 2018, Alimissis et al. used artificial neural networks and MLR methods to model the spatial air pollution changes in the Athens metropolitan area of Greece. It shows that the ANN model has superior performance over the MLR scheme and can simulate complex air pollution more effectively. Spatial variability. It also shows that an important factor affecting the predictive ability of the ANN model is the optimal choice of network architecture and air quality monitoring network density [14]. A considerable disadvantage of the ANN approach is the need for a representative training data set to provide sufficient information to the network to maximize its generalization capabilities. Athira et al. used recurrent neural network (RNN), long-term memory (LSTM), and gated loop unit (GRU) for prediction. The results show that the three models perform relatively well in prediction, and the performance of GRU is slightly higher. On the RNN and LSTM networks [15]. Freeman et al. used deep learning to predict air quality time series with long-term short-term memory (LSTM) recursive neural network (RNN) deep learning to predict 8-h average surface ozone (O3) concentration. Hourly air quality and meteorological data are used to train and predict values for up to 72 h with low error rates. LSTM is also able to predict the duration of continuous O3 overshoot [16] (Fig. 3).

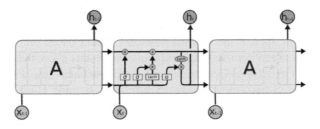

Fig. 3. LSTM

In 2019, Zhao et al. proposed a prediction model based on NARX neural network, and optimized the prediction model according to the nonlinear dynamic description ability of the model. It was compared with CMAQ, NAQPMS air quality numerical model and LSTM statistical model prediction model [17].

4 Prediction Model Based on Combined Neural Network Model

In 2013, Azman et al. conducted a study on API air quality parameters to establish a predictive model combining principal component analysis and neural network. In 2015, Maria et al. used a multi-layer perceptron neural network and clustering algorithm to establish a predictive model. From this, the combined network model has been rapidly developed [18]. In 2016, Mario et al. compared the simple neural network model with the ARIMAX atmospheric prediction model. The results show that the neural network model is slightly better, and the experiment continues to establish a neural network hybrid model based on ARIMAX [19].

In 2016, Zhang used Xi'an as a research area to integrate GIS and BP neural network models to evaluate the accuracy and effect of PM10 concentration prediction of different algorithms and hidden layer neurons. The GIS inverse distance weight interpolation method was used to reveal the spatial distribution of PM10 [20].

In 2018, Stanislaw et al. established an artificial model using an artificial neural network method based on a multi-layer perceptron for multi-directional perturbation and an improved Gaussian model and a radial basis function (RBF) artificial neural network. Research shows that the hybrid model prediction accuracy is better than the traditional model. In the same year, Xin Cheng proposed a new hybrid prediction model based on BP neural network. Reconstruct data in the data preprocessing stage using Variational Mode Decomposition (VMD). Secondly, an improved ant colony algorithm (ACO) hybrid optimization algorithm based on adaptive particle swarm optimization (APSO) algorithm is proposed to optimize the weight and threshold of BP neural network [21, 22]. The experimental results show that the proposed hybrid prediction model (VMD-APSOACO-BP) has higher prediction accuracy. Wang Ping optimized the weight and threshold of BP neural network by two improved whale algorithms, and predicted the concentration of atmospheric pollutants at multiple sites and small riversides in Shenyang. The results show that the improved whale algorithm optimizes the BP neural network and the prediction results are more accurate. In the same year, Qiao Junfei et al. proposed a PM2.5 prediction method based on T-S fuzzy neural network, and applied the soft measurement model based on T-S fuzzy neural network to the actual environment. The experimental results show that the method can predict PM2.5 in real time. Compared with other methods, the PM2.5 prediction method based on T-S fuzzy neural network has better training effect and higher prediction accuracy. Han Wei et al. proposed a hybrid model method based on neural network: spatial and temporal mining and analysis of historical air pollutant data and atmospheric data by means of fully connected neural network method combined with long-term and short-term memory network (LSTM) method. Compared with the traditional single model method, the combination of full connection and LSTM neural network methods can not only get rid of the limitations of the single model feature space, but also improve the prediction accuracy, and have greater applicability and operability. Georgia Miskell et al. proposed a new optimal combination model based on CEEMD (complementary set empirical mode decomposition), PSOGSA (particle swarm optimization and gravity search algorithm), PSO (particle swarm optimization) and combination, which effectively solved blind combination prediction. problem. The proposed model has the highest prediction classification accuracy, more powerful prediction performance, smaller prediction error and better generalization ability.

In 2019, Wu et al. proposed a city quality prediction model based on variational mode decomposition, sample entropy and LSTM neural network. The variational mode decomposition (VMD) was used to decompose the original AQI sequence into different sub-families with different frequencies. Then, applying sample entropy (SE) recombines the sub-series to solve the problem of excessive decomposition and computational burden. Next, establish a long-term short-term memory (LSTM) neural network, predict the new sub-series by accumulating the predicted values of each sub-series, and then obtain the final AQI prediction, which proves that the proposed model has a higher correct AQI-like prediction rate. In the same year, Wang et al. proposed a combined forecasting structure based on the L1 specification, which includes analysis, prediction and evaluation. First, the raw data is broken down into several components. Each component is then expanded into a matrix time series by phase space reconstruction. Then, a weighted combination of prediction results of the three models is performed using the prediction module based on the L1 norm to determine a final prediction result, and the process parameters are optimized using a multi-tracker optimization algorithm. In addition, the comprehensive fuzzy evaluation is used to qualitatively analyze the air quality. Taking the daily pollution sources of three cities in China as an example, the validity and effectiveness of the established combination prediction structure are verified. The results show that the architecture has great application potential in the field of air quality prediction.

5   Conclusion

For the air quality data prediction model, it has been proposed since the 1990s, but due to the hardware conditions at that time, the large-scale development of the neural network model was limited, and it was only after 2005 that it developed rapidly. The method based on neural network model for prediction is largely determined by factors such as the type of sample data and the logical structure of the model.

In the prediction of air quality data, ANN mainly predicts the future time interval according to historical data. In the past few years, the most commonly used methods at home and abroad are mainly static neural network, dynamic neural network and combined neural network. Class, and considering the correlation factors of the impact factor screening of predictive objects, many efficient training algorithms are also proposed. The ANN technology is also widely used in the field of air quality data prediction. However, with the rapid development of information technology, a large number of air quality data has increased sharply in PB. The era of big data is coming, and a new test is put on the ANN air quality prediction model, which can not only satisfy the prediction accuracy of a small number of samples. It is necessary to focus on the 4 V features that meet the basic requirements of big data, to be deeper in the number of layers in the ANN, to be shorter in processing time, and to conduct in-depth research on the direction of high availability in real use.

References

1. Thiede, L.A., Parlitz, U.: Gradient based hyperparameter optimization in Echo State Networks. Neural Netw. Official J. Int. Neural Netw. Soc. **115**, 23–29 (2019)
2. Jiang, H.: Research on air pollution prediction technology based on neural network. Nanjing University (2014)
3. Zhao, M.: Research on Atmospheric Environment Prediction Based on Data Mining Technology. Beijing Jiaotong University (2017)
4. Lin, K., et al.: Air quality prediction based on random forest and neural network. J. Qingdao Univ. (Eng. Technol. Ed.) **33**(2) (2018)
5. Pu, G., Liu, Y.: Prediction of ambient air quality based on improved neural network. Comput. Technol. Dev. **28**(09), 181–184 (2018)

6. Tong, L., Alexis, K.H.L., Kai, S., Jimmy, C.H.F.: Time series forecasting of air quality based on regional numerical modeling in Hong Kong. J. Geophys. Res. Atmos. **123**(8), 4175–4196 (2018)
7. Kang, B., Dang, X.: Air quality prediction system based on deep neural network. J. Harbin Univ. Commer. (Nat. Sci. Ed.) (03), 1–5 (2019). https://doi.org/10.19492/j.cnki.1672-0946. 2019.03.015
8. Hao, X., Zhang, W., Zhu, J.: Prediction of air quality based on fuzzy time series SVR model. J. Lanzhou Univ. Arts Sci. (Nat. Sci. Ed.) **33**(04), 17–23+39 (2019)
9. Zhang, Q., Xu, Z., Zhao, K.: Prediction of pollution source data based on Elman neural network. J. S. China Univ. Technol. (Nat. Sci. Ed.) **37**(5), 135–138 (2009)
10. Chen, L.: Wavelet analysis and support vector machine applied to air pollution prediction. J. Xi'an Univ. Sci. Technol. **30**(6) (2010)
11. Liu, J.: Research on Temporal and Spatial Variation Law and Evaluation Prediction Model of Air Pollutants in Beijing. University of Science and Technology Beijing (2015)
12. Alimissis, A., Philippopoulos, K., Tzanis, C.G., Deligiorgi, D.: Spatial estimation of urban air pollution with the use of artificial neural network models. Atmos. Environ. **191**, 205–213 (2018)
13. Athira, V., Geetha, P., Vinayakumar, R., Soman, K.P.: DeepAirNet: applying recurrent networks for air quality prediction. Procedia Comput. Sci. **132**, 1394–1403 (2018)
14. Freeman, B.S., Taylor, G., Gharabaghi, B., Thé, J.: Forecasting air quality time series using deep learning. J. Air Waste Manage. Assoc. **68**(8), 2018 (1995)
15. Zhao, Q., Qiu, F., Yang, J.: Application of NARX neural network model in the prediction of ambient air quality in Kunming. Environ. Monit. China **35**(03), 42–48 (2019)
16. Cortina–Januchs, M.G.: Development of a model for forecasting of PM10 concentrations in Salamanca, Mexico. Atmos. Pollut. Res. **6**(4), 626–634 (2015)
17. Catalano, M.: Improving the prediction of air pollution peak episodes generated by urban transport networks. Environ. Sci. Policy **60**, 69–83 (2016)
18. Zhang, P., He, L., Zhang, T., et al.: Prediction and spatial distribution of PM10 concentration based on GIS and BP neural network model. Environ. Sci. Manage. **41**(5), 39–43 (2016)
19. Xin, C.: Analysis and prediction of Lanzhou air pollution index based on hybrid model of data preprocessing and machine learning. Lanzhou University (2018)
20. Wang, P.: Application of Intelligent Optimization Algorithm in Environmental Data. Shenyang Aerospace University (2018)
21. Qiao, J., Cai, J., Han, H.: Research on PM2.5 prediction based on T-S fuzzy neural network. Control Eng. **25**(3) (2018)
22. Miskell, G., Pattinson, W., Weissert, L., Williams, D.: Forecasting short-term peak concentrations from a network of air quality instruments measuring PM 2.5 using boosted gradient machine models. J. Environ. Manage. **242**, 56–64 (2019)

Real Estate Spatial Price Distribution in Xining from the Perspective of Big Data

Hongzhang Zhu[1], Lianyan Li[1(✉)], Xiaobin Ren[2], Yangting Fan[3], and Xiaoliang Sui[1]

[1] School of Civil Engineering, Wuhan University, Wuhan 430000, China
Lianyanli@whu.edu.cn
[2] GNSS Research Center, Wuhan University, Wuhan 430000, China
[3] China Construction Eighth Engineering Bureau Zhejiang Construction Co., LTD., Shanghai, Zhejiang, China

Abstract. Studying the spatial distribution of real estate price can not only help consumers to choose the suitable real estate, but also help urban planners to make better decisions. Under the background big data, a new perspective of research on urban real estate prices is produced. Based on the real estate parameters of ordinary residential buildings in Xining city, ArcGIS software is used to conduct the nearest neighbor distance analysis, and it is found that the residential buildings in the area presented significant agglomeration. Then, Moran's I index is selected for spatial autocorrelation analysis, which proves the positive spatial correlation of real estate prices in Xining city. Finally, the spatial distribution map of real estate prices is obtained by fitting the real estate prices in the research area with the statistical analysis of land, which finds the area of Chengxi is the center of local housing price, and the price is decreasing gradually around other areas. In addition, the main factors affecting housing prices in four areas can be obtained by analyzing the distribution figure, which gives a significant reference for residents and policy planners.

Keywords: GIS · Xining city · Real estate price spatial pattern · Geostatistical analysis

1 Introduction

As it well known to all, the real estate price is increasingly recognized as a serious, worldwide public concern, it had been risen continuously, which seemed to over the maximum payment ability of most people [1]. Numerous studies show that the housing price can even increase by over ten percent every year, which is obviously unnormal development tendency especially between 2003 and 2014 [2]. It is notable that the government keep promoting the concept of "houses are for living" since 2017, government has paid more attention to adjust housing prices and promote the steady development of the real estate industry [3].

In the past decades, a number of researchers have sought to determine the mechanism of the change of housing price. A traditional regression model for real-estate price index must be mentioned in the field of real estate modeling, which is conducted

M. Atiquzzaman et al. (Eds.): BDCPS 2019, AISC 1117, pp. 91–99, 2020.
https://doi.org/10.1007/978-981-15-2568-1_14

in taking advantage of the data of repeat sales and the standard techniques of regression analysis to obtain a price index. One of the main obstacles is that the application of the model can produce erratic results due to the instability of the index estimated by the chain method [4, 5]. Additionally, the selection of evaluation indicators has been a largely under explored domain [6].

The purpose of this paper is to provide a new insight to consider the mechanism of the spatial price distribution of real estate. In the context of big data and internet technology, using ArcGIS software to deal with the original data, then research on the distribution form and correlation of real estate price in geographical space. The model proposed can not only reduce the subjective influences, but also make full use of the characteristics of the original data.

2   Methods

2.1   Global Spatial Autocorrelation Analysis

Spatial autocorrelation analysis [7, 8] can determine whether a variable is related. Moran's I is a commonly used tool to obtain the relationship by using spatial correlation tools. When it closes to 1 means agglomeration, closes to -1 means discrete. The Moran's I index can be calculated as follows.

$$I = \frac{\sum_{i=1}^{n} \sum_{j=1}^{m} w_{ij} (x_i - x_j) (x_j - x_m) / \sum_{i=1}^{n} \sum_{j=1}^{m} w_{ij}}{\sum_{i=1}^{n} (x_i - x_m)^2 / n} \tag{1}$$

Where, i is the ith pixel, j is the jth neighboring pixel, w_{ij} is the corresponding coefficient term, when j is one of the four nearest neighbors of the i pixel, the value of w_{ij} is 1, otherwise its value is 0. x_i is the value of the pixel i. x_j is the value of the neighboring pixel of the pixel j, x_m is the mean of the grid pixel.

Generally, the standardized statistic $Z(I)$ can be used to test the significant level of spatial autocorrelation.

$$Z(I) = \frac{(I - E(I))}{\sqrt{Var(I)}} \tag{2}$$

Where, $E(I)$ is the theoretical expectation, $E(I) = (-1)/((n-1))$, $Var(I)$ is the theoretical variance of the Moran index, it can be expressed as follows.

$$Var(I) = E(I^2) - E(I)^2 = \frac{n^2 S_1 - n S_2 + 3n S_0^2}{S_0^2 (n^2 - 1)} \tag{3}$$

In Eq (3), S_0 is the spatial weight matrix element, $S_1 = \sum_{i=1}^{n} \sum_{j=1}^{n} (w_{ij} + w_{ji})^2$, $S_2 = \sum_{i=1}^{n} (w_{i*} + w_{*i})^2$.

2.2   Local Spatial Autocorrelation Analysis

Local spatial autocorrelation analysis can figure out the trend of a subunit obeying the whole unit. The most commonly used indicator is Local Moran's I [9].

$$I = \sum_{j=2}^{m} w_{ij}(x_i - x_m)(x_j - x_m) \tag{4}$$

First, load the elements in the database with ArcGIS software, then use map clustering to analyze the difference in house prices between each region and surrounding areas. Additionally, "HH" means that the average value of the area and its surrounding area is higher than the average price of all the houses, "LL" means that the average value of the area and its surrounding area is lower than the average price of all the houses, "HL" means the house price is over average price in the area, but the housing price of its surrounding area is lower than the average price, "LH" indicates that the residential price in the area is lower than the average price, but its surrounding area average is higher than the average price.

2.3   Geostatistical Analysis

Geostatistical analysis uses the variogram as the main tool to study the science of spatial phenomena that are both random and structural, or spatially related and dependent [10].

The second-order mixed center moment of the random variables of the regionalized variable at the spatial point x and $x + h$, defined as the autocovariance function of $Z(x)$.

$$Cov(Z(x), Z(x+h)) = E[Z(x)Z(x+h)] - E[Z(x)][Z(x+h)] \tag{5}$$

$$C(h) = \frac{1}{N(h)} \sum_{i=1}^{N(h)} [Z(x_i) - \bar{Z}(x_i)][Z(x_i+h) - \bar{Z}(x_i+h)] \tag{6}$$

Where, $Z(x)$ is the regionalized random variable, h is the spatial separation distance of two sample points, $Z(x_i)$ is the sample value of $Z(x)$ at the spatial point x_i, and $Z(x_i + h)$ is the sample value of $Z(x)$ deviating from h at x_i, $N(h)$ is the total number of sample points when the distance is h. $\bar{Z}(x_i)$ and $\bar{Z}(x_i+h)$ are the average number of samples at $Z(x_i)$ and $Z(x_i + h)$ respectively.

$$\bar{Z}(x_i) = \frac{1}{n} \sum_{i=1}^{n} Z(x_i) \tag{7}$$

$$\bar{Z}(x_i + h) = \frac{1}{n} \sum_{i=1}^{n} Z(x_i + h) \tag{8}$$

Then, the discrete calculation formula of the variogram $r(h)$ is seen below.

$$r(h) = \frac{1}{2N(h)} \sum_{i=1}^{N(h)} [Z(x_i) - Z(x_i + h)]^2 \tag{9}$$

Therefore, as for different separated distance h, its corresponding values of $C(h)$ and $r(h)$ can be calculated subsequently. It is possible to directly show the spatial variability of the regionalized variable $Z(x)$.

3   Research Data

3.1   The Research Area

Xining is located in the northeast of Qinghai province, it's a fascinate city at the intersection of Huangshui river and three tributaries, which is high in the southwest and low in the northeast. Furthermore, the built-up area of Xining's downtown area are over 150 km^2, building areas are concentrated in the "cross" shape of the narrow urban area. In Fig. 1, it can be inferred roughly that Chengxi district has more convenient traffic because it is located in the intersection area of the north-south and east-west trends of the city, its economy may be boom as well.

Fig. 1. Study area

3.2   Selection of Research Objects

In this paper, ordinary residential buildings in Xining city are taken as research objects, each region selects communities that are currently being sold or sold recently as a key research object. Due to the large research area, the houses with higher representativeness in a small range are selected as the research objects, while the other houses with similar prices and geographical proximity are not discussed here (Table 1).

The used data are from government statistical bulletin, Qinghai Provincial Yearbook, Real Estate Statistics Bureau, Lianjia.com, Dianping.com, Mobike.com, and Baidu Map POI. Data such as the name and price of ordinary residential buildings for sale in research areas of Xining City are selected for preliminary processing.

Table 1. The selected objects

Urban areas	The research objects
Chengbei district	Greenland International Flower Capital, Country Garden, Jinshui Lake, Chengbei International Village, Ziheng Dijing Garden, Ningrui Water Town, Hengda Mingdu, Qinghai Northwest City, Ding'an Famous City, Xining Kangmei Chinese Medicine City, Yingji Ziyu Yuting, China De Chaoyang Oasis Home, Jinzuya Garden Phase II
Chengzhong district	Guanghui Jiujin Garden, Xiangge Special Zone, Shangri-La City Flower, Xining Red Star Tianplatin, Nanmen Daye World, Xinheng International Center Complex, Tianqiao Xiangfu, Rongcheng, Holiday Sunshine, Hailiang Commercial Plaza, Zhongfayuan Times Square, the first city of Rongfu
Chengxi district	Langyue Xintiandi, Times Shenghua, Wanfang City, Xiadu Jingyuan, Limeng Shangdu, Jinxiu Jiangnan, Hanlin Huating No. 5
Chengdong district	Heng Tai Da Guan Tianxia, Hailiang Daduhui, Hong Kong Oriental Garden, Pantai Yishan County, Ronghao Garden, Zhonghui Zijincheng, Yuezhou International Plaza, Hailiang Commercial Plaza, Shangdong International, Sanhe Cuizhen Star City, Hai Liangdaduhui, Baili Haoting Phase II

4 Results and Discussions

In this part, the procedures of applying ArcGIS can be divided into several parts. First, input the original image and attribute value, the attribute values are mainly extracted from the collected data. Second, conduct raster classification and raster calculator. Finally, calculate tabulate area, create a buffer with the attribute value as the condition, and obtain the area chart. The basis for implementing the above steps is building a database using the ArcCatlog function module. Then, vectorize the roads and houses of research areas, the results are shown below (Figs. 2 and 3).

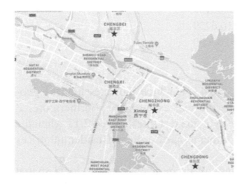

Fig. 2. Base map of research area

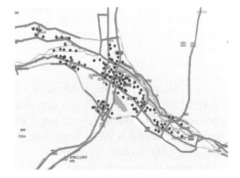

Fig. 3. The vectorization map

The houses in the specific area are distributed in discrete points. In order to further explore the distribution law of real estate prices, the location elements of residential buildings are processed with the analysis mode tool in ArcGIS10.1, the average nearest neighbor distance analysis method is conducted as well. The results are shown in Table 2.

Table 2. Analysis on the nearest neighbor distance of residential buildings in Xining City

Parameters	Values
Average observation distance/m	16.04374
Average expected distance/m	18.36823
Nearest neighbor index	0.73013
Z	−4.68391
P	0.00000

Table 2 indicates that the nearest neighbor index of residential samples is about 0.73, which means that the residential buildings are distributed in agglomerated form. In addition, the test value of Z is approximately −4.68391, which is clearly less than the value (−2.58) at the 1% significance level. P's value is zero, indicating that the probability of this random probability distribution is zero, which all prove the agglomeration characteristics of the residential space distribution are extremely significant.

Subsequently, integrate and aggregate big data from various databases again, only suitable and complete data can be input into the software, the spatial correlation results are shown as follows.

Table 3. Spatial correlation analysis results

Parameters	Values
Moran's I	0.4371
Expected index	−0.00376
Variance	0.001023
Z	16.6739
P	0.00000

As is shown in Table 3, Moran's I index is 0.4371, which indicates that the geographical location is positively correlated with the real estate price. Furthermore, Z score is 16.6879, which is much larger than the two-side test threshold of 2.17 when the confidence interval is 99%. Therefore, the agglomeration of ordinary residential prices in Xining City is quite apparent.

In order to research on the spatial heterogeneity in the research area, local spatial autocorrelation analysis is applied, and Map clustering analysis can suggest the degree of differences on a whole, the processing result is shown below (Fig. 4).

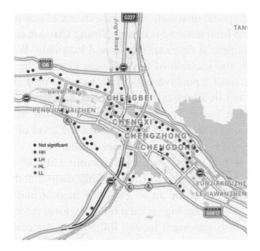

Fig. 4. The graph of spatial association

The red dots appear in the vicinity of the Chengzhong district and the Chengxi district, which imply that the spatial agglomeration in these two areas is pretty significant. Most of the areas showing significant agglomeration are concentrated near the main roads, and the surrounding service facilities are relatively complete as well. It also demonstrates that the impact of traffic on housing prices is quite prominent, which also provides a theoretical basis for determining the factors affecting real estate prices. Then, use the geostatistical analysis tool directly to select Kriging interpolation. Kriging method requires the data to follow a normal distribution, using Box Cox to let all data conform to the normal distribution [11]. Then, calculating the variogram, fit the model, create the matrix, and make the prediction. The result is as follows.

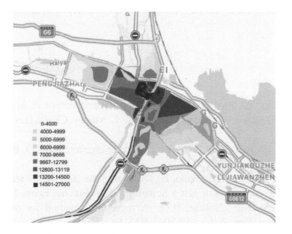

Fig. 5. The graph of spatial association

Figure 5 shows the spatial distribution characteristics of real estate prices in Xining. Some buildings with the high housing prices in Xining City are concentrated in the area of the main roads, commercial districts, school and hospitals. Which indicates that the degree of convenience, the location of the business district, the availability of education, and medical care have a positive impact on housing prices. For example, banks, supermarkets, stores, and living markets in Chengxi district are consistently clustered near the west of Kunlun Road. The concentration in space makes an extreme high spatial autocorrelation of the region, which improves the level of local service facilities and promotes the agglomeration of house prices further.

The distribution of real estate prices in Xining City is in a state of central agglomeration, mainly in the vicinity of Chengzhong district and Chengxi district, and gradually decays to the surrounding areas. People's income and consumption level in the vicinity of these two areas are higher than other regions, indicating that income and consumption level are also important factors influencing the real estate price.

The change of real estate price in Xining City is quite flexible. In other words, the price of the commercial area has changed greatly, while far from the central area, the price decline is less flexible, which is also closely related to the central effect of urban development.

5 Conclusion

This paper analyzes the spatial pattern of the ordinary residential housing prices in the Chengxi district, Chengzhong district, Chengbei district and Chengdong district of Xining City. The ArcGIS software is used to quantitatively analyze the spatial auto-correlation and the trend of housing prices in the research area. The research suggests that the housing prices in Xining City show a significant concentration in space, and the housing prices show a trend of decreasing from Chengxi district to the surrounding area, which indicates that housing prices are significantly affected by the level of education and medical care, the degree of convenient transportation and infrastructure construction.

References

1. Fougere, D., Lecat, R., Ray, S.: Real estate prices and corporate investment: theory and evidence of heterogeneous effects across firms. J. Money Credit Bank. 51(6), 1503–1546 (2019)
2. Glaeser, E., et al.: A real estate boom with Chinese characteristics. J. Econ. Perspect. 31(1), 93–116 (2017)
3. Bianchi, D., Guidolin, M., Ravazzolo, F.: Dissecting the 2007–2009 real estate market bust: systematic pricing correction or just a housing fad? J. Financ. Econ. 16(1), 34–62 (2018)
4. Bailey, M.J., Muth, R.F., Nourse, H.O.: A regression method for real-estate price-index construction. J. Am. Stat. Assoc. 58(304), 933–942 (1963)
5. Li, H., et al.: Analyzing housing prices in Shanghai with open data: amenity, accessibility and urban structure. Cities 91, 165–179 (2019)

6. Wang, W.C., Chang, Y.J., Wang, H.C.: An application of the spatial autocorrelation method on the change of real estate prices in Taitung City. ISPRS Int. J. Geo-Inf. **8**(6) (2019)
7. Yu, S., et al.: Spatial patterns of potentially hazardous metals in soils of Lin'an City, Southeastern China. Int. J. Environ. Res. Public Health **16**(2) (2019)
8. Rousta, I., et al.: Analysis of spatial autocorrelation patterns of heavy and super-heavy rainfall in Iran. Adv. Atmos. Sci. **34**(9), 1069–1081 (2017)
9. Zhang, Z., et al.: Spatial distribution characteristics of forest soil nutrients in Jiangxi Province based on geostatistics and GIS. Res. Soil Water Conserv. **25**(1), 38–46 (2018)
10. Choi, B.G., et al.: A study on the satisfaction analysis on officially assessed land price using time seriate geostatistical analysis. J. Korean Soc. Surv. Geod. Photogramm. Cartogr. **36**(2), 95–104 (2018)
11. El Hamidi, M.J., et al.: Spatial distribution of regionalized variables on reservoirs and groundwater resources based on geostatistical analysis using GIS: case of Rmcl-Oulad Ogbane aquifers (Larache, NW Morocco). Arab. J. Geosci. **11**(5) (2018)

Design of UAV Bird-Driving Linkage System

Xiangguo Lin[(✉)]

Department of Communication, Civil Aviation College,
Guangzhou 510640, China
linxiangguo@caac.net

Abstract. Based on the analysis of the bird situation in airport and the means of bird repelling, a kind of UAV bird repelling linkage system is proposed. The system is mainly composed of UAV, ground bird drive equipment and monitoring center. Unmanned aerial vehicle first finds the bird situation through visual capture and voice recognition, and then reports the bird situation to the monitoring center through the Internet of Things. After the analysis of the monitoring center, the ground bird-repelling equipment is controlled to select sound, laser, gas gun and other means to cooperate with UAV to complete the bird-repelling task.

Keywords: UAV · Driving-birds · GPRS

1 Introduction

In recent years, with the development of the aviation industry, the number of flights has gradually increased. Aviation safety accidents occur from time to time, especially the impact of birds on spacecraft is one of the major accidents affecting aviation safety. Therefore, how to effectively prevent the occurrence of bird strikes has become an important work of airport staff. Practice has proved that driving birds away from aircraft is an effective means. According to the statistics of the World Civil Aviation Organization, about 80% of bird strikes occur at airports and their surrounding areas. During the take-off, landing and low-altitude and ultra-low-altitude flight stages, the proportion of bird strikes occurring at altitudes below 39.4 m is as high as 59.2% [1]. The traditional means of bird repelling mainly include setting up scarecrow, tweeter, shooting gun, spraying bird repellent, firing guns, and placing bird traps. However, these bird-repelling methods have a single function, which requires a lot of manpower, limited scope of action and lack of pertinence, making the work of bird-repelling blindly, especially for the high altitude areas above 80 m and below 200 m, the effect of bird-repelling is not obvious [2, 3]. Therefore, the airport urgently needs a new means of bird repelling to meet the requirements of bird repelling in the high altitude areas above 80 m and below 200 m and outside the airport fence.

In this regard, DeTect Corporation of the United States and Sicom System Company of Canada have developed Merlin and Acipiter radar bird detection systems based on radar respectively. In 2011, Robert et al. [4] combined radar technology with visual analysis and auditory judgment to improve the accuracy of bird monitoring data. In recent years, the rapid development of UAV has provided a new technical means for

© Springer Nature Singapore Pte Ltd. 2020
M. Atiquzzaman et al. (Eds.): BDCPS 2019, AISC 1117, pp. 100–105, 2020.
https://doi.org/10.1007/978-981-15-2568-1_15

the problem of bird repelling in airports [5–7]. In the early stage, UAV bird repellent mission was mainly used in military airport bird repellent mission, as well as farmland, forestry, fishing grounds and other areas [8, 9]. Compared with the traditional bird-driving methods, UAV has the advantages of strong maneuverability, wide range, high efficiency and fast response. It can also carry different bird-repelling equipment in the actual situation of the airport, and the way of bird-repelling is very flexible.

Aiming at the problems existing in the management of avian repellent in airport, this paper designs a set of linkage system which is suitable for avian monitoring, risk assessment of avian aircraft and avian pest removal in airport by using UAV as carrier, combining with Internet of Things technology and connecting with ground avian repellent equipment and monitoring center.

2 Working Principle of the System

The system is mainly composed of UAV, ground bird drive equipment and monitoring center. The functional block diagram of UAV bird-driving linkage system is shown in Fig. 1. The UAV is a hybrid formation of fixed-wing and rotorcraft UAVs with the shape of bionic eagle. Each UAV is equipped with cameras, ultrasound, laser generators, audio generators, flash generators and other bird-repelling equipment. Ground bird repellent equipment mainly includes gas gun, titanium thunder gun, voice bird repellent, bird detection radar and other equipment.

The workflow of the system is as follows:

(1) Regional division. Taking the airport as the center, a three-dimensional coordinate system is established, which is divided into early warning area, driving, maneuvering and standby subspace.

(2) Bird detection. It mainly consists of fleet patrol (carrying drone), radar detection and ground surveillance camera. When bird information is detected, bird information is transmitted to the surveillance center through ZigBee and GPRS.

(3) Bird situation analysis. After receiving the bird information, the monitoring center intelligently analyses and identifies the bird information through the server, and then notifies the ground bird-repelling equipment and the UAV to make the corresponding bird-repelling action.

(4) Bird drive. After receiving the information of bird repelling, ground equipment and UAV make corresponding bird repelling actions according to the command of the monitoring center.

(5) Evaluating the effect of bird repelling. Return to step 2, evaluate the effect of bird repellent, and determine whether to proceed with step 4.

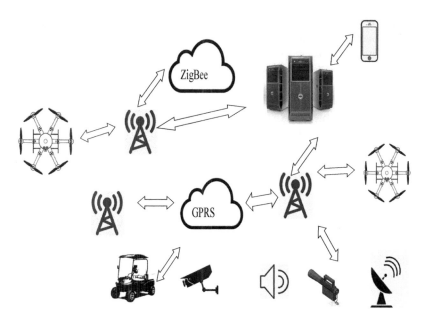

Fig. 1. System schematic diagram

3   Bird-Driving UAV

Using STM32F404ZGT6 as MCU, the UAV uses video monitoring module to detect bird situation on the spot, uses audio module to output various sound combination ultrasonic module to drive birds, and communicates with ground bird driving equipment through GPRS to feedback bird information. Airport staff can monitor the situation of birds in and around the airport at any time through hand-held terminals and operation panels, and choose different ways to combine bird driving to achieve the best effect of bird driving.

STM32F404ZGT6 chip has powerful built-in resources. It integrates FPU and DSP instructions. It has 192 KB SRAM, 1024 KB FLASH, 12 16-bit timers, 2 32-bit timers, 2 DMA controllers (16 channels), 3 SPI, 2 full-duplex I2S, 3 IIC, 6 serial ports, 3 12-bit ADC, 2 12-bit DAC, 1 Camera interface, 112 universal IO ports, etc. It is widely used in security systems, video systems, power systems, instrumentation systems and other fields. The UAV is powered by DC 12 V lithium battery, while the working voltage of STM32F404ZGT6 chip is DC 3.3v, which needs to be converted to 3.3 V DC drive MCU through AMS1117 chip.

Video monitoring module collects bird information through CCD image sensor and passes it to STM32F4 processor after processing. STM32F4 processor can adjust the observation angle of CCD image sensor by controlling the platform at any time to achieve different angles of observation, and then monitor bird situation in an all-round way.

Audio bird repellent module makes birds feel nervous or fearful by making natural enemies'voices and noises, thus leaving the corresponding airspace. After the audio signal is output by STM32F4 processor, it needs to be amplified by voice amplifier circuit. This system uses integrated operational amplifier 1/4LM324 to amplify the voice of MCU. The schematic diagram of the voice amplifier circuit is shown in Fig. 2. Then the power amplifier circuit is output to the loudspeaker to emit bird repellent sound. The core device of the power amplifier circuit is composed of a LM4755 voice power amplifier. The circuit is shown in Fig. 3.

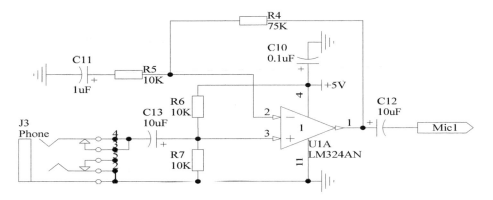

Fig. 2. Audio amplifier circuit

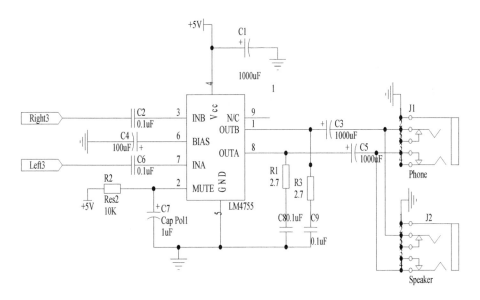

Fig. 3. Power amplifier circuit

Ultrasound module stimulates the nervous system of birds by playing ultrasound, which makes the physiology of birds disordered and escapes from related fields. Ultrasound frequency is generally above 20 kHz, which exceeds the hearing frequency range of human ear. Therefore, it is harmless to human body and is not limited by region. Ultrasound frequency can be continuously changed in use to reduce the adaptability of birds.

4   Ground Bird Removal Equipment

The ground bird-repelling equipment mainly includes bird-repelling patrol car, gas gun, titanium thunder gun, voice bird-repelling device, bird-repelling spray and so on. This system uses the airport Bird Monitoring System Based on three-coordinate search radar, millimeter-wave high-resolution observation radar and optical auxiliary equipment proposed by Zhao et al. [5], to search for bird information, and then uploads it to the monitoring center through ZigBee. Then the monitoring center combines the image taken by the ground surveillance camera to analyze, and then notifies nobody. The machine confirms the situation of birds and chooses the appropriate way to drive birds according to the different situation of birds.

Titanium thunder gun is a kind of bird repelling method suitable for airport. It uses electric control to shoot, blast in the air, sound, light, vibration and wave impact [10]. Titanium thunderbolts can be fired at altitudes of 100–150 m, which can quickly and controllably track the flying explosion of birds. The sound and light effects of the explosion make the birds frightened at a distant location, especially suitable for the needs of flying birds at high altitudes at airports.

5   Conclusion

In this paper, a linkage bird-driving system based on UAV is built, which uses Doppler radar to detect bird situation. When bird situation is found, UAV maneuverability is used to track birds, and non-lethal means such as ultrasound and voice are used to drive birds out of the control airspace.

References

1. Liang, S., Wang, W., Gao, L., et al.: Analysis of the relationship between the birds flying height and guard against the bird striking in the civil aviation airport. J. Saf. Environ. **16**(1), 104–109 (2016)
2. Mingli, N.W.: Comparison of several kinds of bird-repelling ways. Agric. Eng. **4** (Supplement 1), 60–61 (2014)
3. Nohara, T.J., Weber, P., Premji, A, et al.: Affordable avian radar surveillance systems for natural resource management and BASH applications
4. Rasmussen, S.J., Shima, T.: Tree search algorithm for assigning cooperating UAVs to multiple tasks. Int. J. Robust Nonlinear Control **18**(2), 135–153 (2008)
5. Sbihi, A.: A best first search exact algorithm for the multiple-choice multidimensional knapsack problem. J. Comb. Optim. **13**(4), 337–351 (2007)
6. Zong, Q., Wang, D., Shao, S., et al.: Research status and development of multi-UAV coordinated formation flight control. J. Harbin Inst. Technol. **49**(3), 1–14 (2017)

7. Lin, Y.: Domestic airports equipped with unmanned remote-control cars driving birds. Transp. Transp. (3), 45–47 (2006)
8. Beason, R.C., Nohara, T.J.: 3-D radar sampling methods for ornithology and wildlife management [DB/OL], 15 December 2011
9. Zhao, J., Wei, Q., Zhao, H.: Research on an airport avian information monitoring system. Radio Eng. China **41**(10), 21–23 (2011)
10. Jiangchao, Cheng, Y., Wang, Y.: A New Laser UJ Laser J. (04) (2002)

Algorithm Design Based on Multi-sensor Information Fusion and Greenhouse Internet of Things

Rui Huang[1,2], Shuaibo Peng[1,2], Wendi Chen[1,2], Shan Jiang[1,2], Zhe Wu[1,2], Jiong Mu[1,2(✉)], and Haibo Pu[1,2(✉)]

[1] College of Information Engineering, Sichuan Agricultural University, Yaan 0086-625015, China
{jmu,puhb}@sicau.edu.cn
[2] Key Laboratory of Agricultural Information Engineering of Sichuan Province, Yaan 0086-625015, China

Abstract. From the last century, many countries had costed a large amount of energy and material fund to study application of multi-source information fusion technology. At present, whether it is military or civilian, this new technology is widely used, indicating the importance of new technologies. This paper designs and implements a multi-sensor data fusion algorithm combining Kalman filter, Euclidean distance formula and multi-cluster statistical technique. The algorithm can better achieve data fusion and reduce data uncertainty and error caused by various errors such as temperature, humidity and light. We conducted experiments in the cucumber greenhouse in March. The results show that the algorithm is used to process the greenhouse data, which effectively optimizes the decision-making and adjustment basis of the system environmental parameters, which is beneficial to the economic benefits of high greenhouse production.

Keywords: Multi-sensor · Information fusion · Data association · Data decision making · Greenhouse Internet of Things

1 Introduction

Currently, the primary tool for obtaining information is a sensor that functions like human vision, hearing, wake-up, touch, and the like. The sensor mentioned here is a sensor under the broad concept, which can be understood as any way to obtain relevant environmental information [1]. The fusion mentioned here refers to the extraction, analysis, processing and judgment of multidimensional information. Due to its enormous advantages and potential, multi-sensor information fusion technology has gradually played a huge role in military, civil, management, industrial and other fields, making the intersection of a series of new disciplines possible. Multi-sensor information fusion technology has gradually become a new modern information processing model [2].

The greenhouse constructed by the Internet of Things, as an information-efficient and efficient modern agricultural production facility, has achieved basic parameter collection and basic control, such as CO_2 concentration, illuminance, pH, temperature

© Springer Nature Singapore Pte Ltd. 2020
M. Atiquzzaman et al. (Eds.): BDCPS 2019, AISC 1117, pp. 106–115, 2020.
https://doi.org/10.1007/978-981-15-2568-1_16

and humidity, nitrogen, phosphorus and potassium collection and control. It realizes various functions such as automatic watering, ventilation and remote monitoring. However, because the greenhouse Internet of Things is a typical multi-sensor system with multi-point layout, the sensors involved are not only various but also numerous.

2 Multi-sensor Information Fusion Technology

2.1 Basic Concepts and Characteristics

Raw data layer, feature abstraction layer and decision layer processing are often used for information fusion [3]. Information fusion is a method for preprocessing, data correlation, prediction estimation as well as fusion of different levels and different integrated multi-source information [4]. In order to obtain high quality data or information, multi-sensor data is integrated, combined and graded [5]. Due to sensor performance and external environmental interference, the data received by the sensor is uncertain. Comprehensive processing of data from multiple sensors yields more accurate predictions and estimates [6].

2.2 Data Fusion Structure

Data fusion is often used to improve the performance of multi-sensor intelligent detection systems and to reduce the loss of overall information or individual sensor detection information.

Parallel multi-sensor data fusion means that the data output by all sensors is simultaneously input to the internal fusion center. As shown in Fig. 1, each sensor is independent of each other. The Fusion Processing Center will use appropriate methods for all types of data to synthesize the fusion results of the final output. Therefore, in the case of parallel fusion, there is no interaction between the outputs of all the sensors.

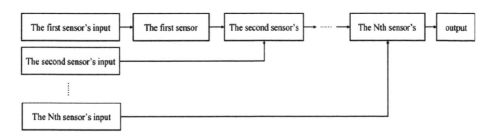

Fig. 1. Data fusion serial structure

The multi-sensor information fusion of the series-parallel hybrid structure, as shown in Fig. 2. It can be connected in series and then in parallel, or in parallel and then in series.

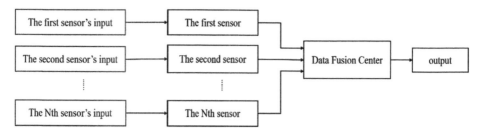

Fig. 2. Data fusion parallel architecture

3 Research on Data Association Technology

The core of multi-sensor multi-target tracking technology is data association. Target tracking technology faces significant challenges due to noise, complexity of the target environment, electromagnetic interference, spurious measurements, and sensor factors. Reasonable handling of the relationship between measurement information and target sources is the key to solving the problem of target tracking technology. A process of determining whether to derive a measurement from a target source based on measurement information received by the sensor is a data association technique. The correlation results determine the performance of the target tracking algorithm.

A multi-hybrid multi-target tracking joint probability data association algorithm is proposed to solve multi-target tracking problems in multi-objective and multi-clutter interference environments. Handling multi-target data interconnects in clutter environments plays a decisive role in the overall multi-target tracking system [7].

For different application backgrounds, many scholars have proposed an improved probabilistic data association algorithm [8–10]. Han Hong et al. [11, 12]. They solved the problem of inconsistent observation space of heterogeneous sensors. Tu Yongjun et al. [13]. proposed an improved data association algorithm based on fuzzy clustering analysis, which solved the problem that it is difficult to obtain correct data association results when the attribute differences of all targets in the target set are small. At present, a complete understanding to improve the scientific, accurate, responsiveness and decision-making risk reduction process of system planning and decision-making is developed [14].

4 Research on Multi-sensor Data Decision

4.1 Introduction

Multi-sensor data decision technique is a process of combining information obtained by multiple sensors and determining target attributes, characteristics, and types. Especially when multiple targets occur at the same time, more decision rules are required. At present, multi-sensor data decision technology has received widespread attention, and many data decision theory and algorithms have emerged. However, due to the limitations of data attributes and data types, multi-sensor data decisions have not yet formed a unified theoretical framework and unique algorithm classification. In the field of multi-sensor data decision making, mainstream decision-making methods include statistics, classical reasoning, Bayesian inference, template methods, voting methods, adaptive neural networks and evidence theory.

4.2 Decision Method Based on DS Evidence Theory

Since the theory of direct evidence is not limited by factors such as prior information, it can be combined with other theories to solve practical problems. The most important feature of evidence theory is that information can be divided into support interval, trust interval and uncertainty interval to explain the uncertainty of information. In terms of flexibility, Beyonce and other methods require complete prior knowledge, conditional probability knowledge and other constraints, and the flexibility is poor. Evidence theory can effectively combine compatible event propositions or mutually exclusive event propositions with greater flexibility.

4.3 Problems in DS Evidence Theory

The main problem of DS evidence theory is the conflict with traditional cognition, that is, the source of evidence has little confidence in the focus elements in the set, but it is almost certainly supported. When the evidence conflict coefficient K is too large, the DS evidence comprehensive formula cannot reasonably combine the evidence acquisition. That is, when the evidence conflict is too large, the DS evidence theory is not ideal. Poor robustness, that is, when the basic probability distribution function of the focus element of an evidence source changes slightly, when the evidence conflict coefficient is large, the combined result of multi-source evidence will change significantly. When the support of the focus element in the evidence is zero, the combined result of the focus element remains zero, no matter how the following evidence supports the focus element.

At present, most scholars have made great progress in dealing with evidence without changing the theoretical model of DS evidence [15]. At the same time, some scholars have tried to modify the combination rules to achieve some visible effects [16].

5 Research on Greenhouse Internet of Things Data Fusion Algorithm

5.1 Algorithm Design

Greenhouse Internet of Things is a typical multi-sensor system. It uses computer technology to obtain sensory information and data analysis to obtain more accurate measurements of the measured object. Decision-making and evaluation to imp rove the accuracy and comprehensiveness of information, reduce information uncertainty and so on. The basic process of system data fusion is shown in Fig. 3.

It can be seen from Fig. 4 that the control system or the IoT sensor node performs A/D conversion to obtain a digital signal of the sensor measurement data, and then

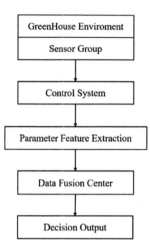

Fig. 3. Data fusion process

performs parameter feature extraction or data preprocessing to eliminate noise or interference signals during parameter acquisition, and finally Send to the data fusion center to complete the data fusion and output the fusion/decision result execution. In this study, the data fusion algorithm combining Kalman filter and Euclidean weighted average is used to process the greenhouse IOT data fusion. The Kalman filter algorithm can be well applied to the fusion processing of sensor redundant data information in a relatively stable linear dynamic environment of systems such as the greenhouse of the Internet of Things. After the Kalman algorithm is used to fuse the data collected by the N sensors, both the current state estimation of the system and the future state of the system can be obtained, so that the feature value or the measured value itself can be extracted from the data. The basic mathematical model of the Kalman algorithm consists of Eqs. (1) and (2).

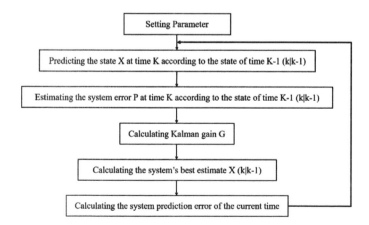

Fig. 4. Kalman algorithm flow

The equation of state at system k:

$$X(k) = A \times X(k-1) + B \times U(k) + W(k) \tag{1}$$

System observation equation:

$$Z(k) = H \times X(k) + V(k) \tag{2}$$

Where: $X(k)$ represents the observation vector at system k; $Z(k)$ represents the system measurement is m order vector; $U(k)$ is the control of the system at time k, if the collected value does not exceed the set threshold If there is no control, it is 0; A, B are $n \times n$-order matrix in multi-model systems, which is the basis system parameters for the algorithm to predict the state variables; H is the measurement system parameter, for multi-parameter measurement system, H is the $m \times n$ order measurement matrix, which converts m-dimensional measurements to n-dimensional, corresponding to state variables; $W(k)$ represents process noise, which is considered Gaussian white noise in greenhouse environment; $QV(k)$ denotes measurement noise, which is considered to be Gaussian white noise in the greenhouse environment, and its covariance is R.

According to the formula (1), the state is further predicted. Using the state-predicted value $X(k-1|k-1)$ at $k-1$ time, the optimal predicted value $X(k|k-1)$ at time K is obtained as

$$X(k|k-1) = A \times X(k-1|k-1) + B \times U(k) \tag{3}$$

The covariance $P(k|k-1)$ corresponding to $X(k|k-1)$ is

$$P(k|k-1) = A \times P(k-1|k-1) \times A' + Q \tag{4}$$

The optimal estimate $X(k|k-1)$ of the system at time K is obtained by Eqs. (2) and (3).

$$X(k|k) = X(k|k-1) + G(k) \times [Z(k) - H \times X(k|k-1)] \tag{5}$$

In order for the Kalman algorithm to continuously run iterations, each iteration needs to update the covariance $P(k|k)$ of the current K moment $X(k|k)$:

$$P(k|k) = I - G(k) \times H \times P(k|k-1) \tag{6}$$

$$G(k) = P(k|k-1)H' / [H \times P(k|k-1)H' + R] \tag{7}$$

In Eqs. (3) through (7): $P(k-1|k-1)$ is the covariance corresponding to $X(k-1|k-1)$; A' is the transposed matrix of A; $G(k)$ Representing Kalman Gain is the intermediate result of filtering.

If there are m sensors for a parameter and there are n data for each sensor disability, the data collected by the kth sensor can be expressed as vector $X_k = (x_{k1}, x_{k2}, \cdots, x_{km})T, k = 1, 2, \cdots, m$, the size of the data difference collected by each sensor can be determined by many methods. In this study, the distance between the distances X_1 and X_i is expressed by the Euclidean distance formula $d_{li} = [(x_l - x_i)(x_l - x_i)]^{\frac{1}{2}}$. The smaller the data deviation, the smaller the Euclidean distance matrix of all sensors.

$$
D = \begin{bmatrix} d_{11} & \cdots & d_{1n} \\ \vdots & \ddots & \vdots \\ d_{m1} & \cdots & d_{mn} \end{bmatrix} \tag{8}
$$

Multi-sensor data can then be merged into the sensor group. According to the clustering method in multivariate statistics, data fusion can be performed in an orderly manner. The fusion formula is as follows:

$$
f(\chi_{ik}, \chi_{jk}) = \frac{c(x_{ik} + x_{jk}) + (c-1)^2 x_{ik} x_{jk}}{1 + c^2 - (c-1)^2 (x_{ik} + x_{jk} - 2x_{ik} x_{jk})} \tag{9}
$$

In formula (8) and formula (9): $f(\chi_{ik}, \chi_{jk})$ is the fusion function; C is the fusion cophenetic coefficient, taking 0–1.

5.2 Algorithm Verification and Analysis

To verify the algorithm, the dominant environmental factors of early spring stubble cucumber in greenhouse in March were selected as the experimental subjects, at which time the plants had a large growth and produced more melons. Optimum temperature is 25–32 °C during the day and the illumination time is as long as possible. The relative humidity requirement of cucumber is different in different growth stages, and the relative humidity required in the fruit stage is 70%–80%. A group of environmental data of temperature, humidity and illumination were collected.

In this study, three sensors of temperature, humidity and illumination were selected. Eight sensors are provided for each type to collect relevant data between 11:00 and 18:00 in the greenhouse (Table 1).

Table 1. Comparison of environmental conditions in cucumber greenhouses at different time

Time slot	Temperature (°C)	Humidity (%)	Illuminance (lx)
11:00	23.1	72.89	1420
12:00	23.85	73.10	1500
13:00	25.20	77.42	1750
14:00	26.33	78.32	2150
15:00	26.82	75.24	2232
16:00	25.42	75.12	2120
17:00	25.30	74.86	1950
18:00	24.74	73.12	1780

The data collected by the greenhouse at 15:00 is processed using a data fusion algorithm. The data fusion results are shown in Table 2.

Table 2. Data fusion results

Serial number	Temperature (°C)	Humidity (%)	Illuminance (lx)
1	27.36	72.12	2420
2	27.45	73.54	2321
3	27.82	72.11	2245
4	26.34	74.21	2357
5	26.54	74.35	2360
6	25.41	74.25	2450
7	26.11	73.82	2420
8	26.72	73.12	2415

When m = 2, k = 7, the resulting Euclidean distance formula is

$$D = \begin{bmatrix} 0 & 29.12 & 28.99 & 28.96 & 28.78 & 29.01 & 28.69 & 28.55 \\ 0 & 0 & 28.33 & 27.49 & 28.36 & 28.47 & 28.45 & 28.60 \\ 0 & 0 & 0 & 28.61 & 28.43 & 28.61 & 27.85 & 28.84 \\ 0 & 0 & 0 & 0 & 28.29 & 27.94 & 28.37 & 28.48 \\ 0 & 0 & 0 & 0 & 0 & 28.39 & 27.82 & 28.73 \\ 0 & 0 & 0 & 0 & 0 & 0 & 0 & 28.46 \\ 0 & 0 & 0 & 0 & 0 & 0 & 0 & 0 \end{bmatrix} \quad (10)$$

The fusion algorithm formula (9) gives a temperature value of 28.46, similar to get a humidity value of 78.89%, illuminance is 2584 (lx). The relative error between the data fusion result and the actual measured value is shown in Table 3.

Table 3. Comparison of data fusion results and relative errors

Parameter	Measured value	Fusion algorithm		Arithmetic averaging	
		Measured value	Relative error (%)	Measured value	Relative error (%)
Temperature (°C)	28.46	28.52	0.21	28.90	1.54
Humidity (%)	78.89	78.50	0.62	79.52	0.79
Illuminance (lx)	2584	2590	0.23	2506	3.01

As can be seen from Table 3, the data obtained in Sect. 5.2 of the data fusion algorithm is closer to the measured value, and the fusion result of simple data with relative error less than the average algorithm improves the accuracy and precision of the data in the greenhouse. As a result, the data fusion results with the common arithmetic mean algorithm fusion results are compared, the results show that the relative error of the data fusion algorithm designed to significantly less than the arithmetic mean algorithm, accurately reflect the actual status of greenhouse, improves the precision and accuracy of data acquisition, for in the greenhouse environment to save energy and improve the accuracy of the greenhouse control has important value.

6 Conclusion

This paper mainly studies measurement preprocessing, data correlation, data decision and information fusion and new ideas and algorithm for improving the performance of the corresponding algorithms are proposed in the multi-sensor information fusion technology. The algorithm can better achieve data fusion and reduce data uncertainty and error caused by various errors such as temperature, humidity and light. We conducted experiments in the cucumber greenhouse in March. The results show that the relative error of illumination is significantly smaller than the arithmetic average algorithm, and the decision-making and adjustment basis of the system environmental parameters are optimized, which is beneficial to the economic benefits of high greenhouse production. There is no unified framework at home and abroad to establish a multi-sensor information fusion model, which generally needs to be analyzed according to the actual situation. Establishing a reasonable multi-sensor fusion model is a problem that needs further study. The main information of this paper is integrated in the greenhouse network application and related algorithm design, but it is not suitable for other problems, which deserves further research and discussion.

7 Conflict of Interest

The authors confirm that this publication does not have any known conflicts of interest.

Acknowledgments. Thanks to the support from the Scientific Research Project of Sichuan Provincial Department of Education: Research on New Agricultural Internet of Things Intelligent Management System Based on Zigbee Technology (project number: 17ZB0336).

References

1. Zhou, Y.Q., Hong, X.Z.: Multi-sensor information fusion technology. Telem. Remote Control **1**, 16–22 (1996)
2. Si, X.C., Zhao, L.J.: Anti-radiation missile anti-bait lure technology research. Proj. Guid. **26**(S7), 550–553 (2006)
3. Jianwei, L., Qingchang, R.: Study on supply air temperature forecast and changing machine dew point for variable air volume system. Build. Energy Environ. **27**(4), 29–32 (2008). (in Chinese)
4. He, Y., Guan, X., Wang, G.H.: Research and prospects of multi-sensor information fusion. J. Astronaut. **26**(4) (2005)

5. Lucien, W.: A European proposal for terms of reference in data fusion. In: Commission VII Symposium "Resource and Environmental Monitoring", Sept. 1998; Budapest, Hungary. **XXXII**(7), pp. 651–654 (1998)
6. Solaiman, B., et al.: Information fusion: application to data and model fusion for ultrasound image segmentation. IEEE Trans. Biomed. Eng. **46**(10), 1171–1175 (1999)
7. Wang, F.C., Huang, S.C., Han, C.C.: Multi-sensor information fusion and its new technology research. Aeronaut. Comput. Technol. **39**(1), 102–106 (2009)
8. Guo, H., Zhang, X., Xia, Z.: Target tracking based on frequency spectrum amplitude 1. J. Syst. Eng. Electron. **17**(3), 473–476 (2006)
9. Yan, F., Zhu, X.P.: An improved multi-sensor multi-target tracking joint probability data association algorithm. J. Syst. Simul. **19**(20), 4671–4675 (2007)
10. Li, J.W., Wang, S.Z., Wan, H.Y.: Research on data association algorithm based on Markov chain Monte Carlo method. J. Wuhan Univ. Technol. **31**(6), 1045–1048 (2007)
11. Han, H., Han, Z.Z., Zhu, H.Y., et al.: Heterogeneous multi-sensor data association algorithm based on fuzzy clustering. J. Xi'an Jiaotong Univ. **38**(4), 388–391 (2004)
12. Li, J., Gao, X.B.: A fuzzy clustering data association method based on sensor weighting. Chin. J. Electron **35**(12A), 192–196, 184 (2007)
13. Tu, Y.J., Huang, G.M., Li, J.H.: An improved data association algorithm based on fuzzy clustering analysis. Radar Confront. **1**, 22–24, 46 pages (2008)
14. Chen, W.H., Ma, T.H.: Research and development of multi-sensor information fusion technology. Sci. Technol. Inf. Dev. Econ. **16**(19), 212–213 (2006)
15. Yager, R.R.: On the Dempster-Shafer framework and new combination rules. Inf. Sci. **41**(2), 93–137 (1987)
16. Josang, A., Daniel, M., Vannoorenberghe, P.: Strategies for combining conflicting dogmatic beliefs. In: Information Fusion, Sixth International Conference of IEEE (2003)

Innovation of Office Management Work in Colleges and Universities Under the Background of "Internet +"

Yanping Wang[✉]

Jilin Engineering Normal University, Changchun, China
76772664@qq.com

Abstract. In the era of "Internet +", vigorously promoting the construction of electronic school affairs in colleges and universities has become the important task of changing management functions, working methods and style, further improving work quality and efficiency, and establishing efficient, coordinated and standard educational management system. This work put forward some opinions and countermeasures on how to carry out the office management work better under the background of "Internet +".

Keywords: Office management work · "Internet +" · Function · Innovation

1 Introduction

In order to improve the level and efficiency of office management work, the office management in colleges and universities needs to strengthen the application of network technology, improve the informatization level of office management, and construct a new mode of collaborative office management. "Internet +" refers to the process of diffusion, application and deep integration of the new generation of information technology (including mobile Internet, cloud computing, Internet of things and big data) in various sectors of economic and social life [1]. If colleges and universities can actively explore the new mode of office management work under the background of "Internet +", its management level and effectiveness will be effectively improved [2].

2 The Significance of Offices in Colleges and Universities

2.1 Service Function

Service is the main and the primary responsibility of offices in colleges and universities. Firstly, it should serve the leadership, such as handling daily affairs, collecting information and providing reference for leadership decision-making. Secondly, it should serve government institutions. The office should play a good role as a link to provide communication and coordination services for functional departments and for leadership, and do a good job of document audit and collation [3]. Thirdly, it should serve teachers, students and staff, such as doing a good job of document management, file management, printing management and other matters to effectively help them solve the problems encountered in learning, scientific research, life and all aspects.

2.2 Coordination Function

If colleges and universities are regarded as a living body, then the office is the central neuron, the meeting point of all kinds of opinions output and feedback and the integrated management organization that communicates and coordinates up and down, and it undertakes the coordination function of comprehensiveness and integrity. Only by coordinating all kinds of relations can all contradictions and problems that arise in the process of various departments work in colleges and universities be solved and the situation of bickering and prevarication be avoided, so that school leaders are able to concentrate their efforts on personnel training and scientific research to achieve the purpose of serving teaching and scientific research.

2.3 Supervision Function

Supervision is an indispensable administrative means in various organizations. Whether the task is completed, what is the quality of the completion, whether the goal is achieved and whether the plan is achieved can not be separated from supervision. Therefore, the office must play the role of supervision to promote subordinates to actively complete the task. At the same time, through the in-depth investigation of practice in the process of supervision, problems can be found and the deviation can be corrected, so as to improve the work efficiency.

3   Problems Existing in the Office Management Work in Colleges and Universities

At present, the management systems of colleges and universities are all local development faced to department. Since there is no shared application architecture and shared technical architecture, isolated island of information and applications must exist [4]. What is more, since the work flow of each department only considers the offices within the department, it is impossible to achieve integrated management among departments and to achieve collaborative work among various departments. The specific aspects are as follows:

Firstly, development lacks unified planning. The construction of office management work mode in colleges and universities should be an organic integrity. However, due to the discontinuity of informatization construction caused by the inconsistency of investment in education, and the relatively loose relationship among departments, the construction of campus informatization lacks unified planning. Therefore, it is difficult to process information at a higher level, such as information mining and decision support. For different application systems, users need to log in to access separately and lack unified access resource and application interface. In the face of a wide range of applications, users are difficult to find what they need, and the degree of humanization is low.

Secondly, information lacks effective sharing. The office management work in colleges and universities should be an organic integrity. However, since colleges and universities are in the reform period of educational reform, the application system is developed by different people in different periods, and lacks the overall system planning. The data sharing among application systems still depends on the backward and inefficient information transmission mode, which makes it difficult for colleges and universities to share the affairs information effectively [5]. In addition, although individual systems run on computers connected to campus network, their own operation mode is personal computer mode, which is difficult to share information with other systems. Lacking of effective information sharing can have a significant impact on the efficiency and accuracy of the whole campus network application system.

Thirdly, application lacks effective integration. The office application system of colleges and universities should be an organic integrity. In addition to the above reasons, since the application system may be developed by different software platforms and the application access interface lacks unified planning, application systems lack integration [6]. Therefore, the same user may need different password and even different identity when entering into the different application system of campus network. Application systems can not directly access each other's data and functions and sometimes need human processing, which is lack of effective integration.

4   Optimization of Office Management Work in Colleges and Universities Under the Mode of "Internet +"

4.1   Constructing Electronic Platform of School Affairs

The level of administrative decision-making in colleges and universities directly affects the vitality of administrative management and the smooth realization of administrative objectives. Under the mode of "Internet +", the electronic platform of school affairs will lay the foundation for the administrative leaders of colleges and universities to make scientific decisions by providing valuable and high-quality information.

Firstly, conceptual guidance. It is necessary to raise the leaders' understanding of the electronic platform of school affairs. The information construction of some colleges and universities is still in its infancy, and it will take long time to realize informatization [7]. If leaders of colleges and universities do not pay enough attention to the construction of electronic school affairs, the large amount of capital investment needed for the construction can not be guaranteed, and the construction process can not be accelerated. Therefore, it is necessary to improve the leaders' understanding of the necessity to construct the electronic school affairs and to improve the efficiency and management level of the office.

Secondly, system guarantee. It is necessary to strengthen the system construction, the standardized management, the coordination management system and the safety guarantee system of electronic school affairs. Colleges and universities should actively formulate and perfect the relevant systems and information security measures in the process of constructing electronic school affairs in order to ensure the smooth and safe circulation of campus information data [8]. At the same time, it is necessary to establish and improve the mechanism of information feedback and response, improve the efficiency of office work, promote the scientific decision-making of leaders, and speed up the process of democratic management in colleges and universities.

Thirdly, material guarantee. The overall popularization and application level of information technology in some colleges and universities still needs to be further improved, and it is urgent to further strengthen the infrastructure construction which is matched with electronic school affairs. It is also necessary to speed up the construction of information network facilities, strengthen the promotion of collaborative technology and speed up the construction and application of the technical standard system of electronic school affairs.

Fourthly, personnel cultivation. The management of electronic school affairs system and the working ability of users in colleges and universities are directly related to the operation efficiency of electronic school affairs. Therefore, colleges and universities should actively carry out all kinds of training related to information network knowledge at all levels and systematically using skills in order to cultivate managers to set up awareness of serving education and teaching, teachers and students' lives and the development of colleges and universities [9]. At the same time, all the teaching staff in colleges and universities should keep up with the reform pace of administration and management more quickly.

4.2   Improving Systematic Management Work

Office in colleges and universities is different from other business departments, since it is responsible for comprehensive affairs up to the top management departments and down to the basic affairs of teachers, students and staff. Although it does not directly participate in the specific business work, the wide scope of its work determines that it exists in all kinds of management work, which is of great significance to the development of higher education [10]. Therefore, in the process of carrying out all kinds of management work, the following work should be improved:

Firstly, humanized management. Colleges and universities should adhere to people-oriented and carry out the humanized management, which is to respect, understand, care for and meet the legitimate needs of each staff, and to create a harmonious working atmosphere. Colleges and universities should provide more professional training opportunities to help everyone find a correct career positioning and expand the space for development. At the same time, the office equipment should be updated step by step, so that the staff can get rid of the tedious affairs to maximize the creativity of people and finally improve the overall work efficiency of the office.

Secondly, institutionalized management. According to the development needs and reality of colleges and universities, a series of rules and regulations of office management should be formulated, so that every work in the office of colleges and universities has rules to follow. Colleges and universities should also strengthen people-oriented consciousness, establish competition mechanism and policy that can both restrain and encourage, create a working environment in which people can make the best use of their talents, encourage and mobilize the initiative, enthusiasm and creativity of staff. It is necessary to standardize the procedures and clarify the order, specific steps and basic requirements, especially not to bypass the rank or prevaricate [11].

Thirdly, performance evaluation. It is very important to carry out scientific performance evaluation and reasonable salary and welfare optimization to improve the enthusiasm of office staff. In the process of the performance evaluation reform of office staff, it is of great importance to seek opinions and suggestions from staff of the content and standard, the implementation of the system, the types of evaluation and the evaluation methods. In this way, not only the democratic atmosphere of the organization will be improved, but also the management leaders of the office will be widely accepted, which is beneficial to the further development of management. Most importantly, the performance evaluation standard comes from the staff themselves, therefore, its persuasive and binding force are stronger. In addition, while formulating a reasonable performance evaluation system, colleges and universities should also implement a more reasonable salary and welfare system, which is also an important factor to stimulate the enthusiasm of staff.

5 Summary

From the perspective of the whole business of colleges and universities, the management mode of office management work must proceed from the orientation of centralization, sharing and coordination and incorporate the advanced collaborative management concept. It should carry out the timely, efficient, orderly and controllable communication and management to all aspects of the daily work of the office to effectively eliminate the loose and isolated information processing among departments and colleges, effectively improve the level and efficiency of school affairs management and achieve data sharing and business coordination. In addition, it should also do a good job in the construction of electronic school affairs platform, strengthen management and supervision to achieve the overall optimization of office management work.

References

1. Zhijun, G.: Thoughts on innovating the concept of office management in colleges and universities. J. Zhangzhou Teach. Coll. (Philos. Soc. Sci. Ed.) 23(02), 174–176 (2009)
2. Jianping, L.: Research on the innovation countermeasures of office management in colleges and universities - research on the "four key laws" of the office. J. Inner Mongoli Agric. Univ. (Soc. Sci. Ed.) 11(03), 152–153 and 157 (2009)
3. Wentao, X.: Research on the problems and countermeasures of office management work in secondary colleges and universities. J. Northwest Adult Educ. (04), 35–36 and 39 (2011). (in Chinese)
4. Jian, H.: Research on the scientific management work of office in colleges and universities. J. North China Electric Power Univ. (Soc. Sci. Ed.) S2, 289–292 (2011)
5. Bing, L.: Research on flexible management of office in colleges and universities. J. Party School Shengli Oilfield 29(02), 113–115 (2016)
6. Xiaolin, S.: Analysis of the ways to improve the level of office management work in colleges and universities. J. Liaoning Econ. Manage. Cadre Inst. 04, 53–55 (2016)

7. Zhijun, G.: Scientization of office management work in colleges and universities. J. Zhangzhou Teach. Coll. (Philos. Soc. Sci. Ed.) (02), 169–171 (2008). (in Chinese)
8. Hongyu, L., Jing, Z., Hui, Q.: Research on improving the efficiency of office management work in colleges and universities. J. Hebei Agric. Univ. (Agric. For. Educ. Ed.) **16**(04), 27–29 and 33 (2014). (in Chinese)
9. Mimi, C.X.: Modernization of office management work in colleges and universities. J. Yuzhou Univ. (Soc. Sci. Ed.) (03), 126–128 (2001)
10. Qinmei, X.: Research on the office management work in colleges and universities from the perspective of management communication theory. J. Fujian Public Saf. Coll. **32**(02), 104–108 (2018)
11. Bing, L.: Research on the innovation of office management work in colleges and universities under the background of comprehensive reform of education. J. Shengli Coll. China Univ. Pet. **29**(03), 55–58 (2015)

Development and Design of Web-Based Distance Physical Education Teaching Platform

Shoucheng Zhang[(✉)]

Chongqing Real Estate College, Chongqing, China
Shoucheng_Zhang85@haoxueshu.com

Abstract. The distance education platform integrates the functions of network learning, interactive cooperation and teaching management. It has friendly interface and convenient use, greatly reduces the workload of teachers, effectively improves the learning efficiency of students, and is conducive to promoting the transformation process from examination-oriented education to quality-oriented education. Sports distance network teaching platform mainly provides long-distance sports teaching service. It is a system that organically combines physical education network courseware with other long-distance teaching services in schools. In this study, the existing sports distance network education platform mainly takes distance learning system, online question answering system and remote sports real-time interactive system as the core, and manages each part through the platform management system.

Keywords: WEB · Distance · Physical education teaching platform · Design and implementation

1 Introduction

With the continuous development of science and technology, the position and role of computer and computer network are becoming increasingly prominent. Especially in correspondence education, vocational education and continuing education, distance learning has become an important teaching mode [1, 2]. Web-based distance education is not limited by time and space. It has the characteristics of low cost, fast updating of teaching content, full utilization of teaching resources and media, strong interaction of teaching process, and emphasis on personalized learning. Therefore, it has been widely used [3–5]. Because Web-based distance education is attracting more and more attention at home and abroad, many universities and companies are investing a lot of energy in the research of distance education system and teaching platform. But at present, there are many problems in the applied teaching system, such as teaching mode, teaching strategy, the organization of teaching content and so on, which can not meet the requirements of students' personalized learning, and the interactive function of the system is not strong, so it needs to be further improved [6].

How to use modern educational technology to design a high-quality distance education platform is a new topic of current educational technology, which has

© Springer Nature Singapore Pte Ltd. 2020
M. Atiquzzaman et al. (Eds.): BDCPS 2019, AISC 1117, pp. 122–129, 2020.
https://doi.org/10.1007/978-981-15-2568-1_18

attracted extensive attention of educators. At present, the development of distance network teaching in our country is slow [7, 8]. For distance sports teaching, domestic scholars and experts have less research. However, due to the need for sports activities to provide larger venues, and the classification and borrowing of sports equipment is cumbersome and complex, making the school human resources and so on have not been more reasonable use [9]. Therefore, in view of the development situation, it designs and develops a WEB-based distance sports teaching platform here.

2 Analysis of Development Environment and Key Technologies of Sports Distance Education Platform

Distance education platform provides a more vivid teaching environment for teachers and students, and enriches the educational model. First of all, the platform of distance physical education needs to understand the characteristics of flexibility of physical education teaching. Based on this kind of teaching characteristics, it designs the specific teaching methods and contents of the course, as well as the design of course assessment. Course design requires that students' enthusiasm for learning be fully mobilized, and that students' curricula be guided, especially the standardized guidance of action class, be clear and thorough.

2.1 Demand Analysis and Teaching Design

On the premise of the rapid development of distance education, the emergence of sports distance network education platform has become a necessary way to solve the bottleneck of physical education. Sports distance education platform is not only a simple and pure technical system platform, but also an organism composed of sports socialization and social sports demand, sports information and theoretical knowledge, and service elements such as teacher-student interaction, expert support and teaching guidance. With its powerful functions and rich content, it provides an integrated interactive learning environment for all users with learning needs. In order to effectively manage and track students' sports learning and sports training, it is necessary to provide effective teaching means and editing tools for sports curriculum design and teaching content for physical education teaching workers in various ways, and to provide rich system and teaching management functions [10, 11].

Sports design learning process according to the characteristics of motor skill learning. In the design of the sports learning process, it is necessary to fully mobilize the participation of the students' various sensations, strengthen the understanding and memory of the movements, enhance the interest in learning and actively learn the emotional state. The network course can achieve the above goal by designing a web interface with artistic aesthetics, vivid video images, slow presentation of action details and key points, clear language explanation, etc. The knowledge structure of the content of physical distance network course is as follows Fig. 1.

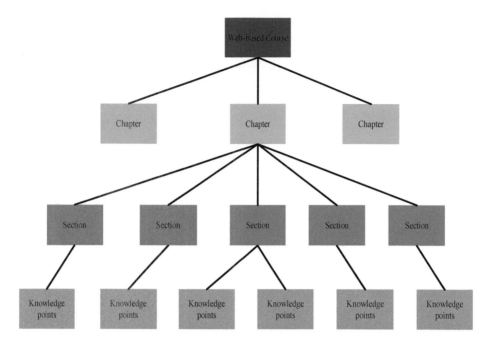

Fig. 1. Knowledge structure map of physical education distance network course content

2.2   Design Criteria of Sports Distance Network Education Platform

The implementation of any project can not be separated from the norms, which mainly play the role of constraints and norms. In the sports distance network education platform, the most used educational technology is technology-oriented, and the personality of people is less reflected. Therefore, to formulate and follow reasonable design norms and make the combination of educational technology and sports humanities become an important breakthrough point in this study. Only by carrying out and implementing the code of conduct can we achieve high-quality platform development results [12]. Generally speaking, sports distance education platform should follow the principles of expansibility, practicability, unity, security and reliability.

(1) Scalability

The research and design of sports remote network teaching platform not only meets the needs of current college physical education, but also lays a good foundation for future platform expansion. The key lies in the sustainability of platform hardware and software selection, and the extensibility and changeability of software programs. Therefore, it is very important to make excellent design in the development stage to make the platform have a good structure. At the same time, it is particularly necessary to distinguish the changing part from the stable part of the platform. The change part is mainly used for the maintenance of the platform after operation, while the stable part is sealed.

(2) Practicality

The practicability includes not only the practicability of functions, but also the practicability of the platform itself in the development process. The purpose of the development of this platform is to promote the reform and innovation of physical education. But we can't blindly seek for perfection in order to seek novelty and difference. Easy to operate and understand, convenient and practical is a very important part of the platform. Moreover, in the process of platform development and design, hardware investment should take into account both sustainable development and practical requirements. No waste can be caused.

3 Design and Implementation of Long-Distance Physical Education Platform

The development and design in this paper follow the criteria of safety, reliability, extensibility and practicability, which is the guarantee of stable and reliable performance of the designed platform. Server - Server is a kind of hardware that can provide services to clients through the network and has strong computing power. Tomcat web software is used in the design and development of this paper. Tomcat is a miniaturized Web application server with free source code. Its latest version is 6.0.18. Because of the advantages of lightweight, Tomcat servers are often used in small and medium-sized systems with fewer users. Figure 2 shows the schematic diagram of the software structure.

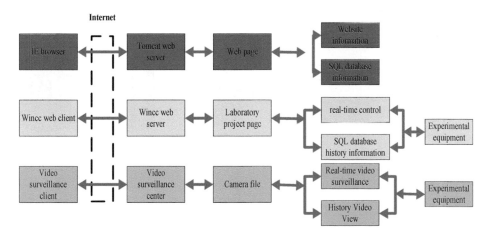

Fig. 2. Software design structure of distance physical education

The purpose of the distance education platform is to provide students with an interactive platform for learning physical education. The system contains various kinds of information, such as students' basic information, teaching videos and homework assessment. This information is stored in the platform's data for users to call and view as needed. Figure 3 is a schematic diagram of the teaching mode of the distance physical education platform. According to the figure, the user sends a request instruction to the server through the platform browser. The server responds to the request in a timely manner and provides feedback to the user.

Database - As we all know, the database system is the support of an information system, which is the core part of the whole information system. The information system designed in this paper chooses the SQL database. Microsoft SQL Server 2003 is developed by Microsoft. It has the advantages of comprehensive function and stable technology. At the same time, it also has good scalability, whether it is a simple personal database or a larger complex enterprise database or even a global site database, SQL can support.

Electronic whiteboard - Electronic whiteboard is a virtual public area based on B/S structure. In the system designed in this paper, it is implemented by Java technology and CSCW technology. In the sports distance education platform, CSCW technology provides a virtual environment for information sharing. In this virtual public environment, teachers and students can learn interactively.

Electronic whiteboard is composed of client and server. Image, audio and other information is stored in the server, users can get these data by means of replication. If these data need to be changed, the change structure will be sent to the server for data updates, and the client will modify the data at any time. Therefore, electronic whiteboard also has the advantage of real-time.

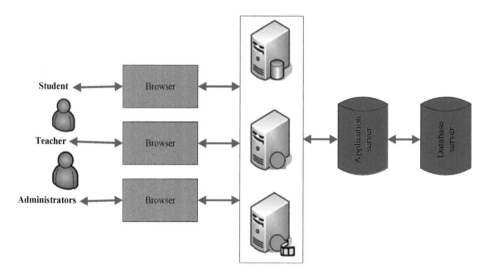

Fig. 3. Schematic diagram of teaching mode

According to the system development and design requirements of the WEB-based remote physical education platform, the platform architecture is divided into five parts. The specific deployment structure is shown in Fig. 4. The system builds an interactive platform for users to learn, communicate and participate in the assessment.

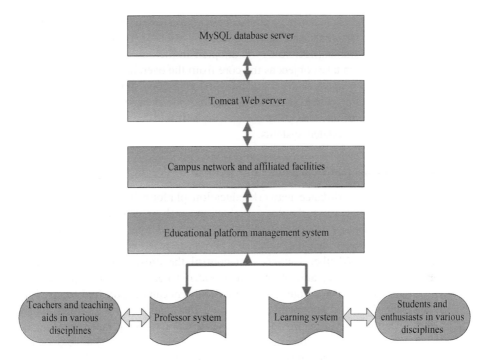

Fig. 4. Schematic diagram of server deployment of distance sports teaching platform

In order to realize a complete distance physical education platform, the system designed in this paper includes three sub-platforms: portal, learning and management. The portal sub-platform is used for publishing information, learning sub-platform is used for user's learning interaction, including course learning and assessment, and management sub-platform is used for the management and maintenance of the whole system to ensure the stability and effectiveness of the whole system. The platform is divided into two parts: the outer network user sub-platform and the inner network user sub-platform. Outer users enter the system mainly through the portal platform, such as the release of various information such as enrollment information, which is fed back to users through the portal platform, while intranet users enter the system through the teacher management system and student channels for the exchange and learning of teachers' curriculum management and students' curriculum.

4   Analysis and Discussion on the Design of Sports Distance Education Platform

4.1   Advantages of Teaching Application in Sports Distance Education

The main function advantages of ontology distance network education are embodied in that it better reflects the role of teachers' leading role and students' principal position, and maximizes the complementary role in and outside the classroom. It can also promote the saving and efficient use of physical education resources, and has prominent effect on the whole teaching and learning-related resource-based learning model. Object-oriented programming technology is adopted in the ontology distance network teaching platform, which takes object as the core from the overall design to the detailed design. It has good openness and expansibility. It is a good platform for follow-up development. It is believed that good economic and social benefits will be achieved after the platform is put into use. At the same time, it will have some reference and practical value for other related systems.

4.2   Main Problems and Deficiencies of Platform in Practical Application

The design of ontology distance network education platform has strict requirements and regulations for software technology evaluation, such as page layout, navigation design, color matching and link format. Whether to provide timely and effective online help, necessary additional plug-ins, necessary human-to-machine or human-to-human interaction functions. Whether learners can control the presentation of multimedia information (such as voice, animation, video, etc.); whether they can provide the retrieval of the course's own resources; whether learners can quickly find the position of knowledge points in the knowledge system and the relationship with other knowledge points; whether they can provide a mechanism for tracking and recording the learning process. As far as possible, the network course platform can be developed into executable files or scripts that can run directly without installation. Therefore, it should be a key point in the development and production of web-based curriculum to formulate the standards of web-based curriculum platform and develop a student-oriented learning platform (web-based learning support environment) which is relatively independent and compatible with specific courses.

Compared with the teaching system of other disciplines, the module function and style of the ontology education remote network platform are too unified. Because the basic construction of campus network teaching is not yet developed, the information technology literacy of teachers and students participating in physical education curriculum needs to be improved, and there is a certain deviation between curriculum design and teaching practice application. In this paper, the knowledge model of sports theory has not been discussed in depth. The establishment of knowledge model needs to refer to all aspects of pedagogical theory. The basic theories of constructivism, ontology, activity theory and educational psychology are all the contents of this paper's instructional design. In this paper, these theories and knowledge models are seldom discussed.

5   Conclusion

The WEB-based distance learning platform provides a new direction for the development of physical education, so that future teaching activities are no longer limited to space and time. This paper introduces a Web-based distance education platform, and introduces in detail the structure and implementation of the individualized teaching subsystem, the Intelligent Question Answering System Based on natural language understanding and the test subsystem. Due to the existence of natural language understanding problems, there are still many problems to be solved in the real intelligence of answering questions and the evaluation of subjective examination questions, which is also the direction of further efforts of distance education platform.

References

1. Chen, Y.C.J., Sinelnikov, O.A., Hastie, P.: Professional development in physical education: introducing the sport education model to teachers in Taiwan. Asia Pac. J. Health Sport Phys. Educ. **4**(1), 1–17 (2013)
2. Zhao, M.J.: A Knowledge-based teaching resources recommend model for primary and secondary school oriented distance-education teaching platform. LNEE, vol. 269, pp. 511–521 (2014)
3. Michael, R.D., Webster, C., Patterson, D., et al.: Standards-based assessment, grading, and professional development of california middle school physical education teachers. J. Teach. Phys. Educ. **35**(3), 277–283 (2016)
4. Liu, J.D., Chung, P.K.: Development and initial validation of the Chinese version of psychological needs thwarting scale in physical education. J. Teach. Phys. Educ. **34**(3), 402–423 (2015)
5. Garmpis, A., Gouvatsos, N.: Design and development of WebLIbu: an innovating web based instruction tool for Linux OS courses. Comput. Appl. Eng. Educ. **24**(2), 313–319 (2016)
6. Cai, J.Y., Zhang, P.P.: The support environment construction for teaching and research of physical education based on emerging information technology. J. Comput. Theor. Nanosci. **14**(4), 2015–2020 (2017)
7. Ward, P., Kim, I., Ko, B., et al.: Effects of improving teachers content knowledge on teaching and student learning in physical education. Res. Q. Exerc. Sport **86**(2), 130–139 (2015)
8. Pill, S., Harvey, S., Hyndman, B.: Novel research approaches to gauge global teacher familiarity with game-based teaching in physical education: an exploratory #Twitter analysis. Asia Pac. J. Health Sport Phys. Educ. **8**(2), 161–178 (2017)
9. Zhou, Ke: Research on Preservice physical education teachers' and preservice elementary teachers' physical education identities: a systematic review. J. Teach. Phys. Educ. **36**(2), 1–29 (2017)
10. Phillip, W., Yung-Ju, C., Kelsey, H., et al.: Teaching rehearsals and repeated teaching: practice-based physical education teacher education pedagogies. J. Phys. Educ. Recreat. Dance **89**(6), 20–25 (2018)
11. Huang, H., Huang, H., Huang, H., et al.: Analysis on the teaching model of physical education in colleges based on interactive teaching method and participatory teaching method. Int. Technol. Manage. **11**, 10–12 (2016)
12. Zadahmad, M., Yousefzadehfard, P.: Agile development of various computational power adaptive web-based mobile-learning software using mobile cloud computing. Int. J. Web-Based Learn. Teach. Technol. **11**(2), 61–72 (2016)

Three-Phase Imbalance Control Method for Distribution Grid Based on SVG and Phase Switch Technology

Yangtian Wang[(✉)]

College of Electrical Engineering and New Energy,
China Three Gorges University, Yichang 443002, Hubei, China
740360384@qq.com

Abstract. With the increasing diversification of unbalanced load in distribution grid in smart city, the three-phase imbalance becomes more and more serious, which has a great impact on the safety of power grid operation and the quality of electric power. To realize the load adjustment in power distribution and transformation for the goal of power loss reduction, this paper firstly systematically analyzes the power loss of each part in the system, and compares the characteristics and balancing outcome between various adjustment technologies on the analysis basis. Then, this paper further comes up with the comprehensive compensation method by combing Static Var Generator (SVG) technology with phase switching technology. This method can not only effectively improve the power loss of low-voltage power lines, but also can realize the optimal balance adjustment in distribution network. The correctness and plausibility of this method is further verified by simulation.

Keywords: Distribution network · Three-phase imbalance · Analysis · Big data · Comprehensive balance control

1   Introduction

With the rapid development of economy in China in recent years, the structure, type and power consumption of the load in distribution network have greatly changed. The electricity load has increased significantly, and the number of distribution transformers has also increased on a large scale, while the imbalance of the three-phase distribution network is becoming more and more prominent. It will lead to the increase in the damage rate of electrical equipment and will affect the electricity quality. The longer the existence of the problem, the more economic losses will be brought, and the decreasing safety and stability will be brought to the operation of distribution network [1]. Therefore, it is necessary to study an effective imbalance control method for distribution grid in smart city.

However, the implementation of big data will make the solution to three-phase imbalance in distribution network possible. The loss and three-phase voltage deviation caused by the unbalanced operation of the distribution transformer are described, and a method of compensation by connecting phase capacitor is proposed in [2]. In [3], it further

© Springer Nature Singapore Pte Ltd. 2020
M. Atiquzzaman et al. (Eds.): BDCPS 2019, AISC 1117, pp. 130–137, 2020.
https://doi.org/10.1007/978-981-15-2568-1_19

analyzes the effect of the three-phase imbalance on the operating capacity of the distribution transformer, and gives the constraints of various types of load operating areas. Big data extraction technology to carry out the basic data management of distribution network to prove the correctness of the analysis and calculation in power electricity line loss of the distribution network was given in [4]. Given the problems of slow data synchronization, insufficient storage resources and slow computing processing in the data analysis of distribution line loss, the scheme based on big data analysis, and improves the high-performance computation of big data in distribution line loss was puts forward [5]. The adjustment workflow for the three-phase load imbalance based on big data technology aiming for the technical loss reduction, and establishes the management system for three-phase load imbalance and technical loss reduction [6]. A new-type SVG is introduced in [7]. Compared with conventional types, SVG can have higher voltage and larger reactive power output current. A control strategy based on SVG compensated reactive power and negative-order current was proposed in [8], which can greatly enhance the stability and electricity power quality of the system.

In the real-life implementation in the distribution network nowadays, there are some limitations when implementing only one device or compensation method to solve the wide-area three-phase imbalance problem. In this paper, simulations under different balance parameters are analyzed, and the compensating results of the electricity quality in the three-phase imbalanced situation are compared. Furthermore, the comprehensive compensation method combined with phase-changing and SVG of the big data is proposed in the paper, which can effectively solve the problem of three-phase imbalance and further lowering the loss. The simulation has proved the correctness and plausibility of the method proposed in this paper.

2   Three-Phase Load Imbalance in Distribution Network

2.1   Danger of Imbalance

Distribution network is an important part of power system, whose electricity quality directly affects its users. The three-phase imbalance is an important index to evaluate the electricity quality supplied by the distribution network. When system is under the imbalance operation, the dangers can be mainly concluded as: (1) It will not only increase the loss of phase line in three-phase and four-wire power supply network, but will also increase neural line losses, thus increasing the overall line losses in the distribution network; (2) The transformer loss will be increased; (3) The utility ratio of the transformer in the distribution network will be decreased; (4) The zero-phase-sequence current will be produced in transformer, leading to hysteresis and vortex loss, and will increase the operation temperature, which will reduce the life span of the transformer; (5) The quality of the supply voltage of the heavy-duty phase in the distribution station area will be seriously reduced [8, 9].

2.2   Calculation for Imbalance

There are many kinds of calculation methods of three-phase imbalance, among which the three-phase current imbalance is relatively simple, thus it is widely used in the real-life engineering implementations. According to the definition in "Corporate Standards in State Grid Corporation of China (Q/GDW519-2010) Operation Procedures in Distribution Network", the following equation can be obtained.

$$\rho_i = \frac{I_{\max} - I_{\min}}{I_{\max}} \times 100\% \tag{1}$$

Where I_{max} is the effective value of the largest current in the three-phase, and I_{min} is the effective value of the smallest current in the three-phase. This calculation method is simple and is with relatively small error. The imbalance of the current in three-phase is calculated based on this method.

3   Three-Phase Imbalanced Governance Process Based on Big Data Platform

Because of the huge number of users and the dispersed operation position in the distribution station area, it is difficult to collect the operation data of the distribution transformer and the heavy workload of manual collection. It leads to the serious problem of three-phase imbalance. Intelligent distribution network monitoring platform provides a big data base for three-phase unbalance control of distribution network. Under the background of big data era, Intelligent distribution network monitoring platform integrates the system information such as EMS(Energy Management System), DAS(Distribution Automation System), GIS(Geographic Information System), PMS (Power Production Management System), marketing, production and repair command platform of distribution network, and electricity information acquisition. It organically integrates the isolated and scattered information into one platform, realizes information sharing, and improving the management of distribution network.

Intelligent distribution network monitoring platform realizes online real-time acquisition of distribution network operation data. The data collected by power information acquisition terminal, DTU(Data Transfer Unit), TTU(distribution Transformer supervisory Terminal Unit), distribution transformer monitoring terminal are uploaded to the main station server. The three-phase imbalance can be identified and analyzed by automatic terminal collection information. The three-phase imbalanced governance process of the distribution area base on big data platform is shown in Fig. 1.

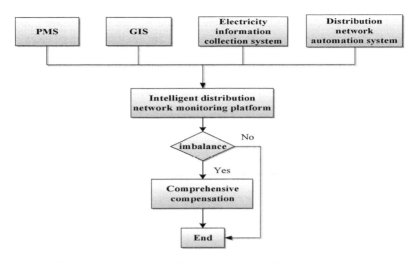

Fig. 1. Three-phase unbalance judgement and treatment process

4   Loss Analysis of Three-Phase Imbalanced System in Distribution Network

4.1   Line Loss at the High-Voltage Side

Assumed that R_g and X_g are the resistance and reactance of the high-voltage line, the calculation equations are shown below:

$$R'_g = k_0 R_0 \tag{2}$$

$$X'_g = 0.0029 f_0 \lg(\frac{d_m}{r_e}) \tag{3}$$

Where k_0 is the skin effect coefficient, R_0 is the direct current resistance, f_0 is the power supply frequency, d_m is the geometry distance between phase lines, and r_e is the effective radius of the line. The overall loss of the line can be calculated as below:

$$p_g = m I_g^2 R' \tag{4}$$

4.2   Transformer Loss

Low-voltage loads in the China distribution network are mainly powered by the electricity in 380 V, thus the motor should be powered by step-down transformer. Compared to the asynchronous motor with same capacity, its loss under rated load is smaller and its effect can still reach up to 99%. In this way, the magnetic saturation is not taken into consideration during the loss calculation in the transformer. Also, the iron-loss fluctuated by the load is ignored. In summary, the circuit and the transformer equivalent circuit are shown in Fig. 2, where R_{b2} and X_{b2} are high-voltage impedances converted from the low-voltage side; R_{bm} and X_{bm} are excitation impedances.

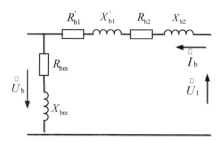

Fig. 2. Transformer equivalent circuit

$$p_b = m[I_b^2(R_{b1}' + R_{b2}) + \frac{U_b^2}{R_{bm}^2 + X_{bm}^2} R_{bm}] \qquad (5)$$

4.3   Loss at the Low-Voltage Side

The highly dependent information for the loss calculation in the distribution network like line switch, equipment account, power network topology and operation data in the power grid can be timely obtained from power grid operation management systems like PMS and electricity information collection system. The classic implicit Z-bus Gaussian method [10, 11] can be expressed as the function below:

$$\mathbf{V} = \mathbf{Y}^{-1}(\mathbf{I} - \mathbf{Y}_{ns}v_s) = -\mathbf{Y}^{-1}\mathbf{I}_L + v_s e_{n-1} \qquad (6)$$

Where e_{n-1} is a unit column vector whose dimension is $(n-1)$; Y_{ns} is the column vector formed by mutual admittance between other bus and slack bus; I_L is the column vector formed by load current at each bus in the distribution network; v_s is the voltage at the slack bus in the distribution system, normally is the flat start and taken as $v_s = v_0 = 1$. Function (6) can be written as follow in the iterative form:

$$\mathbf{V}_{(k+1)} = -\mathbf{Y}^{-1}\mathbf{I}_{L(k)} + v_s e_{n-1} \qquad (7)$$

Then the active power loss at the low-voltage side is:

$$P_1 = \sum_{l=1}^{k_l} R_l I_l^2 \qquad (8)$$

Where k_l is the number of total branches in distribution network, and I_l is the resistance of the lth branch.

5 Comprehensive Compensation Method and Simulation Verification

5.1 Establishment of Comprehensive Compensation Method

In the real practice, the practical load balance adjustment and reduction of line loss at the low-voltage side can be realized by phase switching. However, the optimal balance in load distribution cannot be realized by this method. The compensation effect for SVG technology is good and can realize the smooth adjustment, while have little improvement on line loss at the low-voltage side. In this way, this paper comes up with the comprehensive compensation method by combing phase-switching technology with the SVG technology, which can not only effectively improve the line loss at the low-voltage side, but can also realize the optimal adjustment for the balance in distribution network. The system can be shown in Fig. 3.

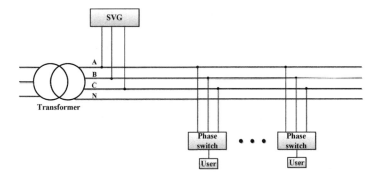

Fig. 3. Comprehensive compensation system

5.2 Simulation Verification for Balance Effect

(1) Effect of phase-switching

For the convenience to conduct analysis, the load number at each phase at the initial condition is set as 6 in this paper. According to the table, loads between phase A and phase C are imbalanced, whose imbalance is up to 0.57.

According to the acquired data, the optimal operation of phase switching can be obtained. When the phase switching is conducted for 3 times, the optimal control can be realized. The balance after phase switching is 0.18. Detailed data are shown in Table 1.

(2) Effect of SVG adjustment

For the convenience to conduct analysis, the initial data in Table 1 is used to conduct the adjustment control during SVG adjustment simulation. Results are shown in Table 2. According to the Table 2, the imbalance in transformer and the high-voltage line is lowered to 0.03, while there is little improvement in the imbalance at the low-voltage line. The reduction for overall loss is not obvious.

Table 1. Data results after phase switching adjustment

Name	Phase A	Phase B	Phase C	Imbalance	Loss (KW)
Before phase switching	560.9	400.7	240.2	0.57	37.5
After phase switching	445.2	400.4	362.5	0.18	31.3
Phase switching times commutations	1	1	1	–	–

Table 2. Data results with SVG compensation method

Name	Phase A	Phase B	Phase C	Imbalance	Loss (KW)
High-voltage side and transformer before compensation	21.4	17.4	13.9	0.35	2.9
Low-voltage side before compensation	560.9	400.6	240.2	0.57	34.65
High-voltage side and transformer after compensation	16.3	16.1	15.8	0.03	1.87
Low-voltage side after compensation	559.8	400.9	240.4	0.57	34.5

(3) Comprehensive compensation results

Table 3 is the data results after the implementation of proposed comprehensive compensation method in this paper. According to the data, the imbalance at the high-voltage line is lowered to 0.025, and the imbalance at the low-voltage line is lowered to 0.17. The overall loss in the system is lowered to 30.2 kW. In this way, the effect of the comprehensive compensation method is more obvious.

Table 3. Data results of comprehensive compensation method

Name	Phase A	Phase B	Phase C	Unbalance	Loss (KW)
High-voltage side and transformer before compensation	21.4	17.4	13.9	0.57	2.9
Low-voltage side before compensation	560.9	400.6	240.2	0.57	34.6
High-voltage side and transformer after compensation	16.2	16.0	15.8	0.025	1.8
Low-voltage side after compensation	441.1	400.1	362.2	0.17	30.2

6 Conclusion

Based on the theory analysis above, this paper proposes a comprehensive compensation method by combining phase switching method and SVG technology. According to the simulation results, compared to mono-compensation technology, the comprehensive compensation method can lower the imbalance at the high-voltage side to 0.025, and the imbalance at the low-voltage side to 0.17. At the same time, the loss can be further decreased, which has prominent effect.

References

1. Haixue, L.: Three-phase Imbalance of The Power System. China Electric Power Press (1998)
2. Yunlong, Y., Fengqing, W.: Additional loss and voltage deviation caused by unbalanced of distribution transformer and countermeasures. Power Syst. Technol. **2004**(08), 73–76 (2004)
3. Zhiqiang,Y., Xia, L.: Study on the influence of three-phase unbalance on the load capacity of distribution transformer. Electric. Measur. Instrum. **55**(8) (2018)
4. Wei, L.: Analysis of distribution line loss based on big data. Electron. Technol. Softw. Eng. **2017**(14), 186–188 (2017)
5. Lihua, S., Mu, H., Meng, Qingqiang: Solution for high performance of distribution network line loss big data. Comput. Modernization **2016**(12), 42–46+50 (2016)
6. Yuan, R., Chengjun, H., Weiwei, Z., Wenshi, H.: Comprehensive research on three-phase unbalance analysis and load adjustment based on big data. Autom. Instrum. **11**, 166–169 (2016)
7. Qian, D., Hu, C., Zhu, M., Liu, Y., Zheng, C.: Research on three-level SVG in the distribution network system. In: IEEE Conference on Industrial Electronics and Applications (ICIEA), Siem Reap, pp. 1112–1116 (2017)
8. Yue, G., Li, X., Sui, X.: The research of SVG based on the unbalanced power system. In: 2016 International Symposium on Computer, Consumer and Control (IS3C), Xi'an, pp. 866–869 (2016)
9. Hengfu, F., Wangxing, S.: Research on the method for real-time online control of three-phase unbalanced load in distribution area. In: Proceedings of the Chinese Society for Electrical Engineering, vol. 35(9), pp. 2185–2193 (2015)
10. Chen, T.H., Chen, M.S., Hwang, K.J., et al.: Distribution system power flow analysis-a rigid approach. IEEE Trans. Power Delivery **6**(3), 1146–1152 (1991)
11. Chiang, H.D., et al.: Convergence/divergence analysis of implicit Z-bus power flow for general distribution networks. In: IEEE International Symposium on Circuits & Systems (2014)

Design of Factory Environmental Monitoring System Based on Internet of Things

Guang-qiu Lu$^{(\boxtimes)}$, Jia-hui Liu, Hai-qin Wang, and Yi Feng

Qingdao Binhai University, Qingdao, China
156360093@qq.com

Abstract. Factory environmental issues have always been the focus of attention, the factory is a highly populated place, the health and safety of workers is very important. The dust in the air can cause respiratory diseases, and the factory exhaust gas is a serious hazard to human health and can easily lead to fires. Based on the above problems, this paper designs a factory environment monitoring system based on Internet of Things, the design of the system uses 4G wireless communication technology, sensor technology, image acquisition technology, MySQL database design technology, dynamic web page development technology, WeChat public number development and other key technologies. The system can monitor a variety of environmental indicators in the factory and has good practical value.

Keywords: Internet of Things · Environmental monitoring · 4G · WeChat public number

1 Introduction

Environmental issues have become a global problem. The environmental problems in China's factories are particularly acute, which seriously endanger the health of workers. Environmental monitoring in factories is extremely urgent [1]. This paper designs an environmental monitoring system for the environmental pollution problem of the factory. The system can realize automatic collection, storage, query, alarm and other functions of environmental data, which can provide a practical and reliable solution for factory environmental monitoring.

2 The Design of Overall Scheme and Hardware Structure of the Environmental Monitoring System

2.1 The Design of Overall Scheme of the System

The main function of the system is to send the data collected by the hardware platform to the server through the communication module, and then establish a complete real-time information collection system through the development of the client. The factory environment data monitoring system designed in this paper uses 4G wireless communication technology, image acquisition technology, data preprocessing technology,

M. Atiquzzaman et al. (Eds.): BDCPS 2019, AISC 1117, pp. 138–146, 2020.
https://doi.org/10.1007/978-981-15-2568-1_20

sensor technology, dynamic web page development technology, and WeChat public number development technology. Firstly, the hardware development platform collects environmental data such as illumination intensity, smoke concentration, temperature and humidity, formaldehyde concentration, PM2.5, etc. The system uploads the collected environmental data to the cloud server through the 4G communication module, and the cloud server processes the received data and stores it in the My SQL database. Users can query the monitored environmental data at any time by accessing the website or Wechat public number. The monitoring system also sets the threshold alarm function, when the indoor temperature is too high and the smoke concentration is too high, the camera takes a picture and transmits the picture to the cloud server, the system can alarm the multiple preset numbers by SMS. The overall framework of the system is shown in Fig. 1.

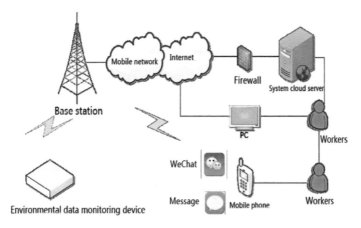

Fig. 1. Overall framework of the system

The functions of the system are described as follows:

(1) Users can view real-time environmental information by visiting webpages and Wechat public numbers through mobile phones and other terminal devices, such as PM2.5, temperature and humidity, light intensity, etc.
(2) The system can realize the visual monitoring environment through the image information collected by the image acquisition module.
(3) When the environmental data is abnormal and exceeds the set threshold, the system will alarm by short message.
(4) The system has the function of Querying Historical data. Users can query environmental data and picture information a few days ago and analyze the recent environmental situation.

2.2   The Design of Hardware Structure of the System

The hardware design of this system is based on the purpose of low cost and high performance. The aim is to design a hardware data acquisition platform which can meet the actual needs and is cheap [2]. The hardware part of the system consists of a main control module, a 4G wireless communication module, an image acquisition module, and various sensor modules. The sensor module includes temperature and humidity sensor, formaldehyde sensor, smoke sensor, dust sensor and light intensity sensor. The overall structure of the hardware part of the system is shown in Fig. 2.

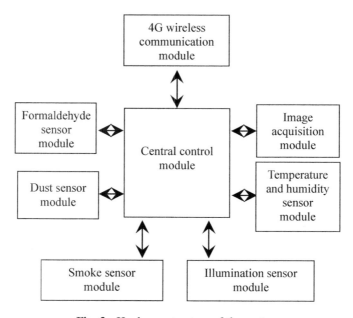

Fig. 2. Hardware structure of the system

The various sensor modules, image acquisition modules, and communication modules are coordinated and unified through the central control module, and the overall structure is clearly visible. The main control module controls the smoke sensor module, the formaldehyde sensor module, the dust sensor module, the illumination sensor module, and the temperature and humidity module to collect data on smoke, formaldehyde, dust, light intensity, temperature and humidity. The system sets the threshold value of environmental data [3]. When the value of the measured data is higher than the threshold value, the system immediately alarms by short message and triggers the camera to take photos. The captured image information is uploaded to the database of the cloud server of the system through the 4G communication module. The manager can make better and safer processing measures according to the picture information. The hardware selection of each module of the system is shown in Table 1:

Table 1. Hardware of each module of the system

Module name	Chip name
Master module	STM32F103RCT6
Communication module	EC20
Camera module	PTC08
Temperature and humidity module	DHT22
Formaldehyde module	ZE08-CH2O
Dust module	GP2Y1010AU0F
Illumination intensity module	GY-30
Smoke Sensor Module	MQ-2

The main technical indicators of the system are as follows:

Working voltage: DC+5 V;
Working current: ≤ 450 mA;
Working temperature: -10–50 °C;
Weight: about 200 g;
Temperature resolution: 0.1 °C;
Humidity resolution: 0.1% RH;
Formaldehyde gas resolution: ≤ 0.01 ppm;
Light intensity resolution: 1 lx;
Minimum dust particle detection value: 0.8 μm;
Maximum pixel size: 200w;
Image resolution (adjustable): 320×240;
Image format: standard JPEG/M-JPEG;

The system hardware is shown in Fig. 3. The hardware system consists of an STM32F103 master module, a 4G communication module, a camera and sensors for temperature, humidity, light, smoke, formaldehyde and dust. The sensor is responsible for collecting environmental data, and the 4G communication module is responsible for uploading the data to the cloud server.

Fig. 3. System physical map

3   The Design of System Software

3.1   The Design of Server-Side Software

The TCP console program of this system is developed in Eclipse development environment using Java programming language [4]. The system adopts the communication mode based on TCP protocol, and realizes the communication between client and server through Socket programming. The TCP console workflow is shown in Fig. 4.

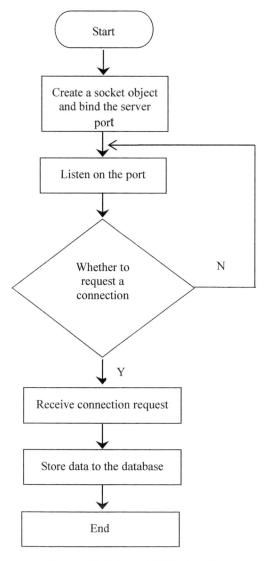

Fig. 4. Workflow of TCP Console

3.2   The Design of Database

The system needs to store the data received from the data collection end, which requires the use of the database [5]. Considering the actual needs of the system and the impact of development costs, this system uses My SQL database to process the environmental data and picture information received from the collector. For the convenience of management, two data tables are established. The Table 2 TSD is used to store the data collected from the sensor, and the Table 3 TID is used to store the collected picture data.

Table 2. TSD data

Field name	Data type	Remarks
seq	int	Primary key
deviceid	varchar	device ID
temp	varchar	temperature
hum	varchar	humidity
illu	varchar	illumination
pm	varchar	dust
forma	varchar	formaldehyde
smo	varchar	smoke
updatetime	Datetime	data update time

Table 3. TID data

Field name	Data type	Remarks
seq	int	Primary key
deviceid	varchar	device ID
image	blob	image
updatetime	Datetime	data update time

3.3   The Design of Data Monitoring Software

(1) The Design of Dynamic Web Page

The Web server program designed in this system is developed by using Eclipse platform. People can visit the web pages by mobile phones, computers and other terminal devices [6]. In this way, we can easily see the environmental data of monitoring points, and we can further understand the specific situation of the current environment through picture information.

The development of Java Web project in this system is accomplished by using JSP +Servlet+Java Bean. In this mode, Servlet is responsible for processing user requests, JSP is responsible for data display and Java Bean is responsible for encapsulating data [7]. The system establishes a connection between JDBC and My SQL database, executes SQL statements to increase, delete, modify and query operations, and then processes the execution results of the SQL statements.

The design of the web page includes the design of the Java program for the dynamic page and the design of the HTML program for the static page. The system has designed three JSP pages, including the factory environment data monitoring center home page (DataCenter. jsp) and the historical environment data query page (HisData. jsp), historical image query page (HisPicture. jsp). Because the environment data acquisition terminal establishes a long TCP connection with the cloud server, according to the data acquisition interval set by the system, the data received by the cloud server will always be updated, and every time the data is stored, the JSP page will be updated. Therefore, when people query environmental data, JSP pages display the latest environmental data [8]. The process of information interaction between dynamic web pages and databases in this system is shown in Fig. 5.

Fig. 5. Information Interaction Process between Web Page and Database

(2) The Development of Wechat Public Platform

Wechat public platform is widely used at present. Users can receive environmental data by sending instructions on Wechat public platform [9]. Because of the convenience of Wechat's public platform, this paper chooses it as the monitoring software.

The WeChat public account needs to be registered on the WeChat public platform official website. The registration process is mainly divided into the following steps [10]:

(1) Log in to the WeChat public platform official website;
(2) Select the registered account type, fill in the email and email verification code, and set the password and other basic information;
(3) Log in to the mailbox, confirm that the email has been sent to your email, check the email, and activate the WeChat public platform account;
(4) Fill in the registration information;
(5) Enter the public number information page, fill in the account name, location, function introduction and other related information.

The process of viewing the factory environmental data collected by the system through the WeChat public account can also be said to be the process of communication between the WeChat server and the cloud server of the factory environmental monitoring system. The specific information interaction process is as follows:

First, users need to fill in the basic configuration of the server in the Wechat public platform: URL (cloud server address), Token (which consists of 3–32 English or Chinese characters), Encoding AESKey (message encryption and decryption key, which is generated randomly by the system).

Then, the Wechat server will send GET requests to the URL, and the GET requests will carry four parameters: signature (Wechat Encrypted Signature), timestamp, nonce (Random Number), echostr (Random String).

Finally, according to the access rules, the system carries out the access program on the server side. In the first step, the system ranks the three parameters of token, timestamp and nonce in dictionary order. In the second step, the system spliced the three parameter strings into a string and encrypted them with sha1. In the third step, the system compares the encrypted string with the signature, and identifies whether the request is from the official WeChat server according to the comparison result. If the comparison result is consistent, the verification is successful. After successful URL validation, the Wechat server will send POST requests to the cloud server. The cloud server will parse the data packet effectively. The factory environment data can be obtained by the get () method of JavaBean. The process of verification and communication between the Wechat server and the cloud server is shown in Fig. 6.

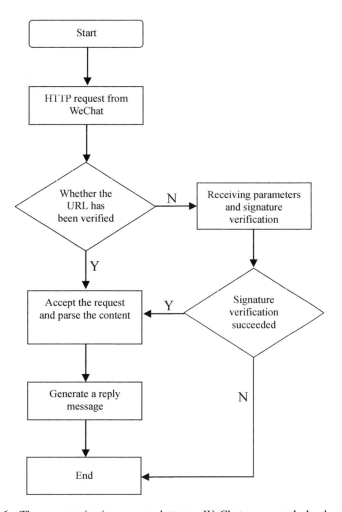

Fig. 6. The communication process between WeChat server and cloud server

4 Summary

This paper focuses on the important field of the development of the Internet of Things, and designs a factory environmental monitoring system for the problem of factory environmental pollution. With this system, users can inquire the sensor data and pictures in real time through the website or Wechat public number; at the same time, when the sensor data exceeds the set threshold, the system automatically alarms by short message, and the environmental monitoring system supported by the Internet of Things technology achieves better environmental monitoring, which lays a foundation for the environmental management of factories.

References

1. Dan, L., Bao, Z., Yang, Y.: Design of workshop environment monitoring system based on internet of things. In: International Symposium on Computational Intelligence & Design (2017)
2. Tang, J., Dong, T., Li, L., et al.: Intelligent monitoring system based on Internet of Things. Wirel. Pers. Commun. **15**, 1–17 (2018)
3. Han, Z., Wu, Z., Lin, S., et al.: An intelligent household greenhouse system design based on Internet of Things (2018)
4. Shen, W.Z., Wang, S.Y., Eltieb, N.E.B., et al.: Design of greenhouse environment control system based on Internet of Things. J. Northeast Agric. Univ. (Engl. Ed.) **25**(71)(2), 56–66 (2018)
5. Peng, H., Bohong, Z., Qinpei, K.: Smart City Environmental Pollution Prevention and Control Design Based on Internet of Things (2017)
6. Tu, Z.X., Hong, C.C., Feng, H.: EMACS: design and implementation of indoor environment monitoring and control system. In: IEEE/ACIS International Conference on Computer & Information Science (2017)
7. Wang, H., Zhang, T., Quan, Y.: Rencai dong research on the framework of the environmental Internet of Things. Int. J. Sustain. Dev. World Ecol. **20**(3), 199–204 (2013)
8. Niu, Z., Chen, J., Xu, L., Yin, L., Zhang, F.: Application of the environmental Internet of Things on monitoring PM2.5 at a coastal site in the urbanizing region of southeast China. Int. J. Sustain. Dev. World Ecol. **20**(3), 231–237 (2013)
9. Su, X., Shao, G., Vause, J., Tang, L.: An integrated system for urban environmental monitoring and management based on the environmental Internet of Things. Int. J. Sustain. Dev. World Ecol. **22**(03), 205–209 (2013)
10. Sun, C.: Application of RFID technology for logistics on Internet of Things. AASRI Procedia **1**, 106–111 (2012)

A Study of Teaching Model Construction Under Big Data Background-Taking 《Computer Information Technology》 for Example

Jun Zheng[(✉)]

Network Information Center, Baotou Teachers' College, Baotou, China
zhj@bttc.edu.cn

Abstract. The advent of the era of big data has driven rapid development of the educational industry, and meanwhile proposed higher requirements for informatization of education and teaching. As for modern education, the concept of big data must be introduced into the daily education and teaching. Teaching objectives, teaching model, and pre-class, in-class and after-class teaching activity models under the big data background were designed in this paper, and a teaching case was designed by taking the Computer Information Technology course for example. Teaching data were collected through the big data teaching platform, followed by an intelligent analysis. In the end, the new-type teaching model was constructed and implemented under the big data background. This teaching model has exerted a promoting effect on educational informatization and provided a big data-based classroom teaching model which can be used for reference.

Keywords: Big data · Education · Reform

1 Introduction

With the continuous development of network technology, rapid promotion of information technology in various industries and appearance of cloud computing and big data technology since the 21st century, enterprises, industries and even countries are under a brand-new competition pattern. The advent of the era of big data has driven the development of various industries, and the educational field is also under its deep influence. Inundated with big data, the educational field includes a great number of educational and teaching resources, student behavioral habit data and big data concept-based education and teaching modes [1–3]. The talents cultivated by the traditional teaching mode can't adapt to the demand for rapid socioeconomic development and can be easily eliminated in their future work. In order to transport more applied talents conforming to the modern development demand, the big data concept must be introduced into daily teaching by following the innovative development in the era closely. Permeating the advanced big data technology into the educational field is a future development tendency of educational informatization and will surely generate an enormous effect on profound reform of the educational field. Therefore, a teaching

M. Atiquzzaman et al. (Eds.): BDCPS 2019, AISC 1117, pp. 147–156, 2020.
https://doi.org/10.1007/978-981-15-2568-1_21

reform scheme of studying and constructing a new teaching model under the big data background, was proposed in this paper.

2 Educational Big Data

"Big data" is spread in numerous social fields and become more and more acquainted in life, study, society, commerce, education, etc. However, it has not a long history. In 1998, Big Data Processing was published on the Science Magazine, when "big data" was put forward for the first time [4].

The concept of educational big data refers to all kinds of behavioral data directly generated in all daily teaching activities in general, while educational big data in narrow sense refers to all kinds of learning behavioral data of students, which mainly derives from online learning platforms and educational management systems [5]. Educational big data can help to find correlations, diagnose existing problems, predict development tendencies, improve educational quality, promote educational fairness, realize individualized learning, optimize allocation of educational resources and assist in scientific decision making in education from numerous educational data, which is enormous in quantity, diversified in types and low in value density. A data model can be constructed by technical means to comprehensively record and collect students' behavioral data. The construction of educational big data environment can be promoted through data acquisition, intelligent data analysis, real-time learning data analysis, real-time pushing of resource data and dynamic feedback result, so as to complete construction and implementation of new-type teaching models under the big data background in the new era [5]. Only in this way can students perceive the interactive, intelligent, data-based and dynamic teaching process and can the education develop toward informatization and intelligent directions rapidly [6].

3 Teaching Model Design Under Big Data Background

3.1 Modeling of Big Data Environment

Many current systems and platforms collect a large quantity of learning process data of learners, analyze application effects of specific class examples and track and test teaching effects through backgrounds. The big data environment in this study is shown in Fig. 1:

The architectural analysis of big data environment is divided into two layers, where the first layer includes big data analysis and decision-making analysis service, based on which coprocessing function is completed. Hadoop technology, a distributed system infrastructure developed by Apache company, is adopted in the system with very simple architecture and it can easily develop a distributed program. Hadoop supports reliable and highly efficient distributed processing of a large quantity of data. Large-scale data are saved by Hadoop using many system clusters so as to form a unified storage layer and unified interface layer. Teaching resources of teachers and learning data of students, which are generated in the education and teaching process, are

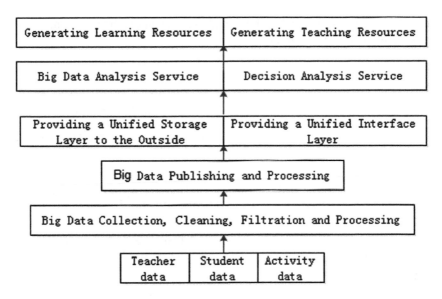

Fig. 1. Big data environment

collected and saved in My SQL. The database can save the data in different servers to generate an individualized database of students. Individualized learning diagnostic reports can be formed by collecting, cleaning and filtering individualized data of all students so that students change their learning behaviors, transform their emphasis points and realize individualized learning on well-founded basis. The technology can realize collection, mining and analysis of dynamic learning data, present the whole learning process and effect data of the students and accurately master learning conditions, and moreover, based on data decision-making, it is convenient for teachers to timely adjust their teaching and transform the past actual situation which depended on teachers' teaching experience.

The second layer is embodied by data sources, including abundant learning resources of students and teaching resources of teachers, and teacher data, student data and activity data generated in the classroom activities are collected, saved, analyzed and processed via the big data analysis technology. Underlying data as educational bog data must be sliced according to disciplinary knowledge points and are associated with a certain number of topics. They can realize preview before class and in-class test of the students and provide a support for teachers to select proper learning starting points.

3.2 Design of Teaching Objectives

Through the teaching model under the big data background, teachers utilize big data technologies to support their own wisdom, generate the "photosynthetic effect" between "wisdom" teaching of teachers and "wisdom" learning of students and reach an excellent effect on wisdom sparking. The teaching model under the big data background can facilitate the generation of students' autonomous ability, cooperative ability, thinking ability, innovation ability and intellectual capability, and the design of teaching objectives is shown in Fig. 2.

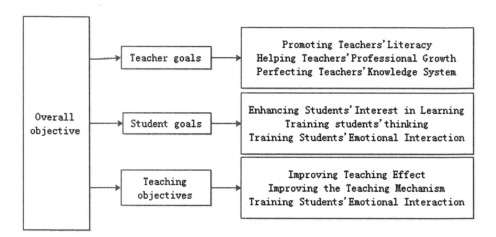

Fig. 2. Design of teaching objectives

1. Teacher layer. Improvement of professional qualities is what each teacher pursues. The teaching model under the support of big data should pay more attention to generation of teaching wisdom and cultivation of informatization qualities among teachers and assist the educational informatization in driving into a high-speed channel so that teacher's knowledge system can be constructed and perfected, and a systematic knowledge system can be used to carry out teaching and improve teachers' teaching ability.
2. Student layer. Students' learning interest is a problem which should be considered firstly in the teachers' teaching process. Only with interests will there be space for progress and improvement. The teaching process should be made as interesting as possible in order to develop students' intelligence, enhance their learning motivation, train their thinking ability, enhance student-teacher emotional interaction and realize interest-based learning and wisdom learning.
3. Teaching layer. The evolution of teaching mechanism into teaching mechanism needs the cooperation of the informatization environment. The generation of teaching mechanism aims at improving the teaching effect. Improving intelligent level of teaching, perfecting the teaching wisdom mechanism and promoting the burst of teaching wisdom relies on data decision-based teaching.

3.3   Design of Teaching Model

"Model" was introduced by Americans Joyce and Will into the education and teaching field for the first time. In opinion of professor Joyce, teaching model is namely learning model. While helping students to acquire basic knowledge, master basic skills and methods, adjust their self-values and divergent thinking, the teachers are importing more highly efficient learning methods to the students. The fundamental goal of the teaching model is to provide learning skills and methods for students better so as to improve their learning efficiency [7].

Simply speaking, teaching model refers to that teachers establish a teaching activity-centered teaching framework according to their own teaching cognition on the ideological basis of related teaching theories. With this teaching framework, teachers can master teaching progress and relation and convergence degree between various factors so that the whole teaching process can become more orderly with stronger operability [8]. The teaching model design under the support of educational big data is as shown in Fig. 3:

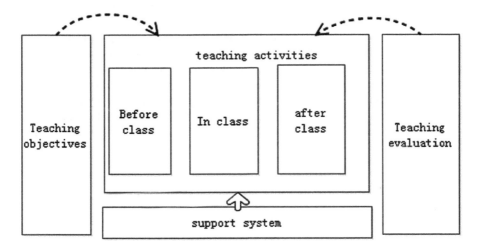

Fig. 3. Teaching model flowchart under the support of educational big data

Teaching activity under the big data background is a gradual generation process [9], including three links: pre-class, in-class and after-class. Teaching evaluation includes online evaluation and offline evaluation, where the former runs through all teaching links and evaluation data are mainly procedural data collected in classroom, and the latter is the evaluation of learning performance, including evaluation of test questions and preview before class, etc. with the emphasis laid on self-evaluation and evaluation of learning performance and classroom performance of students and giving students a complete result evaluation. Teaching objective is both starting point and end point of all teaching activities and it is correlated with educational goal and training objective [10, 11]. Different from educational objective and training objective, teaching objective mainly refers to the implementation direction of teaching activities and expected result. Teaching objectives guide the implementation direction and flag of teaching activities. Teaching evaluation is an important index and improvement strategy. According to the teaching evaluation, teaching activities need continuous modification and perfection.

The supporting system includes adaptative learning system, adaptive learning technology, learning resources and mobile terminal. The supporting system can push preview contents before class to students via smart client end so that students can construct new knowledge according to their own demand and experience, and fuse, transform and recombine old knowledge and new knowledge better. Teaching activity includes pre-class, in-class and after-class activities, and the detailed model design is as below:

3.3.1   Preparation Before Class

The learning analysis report is generated based on the pretesting data collected through the cloud diagnostic analysis system. The teacher analyzes learning conditions and new knowledge features in textbook by combining the report, completes compilation of learning resources like electronic textbook resources, small videos or guided learning plan and pertinently releases learning resources, and the students communicate and discuss with each other over new knowledge before class and timely give feedbacks about preview; the teacher should complete the corresponding teaching design according to feedbacks given by students about preview before class, formulate teaching objectives and set important and difficult teaching points so as to preset the teaching process; the teaching situation is set, followed by the importing of new lesion, e.g. the lead-in link is completed in forms of micro-video, figure story or minigame, and in this link, teachers and students communicate, share and interact with each other so as to make better preparation for the in-class link. The concrete before-class link is shown in Fig. 4.

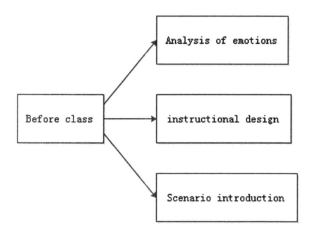

Fig. 4. Before-class link

3.3.2   In-Class Interaction

Firstly, the students are guided to reflect upon problems through the teaching situation designed before class, and new knowledge is introduced in problem-driving-classroom way to motivate students' interests in learning new knowledge and improve their abilities of thinking.

Secondly, the teacher pushes learning tasks and utilize task-driven method and group cooperative learning to facilitate students to get a deep understanding of knowledge, explore new knowledge and cultivate their cooperative inquiry ability. The teacher should guide students to perform communication and interactive activities in class, find common problems and solve them. In class, the teacher should motivate students' initiative and stimulate their learning interests. Students will interact with each other in class and the teacher will formulate teaching pace according to students' learning data generated in class.

Thirdly, the teacher can analyze data results and timely adjust the teaching plan according to teaching data generated in class, carry out, extend and improve the teaching links and upgrade task resources. The students will discuss and communicate with each other again and promote internalization of knowledge.

In the end, the exercise compilation function of the system is utilized to test students' knowledge. The lesson will be analyzed and evaluated according to individual test results of the students, and this link can realize combinations in various forms, including self-evaluation of students, mutual evaluation between students and teacher evaluation. The reward mechanism in the system can be utilized to complete the evaluation link. The specific in-class link is shown in Fig. 5.

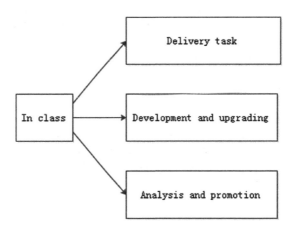

Fig. 5. In-class link

3.3.3 After-Class Improvement

The teacher pushes individualized resources and assignments to students through the mobile terminal, and the students complete their assignment, chase leaks and supply deficiencies. The students will submit the assignment after completing them and the teacher will check and give feedbacks and remedy individualized puzzling questions of students. Meanwhile, students can communicate and interact with each other after class and summarize and reflect upon their shortcomings. The teacher will provide students with individualized guidance so as to improve and give feedbacks over their learning results as shown in Fig. 6.

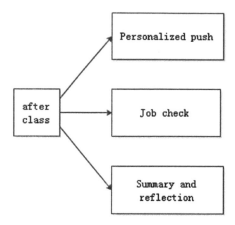

Fig. 6. After-class link

4   Teaching Case Design of 《Computer Information Technology》 Under Big Data Background

As a public compulsory course, Computer Information Technology takes Flash teaching as the main content. Being able to fuse multiple media, Flash, a vectorgraph editing and interactive animation production software, has been extensively applied to numerous aspects and fields like advertising, mathematics, titles, animation, game and greeting card, and moreover, it is also one of indispensable tools for webpage material making. Through this course, students can master basic use methods and common operations of Flash proficiently and initially gain the interactive animation production abilities so as to satisfy daily work and living requirements. The main teaching contents include: 1. Basic knowledge of animation and introduction of basic functions of Flash; 2. Graphing; 3. Graph editing; 4. Texts; 5. Layers; 6. Elements; 7. Flash animation. Taking Computer Information Technology as a practical course, a typical teaching case under the typical classroom teaching model in the big data background is designed in this paper.

4.1   Preparation Before Class

Learning contents will be arranged and student groups will be divided before class. The teacher will decide teaching contents according to teaching objectives and student conditions before class, design teaching programs in accordance with their aptitudes and propose the overall task object. The teacher will divide student groups according to their performances in the leading course with excellent students driving poor ones, so as to guarantee active group learning atmosphere. The general program task is divided and refined into several knowledge points. The teacher will reasonably divide small tasks and corresponding knowledge points in each course, select proper video learning contents on the learning platform according to the knowledge points, arrange pertinent thinking-triggering topics and assignments and distribute to student groups one week before the class.

4.2 In-Class Implementation

Before the classroom teaching, the teacher check students' learning conditions through the big data technology of the teaching platform including whether the students have finished the assignments and learning hours, etc. The teacher will answer questions raised by students via the teacher-student interactive exchange module on the platform. Difficult or common questions can be left in the classroom for centralized discussion. The teacher will master students' condition according to assignments and test results and adjust teaching progress on this basis. In the teacher-guided and student-oriented way, reflection questions arranged in the last week and knowledge points in this lesson will be solved and summarized, followed by a discussion over difficult or common questions. Student groups can present their own programs, raise their questions like technical bottlenecks or seek for help from other groups.

4.3 After-Class Improvement

After the class, students will study video contents arranged by the teacher through the learning platform and complete their reflection questions and assignment. The students with spare energy can select other learning contents according to their own hobbies and interests. Once a puzzling question is encountered, it can be raised through the teacher-student interactive communication module so as to initiate a discussion. The group cooperative program should be completed through division of labor by actively using laboratory resources of the school, and discussions can be conducted both in groups and between groups. In addition, assignments can be submitted through the platform and the test can be regularly carried out, etc. Furthermore, students can check their assignment and testing results, understand their own shortcomings and promote the improvement of their independent consciousness through the platform.

5 Conclusion

The teaching model under the big data background facilitates the transition of "teacher-centered" teaching model into "student-centered" teaching model based on the traditional reflective teaching by taking full advantages of innovation and development of big data technology in the information-based teaching as well as resources, platforms and technologies under the background of internet and big data. The practice indicates that this model is conducive to motivating students' learning interests and promoting the development of thinking pattern, knowledge innovation and improvement of comprehensive qualities of students with remarkable effects. The proposition of this model has exerted a certain promoting effect on the development of educational informatization, provides a theoretical learning basis and practical scheme for the development of wisdom education and promotion of creative teaching model and supplies a big data-based classroom teaching model which can be taken for reference by other schools.

Acknowledgements. This work was supported by Teaching Reform Project of the Educational Administration Department of Baotou Teachers' College (BSJG16Y011).

References

1. Zhao, Y.: The development of network education in the age of big data. In: Proceedings of the 2019 International Conference on Pedagogy, Communication and Sociology (ICPCS 2019) (2019)
2. Geisseler, D., Scow, K.M.: Long-term effects of mineral fertilizers on soil microorganisms-a review. Soil Biol. Biochem. **75**(1), 54–63 (2014)
3. Zhao, M., Zhao W.: Practical research on improving education targeting based on big data mining. In: Proceedings of the 2018 International Workshop on Education Reform and Social Sciences (ERSS 2018) (2019)
4. Zhong, C., Cao Z.: Study on student growth tracking system based on educational big data. In: Proceedings of the 3rd International Conference on Mechatronics Engineering and Information Technology (ICMEIT 2019) (2019)
5. Simović, A.: A big data smart library recommender system for an educational institution. Libr. Hi Tech **36**(3), 498–523 (2018)
6. Special issue on educational big data and learning analytics. Interact. Learn. Environ. **26**(5) (2018)
7. Ruiz-Gómez, J.L., Martín-Parra, J.I., González-Noriega, M., Redondo-Figuero, C.G., Manuel-Palazuelos, J.C.: Simulation as a surgical teaching model. Cir. Espanola **96**(1), 12–17 (2018)
8. Slade, D.G., Martin, A.J., Watson, G.: Developing a game and learning-centred flexible teaching model for transforming play. Phys. Educ. Sport Pedagogy **24**(5), 434–446 (2019)
9. Sodero, A., Jin, Y.H., Barratt, M.: The social process of Big Data and predictive analytics use for logistics and supply chain management. Int. J. Phys. Distrib. Logist. Manag. **49**(7), 706–726 (2019)
10. Gelfman, D.M.: Changing the learning objectives for teaching physical examination at the medical school level. Am. J. Med. (2019)
11. Jia, F.-L., Jing, Y.-P.: Study on the development of innovative experiment teaching model with autonomous learning under the background of big data. In: DEStech Transactions on Economics, Business and Management (2018)

Discussion on Compiling Principle Teaching Mode Based on Modularization

Hong Zheng[✉], Jianhua Li, Weibin Guo, and Jianguo Yang

School of Information Science and Engineering, East China University
of Science and Technology, Shanghai 200237, China
zhjenny@tom.com

Abstract. Compiling principle" is an important professional course for computer science and technology specialty. It aims at introducing the general principles and basic methods of compiling program construction. It plays an important role in the major teaching system. The course is very theoretical and practical, and it is difficult to master the knowledge of compilation principle completely and comprehensively. A mixed teaching mode is put forward to improve students' abilities in teaching effect of compiling programming based on constructivism modularization in the paper.

Keywords: Compiling principle · Teaching mode · Constructivism

1 Introduction

Compiling principle is an important subject for the teaching program of computer major in higher education with high requirement of project practice from team-spirited students who is not only equipped with the capability of software developing but also the overall managing concept of software project. While many courses are directly and indirectly related to software or program design and implementation such as the language courses, and fundamental principle courses such as Compile Principle. It is difficult to master the knowledge of compilation principle completely and comprehensively. The teacher needs to explore the improvement of teaching methods to achieve the students' full awareness of the importance of the course through altering the tradition mode centering on the teacher in the form of positive learning instead of negative one with availability of learning sources, which can generate a good effect with modular organization of compilers.

The paper makes an introduction to the center of constructivism in Sect. 2. Section 3 discusses how to apply the theory and the method to the compiling programming of teaching practice. Section 4 concludes several problems with the teaching effect of the course.

M. Atiquzzaman et al. (Eds.): BDCPS 2019, AISC 1117, pp. 157–164, 2020.
https://doi.org/10.1007/978-981-15-2568-1_22

2 Constructivism

Constructivism is often mentioned in contrast to the behaviorism, and at once viewed as a furtherance of cognitivism [1]. In teaching and learning practices, behaviorism "centers on students' efforts to accumulate knowledge and on teachers' efforts to transmit it [2, 3]. It therefore relies on a transmission, instructional approach which is largely passive, teacher-directed and controlled". On the other end of the spectrum, constructivism stresses that knowledge is seen as relative, varying according to time and space. Constructivism basically means that as we experience something new, we understand it with the aid of our past experiences. A constructivist views learners as actively engaged in making meaning, and constructivist learning design concentrates on what learners can analyze, investigate, collaborate, share, build and generate based on what they already know, rather than what facts, skills, and processes they can parrot [4]. At the core of the theory lies the emphasis of students' initiative in knowledge and construction of it. The theory requires that the knowledge is acquired not through the teacher but the learner in certain background and culture, assisted by others (including teachers and classmates) and necessary sources through meaning construction. Learning is a process of constructing inner psychological aspects including structural knowledge and non-structural experience. The student is the subject of information processing, positive in meaning construction instead of negative in reception of the outside stimulation. The teacher is just a helper to produce progress rather than a major role. Learning environment consists of background, cooperation, discussion and meaning construction [5].

The teaching method, corresponding to the constructivism and it's learning environment, is summarized as the principle that the purpose of effective construction of current knowledge is achieved through the advantage of background, cooperation, discussion by the initiative student in the center surrounded by the teacher as organizer, instructor, helper and promoter. The knowledge in the book derives not from the content passed by the teacher but object of meaning construction [6, 7]. Media is used no longer as a means to pass the methods but the exploring cognitive instrument for initiative learning to create background, cooperation, discussion.

3 Compiling Programming Design Teaching Method Under Constructivism

Characteristics of constructivism are summarized as four: context, collaboration, conversation and construction [8]. One basic notion of constructivism is that learners actively construct their knowledge rather than receive ideas preached by teachers. Further, constructivist learning is based on learners' active participation in problem-solving and critical thinking in learning activities. The activities should be "situated" in real-world contexts, including the physical, cultural and social issues. The collaboration among learners via conversation (in a broad sense) should be encouraged. As a result, the knowledge is constructed as a natural product of the learners' mind.

However, these constructivist principles are merely some theoretical assumptions. They do not provide a model for implementation from a general learning paradigm to software design.

On the one hand, Constructivism recognizes the learner as the cognitive subject of initiative meaning construction; on the other hand, the guiding role of the teacher can not be neglected for his aid, instruction and promotion.

Therefore, the teaching process involves not only the initiative participation of the student but also the guidance of the teacher. The student's positive thinking and exploration are conducted under the instruction of the teacher whose major task is to arouse his interest in learning, helping them form the learning motive, create the various backgrounds suitable for the teaching, give hint to the relationship between new and learned knowledge, assist transferring of the knowledge, organize cooperative learning in order to lead the student deeper and provide the timely guidance and help.

Under constructivism, our compiling programming adopts scaffolding instruction, anchored instruction and random access instruction. The practice has proved that the method has aroused the learning interest thus enabling the student to complete the meaning construction gradually under the guidance of the teacher creating a certain background.

A. Scaffolding Instruction Conceptual Framework

Zone of proximal development means the difference between the independent level of the student and the potential one tapped by the teacher when solving the problem [9]. The meaning construction is complete on the basis of zone of proximal development with learning as the theme through establishing framework, instruct the student to certain background to allow independent exploration and group negotiation.

Firstly, establish a project group. The standard can be the knowledge level, capability and interest. Each group needs to appoint a leader to be in charge of allocation, arrangement and communication during a project. Secondly, each group should decide on a project. The source can be the group interest and knowledge or the options proposed by the teacher. For example, (1) Add PL/0 language in the form of '/* ...... */' like C language. Note/comment, (2) Add conditional and exit sentences with else clauses to PL/0 and so on. Generally, More than one task is required to master the methods of different stages from demand analysis, system design to code testing, accompanied by the structured developing method or object-oriented one.

Thirdly, allocate the tasks for the members to ensure the responsibility on the person. The principle is that each member should not only be responsible for his own task but also for the cooperative one which is relatively independent for the individual. The leader makes allocation of tasks for his members in different stages of software developing. For example, in the stage of feasibility research, the members can be allocated to conduct research on economic, technical feasibility for a certain proposal and in the stage of requirement analysis, the members can be allocated to conduct such tasks as symbol tables, function tables, structure design, lexical analyzer, syntax analyzer, intermediate code generator and software requirement specification on the principle of exposing them to enough training.

Fourthly, launch a exploring learning. The difficulty of practical subject exploration requires the teacher to supervise for the timely fulfillment of the plan through regular reports of the group over the progress or the period reports. It is certain that time may prevent some groups from fulfilling all the functions of the selected project but not from the parts of it on the guiding principle of Compiling principle as shown Figs. 1 and 2.

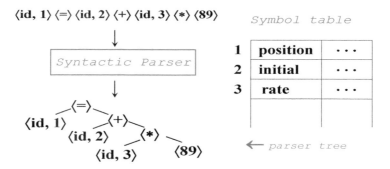

Fig. 1. Syntactic parser and symbol table

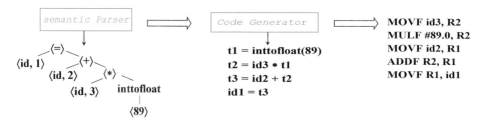

Fig. 2. Semantic parser and code generator

Fifthly, make summary report. Make brief report of the task fulfillment to the whole class with PowerPoint in the limited time after integration of works from each member. The report can not only clearly acquaint the teacher with the fulfillment situation but also urge the group to achieve task of high quality. Meanwhile, it makes possible for mutual learning, mutual improvement, enhancing of interest and confidence for the student by the encouragement form the teacher.

B. Anchored Instruction

Anchored Instruction is also called "teaching based on the question". "Anchor" means to define the real events or questions in the actual situation. If these questions were able to draw forth basic concept and principle that are closely related with the study subject, students can be motivated to learn and explore initiatively [10]. Teachers should gradually hone students on their abstract and logic thought.

The key but also the difficult point in Compiling principle is "modular", which is always considered as an abstract concept and hard to learn. Therefore, Anchored Instruction in the teaching could contribute to a successful understanding of the modular software development in the real situation and to an effective solution on large-scale and complex problems with "modular".

Before teaching the principle of modular design, teachers should ask the actual problems as follows: the modular design of Compiling System(CS), whose functions include: the books warehousing, borrowing and restore, and appointment; the inquiry of books and the borrowing, the establishment of staff jurisdiction and so on, bringing in its wake the digital management of library. In order to achieve this goal, students should, by connecting with the actual situation, divides the CS into five modules including Lexical Analyzer module, Syntax Analyzer module, Semantic Analyzer module, Source Code Optimizer and Intermediate Code module module, and then gradually Target Code Generator module in each module in accordance with the functional requirement. Take reader management module for example, we can divide this module into two sub modules—the reader information input sub module and the reader information renewal sub module, and then further analyze their realization in actuality, gaining a step-by-step approach of complex question by the method called "divides and rules". At last, input the complete procedure codes on the computer and demonstrate the process and result.

The entire teaching process should be launched naturally linking to several cognitive comprised of the construction of principle situation and the significance, the cooperation and the conversation. Students can gain a keen perception of modular concept and principle as well as the modulation software development thought from the top to the bottom through their feelings and experience on the real question and the significance construction of their knowledge.

C. **Random Access Instruction**

Another key of software development is the software testing. This part of course content provides students with a comprehensive knowledge and understanding about software testing through Random Access Instruction.

For example, White Box Testing which adopts several methods of coverage testing content of program process diagram in Fig. 3.

```
var n:integer;
procedure movement(u:integer; v:integer);

    begin
    u:= Boolean;    /* error */
    v:= true;   /* error */
    write(u,v)
    end;

begin
    read(n);
    call movement(1,3)
end..      /* error */
```

Fig. 3. A test case

Design 1: Determination Coverage

① The main feature: The determination coverage is also called branch coverage, which requests enough design of testing subjects, enabling in the procedure each determination to have one time is the true value at least, has one time is the false value, namely: In procedure each branch is carried out one time at least.
② Subject design:

(1) if read 1 and 3, write (1, 3)
(2) if read 1.0 and 3.0 error /* They are not integer numbers */
(3) if read a and b error /* They are not integer numbers */

③ Merit: The determination coverage has testing ways nearly one time than those of the sentence coverage and also has a stronger testing ability. The determination coverage, as simple as the sentence coverage, is able to be possible to obtain the testing subjects without subdividing each determination.
④ Shortcoming: It is very common that the majority of determination sentence is comprises of many logical combination of condition (for example, determines in sentence contain AND, OR, CASE), it is inevitable to omit parts of the testing ways, if we merely judge its final outcome but neglect the value situation of each condition.

Design 2: Combination Coverage

① The main feature: It requests enough design of testing subjects, enabling all possible combinations of the condition result in each determination to present at least one time.
② Subject design:

(1) if read 1 and 3 write (1, 3)
(2) if read 1.0 and 3 error /* There is a real number*/
(3) if read 1 and b error /* b is not an integer number*/

③ Merit: Multi-condition coverage criterion satisfies determination/condition coverage criterion. Changed determination/condition coverage requests enough design of testing subjects, enabling all possible results of each condition in the determination and each determination itself to appear at least one time, and each condition to be demonstrated and affect the determination result alone.
④ Shortcoming: It has linearly increased the quantity of testing subjects.

In addition, there also have the condition coverage, methods coverage testing and so on, all of which relate to the common question in the developmental process of White Box Testing: how to discover the duplicated movements in the algorithm, which ways involved in these coverage methods, how to determinate the ways in White Box Testing? In what situation the coverage method should continue or stop? Or which coverage method should be selected to realize the test and so on. No matter which situation the students choose to study, it can achieve the goal of finding the problem, solving the problem, and significance construction.

4   Conclusion

The experiment of compiling principle is very difficult. Some compilers implement program code as many as 100,000 instructions. Therefore, we must avoid directly letting students analyze and design compilers. We should design experiments according to the compiling process of high-level programming language, and guide students to carry out experiments with the help of experimental teaching materials of compiling principles of the system. At first, we can use compiler model tools of object-oriented programming language to generate lexical analysis and grammar analysis programs of simple code, so that students can understand the program structure of code; on this basis, we can add semantic actions to understand the code generated by virtual machine; finally, we can expand the teaching model and construct a complete compiler at different levels. In the part of comprehensive experiment, teachers can set up procedural generations. Code framework, even provide source code to hollow out key modules and guide students to complete the core. The compilation and debugging of heart code can guide students to complete the course design in a directional way. In this way, they can devote more enthusiasm and energy to the practical development of depth and energy. Extensive curriculum practice promotes further mastery of theory.

Each teaching method has its limitation, for instance, the anchored instruction is question-oriented. If used improperly, it will affect the student's systematic knowledge. In the field research, it is not always to adapt a single teaching method but flexible combines two or more methods in view of the teaching duty. It is improper to lay stress on students while neglecting the leading role the teachers play in the teaching process.

Acknowledgements. We are pleased to acknowledge Reform and Construction of Undergraduate Experimental Practice Teaching in East China University of Science and Technology; the National Natural Science Foundation of China under Grant 61103115; the special fund for Software and Integrated Circuit Industry Development of Shanghai under Grant 150809; the "Action Plan for Innovation on Science and Technology" Projects of Shanghai (project No: 16511101000). The authors are also grateful to the anonymous referees for their insightful and valuable comments and suggestions.

References

1. Steele, M.M.: Teaching students with learning disabilities: constructivism or behaviorism? Curr. Issues Educ. **8** (2005)
2. Wu, Q.Q., Cao, L.C.: Teaching mode of operating system course for undergraduate majoring in computer sciences. In: Proceedings of 4th International Conference on Computer Science & Education, pp. 1412–1415 (2009)
3. He, Y.: The course choice between C language and C++ language. In: Proceedings of 4th International Conference on Computer Science &Education, pp. 1588–1590 (2009)
4. Richardson, V.: Constructivist pedagogy. Teach. Coll. Rec. **105**(9), 1623–1640 (2003)
5. Aho, A.V., Chen, S.Q.: Teaching attempt under the leading of learning theory of constructivism. Educ. Explor. **9** (2002)
6. Chen, Y.-Y.: The research and fulfillment of "Introduction of Computer Science" course reform in education. Softw. Lead Publ. Educ. Tech. (5), 24–25 (2015)

7. Wang, Y., Jiang, H.-Y.: The research and fulfillment of the calculator introduction course reform in education in new situation. Comput. Knowl. Tech. (7), 5391–5392 (2010)
8. Marin, N., Benarroch, A., Jiménez Gómez, E.: What is the relationship between social constructivism and Piagetian constructivism? An analysis of the characteristics of the ideas within both theories. Int. J. Sci. Educ. **22**(3), 225–238 (2000)
9. Wertsch, J.V.: The zone of proximal development: some conceptual issues. New Dir. Child Adolesc. Dev. **1984**(23), 7–18 (1984)
10. Bransford, J.D., et al.: Anchored instruction: why we need it and how technology can help. In: Cognition, Education, and Multimedia, pp. 129–156. Routledge (2012)

Design and Fabrication of Bionic Micro-nano Flexible Sensor

Hengyi Yuan[(⊠)]

Jilin Engineering Normal University, 3050 Kaixuan Road,
Changchun 130052, China
yuanhengyi@163.com

Abstract. With the popularity of intelligent devices, wearable electronic devices show great market prospects. Capacitive flexible pressure sensors with sandwich structure are designed and manufactured, and their performances are studied. A novel graphene (GR)/PEDOT:PSS multi-component mixed ink was prepared based on solution blending method using silver nanowires as electrodes and PDMS as flexible substrates. The GR/PEDOT:PSS multi-component mixed ink was used as conductive material. The different GR dosage pairs were analyzed by means of SEM and electrical testing platform. At the same time, PDMS films with micro-nano structure and planar structure were prepared by using ground glass and smooth glass as templates for flexible substrates respectively. Then AgNWs/PDMS composite electrodes were prepared by spraying method. With another layer of PDMS as dielectric layer, the two electrodes were encapsulated face to face to obtain capacitive flexible pressure sensor. Finally, the influence of electrode Micro-Nanostructure on device performance was systematically studied. The results show that the sensitivity of AgNWs/PDMS composite thin film sensor with Micro-Nanostructure is 1.0 kPa-1, while that of plane AgNWs/PDMS composite thin film sensor is 0.6 kPa-1. It can be concluded that flexible substrates with Micro-Nanostructure can significantly improve the sensitivity of the device.

Keywords: Tactile sensor · PDMS · Micro-nano structure · GR/PEDOT:PSS

1 Introduction

With the development of science and technology, the emergence of intelligent electronic products is also changing with each passing day, and gradually to the direction of flexibility, miniaturization, integration, multi-function, wearability and low cost. Flexible electronic devices have the advantages of unique extensibility, high efficiency and low cost. In recent years, they have shown a trend of rapid development. Potential. Biomimetic micro-nano flexible sensors have the advantages of unique ductility, high efficiency and low cost. In recent years, they have shown a trend of rapid development [1]. Research on flexible electronic devices that can withstand various deformation such as tension, compression, bending and distortion has become one of the research focuses in the fields of physics, chemistry, mechanics, electronics and so on. Extensible electronic devices have many important and emerging applications, such as flexible

© Springer Nature Singapore Pte Ltd. 2020
M. Atiquzzaman et al. (Eds.): BDCPS 2019, AISC 1117, pp. 165–172, 2020.
https://doi.org/10.1007/978-981-15-2568-1_23

displays, flexible electronic skin, intelligent surgical gloves, health monitoring equipment and flexible energy systems. With the development of wearable electronic devices, it will become a new fashion to wear intelligent electronic products like jewelry or clothes. Tactile sensor, as a frontier field of current scientific research, can imitate the tactile perception function of human interaction with the external environment. It is the nerve endpoint of the Internet of Things and the core component of assisting human multi-directional perception of nature and itself.

In the field of flexible sensors with uniform micro-nano structure, Rogers et al. [2] fabricated wavy silicon wafers with micro-nano thickness by silicon thinning process to improve the flexibility of devices, which initiated the main method of early fabrication of flexible electronic devices. Many researchers have introduced wavy microstructures into flexible pressure sensors to fabricate wavy microstructures on polymer surfaces by mechanical pre-stretching or temperature control. Tang et al. [3] Silver magnetron sputtering (Ag) on the surface of pre-stretched PDMS substrate was used as conductive layer by vacuum deposition. After releasing prestressing force, two electrodes with wavy microstructures were prepared. The two electrodes were separated by a deformable dielectric layer composed of carbon nanotubes (CNT)/polydimethylsiloxane. In order to simplify the fabrication of flexible pressure sensors, Chen et al. [4] designed a piezoresistive flexible sensor based on pyramid micro-structure array. Experiments show that the sensor has high sensitivity (15.1 kPa-1), low detection limit (0.2 Pa), fast response time (0.04 s) and is not affected by temperature. It has good application potential in wearable human health monitoring system. Microcolumn/nanocolumn structure is also a common regular geometry. Suh et al. [5] Silicon templates with nanopore were fabricated by photolithography and ion reactive etching technology. Then the flexible substrates with nano-pillar structure were copied by pouring polymer solution after solidification. Then conductive metal platinum was sputtered on the surface of flexible nano-pillars. Finally, the flexible structures with interlocking nano-pillars were constructed. Pressure sensor. Because of its high aspect ratio, the interlocking structure of nano-columns can not only sense the resistance change under positive pressure, but also accurately detect the shear and torsion deformation. In addition, the sensor has high stability (> 10000 cycles) and excellent on/off conversion characteristics. It can monitor human heartbeat, bouncing droplets and other subtle external stimuli. Bao et al. [6] combined pyramid micro-structure with micro-column structure, designed and manufactured a super-thin flexible pressure sensor, which greatly enhanced the signal-to-noise ratio (SNR) of signals obtained from human body by using the amplification effect of micro-column interface. The flexible pressure sensor has excellent stability after long-term cyclic testing (loading and unloading more than 3000 times at 10 kPa). The sensitivity of the sensor is kept at 0.56 kpa-1. It can accurately detect and distinguish weak jugular vein pulse signals of healthy people and patients with heart disease, and is conducive to rapid diagnosis of cardiovascular diseases. Chen et al. [7] A high sensitivity flexible pressure sensor was fabricated based on the principle that the contact resistance between the gold micro-column array (the diameter of the cylinder is 20 um) and the conductive polymer film varies under different static pressures. The high sensitivity of the sensor is mainly derived from the power law relationship between the contact resistance and the pressure. The pressure sensor has adjustable sensitivity from 0.03 to

17 kPa-1 in the pressure area below 1 kPa, and its performance even exceeds the ability of human skin. It has potential application in static detection of small size objects [8].

2 Experiment

2.1 Materials and Instruments

The graphene (flake diameter: 10 um) and conductive polymer PEDOT: PSS were purchased from Sigma Aldridge (Shanghai) Co., Ltd. The high-efficiency Silver conductive adhesives were High Performance Silver Paste 16031 of Shanghai Sango Biotechnology Co., Ltd. and anhydrous ethanol was purchased from China Pharmaceutical Chemical Reagent Co., Ltd. The product specifications are analytical purity (AR). Commercial PMDAODA type polyimide tape (50 um) is used for flexible substrates and SYLGARD 184 (10:1 mass ratio of main agent to curing agent) is used for polydimethylsiloxane. Sartorius BS110S Precision Electronic Balance of Beijing Saidolis Balance Co., Ltd; FS-110 Type Ultrasound Dispersion Instrument of Shenyang Ultrasound Technology Center; UT71C Digital Multimeter of Ulide Technology (China) Co., Ltd.

2.2 Sensitive Mechanism of Sensors

Composite materials are used as strain sensitive materials. Carbon nanotubes have good conductivity and mechanical properties [9]. In order to make the sensor have a larger strain measurement range, PEDOT:PSS is used as a conductive polymer to enhance the conductivity of carbon nanotubes network. When the sensor is in tension state, the contact resistance of graphene increases; when the sensor is in recovery state, the contact resistance of graphene decreases, so the sensor has good response and recovery characteristics [10].

2.3 Fabrication of Flexible Tensile Strain Sensor

60 mg graphene was dissolved in absolute ethanol and treated by ultrasonic oscillation for 1 h to make it evenly dispersed in ethanol. The CNT-PEDOT:PSS nanocomposites were prepared by adding 5 ml PEDOT:PSS solution (mass fraction 1.7%) to the solution of carbon nanotubes and then ultrasonic oscillation for 1 h. The fabrication process of flexible stretchable strain sensor is as follows: AZ4620 photoresist is spin-coated on the surface of silicon wafer using clean silicon wafer as substrate material at a rotating speed of 2000 r/min to obtain sacrificial layer adhesive with thickness of 50 micron; PDMS film with thickness of 500 micron is spin-coated on the surface of sacrificial layer adhesive. The PDMS film was solidified by heating at $60 \sim C$ for 3 h; the photoresist was dissolved in acetone solution to realize the stripping of PDMS; the surface of PDMS was hydrophilic treated by oxygen plasma to make the surface of PDMS have hydrophilic groups such as hydroxyl group and carboxyl group; the composite material was sprayed. The strain sensitive layer is obtained on the surface of

PDMS, and then the strain sensitive layer is rotated and coated on the surface of PDMS to obtain a stretchable strain sensor. The preparation process is shown in Fig. 1.

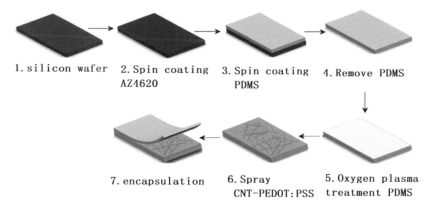

1. silicon wafer   2. Spin coating   3. Spin coating   4. Remove PDMS
                        AZ4620   PDMS

7. encapsulation   6. Spray   5. Oxygen plasma
                    CNT-PEDOT:PSS   treatment PDMS

Fig. 1. Fabrication process of flexible strain sensor

2.4   Preparation of Conductive Ink

PEDOT: PSS/graphene multi-component composite ink was prepared by solution blending method. PEDOT: PSS was weighed by a precise electronic balance of 1 g. Graphene was weighed according to different concentration. Graphene was added to 2 mL anhydrous ethanol. After ultrasonic dispersion for 20 min, PEDOT: PSS was added to stir for 20 min, and the prepared ink was prepared. The precursor ink is heated to volatilize the absolute ethanol in the precursor ink, so that the prepared ink meets the printing requirements. The heating temperature is 40 C and the heating time is 15 min. The preparation process of the ink is shown in Fig. 2. The sensor is prepared by taking a certain amount of self-made multi-component composite ink and placing it in it. In the printing needle of dispensing machine, vacuum treatment was carried out in the vacuum drying box for 5 min to remove the air bubbles in the slurry. The inner diameter of the needle used in the printing process was d = 210 um. The pressure of dispensing machine is 6psi (1psi = 6.89476*103 Pa, the same below). Patterned inkjet printing is carried out on PI film. After printing, the heat plate is cured at 90 C for 60 min. The cured sensor is packaged with PDMS.

Testing characterization of multi-component composite ink sensor The micromorphology of multi-component ink and sensor was characterized by tungsten filament scanning electron microscopy (SEM). In order to facilitate the electrical performance test of the device, copper wires were drawn from both ends of the device as test wires, and silver paste was used between the copper wire and the ink. Fixed. Connect the copper wire to the two test chucks of the multimeter separately. Record the resistance change of the sensor with the change of bending angle during the process. SEM images of PEDT: PSS composite ink prepared with different graphene blending ratio. Graphene blending ratio is 0 wt%, 6 wt%, 8 wt%, 10 wt%, 12 wt%, 14 wt%, 16 wt%, 18 wt%, respectively. From Fig. 3(a) It can be seen that the surface of PEDOT: PSS ink

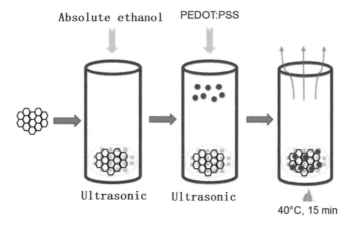

Fig. 2. Process chart for preparation of multi-component composite ink

without graphene is smooth, the particle size of conductive polymer in ink is uniform, and the particle size of conductive polymer is nano-scale. It can be seen that the diameter of graphene sheet is mainly distributed in the range of 10–20 um, and there is a small amount of agglomeration in PEDOT: PSS with 18 wt% graphene content. The results show that with the decrease of graphene content, the dispersion of graphene in composite ink is greatly improved, and the adhesion between flake graphene is weakened. Graphene nanomaterials are coated with PEDOT: PSS conductive ink. Most of the graphene doped is evenly dispersed in PEDOT: PSS conductive ink, and there is no large amount of graphene. The agglomeration of PEDOT: PSS showed that the adhesion and agglomeration between fluffy graphene could be weakened by using absolute ethanol as solvent to disperse graphene, which was beneficial to uniform dispersion in PEDOT: PSS conductive ink, and the addition of graphene could increase the viscosity of PEDOT: PSS ink, so that the prepared multi-component composite ink could respond to the agglomeration. In direct ink-jet printing process, the flexible strain sensor is printed with multi-component composite ink with different graphene blending ratio. The success rate of the sensor is counted. It is found that the printing effect of multi-component ink with graphene thinning ratio between 8 wt% and 10 wt% is better.

The conductive ink obtained from the above experiments was added to the spray gun, and the flexible substrate was placed on the heating table. By adjusting the temperature of the heating table to 100 C, the conductive film could be obtained on the bottom flexible substrate by spraying the conductive ink. Then the dense and highly conductive silver nanowire film could be obtained by annealing at 120 C for 2 h. Two smooth PDMS films of the same size were prepared as dielectric layers of the two sensors. Then two PDMS composite film electrodes were encapsulated face-to-face according to the sandwich structure, and electrode leads were pasted at both ends of the sensor to obtain a capacitive flexible pressure sensor.

3 Results and Discussion

Characterization Fig. 3 shows the planar and micro-nano structures of PDMS thin films respectively. It can be seen from the figure that the surface of PDMS thin films prepared by using smooth glass as template is fairly flat. In Fig. 3, the surface of PDMS thin films prepared by using ground glass as template has many convex microstructures, of which the height of convex is about 4.5 um. It is important to study the influence of PDMS with these two structures on sensor performance. In order to better study the influence of micro-nano structure on sensor performance, two kinds of sensors with planar structure and micro-nano structure were tested. Figure 4 shows the continuous response curve of the planar flexible pressure sensor to 333 Pa pressure, and Fig. 5 shows the continuous response curve of the micro-nano flexible pressure sensor to 333 Pa pressure. It can be seen from Fig. 6 that the sensors with these two structures have good stability and repeatability, but the sensors with micro-nano structure have higher response than those with plane structure.

In this paper, a GR/PEDOT:PSS multi-component composite ink was prepared. Resistance-type flexible strain sensor was fabricated by direct-writing inkjet printing (DIW). The multi-component ink was used as bending resistance material. The influence of graphene content in composite ink and ratio of depth to width on the sensitivity of sensor. The method of ultrasonic dispersion of graphene with absolute ethanol can make graphene more uniformly distributed in PEDOT:PSS. With the increase of printing rate, the linewidth of flexible sensor decreases obviously, and there is no edge spreading phenomenon. With the increase of multi-component, the linewidth of flexible sensor decreases obviously. The sensitivity of the device decreases with the increase of graphene content in the composite ink. The resistance change rate (R/R0) of the 8 wt% graphene content sensor is 1.045 at the highest level, and the bending test range is 0 to 40°. With the increase of graphene content, the service life of the device increases obviously. Graphene is the most important component in PEDOT:PSS polymer.

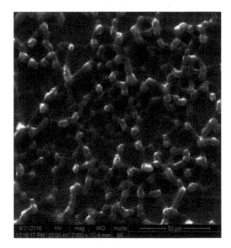

Fig. 3. Electron Microscope with or without Micro-nanostructure

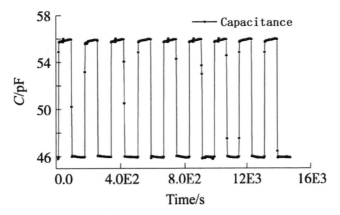

Fig. 4. Response Curve of Planar Structure Sensor Fig

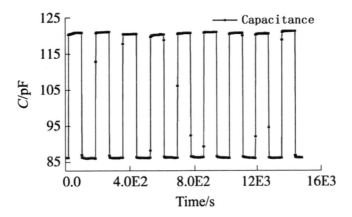

Fig. 5. Response Curve of Micro-nanostructure Sensor

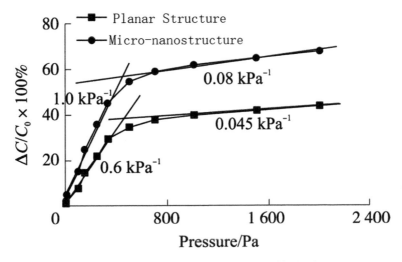

Fig. 6. Sensitivity comparison curves of two kinds of sensors

The optimum mixing ratio is 12 wt%. Increasing the depth-width ratio of the sensor can effectively improve the resistance change rate ($\Delta R/R0$) of the sensor. The maximum resistance change rate is 3.414, and the bending test range is 0–80 which effectively enlarges the measuring range of the bending angle. In view of the above research, the practical application of the flexible strain sensor can be improved by increasing the bending angle. The method of aspect ratio of devices is used to improve the sensitivity of devices. a flexible capacitive pressure sensor with sandwich structure was fabricated by a simple and low-cost method. The flexible PEDOT:PSS/graphene multicomponent composite ink PDMS film with microstructures and the flexible PEDOT:PSS/graphene multicomponent composite ink PDMS film with planar structure were fabricated by template replication. The performance of the flexible pressure sensor with these two structures is compared. The sensitivity of the composite thin film sensor with Micro-Nanostructure is 1.0 kPa-1, while that of the composite thin film sensor with planar structure is 0.6 kPa-1. The results show that the sensitivity of the flexible sensor is greatly influenced by the micro-morphology of the electrode surface. When the PEDOT:PSS/graphene composite ink film electrode surface has micro-structure, the sensor has higher sensitivity. When the surface of the electrode is relatively flat, the sensitivity of the sensor is relatively low. In view of the above research, in the practical application of sensors, the method of fabricating Micro-Nanostructures on the surface of electrodes can be used to significantly improve their sensitivity.

Acknowledgments. Foundation project: 2018 Key topic of Educational Science Planning in Jilin Province, Research on the training mode of innovative engineering talents under the background of Emerging Engineering Education (ZD18089).

References

1. Gong, S.Y., et al.: Tattoolike Polyaniline microparticle-doped gold nanowire patches as highly durable wearable sensors. ACS Appl. Mater. Interfaces. **7**(35), 19700–19708 (2015)
2. Bandodkar, A.J., Jeerapan, I., Wang, J.: Wearable chemical sensors: present challenges and future prospects. ACS Sens. **1**(5), 464–482 (2016)
3. Amjadi, M., Kyung, K.-U., Park, I., et al.: Stretchable, skin-mountable, and wearable strain sensors and their potential applications: a review. Adv. Funct. Mater. **26**(11), 1678–1698 (2016)
4. Jiang, H.Q., Sun, Y.G., Rogers, J.A., Huang, Y.G.: Appl. Phys. Lett. **90** (2007)
5. Cui, J., Zhang, B., Duan, J., Guo, H., Tang, J.: Sensors **16** (2016)
6. Zhu, B.W., Niu, Z.Q., Wang, H., Leow, W.R., Wang, H., Li, Y.G., Zheng, L.Y., Wei, J., Huo, F.W., Chen, X.D.: Small **10**, 3625 (2014)
7. Park, J., Lee, Y., Hong, J., Ha, M., Jung, Y.D., Lim, H., Kim, S.Y.: KoH. ACS Nano **8**, 4689 (2014)
8. Pang, C., Lee, G.Y., Kim, T.I., Kim, S.M., Kim, H.N., Ahn, S.H., Suh, K.Y.: Nat. Mater. **11**, 795 (2012)
9. Pang, C., Koo, J.H., Nguyen, A., Caves, J.M., Kim, M.G., Chortos, A., Kim, K., Wang, P.J., Tok, J.B.H., Bao, Z.A.: Adv. Mater. **27**, 634 (2015)
10. Shao, Q., Niu, Z.Q., Hirtz, M., Jiang, L., Liu, Y.J., Wang, Z.H., ChenX, D.: Small **10**, 1466 (2014)

A Parallel AES Encryption Algorithms and Its Application

Jingang Shi$^{(\boxtimes)}$, Shoujin Wang, and Limei Sun

School of Information and Control Engineering, Shenyang Jianzhu University,
Shenyang, Liaoning, China
shijingang@163.com

Abstract. With the rapid development of the Internet technology, data security is becoming more and more important. Data encryption is an important means to protect data security. AES is an important algorithm for encrypting data. However, when the amount of data needed to be encrypted is large, the traditional AES algorithm runs very slowly. This paper presents a parallel AES encryption algorithm based on MapReduce architecture, which can be applied in large-scale cluster environment. It can improve the efficiency of massive data encryption and decryption by parallelization. And the paper designs a parallel cipher block chaining mode to apply AES algorithm. Experiments show that the proposed algorithm has good scalability and efficient performance, and can be applied to the security of massive data in cloud computing environment.

Keywords: AES algorithm · MapReduce · Cloud computing

1 Introduction

With the wide application of computer technology, data security is also more and more widely concerned by people [1]. Encryption technology of massive data is an effective means to ensure data security. Block cipher algorithms has high security and fast encryption speed, so it is widely used in the security of massive data system. However, the traditional block cipher algorithms do not support large-scale parallel computing. For the rapid growth of massive data applications in cloud computing environment, the traditional block cipher algorithm is increasingly unable to meet the needs of applications. Therefore, more and more attention has been paid to the research of parallel block ciphers for massive data [2].

Wu gives a detailed description and analysis of the research of block cipher [3]. Qing firstly puts forward the basic theory and requirements of parallel cryptosystem construction, which lays a solid theoretical foundation and construction principles for parallel cryptosystem [4]. A parallel working mode of block cipher can achieve linear acceleration ratio without changing the cryptographic characteristics of the original block cipher algorithm [5]. Yang proposes a concurrent design of S-box structure, which can effectively improve the encryption performance of block ciphers [6]. Lee proposes an accelerated block cipher algorithm based on GPU [7], which contributes to the research of high performance computing and cryptography. The above-mentioned researches put forward effective working mode and parallel strategy for block cipher

© Springer Nature Singapore Pte Ltd. 2020
M. Atiquzzaman et al. (Eds.): BDCPS 2019, AISC 1117, pp. 173–179, 2020.
https://doi.org/10.1007/978-981-15-2568-1_24

system, which can provide a good reference and inspiration for large-scale parallel research of block cipher in cloud computing environment in this paper.

Aiming at the problem of parallel cryptographic algorithm for massive data, this paper proposes a parallel implementation mechanism of AES block cipher based on MapReduce architecture, which can realize large-scale parallel encryption and decryption of massive data.

2   MapReduce

MapReduce [8, 9] is a distributed computing framework first proposed by Google that can process massive data concurrently in large computer clusters. It is a simplified distributed programming model. Map and Reduce are two basic operations of the model. Initially, data sets are partitioned and stored in distributed file systems. Users process the initial data key/value pairs by rewriting the Map function, resulting in a series of intermediate key/value pairs, and use a rewritten Reduce function to aggregate the intermediate key pairs with the same key value, and finally output the results.

Runtime system of MapReduce architecture resolves the distribution details of input data, tasks scheduling and errors handling. Hadoop framework [10, 11] is an implementation of MapReduce architecture using Java.

3   Parallel AES Based on MapReduce

In this section we describe in detail the parallel encryption and decryption process of AES algorithm. Massive plaintext data that need to be encrypted are stored in distributed file system. It is stored in the block form (for example, 64M per block). In order to ensure the robustness of data, every block of data has a copy.

The process of MR-AES encryption is divided into three stages: Map, Shuffle and Reduce. The whole process is shown in Fig. 1.

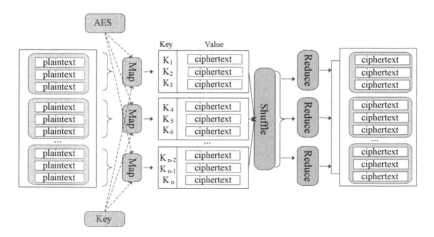

Fig. 1. Parallel AES algorithm MR-AES based on MapReduce architecture

In Map stage, the system first divides the plaintext files into blocks and then generates *<key, value>* pairs according to block size of block cipher algorithm. The key is formed by the offset of plaintext relative to the whole file. The value is a 64-bit long plaintext block. Then for each *<key, value>* pair, block cipher algorithm is invoked to encrypt the current plaintext block m_i into c_i. According to the input key, the compound key K_i of the current cipher text is generated, which is composed of the block number of the file block and the plaintext group number. Finally, each Map task outputs a pair of intermediate values $<K_i, c_i>$.

The Shuffle phase allocates the $<K_i, c_i>$ pairs, assigns the same group value pairs in K_i to the same group, sorts all values in the same group according to offset in K_i, and finally transfers the same group of data to a single Reduce node.

Each Reduce task processes the list with the same set of values. Since the plaintext block has been encrypted into the corresponding cipher text, the Reduce task only needs to be arranged in the order of offsets within the group, and the cipher text can be output directly to the distributed file system.

The decryption process of parallel block cipher is similar to that of encryption, so it is not described in detail.

4   Cipher Block Chaining Mode MR-CBC Based on MapReduce

It is necessary to hide the statistical characteristics and data formats of plaintext by using appropriate working modes in order to improve the overall security and reduce the chances of deletion, replay, insertion and forgery.

In theory, the cipher block chaining CBC mode cannot achieve parallel operation, because when encrypting plaintext of each packet, we need to know the cipher text encrypted by the previous group of plaintext.

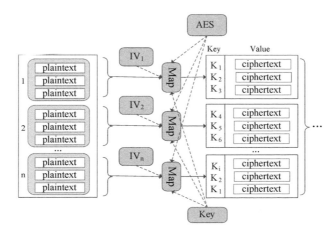

Fig. 2. Parallelization of MR-CBC mode

In MR-CBC mode based on MapReduce, we improve the basic CBC mode by introducing an initial vector *IV*, an initial subvector IV_i and a subvector generation algorithm. The parallelization of MR-CBC mode is realized by the initial subvectors.

4.1   Parallelization of MR-CBC Mode

Suppose that plaintext files are divided into *n* file blocks for storage in the distributed file system, as shown in Fig. 2. In MR-CBC mode, each file block starts a Map task. The system assigns an initial subvector IV_i to each Map task. Then in each Map task, the system encrypts or decrypts the data in common CBC mode according to the specific block cipher algorithm (such as AES), the *key* and the initial subvector IV_i. Then the results generated by each Map task are output to the distributed file system DFS through Shuffle and Reduce processes.

4.2   Subvector Generation Algorithms of MR-CBC Mode

When the amount of data in plaintext file is very large, the number of blocks in plaintext file will be very large, so there are many initial subvectors needed. If each initial subvector is generated independently, not only the task is very heavy, but also it is not easy to store and memorize them.

The message digest method performs a MapReduce tasks on plaintext file blocks, as shown in Fig. 3. We start a Map task for each block of plaintext file, call message digest algorithm (such as MD5 algorithm) in Map task, and get a message digest value of the block. Then in the Reduce task, we do a XOR operation between the digest value of a plaintext, the initial vector *IV* and the digest values of all previous plaintext blocks. Then an initial subvector of the current block is generated. Finally, all the initial subvectors IV_1, IV_2, ..., IV_n are generated.

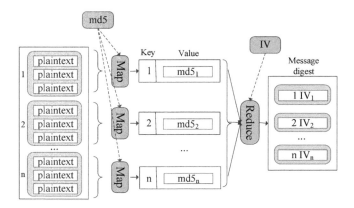

Fig. 3. Parallelization of MR-CBC mode

5   Experiments

In order to evaluate the performance of the parallel AES algorithm, we compare the parallel block cipher based on MapReduce architecture with the traditional block cipher through experiments in this section.

The running environment of the experiment is that the cluster contains 10 server nodes, each node has 4 cores with 2.00 GHz Intel Xeon (R) CPU and 4G memory, Linux Red Hat 5.1 operating system and Hadoop version 1.1.0.

5.1   Performance Evaluation of Basic Block Ciphers

We compare the performance of the traditional AES algorithm with the parallel AES algorithm (MR-AES) based on MapReduce in this section.

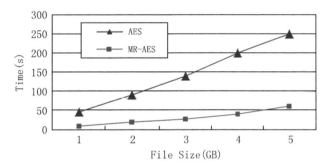

Fig. 4. Parallelization of MR-CBC mode

We compare the encryption performance of traditional AES algorithm with that of parallel AES algorithm in Fig. 4. The horizontal axis represents the amount of data of encrypted files, and the vertical axis represents the time spent in encrypting. The curve AES is the performance curve of encrypting large files using traditional block encryption. The MR-AES is the performance curve of using the AES algorithm based on MapReduce architecture. The number of cluster nodes used in MR-AES algorithm is 5. The experimental results show that the time spent by the two encryption methods

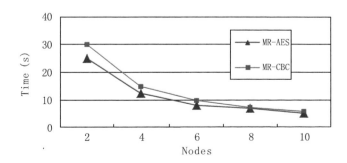

Fig. 5. Parallelization of MR-CBC mode

varies linearly with the amount of encrypted data, and the time spent by the MR-AES algorithm using five node clusters is slightly more than one fifth of the time spent by the AES algorithm.

5.2   Impact of Cluster Nodes

We make an experimental analysis of the impact of the number of cluster nodes on parallel block ciphers in this section. In Fig. 5, it shows the experiment of encrypting 1 GB files. The number of cluster nodes is from 2 to 10. It can be seen from the figure that with the increase of the number of cluster nodes, whether it is the basic parallel block cipher algorithm MR-AES or the CBC parallel working mode MR-CBC, the time spent on encryption decreases linearly with the increase of the number of cluster nodes. Therefore, the parallel block cipher algorithm based on MapReduce has good scalability.

6   Conclusions

This paper proposes a parallel implementation mechanism of block cipher based on MapReduce architecture, and designs a corresponding parallel working mode MR-CBC. The parallel block cipher mechanism proposed in this paper can realize parallel encryption and decryption of block ciphers. Experiments show that the block cipher based on MapReduce architecture has good scalability and performance, and can be applied to the security of massive data in cloud environment.

Acknowledgements. This work was supported by the National Natural Science Foundation of China (No. 61702345).

References

1. Fernandes, D.A.B., Soares, L.F.B., Gomes, J.V., et al.: Security issues in cloud environments: a survey. Int. J. Inf. Secur. **13**(2), 113–170 (2014)
2. Ahuja, S.P., Komathukattil, D.: A survey of the state of cloud security. Netw. Commun. Technol. **1**(2), 66–75 (2012)
3. Wu, W.L., Feng, D.G.: The State-of-the-art of research on block cipher mode of operation. Chin. J. Comput. **29**(1), 21–36 (2006)
4. Qing, S.H.: Construction of parallel cryptographic systems. J. Softw. **11**(10), 1286–1293 (2000)
5. Yin, X.C., Chen, W.H., Xie, L.: Parallel processing model of the block cipher. J. Chin. Comput. Syst. **26**(4), 600–603 (2005)
6. Yang, J., Ge, W., Cao, P., et al.: An area-efficient design of reconfigurable s-box for parallel implementation of block ciphers. IEICE Electron. Express **13**(11), 1–9 (2016)
7. Lee, W.K., Cheong, H.S., Phan Raphael, C.W., et al.: Fast implementation of block ciphers and PRNGs in maxwell GPU architecture. Cluster Comput. **19**(1), 335–347 (2016)
8. Dean, J., Ghemawat, S.: MapReduce: a flexible data processing tool. Commun. ACM **53**, 72–77 (2010)

9. Dean, J., Ghemawat, S.: MapReduce: simplified data processing on large clusters. Commun. ACM **51**(1), 107–113 (2008)
10. White, T.: Hadoop the Definitive Guide. O'Reilly, USA (2009)
11. Landset, S., Khoshgoftaar, T.M., Richter, A.N., et al.: A survey of open source tools for machine learning with big data in the Hadoop ecosystem. J. Big Data **2**(1), 1–36 (2015)

Identifying Influence Coefficient Based on SEM of Service Industry

Yawei Jiang[1], Xuemei Wei[2(✉)], and Fang Wu[1]

[1] Quality Research Branch, China National Institute of Standardization, Beijing, China

[2] National Science Library, Chinese Academy of Sciences, Beijing, China
583451349@qq.com

Abstract. Nowadays, service industry represents the main engine of China's economic development. By using the structural equation model (SEM), we can identify the influence coefficient of different factors of consumer satisfaction, which can help to better promote high-quality development of service industry. In this paper, we find that consumers pay more attention on brand image and perceived quality and care less about expected quality and perceived value of our survey of six service industries, which are express delivery, express hotel, aviation, auto insurance, online travel and mobile communication. The results suggest that we can promote the quality development of service industry by strengthening brand building, enhance brand characteristics and individuality, and improve customer perceptions of different dimensions of service industry, such as service environment, service personnel, service personalization and service professionalism.

Keywords: Consumer satisfaction · Service industry · Service quality

1 Introduction

Nowadays, the contribution of China's service industry to economic growth is improve continuously. In 2018, the service industry accounted for 52.2% of GDP, contributing 59.7% to national economic growth, driving national GDP growth by 3.9% points, and continuing to play the role of "stabilizer" for economic growth. Therefore, it is necessary to find the right way to improve customer satisfaction and identify the key factors to promote high-quality development in service industry.

From the research in recent years, Ahrholdt studied the interrelationships between customer delight and customer satisfaction to consumer loyalty, and measured their marginal effect respectively [1]. Altinay proved that interactions between older consumers and other consumers would impact the customer satisfaction and social welfare directly [2]. Lynn analyzed the net effects of tipping and no-tipping systems to restaurant customer satisfaction [3]. Meesala computed patient's satisfaction from healthcare industry, evaluate path coefficients as well as direct and indirect effects of critical factors [4]. Park explained the tourist satisfaction with current satisfaction and future behavior, and found that visitors' expectation, tourist season as well as first trip would have an important impact on satisfaction factors [5]. Longyu considered

© Springer Nature Singapore Pte Ltd. 2020
M. Atiquzzaman et al. (Eds.): BDCPS 2019, AISC 1117, pp. 180–185, 2020.
https://doi.org/10.1007/978-981-15-2568-1_25

residents' requirements and satisfaction for sustainable urban development, combined the subjective and objective variables to observe urban sustainable development quality [6]. Huaman evaluated the consumer satisfaction of online medicine service and discussed the positive and negative effects of brand image, expectations, perceived quality and perceived value on customer satisfaction [7]. Li Kexin calculate customer satisfaction from supermarket and find out factors, which play important role in it [8]. Zhou Zhiying use SEM to evaluate main factors affecting customer satisfaction of bicycle sharing [9]. Wu Bing introduce APP experience, cycling experience and social factors into ACSI model and evaluate consumer satisfaction of shared cycling [10].

The current research lacks an overall study of factors affecting service satisfaction, by using SEM in six service industries, we can get some common guidance of quality improvement in service industry.

2   Customer Satisfaction Model

The SEM is show in Fig. 1.

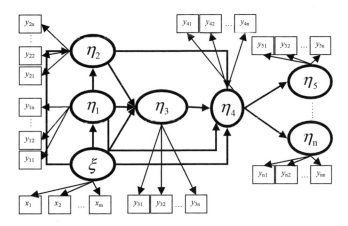

Fig. 1. SEM model

Where, circles and rectangles respectively indicate the latent variables and the observable variables. And the arrows shows the influence relationship as well as the influence direction between them.

And the mathematical model of SEM is shown as follows:

$$\eta = \mathbf{B}\eta + \Gamma\xi + \zeta \tag{1}$$

In the model, η and ξ respectively denote the endogenous and exogenous latent variables; B is interrelationship between endogenous latent variables; Γ is the relationship between exogenous and endogenous latent variables; ζ indicates the unexplained factors in the equation.

Then, we get the mathematical equation as follows:

$$X = \Lambda_x \xi + \delta \tag{2}$$

$$Y = \Lambda_y \eta + \varepsilon \tag{3}$$

Where, X and Y represent the vector composed by exogenous indicators and endogenous indicators; Λ_x reflects the interrelationships between exogenous and endogenous indicators; Λ_y reflects the interaction between endogenous index and endogenous latent variable.

Then, based on SEM, we build the customer satisfaction model of service industry (Fig. 2).

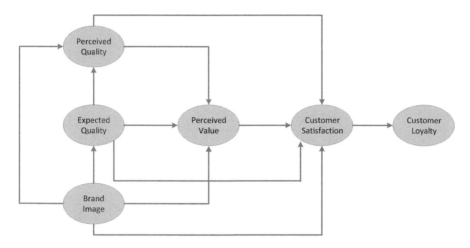

Fig. 2. Customer satisfaction model of service industry

This model is composed by six structural variables. The brand image indicates the overall brand impression and brand characteristics from consumer perspective. The expected quality indicates customers' overall expected to the service quality. The perceived quality is customer's evaluation of different aspects from service quality. The perceived value means customer's evaluation of cost performance. The customer satisfaction contains customer's overall satisfaction, customer's satisfaction compared with expected service, ideal service and other brands service. The customer loyalty is the combination of customer's repurchase willingness and recommendation of the service.

3   Results

We investigated six service industries, and they are expresses delivery, express hotel, aviation, auto insurance, online travel and mobile communication. Through online and telephone surveys, we collect questionnaires from 250 cities across the country, and each industry has received 2000 effective questionnaire at least. Then we use the customer satisfaction model to calculate the influence coefficient of the six structural variables, and the results are shown in Table 1.

Table 1. The influence coefficient of consumer satisfaction in service industry

Industry	Brand image	Expected quality	Perceived quality	Perceived value
Expresses delivery	0.63	0.27	0.60	0.19
Express hotel	0.68	0.30	0.55	0.19
Aviation	0.64	0.27	0.51	0.24
Auto insurance	0.61	0.25	0.54	0.18
Online travel	0.50	0.22	0.51	0.17
Mobile communication	0.68	0.13	0.59	0.29

As can be seen from the Table 1, brand image and perceived quality play a significant role in consumer satisfaction where their average influence coefficients during the six service industries are 0.62 and 0.54. Expected quality and perceived value are not critical to customers, their average influence coefficients are 0.23 and 0.21. Moreover, there is a little bit different of consumer's concerns in different industries. From the calculated results, in express delivery, express hotel, aviation, auto insurance and mobile communication industries, brand image has the biggest impact to customer satisfaction. It indicates that consumers care more about brand when they purchase service. However, in the online travel industries, perceived quality has the largest influence coefficient, that means a slight adjustment in online travel services can have a major impact on customer satisfaction.

The influence coefficient of the variables on the brand image of different industries is shown in Table 2.

Table 2. The influence coefficient of the variables on the brand image

Industry	Brand impression	Brand characteristics
Expresses delivery	0.33	0.30
Express hotel	0.36	0.32
Aviation	0.33	0.31
Auto insurance	0.31	0.29
Online travel	0.25	0.25
Mobile communication	0.39	0.29

It can be seen from Table 2 that brand impressions have a greater impact on customer satisfaction than brand characteristics, reflecting the daily propaganda of the brand and the construction of the brand image is essential. That is to say, making consumers have a good impression of the brand plays an important role in improving customer satisfaction. Of course, we must also pay attention to the importance of brand characteristics. Compared with the brand impression, the influence coefficient of brand characteristics is slightly smaller, but the influence of brand characteristics on customer satisfaction is still critical. Therefore, it is necessary to strengthen the research and construction of brand characteristics and create a brand with distinctive personality, which is also important to meet the needs of consumers and improve customer satisfaction.

In addition, the perceived quality variables of different industries have slightly different factors affecting customer satisfaction.

In addition, different service industries have different perceived quality variables, and the impact of these variables on customer satisfaction in the industry is summarized as follows. As for express delivery services, service timeliness and service attitudes have a large influence coefficient; for express hotel service, hotel facilities and hotel environment have a greater impact on satisfaction; for airline service, cabin equipment and crew are more influential; for auto insurance, meeting the needs of customers and the convenience of insurance claims procedures are the key influencing factors; for online travel, the comprehensive information and the rationality of tourism product design are important to customer satisfaction; for mobile communication, network signal and call quality are key factors.

4   Conclusion

Based on SEM, we build customer satisfaction model of service industry and calculate customer satisfaction from six different service industries. We find out that brand image and perceived quality are the key factor to consumer satisfaction. In the six investigated industries, consumers that purchase expresses delivery, express hotel, aviation, auto insurance and mobile communication service care more about brand image, consumers of online travel pay more attention on perceived quality. Therefore, it is necessary for service enterprises to strengthen their brand building and brand features, as well as to improve their service quality in different aspects, such as consumer environment, service staff, service characteristics and service price. This is important for us to improve the quality of the entire service industry and to promote the high quality development of the whole society.

Acknowledgments. This work is supported by the National Social Science Fund Major Project (Research on Quality Governance System and Policy to Promote High Quality Development), NO. 18ZDA079.

References

1. Ahrholdt, D.C., Gudergan, S.P., Ringle, C.M.: Enhancing loyalty: when improving consumer satisfaction and delight matters. J. Bus. Res. **94**(1), 18–27 (2019)
2. Altinay, L., Song, H., Madanoglu, M., Wang, X.L.: The influence of customer-to-customer interactions on elderly consumers' satisfaction and social well-being. Int. J. Hosp. Manage. **78**(4), 223–233 (2019)
3. Lynn, M.: The effects of tipping on consumers' satisfaction with restaurant. J. Consum. Aff. **52**(3), 746–755 (2018)
4. Meesala, A., Paul, J.: Service quality, consumer satisfaction and loyalty in hospitals: thinking for the future. J. Retail. Consum. Serv. **40**(1), 261–269 (2018)
5. Park, S., Hahn, S., Lee, T., Jun, M.: Two factor model of consumer satisfaction: international tourism research. Tour. Manag. **67**(8), 82–88 (2018)
6. Shi Longyu, X., Tong, G.L., Xueqin, X.: Study progress on the satisfaction degree and evaluation of sustainable development. Acta Ecol. Sin. **39**(7), 2291–2297 (2019)
7. Xu Huaman, X., Yong, L.J., Mingjie, J.: Analysis of influencing factors of consumer behavior in online medicine consumption. E-Commerce Lett. **8**(1), 30–39 (2019)
8. Kexin, L., Zhelei, K., Ruiqi, G., Ruibo, K., Xiang, G.: The model of supermarket satisfaction index. Adv. Appl. Math. **7**(7), 743–748 (2018)
9. Zhiying, Z., Zuopeng, Z.: Customer satisfaction of bicycle sharing: studying perceived service quality with SEM model. Int. J. Logist. Res. Appl. **22**(5), 437–448 (2018)
10. Bing, W., Siqi, C.: A study on customer satisfaction model of shared cycling. Adv. Soc. Sci. **6**(8), 1084–1093 (2017)

Multi-function Bus Station System Based on Solar Energy

Yue Han, Zhongfu Liu[✉], Ke Zhang, and Wannian Ji

College of Information and Communication Engineering,
Dalian Minzu University, Dalian 116600, Liaoning, China
103086817@qq.com

Abstract. At present, solar power is used in many bus stations in China. But most solar panels are installed in a fixed mode, which cannot make the sunlight stay perpendicular to the solar panel in real time. It result in insufficient utilization of solar energy resources. And because most bus stations are not connected to power, there is no real-time display of vehicle movement information and voice broadcast function. Aiming at the above problems, a multi-function bus system based on solar energy is designed in this paper. In the system, the direction of sunlight exposure is collected through the photosensitive sensor, which is transmitted to the STM32 main control system, so that the solar panel is always perpendicular to the sun's rays. The day and night are judged by the system clock, and the working mode is automatically switched accordingly. And the collected solar energy is stored by batteries for use by the bus station system. With the electronic bus stop sign in the system, the GPS positioning information is received by the STM32 main control system, then latitude and longitude information is sent by GSM to the station receiving system. Display bus arrival and voice broadcast are controlled by CPU in the station. Through GPRS and server interaction, the bus coordinate information is transmitted to the mobile phone platform to change the previous phenomenon of passengers blindly waiting for the bus. Through the test, the system can run stably and has high use value.

Keywords: STM32 · Solar tracking · Photoelectric sensor · GSM · GPS

1 Introduction

The development of modern cities has led to a global energy crisis and excessive pollution of the natural environment, so solar energy as a pollution-free renewable energy has been widely used. With the government's advocacy of "Green Travel and Low-carbon Life", more and more citizens choose to travel by bus, and most cities in China also increase investment in bus station facilities distributed throughout the city. However, most of the bus stations are simple bus kiosks [1]. Some bus stations in remote areas are not connected to power, and the bus stop signs are relatively simple. There is no real-time display of vehicle movement information and no voice broadcasting function. How to make the bus station effectively energy saving, and use solar energy power supply, at the same time make the best use of the bus station resources to

© Springer Nature Singapore Pte Ltd. 2020
M. Atiquzzaman et al. (Eds.): BDCPS 2019, AISC 1117, pp. 186–195, 2020.
https://doi.org/10.1007/978-981-15-2568-1_26

better serve the public, is the important content of the bus station construction in the future.

At present, governments around the world strongly sponsor and encourage enterprise schools to carry out solar energy research and development. The design of public facilities is also an important topic of research in various countries. For example, in 2009, Japan launched a bus station called "real-time online", with dynamic bus schedule, billboard and trash bin and other convenient facilities. In addition, the roof of the station is made up of solar panels, which provide power for all the facilities of the station. The rapid development of the public transport system in China has increased the demand for public transport. Solar bus station has become a symbol of urban progress. Although the first solar bus station has appeared in Guangzhou, it has not been popularized nationwide until now. Therefore, the research on solar bus station is of great significance.

The multi-functional bus system based on solar energy designed in this paper mainly includes solar tracking system, battery charging and discharging system, intelligent bus stop display system. Photoelectric tracking and solar Angle tracking are adopted, and the strength of the sun is detected by photodiode to judge the change of working mode on sunny and cloudy days and realize the purpose of tracking the sun. When the CPU detects sunny days, the photosensitivity sensor senses the intensity and incidence direction of sunlight, converts it into an electrical signal and transmits it to the main control system of STM32. When the detection is sunny, the intensity and incidence direction of sunlight are sensed by the photosensitivity sensor, which converts the information into electrical signals and transmits them to the main control system of STM32. After the information is analyzed and processed by STM32, STM32 controls the deceleration motor to rotate so that the solar panel is always perpendicular to the solar rays. When the detection is cloudy, the real-time clock RTC of STM32 is used to calculate the time of year, month and day, and track the sun by calculating the azimuth Angle of the sun according to the function. The collected solar energy is stored by the battery and the charging condition of the battery is tested for the control panel by STC89C52. With the electronic bus stop sign in the system, the GPS positioning information is received by the STM32 main control system, then latitude and longitude information is sent by GSM to the station receiving system. Display bus arrival and voice broadcast are controlled by CPU in the station. Through GPRS and server interaction, the bus coordinate information is transmitted to the mobile phone platform to change the previous phenomenon of passengers blindly waiting for the bus, more intuitive to see the bus.

2 System Scheme Design

The multifunctional solar bus station system focuses on the combination of solar tracking system and diversified bus stations. It is mainly composed of solar automatic tracking system, battery charging and discharging system and intelligent bus stop sign display system. Figure 1 is the overall block diagram of the system.

2.1 Overall Scheme Design of Solar Sun Chasing System

Two deceleration motors are used to control the movement of the solar panels in the horizontal and vertical directions in the system. The energy required for its work is completely from the solar cell power generation and storage battery energy. This product adopts the photoelectric detection and tracking mode and the solar Angle tracking mode, and through the photodiode detection of the sun's strength and strength to judge sunny and cloudy days change the working mode to achieve the purpose of tracking the sun. When the detection is sunny, the photosensitive resistor and the comparison circuit LM324 are used to determine the orientation of the sun, sending an electrical signal to STM32 to drive the rotation of the solar panel. When the detection is cloudy, the real-time clock RTC of STM32 is used to calculate the time of year, month and day, and track the sun by calculating the azimuth Angle of the sun according to the function. Make it can always receive the sun light, greatly improve the utilization of solar energy resources.

2.2 Overall Design of Battery Charging System

Intelligent control of storage charge and discharge, intelligent detection of battery voltage technology, battery overcharge protection and prevent the solar panel backflow is included in the system. With STC89C51 as the core control, the voltage information is collected through ICL7135 and transferred to MCU. When the single-chip micro-computer detects that the voltage at both ends of the battery is greater than 14.8 v, the controller automatically disconnects the battery from the solar panel [2]. If the micro-controller detects that the voltage at both ends of the battery is less than 10.8 v, the controller automatically disconnects the battery from the charging load. The floating charge is 13.2 v [3]. The voltage value collected by ICL7135 is displayed on the 1602 LCD screen.

2.3 Intelligent Bus Stop Sign Display System

GPS, GSM, GPRS, voice broadcast and other technologies are integrated to realize the intelligent function of combining the electronic bus stop sign with voice broadcast. Car terminal CPU reads the information platform (latitude and longitude coordinates, serial number and name of platform), and at the same time GPS location, time, speed and other real-time information, will be effective GPS information compared with the platform of information, calculation, determine the vehicle's current location and

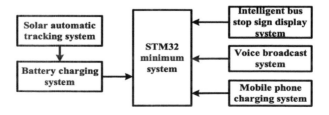

Fig. 1. General block diagram of the system

arrival time, and each other through the GSM communication pass information to stand within the CPU [4], and the GPRS will interact with the server bus coordinate information to the mobile platform, CPU and compare the location, time, speed, and through the speech and LED display to inform the passengers [5].

3 System Hardware Circuit Design

3.1 Solar Tracking System

3.1.1 Microprocessor Circuit

The micro-controller used in the data acquisition and control system is STM32F103C8T6. The micro-controller is a 32-bit micro-controller enhanced by STM32 series micro-controller, with the maximum clock frequency up to 72 MHz.

3.1.2 Photoelectric Detection Circuit Design

The power supply part of this module is composed of chip LM324 and photosensitive resistor, which is powered by 5 V power supply chip. Each chip is connected with four ordinary resistors and a photosensitive resistor, which forms a comparison circuit with operational amplifier. When the light changes, the resistance value of the photosensitive resistor also changes, and the comparison circuit detects the voltage of positive and negative electrodes. When the positive electrode voltage is higher than the negative electrode voltage, the output terminal will output a positive voltage, and vice versa. The output end is connected with the pins of micro-controller PB1, PB2, PB3 and PB4 respectively. After detecting the signal, the pins of micro-controller are analyzed and processed to determine the direction of the sun and control the rotation of the motor. The system is equipped with four comparison circuits to detect the incidence of sunlight in four directions [6]. Figure 2 shows the circuit diagram of photoelectric detection circuit design.

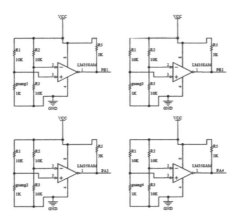

Fig. 2. Photoelectric detection circuit

3.1.3   Rain or Shine Detection Circuit Design

The module is mainly composed of photodiode and operational amplifier LM324. The positive pole of the photodiode is connected to the positive pole of LM324, and the negative pole is connected to the 5 V power supply. The operational amplifier LM324, resistor and photodiode constitute the comparison circuit. As the light intensity changes, the photodiode changes. When the positive voltage of the comparison circuit is greater than the negative voltage, the output is 1; otherwise, the output is 0. The output end is connected to the pin PA5 of the single chip microcomputer, and the single chip microcomputer judges rain or shine through the change of pin level [7].

3.1.4   Clock Circuit

The real-time clock RTC of STM32 is an independent timer and a set of timers that can be counted continuously. The system USES its clock function to adjust its relevant configuration through software, modify the value of the timer to adjust the clock, and use the clock function to judge day and night, and calculate the sun's azimuth on cloudy days. The passive crystal provides the RTC real time clock for the system, which with two capacitors forms the external oscillation circuit.

3.1.5   Motor Drive Circuit

Choose L298N motor drive circuit, the circuit containing two H bridge of high voltage large current full bridge driver, can be used to drive dc motor and stepping motor with two can make control side, in the case of not affected by the input signal is allowed or forbidden to work with a logic voltage input, use internal logic circuit part of the work under low voltage. JGY370 motor is a dc screw reducer motor. The input end is connected to the pin PA1 and pin PA2 of the micro-controller. After the micro-controller determines the direction of the sun, the level of the two pins changes and the motor turns forward and backward to drive the solar panel.

3.2   Battery Protection System Circuit

3.2.1   Design of Voltage Acquisition Circuit

The module USES A 4-bit double-integral A/D conversion chip ICL7135. Which use positive voltage provided by A 5 V power supply and A negative voltage provided by the ICL7660 chip output. Among them, the ceramic capacitor between pins 9 and 10 (small inductance, high resonant frequency and low impedance to high frequency signal) C4 and the ceramic capacitor C3 between pins 7 and 8 play a role in filtering out high-frequency fluctuation interference [8].

3.2.2   Microcontroller Circuit

The micro-controller used in the data acquisition and control system is STC89C52. The micro-controller for the clock frequency up to 11.0592 MHz.

3.3   Intelligent Bus Stop Sign Display System

3.3.1   GSM Circuit

The GSM module circuit is SIM900A chip and its peripheral circuit. It communicates with the micro-controller by means of serial port communication. Therefore, the TX and RX pins of the chip are connected to the PA2 and PA3 pins of the serial port of the main micro-controller.

3.3.2   Rainfall Distribution on 10–12 GPS Circuit

The GPS module circuit is composed of ATGM336H-5 N chip and its peripheral circuit, which supports the single system positioning of BDS/GPS/GLONASS satellite navigation system. GPS communicates with MCU through serial port, so TXD of transmitting end of GPS is connected with RXD of MCU [9]. Figure 3 shows the main circuit diagram of the GPS module.

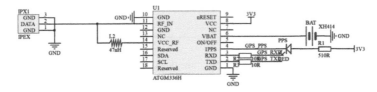

Fig. 3. GPS module circuit

4   System Software Design

The system software is divided into three parts, namely solar automatic tracking system, battery protection system and intelligent bus stop display system.

4.1   Solar Automatic Tracking System

The system software mainly includes rain or shine detection part, Angle tracking part and clock part. All programs are written in C language. After power supply, the system starts to initialize. Day and night can be judged by the clock. If it is day, the system can judge the rain or shine of the weather. If it is dark, the system enters the interrupt waiting state. Figure 4 shows the main program design flow chart.

4.1.1   Rain or Shine Detection Part

As shown in the Fig. 4, after the system is started, the system will judge the photo-electric induction mode when the level of the used pin is high by detecting the rain or shine detection circuit composed of photosensitive diode and operational amplifier. The direction of the sun is determined by the photosensitive resistor, and the level of the pin set by the micro-controller is changed to control the motor rotation until the middle photosensitive resistor detects the sunlight.

When the pin level of the rain or shine detection circuit is low, start the Angle tracking mode, use the RTC real-time clock function of STM32 to initialize the RTC,

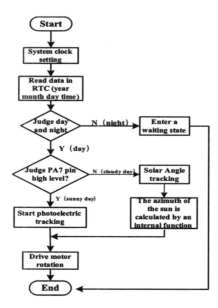

Fig. 4. Main program design flow chart

configure the RTC, modify the count value, re-initialize the RTC configuration time, update the flag position and position information every second according to the interrupt, and finally read the time. The Angle of the sun is calculated by the preset function formula, and the Angle difference is calculated every ten minutes to calculate the rotation time of the motor, so as to achieve the purpose of tracking.

4.2   Battery Protection System

The system software part mainly includes the voltage collection part and the display part. All programs are written in C language. After power on, the system initializes and the display module initializes. When the count overflows, it starts to receive A/D data to determine the switch off of the relay by comparing with the preset voltage value, so as to protect the battery. Figure 5 shows the main program design flow chart.

4.3   Intelligent Bus Stop Sign Display System

The system software mainly includes vehicle GPS positioning, GSM to send positioning information, GPRS and server connection and station receiving information control display and voice module. All programs are written in C language.

4.3.1   Vehicle GPS Positioning

After the system starts, the serial port initialization is first carried out to clear the cache array of GPS information. Read GPS through TXD transmitted information, by looking at the flag bit to determine whether read correctly. After reading correctly, the data will be parsed, and the longitude and latitude will be extracted, and the data will be printed out. Figure 6 is the main program design flow chart of GPS positioning information.

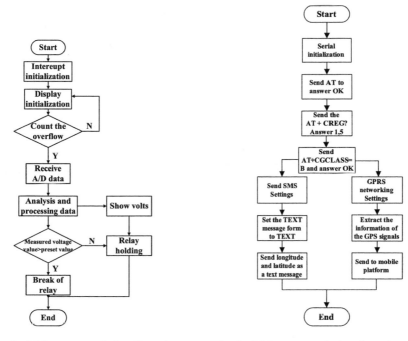

Fig. 5. Main program design flow chart

Fig. 6. Main program design flow chart of GPS positioning information

4.3.2 GSM Sends Positioning Information and GPRS to the Server

After the system starts, the serial port initialization is first carried out, and AT instructions are sent to initialize GSM. The command send AT + CGCLASS = B implements the function of both connecting to the server and sending text messages. The format of sending SMS and GPRS networking are set, and the parsed GPS data is sent to the CPU in the station in the form of SMS, and uploaded to the mobile platform.

4.3.3 Receiving Information

After the system starts, the serial port is initialized, AT instructions are sent, GSM is initialized and SMS format is set. When the longitude and latitude are received, the CPU makes comparison judgment on the longitude and latitude, determines the bus location, and transmits it to the voice control and display module.

5 System Debugging

When testing the function of the electronic bus stop sign system, in order to observe the phenomenon more intuitively, the mobile phone is used as the receiver to verify the process of information transmission. First, GPS and GSM are initialized after power on. After 30 s, the serial port assistant will display the coordinate information of GPS positioning and the longitude and latitude coordinate information after CPU analysis.

Then GPRS connects with the server and transmits the latitude and longitude coordinates to the server. When the serial port assistant shows SEND OK, it means that the upload is successful. In this way, the position can be displayed on the mobile phone platform. GPRS is connected to the server as shown in Fig. 7, and uploaded to the mobile platform as shown in Fig. 8.

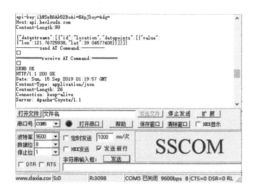

Fig. 7. GPRS connects to the server

Fig. 8. Upload to the mobile platform

Finally, from the latitude and longitude, the CPU determines whether the bus has arrived at the simulated platform. If the coordinates match the GSM coordinates, SMS messages are sent to the station. When the serial port assistant shows station1, it means that the bus has arrived at the station and the SMS is sent successfully [10]. As shown in Fig. 9, GSM sends messages to mobile phones, and as shown in Fig. 10, mobile phones receive messages.

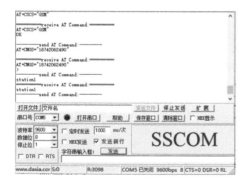

Fig. 9. GSM sends messages to mobile phones

Fig. 10. The phone receives the message

6   Conclusion

In this paper, STM32 single chip microcomputer, AT89C51 single chip microcomputer, GPS positioning module and GSM module are combined to design a multi-functional public transport system based on solar energy. The system is not only of great significance to the utilization of resources, but also through the design of electronic bus stop sign to solve the problem of passengers blindly waiting for the bus, which greatly facilitates the travel of citizens.

Acknowledgements. Fund project: innovation training project support for college students in Liaoning province (201912026158).

References

1. Punz, V., Marzano, V., Simonelli, F.: Guest editorial special issue on models and technologies for intelligent transportation systems. IEEE Trans. Intell. Transp. Syst. **19**(11), 3528–3529 (2018)
2. Krivik, P.: Methods of SoC determination of lead acid battery. J. Energy Storage **15**, 191–195 (2018)
3. Omaiah, V.B., Narayanaswamy, V., Panayan, M., Kannaiyah, J., Ananda, R.G.: Pressure-tolerant electronics and discharge performance of pressure-compensated lead acid batteries under hyperbaric conditions. Marine Technol. Soc. J. **52**(5), 110–117 (2018)
4. Abhishek, B., Kumar, M.A.: GSM-CommSense-based through-the-wall sensing. Remote Sens. Lett. **9**(1–3), 247–256 (2018)
5. Keerthika, C., Singh, M., Tamizharasi, T.: Tracking system for vehicles using GPS, GSM and GPRS. Res. J. Eng. Technol. **8**(4), 453–457 (2017)
6. Xuefu, L., Zhiqin, C.: Dynamic modeling and simulations of a tethered space solar power station. J. Aerosp. Eng. **31**(4), 1–12 (2018)
7. Khan, M.J., Mathew, L.: Lini Mathew: Comparative study of maximum power point tracking techniques for hybrid renewable energy system. Int. J. Electron. **106**(8), 1216–1228 (2019)
8. Ahsanullah, S., Wenguo, X., Samo, K.A., Shiyi, C., Shiwei, M.: Development of a simplified method for the determination of ampere-hour capacity of lead-acid battery. Energy Environ. **29**(1), 147–161 (2018)
9. Derek, C., Xingxin, G.G.: Probabilistic graphical fusion of LiDAR, GPS, and 3D building maps for urban UAV navigation. Navigation **66**(1), 151–168 (2019)
10. Mohan, B., Karthik, N.: Intelligent information passing about road transport using GSM and GPS. Int. J. Syst. Control Commun. **9**(4), 386–391 (2018)

Intelligent City Construction Based on Big Data in the Era of Sustainable Development

Lepeng Chen and Chengjiang Wang[✉]

College of Electricity and New Energy, Three Gorges University,
Yichang 443002, Hubei, China
ccxdxctgu@163.com

Abstract. At present, with the rapid development of computer technology and related network technology as the symbol of the application of the era of big data, and combined with the strategy of sustainable development era, the concept of Intelligent city emerges as the times require. From theory to practice, wisdom city has gradually won the attention of the world. Informatization, networking and technicalization will become the inevitable trend of Intelligent City Construction Based on big data in the era of sustainable development. Taking the construction of Intelligent city based on big data in the era of sustainable development as the research object, this paper starts with the introduction of the importance, significance and achievements of Intelligent city, expounds the relationship between Intelligent city and big data in the era of sustainable development, and studies the construction of Intelligent city based on big data in the appropriate era of sustainable development. This paper tries to analyze the predicament and analysis of the construction model of Intelligent City Based on big data in the era of sustainable development from the evaluation index system of intelligent city, and then tries to put forward the construction model and construction of Intelligent City Based on big data in the era of sustainable development. Discussion measures.

Keywords: Sustainable development era · Big data era · Intelligent city Construction

1 Introduction

In the era of sustainable development, big data plays an important role in the construction of smart cities. In the current process of information network development, the importance of science and technology is becoming more and more obvious. All walks of life need to research and apply new technologies. Only in this way can they occupy their own position in the market, big data. It is the product of high-tech development in the era of sustainable development. It has a strong market potential. Through the analysis of relevant theory and practice, it is found that big data can play an important role in the construction of smart cities. Firstly, the state can understand and grasp the problems in the construction of Intelligent city space through big data technology, take timely measures to deal with these problems; secondly, in the traditional mode, the information channel is single, and the accuracy of data obtained is

© Springer Nature Singapore Pte Ltd. 2020
M. Atiquzzaman et al. (Eds.): BDCPS 2019, AISC 1117, pp. 196–202, 2020.
https://doi.org/10.1007/978-981-15-2568-1_27

limited, which has been unable to match the construction of Intelligent city in the era of sustainable development. Requirements match, the application of large data communication technology can quickly protect the information and sources of smart urban and rural construction data; third, the current speed of information technology is further accelerated, and the use of the Internet is further increased. This provides great convenience for the implementation of big data [1].

This paper briefly describes the current situation of Intelligent city construction based on big data in the era of sustainable development. At present, big data has been widely used in many fields, such as society, education and so on. It has a profound impact on people's lives, and has greatly changed the research methods of scholars. In order to realize the socialist modernization as soon as possible, our country must strengthen the work of building smart cities. In the process of building smart cities, we must strengthen the work of building smart cities. We must use all kinds of advanced technology, especially big data technology, to further improve the scientific nature of Intelligent city construction. At present, many researchers have found that the key research and analysis of big data technology can obtain a lot of research results. Next, it should be noted that big data is not just one kind theory. Only when it is used in practice can it play its role. Big data is a very important tool in the construction of smart cities. Although its role and value have not been fully exploited at present, big data has very broad prospects for development. When the big data theory was put forward, it was not aimed at urban construction. However, from the current point of view, big data will be able to be applied to urban construction and accelerate the development of Intelligent city construction. Big data is very helpful to the construction of urban information sharing platform. The goal has already been achieved. We must strengthen the communication and cooperation between departments. Because of the slow development of urban construction in China, there are many shortcomings in the traditional urban construction work. The generation and extensive use of big data can effectively remedy these mistakes and gradually standardize the construction of Intelligent cities [2].

2 Method

This paper mainly uses literature research, comparative analysis, social network analysis and other related research methods, as follows:

Literature Research Method; By using the method of literature survey, through collecting, sorting out and analyzing the relevant literature on the subject of public service of the intelligent city evaluation index system, the development and changes of the research object in this period are studied. Based on the real situation of the original literature, referring to the research results and experience of predecessors and others, through reasoning and analysis, we can further understand the hot spots and academic frontiers of foreign research, deeply understand the connotation of public service of the intelligent city evaluation index system, and complete the literature review and theoretical summary [3].

Comparative analysis method; Comparative analysis method: comparative analysis method is one of the basic methods of comparative analysis, through a certain index to the same nature of the index evaluation criteria for comparison. This paper mainly carries out statistics on different evaluation indicators of smart cities, compares the public services of smart cities, summarizes their similarities and differences, and analyses their advantages and disadvantages. To further carry out the comparative study of the content, characteristics and modes of network information services; through the study of the current situation of evaluation index public service of smart cities in China, to compare the problems and causes of the service of evaluation index system of smart cities, and to learn from the public service work of evaluation index system of smart cities in other countries.

Social Network Analysis; As a comprehensive and comprehensive analysis method, social network analysis is also called "structural analysis". As its name implies, it is a theoretical method to concentrate on the analysis of social structure and social relations. It is mainly used to explore the dynamic flow and allocation of social relations media, resources, things and locations. Social network analysis explores the organizational structure, individual and network attributes of social network by analyzing the relationship between nodes in the network. The analysis of network includes cohesive subgroup, small group research, small world effect, etc. The analysis of individual social network includes point centrality, proximity centrality and so on. Social network analysis advocates two-way interaction from the perspective of "interaction", which helps to bridge the micro and macro levels. Social network analysis is often used to characterize the relationship between different entities and the mode, structure and function of the relationship. It usually measures the relationship between various things, such as information, resources, and the flow of relationships. Social network analysis method can be used in the following aspects: (1) relationship distance and centrality analysis, including degree, density and centrality analysis. Degree refers to the number of neighboring points in the social network graph: density refers to the degree of close relationship between the points in the network; centrality includes inner centrality and outer centrality, and the more the actor is in the central position of the network, the greater its influence. (2) Small groups, i.e. subgroup analysis, are groups of people in a community that are so closely related that they are combined into a subgroup [4].

3 Experiments

Seven representative indicators of Intelligent cities in China are selected for analysis and sorting. Seven indicators are summarized and classified, and the frequency of indicators is obtained, so as to analyze the key points of the development of Intelligent cities in China at present. In the statistical analysis of indicators, this paper adopts the idea of fuzzy set, replacing the alternative expression of the same indicator with the same one, so as to complete the calculation of the frequency of indicators. The frequency of indicators can be objectively obtained by using the statistical analysis of indicators based on the idea of fuzzy set, which avoids the subjective influence caused by questionnaire survey or less data.

4   Results and Analysis

Seven evaluation indicators of Intelligent cities in China are selected to summarize and sort out, in order to get the focus and frequency of the development of Intelligent cities in China in recent years, and to find out the characteristics of key points and related modes of Intelligent city construction. The seven Intelligent city standard systems are: Shanghai Intelligent City Evaluation Index System 2.0, Ge Jian's Intelligent City Index System, Guo Liqiao's Intelligent City Index System, Zhu Guilong and Fanxia's Intelligent City Evaluation Index System, and the Intelligent City Standard formulated by the SOA Sub-Technical Committee of the National Information Technology Standardization Technical Committee. Quasi-system, Ningbo Intelligent City Evaluation Index System and Nanjing Intelligent City Evaluation Index System [5].

The selection of index system takes into account timeliness and accuracy, and also needs to include research results of research institutes, universities and industry experts, covering a wide range, and these seven are all indicators of the existence of Intelligent city top-level design system, which can better reflect the research and accuracy. Ningbo Intelligent City Evaluation Index System is developed by the research team of Ningbo Intelligent City Construction Standard Development Researcher Joint Advisory Body, Zhejiang University and other famous universities. This index is the first-level indicator system of the standard, including six first-level indicators, including Intelligent infrastructure, Intelligent governance, Intelligent people. The evaluation index system of Nanjing Intelligent City was put forward by Deng Xianfeng of Nanjing Information Center in 2015. There are only two levels of indicators, including five first-level indicators and 18 second-level indicators. The division is simple and clear: Shanghai Intelligent City Evaluation Index System 2.0 is composed of five first-level indicators and 18 second-level indicators [6]. The Shanghai Pudong Institute of Intelligent Urban Development released its research in 2014, which focuses on informatization and citizen perception. Guo Liqiao's Intelligent City Index System is extracted from the book "Research on the Standard System of China's Intelligent Cities" published by Guo Liqiao in October 2013. The index system is in the Department of Housing and Urban Construction Calibration and Architecture. Under the guidance of the Department of Energy Conservation and Science and Technology, the Digital Urban Engineering Research Center of China Urban Science Research Association has organized more than 30 units and institutions, including industries, enterprises and research institutes, and more than 60 experts in various fields. It has carried out the research on the standard system of China's Intelligent cities. The system is a three-level index system with five first-level indicators. Including infrastructure, construction and migration, management and service, industry and economy, safety and operation and maintenance, 19 secondary indicators. This system takes into account the complexity of the urban standardization system, provides specific operational directions in infrastructure and engineering management standards, but ignores the consideration of urban integrity and the causes of urban residents. Su: The Intelligent City Standard System formulated by the SOA Sub-Technical Committee of the National Information Technology Standardization Technical Committee is excerpted from the "Intelligent City Practice Guide–SOA Support Solution" published in July 2013 by the SOA Sub-Technical Committee of the National Information Technology Standardization Technical Committee and the Research Fellow of Electronic Industry Standardization of the Ministry of Industry and Information Technology. Resolve the core problem of Intelligent city: sharing and collaboration. This index has five first-level indicators, namely, foundation, supporting technology, construction management, information security and application. According to classification, the index is characterized by laying particular emphasis on technical elements standards. Ge Jian's Intelligent city index system is extracted from Ge Jian's works in 2016. "Theory and Practice of Intelligent Cities", published in September, set up six first-level indicators, including economic vitality, citizen quality, public management ability, smoothness, environmental friendliness, life happiness index, 32 second-level indicators and more essential oil three-level indicators to evaluate the wisdom of the city: Zhu Guilong, Fanxia. The intelligent city evaluation index system is extracted from Zhu Guilong and Fan Xia's book "Theory and Practice of Intelligent City Construction", which was published in January 2015. The system has four first-level indicators, including social activity level, economic activity level, spatial level, technical level, and 23 second-level indicators (Table 1).

Table 1. Classification table of first-level indicators

Serial number	Proportion	Frequency	Index
1	0.714	5/7	Economic Vitality, Industry and Economy, Social Activities, Intelligent Industry and Intelligent Industry
2		3/7	Subjective Perception of Citizens' Quality, Intelligent Crowds and Intelligent Citizens
3	0.857	6/7	Public Management Ability, Management and Service, Space Level, Intelligent People's Livelihood, Intelligent City Public Management and Service, Intelligent Service Areas
4		3/7	Smoothness, Construction Management and Development of Intelligent City Information Service Economy
5		1/7	Environmental friendliness
6		3/7	Life Happiness Index, Intelligent Environment and Soft Environment Construction of Intelligent City
7	0.571	4/7	Infrastructure, Technology Level, Supporting Technology, Intelligent Urban Infrastructure
8		1/7	Construction and Migration
9		1/7	Safety and Operational Maintenance
10		1/7	Social activities
11		1/7	Basics
12		2/7	Intelligent City Humanities Literacy and Intelligent Humanities
13		1/7	Urban Internet English

As shown in the table above, nearly 86% of the index systems regard public service as a primary indicator, which shows that the focus of the superstructure of Intelligent cities in China is public service, and public service has become the corresponding target of various Intelligent cities. At present, the construction of Intelligent cities in China will be "people-oriented" at the top of the construction, while at the same time, the construction of Intelligent cities in China will focus on public service. The establishment of Intelligent community pilot projects in various places has promoted the process of public service objectives in Intelligent cities based on big data in the era of sustainable development. However, "public service" is mainly aimed at providers, that is to say, this part still emphasizes the "passive connection" in the connection. Citizens "passively" accept the public services of all kinds of Intelligent cities, but does not reflect the importance of "active connection" in it. At the same time, industrial economy and infrastructure ranked second and third, accounting for 71.4% and 57.1% respectively. Industrial economy accounts for a large proportion, mainly because industrial economy is the main manifestation of the vitality of Intelligent cities in China in recent years. The development and test of urban informatization solutions by major operators, product upgrading of various Intelligent machine manufacturing industries, and industry competition within the information technology industry are all the construction of Intelligent cities in recent years. The main manifestation of the situation is that technological enterprises should grasp key technologies in the process of transformation and upgrading of Intelligent cities, operators should stand firm in the promotion of government informatization and other factors, creating a city of all walks of life are exploring Intelligent cities, but these are full of various industry barriers, lack of. Small connectivity and unity [7, 8]. The importance of infrastructure reflects the importance of the basic elements of Intelligent city technology, but the same problem is that it does not reflect the connectivity of technological elements, which easily leads to the problem of each industry fighting independently. Therefore, we have collected the acceptance of public service cases in Aixi Lake streets and towns in Nanchang. Detailed data are shown in the following Fig. 1:

Fig. 1. Statistical chart of acceptance of public service cases in Aixi Lake, Nanchang

5 Conclusion

The emergence of Intelligent city in the era of sustainable development provides a new planning and path for urban transformation. Big data is becoming the "Intelligent engine" of Intelligent city. Whether big data can be used effectively to reflect the comprehensive energy of a city? Effective use of big data in the era of sustainable development to promote the development of Intelligent cities requires not only the construction of a higher level information technology system, top-level design and overall construction, but also the deep involvement of the government, enterprises and the public in personnel training, as well as the reform of the old out-of-date systems and opportunities. System [9, 10]. Only in this way can we explore a suitable model of Intelligent City Construction Based on big data in the era of sustainable development and build a truly intelligent city.

References

1. Fussman: Urban and rural planning and intelligent cities research in the big data era. Urban Constr. Theory Res. (Electron. Ed.) (09), 182 (2018)
2. Xu, W.: Urban and rural planning and intelligent city research in the big data era. Build. Mater. Dev. Orientat. (2), 173–174 (2017)
3. Li, C.: Intelligent city construction in the big data era. Electron. Technol. Softw. Eng. (16), 165–166 (2017)
4. Wen, A., Wang, P.: Innovation of talent information service mode based on big data under the background of wisdom city construction. J. Hebei Softw. Vocat. Tech. Coll. **19**(1), 1–3 (2017)
5. Wei, J., Chen, H., Wu, Z.: Research on intelligent city construction based on big data background. Sci. Technol. Perspect. (20), 6–7 (2016)
6. Wang, K.: Research on intelligent city construction based on big data. Dig. World (3) (2018)
7. Anonymous: Research on intelligent urban construction based on big data. Value Eng. **38**(02), 149–151 (2019)
8. Wang, Z., He, Y.: Research on intelligent city construction based on big data cloud platform. Internet Things Technol. **7**(12) (2017)
9. Anonymous: The construction of intelligent city in the era of internet plus and big data combination. In: Sixth Annual Conference of Yunnan Association for Science and Technology and Red River Basin Development Forum: Topic Two: Construction of Intelligent City in South Yunnan Center (2016)
10. Luo, B., Li, S., Tian, P.: Analysis of new intelligent city construction strategy based on big data. Commun. World (1), 97–98 (2018)

Construction of Supply Chain Information Sharing Mode in Big Data Environment

Jiahui Liu$^{(\boxtimes)}$, Guangqiu Lu, and Haiqin Wang

Qingdao Binhai University, Qingdao 266555, Shandong, China
1193121893@qq.com

Abstract. With the complexity of supply chain network structure and the globalization of supply chain development, information sharing in supply chain becomes more and more important. Supply chain information sharing can not only improve the overall performance of the supply chain, promote cooperation and communication between supply chain nodes, reduce the "bullwhip effect" caused by information non-sharing, but also improve the response speed of supply chain to customer demand, which is conducive to the formation of core competitiveness of enterprises. With the arrival of the era of big data, information sharing has new opportunities and corresponding challenges in supply chain: on the one hand, the development of new technologies such as cloud computing and big data analysis technology has provided new sharing modes and channels for information sharing in supply chain; on the other hand, most of them have to face challenges. According to the storage and processing of massive information caused by massive data in the environment, new challenges are posed to enterprises. Based on the change of information sharing in supply chain under big data environment, it is very important to build a new information sharing mode. Based on the analysis of the advantages and disadvantages of three traditional information sharing modes, this paper constructs an identity-based and role-based information sharing mode using advanced storage platform and data prediction and mining methods in large data environment, which solves the problem of limited data storage and operation speed.

Keywords: Big data · Supply chain · Information sharing · Model construction

1 Introduction

Supply chain information sharing is becoming more and more important to supply chain management. With the rapid development of economy and science and technology, the competitive situation of enterprises has gradually changed from inter-enterprise to inter-supply chain competition [1]. Supply chain management has become the focus and focus of modern management research. In the practice of supply chain management, information sharing among supply chain node enterprises has become an urgent problem to be solved. Under the environment of big data, the supply chain has undergone great changes. Big data will become the next new field for all sectors of society to pursue innovation, competition and productivity. Therefore, it is very important to construct a model of information sharing for supply chain enterprises in the big data environment to adapt to the new environment.

© Springer Nature Singapore Pte Ltd. 2020
M. Atiquzzaman et al. (Eds.): BDCPS 2019, AISC 1117, pp. 203–211, 2020.
https://doi.org/10.1007/978-981-15-2568-1_28

In the field of big data research, the current domestic and foreign research on big data mainly focuses on the big data as a new method. The research focuses on the complexity of big data and related computing models, the description, content construction and semantic understanding of big data, and the key technologies of computing architecture system [2]. However, there are few studies and applications on how big data affects supply chain and what changes it brings to supply chain. The research on information sharing in supply chain is mature both at home and abroad, and there are many achievements on information sharing mode and value sharing. From the existing literature research, it is found that scholars at home and abroad are mostly based on the overall thinking of supply chain, trying to build information sharing model suitable for the whole supply chain is less. Considering big data as a big background, using new technology means to build a new era of information sharing model that can solve the challenges of big data for supply chain is almost not yet available. The value study of information sharing is mostly based on the information provided by one of the sharing parties without considering each of the sharing parties. The mathematical model of this situation is also relatively lacking.

Based on the analysis of the advantages and disadvantages of the three traditional information sharing modes, this paper constructs an identity-based and role-based information sharing mode by using the advanced storage platform and data prediction and mining methods in large data environment to solve the problem of limited data storage and operation speed.

2 Related Concepts

2.1 Large Data

Generally speaking, the so-called big data is the data scale beyond the traditional sense, and it is difficult for general software tools to capture, store, manage and analyze the data. "Big data" is a data set of extremely large size and category that cannot be captured, managed and processed by traditional database tools. Big data environment this article refers to the arrival of big data era, people take all kinds of technical means to deal with large data creates a series of changes, including new big data analysis technology and cloud computing, big data environment for the entire were discussed, and focuses on its as a kind of environmental background, similar to the "information age" "industrial age" [3].

2.2 Supply Chain Information Sharing

Supply chain refers to the core enterprise as the center, based on the related enterprise "third-rate (information, content, capital)" control, starting from the procurement of raw materials, produce intermediate products to the final formation of finished goods and through the enterprise marketing channel will products sold to customers in the hands of the whole process involved in the suppliers, manufacturers, distributors, retailers and ultimately consumers together to form a whole functional network structure [4]. Former members of the supply chain node comparison pay attention to the "c" is closely related to the enterprise, but does not pay attention to the first fourth-rate one stream, some even aware of the importance of information flow cannot for information flow to make a quantitative analysis and evaluation, so the flow of information in a supply chain is not make full use of rise, causing the supply chain to customer demand slow response speed, high inventory cost and bullwhip effect, and other issues, and information sharing is an important mechanism to solve it [5].

2.3   Supply Chain Information Sharing in Big Data Environment

The research on supply chain information sharing is relatively mature, and there are a lot of researches on the definition, impact, value and mode of information sharing. But big data is changing our way of life, entertainment, work. And the emergence of large data brought changes to the supply chain. Many of the original research isn't applicable in the new environment, so we need to study what's the difference between supply chain information sharing and change in the big data environment [6].

3   Information Sharing Mode of Traditional Supply Chain and Its Comparison

3.1   Three Supply Chain Information Sharing Modes

(1) Point-to-point sharing mode
  In this mode, the supply chain nodes carry out end-to-end information transmission from the supplier to the demander through their internal information system. In point-to-point mode, information is provided and acquired in a many-to-many relationship without passing through other data centers or platforms. According to the difference in using information technology, this mode can be divided into EDI and data interface. Value-added network dedicated line between supply chain nodes in EDI mode for information transmission; data interfaces are based on Internet ports, each supply chain node enterprise establishes a special external database, information sharing is based on their external databases without first-hand access to each others' internal databases [7].

(2) Information Centralized Management Model
  Centralized management is to centralize all shared information in a database. In this mode, enterprises operate and access the relevant information shared by their partners. The model can be divided into two modes: third party mode and information platform mode. This model is based on data interface and the supply chain node enterprises can't touch the first-hand information of the internal database of other enterprises [8].

(3) Integrated Information Sharing Model
  This model integrates the point-to-point sharing mode and centralized management model. It emphasizes that the enterprises' information sharing areas can be differentiated. It is generally constructed with the information platform as the core. They are different because of the different levels of cooperation, information acquisition methods, information confidentiality, information processing and

transmission between the two parties. Besides, the difference in information sharing degree caused by different life cycles is unavoidable. Therefore, the differential sharing mode emerges as the times require [9, 10].

3.2 Comparative Analysis of Information Sharing Modes in Traditional Supply Chain

The needs of both sides of information sharing are very clear and can be targeted to plan the data in the format and other aspects in the first mode. At the same time, because of the direct contact with each others' first-hand data, so the enterprise would choose long-term cooperative relationship when choosing partners. Besides the security is also high. But once the partners have problems, enterprise data' security is not guaranteed. The second mode, due to the relaxation of restrictions on information systems, can also be used to expand the scope of enterprises. However, the weakness is that public database supplier need to integrate the information systems of enterprises in the supply chain on the existing basis, or even rebuild them. The cost of information integration is difficult and high. The last mode is feasible and easy to implement because it integrates the first two modes according to the difference in sharing requirements and sharing subjects. At the same time, the rapid development of logistics parks and information platforms throughout the country has also become a fertile ground for the development of this model.

4 Construction of Supply Chain Information Sharing Model Under Big Data Environment

4.1 Data-Information Conversion Model Under Big Data Technology

The supply chain isn't lack of data. The supply chain is lack of an appropriate model to translate huge and diversified raw data into operational information, which is the key for effective supply chain planning. Large data analysis technology has been widely used to transform a large number of raw data into information. It can be described as a data-information model. The model is divided into three stages: data acquisition, data analysis and data utilization to achieve business objectives. Data sources mainly include production promotion data (commodities, prices, sales), release data (increased or reduced commodities), inventory data, consumer data, transportation data. Data Separation and Cleaning: Data exists in many forms, including structured and unstructured, some of which are not suitable for analysis. So suitable data cleaning mechanism should be in place to ensure the large data' quality. Therefore, it is very important to choose the data cleaning and enrichment tools. Data Representation: It is a difficult task to design a database for such a large amount of data. If not well designed, it will lead to serious performance problems. Data representation plays a key role in large data analysis. There are many ways to store data, each of which has its advantages and disadvantages. Analytical data: There are many big data technologies available for forecasting and planning. The key to choosing which technology is decided by the of the enterprise's goal and business plan. Incompatible data formats make the value creation process of big data difficult, so it is urgent to innovate the technology of creating business value from large data.

The model presented below is designed for highly complex systems with huge data, multiple constraints and multiple considerations. It can analyze and provide insights from the system. Secondly, supply chain planning usually has many enterprise goals, such as cost reduction and demand satisfaction, which can be achieved by optimizing technology. The optimization model consists of four components:

(1) Input: The consistent, timely and integrated data of cleaning and quality are the input of the model;
(2) Goals: The model needs to take into account such objectives as minimizing costs, maximizing profits and demand coverage;
(3) Constraints: such as minimizing inventory, capacity constraints, demand coverage constraints, etc.
(4) Output: Based on inputs, objectives and constraints, various plans can be obtained, such as demand plan, inventory plan, etc. Achieving Business
(5) Goals: This stage includes three parts: scheme management, multi-user cooperation and performance tracker.

4.2 Supply Chain Information Sharing Model Under Big Data Environment

Using big data technology to mine useful information from mass consumer data, and in the large data environment, the popularity of cloud computing ensures the storage of mass data, and the capacity limitation is no longer obvious. The following is to construct a new model of information sharing in supply chain under the big data environment with the supply chain without third-party logistics enterprises as a typical representative, as shown in Fig. 1. The new mode of supply chain information sharing in large data environment mainly includes six layers: information sharing main body layer, service layer, data conversion layer, network security layer, cloud storage layer and sharing information release layer. It is summarized as three parts: information sharing participants, sharing information content and large data information sharing platform. The components of this model are described in detail below.

4.2.1 Participants in Information Sharing

The first and the bottom layers are the acts of participants, so they are introduced together. The subject of sharing is the traditional subject of sharing plus consumers, who are not only the provider and publisher of sharing information, but also the users of sharing information.

However, in the process of sharing information, the users of information can not necessarily share all the information provided by the information publisher, and the information provider does not necessarily allow all the information published by the information publisher to be fully shared by all the information sharing subjects. Because each enterprise has different degree of cooperation because of the frequency of business contacts, trade secrets and credibility, the length of cooperation time, the degree of close relationship and so on. According to the close degree of cooperation

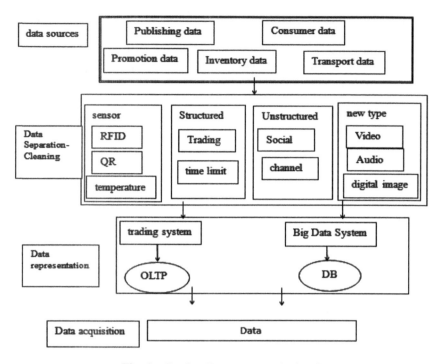

Fig. 1. Get data Down arrow misplaced

among enterprises, the inter-enterprise partnership in supply chain is divided into three levels: important partners, secondary partners and non-partners. However, the three partnerships can be transformed from one another, and important partners may be transformed into secondary or even non-cooperative partnerships. Each subject in the supply chain has its own partners at different levels. According to the difference of the level of cooperation, the types of information they share are also different.

4.2.2   Sharing Information Content

Publishers of shared information are the sources of shared information. They are the main bodies that can provide shared information. Each provides information that can be shared, and enters the shared database of enterprises through enterprise firewall. The information provided by participants in supply chain information sharing can be divided into two categories and three levels. The two major categories are information provided by suppliers, manufacturers, distributors and retailers in shared databases and relevant information released by consumers through various channels. Suppliers and other enterprises from their own internal database will need to share information with the outside world through the enterprise firewall into the enterprise external shared database; Consumers publish their demand for goods, comments and other information through comment sections, social networking sites, media and other channels of various e-commerce websites. These massive, different types and fast updating huge data have entered the Internet. In the process of information sharing, three levels are due to the different levels of cooperation between enterprises and partners and the corresponding permissions are different, so according to the level of cooperation, information sharing can be divided into different levels, which are level 0, level 1 and level 2.

4.2.3 Big Data Information Sharing Platform

Large data information sharing platform is the core part of information sharing mode. It includes network security layer, data conversion layer, cloud storage layer and service layer. The network security layer mainly includes unified authentication system and monitoring system. Unified authentication system is an access control system based on role assignment. Privileges are set according to roles, and different privileges have different levels of shared information. The flow of the unified authentication system is as follows: after a certain authentication procedure, the user can be assigned to the relevant role if he meets the requirements. After entering the role, he can request access permission. The system assigns the corresponding permission to the user according to the type of role, and the user can use the permission and pay a certain fee. After that, we can enjoy the access service of shared information in the shared information database corresponding to its permission level. This unified authentication system has the following characteristics: all users who enter the system can change their roles in a certain way, and at the same time, users have access rights corresponding to that role; The relationship structure of all the roles in the system can be hierarchical and easy to manage and maintain. It can define the role's permission granularity based on the minimum access permission, so as to enhance the security of the system. The people who define roles must be different from other people who play roles, so that roles with different access rights can be mutually restricted and the system can be more reliable. The monitoring system is responsible for the real-time monitoring of the network interface and communication security of the whole large data information sharing platform. The data transformation layer mainly extracts the relevant data from the database and transforms the data-information model constructed in the previous section into useful and understandable information for the access of users with relevant privileges. Cloud storage layer is mainly used to store huge amounts of data, including three large databases: shared information database, consumer database based on large data analysis, Internet database. Shared information database is an integrated database of external shared information database provided by information sharing publishers based on the level of partnership. Enterprises in the supply chain extract enterprise information needed to be shared for publishers of shared information, and enter the external shared information database through firewall, and the database is stored. Store in the cloud of each enterprise. This shared information database is the integration of these databases. Consumer database is based on the information released by consumers through various channels on the network. Available information and related forecast data of various businesses is analyzed by using big data technology. Internet database is a composite database which extracts part of the 0-level shared information from the shared information database and part of the information from the consumer database based on large data analysis. Except a small part of the information enters the Internet database in the consumer database based on big data analysis. the majority enters the shared information database to help enterprises making decisions according to the feedback of consumers. The data between the shared information database and the

Internet database can also be transferred to each other, but most of them are transferred from the Internet database to the shared information database. At the same time, the shared information database only transmits part of the 0-level shared information to the Internet database. The service layer mainly provides evaluation services, personalized services, charging services, access services, upload and download services. The evaluation system is mainly a system in which each information sharing agent in the supply chain enjoys the information sharing platform and gives feedback to its related services. Personalized service and management system is responsible for personalized customization and service settings related to information platform interface. The charging system is based on access. Big data information sharing platform is generally provided by third-party software providers. The cost of the whole system should be collected from the information sharing services enjoyed by the main bodies of supply chain information sharing. Access service is user authentication and has the right to access the corresponding level of information in the shared data after payment. Upload and download service system is mainly provided to the main body of information sharing for publishing and receiving low-level shared information. The system can access the Internet database through the firewall. It can obtain relevant information from the database to provide download service for the information sharing subject, and also can transfer the related data uploaded by the sharing subject to the Internet database safely. This new mode of supply chain information sharing in large data environment mainly combines the new technology means (cloud storage and large data analysis) in large data environment with the traditional integrated information sharing mode, but at the same time it retains the point-to-point information sharing mode, which is suitable for the information sharing needs of different large-scale enterprises.

5   Conclusions

This paper introduces three typical traditional supply chain information sharing modes: point-to-point sharing mode, centralized information management mode and integrated information sharing mode; compares and analyses the advantages and disadvantages of the three modes, and constructs a new mode of supply chain information sharing under large data environment to meet the needs of different enterprises of different scales information sharing needs.

Acknowledgement. Qingdao Bin University science and technology planning research, design and application of automatic material handling system in tire enterprises, Project NO. 2019KY03, 2018/11/22.

References

1. Ji, G., Hu, L., Tan, K.H.: A study on decision-making of food supply chain based on big data. J. Syst. Sci. Syst. Eng. **26**(2), 183–198 (2017)
2. Gunasekaran, A., Papadopoulos, T., Dubey, R.: Big data and predictive analytics for supply chain and organizational performance. J. Bus. Res. **70**, 308–317 (2017)

3. Ma, J., Yang, Y., Chen, C.: Evaluation and case study on information sharing of apparel supply chain. Wool Text. J. **45**(9), 65–71 (2017)
4. Lv, Q.: Supply chain coordination game model based on inventory information sharing. J. Interdiscip. Math. **20**(1), 35–46 (2017)
5. Huang, Y.S., Ho, C.H., Fang, C.C.: Information sharing in the supply chains of products with seasonal demand. IEEE Trans. Eng. Manag. **64**(1), 57–69 (2017)
6. Fosso Wamba, S., Gunasekaran, A., Papadopoulos, T.: Big data analytics in logistics and supply chain management. Int. J. Logist. Manag. **29**(2), 478–484 (2018)
7. Kembro, J., Näslund, D., Olhager, J.: Information sharing across multiple supply chain tiers: a Delphi study on antecedents. Int. J. Prod. Econ. **193**, 77–86 (2017)
8. Yang, C., Huang, Q., Li, Z.: Big Data and cloud computing: innovation opportunities and challenges. Int. J. Digit. Earth **10**(1), 13–53 (2017)
9. Wang, G., Gunasekaran, A., Ngai, E.W.T., et al.: Big data analytics in logistics and supply chain management: certain investigations for research and applications. Int. J. Prod. Econ. **176**, 98–110 (2016)
10. Kache, F., Seuring, S.: Challenges and opportunities of digital information at the intersection of Big Data Analytics and supply chain management. Int. J. Oper. Prod. Manag. **37**(1), 10–36 (2017)

Innovative Experiment Teaching of Computer Basic Course Based on SPOC

Rong Ding[(✉)]

Information Technology Center, Qinghai University,
Xining 810016, Qinghai, China
dingding@qhu.edu.cn

Abstract. The paper analyzed the present situation and the problems about experiment teaching of Computer Basic Course in Qinghai University, and proposing innovative experiment teaching model based on SPOC. This paper summarized the innovative experiment teaching methods with the SPOC platform and the flipped class, in order to stimulate the students' learning interest and improve the students' practical ability, then improve the quality of experimental teaching.

Keywords: SPOC · Computer basic course · Innovative experiment teaching

1 Introduction

The computer basic course is facing a rather awkward situation in the university computer education. Usually this basic course is opened when the students are freshman, but most of them know more about the computer basic knowledge. If the teaching content is too shallow, many students have mastered it. And if the teaching content is too much, but many students do not even grasp the basic knowledge. So how to teach this computer basic course, how to set up the course content, how to use the teaching means and methods, in order to link up this computer course from high school to university?

2 The Current Situation of Computer Basic Course Teaching

The computer basic course in our university is a compulsory public basic course for freshmen of all majors. As the basis of follow-up courses, the main purpose and task of teaching is to enable students to master the basic computer knowledge and necessary skills for better work, study and life in the information society, and enable them can be more skilled in using the operating system and various office software, using the network and other IT means to obtain information and exchange information, then lay the foundation for further study of other computer-related courses in the future.

The course form include theory and experiment teaching. The main contents of the course include computer composition and basic principles, operating system, Windows, Microsoft office, computer network, etc. With the continuous updating and development of computer technology, and according to the actual situation of students, the course is also constantly reforming and changing in teaching content and class hours.

© Springer Nature Singapore Pte Ltd. 2020
M. Atiquzzaman et al. (Eds.): BDCPS 2019, AISC 1117, pp. 212–218, 2020.
https://doi.org/10.1007/978-981-15-2568-1_29

3 Problems in Experiment Teaching Under Traditional Teaching Mode

Because students have different degrees of computer basic knowledge when they are freshman, and the course is very practical, so in the several reforms, the total course hours were reduced from 48 teaching hours to 40 teaching hours, and finally reduced to 32 teaching hours, 16 h each for the theory and experiment teaching, and the teaching content also changed a lot. For example, the data processing and frontier technology have been added, the previous teaching content of Frontpage software is cancelled.

In recent years, the following problems have been summarized through students' responses, questionnaires and teaching experience in teaching.

3.1 The Experiment Teaching Time is Insufficient

The experiment teaching of computer is a practical course for students, which enables them to put the theory knowledge learned in the classroom into practice, so as to improve the students' ability to solve practical problems. However, in the experiment class, most of the teachers first explain and demonstrate the experiment content, and even some teachers use the experiment class to lecture the questions bank in order to improve the passing rate of students' exams, which greatly reduces the time of students' practice. As a result, students can not personally complete the experiment, and cannot improve students' self-conscious learned ability.

3.2 Lack of Interaction in Experiment Teaching

It's wrong to think that the theory class should have interactive teaching while the experiment teaching does not need interactive teaching. And it is also wrong to think the experiment teaching is completed by the students themselves. Some teachers only set up the experiment task to the students in the computer lab, and then turn a deaf ear to the students' practice, not to care about the students' operation level, fail to find the problems in the students' practice in time, and even fail to give targeted guidance to the students with difficult and key points.

3.3 The Experiment Teaching Content Without Bright

The task of each experiment teaching is based on the theory knowledge, but some of the experiment contents are unchanged for many years. No matter what version Windows has been upgraded from XP to, no matter how many upgrades office has undergone, the experiment content is still the basic operation. For example, the experiment of network is to ask students to apply for a free mailbox, and use this mailbox to write to teacher with attachments, it have no any outstanding features and bright spots, and without any innovation and teaching reform.

3.4   Lack of Experiment Teaching Methods

Talking about experiment class, many teachers will think that experiment teaching is student' operation, it's unnecessary to explain the content too much, and there is no need to use teaching methods. Teachers assign experiment tasks, simple demonstration and students practice, which is a single form experiment teaching. However, students can better grasp theory teaching knowledge through experiment class with practical operation. If teachers do not pay attention to experiment teaching methods, then students are in a passive and perfunctory state, so teachers can not improve students' practical ability and innovation ability, hence there no guarantee for the quality of experiment teaching.

3.5   The Form of Experiment Teaching is not Various

The organizational form of the experiment teaching of this course is that nearly 100 students of two classes are in a large computer lab. Teachers are unable to give guidance to 100 students as well as who would wish, and students can only complete the experiment tasks in accordance with the teacher's requirements. Then the experiment teaching lack of creativity and diversity. Because of the problems of weak teacher resources and a large number of students in one class, experiment teaching have not been improved and innovated for a long time.

3.6   The Effect of Experiment Teaching is not Obvious

Based on the single experiment teaching method, dull content, lack of interaction and other reasons, students lack initiative and independent thinking in the practice, and also fail to mobilize students' interest and enthusiasm, resulting in the experiment teaching effect is not obvious, and students pass the examination on the subject bank with constantly many exercises, which fundamentally does not improve students' practical ability and problem-solving ability.

3.7   The Examination of Experiment Teaching is Inaccurate

Although the course examination has been changed from written examination to computer examination, but the question bank has been used for a long time, which leads students to sum up a set of "examination secrets" and "treasure books" by brushing the questions. Most students can cope with the examination freely, while the operation steps, experiment reports and operation norms are only minimum score account for the test results. This situation inevitably leads to students to grasp reliability and authenticity of the practice, which is not conducive to the cultivation of students' practical ability, independent thinking ability, as well as the seriousness of learning and the rigorous attitude to academia.

4 Innovative Experiment Teaching Based on SPOC+ Flipped Classroom

4.1 SPOC+ Flipped Classroom

In recent years, with the calm thinking of MOOC (Massive Open Online Course [1]), experts and scholars have introduced a new concept SPOC (Small Private Online Course [2]). "Small and Private" in SPOC are relative to "Massive and Open" in MOOC. SPOC have restrictive conditions for applicants to select SPOC courses only when they meet the requirements. The number of applicants is generally limited to dozens to hundreds [3].

Flipped classroom is a kind of reform and innovation of traditional classroom, which combines classroom teaching with online teaching through watching video before class to learn knowledge, internalization of knowledge in class [4].

4.2 Experiment Innovation of SPOC and Flipped Classroom

Faced with the problem of experiment teaching, how can teachers improve the teaching quality and students' practical ability under the limited teaching hours? In view of the experiment teaching mode and teaching means, the curriculum combines SPOC with the flipped classroom and carries out innovative experiment teaching from the following aspects.

4.2.1 Define Teaching Objectives and Plan Scientifically

The theory teaching of the computer basic course generally takes four classes as the teaching class nearly 200 students at a time. The experiment teaching usually has two classes of nearly 100 students in a computer lab. In the past, it was considered very difficult to implement blended teaching. Firstly, the selection of MOOC resources is troublesome [5], because there are more and more resources online, but teachers do not have enough energy to find the MOOC resources suitable for students and corresponding to the syllabus of this course in the vast resources.

Therefore, the teaching office reduces its scope step by step. Finally, the teaching materials are determined by the teaching office in the selected textbooks, and the SPOC platform of the textbooks is selected as the online learning resources for students. With online resources, and then a reasonable plan, the syllabus, teaching content, etc. are scientifically and rationally adjusted with the SPOC resources.

4.2.2 Online and Offline Learning to Extend Time and Space

The teaching resources is very abundant on the SPOC platform [6, 7]. Students can master basic concepts and knowledge points through online learning, and preview the course content. Teachers only explain the key and difficult knowledge in theory class. If students do not understand what the teacher has taught, they can also learn repeatedly on the platform and practice in the experiment software environment. SPOC platform enables students to study anywhere and at any time, it also greatly increases the time of study and practice, and expands other learning resources through the platform. The SPOC platform really enables students to "come prepared" and "have no worries about the future". The most important thing is to achieve the time and space extension of learning, which fully solves the problems of insufficient teaching time and insufficient teaching content.

4.2.3 Enriching the Teaching Content, Weeding Out the Old and Bringing Forth the New

Because of students' uneven learning level, for the students whose teaching content cannot be digested and students whose learning ability is strong enough to "eat less", SPOC platform can fully "teach students in accordance with their aptitude". In the experiment teaching, in addition to the office content that must be done in the past, new contents such as "data representation and computing in computer", "working principle of computer", "wide area network communication and mail transmission", "cloud computing and virtual services" have been added [8, 9].

For example, in the "working principle of computer" experiment, students first observe the execution process of instructions through animation demonstration, and then the teacher arranges the students to use desks, aisles, books and so on to launch a special "animation simulation" about CPU, memory and three buses. After fully understanding the working principle, the students talked about and wrote to describe it, then students completed a pleasant experiment in the process of curiosity and novelty.

4.2.4 Activate Classroom Atmosphere and Strengthen Interaction

It's very difficult to realize interaction in the experiment computer lab faced nearly 100 students, so teachers should learn to make full use of various technical means, such as network teaching platform, teaching blog, QQ and Wechat learning group, teaching forum, mailbox and other forms of interactive teaching mode to assist experiment teaching. Using computer teaching software in computer lab [10], the interaction between teacher computer and student can be realized through the management and control of student computer.

For example, teacher can share a student's computer operation process which has the typical mistakes with other students in real time, so that teacher can demonstrate correct steps for students. Students can also be asked to explain the requirements and operation of the experiment. Because students reflected the gap and time difference between theory and experiment course, resulting in "no way to start" in the experiment class. It can also show students' experiment projects, such as PowerPoint, let students finish PPT homework in groups in advance (content is free, such as introducing my hometown, my university, etc.), and select 1 or 2 excellent projects to show in class, which is also a kind of flipped class. In addition, the SPOC platform test [11, 12] can be used to organize two classes of rolling tables competition and inter-group challenges, knowledge points rush to answer the game, unexpectedly, the enthusiasm of students is very high, once the host in order to reward the winner with their own chocolate as prizes, attracted the classroom atmosphere "for its great vibration". The animation similar to "working principle of computer" on SPOC platform can also be simulated on the spot and create vivid and interesting experiment teaching for students.

4.2.5   The Blended Teaching Method is Various and the Effect is Obvious

Make full use of the online learning function of SPOC platform, then how to organize the flipped classroom? Facing the theory teaching class about 200 students and the experiment teaching of nearly 100 students, teachers are really afraid that the flipped classroom will "overturn the river and overturn the sea". In this case, students are well-organized freely according to class and composed of 6–8 students to form a learning mutual aid group. An excellent student is elected as the group leader, and the "group leader responsibility system" is implemented. The group should complete experiments, questions, after-class discussions and group assignments, and group members should help each other. This form effectively solved the contradiction of insufficient teaching time and teachers' inability to tutor all the students, and also fully exercises the students' team cooperation ability.

Under the blended teaching mode of SPOC platform and flipped classroom [13, 14], teaching methods such as lecture, visual demonstration, exercise and task-driven method are adopted, group discussion and individual teaching are also realized in a big class. All of the teaching methods fully mobilize students' initiative and interest. For example, after learning the frontier technology on SPOC platform, some students really took 3D glasses, wearable equipment and VR equipment into the classroom to share with others. That undoubtedly builds a bridge to the knowledge sea for students.

4.2.6   Multiple Evaluation System with Everyone's Participation

It is more difficult to evaluate nearly 200 students exactly and meticulous, which requires teachers to spend a lot of energy and time. Under the blended teaching mode based on SPOC platform and flipped classroom, teachers can first use the statistical data of the platform to evaluate students' situation of access to the platform, participation in discussion and completion of the test. In addition, the proportion of the experiment results has been increased in the usual results, including the final results and the mid-term results of the computer test. But this cannot fully reflect the true level of students, so it adds other means such as teacher evaluation, student mutual evaluation, such as demonstration of experiment projects, demonstration of experiment steps, performance in experiment games, comprehensive evaluation of students' process skills etc. And each student can score for other students as a judge, which fully mobilized the enthusiasm and participation of students, and virtually reduces the workload of teachers too.

5   Conclusion

Compared with the traditional teaching mode, the experiment teaching mode based on SPOC platform and flipped classroom has obvious characteristics and advantages. It undoubtedly mobilizes students' learning initiative and interest, expands knowledge resources for students, and develops students' practical operation ability, problem solving ability, oral expression ability and adaptability. The training effect is remarkable, which is helpful to improve the effect and quality of experiment teaching.

However, it is not easy for SPOC platform and flipped classroom to face the experiment teaching of nearly 200 students [15]. If only one teacher carries out the experiment teaching, it will spend a lot of time and energy to pay attention to each student's online learning and experiment learning. So a teaching team and several assistants are needed to complete the experiment teaching. If teachers carry out small class discussion, it also need a lot of resources such as classroom, computer lab and more teachers.

Therefore, how to better make use of SPOC and flipped classroom blended teaching especially in large class courses with computer experiments? How to evaluate students more comprehensively and effectively? How to ensure the quality of experiment teaching? How to improve the effect of experiment teaching, and how to effectively improve students' practical ability, and cultivate students' independent thinking ability and innovative thinking? These are the problems that will be explored and studied continuously in the teaching process, more and better methods need to be summarized in the future.

Acknowledgements. This work was supported by JY2017015. (Educational and Teaching Research Projects of Qinghai University, 2017).

References

1. Hao, D.: Literature analysis of MOOC research in China. Distance Educ. China **11**, 42–49 (2013)
2. Sang, X.: Reflection on MOOCs fervor. China High. Educ. Res. (06), 5–10 (2014)
3. Kang, Y.: An analysis on SPOC: post-MOOC era of online education. Tsinghua J. Educ. (01) (2014)
4. Qiang, L.: Cold reflection on flipped classroom: empirical study and reflection. E-Educ. Res. **34**(08), 91–97 (2013)
5. Zuona, W.Z., Wu, T.: SPOC - innovation and reflection on MOOC model. Chin. J. ICT Educ. (02), 6–9 (2016)
6. Zhang, D., Zhu, X.: Experimental teaching of computer software courses based on SPOC and multimedia network classroom. Exp. Technol. Manage. **34**(08), 195–198+205 (2017)
7. Fang, S., Dong, L.: Exploration on the teaching model of information engineering basic experiments courses based on SPOC. Lab. Sci. **20**(05), 156–160 (2017)
8. Pan, X., Ang, C., Guo, H., Zhou, G.: Teaching mode of university computer basic course under the condition of educational information technology. Comput. Educ. (04), 5–8 (2017)
9. Yang, W: The research of instructional design model and its supportive strategies on experimental course of modern educational technology based on SPOC. High. Educ. Forum (12), 66–70 (2016)
10. Panbo: application of polar multimedia electronic classroom software in computer teaching. Pract. Electron. (Z1), 22–23 (2016)
11. Hao, J., Guo, X.: Study on electronic experiment teaching in SPOC mode based on Internet+. Res. Explor. Lab. **35**(09), 209–213 (2016)
12. Lin, X., Hu, Q., Deng, C.: Research on the model of innovative ability training based on SPOC. E-Educ. Res. **36**(10), 46–51 (2015)
13. Xu, B., Li, T., Shi, X.: The analysis and educational revelation of learning motivation in MOOC, flipping classroom and SPOC. China Educ. Technol. (09), 47–52+61 (2017)
14. Ding, Y., Jin, M., Zhang, X., Zhang, Y.: Design and implementation process of the teaching model of flipped classroom 2.0 based on SPOC. China Educ. Technol. (06), 95–101 (2017)
15. Zeng, X., Li, G., Zhou, Q., et al: From MOOC to SPOC: construction of deep learning model. China Audiov. Educ. (11), 28–34+53 (2015)

Automated Design Method of Environmental Art Design Scheme Based on Big Data Analysis

Dongyou Wang$^{(\boxtimes)}$

Jilin Engineering Normal University, Changchun 130052, Jilin, China
Tougao_001@163.com

Abstract. Customers' requirements for environmental art design are constantly improving, which undoubtedly brings greater challenges to environmental art designers. There are many excellent environmental design schemes around. If we can develop new design schemes according to existing excellent schemes, it will greatly reduce the workload of designers, and at the same time, it will bring excellent practical experience to customers. In view of the above problems, this paper proposes an automatic design method of environmental art design based on large data analysis. By describing the environmental design works with text, and then according to the input statements (i.e. the description of the desired works), using the generation of confrontation network and sorting algorithm to obtain the best match with the existing environmental design schemes, thus completing the automatic design of environmental art design schemes. Through the simulation of existing environmental design cases such as gardens, residential buildings, entertainment cities and office buildings, the results show that the proportion of the method in this paper meets the requirements of customers reaches 95.1%.

Keywords: Environmental art design · Program automation design · Large data analysis · Generation of countermeasure network

1 Introduction

With the continuous development of economy and the improvement of people's living standards, people's requirements for the surrounding environment are also increasing. Excellent environmental art design gives people a good mood. Environmental art design contains a wide range of contents, such as urban planning, public facilities, landscape construction and other works of art [1]. In environmental art design, it is also a trend to incorporate ecological concepts into the design, which can effectively alleviate the greenhouse effect, ozone depletion, acid rain and other problems, and is conducive to maintaining ecological balance and polluting the green environment [2]. With the rapid development of art design, designers must pay more attention to the cultural and spiritual needs. Designers need to find the roots in the art and design history of the East and the West as a reference basis to bring innovation and inspiration [3]. In [4, 5], we consider applying Chinese traditional culture elements to modern environmental art design. By learning about tea culture, we can integrate tea culture, a traditional culture of our country, into our modern environmental art design, which

© Springer Nature Singapore Pte Ltd. 2020
M. Atiquzzaman et al. (Eds.): BDCPS 2019, AISC 1117, pp. 219–225, 2020.
https://doi.org/10.1007/978-981-15-2568-1_30

greatly improves the surrounding environment of people's living. However, nowadays, environmental art design is manually made, which is inefficient, so it does not meet the needs of the development of information technology today.

Big data analysis is the product of the information age. There are not a few studies and applications on big data analysis. In [6, 7], the author points out that in order to meet the increasing challenges of agricultural production, it is necessary to better understand the complex agricultural ecosystem. This can be achieved through modern digital technology to continuously monitor the physical environment, generating a large amount of data at an unprecedented rate. The analysis of these (large) data will enable farmers and companies to derive value from it, thereby increasing their productivity. In [8], in view of data-intensive situation, the author integrates intelligence into fog intelligent computing to realize large data analysis of intelligent city, so as to protect the future community. In [9], the author applies big data analysis to the medical and health care industry, analyses big data of medical treatment, and shows the prospect of medical and health care. In [10], the author uses random forest algorithm to analyze insurance data. On the basis of insurance business data, this algorithm is used to analyze potential customers. Compared with traditional manual methods, it helps to improve the accuracy of product marketing. The above research shows that the method of large data analysis can effectively deal with complex problems, and is very efficient. We consider using the method of large data analysis to model a large number of existing art designs, and automatically formulate environmental art design plans according to needs.

In order to solve the above problems, this paper uses the combination of large data analysis and software design method to realize the design of automatic environmental art scheme. Through the simulation of 2000 cases, the results show that the proportion of the method in this paper meets the requirements of customers reaches 95.1%.

2   Method

2.1   Design of Automated Environmental Art Scheme

Designing work and artistic creation depend strongly on the author's sensibility and creativity, but they are also repetitive mental activities in some environments, such as gardens, residential buildings, entertainment cities and so on. Most of them are repetitive patterns. Therefore, such artistic creation as environmental art design can be abstracted as a production or production process, as follows:

$$ArtDesign = Generator(f_0, f_1, f_2, \cdots, f_N, r_{x \sim p}) \tag{1}$$

In the formula, f_i is an existing work of art, and $r_{x \sim p}$ is a set parameter.

This paper selected 2000 environmental art design cases. The sorting function is used to sort all the descriptive texts of art works according to search terms, and then the first n are selected. Each artwork has unique text information, corresponding to the only artwork sentence vector. By calculating the sentence vectors of each artwork, comparing the input sentence vectors with each artwork sentence, selecting the first n, we can locate the n most matching artworks. The sentences and input sentences of an artwork are semantically the closest, which determines that the artwork is the most appropriate artwork to the input sentences. Then, according to the most matched works of art, a new design scheme of environmental art is automatically given.

2.2   Generating Antagonistic Networks (GAN)

In order to give the best matching effect, GAN is used to perform matching operation. GAN integrates an adversarial discriminator model (discriminator) skillfully into a generating model (generator), which generates data, and the discriminator judges the authenticity of the data. They compete with each other and promote each other. The generator produces more and more true results, reaching the level of false and true. The discriminator's ability is also getting stronger and stronger, and the discriminant of true and false data is getting stronger and stronger. The GAN training process is a fully automated, non-guided learning process that requires little manual intervention, so it is more suitable for automatic design of environmental art programs. Both the generator G and the discriminator D can use deep neural networks. The optimization process of GAN can be described as a "binary minimax" problem. The objective function is:

$$\min_{G} \max_{D} V(D, G) = E_{xp_{data}(x)}(\log D(x)) + E_{zp_z(z)}(\log(1 - D(G(Z)))) \qquad (2)$$

Where p_z is the distribution of random noise; p_{data} is the distribution of real samples; $E(\cdot)$ is the calculation of expected values.

The generator output layer uses Tanh as the activation function. In addition to the output convolutional layer, the discriminator adds an instance normalization layer and a LeakyReLU layer after the remaining convolution operations. Using LeakyReLU as the activation function in the discriminator, LeakyReLU is an improved version of the ReLU activation function. The expression of the LeakyReLU function is as follows:

$$f(x) = \begin{cases} 0.02x, & x < 0 \\ x, & x \geq 0 \end{cases} \qquad (3)$$

2.3   Large Data Analysis Method

Text big data analysis technology refers to the representation, understanding and extraction of words, grammar, semantics and other information contained in unstructured text strings. Mining and analyzing the facts that exist in it, quantify the feature words extracted from the text to represent the text information. Since each work of art has corresponding text description information, first of all, it is necessary to consider the storage of data. The data storage scheme uses HDFS distributed file system to store large files in streaming data access mode. Named entities such as place names, environmental styles, occupied area, materials and spatial contents are vectorized. Cluster classification method is adopted to classify the whole data set into different categories according to the similarity of objects. The text retrieval technology is implemented on Elastic Search. Eelastic Search provides distributed full-text retrieval through clusters. An Elastic Search cluster can be composed of multiple nodes and can be dynamically increased.

3 Data

This paper uses 2000 samples of existing environmental art design. In the 2000 samples, the sample types include gardens, residential buildings, entertainment cities, office buildings and other environments. Among them, the sample number of garden design scheme is 600, accounting for 30% of the total sample number; the sample number of residential design scheme is 400, accounting for 20% of the total sample number; the sample number of entertainment city design scheme is 500, accounting for 25% of the total sample number; and the sample number of office design scheme is 500, accounting for 25% of the total sample number.

Sample description information includes scale, color (color attributes, color matching, etc.), material (natural and artificial materials; hard and soft materials; exquisite and rough materials), spatial contents (furniture, greening, sculpture, furnishings), structure and structure (wood structure, reinforced concrete structure, etc.). Based on the analysis of large data of existing environmental art design schemes, the automatic design of new environmental art design schemes is carried out according to the characteristics of existing schemes.

4 Results

Result 1: Feature extraction of environmental art design samples

Text description of sample features of environmental art design works mainly includes the name, category, scale, material, spatial contents, structure and structure of the works. Therefore, text feature extraction based on text structure and regular expression is adopted, that is, the corresponding paragraph is found through regular expression. Then Chinese word segmentation and feature extraction are carried out to extract the content of key paragraphs. When extracting, it needs to match the professional thesaurus such as the category Thesaurus of environmental art design and the place thesaurus. If there is a professional lexicon, it will be extracted directly. If there is no professional lexicon, it needs to be extracted according to the grammatical structure of regular expressions. The text feature extraction process is shown in Fig. 1.

Comparing the retrieval time of the collected descriptive documents of works of art, the PC is configured with 4 GB memory and the processor is Pentium (R) Dual-Core CPU E5800 @ 3.20 GHz, among which Eelastic Search contains 3 nodes. The Eelastic Search retrieval method used in this paper is compared with the traditional retrieval method. The retrieval time of the two methods is shown in Fig. 2 under different number of documents.

As can be seen from Fig. 2, the retrieval time based on Eelastic Search is only about 10 ms, which can be regarded as real-time retrieval. At the same time, due to the small amount of data used, the time of Eelastic Search full-text retrieval is basically unchanged, but with the increase of the number of text, the time of traditional retrieval increases linearly. The larger the amount of text data, the better the full-text retrieval of Eelastic Search. This method performs well in the application of large data analysis.

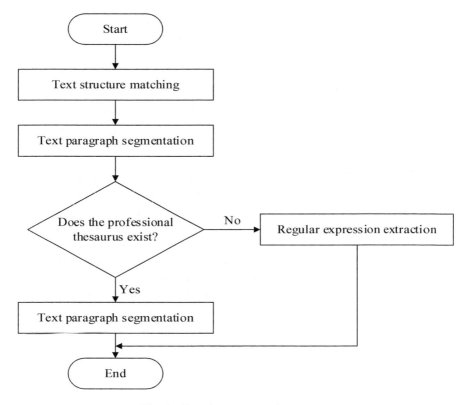

Fig. 1. Text feature extraction process

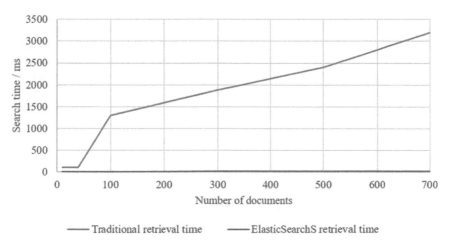

Fig. 2. Comparisons of Eelastic search retrieval time with traditional retrieval time

Result 2: The design accords with the expected accuracy

After retrieving, extracting and training the text description data of artistic works, the matching prediction of 2000 samples is carried out. The predicted results for different types of design schemes are shown in Table 1.

Table 1. Prediction accuracy of different design schemes

	Garden	Residential	Entertainment city	Office building	Average
Accuracy	96%	94.3%	95.2%	94.9%	95.1%

From Table 1, it can be seen that the design scheme obtained by the automatic design method of environmental art design scheme based on large data analysis proposed in this paper has a higher accuracy rate, with an average accuracy rate of 95.1%. Further analysis shows that the accuracy of prediction varies due to the inconsistent number of samples for each type of work. The number of samples of gardens, residential buildings, entertainment cities and office buildings accounted for 30%, 20%, 25% and 25% respectively. Because the model built by generating antagonism network is trained based on sample data, the more sample data, the more features the model learns, so the higher the prediction accuracy of different design schemes is, the more it meets the requirements of environmental art design.

5 Conclusion

Based on the technology of big data analysis, this paper puts forward the research of automatic design method of environmental art design scheme. Firstly, the text information of existing schemes is described, and large files are stored in streaming data access mode by using HDFS distributed file system. Then according to their own text description of the new scheme, through the generation of confrontation network to achieve the optimal matching of the characteristics of the case samples, to achieve automatic design of a new environmental art design scheme. Experiments show that this research has achieved good results, and the research results can provide a reference for large data analysis in the field of environmental art design related business. Of course, the next step of the study can consider extending the method to other areas of automation design, in order to expand the general applicability of the research results in this paper.

References

1. Li, X.: Problems in environmental art design and their individualization. Rural Sci. Exp. (10), 85 (2017)
2. Yang, Q.: Ecological concept in environmental art design. Tomorrow's Fashion (9), 44 (2017)

3. Ning, X., Sitong, L.: Analysis of the reference method perception in environmental art design. Furniture Inter. Design (9), 27 (2016)
4. Xiang, W.: The application of Chinese traditional cultural elements in modern environmental art design. Mod. Hortic. (2), 128 (2018)
5. Zhao, W.: The infiltration and application of tea culture in modern environmental art design. Fujian Tea (2), 104 (2018)
6. Kamilaris, A., Kartakoullis, A., Prenafeta-Boldú, F.X.: A review on the practice of big data analysis in agriculture. Comput. Electron. Agric. **143**, 23–37 (2017)
7. Mao, L., Cheng, W.: Research on the construction of smart agriculture big data platform. Agric. Netw. Inf. **264**(6), 8–12 (2018)
8. Tang, B., Chen, Z., Hefferman, G., Pei, S., Wei, T., He, H., Yang, Q.: Incorporating intelligence in fog computing for big data analysis in smart cities. IEEE Trans. Ind. Inf. **13**(5), 2140–2150 (2017)
9. Lo'ai, A.T., Mehmood, R., Benkhlifa, E., Song, H.: Mobile cloud computing model and big data analysis for healthcare applications. IEEE Access **4**, 6171–6180 (2016)
10. Lin, W., Wu, Z., Lin, L., Wen, A., Li, J.: An ensemble random forest algorithm for insurance big data analysis. IEEE Access **5**, 16568–16575 (2016)

Exploration on the Construction of the English Ecological Classroom in Vocational Colleges from the Perspective of the Internet Plus

Hong Wei[✉]

Shandong Vocational College of Light Industry, Zhoucun, Zibo City 255300,
Shandong Province, China
175203039@qq.com

Abstract. As a new form in the information age, the Internet plus results from the deep integration of the innovation of Internet technologies and various fields of societies, while Cloud computing, big data technology, mobile Internet and Internet of Things have been widely used in education. The application of Internet technology and Internet thinking for education has made Internet and traditional education integrated tightly, which has brought both vitality and challenges to the development of education in China. With the advent of the Internet plus era, the English teaching system in vocational colleges remains in disruption of ecological balance, and the contradiction surges prominently and increasingly among teachers, students and teaching environment, by which the English teaching in vocational colleges is seriously affected. Under this circumstance, the author suggests that the English ecological classroom for vocational colleges should be constructed on the basis of educational ecology and language ecology, and that the ecological elements of teachers, students, teaching resources and teaching environment should be coordinated in order to promote the harmonious and dynamic development of English teaching in vocational colleges.

Keywords: "Internet plus" · Information technology · Ecological classroom · English in vocational college

1 Introduction

Under the background of the Internet Plus, the information technology has penetrated into various areas concerning education and characterized by its real-time, interaction, efficiency and convenience, the Internet has an overturning effect on education and teaching. Information technology such as Internet provides a broad platform and powerful technical guarantee for vocational education in China, and promotes the reforms of traditional teaching elements, including teaching contents, teaching structure, teaching process [1]. The Action Guide for Educational Informatization 2.0 points out that the Internet Plus education should be actively promoted and the information technology and teaching should be deeply integrated [2]. The era of the Internet plus has brought both opportunities and challenges to English education in vocational colleges. When the information technology and English teaching in vocational colleges

© Springer Nature Singapore Pte Ltd. 2020
M. Atiquzzaman et al. (Eds.): BDCPS 2019, AISC 1117, pp. 226–230, 2020.
https://doi.org/10.1007/978-981-15-2568-1_31

are being consolidated, English teaching cannot be positively interacted with the information technology, owing to the application of the traditional teaching mode. The contradiction between teachers, students and teaching environment is becoming more and more serious and the phenomenon of ecological imbalance occurs occasionally, all of which make the harmonious and dynamic development of college English teaching difficult to be realized. As a result, it is significantly urgent to construct an ecological classroom for English in vocational education in response to the development of the Internet plus era.

2 The Connotation and Theoretical Basis of the English Ecological Classroom in Vocational Colleges

2.1 The Connotation of the English Ecological Classroom in Vocational Colleges

The ecological classroom is a new kind of classroom from the perspective of ecology [3]. As an ecological system, the English classroom in vocational colleges includes students, English teachers, teaching forms, teaching modes, teaching evaluation and other ecological factors, where material flows and energy transfers. Furthermore, the English ecological classroom in vocational colleges reconstructs the teacher-student relationship, curricular structures and teaching forms in order to provide a good environment for the development of students. Following the rules of language learning and combining the students' learning standard with the characteristics of teaching in vocational colleges in China, the English ecological classroom incorporates all kinds of elements concerning English teaching in vocational education and coordinates the status of the development for ecological factors in the teaching system, in which way the ecological environment is improved and the dynamic. The harmonious and sustainable development is accomplished for English teaching in vocational education.

2.2 The Theoretical Basis of the English Ecological Classroom in Vocational Colleges

In the early 20th century, ecology formed a comparatively complete theoretical system [4], which was gradually applied to relevant subjects later. In 1976, Lawrence Cremin, L. A of Columbia University put forward the concept of educational ecology, and from then on the theory of education ecology became more and more matured and developed step by step. Based on the principle of ecology, the specialists and scholars began to study various problems in education and teaching.

After the 21st century, the ecological study was extended to the linguistic field. The researchers, from the perspectives of ecological principles and theories, have studied the related issues about language learning [5], second language acquisition and classroom teaching for foreign language, and as a result, formed a new branch of linguistics, which is the ecological linguistics. Focusing on the interaction between language and its environment, the ecological linguists lay emphasis on the dynamic interaction between the process of English learning and internal and external environment. English study occurs in a particularly social and cultural context [6].

3   The Construction of the English Ecological Classroom in Vocational Colleges

Based on the theories of educational ecology and language ecology, by means of informationalized teaching methods, the English ecological classroom in vocational colleges helps to create a relaxing, free and harmonious environment for the students' English learning, which follows the principles of integrity, enlightenment, integrity and expansibility. In this way, English teaching in vocational colleges and English competence for students are enhanced mutually and harmoniously. The author proposes that the following things should be done in order to construct the English ecological classroom in vocational colleges.

3.1   The Function of Guidance for Teachers' Niche Needs to Be Strengthened

When the information technology has a tremendous influence on English teaching, many problems, such as overlap and specialization, appear in the teachers' niche [7], which is an indispensable ecological factor of the English teaching ecosystem in vocational education. English teachers need to combine linguistics with social culture to form the conception of linguistic knowledge and culture, and connect the relevant theories of pedagogy and psychology with the students' profession tightly so as to form a foreign language teaching viewpoint, in accordance with the learning characteristics of students in vocational colleges.

In the era of the Internet Plus, students learn independently through a variety of network resources, and sometimes they can complete their learning tasks without attending classes. Under the influence of the niche overlap, the teachers' role as "preaching, teaching and guiding for the perplexed" has been gradually overshadowed; on the contrary, the teachers play the role as guides [8], participants or promoters in the teaching ecosystem. In consequence, it is required for English teachers in vocational colleges to pay close attention to the change of their niche, master the basic knowledge and theories about information technology, use network technology tools and software skillfully, implement the activities for teaching designs with the help of Internet resources, and promote the integration between information technology and curriculum. What's more, the teachers needs to learn how to use information resources tools and technology to solve the issues occurring in English classrooms in vocational education for the sake of adjusting the path for professional development and adhering to the development trend of educational informationization.

3.2   The Function of Subjectivity for Students' Niche Needs to Be Enhanced

In the English teaching ecosystem of vocational education, the force produced by the Internet promotes the adjustment and change of ecological factors in the teaching system, while the innovation of teaching technologies impacts the ways how students learn, the methods that students use to learn, and what students learn. As an important ecological factor, students are acted as the consumers and decomposers of knowledge in the English teaching ecosystem and ultimately give the energy produced back to the society. The students are required to change the passive, perfunctory and stubborn learning methods which were used in the past and accomplish exploratory, cooperative and autonomous learning by means of information technologies and abundant network teaching resources [9]. They are able to finish the tasks arranged by teachers on the network platform voluntarily before class, cooperate with members of groups in class, and complete the tasks independently to develop their own individualized learning abilities after class [10]. The English ecological classroom in vocational colleges aims to improve students' competence for English application for the workplace in the future and to accomplish the harmonious and unified development between students and the society.

3.3   The Three-Dimensional Ecological Teaching Resources Needs to Be Designed

Having Teaching resources is the premise and basis of material flow and energy transmission in the English teaching ecosystem. To construct the English ecological classroom in vocational colleges, the three-dimensional English teaching resources can not be separated. English teaching resources in vocational colleges include micro resources, teaching contents and extensive curricular resources. Curricular micro resources consist of English micro lessons, communicative tasks in daily life and English dialogues etc. Teaching contents mean how the teachers integrate teaching materials and design the subjects. Curricular extensive resources are as follows: massive online courses, English contests, classic English movies and songs, articles related to British and American cultures and celebrity speeches, etc. Three-dimensional and diverse ecological English teaching resources are of great benefit to the development of students' niche and the harmony of the overall teaching ecosystem.

3.4   The Informationized Ecological Teaching Environment Needs to Be Established

The teaching environment is a comparatively complex ecological factor in the teaching ecosystem, which consists of the ones generated before class, in class and after class. To optimize the teaching environment can keep the dynamic balance of the whole teaching ecosystem. In the era of the Internet plus, the digital campus is fully established, the intelligent classroom is totally set up and the network teaching platform is completely constructed. Those provide a guarantee for the construction of the English ecological classroom in vocational colleges. These App software in cell phones, man-machine conversation simulation systems, analogue simulation software and other relevant technologies provide a "natural and real" language context for English learning in vocational colleges so as to improve students' English competence for workplace application. Information technologies have penetrated thoroughly and totally into the English teaching environment in vocational colleges, optimizing the teaching effects and vitalizing the teaching ecology.

4   Conclusion

It is of great significance to construct the English ecological classroom for the development of English teaching in vocational colleges. Connected with the English teaching practice, this kind of classroom is able to provide a theoretical basis for solving the ecological imbalance under the background of the Internet plus. At the same time, it makes the whole English teaching system more optimized and helps to enhance the establishment of the harmonious teacher-student relationship and improve the efficiency of English teaching.

References

1. Aldridge, J.M., Laugksch, R.C., Fraser, B.J.: School-Level Environment and Outcomes-Based Education in South Africa. Learning Environ. Res. **1**, 123–147 (2006)
2. Guowen, H.: The rise and development of ecological linguistics. Foreign Languages China (1), 8–12 (2016). (in Chinese)
3. Browy, W., Smith, J.E.: Ecology and Development in Classroom Communication. Linguist. Educ.: Int. Res. J. **19**, 149–165 (2008)
4. Priestley, M., Biesta, G., Robinson, S.: Teacher Agency: An Ecological Approach. Bloomsbury Publishing, London (2015)
5. Boylan, M.: Ecologies of participation in school classrooms. Teach. Teach. Educ. **26**, 61–70 (2008)
6. Gass, S., Mackey, A.: Data Elicitation for Second and Foreign Language Research. Foreign Language Teaching and Research Press, Beijing (2011)
7. Liu, C.: A Study on College English Classroom Ecology in the Context of Informatization. Xingsi Books Publication Press, Guangzhou (2014). (in Chinese)
8. Kramsch, C.: Teaching foreign languages in an era of globalization: Introduction. Mod. Lang. J. **98**(1), 296–311 (2014)
9. Porto, M.: Ecological and intercultural citizenship in the primary English as a foreign language (EFL) classroom: an online project in Argentina. Cambridge J. Educ. **46**, 395–415 (2016)
10. Shu, X.: An empirical study on a flipped classroom in open university teaching based on an ecological perspective: a case study on a translation theory and practice course. Asian Assoc. Open Univ. J. **10**(1), 53–63 (2015)

Analysis and Enlightenment of New Media Editor Talent Demand Based on Python

Yue Wang[✉], Jieping Liu, and Guangning Pu

Institute of E-Commerce Operations, Chengdu Neusoft University,
Chengdu, China
wang-yue@nsu.edu.cn

Abstract. The cultivation program of new media editor talents must be conducted based on the needs and requirement analysis of its labor market. This article mainly uses Python to crawl and clean data from four mainstream recruitment websites, such as Lagou.com, Zhaopin, 51job.com, and Liepin.com. It has gained 5,381 pieces of data highly correlated with "new media editors". Through the analysis of the data, the distribution of the demand for new media editors in different cities, industries, work experience and education levels and salary levels were discovered. Through the jieba segmentation, new media editor position requirements were explored in Business Skills, IT Skills and Calibre. In accordance with the job descriptions, the talent training program for college-related majors under the new media era was constructed, as so a certain basis to be provided for implementing the talent cultivation in the future.

Keywords: New media editor · Talent needs and requirements · Python · Jieba

1 Introduction

With the development of digital technology, computer network and mobile Internet, micro-blog, WeChat, APP client and other channels emerge, and New Media has gradually penetrated into people's daily life. In the new media era, content is the first priority in communication. The demand for new media editors is increasing, and this positon is becoming more and more prominent to today's organizations. It is obvious that the cultivation of new media editors is imminent. The authors believe that the training program of new media editors must be demand-oriented. Only if based on the analysis of current talent needs and requirements, can the target learners clearly acknowledge the talent demand situation of new media editors, and engage themselves in new media editing learning with clear and specific goals. Therefore, the training program can be more effective [1–3].

2 Data Collection

2.1 Identify Data Crawling Websites

According to the market share of recruitment websites and the number of new media editor positions, this article selected 4 mainstream recruitment websites including

© Springer Nature Singapore Pte Ltd. 2020
M. Atiquzzaman et al. (Eds.): BDCPS 2019, AISC 1117, pp. 231–237, 2020.
https://doi.org/10.1007/978-981-15-2568-1_32

Lagou.com, Zhaopin, 51job, and liepin. Among them, Zhaopin and Lagou.com are dynamically loaded by Ajax; 51job and liepin are statically loaded by Html [4–6].

2.2   Get the Urllist of New Media Editor Jobs

The recruitment websites which are statically loaded by Html, this article uses requests library, xPath library and re library to get the href attribute (url) of the <a> tag of new media editor job. The recruitment website which is dynamically loaded by Ajax, this article gets the href attribute (url) by grabbing the json file in the network. Then, save all those href attribute (url) as a list type whose name is urllist.

2.3   Get the Detail Information of New Media Editor Jobs

Through the for-loop, request all url in the urllist to the server again by requests library, and get details information of new media editor jobs. Then use xPath library to phrase the selected pages, then to obtain all detailed position information, which includes position links, names, locations, salaries and job descriptions.

2.4   Data Cleaning and Standardization

Clean and standardize data from different recruitment websites and different formats, such as deleting duplicate values by position links, unifying salary units into thousands per month, and classifying education, working experience, enterprise scale, and enterprise industry, etc. Then, save the data as a csv file through the pandas library.

3   Data Analysis and Processing

According to the recruitment information of 5381 new media editors, the national wide average salary level of new media editor position is about 7.59 k. According to the average salary data of 2018 workers published by the National Bureau of Statistics, in 2008 the average annual salary for employees of enterprises above the designated size is 68,380 RMB and the average monthly salary is 5.70 k. Obviously, the average salary of new media editors is higher than the average salary level of the above-mentioned enterprises.

3.1   New Media Editors' Talent Demand and Wage Distribution in Different Cities and Different Industries

The following quantity demanded is referred to as Qd.

According to the data analysis from Fig. 1, the geographical distribution of new media editor job demand is very wide. There is demand for new media editor position in more than 140 cities in China. The top 15 cities with high demand are analyzed and found that the demand is mainly distributed in the first-tier cities with developed economies. Especially, Beijing, Shanghai, and Guangzhou have very high needs for this position. The total quantity of demanded in these three cities exceeds 46% of the

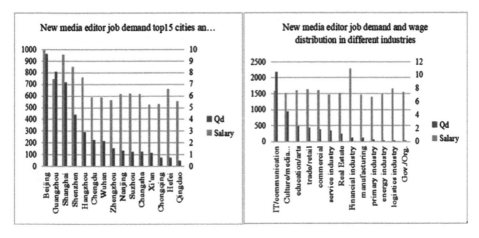

Fig. 1. New media editor job demand and wage distribution in different cities and different industries

whole country. From the perspective of the industry distribution of the enterprises affiliated with the new media editor positons (the industry division standard is unified as the division standard of Zhaolian recruitment), the highest needs for new media editor positons is IT related industry, it takes more than 40%. The industry of Culture, media, entertainment, sports takes the second place, accounting for 17.38%. Obviously, the demand for new media editors in the Internet-related industries is still the strongest, but it can be seen that in the new media era, other industries such as government, energy, environmental protection, agriculture, and etc. also have the demand for new media editor talent. And in the future the demand for new media editors in such industries should not be underestimated.

In terms of salary, the salary ratings of new media editors in Beijing and Shanghai is absolute an advantage, both exceeding 9.5 k/month, and then followed by Shenzhen, it's about 8.5 k/month. While the salary level in Guangzhou, the second highest demand, has a fault, only 7.48 k/month, even lower than the salary level of 7.6 k/month in Hangzhou. The salary levels of the other first-tier cities mentioned above has no much difference, and the salary level fluctuates within the range of 5.26–6.58 k/month. Besides, in addition to the financial industry, which has achieved an absolute advantage of 11 k/month, the average salary of new media editors in other industries has not shown a significant gap, ranging from 6.65 to 7.94 k/month.

3.2 New Media Editors' Talent Demand and Wage Distribution in Different Work Years and Different Academic Degrees

As an emerging profession, the New Media Editor has low requirements for work experience. There are more than 30% of the jobs have no requirements for work experience. Additionally, the demand for more than five years of work experience is only 1.88%. In terms of salary, the salary of newcomers in one year can exceed 6 k/month, and the salary of new media editors with 3–5 years of work experience can

Table 1. New media editors' talent demand and wage distribution in different work years

Work years	Qd	The percentage	Wage (k/month)
No limitation for experience	1645	30.57%	6.84
0–1 yrs	837	15.55%	6.16
1–3 yrs	2088	38.80%	7.81
3–5 yrs	710	13.19%	10
Above 5 yrs	101	1.88%	15.67

Table 2. New media editor demand and wage distribution in different academic degrees

Academic levels	Qd	The percentage	Wage (k/month)
Below college diploma	66	1.23%	6.07
College diploma	2773	51.53%	6.67
Bachelor degree	2517	46.78%	8.62
Master degree & above	25	0.46%	11.03

reach 10 k/month. The salary of new media editors with more than 5 years of work experience can exceed 15 k/month (Table 1).

From the perspective of the demand distribution of academic qualifications, almost 99% of the job requirements are concentrated in college and undergraduate degrees. In other words, the new media editing position does not have high requirements for academic qualifications, and the candidates with college diploma or bachelor degree can apply for this job (Table 2).

3.3   Key Vocational Skill Requirements for New Media Editor Positions

In this study, the jieba library was used to classify the word descriptions of New Media Editor from the 5,381 results obtained. Through multiple optimizations, it can eliminate the unrelated high-frequency phrases (such as responsibility, possession, work, and so on). And the core information of position description is classified manually according to Business Skills, IT Skills and Calibre, as shown below [7, 8] (Table 3).

As can be seen from the above figure, regardless of the common skills of "text editing ability", "creative innovation ability", "operation planning ability", etc. in the Business Skills column, "data analysis and processing capabilities" has been listed as well. It indicates that the new media editors are not only going to complete the "content", but also should focus on the operation data and its results. Therefore, they can revise the "content" through the analyzed results.

Table 3. Key words of new media editor

	Key words of new media editor		
	Business skills	IT skills	Calibre
New media editor	Text editing, original thinking ability, integration, topic maker, material collection, creative innovation, operation planning, event planning, copywriting promotion skills, keyword optimization, data analysis and processing, official account operation and maintenance, online community maintenance, user needs analysis	Typesetting editing software Office software Photoshop Video editing	Communication, responsibility, teamwork, learning skill, expression skill, passionate and enthusiastic, acuity, active mind, execution, critical thinking ability

Besides, in the IT Skills section, in addition to necessary ability of using "typesetting editing software" and "office software", how to use "PS" and "video editing" is also required. It can be seen that in the era of the Internet, the requirements of the new media editing talents are more comprehensive, not only to complete the traditional text editing and typesetting work, but also to have certain image processing and video editing capabilities.

In addition, basic professionalism is also the main point in corporate recruitment. Unlike most other positions, new media editors are required to be highly sensitive, which requires them to be able to respond quickly to hot spots and grasp the hot trends keenly.

4   Enlightenment

Referring to the previous refinement and analysis of the recruitment information of new media editors, this paper builds a plan for new media editor talent cultivation [9, 10] (Table 4).

As from the 5381 recruitment information, it can be found that the professional requirements for new media editor recruitment are mainly concentrated in the majors of Chinese literature, journalism, advertising media, marketing and e-commerce. At present, the training of the above-mentioned majors offered by domestic universities are not mainly directed at cultivate new media editors, and their curriculum setting is not targeted on this position as well. For example, the curriculum in traditional Chinese literature and journalism majors are more oriented towards text and editing skills, it will make more sense, if new media oriented and data analysis courses can be added. The e-commerce major may prefer to develop operational planning capabilities, and can add copywriting writing training course.

Table 4. A plan for new media editor talent cultivation

Course type	Develop skills	Opening a course
Business skills course	Operation planning, event planning, copy promotion, public operation and maintenance, community maintenance	New media operations
	Text editing, original ability, integration ability, topic manufacturing, material collection, keyword optimization	Copywriting
	Data analysis and processing, user needs analysis	Business data analysis and application
	Creative innovation	Innovative thinking training and practice
It skills course	Office software	Basic computer science
	Photoshop	Visual design
	Typesetting editing software	New media operation training
	Video editing	Multimedia technology and application
All-round development course	Communication and expression skill	Communication and speech

5   Conclusions

This article uses the Python and jieba library as tools to crawl and segment the data obtained from job description of new media editors. It comprehensively analyzes the needs and salary level of new media editors in different industries, cities. It also discusses the relationship between salary levels with candidates' different education levels and work years. The research results show that there is strong demand for new media editors, the salary level is high, and it has bright developing prospects. Based on the research results, this paper also constructs a talent training program for college-related majors in the new media era, which provides a basis for the future implementation.

References

1. Tingting, Z.: Editors' gorgeous transformation in the "Internet +" era—on the role of new media editors. Jin Media **1**, 138–139 (2016). (in Chinese)
2. Botao, Z.: Analysis of the similarities and differences of news editors between new media and traditional paper media. Journal. Res. Guide **10**(19), 175–177 (2017). (in Chinese)
3. Xueyan, W.: Discussions on the similarities and differences of news editors between new media and traditional paper media. News Window **1**, 54–55 (2016). (in Chinese)

4. Zhongyi, H., Ya, l, Jiang, W., Yizhen, Z., Yang, Z.: Analysis and enlightenment of business intelligence talents based on recruitment information. J. Inf. Resour. Manage. **9**(3), 111–118 (2019). (in Chinese)
5. Yanhong, J., et al.: Beijing area accounting talent market demand analysis based on industry and job demand differences. Acc. News (7), 36–39 (2019). (in Chinese)
6. Tao, W.: Pyhton-based software technology talent recruitment information analysis and implementation - use 51job.com as an example. Fujian Comput. (11), 118–119 (2018). (in Chinese)
7. Xiaoli, W.: Discussions on the basic qualities of new media editors. J. Sci. Consult. (Technol. Manage.) (44), 47–48 (2013) (in Chinese)
8. Xuefei, K., Xiaoyang, W., Tengfei, L.: Basic qualifications of sci-tech periodicals editors under the new media environment. Journal. Res. Guide **8**(18), 29–30 (2017). (in Chinese)
9. An, S.: The path choice for strengthening the training of new media talents in colleges and universities. News Outpost **8**, 37–39 (2019). (in Chinese)
10. Yanli, W.: Analysis of the capabilities of new media editors. Journal. Res. Guide (02) (2018). (in Chinese)

A Fast and High Accuracy Localization Estimation Algorithm Based on Monocular Vision

Junchai Gao[1], Bing Han[2], and Keding Yan[1(✉)]

[1] School of Electronic Information Engineering, Xi'an Technological
University, Xi'an 710032, China
gaomuyou@126.com
[2] Xi'an Technological University, Xi'an 710032, China

Abstract. In the process of mobile robot SLAM based on monocular vision, the data association between the landmark and the imaging feature point is vital to localization accurately. RANSAC can remove mismatching points, and estimate localization robustly. But it chooses the correct model by random sample without efficient, and doesn't distinct between inlier points for localization, which decreases localization accuracy. For the above questions, a fast and high accuracy localization estimation algorithm is proposed. A approximate localization model is predicted with the KF filter, which accelerates initial correct localization model selection. The consistent set is obtained based on geometric distance threshold, and variance weighted BA optimization is used to improve the accuracy of SLAM localization estimation. The numerical simulations show that the proposed localization estimation method is effective.

Keywords: Monocular vision · Localization estimation · Kalman filter · Variance weighted bundle adjustment

1 Introduction

That the correct matching pairs between landmark and the imaging feature points in the image is the key of the SLAM method for mobile robots [1]. So, it is key to find out correct matching pairs based on the data association characteristics.

The ICNN [2] (Individual Compatibility Nearest Neighbor, ICNN) and JCBB [3] (Joint Compatibility Branch and Bound, JCBB) data association method would still exist mismatching pairs because of the repeated texture and noise, which will lead to the inaccuracy or error of localization estimation. RANSAC [4, 5] (RANdom SAmple Consensus, RANSAC) is a robust data association method of statistical model that takes into account the rejection of mismatching pairs. It uses random sampling to find the maximum consistent set to reject the mismatching pairs (outlier point), moreover, this method assumes that correct association data model is the main model. In the static environment, the mismatching pairs is random, and its model is not major, which satisfies this assumption. However, the RANSAC data association method no longer differentiates the internal points, so the accuracy of mobile robot localization estimation

M. Atiquzzaman et al. (Eds.): BDCPS 2019, AISC 1117, pp. 238–245, 2020.
https://doi.org/10.1007/978-981-15-2568-1_33

is affected. Moreover, the localization model requires more times sampling to estimate, so the estimation efficiency is not high.

For the data association errors between landmarks and imaging feature points in the SLAM, on the basis of RANSAC data association, a fast and high accuracy localization estimation algorithm is proposed. The variance weighed BA [6, 7] (Bundle Adjustment, BA) combined with KF [8] (Kalman Filter, KF) to reject mismatching pairs. The variance weighed loss function is given, which not only rejects the mismatching pairs in a reliable way, but also distinguishes inlier points of the data association effectively. Combined the KF prediction equation, the efficiency of data association is improved. Finally, the robust localization estimation of mobile robot can be realized based on the BA. Through experimental simulation and analysis, the proposed method can estimate the mobile robot localization robustly and effectively.

2 Localization Model Estimation Algorithm Based on RANSAC Data Association

2.1 The Localization Algorithm Principle Based on 3D-2D Matching Points

If there are mismatching pairs or multi-model among the matching points, the RANSAC data association algorithm could obtain the main association model by repeatedly random selecting a sample among the matching points, which is the localization model of mobile robot [8]. The process is as following:

Firstly, the smallest sample is assumed to be the inlier points set of the association localization model. If the localization model of the mobile robot is solved with the direct linear method, six pair matched points are needed.

Then, the localization model of mobile robot is used to test and decide for other matching points, if it conforms to the estimated localization model of mobile robot or has small error, it is judged as the consistent inlier point of the localization model, and all the inlier points of consistent sets are used to estimate the localization model of mobile robot again based on least square method.

Finally, select the localization model by the number of inlier points contained in the consistent sets, discard the localization model of too few inlier points, and obtain the localization model with the maximum number consistent set.

According to the above steps, the localization model of mobile robot is estimated.

2.2 The Analysis of Localization Model Estimation Algorithm

(1) **The Computational Complexity Analysis**: For the monocular vision SLAM, the 3D landmark is associated with the 2D feature imaging points by projective imaging model. Here the internal parameters of the imaging model are invariant and known, the external localization parameters of the imaging model have 6 degrees of freedom, including 3 translation and 3 rotation parameters. According to the n point localization principle of PnP [9, 10] (Perspective-n-Point, PnP), at least six pairs of correct data association points are needed as a sample to uniquely

determine the localization model of mobile robots. For the convenience of estimation, the localization matrix model can be directly established combined with internal imaging parameters and external localization parameters, which is a 3 × 4 matrix, without the need to estimate specific external parameters. A sample needs at least six pairs of points to uniquely determine the location matrix model.

When there are mismatching pairs between the landmark and the feature point, the sampling times of obtaining a correct localization estimation model under a certain probability is non-linear with the size of a sample under a certain proportion of inlier points. Assuming that the proportion of the association inlier points is $w = 0.45$, the probability of obtaining a good observation sample is $z = 0.98$, when the size of the observation sample is n, the sampling times of an correct observation sample is K. the specific corresponding values between them are shown in Table 1:

Table 1. Relationship between sampling size and sampling time

n	1	2	3	4	5	6	7	8	9	10
K	6	18	41	94	210	469	1045	2325	5168	11487

For the localization matrix model of monocular vision SLAM mobile robot, the size n of observation sample J is at least 6. As can be seen from Table 1, its sampling times K is about 78 times that of $n = 1$. Therefore, it is necessary to reduce sampling times and improve the efficiency of obtaining a good observation sample.

(2) **The Loss Function Analysis:** The localization model least square estimation with the consistent sets based on the RANSAC data association is actually minimization of the following loss function:

$$C = \sum_{i=1}^{n} \rho\left(\frac{e_i^2}{\sigma^2}\right) \tag{1}$$

here:

$$\rho\left(\frac{e_i^2}{\sigma^2}\right) = \begin{cases} 0, & \frac{e_i^2}{\sigma^2} < T \\ s > 0, & \frac{e_i^2}{\sigma^2} \geq T \end{cases} \tag{2}$$

Formula (1) is the loss function of RANSAC data association algorithm. The inlier points loss is the same. If T is large enough, each matching data will be an inlier point, the loss function will be not working. So, in order to reduce the difficulty of the distance threshold selecting, it is necessary to evaluate the inlier point set differently.

3   Fast and High Accuracy Localization Estimation Algorithm Based on KF and Weighted BA

Aiming at the problems of high computational complexity, and inlier points withnot distinguished, resulting in localization estimation ineffective and low accuracy, introducing KF motion equation to select the approximately correct localization model quickly, and obtain the consistent set with 3σ limit criterion. The inlier points is weighted, and localization model is estimated based on BA with iterative optimization, which effectively measures the quality of inlier points for localization estimation and reduces the over-dependence on distance threshold.

3.1   The Consistent Set Selected Based on KF Prediction

In order to reduce the sampling times of acquiring a correct sample by RANSAC, and reasonable reject outlier points, the consistent set selected based on KF prediction is designed, the specific steps are as following:

(1)   The mobile robot's motion equation was used to predict the current position and pose, supposing its motion equation is uniformly accelerated motion:

$$r_k = f(r_{k-1}, u_{k-1}) + w_{k-1} \tag{3}$$

$$\overset{...}{r_k} = w_{k-1} \tag{4}$$

Here, r_{k-1} is the state at $k-1$ moment, r_k is the state at k moment, w_{k-1} is the motion noise. u_{k-1} is the input, $f(\cdot)$ is the motion function, acceleration α is measured with IMU.

Then, the mobile robot's current position and pose $r_{k|k-1}$ is predicted with the motion Eq. (3), and initial localization is obtained.

(2)   Suppose the landmark set is M_i, the corresponding feature point data set is $z^{IC} = \{z_1, z_2, \cdots, z_i, \cdots, z_n\}$, and the association data set is $X = \{M, z^{IC}\}$, in order to selecting consistent set for $r_{k|k-1} = [R_{k|k-1} \quad T_{k|k-1}]$, a landmark M_i is re-projection to $m_{i,k|k-1}$ at the monocular vision state $r_{k|k-1}$ as formulas (5) and (6), and re-projection error e_i is calculated with matching actual imaging point z_i as formula (7).

$$\begin{bmatrix} x^c_{M_I,k|k-1} \\ y^c_{M_I,k|k-1} \\ z^c_{M_I,k|k-1} \end{bmatrix} = R_{k|k-1} \begin{bmatrix} x^w_{M_I,k|k-1} \\ y^w_{M_I,k|k-1} \\ z^w_{M_I,k|k-1} \end{bmatrix} + T_{k|k-1} \tag{5}$$

$$m_{i,k|k-1} = \begin{bmatrix} u \\ v \end{bmatrix} = \begin{bmatrix} f_u \frac{x_{M_I,k|k-1}}{z^c_{M_I,k|k-1}} + u_0 \\ f_u \frac{y_{M_I,k|k-1}}{z^c_{M_I,k|k-1}} + v_0 \end{bmatrix} = h_k(x_{k|k-1}, M_i) \tag{6}$$

$$e_i = z_i - m_{i,k|k-1} \tag{7}$$

Where, f_u and f_v are the focal lengths of monocular vision, u_0 and v_0 are the main points of monocular vision, all in pixels. $x^w_{M_i,K}$, $y^w_{M_i,K}$ and $z^w_{M_i,K}$ are respectively the coordinates of landmark M_i under the world coordinate system. $x^c_{M_i,K}$, $y^c_{M_i,K}$ and $z^c_{M_i,K}$ are respectively the coordinates of landmark M_i under the mobile robot coordinate system of current initial pose.

(3) Suppose the re-projection error $e \sim N(0, \sigma^2)$, so $e^2 \sim \chi^2_2$. Random variable χ_2 probability distribution of less than $\alpha = 95\%$ is:

$$F_m(\alpha) = \int_0^\alpha \chi^2_2(\xi)d\xi \tag{8}$$

Then distance threshold T:

$$T = \sqrt{F_m^{-1}(\alpha)\sigma^2} = \sqrt{5.99}\sigma \tag{9}$$

Here, the σ is the imaging error.

According to formula (9), re-projection error less than the distance threshold T is the consistent set.

3.2 Inlier Points Weighted for Localization Estimation Based on BA with Iterative Optimization

For the consistent set, the loss of the inner points are distinguished according to that they fit with the association model, the loss of the outer points are the same:

$$C_{map} = \sum_{i=1}^n \rho_{map}\left(e_i^2, T^2\right) \tag{10}$$

Here:

$$\rho_{map}\left(e_i^2, T^2\right) = \begin{cases} e_i^2, & e_i^2 < T^2 \\ T^2, & e_i^2 \geq T^2 \end{cases}$$

Because the effect of the inlier points for the localization estimation is changing with the loss, so they are weighted in formula (11):

$$w_i = \begin{cases} 1 & e_i^2 \leq \sigma^2 \\ \sigma/|e_i| & \sigma^2 \leq e_i^2 \leq 5.99\sigma^2 \\ 0 & 5.99\sigma^2 < e_i^2 \end{cases} \tag{11}$$

With the weighted inlier points, we can solve the high accuracy localization parameters based on BA [8] using the iteration optimization method.

4   Simulation Experiments and Analysis

The test data used in the experiment is derived from the Mikolajczyk dataset(matched feature points are obtained with SURF algorithm), there are 6 group real test data used, and there are 3 frame images in every group, the first frame image of every group is shown in Fig. 1.

|(a)blur group 1|(b)blur group 2|(c)Viewpoint group 3|

(d)Zoom+rotation group 4   (e) Zoom+rotation group 5   (f) Light group 6

Fig. 1. Test data

And 54 group feature points of simulated data from real test data (standard deviation is 0.1, 0.5 and 1.0, outlier points has a percentage of 5,10 and 20) are generated.

For the real and simulated data, the first two frames image used 2D-2D matching to estimate position and initialize the local map. The third frames image used 3D-2D matching to estimate position, and to verify the effectiveness of the robust localization estimation algorithm based on KF and weighted BA. The processing real-time performance, localization accuracy of the algorithm are discussed between proposed algorithm and RANSAC algorithm. All algorithms realized with MATLAB.

4.1   Processing Efficient Analysis

To verify processing performance of the fast and accuracy localization estimation algorithm, its runtime is compared with RANSAC algorithm. Each group data is tested 10 times, the average runtime is as shown in Fig. 2.

From Fig. 2, it can be seen that the average processing speed of the proposed algorithm is increased greatly than the RANSAC algorithm.

4.2   Accuracy Analysis

This paper compares the localization accuracy under different actual scene, evaluates the accuracy with mean and standard deviation of the feature point re-projection error

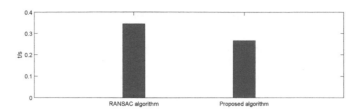

Fig. 2. Comparison of average runtime

under the localization estimation. The results of the comparison between RANSAC and the proposed algorithm are shown in Table 2.

Table 2. Accuracy comparison of two algorithms

Different group	RANSAC algorithm		Proposed algorithm	
	Mean	Standard deviation	Mean	Standard deviation
Blur group 1	0.402	0.316	0.201	0.978
Blur group 2	1.811	2.035	0.403	0.247
Viewpoint group 3	20.139	23.173	0.295	0.204
Zoom+rotation group 4	2.598	3.413	0.184	0.115
Zoom+rotation group 5	3.012	3.564	0.197	0.108
Light group 6	56.071	69.248	0.957	0.061

From Table 2, it can be seen that for different test scene, not only blur and zoom, but also light changing, the mean and standard deviation of re-projection error are relatively low with proposed algorithm, while the estimation error of the RANASC algorithm is always bigger than the proposed algorithm. So, the localization accuracy of proposed algorithm is good.

4.3   Performance Comparison of Two Algorithm

In order to compare the two algorithm visually, the feature points matching results of images is as shown in the Fig. 3. Data association with the RANSAC and proposed algorithm are shown in Figs. 4 and 5.

Fig. 3. The matching points   **Fig. 4.** RANSAC result   **Fig. 5.** Proposed algorithm result

It can be seen from Fig. 4, there are still some outlier points in the inlier point set obtained with RANSAC, its the processing effect is not good. While the proposed algorithm can reject all of the outlier point, and has a good robustness. The robustness of proposed algorithm is better than the RANASC algorithm, and the changing trend of correct matching point is consistent.

5 Conclusion

In this paper, the problem of localization estimation for SLAM with mismatching data in monocular vision is studied. In the case of mismatching data, combined with the prediction of motion equation of mobile robot, a fast and high accuracy localization estimation algorithm based on KF and weighted BA is proposed, to improve the efficiency of localization estimation. The loss function of weighted BA improves the accuracy of localization estimation. Through the simulation experiments, it is shown that the proposed localization estimation algorithm can reliably reject the mismatching point, tolerate influence of noise data, and provide a reliable guarantee for SLAM localization estimation.

Acknowledgments. This work was supported by Shaanxi Province Key Research and Development program (Program No. 2018GY-184), and supported by the Program for Innovative Science and Research Team of Xi'an Technological University.

References

1. Lee, T., Kim, C., Cho, D.D.: A monocular vision sensor-based efficient SLAM method for indoor service robots. IEEE Trans. Ind. Electron. **66**(1), 318–328 (2018)
2. Yu, H., Ji, N., Ren, Y., et al.: A special event-based K-nearest neighbor model for short-term traffic state prediction. IEEE Access **7**, 81717–81729 (2019)
3. Kaess, M., Dellaert, F.: Covariance recovery from a square root information matrix for data association. Robot. Auton. Syst. **57**(12), 1198–1210 (2009)
4. Hossein-Nejad, Z., Nasri, M.: A-RANSAC: adaptive random sample consensus method in multimodal retinal image registration. Biomed. Signal Process. Control **45**, 325–338 (2018)
5. Hossein-Nejad, Z., Nasri, M.: An adaptive image registration method based on SIFT features and RANSAC transform. Comput. Electr. Eng. **62**, 524–537 (2017)
6. Li, Y., Fan, S., Sun, Y., et al.: Bundle adjustment method using sparse BFGS solution. Remote Sens. Lett. **9**(8), 789–798 (2018)
7. Ovechkin, V., Indelman, V.: BAFS: bundle adjustment with feature scale constraints for enhanced estimation accuracy. IEEE Robot. Autom. Lett. **3**(2), 804–810 (2018)
8. Karrer, M., Schmuck, P., Chli, M.: CVI-SLAM—collaborative visual-inertial SLAM. IEEE Robot. Autom. Lett. **3**(4), 2762–2769 (2018)
9. Ruchanurucks, M., Rakprayoon, P., Kongkaew, S.: Automatic landing assist system using IMU + PnP for robust positioning of fixed-wing UAVs. J. Intell. Robot. Syst. **90**(1–2), 189–199 (2018)
10. Morgan, I., Jayarathne, U., Rankin, A., et al.: Hand-eye calibration for surgical cameras: a procrustean perspective-n-point solution. Int. J. Comput. Assist. Radiol. Surg. **12**(7), 1141–1149 (2017)

Reasonable Deployment of Computer Technology and Enterprise Human Resources

Juan Chen[1], Chaohu He[1,2(✉)], and Dan Zhang[3]

[1] Graduate School, Cavite State University, 4122 Indang, Cavite, Philippines
holy921@163.com
[2] Yunnan Technology and Business University,
Kunming 651701, Yunnan, China
[3] Graduate School, Jose Rizal University,
80 Shaw Blvd, Mandaluyong, Metro Manila, Philippines

Abstract. With the development of society, human resources has become the core of today's society, reasonable management and development of human resources can improve the efficiency of enterprises and promote the development of the economy. Human resource management system is a modern way of human resource management using information and computer technology to achieve the best integration of human resources and efficient management. According to the needs of enterprises, this paper takes information management as the basis and personnel management as the goal to collect, store and modify the information of personnel of various enterprise organs, and timely reflect the information of personnel deployment and personnel training. This paper adopts B/S mode in the implementation of the system, J2EE development framework, unified modeling language to analyze the system, and Oracle server for the database. Through testing, the system has the advantages of flexible structure, reusability, easy maintenance, good expansibility and friendly interface. The research of this paper realizes the information-based management of information mobilization and training of enterprise personnel, records the dynamic of each employee in the unit, and has certain flexibility. Therefore, the research in this paper is feasible in both technical and practical aspects and has certain practical significance.

Keywords: Computer technology · Enterprise human resources · Personnel deployment · J2EE development framework

1 Introduction

Resources are the most precious wealth in the society, and human beings are the core of the society, so the management of human resources has become the top priority [1]. Effective management of human resources can promote economic development, consolidate social stability and accelerate the pace of social development. The human resource management system improves the work efficiency of the management department, improves the technical content of the department's human resource management, accurately positions the roles of personnel, and effectively manages personnel information [2]. Starting from the lofty realm of saving resources, this paper

M. Atiquzzaman et al. (Eds.): BDCPS 2019, AISC 1117, pp. 246–252, 2020.
https://doi.org/10.1007/978-981-15-2568-1_34

proactively pays attention to the changes in the internal and external environment of the organization, such as technological update and employees' mentality, and explores challenges according to the needs of the organization's development [3]. In order to improve the work efficiency of various organs, reasonably allocate and use high-quality resources, and achieve the "people-oriented" human resource management goal of organs, it is very necessary to develop a set of advanced human resource management technologies and programs for the fine management of personnel [4].

In the last century, foreign enterprises took the lead in adopting advanced human resource management technologies and programs to allocate and manage human resources [5]. To put into huge cost to implement the standardization of the human resource management and information technology, its main reason is that expected to apply human resources to the best economic benefit, but also because the coming of knowledge economy, the concept of human capital has been formed, the importance of human capital as more land, plant, equipment and funds, and even beyond, in addition to this, people is the carrier of knowledge, in order to effectively use knowledge, knowledge to maximize utility, it need proper human resources management, can use of human resources [6]. China's human resource management system lags behind advanced countries. After entering the 1990s, most enterprises in China have computers, but the popularity of computers in human resources departments is still not enough [7]. The human resource management system of enterprises is not developed very much. Most of them develop by themselves or entrust small software companies to develop, and some of them directly use EXCEL for employee information management [8]. It was not until the end of 1990s that foreign advanced HRM concepts began to be widely accepted in China. Driven by the process of economic marketization and the rapid development of the Internet, enterprises paid more and more attention to information construction, and HRM technologies and programs became one of the core contents of enterprise informatization [9].

Based on the above background, this paper makes an in-depth study on the internal operation mechanism of Struts framework and Hibernate technology, and integrates it on this basis, aiming to develop a human resource management technology and solution with flexible structure, high reusability, easy maintenance, good expansibility and multi-layer structure by utilizing J2EE technology [10]. To realize the full use of computer technology by various organs and administrative units to reasonably manage employees' files, personnel transfer, education and training management and system management.

2 Methods

2.1 Feasibility Analysis

Based on the development of existing technologies, the conditions for the development of human resource management technologies and programs have been very mature. In terms of hardware, the speed of computer hardware is no longer a problem. Large capacity and high speed hard disks are very common, and the improvement of network speed lays a solid foundation for the operation of the system. From the perspective of software, database technology has been quite mature, and the processing capacity is very strong. At the same time, based on Struts framework for development, data persistence operation using Hibernate framework. Struts, as an open source framework of MVC design pattern, divides view, model and control well, reduces complexity and improves the reusability of code. With the support of these technologies, there is no technical risk to be taken for our successful research on human resource management technologies and programs.

2.2   Introduction to Struts and Hibernate Frameworks

(1) Struts framework

Struts was originally built as part of the Apache Jakarta project. The framework is called Struts to remind us of the foundations that support our houses, buildings, Bridges, and even when we're walking on stilts. This is also an excellent description of Struts's role in developing Web applications. When software engineers use Struts to support every layer of a business application, its purpose is to help reduce the time we spend developing Web applications using the MVC design model.

Struts is an Open Source project from the Apache foundation Jakarta project group. At its core is an mvc-style controller that helps Java developers develop Web applications with J2EE. Struts includes the following four core components: Action-Servlet, ActionClasses, ActionMapping, and ActionFormBean.

(2) Hibernate framework

Hibernate is an open source object-relational mapping framework that encapsulates JDBC (Java Data BaseConnectivityjava database connection) in a very lightweight way, allowing Java programmers to manipulate databases with object programming thinking at will. Hibernate can be used in any situation of using JDBC, not only in Java client program, but also in Servlet/JSP Web application. The most revolutionary is that Hibernate can replace CMP in the J2EE architecture of application EJB and complete the task of data persistence.

Hibernate is based on the idea of object/relational mapping technology, enabling developers to manipulate and manage relational databases in an object-oriented manner. In Java Web applications based on the MVC design pattern, Hibernate is used as the data access layer or persistence layer of the application. When the system is released, according to the database platform selected by the user, modify the configuration file of Hibernate, which can easily achieve the database transplantation of the system, Hibernate can greatly reduce the workload of operating the database. In addition, Hibernate can make use of the proxy mode to simplify the process of loading classes, which will reduce the code needed to be written when extracting data from the database by using Hibernate, reduce the difficulty and complexity of software development, improve development efficiency, and thus make the development of enterprise-class applications easier and more powerful.

3   Experiments

3.1   System Framework Design

Based on the above requirements, this system mainly applies JSP+struts+hibernate technology to design a small human resource management system, which mainly includes four modules: personnel files, personnel deployment, education and training, system management, etc. The system framework is shown in Fig. 1 below.

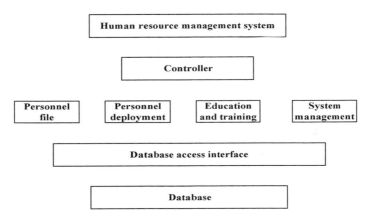

Fig. 1. System frame diagram

The whole system is composed of four function modules, this paper mainly to the personnel files of information management system of worker individual information maintenance, personnel deployment module of personnel transfer management and education training module of the training records maintenance system function modules of the detailed design as an example, and mainly from the logic to handle business processes, business process design class diagram and system data concept model from three aspects.

3.2   Database Conceptual Structure Design

In database design, the data storage in the data dictionary obtained by system analysis should be analyzed first, and then the relational mode of the system can be obtained. The entity – relation graph (ER graph for short) can be used to describe the data structure. ER diagram is composed of entity, attribute and relation. The entity includes employee, file, training, contract, department, department level and certificate. The employee is hired and signs the employment contract. The employee has the file in the unit and divides it into a certain level department.

4   Discussion

4.1   Realization of Design Scheme

This system USES Java development environment JDK1.5, the web server USES Apache tomcat5.5, the database USES Oracle, and the development tools are Eelipse5.5 and MyEelipse6.0 plug-ins. Add the Struts, Hibernate, and junit jars to the developed Javaweb project, and the human resource management system interface is shown in Fig. 2.

Fig. 2. Interface diagram of human resource management system

4.2   Implementation of Struts and Hibernate

The Struts and Hibernate framework is adopted to build the system. Each module in the system has its own configuration document to control the process of the module. In this way, except for the different business logic implementation of each module, the Struts and Hibernate architectures are adopted to implement the module with the same technology and process.

4.3   Application of Hibernate Technology in Model Layer

The main task of the model layer is to save the transaction logic code and data access code, which is the core of the whole system application. Our system implements data persistence through Hibernate. In terms of user login, specific information of the user can be obtained from the database according to the user name and password, including geographical information, user permissions, etc. The configuration information for Hibernate is shown in Table 1 below.

Table 1. Hibernate configuration information table

Database connection user name	Database connection url	Underlying database dialect
Localhost: 8080	acle:// 222.195.151.11:1433/	org.hibernate.dialect. DB2Dialect
Localhost: login	acle:// 222.195.151.11:1376/	org.hibernate.dialect. OracleDialect
Localhost: Bing	acle:// 222.195.151.11:1132/	org.hibernate.dialect.H2Dialect
Localhost: ADS	acle:// 222.195.151.11:1243/	org.hibernate.dialect. IngresDialect

4.4   System Test

The testing of web-based application system is an important and challenging task, which is different from traditional software testing. It not only needs to detect and verify whether the system runs according to requirements and design requirements, but also needs to test whether the system runs normally in different clients, and more importantly, it also needs to test security, pressure load and other aspects from the perspective of the end user. In the development process of this system, unit test, integration test, performance test and acceptance test are carried out. This chapter will focus on unit testing with the Java open source testing tool Junit. Test login and exit, personnel file information management, employee personal information maintenance, personnel transfer management and core system Settings were conducted for 2000 times respectively, and the number of bugs in each plate was 5 times, 9 times, 16 times, 3 times and 34 times respectively. The test results are shown in Fig. 3.

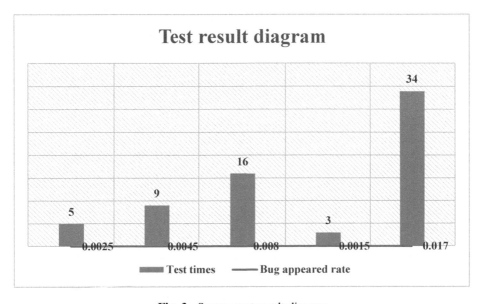

Fig. 3. System test result diagram

In the process of web system development, system testing is a complex and tedious thing. We use Junit to carry out unit testing of the system, which helps us locate the error location as soon as possible, greatly improves the development efficiency, and also reduces the chance of bugs in the program.

5 Conclusions

Based on the combination of Struts and Hibernate, this paper develops a multi-layer human resource management system, which combines the advantages of Struts and Hibernate and reasonably applies them in practical projects. Junit is used to test the system, and the result is satisfactory.

In particular, according to the introduction of the above can be concluded that the Struts is a good way of MVC architecture, coordinate it by putting a set of classes, the Servlet and tag library combining in a unified framework, the MVC pattern is the separation of business logic and display logic ability to play to the extreme, but Hibernate as excellent persistence layer implementation, provides a powerful, high performance ORM mechanism, greatly reduces the Struts model part of the difficulty of development. The research of this paper has realized the functions of personnel file management, personnel deployment management, education and training management, system management, etc. with the help of computer technology, to meet the needs of saving social resources.

References

1. Kianto, A., Sáenz, J., Aramburu, N.: Knowledge-based human resource management practices, intellectual capital and innovation. J. Bus. Res. **81**, 11–20 (2017)
2. Guest, D.E.: Human resource management and employee well-being: towards a new analytic framework. Hum. Resour. Manag. J. **27**(1), 22–38 (2017)
3. Presbitero, A.: How do changes in human resource management practices influence employee engagement a longitudinal study in a hotel chain in the Philippines. J. Hum. Resour. Hosp. Tour. **16**(1), 56–70 (2017)
4. Chowhan, J., Mann, F.P.S.: Persistent innovation and the role of human resource management practices, work organization, and strategy. J. Manag. Organ. **23**(3), 456–471 (2017)
5. Belizón, M.J., Morley, M.J., Gunnigle, P.: Modes of integration of human resource management practices in multinationals. Pers. Rev. **45**(3), 539–556 (2017)
6. Chow, J.: Internet-based computer technology on radiotherapy. Rep. Pract. Oncol. Radiother **22**(6), 455–462 (2017)
7. David, B.: Computer technology and probable job destructions in Japan: an evaluation. J. Jpn. Int. Econ. **43**, 77–87 (2017)
8. Wright, F.D., Conte, T.M.: Standards: roadmapping computer technology trends enlightens industry. Computer **51**(6), 100–103 (2018)
9. Alkhawaldeh, M., Hyassat, M., Alzboon, E., et al.: The role of computer technology in supporting children's learning in Jordanian early years education. J. Res. Child. Educ. **31**(6), 1–11 (2017)
10. Hao, B., Rong, Z., Chen, S., et al.: Research on concurrency comparison modeling based on calculus in future computer technology. Cluster Comput. **23**(5), 1–12 (2018)

Strategies for Improving College Counselors Ideological and Political Education Capability by Using New Media Against the Background of Internet+

Xin Zhou and Fang Ding[✉]

Kewen College of Jiangsu Normal University, Xuzhou 221116, Jiangsu, China
zhouxin@jsnu.edu.cn

Abstract. Nowadays, the rapid development of new media, leads a great revolution in college students that affects their way of thinking, learning and living. In 2017, Ministry of Education of the People's Republic of China proposed that universities should play a more active role in promoting internet education. In Chinese universities, counselors have close relationships with students. They meet with students almost every day and are responsible for them, which plays an important role in the growth of students. Meanwhile college students like to understand and comment the news that happens every day through the Internet. They would like to express their feelings and record their lives through social media also. For counselors, the traditional method of ideological and political education for college students must be altered immediately. Under the background of Internet+, college counselors shall make better use of the new media as the most important tool to operate the management of student affairs, establish a great fascinating and influential brand of the new media, and that can make tremendous contributions to the ideological and political education which would enter a new era.

Keywords: Internet+ · Counselor · Ideological and political education · New media · Network education

1 Introduction

In December 2017, MOE of China released the outline of implementation of the project to improve the quality of ideological and political education in universities, proposing to build a "ten systems for educating students". Nowadays, the rapid development of new media [1] has an increasing impact on college students' learning style, way of thinking and way of life. Therefore, college counselors' ideological and political work is faced with new opportunities, new tasks and new challenges. Some domestic colleges and universities, is actively applying all kinds of new media tools to carry out ideological and political education work, but because of the new media tools development changes quickly, new media education work running mechanism and working mode is still at groping stage, most of the counselor team in the new media tools on the unique status and advantage to explore, in the aspect of theory research is in its infancy.

M. Atiquzzaman et al. (Eds.): BDCPS 2019, AISC 1117, pp. 253–259, 2020.
https://doi.org/10.1007/978-981-15-2568-1_35

This topic selected topic unique perspective, starting from the identity of the counselor, around the counselor skills upgrading, the kernel in the spirit of new media, brand construction and management system for the three points, to explore the use of new media to carry out the ideological work of the operating mechanism and working mode, for China's colleges and universities carry out the work of ideological instruction through new media path construction provides the certain model reference and theoretical support.

2 Explore the New Path of Counselors Ideological and Political Education from the Perspective of New Media

The campus spiritual culture is the core of the new media culture. Building a campus spiritual culture full of positive energy can better help college students grow up. The construction of a strong campus new media brand can attract students' great interest in the propaganda content, so that students can keep their attention to the official new media platform [2]. Improving the management of new media platforms on campus can make all kinds of new media platforms develop more rapidly, and also make the content released by these new media platforms positive and full of energy.

2.1 Transform the Spiritual Core of Campus New Media and Build a Good Campus Spiritual Atmosphere to Guide the Value Orientation of College Students

Campus spiritual culture is the core, essence and soul of campus new media culture. Building a campus spiritual culture full of positive energy can better help college students grow up. Building a strong new media brand on campus can attract students' attention and interest in the propaganda content [3]. Sound management system can ensure the healthy growth of new media on campus.

Build the theme of campus spirit with the socialist core value system.

Socialist core value system is the spiritual banner of socialist China and should be the banner guiding the development direction of new media cultural spirit on campus. Core socialist values play a very important role, which can "better build the Chinese spirit, Chinese values and Chinese strength, and provide spiritual guidance for the people". Such values are closely related to everyone's study and life. In particular, the core socialist values for the personal level of the value standards, "patriotism, dedication, integrity, friendly", it points out that everyone should abide by the moral standards, is also the counselor should focus on cultivating college students have a good ethics [4]. Therefore, in the concrete measures to build the core of campus spirit, we can adopt the following ways, including: adjust the layout of the official website of the school to build a distinctive moral education plate; Intensify the development of official mobile phone client and WeChat public number and other new media tools to launch online ideological and political education [5]. Counselors and teachers should make good use of popular new media tools, such as Weibo and Douyin, to carry out some online interactions with students, walk into students' lives, and promote positive social energy around them as peers.

Constructing the core content of new media culture on campus through the university spirit. University spirit is the core of campus culture, including the spirit of freedom, the spirit of science and humanity, and the spirit of innovation Motto of a university can be regarded as the university spirit that this university focuses on, for example, Tsinghua University's "unremitting self-improvement and social commitment", Fudan University's "erudite and resolute, asking questions and thinking closely". Taking the school motto of Kewen college of Jiangsu normal university as an example, "Chong DE Shang Neng, Ming Zhi Du Xing" exactly hopes to cultivate students to become versatile and practical talents who have the combination of knowledge and practice. Therefore, the content of new media release should be in line with the positive orientation of campus spirit and cultivate students' modern university spirit of science, innovation and humanities [6]. Such as the recent online buzz "WeChat business phenomenon", the counselor can through combing the origin and evolution of WeChat business phenomenon, help students identify means of fraud, the harmfulness of correct cognition "borrowed" campus eliminate something for nothing, and rich fantasy, guide students to learn by own efforts and scientific business, can we truly achieve wealth accumulation and life ideal.

Take hot events as a window into the construction of new media culture. Many hot topics on the Internet, new things around, because of the interest of the college students, and thus caused a wide range of attention [7]. For these topics that are likely to cause widespread controversy, counselors should maintain their sensitivity to the issues, stand firm and correct the direction, and guide and educate students with the help of new media tools. For example, the topic of "black areas" is common. Students from different places of origin cause blind discrimination and disputes because of their different hometowns and provinces. More and more foreign students are coming to campus, especially the number of black students is increasing in recent years. Guide students to treat the increase of campus halal restaurants rationally, learn to respect the religious beliefs of different ethnic groups, and consciously maintain ethnic unity; And the civilized etiquette in the intercourse between the sexes, including rational treatment of homosexuality and AIDS and other issues. Counselors should firmly grasp hot issues, scientifically apply socialist core values, moistening things silently to guide students to distinguish right from wrong, surf the waves, and help them establish correct values.

2.2 Transform the Campus New Media Culture Market and Build an Influential Brand of Campus New Media Culture

At present, the new media culture market of colleges and universities is in a period of rapid development. The official new media platform of many colleges and universities has only completed the construction from scratch, with the lack of innovation in form, single release content and slow update, which is difficult to attract students' interest in reading and cannot be aggregated into an influential new media brand. The official new media platforms of many universities have only completed the construction from scratch, and they follow the rules in content and form without innovation. For example, in terms of media content, they simply copy the news on the college news website and release the news with a single content, which seems to be full of "official taste".

Moreover, they update the news slowly and lack original content, which is not suitable for the daily life of college students, hard to attract students' interest in reading and arouse the resonance of college students [8]. In the form of media, the structure is old and rigid, and the layout of the old school is not new, which makes it difficult to accurately feel the interests and needs of contemporary college students, and fails to give full play to the advantages of the new media platform as a lively and free platform. Therefore, the transformation of the new media culture market on campus and the building of an influential brand of new media culture on campus should be mainly carried out from the two aspects of content and form.

To transform content, counselors should keep pace with the era and get close to students' life. In addition to the release of authoritative information, the official new media platform should focus on displaying colorful campus life, shape the unique campus cultural spirit, and build a bridge of emotional communication between teachers and students [9]. In particular, the official WeChat public number cannot restrict the traditional administrative news, department dynamics and scientific research results display, so it is similar to the old campus media channels, including the content on the official website, and cannot reflect the natural advantages of the new media platform itself. For example, Shaanxi normal university's official WeChat ID "Shan Star" means "Shan" and "Star" is the same as "normal university". Second, the new media platform is mainly original in the release of content, actively adhere to the network hot events, so as to spread, to attract students' interest in reading and attention. For example, the article "just met you – monologue of a cherry blossom" published on the official WeChat of Wuhan university in March, and the article "Sichuan southwest aviation vocational college", "the little brother of the flight attendant who gets angry with Douyin is also our schoolmates" [10]. These articles can successfully capture students' interest in reading and make them willing to pay attention to the content published by the school.

From the form transformation, we should pay attention to the application development. For example, At Tsinghua APP has launched the function of "finding self-study room in real time", which has been favored by the majority of students. This function can roughly calculate the number of students in the self-study room in real time and refresh the data in real time by monitoring the wireless network connected by the client in each teaching building, successfully solving the difficult situation of one self-study room on campus. I Fudan APP, the official mobile phone client of Fudan university, not only provides common functions such as mobile welcome service, school bus service and map service, but also provides convenient service functions such as campus lecture information inquiry and shuttle bus inquiry, providing practical and convenient mobile services for students, faculty and staff. In addition, Tianjin university cooperated with Tencent in May this year to use "WeChat campus card" as the application entrance of smart campus to realize mobile WeChat scanning code to enter the library, and support mobile phone to reserve seats in self-study room, borrow and return books and other functions, greatly facilitating students' daily life. With these measures close to students' daily life, it is natural to develop students' dependence on the official APP and realize the role of new media platform as a bridge of communication. Through the transformation of the form of the new media platform, with the official mobile phone client as the main, supplemented by microblog and WeChat, the three vehicles jointly build the featured brand of the new media of colleges and universities, form a strong brand effect, and build a broad influence among students.

3 Improve the Management System of New Media on Campus

New media centers in colleges and universities have been established for a short time, and their functions and responsibilities are still in the exploratory stage. In the face of major online public opinions, it is inevitable to be in a panic. In the new media market, all kinds of micro-blogs and WeChat public accounts set up by students themselves have sprung up like bamboo shoots after a spring after a spring after a spring after a spring after a spring after a spring after a spring after a spring after a spring after a spring after a spring after a spring after a spring after spring [11]. For example, QQ vindicate wall of many universities has a strong attraction and influence on college students through publishing vindicate content, lost property and other information. But these QQ Numbers, which purportedly carry an official name, are actually created by individuals who register and publish content that cannot be approved by the authorities. Many Confession Wall, in order to catch the eye, without the consent of the parties will be published by the love photos, revealing students' privacy, to students caused some trouble, is a new type of "human flesh search. At the same time, there are some false advertisements in the release content, which can deceive students and cause certain economic losses to students. Many universities also give different support to new media departments, which lead to many new media departments with limited funds and difficult to exert their efforts. At the same time, there is a lack of echelon construction for the management personnel of the new media department, the counselors hold several jobs, rely too much on student groups such as the members of the student union, and the reward and punishment mechanism is not perfect.

Colleges and universities should establish or perfect full-time new media departments to manage the operation of new media on campus [12]. At the same time, the new media alliance on campus was established to establish a three-level management echelon of the school, college and department, strengthen the supervision of the new media on campus, and form a new media operation and management order with responsible leadership and division of labor. In addition, colleges and universities should establish the management system and moral code for new media as soon as possible, and establish the operating rules from the system level. In particular, in the face of network public opinion problems, major network public opinion emergency plan should be set up as soon as possible to form a complete quick handling mechanism. To strengthen the campus new media management team construction, should include: clear main responsibility consciousness; Standardize the work flow of daily operation; Regular training and exchange business improvement; Strengthening foreign exchanges and cooperation; Improve the award evaluation and punishment mechanism, etc. Therefore, under the correct leadership of the university party committee, standardize the operation of new media on campus, promote the new situation of new media on campus, and comprehensively enhance the influence and brand image of university publicity [13]. When it comes to content, counselors should pay special attention to sensitive topics. Colleges and universities should establish the management system and ethics of new media as soon as possible, and establish the operating rules from the system level. In particular, in the face of network public opinion problems, major network public opinion emergency plan should be set up as soon as possible to form a complete quick handling mechanism. For example, on September 27, 2018, the official Weibo account of Shijiazhuang institute of engineering and technology posted a message insulting revolutionary martyrs at noon. In the afternoon, the communist youth league committee of the school found the offending message and deleted it, and the official Weibo account issued an apology the next day [14]. Although the school responded to the crisis quickly and took measures in time, it also exposed the bad management of the new media platform, the lack of training and education for the student operation team, the weak sense of responsibility of the students involved and other prominent problems, which had a very bad impact on the Internet.

To strengthen the campus new media management team construction, should include: clear main responsibility consciousness; Standardize the work flow of daily operation; Regular training and exchange business improvement; Strengthening foreign exchanges and cooperation; Improve the award evaluation and punishment mechanism, etc. For example, Southwest jiaotong university promotes the communication among new media platforms on campus by offering training courses on new media content and regular discussion salons, and comprehensively improves the skills and qualities of new media platform operators. The school also exchanges with other universities in Sichuan and participates in national high-level new media BBS and seminars to expand the vision pattern of new media managers. Southwest jiaotong university attaches great importance to the construction and development of new media platforms on campus, especially in terms of development funds. The propaganda department of the party committee of the university budgets 100,000 yuan each year from special funds for cultural construction to support the construction of new media on campus [15]. And at the end of each year, through holding the campus form such as the annual meeting of the new media, so as to the selection for New Year performance outstanding campus media units and individuals, in order to support the construction of campus of new media, rapidly improve the development of new media on campus, firmly attract new media dependency on campus, students start in the new situation and effective to carry out the ideological work of the new situation.

4   Conclusion

The rapid development of new media is not only a challenge but also an opportunity for counselors to carry out ideological and political education. Counselors can learn and skillfully use new media tools, make use of new content and interesting forms, establish brand effect of new media for counselors, and attract the attention of students around, which is conducive to the development of ideological and political education and the growth of college students. At the same time, to strengthen the improvement of management system and the construction of management echelon, colleges and universities can certainly open up the new situation of student education management through new media.

Acknowledgements. This work was supported by The Special Research Project of Philosophy and Social Science Research in Colleges and Universities in Jiangsu Province 2017 (project number: 2017SJBFDY269).

References

1. Gone, C.B.: Introduction to New Media. China Radio Film & TV Press, China (2016)
2. Huang, X.L.: Study on the Educational Function and Realization of New Media on Campus. MS., Central China Normal University, China (2015)
3. Huang, J.P.: Study on Campus Culture Construction in Universities Under the New Media Environment. MS., Southwest University of Political Science and Law, China (2012)
4. Chang, H.: Research on Campus Media Integration and Development in the Context of New Media. MS., Hunan University, China (2016)
5. He, W.X.: Research on the Application of New Media in Ideological and Political Education of College Counselors: Think Tank Era, No. 35, pp. 74–75 (2019)
6. Polatu: Utopia. The Commercial Press, Beijing (1986)
7. Aristotle: Political Science. The Commercial Press, Beijing (1997)
8. Rousseau, J.-J.: The Social Contract Theory. The Commercial Press, Beijing (1997)
9. Locke, J.: The Theory of Attacking the House. The Commercial Press, Beijing (1996)
10. Kant, I.: Principles of Moral Metaphysics. People's Publishing House, Shanghai (1986)
11. Rawls, J.: Political Liberalism. Yilin Press (2000)
12. Wang, H.: The Logic of Politics: The Political Principles of Marxism. People's Publishing House, Shanghai (1994)
13. Wang, P.: Reformation of the Foundation. Peking University Press, Beijing (1995)
14. Fang, N.: Political Analysis Course. Capital Normal University Press, Beijing (1995)
15. Wang, Y.: Quadruple International Reform: History and Theory. People's Daily Press, Shanghai (1997)

Application of Machine Learning in the Evaluation Model of Scientific Research Performance of Teachers

Zijiang Zhu[1], Weihuang Dai[2(✉)], Junhua Wang[1], Jianjun Li[1],
and Ping Hu[3]

[1] School of Information Science and Technology, South China Business College
of Guangdong University of Foreign Studies, Guangzhou 510545,
People's Republic of China
[2] Human Resources Department, South China Business College of Guangdong
University of Foreign Studies, Guangzhou 510545, People's Republic of China
weihuangd2002@126.com
[3] School of Economics, South China Business College of Guangdong University
of Foreign Studies, Guangzhou 510545, People's Republic of China

Abstract. Due to the small number of scientific research evaluation indicators of college teachers, the mathematical model of evaluation is uncertain and the subjective process of the evaluation process is too strong, this paper proposes a model of university teachers' scientific research performance evaluation based on machine learning. The evaluation model uses the neighborhood rough set to reduce the index of the evaluation index and form the decision table as the input data of the support vector machine algorithm, which reduces the dimension of the sample data and improves the training speed of the evaluation model. The particle swarm optimization algorithm is used to optimize the parameter search of the support vector machine algorithm, which improves the prediction accuracy of the evaluation model. Finally, the feasibility and practicability of the scientific research performance evaluation model based on machine learning are proved through experiments.

Keywords: Machine learning · Evaluation model · Neighborhood rough set · Support vector machine · Particle swarm optimization

1 Introduction

At present, the scientific research level of colleges and universities has become one of the important indicators of the comprehensive strength of colleges and universities. As the main force of school research work, teachers need a fair, objective and effective evaluation method to make a reasonable evaluation of teachers' scientific research ability.

For the evaluation of the scientific research ability of college teachers, the peer review method is mainly adopted in foreign countries. As early as 1993, the United States established a peer-reviewed research evaluation system [1, 2]. The UK university research evaluation system mainly focuses on the evaluation of disciplines, and its core

M. Atiquzzaman et al. (Eds.): BDCPS 2019, AISC 1117, pp. 260–270, 2020.
https://doi.org/10.1007/978-981-15-2568-1_36

is also peer review [3]. The research evaluation system of Japanese universities is based on self-evaluation and third-party evaluation [4]. Although the peer review method can more accurately evaluate the teacher's scientific research ability, it has the disadvantage of strong subjectivity. For the evaluation of scientific research in Chinese universities, the quantitative analysis method of scientific research results is mainly used [5–8]. Tang et al. [5] proposed a quantitative index evaluation system that includes scientific research papers, projects, and achievements. Han et al. [6] proposed a teacher performance evaluation system based on classification, which assigns different weights to different disciplines and different disciplines. At present, there are not many researches on the evaluation of the research ability of applied college teachers. Zhang et al. [8] used the total amount and increment of scientific research projects, papers, and achievements as the evaluation indicators of teachers' scientific research ability for secondary units.

From the point of view of evaluation methods, there are still a lot of imperfections. For example, there is a certain difficulty in the distribution of weights on the secondary indicators of scientific research evaluation. Most of them are weighted according to the experience of experts, which will lead to certain errors between the prediction results and the actual results. Therefore, this paper proposes a model of university teachers' scientific research performance evaluation based on machine learning, which makes the evaluation of teachers' scientific research ability more accurate and closer to the actual research.

2   Introduction to Machine Learning Relevant Algorithms

2.1   Neighborhood Rough Set (NRS)

Hu et al. [9] used the concept of spherical neighborhood in topological space to define the neighborhood decision system and constructed a numerical feature attribute selection algorithm based on the Neighborhood Rough Set (NRS) model. The basic concepts of a neighborhood decision system are as follows: a given sample set $U = \{x_1, x_2, \cdots, x_n\}$, A is the real number feature set used to describe U, D is the decision attribute. If A generates the family relationship in the domain, then $NDS = (U, A \cup D, N)$ is a neighborhood decision system. A is also called the condition attribute, N is the equivalence class of D to divide U [10].

For $\forall \subseteq A$, then the decision attribute D is defined as follows for the upper and lower approximation sets of B, as shown in Eqs. (1) and (2).

$$\overline{N}_B D = \bigcup_{i=1}^{N} \overline{N}_B X_i \qquad (1)$$

$$\underline{N}_B D = \bigcup_{i=1}^{N} \underline{N}_B D X_i \qquad (2)$$

The attribute dependency of the decision attribute D on the subset condition attribute B can be defined as follows, as shown in Eq. (3).

$$k_D = \gamma_B(D) = \frac{|Pos_B(D)|}{|U|} \tag{3}$$

If $a_i \in B$ (i refers to the i condition attribute, and then the importance of the condition attribute a_i to the decision attribute D is calculated as shown in Formula (4). If $\{B_1, \cdots, B_i, \cdots, B_k\}$ is all the reductions of the neighboring decision system, the core definition of the decision system is shown in [11] and formula (5).

$$Sig(a_i, B, D) = \gamma_B(D) - \gamma_{B-\{a_i\}}(D) \tag{4}$$

$$Core = \bigcap_{i=1}^{k} B_i \tag{5}$$

2.2   Support Vector Machine (SVM)

Support Vector Machine (SVM) has certain advantages in solving small-scale samples, nonlinear separable and high-dimensional space recognition, and it is now more and more widely used [12]. In solving the dual problem, the dot product operation in the case of linear separability is replaced by the kernel function satisfying the Mercer theorem. The kernel function is shown in Eq. (6). The mapping of kernel functions to high-dimensional feature spaces is the solution to the dual problem of Eq. (7).

$$K(x_i, x_j) = \Phi(x_i)\Phi(x_j) \tag{6}$$

$$\begin{cases} maxQ(a) = \sum_{i=}^{l} a_i - \frac{1}{2}\sum_{i=i}^{l}\sum_{j=1}^{l} a_i a_j y_i y_j K(x_i, x_j) \\ s.t. \begin{cases} \sum_{i=1}^{l} a_i y_j = 0 \\ 0 \le a_i \le C \end{cases}, i = 1, 2, \cdots, l \end{cases} \tag{7}$$

In Formula (10), C is a penalty factor, indicating the degree of tolerance for misclassification of samples. The larger C, the less likely it is to tolerate the misclassification of the sample. Set $a^* = [a_1^*, a_2^*, \cdots, a_l^*]^T$ is the solution to Formula (10), x_r and x_s are a random pair of support vectors in the two categories. The optimal classification function can be obtained, as shown in Eq. (8).

$$f(x) = sgn\left(\sum_{i=1}^{l} a_i^* y_i K(x_i, x) + b^* - \frac{1}{2}\sum_{i=1}^{l} a_i^* y_i (x_r + x_s)\right) \tag{8}$$

2.3   Particle Swarm Optimization (PSO)

Parameter selection is an important issue in support vector machines, and it is essentially a continuous optimization process. The penalty parameter c and the function parameter g have multiple sets of solutions, there is no fixed value, and different parameters are set to make the established model have different prediction precision and error. This paper proposes to use the particle swarm optimization support vector machine method to establish a teacher scientific research ability evaluation model. The global optimal parameter combination is obtained by iterative search penalty parameter c and kernel function parameter g, and the optimized evaluation model is established to make the model have higher prediction accuracy [13].

In order to improve the convergence performance of the particle swarm optimization algorithm, inertia weight parameters are introduced in the original particle swarm optimization algorithm. The inertia weight w plays the role of weighing local optimal ability and global optimal ability. For the calculation of inertia weight w, use the linear decreasing weight method, see Formula (9).

$$w^{(t)} = (w_{ini} - w_{end})(T_{max} - t)/T_{max} + w_{end} \qquad (9)$$

Among them, t is the search iteration times, T_{max} is the maximum evolution algebra, w_{ini} is the initial maximum inertia weight, and w_{end} is the inertia weight value when evolving to the maximum algebra.

3 Construction of Teacher's Scientific Research Performance Evaluation Model Based on Machine Learning

3.1 Evaluation Index System

This paper constructs a scientific research evaluation index system based on the evaluation standards of scientific research performance of most engineering universities and the related literatures at home and abroad. The indicator system mainly has five first-level indicators, and 30 second-level indicators are used as attributes of teacher scientific research ability evaluation. The hierarchical structure of the evaluation index system is shown in Table 1.

Table 1. Research performance evaluation index

Primary indexes		Secondary indexes	
No.	Index	No.	Index
S_1	Research project	S_{11}	National key project
		S_{12}	National key project
		S_{13}	National key project
		S_{14}	Provincial key project
		S_{15}	Provincial key project
		S_{16}	General provincial project
		S_{17}	Department level project
		S_{18}	Horizontal project
		S_{19}	School project

(continued)

Table 1. (*continued*)

Primary indexes		Secondary indexes	
No.	Index	No.	Index
S_2	Research papers	S_{21}	Special journal paper
		S_{22}	Grade A paper
		S_{23}	Grade B_1 paper
		S_{24}	Grade B_2 paper
		S_{25}	Grade B_3 paper
		S_{26}	Grade B_4 paper
		S_{27}	Grade B_5 paper
		S_{28}	Grade C paper
S_3	Publishes	S_{31}	First-level publishing house
		S_{32}	Second-level publishing house
		S_{33}	Third-level publishing house
		S_{34}	Other publishing houses
S_4	Patent	S_{41}	Invention patent
		S_{42}	New utility patent
		S_{43}	Appearance design patent
S_5	Awards	S_{51}	National award
		S_{52}	Provincial award
		S_{53}	Department-level award
		S_{54}	School-level award

3.2   Evaluation Model of PSO Optimized NRS + SVM

According to the evaluation data obtained from the second-level indicators in Table 1, the scientific research evaluation index attribute data set is constructed. After the data is pre-processed, the NRS is used to reduce the index attribute, that is, the index attribute is reduced. The reduced attribute subset data is constructed into a decision table, and an evaluation model is established as input data of the SVM algorithm. In the evaluation model training process, the optimal parameters are obtained by PSO optimization SVM, and the prediction accuracy of the evaluation model is improved. The logical block diagram of the PSO optimization NRS + SVM teacher research ability evaluation model is shown in Fig. 1.

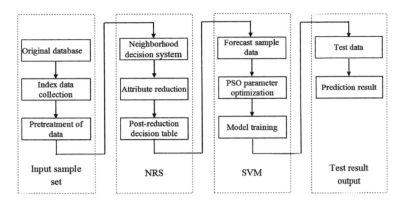

Fig. 1. Evaluation logical diagram of using PSO to optimize the NRS + SVM scientific research

3.3 Indicator Attribute Value Reduction

In this paper, the neighborhood rough set attribute reduction method is used to reduce the indicators in the scientific evaluation index system. An optimal reduction of the teacher's scientific research ability evaluation index set is obtained by setting the neighborhood size and selecting the attribute reduction greedy algorithm, and constructing a new decision table as the input data of the evaluation model.

A. Pretreatment of Index Set Attributes

In order to improve the reliability of the reduction results and reduce the impact of different index data dimensions on the reduction results, the sample data is first normalized so that all values are normalized to the interval [0, 1]. Before performing the range normalization, first remove the indicator data whose attribute values are all 0, so as not to affect the accuracy of the model. The range normalization process is shown in Eq. (10).

$$x_{ij} = \frac{x_{ij} - \min_{1 \leq k \leq n} x_{kj}}{\max_{1 \leq k \leq n} x_{kj} - \min_{1 \leq k \leq n} x_{kj}}, i = 1, 2, \cdots, n; \ j = 1, 2, \cdots, p \qquad (10)$$

Among them, $\min_{1 \leq k \leq n} x_{kj}$ is the minimum value of the variable x_j attribute in the column, and $\max_{1 \leq k \leq n} x_{kj} - \min_{1 \leq k \leq n} x_{kj}$ is the range of variable x_j attribute in the column.

B. Forward Numerical Attribute Reduction based on Neighborhood Model

According to Hu et al., the neighborhood rough set forward search attribute fast reduction algorithm [10]: If there are D samples in U, under the selected subset E of attributes, the positive domain sample set is S, and the attribute dependence of D on E is k. After r is added, the positive domain sample of $(U-S)$ on $(E + r)$ is s, and the number is n. And the dependence of D on $(E + r)$ is $(k + n)/N$. The specific algorithm is shown in Table 2.

Table 2. Fast reduction algorithm for neighborhood rough set forward search attribute

Algorithm
Begin

Enter: decision table $< U, C, D, V, f >$

(1) Initialize the core attribute collection $red = \Phi$ and the sample set $smp\_chk = U$ to be tested.

(2) while $smp\_chk = U$

        For each $k_i \in (C - red)$

          $DT_i = < U, red \cup k_i, D, V, f >$;

          Initialize $POS_i = \Phi$;

            For each $a_j \in smp\_chk$

              Calculate the neighborhood domain $\delta(a_j)$ of a_j in DT_i;

              if $\delta(a_j)$ the decision attribute D of each sample is the same

                $POS_i = POS_i \cup a_j$;

              end if

            end for

          end for

          Find the maximum POS_i and the corresponding k_i;

          if $POS_i = \Phi$

            $red = red \cup k_i$;

            $smp_{chk} = smp_{chk} - POS_i$;

          else

              Exit the while loop;

          end if

      end while

(3) return red;

Output: red
End

3.4 Particle Swarm Optimization SVM Algorithm

The particle swarm optimization SVM algorithm is used to find the optimal parameters of SVM, further optimize the scientific research ability evaluation model, and improve the prediction accuracy of the model. The main factor affecting the accuracy of the model is the parameters in the SVM algorithm. The essence of parameter selection is a process of continuous optimization. In view of the fact that the parameters are not globally optimal, this paper uses particle swarm multi-objective optimization algorithm, knot support vector machine to find the optimal parameters c and g, improve the evaluation prediction accuracy, and optimize the teacher scientific research ability evaluation model.

In the process of optimizing SVM by PSO algorithm, the fitness function is the performance of particle initialization variables in the prediction accuracy, which is used to measure the performance of each particle [12]. The fitness function of the particle is shown in Formula (11).

$$fitness = f(i_d) - p\frac{m_C}{m_A} \tag{11}$$

In Formula (11), m_C refers to the number of variables in this particle, $f(i_d)$ is the particle's positive fraction; m_A is the total number of variables collected by the system; p is the adjustment parameter. Set according to system variables and prediction accuracy requirements to balance the maximum positive fraction and the number of variables.

4   Experiment Process and Results Analysis

4.1   Experimental Data Acquisition and Pretreatment

The experiment selected representative research 12 (T_1–T_{12}) scientific research activity indicator data as sample data set, each data contains 29 indicator data, and the T_1–T_3 teacher's decision attribute value is 4. The decision attribute value of the T_4–T_6 teacher is 3, the decision attribute value of the T_7–T_9 teacher is 2, and the decision attribute value of the T_{10}–T_{12} teacher is 1. 80 teachers were randomly selected from the database of scientific research information platform as the data set of model test. The preprocessing process of all data is the same. This paper only describes the processing of sample dataset.

After the data set is subjected to the range normalization, the index attribute reduction is performed according to the algorithm in Table 2. In this paper, the neighborhood initial value $\delta = 0.15$ is set based on experience. According to the obtained reduction result, the accuracy of the prediction is obtained, and a specific step size solution is set, and the minimum reduction result subset which can best represent the original evaluation index is obtained. In the experiment, when the classification accuracy reaches the highest, the corresponding neighborhood $\delta = 0.27$. The importance of the correspondence between the attributes of the index set obtained by the algorithm is shown in Table 3.

Table 3. Relative importance of index attribute reduction (%)

Index attribute	Importance	Index attribute	Importance	Index attribute	Importance
S_{11}	28.72	S_{17}	50.12	S_{26}	57.15
S_{12}	39.21	S_{18}	29.03	S_{33}	27.81
S_{13}	35.76	S_{23}	35.33	S_{34}	48.24
S_{15}	42.80	S_{24}	53.61	S_{42}	35.06
S_{16}	46.43	S_{25}	60.71	S_{43}	28.49

4.2   Experiment Simulation

In this paper, the cross-validation iterative method and the particle swarm optimization algorithm are used to simulate the experiment. The better parameter finding method is determined by comparison, and the advantages of the proposed particle swarm optimization NRS + SVM model are verified. The simulation experiment used 12 sets of sample data to train the NRS + SVM model. The scientific research performance data of 80 teachers were randomly selected from the scientific information platform database for data preprocessing and used as test data.

A. Cross-validation Iteration Method

The SVM's penalty parameter c and kernel function parameter g search range are set to $[2^{-10}, 2^{10}]$, and the pace length range is 0.5. After experimental simulation, the results of the teacher's scientific research ability test and the actual results are shown in Fig. 2.

As can be seen from Fig. 2, only 6 of the 80 teachers have incorrect test results, and the model's predictive ability is basically sufficient, with an accuracy rate of 92.5%.

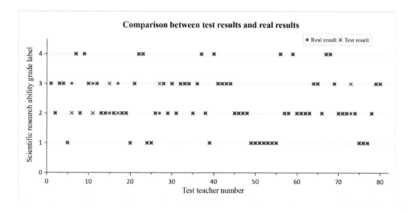

Fig. 2. NRS + SVM model test results

B. Particle Swarm Optimization

The parameters for setting the PSO are: the number of individuals in the population $N = 28$, inertia factor $w = 0.6$, acceleration constant, dimension $D = 2$, and the maximum iteration time is 300. The penalty factor c and kernel function parameter g are obtained by iterative optimization when the termination condition is satisfied. After experimental simulation, the results of the teacher's scientific research ability test and the actual results are shown in Fig. 3.

As can be seen from Fig. 3, only one of the 80 teachers had incorrect test results, and the prediction accuracy of the model reached 98.75%. The experimental results show that the NRS + SVM model optimized by particle swarm optimization algorithm has higher prediction accuracy.

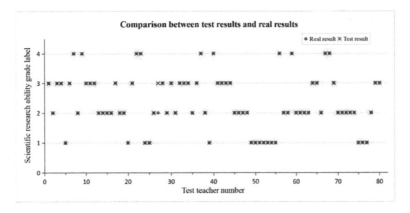

Fig. 3. Test results of PSO optimized NRS + SVM model

4.3   Analysis of Experiment Results

For the main parameter penalty factor c and kernel function parameter g which affect the performance of SVM, the PSO-optimized SVM algorithm is used to find the global optimal parameters, and the new decision table data formed by reduction is used to optimize the established evaluation model. The feasibility and effectiveness of the combination of PSO algorithm and SVM algorithm are proved by the comparison of accuracy.

5   Conclusion

In this paper, the decision table data formed by reduction is selected as the input of the SVM algorithm, and the appropriate parameters are selected for modeling in the SVM algorithm modeling, and the model is established with the SVM algorithm without the reduced sample data. In both cases, comparative analysis was carried out on the prediction accuracy, model training time and prediction time. The experiment proves that particle swarm optimization SVM algorithm can improve the accuracy of teacher scientific research ability evaluation and prediction, and shows that PSO optimization SVM algorithm has better applicability in teacher scientific research ability evaluation model.

The combination of NRS, PSO and SVM in the evaluation model of teachers' scientific research ability not only reduces the scale of sample data, improves the training speed of the model, but also improves the accuracy of classification. However, the combination of machine learning methods has a certain impact on the time complexity of the evaluation model. In the application of the theoretical project, the feasibility and accuracy of the theoretical results are first satisfied, and the time complexity for better reducing the classification needs further study in the future.

Acknowledgements. This work was supported in part by a grant from the philosophy and Social Sciences Project of Guangdong Province (No. GD18XYJ19).

References

1. Wang, X.S.: An analysis of the evaluation mechanism of scientific research performance of college teachers in the United States. Coop. Econ. Sci. **23**, 40–41 (2013)
2. Xu, D.C.: Characteristics of American teacher evaluation system and its inspiration to China. J. Bingtuan Educ. Inst. **22**(6), 38–43 (2012)
3. Liu, X.K., Liang, X.: Evaluation of university scientific research level and its enlightenment to China. Contemp. Educ. Sci. **23**, 51–55 (2014)
4. Han, X.J., Gao, J.: Concept, contents and methods of academic evaluation for university teachers in Japan. High. Educ. Dev. Eval. **29**(2), 64–71 (2013). 107
5. Tang, W.S., et al.: A research on the scientific evaluation quantification system in local universities. Sci. Res. Manage. **37**(S1), 324–327 (2016)
6. Han, X.L., et al.: A research on the performance evaluation of college teachers based on the view of classification and type-discrimination. J. Chongqing Univ. (Soc. Sci. Edn) **20**(1), 114–119 (2014)
7. Wei, J., Tang, J.K.: Research on performance appraisal of college teachers based on fuzzy comprehensive evaluation. Jiangsu High. Educ. **6**, 100–103 (2014)
8. Zhang, B., Ye, X.: Study of scientific research evaluation of application-oriented university. J. Beijing Union Univ. **29**(1), 10–13 (2015)
9. Hu, Q.H., et al.: Numerical attribute reduction based on neighborhood granulation and rough approximation. J. Softw. **19**(3), 640–649 (2008)
10. Wang, L.N., et al.: Supervised neighborhood rough set. Comput. Sci. **45**(8), 186–190 (2018)
11. Zhao, R.Z., et al.: Extraction of decision rules of rotor faults by means of neighborhood rough set combined with fisher discrimination approach. J. Lanzhou Univ. Technol. **45**(1), 43–48 (2019)
12. Li, C.B., et al.: Short-term wind power prediction based on data mining technology and improved support vector machine method: a case study in northwest China. J. Clean. Prod. **205**(12), 909–922 (2018)
13. Man, C.T., Liu, B., Cao, Y.C.: Application of particle swarm and genetic algorithm optimization support vector machine. Inverse Prob. Sci. Eng. **6**, 87–92 (2019)

Spatial Pattern of Talents and the Evolution of the Situation of Chinese "Heihe-Tengchong Line" Under the Background of Big Data

Hui Yao[1]([⊠]) and Lei Zhang[2]

[1] Faculty of Education, Yunnan Normal University, Kunming, China
yaohui.11@163.com
[2] School of Tourism and Geographical Science, Yunnan Normal University, Kunming, China

Abstract. In this paper, the education level and education structure of the population of the whole country and the provinces on the northwest and southeast sides of the Heihe-Tengchong Line are systematically analyzed based on the national census data of 1964, 2000 and 2010 and the relevant data of economic and educational development of provinces. The correlation between population education level and education structure of the provinces on the northwest and southeast sides of the Heihe-Tengchong Line and regional economic development and regional higher education development is analyzed, and the factor matching and structural similarity are evaluated, so as to provide a basis for scientific regulation and control of the evolution of regional talent spatial pattern.

Keywords: Heihe-Tengchong Line · Talent flow · Spatial pattern · Geography of education

1 Introduction

In 1935, Mr. Hu Huanyong published the article "The Distribution of Population in China: With Statistics and Maps" in the Acta Geographica Sinica. On the basis of county-level unit-scale population data in all regions of China, he drew the first population distribution map and population density map of China, and put forward the famous population geographic demarcation line between Aihui in Heilongjiang Province and Tengchong in Yunnan Province, also known as Heihe-Tengchong Line [1]. Heihe-Tengchong Line quantitatively depicts the spatial characteristics of China's population for the first time, revealing the basic fact that the population density of the southeastern half and northwestern half differs greatly. More than 80 years later, many scholars of geography, economics, demography and other disciplines have conducted in-depth research on Heihe-Tengchong Line, which has generated intense academic exploration and contention in its population, economic and geographical studies [2]. Some scholars have realized what differences exist in the population structure of the southeast half and northwest half of the Heihe-Tengchong Line, especially in the

M. Atiquzzaman et al. (Eds.): BDCPS 2019, AISC 1117, pp. 271–281, 2020.
https://doi.org/10.1007/978-981-15-2568-1_37

educational structure of the population (talent structure), and the basic trends and future trends of this difference and their impacts on regional development.

2 Multi-factor Interpretation of the Basic Pattern of Heihe-Tengchong Line in Existing Literature

Lu Dadao academician has clearly pointed out that the stability of Heihe-Tengchong Line will exist for a long time. Due to the influence of social environment, historical environment, natural environment and economic conditions, the spatial and temporal evolution characteristics of population distribution in China are different, the eastern situation is becoming more and more obvious. There are differences in population pattern, economic pattern and educational pattern between northwest and southeast regions of China divided by Heihe-Tengchong Line. Moreover, these factors interact and influence each other, which forms the basic situation of the current regional pattern and also affects the future evolution trend of this pattern to a large extent.

According to the existing research, Heihe-Tengchong Line, as the most important population demarcation line in China, remains basically stable. The population of 94:6 on both sides has been relatively stable, but the population of the southeast half continues to decline slightly. Thanks to the higher natural growth rate, the northwest half has a higher population growth rate [3]. In 1990, Mr. Hu Huanyong used the data of the third census to compare the population distribution on both sides of the Heihe-Tengchong Line. The results showed that although the western economy developed rapidly after the founding of new China, the population proportion of the southeast half and northwest half did not chang much [4]. In addition,many scholars have studied the spatial distribution of China's population from multiple perspectives and by various means. The results all prove that the spatial distribution pattern of China's population is relatively stable. From the perspective of the influencing factors of population spatial distribution, although the influence of economic development factors on population spatial distribution is increasing in general, the natural geographical background conditions such as the three steps still have an important impact on population distribution in China. The influence of economic development factors on population spatial distribution is increasing, while the influence of natural environment factors and social and historical factors is decreasing and the institutional constraints are weakening [5]. The spatial consistency of China's population distribution and economic development has obvious regional differentiation characteristics, but the interprovincial consistency is shrinking, and the population agglomeration has a strong economic orientation [6].With the increasing population in China in recent years, although the population pattern of Heihe-Tengchong Line is basically stable, the change of its population structure, especially the regional talent pattern and its evolution caused by the change of population education structure, deserves attention. In the relationship between regional higher education level structure and population employment structure and regional economy, higher education level structure should actively adapt to the economic structure and optimize the adjustment on the basis of considering the changes of employment structure, especially the education structure of the employed population [7].

3 Research Hypothesis and Process of the Talent Pattern of Heihe-Tengchong Line

3.1 Research Hypothesis

(1) *Regional hypothesis of research*

Heihe-Tengchong Line crosses various geographical units and administrative regional units and is a "virtual boundary". In the regional hypothesis of talent pattern research:

- China mainland is taken as the regional scale, and provinces are taken as the regional analytical scale.
- Consolidation is carried out according to the spatial location of each province in China. The northwest side of Heihe-Tengchong Line includes six provinces of Qinghai, Tibet, Xinjiang, Gansu, Inner Mongolia and Ningxia, while the remaining 25 provinces are in the southeast.

(2) *Elemental hypothesis of research*

- Talent and talent structure. The concept of "talent" used in the study refers to regional human capital in connotation. The study assumes that it is the number of population with higher education in the region. The talent structure refers to the educational level structure of the population receiving higher education in the region.
- Spatial and temporal pattern of talents. In terms of time, the number and structure of talents in a certain region change with time, and show a curve with certain change rule or change characteristics. Spatially, the layout focus of number and structure of multi-regional talents has shifted, showing a certain regional or directional trajectory.
- The motivation mechanism of talents. The change and distribution of talents affect the gap between regional development level and inter-regional development level, which is realized through the interaction between regional talent status and regional politics, economy and society.

(3) *Viewpoint hypothesis of research*

- The distribution of regional talents is restricted by the level of regional economic and social development, and its scale and structure may change regularly at different stages of regional development.
- The distribution of regional talents has a direct impact on the level of regional economic and social development. The change of its scale and structure can promote the improvement of regional economic and social development. The goal of narrowing the development gap between regions can be achieved by regulating and controlling the regional talent flow and influencing its layout.
- There is a latent or implicit correlation between the distribution of regional talents and the development factors of regional politics, economy and society [8–10]. In the process of regional coordinated development, a reasonable structure of talent scale is needed to match it.

3.2   Data Sources

The data used in the study mainly come from the following sources: (1) the second census of 1964, the fifth census of 2000 and the sixth census of 2010 of the People's Republic of China; (2) the statistical yearbook of China's education in 2001 and 2011; (3) the statistical yearbook of China in 2001 and 2011.

3.3   Indexes and Interpretation

In this study, the factor matching index and structural similarity index are selected to make a comparative analysis of the factors and structures of the whole country and the two sides of Heihe-Tengchong Line. The index explanations and calculation formulas are as follows.

(1) *Factor matching index*

Factor matching index refers to the measurement index of economic, social and educational factors of regional development with dominant correlation or the measurement index that there are fitting or differences between specific indicators and indexes, similar to correlation analysis. If the factor matching index is μ, there is:

$$\mu = \frac{\sum_{i=n}^{3} |x - x'|}{n} \tag{1}$$

In formula 1, n is the number of regions, x' and x are the index values of a group of matching factors in the region after elimination of dimension. And the smaller the value is, the higher the matching index of the group of factors is. M ≥ 0, and the smaller the value, the higher the matching index of the group of factors.

(2) *Structural similarity index*

Structural similarity index is used to quantify the index of whether the two normally correlated systems are structurally similar. If the structural similarity index is ϕ, then there is

$$\gamma = \sum_{i=n}^{3} \left| \frac{1}{n} (x_{ij} - y_{ij}) \right| \tag{2}$$

In formula 2, x_{ij} is the index value of the single factor level after eliminating the dimension of the j-th structural factor in i-th region, and n is the number of regions. $\gamma \geq 0$, and the smaller the value is, the higher the matching index of the group of factors is.

4 The Basic Trend of the Evolution of the Talent Pattern of Heihe-Tengchong Line

4.1 The Proportion of the Population with Higher Education Qualifications in the Total Population

According to the second, fifth and sixth national censusdata, in 1964, the percentage of the population with higher education qualifications in the total population was low, with only 0.4356%. The difference between urban and rural areas was 38.32 times, and the gap was significant. By 2010, the percentage of the population with higher education qualifications in China's total population had increased from 0.4356% to 12.8657%, an increase of about 29.5 times, with an average annual growth rate of 7.81%. Among them, the percentage of urban population with higher education qualifications increased from 2.6096% in 1964 to 16.7217% in 2010, about 6.4 times, with an average annual growth rate of about 4.21%. The percentage of rural population with higher education qualifications increased from 0.0681% in 1964 to 2.0589% in 2010, about 30.2 times, with an average annual growth rate of about 7.87%. In 2010, the difference between urban and rural areas in the percentage of the population with higher education qualifications was 8.12 times, which was significantly smaller than that in 1964 (as shown in Table 1).

Table 1. Higher education level of population in northwest and southeast of Heihe-Tengchong Line (1964, 2000 and 2010)

Years	Regional scope	The Percentage of the population with a college degree or above on total population (%)		
		Overall	Town	Countryside
1964	The whole country	0.4356	2.6096	0.0681
	Northwest provinces	0.4388	2.2308	0.1190
	Southeast provinces	0.4155	2.6331	0.0654
2000	The whole country	3.8057	*	*
	Northwest provinces	3.9603	*	*
	Southeast provinces	3.7952	*	*
2010	The whole country	12.8657	16.7217	2.0589
	Northwest provinces	13.1626	18.9473	2.4991
	Southeast provinces	12.8450	16.5862	2.0242

Note: *is the data missing.

4.2 Education Level Structure of Population with Higher Education Qualifications

In 2000, among the higher education population in China, the proportion of the population with junior college degree was the highest, 65.8460%, followed by the population with bachelor degree, and the proportion of the population with graduate degree was the lowest, only 2.0080%. By 2010, in the proportion of junior college students, undergraduates and postgraduates in the hierarchical structure of higher education population, the junior college degree level decreased, about 57.9604%. The undergraduate degree level and graduate degree level increased to 38.5435% and 3.4962%, respectively, up by 19.90% and 74.11% respectively. In China, the focus of education level of population with higher education is shifting from junior college degree level to undergraduate degree level, and the population education level structure is evolving to higher education level (As shown in Table 2).

Table 2. The educational hierarchy of the population with university degree or above on the northwest and southeast sides of Heihe-Tengchong Line (2000 and 2010)

Years	Regional scope	The educational hierarchy of the population with higher education academic degrees (%)		
		Junior college students	Undergraduate	Graduates
2000	The whole country	65.8460	32.1460	2.0080
	Northwest provinces	71.9409	27.2173	0.8418
	Southeast provinces	65.4159	32.4938	2.0903
2010	The whole country	57.9604	38.5435	3.4962
	Northwest provinces	63.0889	35.0454	1.8657
	Southeast provinces	57.5887	38.7970	3.6143

5 Regional Relevance Analysis on the Evolution of the Talent Pattern of Heihe-Tengchong Line

5.1 Matching Analysis of Economic Factors of the Whole Country and on Both Sides of the Heihe-Tengchong Line

(1) *Matching analysis of talent factors and economic factors on the whole country and both sides of the Heihe-Tengchong Line*
    In 2000, the ratio of per capita GDP of the whole country and the provinces on the northwest and southeast sides of Heihe-Tengchong Line was 0.3681:0.2560:0.3759. The level of per capita GDP of the provinces on the southeast side of Heihe-Tengchong Line was higher than that of the northwest and the whole country. By 2010, the ratio had changed to 0.3475: 0.3017: 0.3507. The per capita GDP of the whole country and the provinces on the southeast side of the Heihe-Tengchong Line decreased by 5.59% and 6.69% respectively compared with 2000, while that of the provinces on the northwest side increased by 17.86% compared with 2010. The gap of per capita GDP between the northwest provinces of the Heihe-Tengchong Line and the whole country and the southeast provinces of Heihe-Tengchong Line has been narrowing continuously (as shown in Table 3). In 2010, the matching index (μ) between the per capita GDP on the whole country and provinces on the northwest and southeast sides of the Heihe-Tengchong Line and the proportion of the regional

population receiving higher education was 0.1730. By 2010, the matching index was 0.0737, which was reduced by about 57.40% compared with 2000. During this period, the matching between the per capita GDP of the whole country and the provinces on the northwest and southeast sides of Heihe-Tengchong Line and the proportion of regional higher education population continued to rise, and the gap between the proportion of higher education population and the state of economic development combination between regions significantly narrowed.

Table 3. Gross regional domestic product and industrial structure of the northwest and southeast sides of Hu Huanyong Line (2000 and 2010)

Years	Regional scope	Per Capita GDP (yuan/person)	Industrial structure(%)		
			Primary industry	Secondary industry	Tertiary industry
2000	The whole country	7701	15.2702	47.0985	37.6313
	Northwest provinces	5356	21.6837	41.9568	36.3595
	Southeast provinces	7864	14.9665	47.3420	37.6915
2010	The whole country	32792	9.2744	50.3533	40.3723
	Northwest provinces	28472	12.6564	51.1785	36.1611
	Southeast provinces	33094	9.0711	50.3037	40.6254

(2) *Similarity analysis of talent structure and economic structure on the whole country and both sides of the Heihe-Tengchong Line*

From 2000 to 2010, the industrial structure of primary industry, secondary industry and tertiary industry in China changed from 0.1527: 0.4710: 0.3763 to 0.0927: 0.5035: 0.4037. The proportion of primary industry declined and the output value of secondary and tertiary industries increased. The evolution trend of the three industrial structures in the southeast provinces of Heihe-Tengchong Line was similar to that in the whole country, but for the three major industrial structures of the northwest provinces, the proportion of the primary and tertiary industries reduced, while the proportion of the secondary industry increased. In 2000, the similarity index (γ) of the economic industrial structure on the whole country and the northwest and southeast provinces of Heihe-Tengchong Line and the educational level structure of regional population was 1.0085. In 2010, the similarity index was 0.9842, which was reduced by about 2.4115% compared with 2000. During this period, the similarity of the economic industrial structure on the whole country and the northwest and southeast provinces of Heihe-Tengchong Line and the educational level structure of regional population continued to rise, and the gap between the hierarchical structure of regional population receiving higher education and the combination state of economic industrial structure significantly narrowed.

5.2 Similarity Analysis of Educational Structure on the Whole Country and Both Sides of Heihe-Tengchong Line

(1) *Matching analysis of talent factors and education factors on the whole Country and both sides of the Heihe-Tengchong Line*

In 2000, the ratio of the number of students enrolled in higher education per 10,000 people of the whole country and northwest and southeast provinces of the Heihe-Tengchong Line was 0.3563: 0.2801: 0.3636. The number of students enrolled in higher education per 10,000 people in the southeast provinces of the Heihe-Tengchong Line was higher than that in the northwest and the whole country. By 2010, the ratio had changed into 0.3572: 0.2790: 0.3638. The number of students enrolled in higher education on the whole country and the southeastern provinces of the Heihe-Tengchong Line increased by 0.2790% and 0.0408%, respectively, compared with 2000. The number of students enrolled in higher education in the northwestern provinces of the Heihe-Tengchong Line decreased by 0.4078% compared with 2010. The gap between the number of students enrolled in higher education in the northwest provinces of Heihe-Tengchong Line and the whole country and the southeast provinces continues to widen (as shown in Table 4).

Table 4. The scale and hierarchical structure of higher education on the northwest and southeast sides of Heihe-Tengchong Line (2000 and 2010)

Years	Regional scope	Number of students per 10,000 (person)	Hierarchical structure of higher education (%)		
			Junior college	Undergraduate	Graduate
2000	The whole country	145	47.6141	47.9185	4.4674
	Northwest provinces	114	47.3803	50.0962	2.5235
	Southeast provinces	148	47.6267	47.8012	4.5721
2010	The whole country	219	43.7137	51.0207	5.2655
	Northwest provinces	171	43.3704	52.7103	3.9192
	Southeast provinces	223	43.7322	50.9301	5.3378

In 2010, the matching index (μ) between the number of students enrolled in higher education per 10,000 people on the whole country and northwest and southeast provinces of Heihe-Tengchong Line and the proportion of the regional population receiving higher education was 0.1249. By 2010, the matching index was 0.1193, which reduced by 4.4931% compared with 2000. During this period, the matching between the number of students enrolled in higher education per 10,000 people on the whole country and northwest and southeast provinces of Heihe-Tengchong Line and the proportion of the regional population receiving higher education continued to rise, and the gap between the proportion of the regional population receiving higher education and the combination state of economic development significantly narrowed.

(2) *Similarity analysis of talent structure and educational structure on the whole country and both sides of Heihe-Tengchong Line*

From 2000 to 2010, the hierarchical structure of junior college, undergraduate and graduate degrees of the number of students enrolled in higher education nationwide evolved from 0.4761:0.4792:0.0447 to 0.4371:0.5102:0.0527. The proportion of junior college degree level declined, while the proportion of undergraduate degree level and graduate degree level increased. The evolution trend of the junior college, undergraduate and graduate degree level structure of the number of students enrolled in higher education on both sides of Heihe-Tengchong Line was similar to that of the whole country. In 2000, the similarity index (γ) between the education level of students enrolled in higher education on the whole country and northwest and southeast provinces of Heihe-Tengchong Line and the education level structure of regional population receiving higher education was 0.4039. By 2010, the matching index was 0.3188, which was reduced by about 21.06% compared with 2000. During this period, the similarity between the education level of students enrolled in higher education on the whole country and northwest and southeast provinces of Heihe-Tengchong Line and the education level structure of regional population receiving higher education continued to rise. The regional development of higher education service, especially the improvement of regional talent educational degree hierarchy, had obvious effect. The gap between the level structure of regional population with higher education and the combination state of economic and industrial structure significantly narrowed.

6 Main Conclusions and Discussions

Based on the national census data of 1964, 2000 and 2010 and the relevant data of economic and educational development of provinces, the education level and education structure of the population of the whole country and the provinces on the northwest and southeast sides of the Heihe-Tengchong Line are systematically analyzed. The following conclusions are drawn:

- The population of the provinces on northwest side of Hu Huanyong Line in China is more highly educated than that of the provinces on southeast side. The rural population of the provinces on northwest side is much more educated than that of the provinces on southeast side. The degree of higher education of urban population has changed from less than that of the provinces on southeast side in 1964 to more than that of the provinces on southeast side in 2010.

- The evolution rate of "junior college degree to graduate degree" in the population of the northwest provinces of Heihe-Tengchong Line is higher than that of southeast provinces.
- The matching between per capita GDP of the provinces on the northwest and southeast sides of the Hu Huanyong Line and the proportion of the regional population receiving higher education has continued to rise, and the gap between the proportion of the population receiving higher education and the combination state of economic development has narrowed significantly. The similarity between the economic industrial structure and the educational level structure of the regional population receiving higher education has continued to rise, and the gap between the level structure of population receiving higher education and the combination state of economic and industrial structure among regions has narrowed significantly.
- The matching between the number of students enrolled in higher education per 10,000 people in the whole country and the northwest and southeast provinces of the Heihe-Tengchong Line and the proportion of the population receiving higher education in the region has continued to rise, and the gap between the inter-regional proportion of the population receiving higher education and the combination state of economic development has significantly narrowed. The similarity between the educational level structure of the number of students enrolled in higher education and the educational level structure of regional population receiving higher education has continued to rise. The gap between the educational level structure of regional population receiving higher education and the combination state of economic and industrial structure has significantly narrowed.

Acknowledgments. The research was influenced by the National Natural Science Foundation of China, Study on the Time and Space Structure of China's Compulsory Education, Early Warning and Balanced Development Countermeasures System (Project No. 41671148), and Yunnan Province Philosophy and Social Science Planning Project Equilibrium of Educational Resources in the Process of Urban-Rural Integration in Yunnan Province Research" (Project No.: QN2016018).

References

1. Huanyong, H.: The distribution of population in China, with statistics and maps. Acta Geogr. Sinica **2**(2), 33–74 (1935)
2. Dadao, L., Wang, Z., et al.: Academic contention on "whether Heihe-Tengchong Line can break through". Geogr. Res. **35**(5), 805–824 (2016)
3. Qi, W., Liu, S., Zhao, M.: The stability of "Heihe-Tengchong Line" and the difference of population gathering and dispersion patterns on both sides. Acta Geogr. Sinica **70**(4), 466–551 (2015)
4. Huanyong, H.: Population distribution, regionalization and prospect in China. Acta Geogr. Sinica **45**(2), 139–145 (1990)

5. Li, J., Dadao, L., et al.: Spatial differentiation and change of population on both sides of Heihe-Tengchong Line. Acta Geogr. Sinica **72**(1), 148–160 (2017)
6. Yang, Q., Wang, Y., et al.: Analysis of spatial-temporal coupling characteristics of population distribution and socio-economic development in China from 1952 to 2010. J. Remote Sens. **6**, 1424–1434 (2016)
7. Liu, L., Yao, H.: Adaptability evaluation of hierarchical structure of higher education and population employment structure in China. J. Yunnan Normal Univ. (Philos. Soc. Sci. Ed.) **47**(6), 134–139 (2015)
8. Luo, S., Wang, A., Gao, R.: Analyses of factors of high-level falents' migration and significance of establishing anti-selection mechanisms among regions. Scientia Geogr. Sinica **29**(6), 779–786 (2009)
9. Zhang, B., Ding, J.: Spatial distribution and change trend of high- level talent in China, 2000–2015. J. Arid Land Resour. Environ. **33**(2), 32–36 (2019)
10. Zhang, J., Debin, D., Jiang, H.: Regional differences of S&T talents in Jiangsu province. Scientia Geogr. Sinica **31**(3), 378–384 (2011)

Smart Tourism Management Mode Under the Background of Big Data

Leilei Wang[1] and Heqing Zhang[2(✉)]

[1] Guangzhou Panyu Polytechnic, Guangzhou, China
Leileiwang@gzpyp.edu.cn
[2] Sino-French College of Tourism, Guangzhou University, Guangzhou, China
lyzhq8007@gzhu.edu.cn

Abstract. With the rapid development of the Internet of things and communication technology, the arrival of the era of big data, the construction of smart tourism is no longer an impossible slogan, the significance of this paper is to respond to the current development of China's tourism strategic goals, high-quality development of smart tourism industry. The main purpose of this study is to study the management model of smart tourism and promote the development of China's tourism industry. Based on big data era as the background, on the theory of artificial colony algorithm support, elaborated profoundly the big data and the connotation of the wisdom of tourism, by analyzing the wisdom tourism development ideas, to build with the wisdom of the big data platform based on forecast and feedback service platform, and puts forward the implementation wisdom tourist service platform construction mode and path. The research results of this paper show that smart tourism has become the mainstream development mode under the current background, especially the smart tourism supported by big data, which has optimized the allocation of tourism resources, improved the management efficiency and quality of tourism, and promoted the development of China's tourism economy.

Keywords: Big data era · Smart tourism · Artificial swarm algorithm · Tourism resource allocation

1 Introduction

Social development has promoted the arrival of the era of big data information, which has also brought great changes to various industries in China's social and economic development. The arrival of big data has changed the traditional operation and management mode of tourism industry [1]. In the development process of China's tourism industry, a large number of data will be involved in the application of information. If these large amounts of data cannot be timely collected and collated, great disadvantages will be brought to the tourism industry. Thus, tourist experience is affected [2]. Therefore, it is necessary to improve the big data processing of tourism industry, actively adopt modern Internet and high and new technologies, and build a perfect information processing platform.

© Springer Nature Singapore Pte Ltd. 2020
M. Atiquzzaman et al. (Eds.): BDCPS 2019, AISC 1117, pp. 282–288, 2020.
https://doi.org/10.1007/978-981-15-2568-1_38

The combination of big data, cloud computing and other technologies with traditional industries is increasingly close, which promotes the development and transformation of various industries in China [3]. For tourism development in our country, the application of large data effectively change the current tourism business, tourism management difficulty higher and tourism resource configuration is not reasonable, greatly improve the tourism management work efficiency and quality, optimize the tourism resources configuration, steady and healthy development of tourism industry in our country made great contributions in the [4, 5]. Smart tourism in the context of big data makes use of efficient Internet and data platform to carry out tourism management according to customers' needs, enhance customers' satisfaction with tourism experience, and promote the healthy and sustainable development of China's tourism industry [6].

This paper argues that usually involves two main participants in the process of tourism, the tourism information and services provider, tourism consumers, but China's current big data technology development can be achieved should the two parties involved in the course of travel, for online travel companies marketing strategy to provide more accurate solid technical support and data reference [7]. From another perspective, big data network is like the link between tourism supply chain and consumption chain, successfully connecting the two chains. In the process of tourism, there is a very close relationship between all parties. If the characteristics of one part change, different reactions will be made and different information will be given in the whole process, which can be regarded as a driving force. Big data technology can help tourism enterprises understand and apply this rule [8]. In the whole process of tourism and the whole chain, in-depth analysis of data is the best way to collate and integrate enterprise data and more accurately identify the information needed by enterprises and the information helpful to enterprises [9]. Furthermore, it can provide consumers with the information and services they really need, so as to make more potential users involved in the supply chain of tourism services. In addition, tourism suppliers can better understand and penetrate the tourism industry, accurately develop the vast potential customer groups, and thus create rich value [10].

2 Methods

2.1 Swarm Algorithm

Artificial colony algorithm is a kind of swarm intelligence algorithm based on multi-objective function optimization, in recent years have been applied in many ways, this is the first time that the algorithm is introduced into the search for wisdom travel patterns, application, and in theory and application on made adequate preparations, believe that will play an important role. Artificial colony algorithm is a kind of heuristic algorithm, it is produced by the intelligent foraging behavior of bees and nature, natural food is looking for three members of the hive group behavior of the division of labor cooperation, the three bees include hiring reconnaissance bee bee, bee, hire bee according to memory and special dance of the bees for honey and find food and watching the bees share the recent information, watching the bees in hiring near honey bees find the search of a better food source, and random search for a new scout bees honey.

In this study, the use of artificial colony algorithm is based on a single search equation, puts forward a kind of artificial colony algorithm based on double search equation, and apply it in the wisdom of tourism management mode analysis, the income of each local optimal value is the various types of hot issues, the global optimal value that corresponds to the scenic spots in the most successful tourism management mode, if a negative value is the failure model of tourism management. In the population of artificial swarm algorithm, the number of hired bees is equal to the number of watching bees and the number of solutions in the artificial swarm algorithm. During initialization, all food sources are regarded as solution vector RX_i and RN feasible solutions are generated randomly. The position of each feasible solution RX_{ij} can be expressed as follows:

$$RX_{ij} = RX_{k,j} + \beta(RX_{u,j} - RX_{k,j}) \tag{1}$$

In the artificial swarm algorithm, the quantity and quality of nectar sources related to each food source and the fitness value of the correlation solution fit_i correspond to each other. Three kinds of bees search for nectar sources iteratively in the algorithm, and discard the nectar sources that cannot meet the conditions. Otherwise, choose the nectar sources according to the probability of the fitness value of the corresponding nectar sources.

2.2   New Search Equation

In artificial colony algorithm, three kinds of bees are repeating their search process, to find the global optimal, but three bees are in the process of the optimization of local optimum and global optimal balance problems, problems are often fall into local optimum and lead to a long time can't convergence, many researchers have studied, have also made some progress, the artificial colony algorithm is improved representative GABC algorithm, GABC algorithm of artificial colony algorithm improvement mainly reflects in the improvement of search formula:

$$RV_{ij} = RX_{i,j} + \theta_{i,j}(RX_{best,j} - RX_{i,j} + \omega_{i,j}(RX_{i,j} - RX_{k,j}) \tag{2}$$

According to this study of network need emergency prediction based on artificial swarm algorithm combining particle swarm optimization algorithm and on the basis of colony dynamic characteristic of the individuals in the high dimension space, and watching the bees to hire swarm presented a search formula, due to the scout bees is a random search, so scout bees use employed bees search formula proposed in this paper, employed bees and see bee search the corresponding formula is as follows:

$$RV_{i,j} = RX_{i,j} + \rho_{i,j}(RX_{best,j} - RX_{r,j} + \lambda) + \mu_{i,j}(RX_{i,j} - RX_{k,j}) \tag{3}$$

In this improved algorithm, $\rho_{i,j}, \mu_{i,j}$ can ensure that when searching network emergencies with this algorithm, it can quickly search all possible emergencies and obtain the most possible emergencies in the global optimal.

3   Experimental

Based on big data platform of Jia tourism management model, to be able to offer the latest tourist route, the most preferential qiao ticket information, the most reasonable travel advice and detailed travel information. Relying on the intelligent tree platform, the smart tourism management model classifies and integrates tourism related information, and provides tourists with personalized travel routes, pre-purchase tickets, visa services, air tickets and hotels, travel insurance, car hailing, travel records, travel blog writing and other services. Real-time feedback information to the tourists, the current regional tourism traffic, tourism accommodation, tourism catering, tourism scenic spot status quo, tourism flow direction and other information, for tourists' reference, provide practical Suggestions for tourists to adjust.

This article will wisdom tourism management model of the platform on the basis of different interest subjects, divided into four subsystems platform, respectively, for the government tourism doors platform, tourists, tourism enterprise platform and community platform, although different between each platform is only part of the face group, but its implementation are common on the basis of tourism activities of big data.

In terms of the development of smart tourism, this paper will randomly interview some relevant groups, so as to understand the impression and evaluation of smart tourism in consumers' mind under the background of big data. This paper also included the operation status of big data platform into the investigation scope, and planned to take this as an important basis for formulating measures.

4   Discuss

4.1   Problems Existing in Smart Tourism Under the Background of Big Data

(1) Insufficient construction of basic information facilities
     The result of crowd honeycomb algorithm shows that China's smart tourism is still in the initial stage, so the basic information facilities are not perfect. Wireless network is widely used in people's life, and has a huge impact on people's life, many tourist attractions and tourism enterprises have begun to carry out wireless network coverage, but not fully meet the needs of intelligent tourism, there is still a lot of room for improvement. In the process of the experiment, the author randomly interviewed 104 people from different groups and 87 pieces of valid data. The interview results are shown in Table 1 below. Such survey results show that big data still fails to play its maximum role in the smart tourism industry.

(2) The resource value of big data has not been fully explored
     At present, the basic database of smart tourism is not very perfect, and the data cannot be Shared and open, and a large amount of data information cannot be collected, and the advantages of big data cannot be played and recognized. Although in the context of big data, many big data platforms and enterprises begin to attach importance to data information collection and hope to take it as their competitive advantage, the collection, analysis and processing of data are not

Table 1. Interview results on the development of smart tourism

The serial number	Question	Satisfied number	Proportion
1	Are you satisfied with WIFI coverage in the scenic area	45	51.72%
2	Whether the scenic area text slogan translation wisdom	56	64.37%
3	Whether the scenic area functions intelligently	67	77.01%
4	Whether there are more intelligent machines in the scenic area	71	81.61%
5	Whether the parking area intelligent	43	49.43%
6	Whether the scenic area settlement system is intelligent	59	67.82%

timely, which leads to the failure of data to provide better help for smart tourism. In short, the lack of sharing and openness of big data will lead to the fact that the resource value of big data cannot be fully explored, which is not conducive to the modernization of enterprises. Similarly, 23 data management platforms with big data management qualifications were randomly interviewed in this study. Through the interviews, it was found that although big data has invested a lot in social networking, transportation, education and medical industry, smart tourism has not aroused great interest. The interview results on big data platforms are shown in Fig. 1 below.

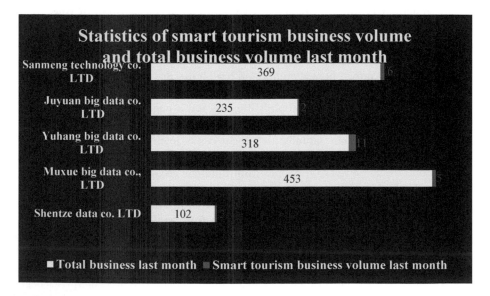

Fig. 1. Statistics of smart tourism business volume and total business volume of each big data platform last month

4.2 Development Path of Smart Tourism Under the Background of Big Data

(1) Build a complete intelligent service system with the help of big data

Tourism, as the core service industry in China, has made great contributions to China's economic growth and the construction of social and cultural system. Smart tourism is the inevitable trend of China's tourism development under the background of big data. In order to meet the needs of tourists and enhance their tourism experience, it is necessary to establish a perfect intelligent service system with the help of big data to publicize and manage the network information platform. Tourists can book travel and choose services on the Internet, and tourism enterprises can also provide corresponding services according to tourists' preferences to meet their personalized needs.

(2) Strengthen the construction of smart business system in smart tourism

The rapid development of tourism also promotes China's economic development and adjustment of social and economic structure. With the help of big data, the integration of smart tourism services is established to strengthen the connection between tourism enterprises and hotels, scenic spots and transportation, so as to fully explore the economic efficiency in tourism. Strengthen the establishment of intelligent business system, ensure the efficient operation of the tourism management system, reduce the cost of tourism management, select the best solution in the integrated catering and entertainment services, and conduct reasonable data screening and analysis. In addition, the business system of smart tourism can be adjusted according to tourists' psychology, so as to meet tourists' service demands and increase tourism economic benefits.

(3) Promote the intelligent management level of tourism management enterprises

With the rapid development of tourism in our country, the number of visitors has increased dramatically, and brought great pressure to tourism management, in order to enhance the effectiveness of tourism management, improve the satisfaction of tourists, need to strengthen the tourism management wisdom, wisdom through tourism management to ensure the quality and level of tourism service, timely adjust the problems appeared in the process of tourism services, strengthen the improvement of the safety management system. According to the information and data access in the intelligent management system, we can understand the needs of tourists, so as to adjust the original tourism management system accordingly and innovate the tourism management system synchronously.

5 Conclusion

Based on the background of the era of large data, elaborated profoundly the big data and the connotation of the wisdom of tourism, on the theory of the swarm algorithm support and more targeted optimization, by analyzing the wisdom tourism development ideas, to build with the wisdom of the big data platform based on forecast and feedback service platform, and puts forward the implementation wisdom tourist service platform construction mode and path. The research of this paper shows that smart tourism supported by big data can better meet people's demands for tourism activities, efficiently allocate tourism resources and enhance people's tourism experience. At the same time, the tourism enterprises should seize this opportunity, through constructing a high level of data platform, the implementation of intelligent management model, improve the tourism enterprise's management and service level, to achieve the purpose of tourism management wisdom, to better meet the needs of tourists, let the people really feel under the background of big data for the development of tourism industry, promote the development of wisdom tourism better.

References

1. Ruíz, M.A.C., Bohorquez, S.T., Molano, J.I.R.: Colombian tourism: proposal app to foster smart tourism in the country. Adv. Sci. Lett. **23**(11), 10533–10537 (2017)
2. Perfetto, M.C., Vargassánchez, A.: Towards a smart tourism business ecosystem based on industrial heritage: research perspectives from the mining region of Rio Tinto, Spain. J. Herit. Tourism **13**(3), 1–22 (2018)
3. Jovicic, D.Z.: From the traditional understanding of tourism destination to the smart tourism destination. Curr. Issues Tourism **12**(5), 1–7 (2017)
4. Tripathy, A.K., Tripathy, P.K., Ray, N.K., et al.: iTour: the future of smart tourism: an IoT framework for the independent mobility of tourists in smart cities. IEEE Consum. Electron. Mag. **7**(3), 32–37 (2018)
5. Bryson, M., Castell, S.D.: Queer pedagogy: praxis makes im/perfect. Can. J. Educ. **18**(3), 285–305 (2017)
6. Wu, M.Y., Huang, K.: Appraising netnography: its adoption and innovation in the smart tourism era. Tourism Tribune **29**(12), 66–74 (2017)
7. Mekawy, M.A., Croy, G.: Smart tourism investment: planning pathways to break the poverty cycle. Tourism Rev. Int. **18**(4), 253–268 (2015)
8. Boes, K., Buhalis, D., Inversini, A.: Smart tourism destinations: ecosystems for tourism destination competitiveness. Int. J. Tourism Cities **2**(2), 108–124 (2018)
9. Han, D.I.D., Dieck, M.C.T., Jung, T.: Augmented Reality Smart Glasses (ARSG) visitor adoption in cultural tourism. Leisure Stud. **4**, 1–16 (2019)
10. Choi, I.Y., Ryu, Y.U., Kim, J.K.: A recommender system based on personal constraints for smart tourism city. Asia Pacific J. Tourism Res. **6**, 1–14 (2019)

Optimization of Support Capability Under Emergency Conditions

Ou Qi[1(✉)], Lei Zhang[2], Wenhua Shi[3], and Yanli Wang[4]

[1] Army Academy of Amored Forces, Changchun, China
haikuotiankongru@163.com
[2] Changchun Military Representative Office of Shenyang Military
Representative Bureau of Ground Force, Changchun, China
[3] The Army of 95795, Guilin, China
[4] PLA 32256 Troops, Beijing, Yunnan, China

Abstract. China is a country with frequent natural disasters, and the army is an important force to ensure the safety of people's lives and property. Non-war military action is a kind of activity in which military forces are reasonably used to safeguard people's life and property security under sudden disasters without changing the basic structure and functions of the army. It mainly includes anti-terrorism and stabilization operations, rescue and disaster relief operations, emergency prevention operations, peacekeeping operations, urban rescue operations and so on.

Keywords: Support capability · Emergency condition · Optimization

1 Introduction

The sudden outbreak of non-war military operations has strong uncertainty in time and region [1]. In order to save costs, less equipment is stored in peacetime, and a large number of emergency equipment is temporarily raised when disasters occur [2–5]. It is the main body of support equipment for non-war military operations. Emergency equipment has the following characteristics: (1) Most of the equipment belongs to non-organizational equipment, which is not uniform in type, poor in versatility and difficult to implement support; (2) the intensity of loss in the early stage of the mission is high and the pressure of support is high; the amount of loss in the late stage of the mission is sharply reduced, and the amount of equipment used in the later stage of the operation is small; (3) the difference of peacetime and wartime management is large, and the source of equipment is large. Dispersion, spare parts delivery pressure is high.

Emergency financing equipment, as the main equipment for non-war military operations, its support efficiency is very important for the implementation of Non-war Military operations, which is a complex issue involving support personnel, equipment and command [6–9]. In order to effectively respond to non-war military operations, each grass-roots unit has a special emergency equipment support capacity training program [10–13]. Whether the scheme is reasonable or not, and whether it can meet the requirements of Non-war Military operations, how to evaluate is an urgent problem to

© Springer Nature Singapore Pte Ltd. 2020
M. Atiquzzaman et al. (Eds.): BDCPS 2019, AISC 1117, pp. 289–292, 2020.
https://doi.org/10.1007/978-981-15-2568-1_39

be solved. Index optimization is the key link to realize the evaluation of equipment support capability.

Factor analysis, through the analysis of the relationship between potential variables and explicit variables, can achieve the removal of correlation between factors, and keep the indices independent or relatively independent. Structural equation modeling uses a set of multivariate analysis processes, which can stabilize the evaluation results. In order to solve the problem of unstable evaluation results caused by the independent relationship of evaluation factors in current research, factor analysis and structural equation modeling method are used to model and analyze the equipment raised by non-war military operations in emergency, and index system is established to provide technical support for the implementation of Non-war Military operations.

2 Index System Construction

The indicators of completeness must cover the main key factors that affect the supportability of equipment. The lack of key factors that affect the supportability should not lead to too large deviation between the evaluation results and the actual results under the condition that the evaluation methods and data are correct.

Indicators in the self-consistency system should constitute a whole describing the evaluation objects. Each evaluation sub-index supports each other and completes the evaluation of the evaluation objects. The contradictions among the indices in the system should be avoided.

Feasibility evaluation index must be available in practice, and evaluation can be carried out in practical application. Especially, it is necessary to avoid the construction of index system being too large, and the evaluation results are difficult to meet the consistency test, which leads to the evaluation in "data acquisition-sub-index calculation-consistency test-data acquisition". It runs in a dead cycle.

On the basis of equipment support index system for military operations in war and considering the characteristics of emergency equipment for non-war military operations, a preliminary index system for emergency equipment support for military operations in Non-war based on personnel support capability, operational support capability and equipment utilization capability is constructed to improve the description accuracy of the index system.

3 Model Initialization

The support capability of emergency equipment in non-war military operations is the general objective of index optimization, the purpose of optimization is to make the new index have better support for the overall objective; the second level index is the sub-objective under the overall objective, but the variable can not be obtained by observation experiment, and it is the potential variable of the system; the third level index can be seen through observation. The process of optimization is to analyze the relationship between potential variables and explicit variables of the system, and put forward the essential factors of the system based on the data given by experts. Assume

$$X' = (X_1, X_2, \ldots, X_p) \tag{1}$$

is $P * 1$ vector, the mean of which is

$$\mu' = (\mu_1, \mu_2, \ldots, \mu_p) \tag{2}$$

Then the relationship between variables can be written as follows:

$$\underset{(p \times 1)}{X} = \underset{(p \times 1)}{\mu} + \underset{(p \times m)(m \times 1)}{AF} + \underset{(p \times 1)}{\varepsilon} \tag{3}$$

Error Elimination Ability for Non-War Military Operations:

$$X_6 = 0.114u_{27} + 0.312u_{28} \tag{4}$$

Factor analysis condition setting. By calculating the correlation between potential variables and explicit variables, the system is de-redundant and dimension-reduced, and the common factors of high-correlation variables are extracted, and the essential factors of the system are composed of relatively independent variables.

Setting up the condition of structural equation modeling. Through structural equation modeling, the explanation variance between the essential factors and the original variables of the system is calculated to verify the support degree of the essential factors to the system. If the explanation variance is too low, it indicates that the redundancy and dimensionality reduction are excessive, factor analysis needs to be carried out again; the correlation between the essential factors is calculated, and the essential factors with too high correlation are introduced. One step is factor analysis to achieve decorrelation and ensure the independence of the essential factors.

4 Conclusion

Aiming at the problem of insufficient quantification of the index of emergency equipment support capability in Non-war Military operations, based on the evaluation index of equipment support capability in war military operations and combined with the characteristics of emergency equipment support in Non-war Military operations, factor analysis is carried out on the index of emergency equipment support capability. Common factor extraction, factor load determination and index are adopted. Weight normalization establishes the weight of the new index system. By explaining variance analysis, the comprehensiveness of index selection and the accuracy of weight determination are verified, which solves the problem that the validity of traditional index system is difficult to verify.

References

1. Kumar, R., Kumar, M., Singh, A., Singh, N., Maity, J., Prasad, A.K.: Synthesis of novel C-4'-spiro-oxetano-α-L-ribonucleosides. Carbohydr. Res. **445**, 88–92 (2017)
2. Salame, P.H., Prakash, O., Kulkarni, A.R.: Colossal dielectric response of Mott insulating, nanocrystalline, T'-type Sm2CuO4 ceramics. Ceram. Int. **43**(16), 14101–14106 (2017)
3. Durey, A., Kang, S., Suh, Y.J., Han, S.B., Kim, A.J.: Application of high-flow nasal cannula to heterogeneous condition in the emergency department. Am. J. Emerg. Med. **35**(8), 1199–1201 (2017)
4. Purkayastha, A., Frontera, A., Ganguly, R., Misra, T.K.: Synthesis, structures, and investigation of noncovalent interactions of 1,3-dimethyl-5-(4'/3'-pyridylazo)-6-aminouracil and their Ni(II) complexes. J. Mol. Struct. **1170**, 70–81 (2018)
5. Jokisaari, A.M., Naghavi, S.S., Wolverton, C., Voorhees, P.W., Heinonen, O.G.: Predicting the morphologies of γ' precipitates in cobalt-based superalloys. Acta Mater. **141**, 273–284 (2017)
6. Ashraf, A.R., Akhter, Z., Simon, L.C., McKee, V., Dal Castel, C.: Synthesis of polyimides from α, α'-bis(3-aminophenoxy)-p-xylene: spectroscopic, single crystal XRD and thermal studies. J. Mol. Struct. **1160**, 177–188 (2018)
7. Amirkhani, A., Papageorgiou, E.I., Mosavi, M.R., Mohammadi, K.: A novel medical decision support system based on fuzzy cognitive maps enhanced by intuitive and learning capabilities for modeling uncertainty. Appl. Math. Comput. **337**, 562–582 (2018)
8. Xiao, S., Zhang, X., Zhang, J., Wu, S., Wang, J., Chen, J.S., Li, T.: Enhancing the Lithium storage capabilities of TiO 2 nanoparticles using delaminated MXene supports. Ceram. Int. **44**, 17660–17666 (2018)
9. Subasi, O., Di, S., Bautista-Gomez, L., Balaprakash, P., Unsal, O., Labarta, J., Cristal, A., Krishnamoorthy, S., Cappello, F.: Exploring the capabilities of support vector machines in detecting silent data corruptions. Sustain. Comput.: Inf. Syst. **19**, 277–290 (2018)
10. Järvenpää, E., Lanz, M., Siltala, N.: Formal resource and capability models supporting re-use of manufacturing resources. Procedia Manuf. **19**, 87–94 (2018)
11. Balis, B., Brzoza-Woch, R., Bubak, M., Kasztelnik, M., Kwolek, B., Nawrocki, P., Nowakowski, P., Szydlo, T., Zielinski, K.: Holistic approach to management of IT infrastructure for environmental monitoring and decision support systems with urgent computing capabilities. Future Gener. Comput. Syst. **79**, 128–143 (2018)
12. Adamides, E., Karacapilidis, N.: Information technology for supporting the development and maintenance of open innovation capabilities. J. Innovation Knowl. (2018)
13. Getnet, H., O'Cass, A., Ahmadi, H., Siahtiri, V.: Supporting product innovativeness and customer value at the bottom of the pyramid through context-specific capabilities and social ties. Ind. Mark. Manag. **83**, 70–80 (2018)

Coordinated Development of Coal and Water Resources Based on Big Data Analysis

Yu-zhe Zhang, Xiong Wu[✉], Ge Zhu, Chu Wu, Wen-ping Mu,
and Ao-shuang Mei

School of Water Resource and Environment, China University of Geosciences,
Beijing 100083, China
wuxiong@cugb.edu.cn

Abstract. Human beings have entered the stage of big data in the information age. How to use big data to serve all walks of life is a very worthy of discussion. This article thought that the mine well under - ground real-time monitoring data and mining production, living and ecological water demands as the research object, from the large data set up, management, analysis, and maintain the four aspects, combining the reality of mine work, using the large system decomposition-coordination principle and fuzzy hierarchy comprehensive evaluation method, the multi-objective dynamic programming model for non-linear iterative simulation optimization, determine coal water - coordinate the development of new pattern and the underground water source water coal mining technology, mine Wells under - ground multi-objective efficient mixing technology, realizing the accurate, intelligent mine water utilization. Research method, this paper USES "Knowledgediscovery from GIS" and "Simonett rubik's cube" concept, cyber GIS method is applied to build the mine Wells under - ground real-time monitoring to the big data is discussed, at the same time points out the characteristics of the coal mining situation of real-time, and possible problems are derived, which provide reference for other mines to establish large data.

Keywords: Big data · Coal water resources · Mining production · Cybergis method

1 Introduction

From pre-inspection, survey, exploration to mining of large-scale mines, it usually takes more than 10 years. With the passage of time, the change of understanding, the update of technology, the alternations of ideas and the development of engineering, the participation of different industries will accumulate a large amount of data. If traditional concepts and methods are used to manage, analyze and maintain these data, it is already out of step with the development of The Times and cannot be well served for mine development and comprehensive research [1, 2]. How to establish, manage, analyze and maintain big data is a question worth discussing. In addition, after decades of wild pursuit of rapid economic development in the era of resource indulgence, China's large mine enterprises are gradually transforming from resource development to

© Springer Nature Singapore Pte Ltd. 2020
M. Atiquzzaman et al. (Eds.): BDCPS 2019, AISC 1117, pp. 293–299, 2020.
https://doi.org/10.1007/978-981-15-2568-1_40

comprehensive energy suppliers, which also brings problems to the industry [3, 4]. Large domestic mine enterprises have been faced with many practical problems, such as extensive and scattered traditional management mode, complex and volatile market situation, as well as the increase in scale and volume brought by merger and reorganization. It is urgent to find a breakthrough for upgrading in the aspect of resource standard management innovation [5].

Since human beings have entered the stage of big data in the era of information explosion, any discipline and industry can no longer develop in isolation. Their development is intermingled with a large number of other disciplines and majors, and blended with a variety of previously seemingly unrelated information [6]. It can be predicted that big data will be the new driving force of mine comprehensive information analysis in the next 10 years. Mine comprehensive information analysis will rely more on real-time acquisition from complex and changeable data streams, and more mine comprehensive information analysis will also rely on high-performance computing to become real-time and timely [7]. Some researchers even believe that big data is the "fourth scientific paradigm" after experimental research, theoretical derivation and computational simulation, which will significantly accelerate and promote the pace of innovation in science and engineering [8].

For large data into the mine development and management practice of thinking, to face with some text, through the application of large data intelligent analysis platform and the large system decomposition-coordination principle, forming underground - ground, quality coupling, real-time tracking and adaptive adjustment of forecast and early warning, dynamic allocation, the multi-objective management efficient intelligent module, forming mine Wells under - ground of multiobjective intelligent deployment of technology; A nonlinear multi-objective coal-water dual resource regulation model with dynamic replacement of indicators was established to form a whole-life cycle regulation technology for coordinated development of coal-water [9, 10].

The study in this paper draws the corresponding logging curve by collecting the data needed in the drilling hole to determine the spatial position of the drilling hole and divide the lithology stratification around the drilling hole, so as to analyze the distribution and reserves of coal seam. For the present status of the mine mining technology in China, it is far less than foreign countries, this topic is developed in this context, the development of coal water logging system has certain reference significance, through the research and application of laptop, expect to mine comprehensive development and utilization of coal, mining industry and make a certain contribution.

2 Research Methods

2.1 Big Data Technology Theory

Big data refers to the technology of obtaining, analyzing and processing massive data to obtain valuable information. Through analyzing and processing valuable information, it can provide enterprises and organizations with more effective and accurate decision-making ability. Big data technology is not to use random sampling method for processing, but to analyze and process all the data. It has the following characteristics:

large data volume; Fast processing speed; Various data types; Low value density; A wide range of applications. The "big" of big data originally means a large number and complex types. Therefore, it is particularly important to obtain data information through various methods.

2.2 Big Data Thinking Framework

Combined with the characteristics of mine mining, the idea of establishing a coordinated development system of coal-water mine with the help of big data is to adhere to the principles of accuracy, science, classification and rationality. Identify the elements that need to be collected; Scientific selection of data elements collection means; Reasonable arrangement of data collection procedures. The principle of big data management is to adhere to the number, choice, security and innovation; All kinds of data elements need to be digitized and vectorized. Select the collected data and select the best; All kinds of data collection and transmission equipment should be arranged reasonably and information exchange should be safe and reliable. Where conditions permit, the cloud service model should be actively introduced. Big data analysis is to adhere to the principles of science, rigor, economy and speed; Select software that processes big data to select existing software as far as possible to avoid waste and save costs; According to the actual situation, the establishment of big data maintenance system and norms, in addition, the big data classification, to achieve order.

2.3 Cybergis Algorithm Technology

The utilization process of big data is a process of separating the wheat from the chaff, mining valuable information from the huge data. The value of big data density is low, the data is not its value, only the data of business information excavation is the value of big data, big data technology and artificial intelligence algorithm is combined with the trend of the development and application technology of data, efficient modeling algorithm is the carrier of technology application about big data, the algorithms for mining the most commonly used is cyber GIS algorithm, this method can from different angles the data mining. The GIS algorithm reflects the characteristics of the attribute values of the data in the database, and finds the dependencies between attribute values by expressing the relationship of the data mapping through functions. The functional relationship of GIS algorithm can usually be expressed as follows:

$$b_{sn} = \frac{123}{\eta_b \eta_i \eta_m \eta_\delta \eta_p (1 - \sum \xi_1)} \tag{1}$$

Large system decomposition coordination principle and fuzzy hierarchy comprehensive evaluation method, the multi-objective dynamic programming model for nonlinear iterative simulation optimization, determine coal water - coordinate the development of new pattern and the underground water source water coal mining technology, mine Wells under - ground multi-objective efficient mixing technology, realizing the accurate, intelligent mine water utilization. In addition, the parameters with high sensitivity to energy consumption can be determined through the selection of parameters, and the controllable operation parameters can be optimized. The GIS algorithm can be optimized through the following function equation:

$$\eta_b = 1 - (q_2 + q_3 + q_4 + q_5 + q_6)\% \tag{2}$$

Through the overall thinking framework of big data and the integration and optimization of GIS algorithm, based on the Internet of things + cloud technology, a multi-level and layered intelligent monitoring system of integration of hydrology, geology, environment and ecology of tian-earth-well interaction in the mining area was formed, and the coordinated monitoring of multiple indicators in the mining area was also realized.

3 Experimental

Mining data collection method must be scientific, reasonable and rigorous, in order to achieve twice the result with half the effort. Data collection mainly relies on GNSS (beidou, GPS and GLONASS) to collect first-hand data, determine the coordinate system and leveling model to be used in the mining area, and establish the three-dimensional data relationship of the mining area. After field data collection, indoor data processing is carried out, and the data is adjusted to establish complex graphics of layers, information and elements. GPU is used for processing, and then the cyber GIS with more inclusive and computing capacity is established.

Its concrete steps are generally, the first is for a variety of data collection, carding and comprehensive analysis, narrowing the prospecting target areas, key target areas for geochemical exploration work, according to the geochemical exploration work of information, on the appropriate key sections of geological engineering construction and validation of geophysical-geochemical anomaly, after the completion of the construction of geological engineering logging, the collection of samples for chemical analysis, according to the analysis result, geography relations, rock name and characteristics, mineralization alteration information such as a comprehensive, all above constitute the original big data of yangshan mine area. Secondly, with the deepening of mine work, according to the time information, we arrange and classify various data, discriminate and summarize new discoveries and new understandings, so as to build the current big data of mine. The composition of big data in mining areas should also pay attention to the following three points: first, the premise of using the above data is to select the collected data, eliminate the data with large errors, in a word, to ensure the accuracy of the data; Second, all the data are classified and selected to form the original big data. We must pay attention to the corresponding relationship between the original big data and time, pay attention to the generation of time-varying data, and then form the final big data of mine resources according to the integration of time and space relations. Third, according to mine Wells under - ground real-time monitoring data and dynamic mining production, living and ecological water demand, the application of large data intelligent analysis platform and the large system decomposition-coordination principle, later we will use big data in real time to monitor coal mine exploitation, so as to prevent the mine water disasters or other because of geological and hydrological disasters caused by mining, should be to install a lot of intelligent monitoring equipment, various data real-time transmission, building and mine water transport, store and control function of comprehensive monitoring system and the intelligent analysis platform and mine water monitoring system.

4 Discuss

4.1 Possible Problems of Big Data in Mining Area

Mine big data is a prospective study area, there is no previous experience, the whole construction idea using GIS and Simonett rubik's cube two theoretical concept as a guide, method as far as possible based on the existing equipment and software, can be combined with the GPU graphics and cyber GIS two new methods, in the framework of such a large, we can deduce about possible problems. First, because of the space in the real world is more than a parameters, nonlinear, time-varying instability of the system, all kinds of data, especially the uncertainty of spatial data is an unavoidable question, especially since we now use two traditional methods of statistics and induction, more practical determination of stability of the two methods are shown in Table 1 below, these factors have caused the big data uncertainty, but also want to see, this is the uncertainty about the challenge of making we need continually to mining the value of big data and correct it. It is in this process of constant self-denial that the development of big data of coal water resources in mining areas is promoted. Second, the restriction of collection methods leads to the accumulation of big data errors. The method and equipment of data collection will cause errors, which is an objective fact and cannot be avoided. Therefore, we need to find a new model and a new method to eliminate the accumulation of big data errors as much as possible. Third, network security. The storage and transmission of big data need to use the network or cloud client, which needs to ensure the security of the network but cannot make the network isolated.

Table 1. Stability test results of the two methods

	Ability to predict		Training efficiency		Modelling efficiency
	Predictive value	Output value	Degree	Time consuming	
GIS	5.8	5.27	6	38 s	91%
Simonett	5.5	3.61	37	76 s	66%

In the test of the dip Angle and azimuth Angle of the sampling frequency, we used the host computer software to receive the data and display the corresponding test value, and compared the theoretical value and test value. This method can also test the accuracy of the system. For the calibration test of inclination Angle of the rapid inclinometer, the theoretical inclination Angle of the inclinometer was measured from $0\,°C$ to $90\,°C$, and 10 sets of data were collected. The specific measurement data and error data analysis are shown in Fig. 1 below.

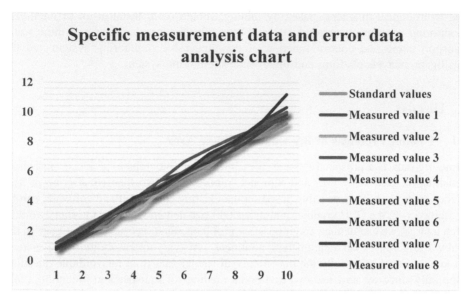

Fig. 1. Analysis diagram of specific measurement data and error data

4.2   Resource Coordinated Development Target Value Optimization

The actual operation of mine big data is an extremely complex thermal system. The change of each parameter of the unit will affect the consumption of the unit and have a significant impact on the economic benefits of the mine big data. When the coal-water unit is in stable operation, it is in the best working condition and its parameters reach the optimal value under the current operating conditions, which is conducive to reducing the coal consumption of the unit and improving the operating economy of the unit. Therefore, it is difficult and key to improve the coordinated development of mine resources to reasonably control the parameters of coal water resource unit under the guidance of big data and optimize them.

Use of big data modeling technology to analyze energy consumption characteristics of coal water, from the mass production run data mining unit running state and the relationship between the parameters of avoiding the traditional complex analysis and calculation, the mechanism of forecast for coal water consumption, the development efficiency of coal water was realized the online monitoring, and through the large data model to optimize the unit operation condition, so as to reduce the effect of the energy consumption of unit.

5   Conclusion

No matter how big the data is or how many types there are, everything in the world has its internal rules and evolution process. We should summarize, sort out, select and analyze the observed data, and use the evolved data to analyze and grasp the rules of things. Big data applications will significantly promote the development of information technology and application of mine mine big data more need to process the data space, liquidity, diversity caused by many factors such as mixing big data complexity, in the future by computing, visualization, and analysis and other methods to reduce the data dimension of research is still full of challenges.

Acknowledgements. This work was supported by National Key R&D Program of China (No. 2018YFC0406400)

References

1. Oparin, V.N., Kiryaeva, T.A., Potapov, V.P.: Methods and models for analyzing methane sorption capacity of coal based on its physicochemical characteristics. J. Min. Sci. **53**(4), 614–629 (2018)
2. Han, L., You, S., Zhang, H., et al.: Modelling of AQI related to building space heating energy demand based on big data analytics. Appl. Energy **203**(109), 57–71 (2017)
3. Li, F., Ran, L., Zhang, Z.: Big data analytics for flexible energy sharing: accelerating a low-carbon future. IEEE Power Energy Mag. **16**(3), 35–42 (2018)
4. Tan, P.L., Koh, E.: Situating learning analytics pedagogically towards an ecological lens. Learn. Res. Pract. **3**(1), 1–11 (2017)
5. Qiang, W., Li, R.: Decline in China's coal consumption: an evidence of peak coal or a temporary blip. Energy Policy **108**(712), 696–701 (2017)
6. Wang, X., Hu, H., Jia, H., et al.: SVM-based multisensor data fusion for phase concentration measurement in biomass-coal co-combustion. Rev. Sci. Instrum. **89**(5), 055–106 (2018)
7. Lu, J., Yun, W., Chen, J.: Detection of tectonically deformed coal using model-based joint inversion of multi-component seismic data. Energies **11**(4), 34–41 (2018)
8. Yang, Z., Gao, X., Sheng, C.: Impact of co-firing lean coal on NO X emission of a large-scale pulverized coal-fired utility boiler during partial load operation. Korean J. Chem. Eng. **34**(4), 1–8 (2017)
9. He, P.F., Kulatilake, P.H.S.W., Yang, X.X., et al.: Detailed comparison of nine intact rock failure criteria using polyaxial intact coal strength data obtained through PFC 3D simulations. Acta Geotech. **6**, 1–27 (2017)
10. Hou, W.: Identification of coal and gangue by feed-forward neural network based on data analysis. Int. J. Coal Prep. Util. **39**(1), 1–11 (2017)

On Diversified Teaching Mode of Comprehensive English in Big Data Era

Yongjia Duan[✉]

Department of Applied Foreign Language, Chengdu Neusoft University,
Chengdu, Sichuan, China
duanyongjia@nsu.edu.cn

Abstract. With the tremendous changes in students' learning styles brought about by the big data the traditional teaching mode of Comprehensive English cannot fully meet the needs of students. Diversified teaching mode, driven by big data, which integrates Comprehensive English with modern information technology, meets the requirements of the new era. Under the current background of big data, teachers should pay attention to the integration of big data and students' individualized education. Through designing teaching from the perspective of students, teachers must create diversified teaching methods and properly organize diversified teaching activities, so as to improve the effectiveness of Comprehensive English teaching.

Keywords: Big data · Comprehensive english · Diversification · Effectiveness

1 Introduction

With the rapid development of information technology and the continuous progress of Internet technology, cloud services, cloud computing and cloud storage of big data have penetrated into all aspects of people's daily life, learning and work, changing people's lives in different aspects tremendously. As a core and major course for English majors, Comprehensive English should meet the needs of the new era by eliminating some outdated teaching methods and modes, and applying big data to teaching practice. In Comprehensive English, teachers can use the big data platform to dig deeply into the teaching mode of Comprehensive English and carry out diversified teaching according to the individual needs of students. Therefore, how to integrate big data technology into Comprehensive English teaching, to further understand students' English learning needs, to realize the close combination of teaching and learning, to improve the efficiency of the teaching of Comprehensive English, and to construct a diversified Comprehensive English teaching mode under the background of big data era to solve the problems existing in the traditional teaching, to further enhance students' learning interest, subjective initiative and autonomous learning ability, all which are the important issues to be solved in teaching of Comprehensive English.

© Springer Nature Singapore Pte Ltd. 2020
M. Atiquzzaman et al. (Eds.): BDCPS 2019, AISC 1117, pp. 300–307, 2020.
https://doi.org/10.1007/978-981-15-2568-1_41

2 The Background of Big Data Era and Overview of Diversified Teaching Mode

Big data is mainly composed of a large number of data collection containing many types and complex structure. It can predict the development trend or law through search, processing, analysis, induction and summary [1]. With the help of data processing and application mode of cloud platform and cloud computing, big data can be integrated, shared and cross-multiplexed to form an aggregation of intellectual resources and knowledge services. Its data processing ability is strong, not only has the characteristics of large data processing volume, but also has a variety of data types and high speed. Through the collection, processing, analysis and digging of data, the essence and connotation of the development of things can be examined and the core things with new values can be acquired.

2.1 The Impact and Challenge of Big Data Era on College English Teaching

The arrival of the era of big data has brought about an unknown impact on the traditional education system. At the same time, it will become a direct driving force for the reform of the current education mode and teaching methods. In the traditional college English teaching, the teacher-centered teaching mode is mostly adopted, which means the content transmission and information transmission in books are usually emphasized. Even nowadays when multimedia-assisted teaching has been popularized in College English teaching, a lot of teachers just use the equipment superficially to transfer and present the contents from blackboard electronically, but they haven't changed teaching in essence.

Big data itself has the characteristics of comprehensiveness, and its data volume is also very powerful. Educators are required to find the required resources from the vast amount of information resources, and to find the connections and impacts among them, so as to analyze the problems and shortcomings in the current education work and effectively improve the effectiveness and individualization of College English teaching. Therefore, there is still a long way to go for the reform and development of College English teaching in the context of big data, which also poses great challenges to college English teachers and students. Firstly, the teachers themselves have traditional and conservative thinking, which, to some extent, makes that they have a certain resistance to change and emerging things, so they cannot accept and adapt all these in a short time. At the same time, some teachers cannot make good use of big data platform to combine their excellent teaching ideas and teaching methods with big data devices and platforms. Secondly, College English teaching in the big data environment involves the renewal and maintenance of equipment. It's possible that colleges and universities don't have enough financial support to upgrade equipment in time and improve their own infrastructure construction. As a consequence, even though they are in the big data environment, they still cannot use the big data platform for teaching and learning, which blocks the breadth and depth of students' learning. Finally, teachers and students' ability to identify massive data information cannot fully meet the needs of the era of big data. Big data provides us with massive information, but not all of them are effective, or useful. We must screen data and information effectively based on the characteristics of the course, so as to avoid being misled by invalid information and spam information.

2.2 The Necessity of Implementing Diversified Teaching Mode in Comprehensive English Course Under the Background of Big Data Era

Big data is a hot word in today's technology and academic field. With the increasing popularity of the Internet and various mobile terminals, it has a tremendous impact on social development and human development. In education, traditional teaching and learning modes are also facing innovation because of the information storm brought by big data. College English teaching is now in the era of big data, and naturally it is impossible to stay out of the changes of the times. Comprehensive English has become a major core course for English majors with its long teaching hours, long span and wide contents. The course directly determines the students' English foundation. As the core course of English majors, it must change with the times. Open-source course platform, MOOC, online sharing courses, the development of various intelligent software and the continuous development of network technology at home and abroad all provide English majors with a ubiquitous learning environment, learning atmosphere and learning resources, which, at the same time, also influence their way of thinking and learning methods implicitly.

The new national quality standard for English majors embodies the people-oriented principle, highlighting the comprehensive quality training and individualized development of students. After 2010, with the globalization of economy and diversification of demand, the society urgently needs diversified talents. Professor Peng Qingdong pointed out that the View of Diversified Talents is the logical starting point for exploring the reform and innovation of English teaching for English majors and formulating national quality standards [2]. Since then, English majors have entered an era of diversified and multi-level innovative foreign language talents training mode. Under the background of big data, with the implementation of the new national quality standard for English majors, the pilot project of comprehensive reform for English majors has been widely launched in China. In the reform of Comprehensive English teaching, how to optimize the teaching mode of teachers and the learning mode of students has become an unavoidable problem in the reform of Comprehensive English teaching.

3 Measures for Innovative and Diversified Development of Comprehensive English Teaching Mode in the Big Data Era

3.1 Creating Diversified Teaching Methods and Organizing Diversified Teaching Activities Properly

The View of Diversified Talents emphasizes the cultivation of students' multiple abilities to create Compound Applied Talents to better meet the needs of society. Diversified teaching forms can break the traditional classroom teaching mode, strengthening the individual needs of students, which are more conducive to stimulate students' interests and dig students' potential. As a result, students' multiple abilities can be cultivated. Using the statistic result of big data, the teaching methods that students are interested in can be found out, so as to increase the ways and resources of students' interest in the teaching process, further enrich the corresponding knowledge, and help students better understand and consolidate the teaching content [3]. In the classroom teaching of Comprehensive English, teachers should change the dull cramming teaching method to construct a student-centered class by using diversified teaching methods, taking individualized development of students as the direction, teaching students according to their aptitude.

Under the current big data background, teachers can use information technology and equipment, such as multimedia devices, personal computers, learning platforms and personal mobile terminals, and use a variety of teaching methods to fully apply students' sensory feelings such as hearing and vision based on the teaching content and students' actual level, not only to make good use of class time, more importantly, to make students fully use the fragmented time after class [4]. All aspects of teaching activities in Comprehensive English should emphasize the combination with new media, breaking the dull traditional class mode [5]. For example, the teaching organization is not limited to traditional classes, and group teaching and group cooperative learning can be adopted [6–9]. At the same time, combined with modern technology, the combination of Online and offline teaching can also be adopted. In this way, all the small time fragments can be utilized to guide students' learning. From teachers' perspective, constant renewal of teaching concept and integration of practicability and innovation into teaching are very effective way to help teachers keep pace with times, which is also helpful for focusing on cultivating students' self-learning ability and initiative learning consciousness to inspire students' innovative consciousness [10]. Teachers can use multimedia to assist teaching, and add information elements to design questions, which makes the classroom more lively and stimulates students' enthusiasm for learning. After class, teachers interact with students through social media such as QQ, Wechat, Blog or network platforms such as electronic reading room and teaching platform to answer questions and solve students' questions. The alternation of classroom teaching and after-class interaction can strengthen and complement the teaching, and effectively cultivate students' language ability and language application ability.

3.2   Integrating English Teaching Resources and Developing Curriculum-Related Diversified Teaching Resources

The arrival of the big data era has brought great impact on College English teaching. In traditional English teaching, teachers occupy the leading position in the classroom. In the big data era of education, students show a trend of diversified development. In terms of knowledge acquisition, students can use the network platform to solve various problems and difficulties encountered in the learning process. Teachers should also encourage students to solve their own problems by themselves. In this process, students' autonomous learning ability has been trained. In the process of Internet popularization, people have gradually enriched their own ways of knowledge acquisition, and can obtain various kinds of information through various channels. Knowledge acquisition is not limited to classroom environment. Teachers' role has also changed from the imparters of knowledge to the guides or organizers of teaching activities. Therefore, the responsibility of teachers is not only to teach students and solve their problems, but also to guide students to better participate in various teaching activities, to learn by themselves autonomously, and improve students' learning ability and information screening ability. Teachers themselves should adapt to this role change as soon as possible. Teachers also need to constantly learn and adapt to the changes brought about by new technologies in the new era.

Colleges and universities should also make full use of the development trend of technology to integrate English teaching resources in Colleges and universities under the background of big data, so as to enable English major teaching to obtain more advantageous resources. In the process of resource integration, we should take the cultivation of students' ability to use English as the main task in our teaching, combining students' professional needs and individualized development, to search valuable network resources for students from diversified perspectives, such as audio and video. For the course of Comprehensive English, English teaching resource database can be constructed, and then the integration and development of teaching resources can be carried out according to the specific use of related knowledge. In view of the construction of various learning resources of textbooks, courseware development, vocabulary database, the key sentence patterns and grammar database, or the construction of related resources of TEM-4(Test for English-major Students-Band four) can be put into practice. The designing of various teaching auxiliary teaching materials with high quality and the development diversified teaching resources related to Comprehensive English are helpful for the teaching and learning, such as MOOC or teaching platform. It is necessary to highlight proper sequence in the process of resource integration and development. Following the principles of initiative and guidance, teachers must dig and screen English information resources to find out useful information and strengthen the guidance of students' curriculum-related learning.

3.3   Establishing Flipped Class and Multi-dimensional Teaching

In order to promote the transformation from knowledge imparting to students' learning ability training, the flipped class can be adopted in Comprehensive English to restore students' dominant position in classroom teaching under the support of big data technology. Flipping class requires students to complete the learning of teaching content before class, and then internalize knowledge through interaction in class. Teachers should strengthen the guidance and instruction to ensure that students can grasp the key points of learning, and be able to break through the difficulties of learning through thinking and then consolidate and apply knowledge. Flipping class learning mode can create an environment for students to think and exercise, and make the classroom a place where students can participate actively. To achieve this goal, teachers need to use big data in advance to find out relevant teaching resources, rationalize and optimize the integration of curriculum content, and make good use of micro-class and micro-vision and other means and technologies. After class, teachers should also strengthen the supervision and management of students, require students to make good use of the network and learning platform to download and learn curriculum-related resources before class, and think about the questions raised by teachers. In the classroom, teachers can organize teaching variously: students can solve problems in groups; teachers can get students' learning situation according to students' understanding for the teaching contents, and then help students better understand the learning content through various forms of communication with students. To establish flipped classroom in Comprehensive English, teachers can provide students with a platform for self-development, learning and performance to form a strong learning atmosphere among students, guide students to complete self-learning, cultivate students' self-learning habits and self-learning ability, and then contribute to the formation of a good English teaching environment.

The teaching of Comprehensive English should also cooperate with the talent training program, integrate modern information technology and teaching theory under the background of big data to complete the dynamic curriculum framework setting. From the perspective of diversified talents cultivation, it is difficult to meet the needs of English talents cultivation in the new era by only using the paper textbooks and CD resources. It is also necessary to break through the situation that all these old materials, taken as carriers, were used to transmit knowledge to strengthen the application of modern information technology and complete the construction of multi-dimensional courses by integrating the teaching tasks and teaching subjects, a variety of carriers of teaching resources in Comprehensive English. As a major course, Comprehensive English needs to play a leading role in the talent training program, breaking through the limitations of "print media", and displaying the advanced nature and scientific nature of the course content in various forms, such as image and video, so as to form a systematic curriculum system.

3.4   Creating a Diversified and High-Quality Classroom Environment

The English teaching environment in the classroom has dramatically changed in the era of big data. The teaching of Comprehensive English should also conform to the trend by creating a diversified and high-quality classroom environment to improve the teaching level. With the help of big data, teaching process digitization can be realized to provide data basis for teachers to teach in accordance with students' aptitude. At the same time, teachers should customize personalized teaching rules based on students' individualized needs, so that teachers can master more data and information related to students' learning. Then teachers can implement teaching through network, micro-vision, teaching platform and other media platforms. To achieve this goal, teachers themselves need to learn continuously, complete the transformation of teaching ideas, and strengthen the learning of information technology in order to better meet the needs of students in the new era. In Comprehensive English class, the existing mature teaching platforms can be utilized to realize the innovation of classroom teaching, such as Mosoink, Moodle and so on, therefore, a good English learning environment for students can be created, and class teaching as well as after-class learning can be combined. In class teaching, teachers can adopt diversified forms of classroom orga-nization, such as dividing students into group. In addition, on the basis of the existing technical equipment, teachers can make the best use of personal and school equipment, such as cloud platform and mobile devices, and study under the organization and guidance of teachers, combining with their own individual needs, in order to effectively improve the quality of classroom teaching.

3.5 Constructing a Diversified Teaching Evaluation System

Evaluation in a course plays an important role in evaluating the teaching effectiveness. An objective and scientific evaluation system is a true feedback to teaching and is essential to the realization of teaching objectives. In the new national quality standard for English majors, it is required to highlight the feedback function of evaluation, to combine formative evaluation with summative evaluation, and to explore scientific diversified evaluation forms. The traditional teaching evaluation mode takes exami-nation results as the main evaluation criterion, which cannot reflect students' learning attitude, interest, feeling and so on. The scientificity of the evaluation results is questioned. Therefore, the evaluation of Comprehensive English for English majors should not only depend on the final exam's results, but also a comprehensive assess-ment of teaching in the whole stages and the whole process. A multi-level and multi-dimensional assessment method should be adopted, and the attitudes and progress of students in the course of learning can also be taken into account. Firstly, the subject of evaluation should be diversified. The subject of evaluation can be teachers, students or groups. The evaluation can be teachers' and students' evaluation, students' mutual evaluation and students' self-evaluation. Teachers guide students to conduct objective self-evaluation and mutual evaluation in the evaluation process, which can help stu-dents correctly recognize their working ability and the correct working attitude, reflect on self-learning behavior, so as to achieve the purpose of complementing each other's strengths and weaknesses, correcting mistakes and conquering shortcomings in time, helping students to adjust learning strategies and methods in time, so as to enhance their work practice ability and promote individual progress. Team evaluation can help students better integrate themselves into team work, realize the importance of team spirit and small group cooperation, and cultivate students' team spirit.

4 Conclusion

The big data era has brought severe challenges to traditional English teaching, but it is both a challenge and an opportunity. Teachers need to learn how to effectively analyze and filter data and implement it in specific teaching. Based on students' foundation, teachers should teach students in accordance with their aptitude so as to create a good learning environment and atmosphere for students. Diversified teaching mode meets the requirements of the new era. Driven by big data, Comprehensive English and modern information technology are closely integrated, which can meet the needs of individualized training for students based on the new national quality standard. Under the background of big data, teachers should pay attention to the integration of big data and students' individualized education. In this process, teachers can change their roles and design teaching from the perspective of students, so as to extend the communication between teachers and students and teaching activities from class to after class, to meet the language learning needs of students at different levels, so as to broaden students' learning time and space by emphasizing diversified evaluation of the course, and eventually effectively improving the effectiveness of integrated English teaching.

References

1. Yu, S.: Teaching design of english writing in higher vocational colleges based on big data. Course Educ. Res. (8), 100 (2018). (in Chinese)
2. Qingdong, P.: On the core competence criteria in the view of diversified foreign language talents: ideological innovation and practical innovation. In: The conference paper of the National Seminar on the Reform and Development of English Major Teaching in Colleges and Universities: Exploring the Reform and Innovation of English Major Teaching Based on the View of Diversified Talents) (2013). (in Chinese)
3. Qiqi, Z.: Research on the construction of multimodal teaching mode in higher vocational English in the information age. J. Suzhou Inst. Educ. **6**, 140–141 (2017). (in Chinese)
4. Hongxin, J.: On the establishment of national standard for teaching quality of English major undergraduates. Foreign Lang. Teach. Res. **46**(3), 456–462 (2014). (in Chinese)
5. Yanlong, Y.: Research on the innovation and informatization reform of college english teaching mode in the big data era. Technol. Enhanc. Foreign Lang. Educ. (4), 56–59 + 84 (2017). (in Chinese)
6. Xiang, C.: Study on the "group cooperation" learning mode of comprehensive english course in the big data era. J. Hubei Univ. Econ. (Humanit. Soc. Sci. Ed.) **4**, 215–217 (2016). (in Chinese)
7. Yun, Z.: The innovation and informatization development of college english teaching mode in the big data era. China J. Multimed. Netw. Teach. **8**, 50–51 (2019). (in Chinese)
8. Chen, W.: Research on the teaching reform of translation course for english majors in the context of big data. In: The conference paper in 2019 International Conference on Arts, Management, Education and Innovation (ICAMEI 2019) (2019)
9. Tingting, D.: Micro communication in English teaching in the era of big data. J. Shandong Agric. Eng. Univ. **34**(12), 5–6 (2017)
10. Yuanyuan, W.: Reflections on reconstructing the ecosystem of college English students in the big data era. J. Heilongjiang Coll. Edu. **36**(9), 128–130 (2017)

A Summary of Domestic and Foreign Education Big Data Literature

Dan Bai and Hongtao Li[(✉)]

School of Liaoning, Dalian University, Dalian 116622, China
Baidan@dlu.edu.cn, 921459057@qq.com

Abstract. As a new technology, big data has been applied to various industries at a rapid speed due to its unique advantages. The education industry has also begun to use big data technology. Recently, domestic and foreign scholars have deepened their research on educational big data. Through the analysis of educational big data literature at home and abroad, this paper explores the concept and role of educational big data, and provides reference and suggestions for the better popularization of big data technology in education.

Keywords: Educational big data, role · Research review

With the popularity of big data, more and more fields are beginning to apply big data technology, and the education field is no exception. Education is extremely important for the development of a country and a nation. If big data can be applied well in the field of education, it will certainly raise the level of education, train more talents, and contribute to the early realization of the rejuvenation of the Chinese nation.

Domestic and foreign scholars' research on educational big data is still immature, and there is no systematic theory. The relevant research involves a wide range of content, but the research depth is not enough, and there is no specific practical application experience. Therefore, it is necessary to carry out educational big data. Deeper research.

This paper summarizes the conclusions of famous scholars at home and abroad on educational big data, further studies its concept and role, and provides suggestions and references for educational services to promote big data better.

1 The Concept of Educational Big Data

If you want to study the concept of educational big data, you have to study the concept of big data first. The first to refer to the term "big data" is McKinsey, a foreign consulting firm. McKinsey believes that "the concept of big data is a data set that exceeds the normal database. Data, which has now spread to every industry, people have a lot of data. Excavation and application will definitely increase the productivity growth and consumer surplus of the enterprise. [1] According to the concept proposed by McKinsey, other companies and scholars also give the definition of big data. Gartner believes that big data is large capacity. The speed of generation is high, and it has many types of information value. At the same time, it is necessary to use new

M. Atiquzzaman et al. (Eds.): BDCPS 2019, AISC 1117, pp. 308–312, 2020.
https://doi.org/10.1007/978-981-15-2568-1_42

processing methods to ensure judgment, insight discovery and optimization processing. [2] Wikipedia concludes that big data refers to the adoption of software tools. Data sets that capture, manage, and process data take longer than tolerable. That is, big data is a large data set with a very large variety of data, and such data sets cannot be paired with traditional database tools. Its data is acquired, managed and processed [3].

Peng et al. conducted in-depth research on the relevant reports on educational big data issued by the US Department of Education, and concluded that the concept of educational big data has a broad and narrow distinction. The broad concept is mainly all behavioral data derived from humans. Most of the data comes from educational activities; narrow educational big data refers to student behavior data, which is mainly derived from the learning platform [4]. Yumin and others concluded that the definition of educational big data contains three meanings: the first meaning, education big data is big data of the education industry, mainly refers to the data collection of the types, dimensions and forms of educational topics; The second meaning is that education big data is the data of the whole process of education. It improves the efficiency of educational decision-making and the level of individualized learning through data mining and learning analysis. The third meaning is mainly a distributed computing architecture, through data sharing. Various support technologies have reached the idea of building and sharing [5]. Xianmin and others found that the so-called educational big data refers to the data set that emerged in the whole process of education and can be tapped according to educational needs, all used for the development of the education industry and can create potential value [6].

Based on the above analysis, this paper combines the views of Yang Xianmin and others that educational big data is a data set that can bring potential benefits to educational stakeholders in educational activities.

2   The Role of Educational Big Data in Education

(1) Change the Teaching Mode

In recent years, the U.S. Department of Education has issued a report on the study of big data, which points out that the current role of big data in education is mainly in two directions: educational data mining and learning analysis. The first is to predict learners' future learning trends by processing big data in education and modeling it. Learning analysis is a comprehensive application of information science and society. According to the processing and research of big data of general education, the theories and methods of various disciplines such as learning are used to explain the major problems affecting learners' learning, evaluate learners' learning behavior, and provide learners with artificial adaptive feedback [4]. From the perspective of teachers and students, Wenxin analyzed the changes brought about by the arrival of big data to teaching: from the perspective of teachers, big data made teaching decisions break through the bounded rationality of human beings, from relying on the existence of teachers' minds to relying on the data analysis of teaching cases, and from the perspective of students, it was concluded that the development of students could be changed from relying on teachers' help and reminders to relying on the analysis of teaching cases. Data analysis of self-learning process [7]. Xing Qiudan and others believe that the emergence of big data can provide online learners with the learning resources they need, as well as online teachers with more quality teaching resources and research basis. Through the collection, collation, analysis, in-depth mining and summary of user data, online education interactive platform can analyze the development trend and hot spot changes in the related education field macroscopically, understand learners' interests at a faster speed, keep up with the pace of the times, update online learning resources in a timely manner, and ensure the real-time and cutting-edge of learning content [8]. Based on the above analysis, we find that the emergence of big data provides a new teaching method for learners and teachers, and promotes the rapid development of education.

(2) Optimize Educational Resources

Educational resources are the foundation of education. How to optimize educational resources is the key to the success of education. Zhongyu and others found that the use of cloud computing, the emergence of big data technology and the changes in user demand for knowledge have enabled the university's resource services to change. Under the impetus of big data, the educational resources of colleges and universities will continue to change in terms of service methods, ways and modes. The services provided by future colleges and universities for teachers and students will be derived from data aggregation and filtering. Secondly, the resource service after the rise of big data will meet the many needs of teachers and students. With the continuous development of information technology, many universities will continue to integrate resource systems under the impetus of digital campuses to maximize the application of resources. [9]. Fengjuan's research shows that the emergence of cloud computing and big data will quickly understand the "travel" of students' operations on the educational resource database to grasp the dynamic demand for learning resources, and also analyze the learners' clicks, downloads and evaluations on learning resources. Such data information defines the "quality teaching resources" objectively, and the acquisition and storage of resources becomes simple. It can also avoid redundant construction of resources and waste of high-quality resources, so that high-quality resources can be shared and utilized to the greatest extent [10]. In summary, the rise of big data provides teachers and students with an opportunity to make full use of educational resources.

(3) Powering the Smart Campus

Smart campus is an important step to improve the education level of our country. The rise of big data has accelerated the pace of smart campus. Lijun believes that big data is a fundamental part of a smart campus. Big data can solve problems: (1) Expand the traditional BI field. Previous statistics, correlation analysis, and trend prediction for large data volumes changed from sampling to full-scale analysis. (2) Analysis of the full amount of data is used as a basis for business process improvement and assessment. (3) Service or product packaging of existing data or data processing capabilities to form data products or data services. The application model of big data on campus can help students discover themselves, discover society, and discover the world [11]. Wu Hao's research found that in the smart campus environment, cloud computing-based big data applications can analyze and predict teachers' teaching behaviors, student learning behaviors, and student personality characteristics, thus providing timely guidance and help for students' physical and mental development. In addition, it can also provide dynamic data of school operation so that school leaders and teachers can keep up-to-date with the latest management and teaching information to help teaching management become more scientific and intelligent [12]. He and others found that the flexible use of big data in the construction of smart campus is very important for improving the construction level and improving the quality of service [13]. Based on the above analysis, we find that big data plays an extremely important role in promoting the advancement of education and provides unlimited vitality for the development of smart campuses.

3   Conclusion

Recent research on big data in education has become a hot topic for many scholars. This paper summarizes the concept and role of educational big data by studying the relevant conclusions of educational big data of a large number of famous scholars, and provides reference and suggestions for improving education level through big data.

Acknowledgements. This work is partially supported by National Social Science Foundation of China No. 16BGL021. The authors also gratefully acknowledge the helpful comments and suggestions of the reviewers, which have improved the presentation.

References

1. Qinhong, S., Fengxian, S.: Data mining and application in the era of big data. Electron. Technol. Softw. Eng. (6), 204–204 (2016)
2. Ji, C., Li, Y., Qiu, W., et al.: Big data processing in cloud computing environments. In: International Symposium on Pervasive Systems, Algorithms and Networks, pp. 17–23. IEEE (2013)
3. Big Data [EB/OL]. http://en.wikipedia.org/wiki/Big. Accessed 28 June 2013
4. Peng, X., Yining, W., Yanhua, L., et al.: Analysis of learning change from the perspective of big data——interpretation and enlightenment of the US report on promoting teaching and learning through educational data mining and learning analysis. J. Distance Educ. (6), 11–17 (2013)
5. Yumin, D., Haiguang, F., Weiyang, L., et al.: A review of educational big data research. China Educ. Informatiz. **19**, 1–4 (2016)
6. Xianmin, Y., Sisi, T., Yuhong, L.: Development of education big data: connotation, value and challenges. Mod. Distance Educ. Res. **1**, 50–61 (2016)
7. Wenxin, L.: The era of big data - classroom teaching will usher in real change. J. Beijing Inst. Educ. **01**, 14–16 (2013)
8. Qidan, X., Jing, J., Zhanhe, D.: Research on online education interaction in cloud computing and big data environment. J. Inf. Resour. Manag. (3), 22–28 (2013)
9. Zhongyu, L., Hailiang, L.: Application of cloud resources in colleges and universities in the era of big data. Mod. Educ. Technol. **7**, 59–62 (2013)
10. Fengjuan, L.: A review of educational application research of big data. Mod. Educ. Technol. **24**(8), 13–19 (2014)

11. Lijun, C.: Cloud computing, big data, mobile Internet - driving the martial arts of the smart campus. China Educ. Informatiz. **20**, 6 (2013)
12. Wei, W.: A review of the application research of big data in education at home and abroad. Chin. Off-Campus Educ.: Late **6**, 83–85 (2017)
13. Yong, M., Zeyu, Z.: Big data innovation and smart campus service. China Educ. Inf. **24**, 6 (2013)

Intelligent Production Logistics System Based on Internet of Things

Haiqin Wang$^{(\boxtimes)}$, Guangqiu Lu, and Jiahui Liu

Qingdao Binhai University, Qingdao 266555, Shandong, China
94198982@qq.com

Abstract. Internet of things technology is the basis of transformation and upgrading of manufacturing industry and it is also the basis of realization of intelligent manufacturing. The introduction of Internet of things technology in the production workshop can promote the transformation of manufacturing and logistics business models, as well as collaborative production scheduling and logistics scheduling. This paper discusses the goal setting of intelligent workshop production logistics system, the structure of intelligent workshop production logistics system and the key technologies to be solved in constructing intelligent workshop production logistics system. On this basis, the paper analyzes the characteristics of production logistics of tire enterprises, puts forward the train of thought of constructing intelligent production logistics system in rubber refining workshop, component workshop, molding workshop and vulcanization workshop of tire enterprises, and finally analyzes the expected effect. Intelligent workshop production logistics system can realize the integration of intelligent manufacturing and intelligent logistics, improve the production efficiency and logistics efficiency of manufacturing enterprises.

Keywords: Internet of things · Production logistics · Intelligent manufacturing

1 Introduction

Internet of things technology is the basis for the transformation and upgrading of the manufacturing industry and also the basis of the realization of intelligent manufacturing [1]. In promoting the implementation of intelligent manufacturing, the Internet of things is expanding from local processes in the industrial field to workshops and factories, from improving quality and efficiency to promoting the transformation of manufacturing and logistics business models. At present, intelligent manufacturing and intelligent logistics are in the process of integration. Intelligent manufacturing, with intelligent workshop as the carrier, realizes end-to-end seamless cooperation in all links of design, supply, manufacturing and service. Intelligent logistics can carry out perception, thinking, reasoning, path planning and decision-making, etc., which is an important link connecting the supply and manufacturing of intelligent workshops and the cornerstone of building an intelligent factory. The Internet of things technology, introduced in the production workshop, equipped with sensors for production resources (RFID, sensor, etc.), the implementation content and content, and the interconnection between people, objects and environment, to the manufacturing process of the real-time

M. Atiquzzaman et al. (Eds.): BDCPS 2019, AISC 1117, pp. 313–319, 2020.
https://doi.org/10.1007/978-981-15-2568-1_43

data acquisition, and on this basis to management of workshop production activities, so as to realize intelligent, digital and networked manufacturing shop, so the reasonable and efficient workshop production logistics scheduling is necessary.

2 Intelligent Workshop Production Logistics System Planning

2.1 Goal Setting of Intelligent Workshop Production Logistics System

According to the production needs of enterprises, the intelligent workshop production logistics system puts forward higher requirements for production performance, among which the main goal is to improve the utilization rate of resources, realize load balance of equipment, shorten production cycle and improve the comprehensive benefits of the intelligent workshop. Because of the conflict among these goals, it is impossible to achieve the best of each goal at the same time, so it needs a balance to be struck among the goals. The constraints of complex scheduling, flexible adaptation and fault handling are also considered. Among them, the complex scheduling demand is that compared with traditional automated workshops, intelligent workshops need to face and adapt to more flexibility, and flexible intelligent workshops require more intelligent fault disposal. For equipment failure, the system needs to adapt to equipment failure, achieve rescheduling, signal safety reset, state restoration, wip abnormal disposal process, etc. For roller way fault, the system should calculate the path accessibility automatically and realize rescheduling. For AGV/RGV fault, the system should implement automatic exit of fault AGV/RGV, automatic introduction of standby AGV/RGV, etc.

2.2 The Framework of Intelligent Workshop Production Logistics System

The intelligent workshop production logistics scheme based on the internet of things adopts a three-tier system architecture: scheduling layer, monitoring layer and execution layer. The scheduling layer carries out production scheduling, path planning and instruction analysis, and issues production and transportation tasks to PLC executive layer. The execution layer receives the tasks from the scheduling layer and transports the wip from the beginning to the end point. The monitoring layer is responsible for data collection and transmission and dynamic monitoring of the production logistics system [2]. The framework of intelligent workshop production logistics system is shown in Fig. 1.

2.3 Key Technical Issues

Key technologies to be solved in constructing intelligent workshop production logistics system include workshop production scheduling theory and model, production scheduling method, path planning algorithm, AGV/RGV scheduling method, buffer dynamic construction method, fault rescheduling method, batch constraint satisfaction algorithm, etc.

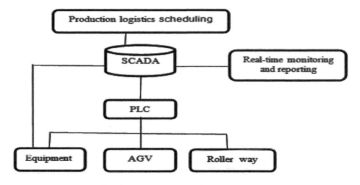

Fig. 1. The system framework.

Based on graph theory and extend event-driven process chain, this paper will model smart workshop production logistics which includes elements, elements of logistics equipment, vehicles, personnel element and logical elements, etc. [3]. Establishing a multi-dimensional integration model of multiple views can help understand workshop production and logistics all-round and also can supply data support for production scheduling and logistics scheduling model.

Based on constraint theory and pull production scheduling idea, this paper will propose a production scheduling method suitable for intelligent workshop considering the production time of each process. Dynamic real-time scheduling method and dynamic rescheduling method are adopted, and combined with dynamic programming model [4]. The optimal assignment of production tasks is carried out according to the demand of intelligent workshop production scheduling.

In order to improve the utilization rate of AGV and transportation efficiency for multiple AGVS on linear reciprocating roller channel, the safety interval is dynamically calculated according to the position of production equipment and the actual position of AGVS [1]. This can ensure that two AGVS are loaded and unloaded separately in their respective regions, so as to improve the utilization rate or transportation capacity of AGVS.

Dynamically establishing and maintaining the buffer zone of wip tray for the production equipment that needs wip tray can reasonable and timely supply wip tray for AGVS [5]. It can ensure continuous supply without congestion at any state and any time. Thus it can ensure the continuity of production In case of equipment failure, the system can reschedule related wip, at the same time, increase the adaptability of the system, allow the fault equipment off line to maintenance. When the equipment goes online, the system automatically accepts the equipment.

According to the intelligent workshop with strict batch constraint requirements, batch constraint was added into the production scheduling and logistics scheduling to ensure the separation of the old batch from the new batch when batch switching and batch separation during mixed batch production [6].

According to the demand of user manual intervention, the adaptability of the system was increased, which allowed manual production task assignment, path planning, etc. during the automatic operation of the system, and automatic path conflict avoidance and load balancing [7].

3 Application of Intelligent Workshop Production Logistics System in Tire Enterprises

3.1 The Characteristics of Production and Logistics Process in Tire Enterprise

The production logistics in tire enterprises has its own characteristics. The tire production process has both the comprehensive characteristics of flow type and discrete type, and the tire components are numerous, each production area is relatively independent, most intermediate materials need to meet a certain parking time.

In the dense mixing process, the materials involved in the logistics transfer include all kinds of raw materials, intermediate finished products and final refined rubber materials. All kinds of raw materials are mainly delivered to storage by forklift and transferred to each floor of the production workshop. After being processed by different equipment, the intermediate finished products are stored in the temporary storage area of the workshop for a certain period of time. After several repeated processing processes and parking, the final refined rubber material is formed and used in subsequent processes.

In the preparation process, the final refined rubber material in the dense mixing process is transported to different equipment by forklift. After different processing processes, different tire billets (raw tire) parts are produced and parked in the material temporary storage area of the preparation process.

In the forming vulcanization process, a variety of different models of tire blank components through the molding equipment assembled into different models of tire blank (birth), after manual or other power transport to different vulcanization equipment vulcanization into different specifications and models of tires. Vulcanized completed tire after several quality inspection procedures, storage for sale.

3.2 Construction of Intelligent Production Logistics for Tire Enterprises

According to the characteristics of production logistics in tire enterprises, the idea of constructing intelligent production logistics system in each workshop of tire enterprise is put forward as follows.

The rubber refining workshop adopts AGV system to realize automatic handling, storage, identification and management of rubber materials in the production line of the mixer [8]. The rubber material produced by mother mixer and final mixer can be folded Automatically. AGVS carry the rubber materials to the temporary storage area of the mother rubber refining and the final rubber refining respectively. The AGV has WMS storage management, which can automatically store and manage original materials according to the specifications and models of the rubber materials. After MES plans to issue production instructions, AGV will send the materials in the temporary storage area of the mother rubber refining to the on-line position of the final rubber refining and recycle the empty trays. The rubber material in the temporary storage area of the final refining is sent to the line position of the calender extruder in the parts of workshop. There is a conveyor line at the bottom of the mixer to realize automatic conveying of film, caching and automatic output of empty tray. The conveying system of rubber refining workshop is shown in Fig. 2.

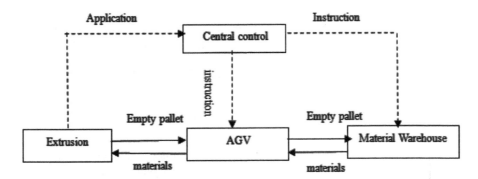

Fig. 2. Conveying system in rubber refining plant

The parts workshop adopts AGV system to realize the rubber material of calendering and extruding production line, as well as manage the empty pallets and empty trolleys [9]. This system can realize move store and identify materials automatically. According to MES plan, AGV will carry the final refined rubber of the required specifications to the position of glue on the corresponding calender outgoing line, and recycle empty pallets. AGV shall move the outgoing calender parts to the temporary storage area for storage and management, and move the empty parts to the corresponding outgoing calender parts and incoming parts.

According to the difference of process between semi-steel tire and all-steel tire factory, this paper suggests adopt different AGV to realize the parts moving and Transporting. Such as tire face parts, semi-steel tire factory adopts i-block wheel to take up the tire, and the forklift AGV carries the platform. The whole steel tyre workshop adopts grille car to carry the tread, and adopts load-bearing AGV straightly to pick up dive lift bearing tread grille car. The load of tread AGV is equal or greater than 4 tons, which can operate in a full range in horizontal direction. They can be the tread grille dense storage, greatly save the floor area.

The forming workshop realizes the intelligent distribution of parts required by the forming machine through AGV system. According to the MES production plan, the tire specifications to be produced are arranged to the corresponding molding machine, and various materials required by the molding machine are also sent to the AGV system. The AGV will transport the inner lining layer, tire, tire side, tire ring, tire body, belt and other parts of the temporary storage area to the waiting position of the molding machine for loading, and transport the corresponding empty trolley back to the temporary storage area for storage.

The AGV system in Vulcanization workshop is mainly used for automatic distribution of embryos to vulcanization machine, as well as mold storage and transportation. Embryo logistics adopts AGV to carry embryos to the side of vulcanizing machine. Generally, embryo trucks can carry 4 or 6 embryos to improve the transfer efficiency. Mould storage adopts high shelf multi-layer form, mold specifications, models and goods can realize intelligent management with AGV system. AGV has its own WMS system, When it receives Mold replacement information, stacking AGV automatically moves to the corresponding mold cargo position fork and pick up the mold tray, sent to the mold assembly area, and send the replaced mold to the corresponding cargo space. Adopting the form of AGV+ shelf, it not only realizes the storage management of the mold, but also realizes the handling of the mold. The lifting height of the stacking AGV is up to 8 meters, and the pallet of high cargo position can be picked up automatically without manual intervention, which is safe and reliable.

4 Conclusions

This paper discusses the goal, structure and key technology of intelligent workshop production logistics system. The application of intelligent workshop production logistics system in tire production workshop shows that the system can realize the integration of intelligent manufacturing and intelligent logistics, improve the production efficiency and logistics efficiency of manufacturing enterprises, and improve the productivity of the workshop.

Acknowledgements. This research was supported by science and technology planning research of Qingdao Binhai University, design and application of automatic material handling system in tire enterprises, Project NO.: 2019KY03,2018/11/22

References

1. Ting, Q., Thurer, M., Wang, J., et al.: System dynamics analysis for an internet-of-things-enabled production logistics system. Int. J. Prod. Res. **55**(9), 2622–2649 (2017)
2. Huang, S., Guo, Y., Zha, S., et al.: An internet-of-things-based production logistics optimization method for discrete manufacturing. Int. J. Comput. Integr. Manuf. **32**(1), 1–14 (2018)
3. Zhang, Y., Guo, Z., Lv, J., et al.: A framework for smart production-logistics systems based on CPS and industrial IoT. IEEE Trans. Ind. Inform. **14**(9), 4019–4032 (2018)
4. Ting, Q.U., Zhang, K., Luo, H., et al.: Internet-of-things based dynamic synchronization of production and logistics: mechanism, system and case study. J. Mech. Eng. **51**(20), 36–44 (2015)
5. Li, J.T., Yuan, H., Zhang, H.B.: Research on submerged logistics carrying AGV used in e-commerce distribution center. Appl. Mech. Mater. **722**, 436–441 (2015)
6. Wan, J., Tang, S., Hua, Q., et al.: Context-aware cloud robotics for material handling in cognitive industrial internet of things. IEEE Int. Things J. **PP**(99), 1 (2017)

7. Chang, F., Liu, X., Pei, J., et al.: Optimal production planning in a hybrid manufacturing and recovering system based on the internet of things with closed loop supply chains. Oper. Res. **16**(3), 543–577 (2016)
8. Zhang, X., Zhou, H., Liu, G.S.: Design of the automatic guided vehicle control system applied to automotive logistics. Appl. Mech. Mater. **644–650**, 381–384 (2014)
9. Aguilar, C.M., Sant Ana, C.T., Costa, A.G.V., et al.: Comparative effects of brown and golden flaxseeds on body composition, inflammation and bone remodelling biomarkers in perimenopausal overweight women. J. Funct. Foods **33**, 166–175 (2017)

Industrial Engineering Major Manufacturing Dig Data Application Analysis Direction Curriculum

Jui-Chan Huang[1], Ming-Hung Shu[2], and Kun-Chen Chung[2(✉)]

[1] Yango University, Fuzhou 350015, China
[2] Department of Industrial Engineering and Management,
National Kaohsiung University of Science and Technology,
Kaohsiung 80778, Taiwan
jackson0323@gmail.com

Abstract. Based on the general application of big data and the increasingly mature background of cloud computing technology, this paper, guided by the knowledge system of industrial engineering, made a preliminary analysis on how Chinese manufacturing enterprises optimize their production management by industrial engineering. Its central idea is that manufacturing enterprises should combine the basic methods of industrial engineering with big data and cloud computing technology to improve the efficiency and reliability of production. In terms of research methods, this paper proposes and improves bp-miv algorithm to rank how big data technology is applied to industrial engineering manufacturing field. On the one hand, the research in this paper provides ideas for the course design of manufacturing based on big data. On the other hand, the improvement of bp-miv algorithm overcomes the defects of sample size and parameter selection of the traditional bp-miv algorithm. Through the improvement idea of "first classification, then calculation of difference", the paper makes a better sorting. All in all, this article made in manufacturing large data meaning, big data meaning and manufacturing technology on the basis of the data, this paper discusses the manufacture large data application analysis is an important developing direction of industrial engineering discipline, is proposed based on industrial engineering professional training manufacture large data application analysis of curriculum system and gives the manufacture process of large data application oriented curriculum content analysis technology.

Keywords: Manufacturing big data · Big data analysis · Industrial engineering

1 Introduction

Nowadays, big data has become a valuable resource. Under the background of big data, data science is based on data, especially the big data as the research object, in order to obtain knowledge and wisdom from the data as the main purpose, in mathematics, statistics, computer science, visualization and the theoretical basis of professional knowledge and so on, with data acquisition, preprocessing, data management and data calculation, etc as the research content of a discipline [1, 2]. The development of iot

© Springer Nature Singapore Pte Ltd. 2020
M. Atiquzzaman et al. (Eds.): BDCPS 2019, AISC 1117, pp. 320–326, 2020.
https://doi.org/10.1007/978-981-15-2568-1_44

engineering technology, data science and big data technology, artificial intelligence technology and other technologies and related industries have promoted the application of big data technology in the manufacturing industry [3].

Many countries' manufacturing development strategies, such as German industry 4.0, American industrial Internet, and made in China 2025, have clearly pointed out that big data is the key technology of the new generation of industrial revolution and manufacturing industry [4]. The platform for action to promote the development of big data issued by the State Council of China in 2015 listed big data in manufacturing as one of the ten key projects. Driven by basic supporting technologies, industrial policies and international competition, China's manufacturing big data is in urgent need of development, and the use of big data technology to enhance the value of manufacturing industry demands a large number of big data talents [5]. Big data talent training is mainly divided into big data application architects and big data application analysts. This paper analyzes and discusses the course setting of manufacturing big data application analysis [6]. It is no exaggeration to say that if China wants to become a manufacturing power, it must have first-class industrial engineering majors and first-class industrial engineering talents. How to adapt to the development of The Times, the needs of the country, training first-class industrial engineering talent, is worth in-depth discussion. In this context, this paper studies the construction of curriculum system for industrial engineering majors [7].

The undergraduate education of big data in China originated in 2016. The Ministry of Education approved the first three universities to start training undergraduate talents in data science and big data technology in 2016 and 32 universities in 2017. The talent cultivation of manufacturing big data application analysis belongs to the talent cultivation of pan-professional data science literacy proposed by literature [8, 9]. As the major education of big data technology has just started, the talent cultivation of pan-major data science literacy has not started yet. The cultivation of pan-professional data science literacy is the deep integration of data science and practical application knowledge, and the cultivation of interdisciplinary talents with both professional background and certain data science literacy. In practice, simplified and integrated courses such as "industrial engineering + big data" can be set, or "micro-majors" that intersect with big data technology can be set, and a series of relatively systematic courses of data science can be set, so as to cultivate dual-majors and multi-talents who highly combine domain knowledge and data thinking [10]. Next, we will analyze the course setting of personnel training for manufacturing big data application analysis from the integrated course mode of "industrial engineering + big data".

2 Research Methods

2.1 Main Methods to Apply Curriculum Design

In the production management driven by market demand based on big data, industrial engineering major manufacturing should pay attention to the following main methods to improve the economy and efficiency in the implementation of manufacturing courses.

(1) Bp-miv algorithm

The professional manufacturing enterprise shall, in the product design and development of link, is fully with big data market feedback and consider products from design to sale, all kinds of problems in the whole process of using, and strive for in the product development phase can solve these problems in parallel algorithm of the whole project is an important content in industrial engineering knowledge system, a weighted formula is:

$$\omega_{firsthand} = \frac{\sum_{i=1}^{N} y_{firsthand,i}}{\sum_{i=1}^{N} y_{firsthand,i} + \sum_{i=2}^{N} y_{firsthand,i} + \sum_{i=3}^{N} y_{firsthand,i}} \qquad (1)$$

Manufacturing enterprises can reduce the waste of resources in product research and development, manufacturing, sales, use, rework and other phenomena and improve product quality and satisfaction through the parallel crossing of these different businesses, system integration and overall optimization.

(2) Optimized simulation technology.

It is necessary for manufacturing enterprises to conduct computer simulation of the production process of products, and analyze the possible technical problems in the manufacturing process through simulation software, so as to shorten the working hours of manufacturing and improve the quality of products.

In other words, when industrial engineering USES intelligent equipment for production, sensor technology is used to collect relevant data of intelligent equipment and upload it to the cloud. Through cloud computing, corresponding parameters and variables of the equipment are analyzed, and then the action and speed of the equipment are dynamically adjusted. By the same analogy, as long as an enterprise applies the concept and principle of industrial engineering and optimizes the algorithm and other systems of cloud computing, it can dynamically adjust all intelligent devices of the enterprise and greatly improve the production efficiency and stability at the same time. The improvement idea of "classify first, calculate difference later" can be expressed as follows:

$$H_j = f\left(\sum_{i=1}^{n} y_{firsthand} - a_i\right) \qquad j = 1, 2, \ldots, l \qquad (2)$$

(3) Big data participates in dynamic control

If China's industrial engineering manufacturing can control labor costs while further improving production efficiency and reducing the risk of labor disputes, it will surely bring huge economic benefits. Therefore, in recent years, China's manufacturing enterprises have begun to use more intelligent equipment to participate in production and manufacturing. In order to manage smart devices, these big data technologies are needed.

Therefore, in the information age, the optimization of production-driven model is also the first step for manufacturers to upgrade their competitiveness. The traditional production-driven model based on prediction can no longer adapt to the current industrial market dominated by differentiated competition. The analysis of market demand depends on the application of big data technology. Through big data mining, the content of different types of demand groups can be accurately analyzed to lay a solid foundation for product differentiation.

To sum up, as the big data and cloud computing technology matures, scope of application of industrial engineering in manufacturing enterprises and greatly broaden the application scenario, the future of our country's manufacturing engineering should be the basic principles of industrial engineering and basic tools flexible new technology applied to lead the production activities, let the big data technology to assist industrial engineering professional course more scientific, make industrial engineering manufacturing more efficient production.

3 Experiment

In order to achieve an effective ranking of importance, this study proposes an improved bp-miv algorithm to overcome the defects of the traditional bp-miv algorithm in overcoming the large sample size and complex parameter selection. Through the improved idea of "first classification, then calculation of difference", the algorithm can perform the ranking better.

Mean Impact Value (MIV) applied by bp-miv algorithm is considered as an indicator to evaluate the importance of impression of related variables on dependent variables. Its symbol represents the direction of correlation, and its absolute Value represents the relative importance of Impact.

The basic process of the improved bp-miv algorithm is as follows:

Step 1: The original sample according to the strain magnitude order into two kinds of P1 sample (N1 optimization of TV, and this) and P2 (class a, N2 samples), for P1 and P2 class samples one by one in the match, the team's match difference as the new sample, this sample size into N1 N2, 20% samples;

Step 2: Training a new sample of neural network. Network initialization, according to the system input and output sequence (X, Y) determine the input layer node number n, the number of hidden layer nodes l, the output layer node number m, initialize the input layer, hidden layer and output layer connection weights between neurons wij, WJK, initialization value a hidden layer, output layer threshold b, given the learning rate and neuron excitation function;

Step 3: The weight trained is the weight of the improved bp-miv algorithm, which is extracted for 10 times in a continuous manner, and the mean value of MIV is taken to eliminate the influence of abnormal conditions.

Step 4: Output the final MIV mean as the result of the experiment.

4 Discuss

4.1 Manufacturing Big Data Analysis

As defined by the authority of the American society of Industrial engineers, Industrial Engineering (Industrial Engineering, IE) is to the people, equipment, energy and information integration system of design, improvement, implementation and control of Engineering technology, it is the integrated use of mathematics, physics, Engineering design and manufacture, the specialized knowledge of social science and technology, combined with the principle and method of Engineering analysis and design, the results of the system are analyzed to determine, prediction, optimization and evaluation. The integrated system can be divided into manufacturing system and service system.

Industrial engineering originates from the improvement of manufacturing problems in the manufacturing industry. Manufacturing industry has always been the main application field of industrial engineering. In the field of industrial engineering, in order to accurately predict the production duration and cost and conduct accurate dynamic real-time scheduling of manufacturing resources according to changes in product demand, status quo of materials and equipment and other factors, it is necessary to analyze manufacturing big data resources. To explore law of equipment operation, product manufacturing quality influence factors and their mutual relations, the prediction quality factor condition, thus effectively for production line of health management, manufacturing quality guarantee, the need to make big data resource analysis, this paper studies done to the above data, large data analysis results are shown in Table 1 below. In other words, when big data technology is mature, the analysis of manufacturing process with industrial engineering technology needs the help of big data technology to improve the value of data resources and industrial application value. It can be said that manufacturing big data application analysis is an important development direction of industrial engineering technology in the era of big data.

Table 1. Big data analysis results

Project	Health management valuation	Manufacturing quality	Production cost valuation
Manufacturing quality factors	Q > 78.3	–	S > 55.9
Equipment running status	Q > 33.6	Z > 64.3	–
Operating condition factor	Q > 56.5	Z > 79.3	S > 72.3
Interrelation factor	–	Z > 67.8	S > 91.6
Predictive quality factor	Q > 89.4	Z > 81.5	S > 55.8

In conclusion, it is necessary and feasible to cultivate manufacturing big data analysis talents relying on industrial engineering majors.

4.2   Course Setting Analysis

Manufacturing big data platform technology includes manufacturing big data platform architecture, manufacturing big data platform network configuration, manufacturing big data platform hardware configuration, manufacturing big data platform support software technology, manufacturing big data commercial products. According to the knowledge module system of manufacturing system engineering, the course system of manufacturing big data application analysis was set up by the course integration mode of "manufacturing system engineering + big data application analysis". The courses of this system include mechanical design foundation, manufacturing engineering technology foundation, automatic manufacturing system, advanced manufacturing technology. This course setting is based on the overall analysis of big data. After analysis, in the course setting, the research of this paper believes that these courses will be scored due to factors in the performance assessment. We set that each student should complete 25 points, which can be regarded as the conditions for graduation. The basic mechanical design foundation and manufacturing engineering technology foundation should occupy 8 points respectively, the automatic manufacturing system should occupy 5 points, and the advanced manufacturing technology should occupy 4 points. The distribution of the assessment points is shown in Fig. 1 below.

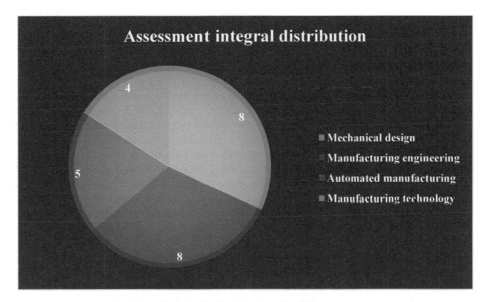

Fig. 1. Assessment integral diagram

Meanwhile, the analysis and application of manufacturing process big data include manufacturing process quality control, product manufacturing process big data mining, quality improvement and control strategy. Equipment operation needs reliability prediction, production equipment operation maintenance and health management, efficiency management, manufacturing process energy consumption optimization.

5 Conclusion

The cultivation of industrial engineering professionals plays an important role in China's manufacturing strategy. In order to adapt to the development of The Times to a greater extent and improve the quality of personnel training, it is necessary to reform the training system of industrial engineering majors by adding big data analysis on the basis of years of practice. Manufacturing big data is an important type and component of manufacturing big data, and manufacturing big data analysis is an important direction of manufacturing big data technical personnel training.

Application in this paper, the research shows that making big data analysis is an important developing direction of industrial engineering discipline and application of "manufacturing system engineering+big data application analysis" mode of manufacturing large data analysis application talents training course system, the original manufacturing system engineering in industrial engineering and information science course, on the basis of courses in big data analysis and manufacturing process of data analysis, the course will greatly accelerate the industrialization of our country.

References

1. Buyurgan, N., Kiassat, C.: Developing a new industrial engineering curriculum using a systems engineering approach. Eur. J. Eng. Educ. **42**(6), 1–14 (2017)
2. Sengupta, D., Huang, Y., Davidson, C.I., et al.: Using module-based learning methods to introduce sustainable manufacturing in engineering curriculum. Int. J. Sustain. High. Educ. **18**(3), 307–328 (2017)
3. Chien, Y.H.: Developing a pre-engineering curriculum for 3D printing skills for high school technology education. Eurasia J. Math. Sci. Technol. Educ. **13**(6), 34–35 (2017)
4. Strauti, G., Dumitrache, V.M., Taucean, I.M.: Entrepreneurial competences in economical engineering curriculum in Romania. Procedia – Soc. Behav. Sci. **238**, 737–742 (2018)
5. Sayin, G.: An engineering curriculum nourished by literature: a dream or a necessity? [Testing Ourselves]. IEEE Antennas Propag. Mag. **59**(3), 128–130 (2017)
6. Murphy, J.F.: Process safety in the undergraduate chemical engineering curriculum. Process Saf. Prog. **38**(1), 3 (2019)
7. Nguyen, A.T., Nguyen, V.P.: Benchmarking industrial engineering programs. Benchmarking Int. J. **25**(4), 1194–1212 (2018)
8. Rodríguez-Prieto, Á., Camacho, A.M., Sebastián, M.Á.: Development of a computer tool to support the teaching of materials technology. Mater. Sci. Forum **903**, 17–23 (2017)
9. Chabalengula, V.M., Mumba, F.: Engineering design skills coverage in K-12 engineering program curriculum materials in the USA. Int. J. Sci. Educ. **39**(3), 1–17 (2017)
10. Chang, K.H., Bassue, J.: Green tricycle design through experiential learning—an open courseware enriching engineering curriculum and entrepreneurship. Comput.-Aided Des. Appl. **14**(6), 1–10 (2017)

Principle of Big Data and Its Application in the Field of Education

Zhuohan Liu$^{(\boxtimes)}$

Department of Telecommunications Engineering and Management,
Beijing University of Posts and Telecommunications, Beijing, China
liuzhuohan@bupt.edu.cn

Abstract. Big data is another disruptive technological revolution in the IT industry after cloud computing and the Internet of Things. Data is growing and accumulating at an unprecedented speed. Big data has been applied in various fields and become an important driving force for social development. To understand the connotation of big data and its key technologies and grasp the significance to the development of education will help us to further study and understand big data, so as to play its important role in promoting the development of education and teaching in China. This paper introduces the concept and background of big data technology and focuses on the analysis and prospect of its application in the field of education to make a comprehensive analysis of the educational application of big data and provide reference for the better application of big data in education.

Keywords: Big data technology · Educational application · Innovative education

1 Introduction

In December 2008, the computer community alliance released a report that outlined the technologies needed to solve big data problems in a data-driven context and some of the challenges they face. EMC corporation held EMC World conference in 2011 and proposed the theme of that year was "cloud computing meets big data". In May 2012, the United Nations released white paper, discusses how to use the Internet to generate large Data promote the Development of the world.

For the status quo of education application of big data, big data in the domestic and overseas in the study of the application of the education is still in its infancy, research into the content of the proved to be more widespread but the depth is not enough, also the lack of specific practical application experience, also need to further strengthen the big data research in education, in order to push the big data as early as possible in specific education practice really exert its advantage and function.

Although big data technology has many applications in the field of education, there are still some problems such as mixed data system, difficult data mining and flawed data conclusions. The application of big data analysis is still in its infancy and still has a long way to go.

© Springer Nature Singapore Pte Ltd. 2020
M. Atiquzzaman et al. (Eds.): BDCPS 2019, AISC 1117, pp. 327–331, 2020.
https://doi.org/10.1007/978-981-15-2568-1_45

2 Analysis of Big Data Technology

2.1 The Concept of Big Data

Big data, or huge amounts of data, massive data, and large data, means that the amount of data involved is so large that it cannot be artificially intercepted, managed, processed, and organized into a form that humans can interpret at a reasonable time, is the people involved in the production and living together all kinds of information, adopting some technical difficult for the traditional database management of special grab and processing the data. "Big data" is not a simple storage of data, it also contains a lot of information after people may be able to create and exploit. Big data is a special technology to analyze data to help us translate into greater value [1]. Practice is the ultimate value of big data. Big data is not "big" but "useful." Big data is not about how to define, the most important thing is how to use it. Value content and mining costs are more important than quantity. For many industries, how to use these large-scale data is the key to winning competition. The biggest challenge is which technologies can better use the data and how big data is applied.

2.2 Major Features of Big Data

Data in the era of big data has the following main characteristics:

(1) Bulky. Relevant data shows that before 2020, the total amount of data will approach 40 ZB, which is a figure hard to imagine. Countries around the world have also set up databases to assist data collection through satellite and other means. China's database is relatively large in scale. At present, some large data sets may reach 10 TB in total. It is not ruled out that one day in the future, the data scale will be measured by PB.
(2) A variety of types. The data comes from various sources. The content includes office documents, texts and pictures in all formats. These are multi-perspective, not only formal data, media news data, timely data, but also with personal emotion data. However, these data broke the previously restricted category of structured data [2].
(3) Quick. Complex data in people's lives, like a double-edged sword, can not only enrich our life but also easy to interfere with our exploration. The advancement of big data lies in its rapid management, integration and real-time processing as much as possible, which is extremely important for our current situation [3].
(4) Data value density is relatively low. The value density is inversely proportional to the size of the total amount of data. The larger the total amount of data, the more data is invalid redundant. In big data, single data may have no value or more useless data, but its comprehensive value is large. How to quickly complete the value of data through powerful machine algorithms "purification" is a difficult problem to be solved in the current big data background.

3 The Main Applications of Big Data in the Field of Education

3.1 Innovate Educational Idea and Educational Thinking

The advent of the era of big data has brought new opportunities and challenges to the transformation of traditional education. It has changed the concept of education and the way of thinking in education. It provides another possibility for education, covering all aspects of classrooms, courses, teacher-student interactions, etc. every aspect. Educational professionals can also apply their core teaching skills to educational data sets and other data sets. More importantly, big data has profoundly changed the way educators' ideas and ways of thinking, making big data smaller and more structured. Education policies can be formulated and implemented with a definite aim, and more practical education and teaching strategies can be formulated. Such as using Moodle set of network learning platform, using the Moodle platform itself provides some forms, using Excel to study samples to carry out the test of Moodle platform log mining, data mining, and interactive evaluation data mining research, in order to explore the network learning activity distribution characteristics, interaction between teachers and students, live network structure characteristics of interaction, for auxiliary teaching on network learning platform to carry out the situation and teachers' use of network to carry out the teaching support service to provide the reference [4].

3.2 Achieve Personalized Education

In the era of big data, learners leave a lot of digital fragments in the process of digital learning. By analyzing these digital fragments, we will find various learning behavior patterns of learners [5]. Big data not only conducts full-sample analysis of large-scale data, but also has general rules. More importantly, it can reflect personality. It can record the changes of each student, and it is convenient for teachers to adjust classroom teaching methods for each student and expand the classroom. Content, reform the traditional classroom mode, make the classroom program and teaching mode more scientific, more in line with the needs of students' individual development. Make the traditional collective education move towards the individualized education of learners [6]. While talking about personality education must mention the current epidemic of Massive Open Online courses (MOOCs) education, MOOCs high education is the main reason of the learning analysis technology and data support for it, with the study analysis and big data technology, high quality teaching, curriculum resources and services through the real objective data was presented.

3.3 Conducive to Educational Decision-Making

Big data is of great value to school management. Various decision-making and control activities in school management, such as the determination of training objectives, the formulation of teaching plans, teaching organization and command, teaching quality control, teaching evaluation, teacher management, student management, etc., are based on a large number of data. And constantly generate a variety of new data, the processing and mining of big data plays a key part in school management. Utilize big data technology, comprehensively collect school management big data. For example, Qian designed a data-based, model-based, problem-oriented provincial education decision-making system [7]. The phased goal of the construction of the system, starting from the user's needs, proposes the construction process and key points of the system design from the bottom up, and gives a series of safeguard measures for the application of the system, which is the research and development of the provincial big data decision system. Providing top-level design ideas, further promoting the scientific decision-making of education and the modernization of education governance.

4 Conclusion

Big data makes it possible to investigate information about student performance and learning pathways without relying on stage test performance [8]. The instructor can analyze what students know and what is the most effective technique for each student. By focusing on the analysis of big data, teachers can Study the learning situation in a more subtle way. Education big data is the need for students to learn personalization and teacher teaching precision. It is the need to quantify the learning process and deeply study the internal mechanism of learning. It is also the need for refined management and decision-making support using data. The development of educational big data should be based on the education big data platform, aiming at solving the actual problems facing the current education [9]. Through the collection, storage and calculation of massive educational data, and analysis and mining, driving education in personalized learning, precision teaching, science Decision-making, education and research, and many other aspects of change and innovation, promote the deep fusion of big data and education. But the application of big data analysis is in its infancy, and it will take several years to mature. Although the existence of big data has proved to be valued, big data is not a panacea for all educational problems. It only provides people with a plan for education. Part of the decision-making reference when solving a problem [10].

References

1. Zhu, A.: A brief analysis of big data technology and its application in the field of education. Digital Communication World (2016). 06-0292-02
2. Zhao, S., Sun, S.: Big data technology and its application in education. Technology Application, pp. 1671–7584 (2016). 03-0064-05
3. Liu, F.: A summary of the research on the application of big data in education. Procedia Comput. Sci. **24**, 08 (2016)
4. Yu, F., Qu, J.: Thinking on "Big Data" and education. China Information Technology Education, August 2017
5. Zhang, Y.: The enlightenment of thinking mode in big data era. Educ. Dev. Res. **21**, 1–5 (2018)

6. Liang, W.: The era of big data - classroom teaching will usher in real change. J. Beijing Inst. Educ. (Nat. Sci. Edn.) **1**, 14–16 (2017)
7. Qian, D., Anni, L.: Design and research of provincial education science decision service support system under the background of big data. Educ. Dev. Res. **38**(5), 68–74 (2018)
8. West, D.M.: Big data for education: data mining, data analytics, and web dashboard. Governance Studies at Brookings, pp. 1–10. Brookings Institution, Washington DC (2016)
9. Zhen, L.: Platform construction and key implementation technology of education big data. Procedia Comput. Sci. **28**, 01 (2018)
10. Picciano, A.G.: The evolution of big data and learning analytics in American higher education. J. Asynchronous Learn. Netw. **03**, 9–20 (2016)

Distributed Accounting Resource Sharing Method Based on Internet

Xuedong Wang[✉]

Ji Lin Engineering Normal University, Changchun, China
1404173368@qq.com

Abstract. As the Internet enters thousands of households, it has spawned emerging network technologies such as big data and cloud computing, constantly shaping people's lifestyles and changing the way people work. This paper analyzes the construction of accounting information resource sharing platform for enterprises in the Internet era, and discusses the positive effects and potential risks brought by enterprise accounting information sharing. The construction of accounting information resource sharing platform can provide enterprises with more advantageous resources, improve the efficiency of accounting information processing, and also expose enterprises to many risks at the technical level, information security level and personnel management level. Based on the distributed algorithm and centralized joint allocation algorithm, the game results of shared information resources are deducted. Based on this, a series of countermeasures and suggestions are proposed for how to prevent the outstanding risks in the construction of accounting information sharing and promote the construction of resource sharing platform. Therefore, the results of this study show that in order to improve the office efficiency of the accounting industry by using network data resources and platforms in the Internet era, it is necessary to start from cultivating human resources, improving relevant laws and regulations, corporate attention and policy support.

Keywords: Internet era · Accounting resource sharing · Distributed resources · Construction of shared platform

1 Introduction

With the continuous development of the social economy and the popularization of Internet technology, the working mode of the accounting industry in modern society has undergone great changes. In the network environment, accounting information will have a greater degree of openness and openness. A large amount of data is directly collected from the internal and external systems of the enterprise through the network, and various agencies and departments inside and outside the enterprise can also directly obtain information through the network according to authorization [1]. In this way, network information realizes resource sharing through mutual access, making up for the shortage of accounting computerization [2], and has become a new field of accounting industry development in the information age. In this era of accounting era to achieve a win-win situation, the trend of accounting resources sharing based on the times is unstoppable.

© Springer Nature Singapore Pte Ltd. 2020
M. Atiquzzaman et al. (Eds.): BDCPS 2019, AISC 1117, pp. 332–339, 2020.
https://doi.org/10.1007/978-981-15-2568-1_46

Distributed processing means that a complex task can be divided into multiple parts that are processed simultaneously by different computers on the network, thereby improving the overall performance of the system [3, 4]. The computer network can realize the data transmission under the geographical distribution to achieve the purpose of data sharing, but it is only the copy of the whole file and lacks the management of the data. Therefore, the researchers envision a "database system + computer network" to achieve a distributed database system, which not only achieves centralized management and sharing of data, but also enables geographical dispersion to be hidden by the system [5].

Under the background of "Internet +", the concept of sharing has gradually penetrated the hearts of the people, which provides a good development space and a large amount of information resources for enterprises. The development of information technology provides a convenient way to share accounting resources, and makes an indelible contribution to the sharing of accounting resources across regions [6]. The establishment of the accounting resource sharing service model will greatly reduce the heavy accounting work of accountants, and enable accountants to participate in the business management decisions of enterprises, provide a large number of services for business management decisions, and realize the development of financial and business integration [7]. The standardization of financial business processing has greatly improved the efficiency of work, but it also puts forward higher requirements for the professional judgment ability of accounting personnel [8]. Financial information is an important core part of the accounting sharing model. It mainly meets the needs of accounting personnel management and economic management. The content mainly includes two aspects: one is to share the information resources of others, and the other is our information resources and others share it. The information resources that can be shared can facilitate the exchange and utilization between various functional departments in the accounting body, so that the modes of management of various enterprises are integrated with each other and ensure the financial data in the financial department and society of the enterprise. Financial status is disclosed and collected [9, 10].

In view of the current innovation trend of the accounting industry and the powerful advantages of Internet big data, this paper analyzes and discusses the distributed accounting resource sharing methods in the current Internet environment, and provides some reference for optimizing resource sharing and promoting accounting innovation.

2   Methods

2.1   Distributed Resource Sharing Scheme

The wide application of Internet technology in production and life has made life more colorful and diversified. Nowadays, all walks of life have gradually felt the advantages of Internet resource sharing platform construction. Under the background of big data era, the construction of resource sharing platform has become the new normal for the development of the accounting industry, which has played a certain impact on the development of enterprises in all aspects. While enjoying the benefits of resource sharing, enterprises also need to pay enough attention to the negative impact of sharing platform.

The Internet era has provided many favorable conditions for enterprises to share accounting resources and provided many channels for enterprises to realize rapid development. First of all, the Internet provides a good information exchange platform for enterprise accounting resource sharing, enabling information resource sharing among various accounting industries and realizing the goal of rapid accounting business processing and solution. Secondly, the Internet has improved the efficiency of accounting business processing to a new level. Through information exchange, enterprises can learn and apply more advanced technologies to help enterprises improve the efficiency of accounting business processing. In addition, data has become increasingly cheap in the Internet era, enabling enterprises to save a lot of money when purchasing information resources. At the same time, cloud services, and the application of information technology to achieve the large capacity data recording, storage and processing, and can guarantee the accuracy, reliability and timeliness of information, both can satisfy the business enterprise internal information resources sharing, also can satisfy the business enterprise external, such as the upstream suppliers, downstream customers, government agencies and other aspects of the information resources sharing.

2.2   Centralized Joint Resource Sharing Allocation Algorithm

The centralized algorithm obtains the related information of other data sets based on dynamic replication and converges to the equilibrium state. Centralized algorithm can effectively change the convergence speed of strategy evolution to reach the equilibrium state of the game. In the Internet with overlapping overlay structure, the set of policies available for the user to choose is K = {0,1,2...,K}, the strategy of data group j is about the share of data group XK of 1 MSP and K SSP. In dynamic replication, gain parameters are used to control the speed of terminal observation and change of SP selection strategy. In the intra-group game, the goal of the terminal is to minimize its own cost function XK. Therefore, the dynamic replication is formulated as:

$$\dot{x}_k^j(t) = \sigma x_k^j(t)\left[\bar{\pi}_j(t) - \pi_k^j(t)\right] \tag{1}$$

The population share of data group j after one round of dynamic replication is updated as follows:

$$x_k^j(t+1) = x_k^j(t) + \dot{x}_k^j(t) \tag{2}$$

According to the dynamic replication of data group j, if the cost of selecting SPK is lower than the average cost, the growth rate of selecting SPK is greater than 0, and the number of terminals selecting SPK will increase. According to formula (2), dynamic replication satisfies xk, then T > 0.

2.3   Distributed Resource Sharing Algorithm

The distributed algorithm obtains EE of the evolving game by obtaining Q value. For distributed algorithm, it represents the current SP allocation and sharing strategy of terminal I in data group j. U represents the next SP selection strategy of terminal I in data group j, and is also known as the control state of terminal I.

The next state of is expressed as:

$$\partial_{i+1}^{i} = u_l^i, l = 0, 1, 2, i \in I \tag{3}$$

PI can be calculated according to the formula to select SPK and f on behalf of terminal I, W and f respectively, where N is all terminals that select SPK. Q selects the initial state of the policy for terminal I. Given the q state, there is a control state U<s.< span="">. The calculation formula of U is as follows:

$$\pi^j(a_l, u_l) = \left[\pi_0^j, \pi_1^j, \ldots, \pi_k^j, \ldots, \pi_K^j\right] \tag{4}$$

The next SP select state UI is then obtained. The Q value of all data group resources distribution and Shared iteration is as follows:

$$Q_{l+1}(a_l, u_l) = (1 - \lambda_l)Q(a_l, u_l) + \lambda_l(\pi(a_l, u_l) + \theta \min Q(a_l, u_l)) \tag{5}$$

Finally, when the values converge, we obtain the optimal shared distribution vector and the optimal control vector and the optimal resource allocation.

3   Experimental

It should not be overlooked that in today's Internet era, the construction of corporate accounting resource sharing system is to use the network to connect the enterprise's customers, raw material suppliers, banks, taxation and other departments. Through the Internet, the various departments can greatly improve the communication between various departments. The efficiency of contact, thereby improving the level of business management and improving the efficiency of accounting work. However, at present, China's big data technology is still not mature at the initial stage, which brings challenges to the development of enterprise accounting information resource sharing. In response to these possible risks, we investigated the attitudes of some employees. The results are shown in Table 1.

First, where is the data coming from and how it is handled. A data scientist in the United States once said in his book: In the past, data was useless after the completion of data collection. For example, the data on the train ticket after the train arrives is useless; when searching for information on the Internet, when the information is searched, its network retrieval command is useless. But in today's era, data information has become commercial capital, and data information can create new revenue. The Internet platform has all the user's usage information. A simple software can know what goods and services are needed, the user's economic situation is good, and the user's preferences

Table 1. Survey of employees' perceptions of accounting information resource sharing risks.

Category		Number of persons	Proportion
Employee ability	Completely unsuitable	5	0.05%
	Trained to be competent	57	0.57%
	Basically adaptable	37	0.37%
	Fully adaptable	7	0.07%
Data security	No risk	2	0.02%
	Smaller	30	0.3%
	Safer	63	0.63%
	Very safe	5	0.05%
Customer acceptance	Almost no	32	0.32%
	Very few	39	0.39%
	More	29	0.29%
Software and hardware device performance	Weak	48	0.48%
	Need to be strengthened	45	0.45%
	Excellent quality	7	0.07%

are. Enterprises store these data information in the cloud, but the cloud back-end is manipulated and maintained by people. Therefore, the information maintenance personnel in the cloud can easily obtain various information of users, so it is difficult to protect the security of data information. After obtaining the user's data information, what to do with the data information and how to use it becomes something that is closely related to all of us. Various information data is processed by a certain program, and the information obtained by the program processing is more objective, but the obtained information is also limited. Once the data provider provides false information, it may have a large negative impact.

Secondly, old users may appear to be unsuitable or reject the sharing of accounting information resources. There may be some users who do not understand and trust the information sharing platform, and do not understand what kind of benefits can be brought to the enterprise through cloud computing and cloud accounting, which will cause greater resistance to the promotion of enterprise accounting informationization.

Finally, you may experience overloaded network traffic. Enterprise accounting information resource sharing system needs to rely on the Internet, which requires enterprises to be equipped with good network equipment, but for now, due to internal and external reasons, there are Internet congestion, resulting in a delay in data transmission, so overloaded networks Transmission will become a key issue in the process of building enterprise accounting information resources.

4 Discuss

4.1 Suggestions for the Establishment of a Distributed Accounting Resource Sharing Platform for Enterprises in the Internet Age

First, an important role in promoting the construction of accounting resource sharing platforms in the Internet era is to provide enterprises with a complete accounting resource information sharing platform, so that enterprises can fully understand the market trends and select resources that meet their own needs. In recent years, more and more enterprises have joined the ranks of accounting information resource sharing platforms, as shown in Fig. 1, but for most enterprises in China, the ones they need to build an information resource sharing platform are needed. Most of the hardware and software in the series come from foreign companies, and China has not achieved much research results in this regard. Therefore, China should invest more resources in this area, vigorously develop related equipment and software, so that more enterprises can use the equipment developed and produced by China independently, avoiding a large amount of capital outflow. For enterprises, starting the construction of accounting information resource sharing will be a long process. In this process, enterprises must invest a lot of human resources and part of the budget to carry out work, which undoubtedly increases the economic burden of enterprises and causes the speed of enterprise development. Slow down. At this time, if we rely solely on the strength of the enterprise to accomplish this important task, it will undoubtedly bring great risks to the enterprise. Therefore, the enterprise should provide full assistance for the construction of accounting information sharing and provide convenience in policy.

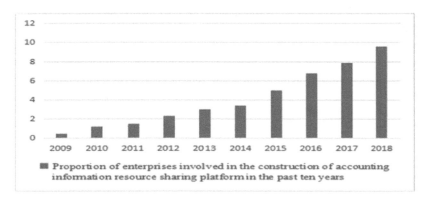

Fig. 1. Proportion of companies involved in the construction of accounting information resource sharing platforms in the past decade

Second, master the core data of the enterprise and build a safe and reliable platform for sharing accounting information resources. Cloud accounting is based on cloud storage and therefore requires a robust security mechanism with robust data information. However, there is also the possibility of information leakage. Therefore, enterprises need to analyze and judge according to the actual situation of the enterprise when using cloud accounting, and whether the more important accounting information should be transmitted into the cloud. Therefore, for the construction of enterprise accounting information resource sharing platform, we must first choose a safe and reliable cloud. Enterprises need to conduct comprehensive comparative evaluation of each cloud storage provider, and choose the cloud with the highest security level to ensure the security of enterprise cloud accounting processing. Sex, companies can flexibly combine business modules according to their actual situation. In addition, in order to ensure the security and reliability of corporate accounting information and prevent corporate accounting information from being leaked, it is necessary to conduct a security level assessment for each person who may be exposed to accounting information and to require identity authentication.

Third, changing corporate institutions has enabled new corporate organizations to leverage the role of the accounting resource sharing platform. At present, most enterprises in China have been using traditional accounting, and enterprises have insufficient understanding of shared accounting. Therefore, it is difficult to promote sharing of accounting information. For this problem, you can take a change to the enterprise to solve. Changing the organization setting can encourage every financial person in the enterprise to feel the efficiency of the sharing of accounting information resources. Enterprises need to configure institutions that are beneficial to the full use of shared accounting efficiency according to their actual situation.

5   Conclusion

With the development and popularization of information technology, enterprises have realized the importance of data information, how to apply information sharing construction to the enterprise management level and how to reduce the risk of information sharing and construction, which has become a management of each enterprise. Those should jointly explore the issues solved. In today's society, under the background of rapid development of information, under the Internet era, information construction is a double-edged sword, which can bring huge benefits to enterprises, and at the same time let enterprises bear a series of risks. Therefore, in the process of building an accounting information sharing platform, enterprises should avoid the risks brought by information sharing as much as possible, make full use of the advantages of information sharing platform, help enterprises expand development channels, and lay a solid foundation for the future development of enterprises.

References

1. Homburg, C., Nasev, J., Plank, P.: The impact of cost allocation errors on price and product-mix decisions. Rev. Quant. Finan. Account. **51**(2), 1–31 (2018)
2. Belval, E.J., Wei, Y., Calkin, D.E., et al.: Studying interregional wildland fire engine assignments for large fire suppression. Int. J. Wildland Fire **26**(7), 32–34 (2017)
3. Liberman, Z., Shaw, A.: Children use partial resource sharing as a cue to friendship. J. Exp. Child Psychol. **159**, 96–109 (2017)

4. Sun, Y., Zhang, N.: A resource-sharing model based on a repeated game in fog computing. Saudi J. Biol. Sci. **24**(3), 687 (2017)
5. Tak, B.C., Kwon, Y., Urgaonkar, B.: Resource accounting of shared IT resources in multi-tenant clouds. IEEE Trans. Serv. Comput. **132**(99), 1 (2017)
6. Matteazzi, E., Menon, M., Perali, F.: The collective farm-household model: policy and welfare simulations. Appl. Econ. Perspect. Policy **39**(1), 004–009 (2017)
7. Telem, D.A., Majid, S.F., Powers, K., et al.: Assessing national provision of care: variability in bariatric clinical care pathways. Surg. Obes. Relat. Dis. **13**(2), 281–284 (2017)
8. Hormiga, E., Saá-Pérez, P.D., Díaz-Díaz, N.L., et al.: The influence of entrepreneurial orientation on the performance of academic research groups: the mediating role of knowledge sharing. J. Technol. Transf. **42**(1), 10–32 (2017)
9. Duro, J.A., Teixidó-Figueras, J., Padilla, E.: The causal factors of international inequality in emissions per capita: a regression-based inequality decomposition analysis. Environ. Resour. Econ. **67**(4), 683–700 (2017)
10. Wang, F., Wang, J.D.: Telehealth and sustainable improvements to quality of life. Appl. Res. Qual. Life **12**(1), 1–12 (2017)

Students' Autonomous Learning in the Era of "Internet+" Big Data

Qunying Li[(✉)]

Department of Textile and Clothing Engineering,
Shandong Vocational College of Light Industry, Zibo 255300, Shandong, China
liqunying2000@126.com

Abstract. Under the social environment of "Internet+", both instructional design and systematic planning organization play an important role in the comprehensive analysis of students' autonomous learning ability. The information age has promoted the formation of many new teaching ideas and methods, but also posed a severe challenge to the traditional teaching methods. Improving the self-learning ability of college students is a trend in the teaching field of colleges and universities at present. Make full use of the Internet, big data and other technical means, explore new teaching links and methods, and build more effective teaching means, to achieve the purpose of improving college students' autonomous learning ability. This research topic is mainly based on the perspective of teachers. Meanwhile, on the premise of the social environment of "Internet+" big data, the three levels of instructional design, classroom organization and assessment are interpreted, so as to discuss students' autonomous learning, self-management and assessment ability.

Keywords: "Internet+" · Big data · Autonomous learning

1 Introduction

According to the field survey of the graduates of higher vocational colleges in Zibo City, among the graduates of higher vocational colleges who will graduate and enter the society, there are a large number of students whose skills obviously need to be improved [1–3]. Even some students do not know how to send mail with attachment materials and how to compress relevant documents. The lack of these basic computer skills has aroused the author's attention and concern. In discussing why students' skills can not meet the needs of social development, the author finds that students' abilities are not only related to their own abilities and social impact, but also inseparable from teachers' teaching methods [4–6]. Therefore, faced with the increasingly developing and deepening social environment of "Internet+" big data, teachers are not only responsible for imparting knowledge, but also for cultivating students' autonomous learning, self-management and assessment ability [7–10].

M. Atiquzzaman et al. (Eds.): BDCPS 2019, AISC 1117, pp. 340–344, 2020.
https://doi.org/10.1007/978-981-15-2568-1_47

2   Research Significance

2.1   The Needs of Students' Acquisition of Skills

For students, if they want to truly grasp the knowledge explained by teachers and turn it into their own use, they must have certain autonomous learning ability, so that they can realize their own views, analysis and absorption of relevant knowledge.

2.2   The Needs of Students' Survival

In the current social environment of the Internet, students have a certain ability of self-learning, which can not only broaden their ways of learning knowledge, but also enable them to better master the computer operation, so that they can timely find and properly solve problems. Only with this kind of self-learning skills can students have basic survival ability.

2.3   The Needs of Students' Development After Employment

In the process of self-learning, if students want to further enhance and strengthen their organizational and managerial abilities, they should not only formulate their own learning goals and plans, but also collect relevant learning resources, and conduct self-monitoring and summary evaluation of the learning process. In fact, this kind of self-learning ability has a very important influence and promotion on students and even on their future work.

3   Reason Analysis

In the actual teaching process, students were guided to explore independently by grouping mode, were guided to carry out group discussions by assigning teaching tasks before class, and were encouraged to discuss and communicate with each other in class. In practice, it is found that there are significant differences in the teaching effect among different groups. Some group members are very active in participating, they are usually able to complete the teaching tasks assigned by the teacher very well, and they can maintain good communication with other members and transform the knowledge explained by the teachers into their own use.

4   Optimization Strategy

4.1   Instructional Design

In fact, under the "project-based" teaching mode, the teaching work should follow the basic idea of "assigning tasks before class, discussing in class and summarizing after class". However, it is not difficult to find from a survey of the current situation of most vocational college students that due to the lack of self-management ability, many students tend to have the behavior of "free-riding" in autonomous learning, so the "assembly line" teaching method can be adopted. The so-called "assembly line" teaching means that the implementation of the next link should be based on the completion of the previous link. That is to say, each student is responsible for one link of the task, for example, the first student is responsible for the collection of relevant information, the second is responsible for the sorting and induction of these information, the third will make the courseware according to the materials, the fourth student will summarize the knowledge points and so on. In this way, this mode can not only strengthen the relationship between team members, but also ensure that each student must complete their own work within the prescribed time, otherwise it will affect the development of the next link, so as to prevent the occurrence of "free-riding" behavior to the greatest extent.

Compared with the traditional teaching mode, the development of this teaching mode must be based on modern electronic media as a basic tool. For example, students often use the Internet, WORD documents, PPT software and other tools to collect relevant learning materials and information. In this process, students' information collection, analysis and induction abilities can be better improved, and meanwhile, their computer software and other operational skills can be further strengthened. In addition, in the discussion and communication with the group members and teachers, their expressive ability can also be enhanced. More importantly, students can greatly improve their ability to analyze and solve problems, and enhance their team cooperation and time management ability through this teaching mode.

4.2   Classroom Organization

The so-called classroom organization mainly refers to the presentation of the implementation results of pre-class tasks in the classroom, which is realized through the explanation of relevant contents and the discussion of cases, which is essentially different from traditional teaching. Classroom organization pays special attention to students' principal position and can be said to be the place of students' thinking collision and interaction. Therefore, teachers should pay more attention to the effectiveness of organizational strategies and stimulate students' interest and participation to the greatest extent.

4.3   Assessment and Evaluation

First of all, the evaluation methods adopted by teachers are more standardized, which not only helps greatly reduce the workload of assessment and reduce the problems faced in the assessment process, but also makes the assessment relatively diversified. In the setting of assessment indicators, it can be set according to different links. There are four assessment criteria for each link, such as incomplete, general, good, very good, etc. In particular, in the specific assessment process, if there are improper or untrue acts, they must be eliminated. For serious violators, their points will be deducted and their behavior will directly link to their final grades. In this way, this method not only helps to ensure the standardization and rationalization of the assessment process, but also helps to ensure the fairness and rationality of the assessment results. At the end of the assessment, teachers can show the results of the assessment. Students who fail to pass the assessment should be punished more, so as to give full play to the encouragement and warning role of the assessment.

Secondly, teachers can actively integrate students' self-assessment, other assessment and teacher assessment into the final assessment in the process assessment. According to the above-mentioned teaching design, the development of each teaching link is based on the high-quality completion of the previous link and the completion quality of the previous link has a great influence on the development of the next link. Therefore, teachers must add students' self-evaluation, other evaluation and teacher evaluation in the assessment. Only in this way can the evaluation be fair and comprehensive. Four assessment criteria can be set in the self-assessment of students, such as incomplete, general, good, and very good. The evaluation only needs to consider the time and quantity of students' completion. However, quality evaluation should be made in other evaluation and teacher evaluation. Other evaluation means that the students in the next link evaluate the completion degree of tasks in the previous link, while teacher evaluation means that teachers conduct all-round evaluation on students' self-learning process. The addition of self-evaluation, other evaluation and teacher evaluation can not only minimize the subjectivity of teacher evaluation, but also help to ensure the accuracy and comprehensiveness of the final evaluation results.

5 Summary

Only by fully mobilizing students' autonomous learning consciousness and making them integrate into teaching activities can students' learning enthusiasm and creativity be stimulated and personalized development be promoted, and then students' exploratory spirit, team cooperation spirit, self-learning consciousness and lifelong learning consciousness be cultivated. Through the teaching process, students are realized that the purpose of university learning is not to graduate but to prepare for their career after graduation.

Acknowledgments. This research was supported by 2018 Zibo City School Integration Development Plan: Public Training Base for Textile and Garment Specialty.

References

1. Zhang, J.F.: On the improvement of college students' self-regulated learning ability from the perspective of big data. J. High. Educ. **13**, 76–79 (2019). (in Chinese)
2. Yao, D.: On the optimizing path of teaching management in applied universities under the background of big data. Learn. Wkly. **23**, 9–10 (2019). (in Chinese)
3. Zhang, R.S.: Design and implementation of educational intelligence platform based on big data. Mod. Electron. Tech. **42**(14), 91–94 (2019). (in Chinese)
4. Wang, Y.Y.: Changes in learning styles of Chinese university students in big data environment. Vocat. Technol. **18**(09), 52–55 (2019). (in Chinese)
5. Luan, C.W.: Application and practice of big data in higher vocational education teaching. Policy Res. Explor. **04**, 83–84 (2019). (in Chinese)

6. Zheng, J., Wang, Y.M.: Smart education model under ideas of "Internet + Education". J. North China Univ. Sci. Technol. (Soc. Sci. Ed.) (4), 76–79 (2016). (in Chinese)
7. Hou, H., Li, K.B., Zheng, W., Ma, L.: Research on the application of educational technology based on "Internet + big data". Chin. Mark. **1**, 219–220 (2018). (in Chinese)
8. Yuan, C.L., Mu, X.Y.: The innovation of instruction model in the background of "Internet + science education". Curric. Teach. Mater. Method **38**(8), 92–98 (2018). (in Chinese)
9. Fan, R.: Study on informatization construction of higher education based on intelligent campus under "Internet+" educational environment. Jiangsu Sci. Technol. Inf. **29**, 79–80 (2017). (in Chinese)
10. Sun, H.L.: Research on modern education technology based on "Internet+" and big data technology environment. Course Educ. Res. **47**, 225–226 (2018). (in Chinese)

Gene Network Cancer Prediction Method Based on Multi-group Algorithm

Ming Zheng and Mugui Zhuo[(✉)]

Guangxi Colleges and Universities Key Laboratory of Professional
Software Technology, Wuzhou University, Wuzhou, China
370505375@qq.com

Abstract. Multi-group-based identification of cancer markers is of great significance to the study of cancer molecular mechanisms, but most of the current work is based on protein-protein interaction data. Therefore, a novel approach based on multi-gene regulatory network and multi-group data is proposed to analyze the molecular mechanisms of cancer and predict biomarkers. Firstly, this method integrates multi-group data, and takes gastric cancer and esophageal cancer as examples to construct cancer-specific networks of gastric cancer and esophageal cancer respectively. Then, weighted co-expression network analysis is carried out on these two networks, and hierarchical clustering module is used to calculate the relationship between the first principal component of the module and all known cancer markers. Finally, cancer-specific modules are screened out. Then, disease-specific biological pathways were extracted and potential cancer markers were identified by similarity assessment. The experimental results show that the specific modules predicted by this method have functional characteristics, and the result of prediction using the correlation coefficient method in the module is more accurate.

Keywords: Cancer · Gene Co-expression Network · Gene expression regulation · Multigroup data

1 Introduction

Complex disease [1] is a kind of disease that does not obey the law of heredity. Its occurrence process involves many complicated biological processes and is controlled by a variety of genetic materials. In recent years, there has been an endless stream of research on cancer markers for complex diseases. With the development of high throughput technology [2], more and more biological data, such as gene expression data [3], somatic cell mutation data [4], protein interaction data [5], have been proved to be applicable to the prediction of cancer markers. After the recognition of cancer pathogenesis in biomedical field has entered the molecular level, researchers in other related fields have gradually carried out a variety of analysis of cancer markers.

Gastric cancer [6] and esophageal cancer [7] are complex diseases. There are no obvious symptoms in the early stage of gastric cancer and esophageal cancer. After diagnosis, they often spread to advanced cancer cells, so the fatality rate is very high. Therefore, pathological studies on gastric cancer and esophageal cancer have been in

© Springer Nature Singapore Pte Ltd. 2020
M. Atiquzzaman et al. (Eds.): BDCPS 2019, AISC 1117, pp. 345–350, 2020.
https://doi.org/10.1007/978-981-15-2568-1_48

progress. A large number of gene expression profiles of gastric cancer have been integrated in the comprehensive database of gene expression [8]. 433 differentially expressed genes in gastric cancer have been extracted, and these genes have been constructed into a co-expression network for network-level analysis. It has been successfully verified that COL1A2 [9] is highly expressed in gastric cancer, and this characteristic does not vary with patient's age, gender, and the change of tumor before and after.

Based on the biological background of gastric cancer and esophageal cancer, a new method was proposed to predict the potential cancer markers in the related pathogenic modules and pathogenic modules in the gene regulatory network [10]. By virtue of the connectivity characteristics of known cancer markers in the network, the maximum-minimum correlation difference between the first principal component of each module and known cancer markers is calculated. Specific modules are selected by ranking the values, and the relationship between biological pathways and modules within the module is analyzed. Five different distance measures are used to identify potential cancer markers in specific modules. The results show that the correlation-based method is more accurate in identifying potential cancer markers than the distance-based method and the correlation-based method.

2 Method

Firstly, abnormal methylation molecules were obtained in two kinds of cancers. Methylation abnormalities include hypermethylation and hypomethylation. Hypomethylation of some DNA leads to transcriptional activation of oncogenes, and hypermethylation of some tumor suppressor genes has been confirmed to be associated with its low expression. In order to construct a more specific network, methylation-specific molecules are selected as the nodes of the network, i.e. the methylation value of the molecules is greater than 0.8 or less than 0.2. These molecules are then used to select the regulatory and phosphorylation relationships under the following specific conditions: if there are genes or microRNAs in two molecules involved in a regulatory or phosphorylation relationship, it must be an abnormal methylation molecule; if there are transcription factors in two molecules involved in a regulatory relationship, the molecule must be in the expression profile. There are expression value vectors. Selected regulatory and phosphorylation relationships are regarded as the links of gene regulatory networks.

After the above steps, the nodes and edges of the two specific networks have been determined, and the network attributes are weightless undirected networks. In order to enhance the tightness and accuracy of the network, the current network is weighted and further adjusted to optimize the connection. Combining the network and the expression spectrum, we calculate the Spielman correlation coefficients of two nodes for each link in the network. If p-value < 0.05 and the Spielman correlation coefficient are absolutely greater than 0.3, then retain the connection in the network and use the Spielman correlation coefficient as the weight of the connection. After this operation is performed on all the edges, the unconsolidated edges are removed. Therefore, if the edges between a node and its neighbors are removed, the node will also be removed as an isolated node. Both networks belong to undirected weighted networks, and the weights of connecting edges are greater than 0.3.

Weighted Gene Co-expression Network Analysis (WGCNA) is generally used to analyze the expression patterns of multiple sample genes, and can be used to find highly related clusters and modules in the network. Topological overlap matrix is used as input of WGCNA. Compared with adjacency matrix, topological overlap matrix adds first-order correlation, which improves the accuracy of network description. The weight of i and j w_{ij} can be shown as Eq. (1) below:

$$w_{ij} = \frac{\sum\limits_{k} a_{ik} a_{kj}}{\min\{\sum\limits_{k} a_{ik}, \sum\limits_{k} a_{jk}\}} \tag{1}$$

The WGCNA module segmentation method for the constructed gastric cancer and esophageal cancer gene regulatory specific network is divided into two steps. Firstly, the network is clustered hierarchically, and then the hierarchical clustering tree is cut dynamically. Before hierarchical clustering, the adjacency matrix of the specific network is transformed into a topological overlap matrix to reduce the noise and false correlation in the original network, and the dissimilar topological overlap matrix is obtained and then hierarchical clustering is carried out. In this experiment, average-linkage method is used for clustering, calculating the distance between each data point of two groups and other data points, and taking the mean of all distances as the distance between two groups. This method is representative for most of the molecules in each cluster and can get representative results. Setting the minimum number of molecules in each module to 30, dynamic hybrid cutting of the two networks is carried out. On the one hand, all modules satisfying the following conditions are found: the number of molecules in the module to meet the set minimum number; removing distant branches from the module; the module is tightly connected by the central molecule. On the other hand, the non-attributable molecule is allocated to its nearest initial module. Finally, gastric cancer was divided into 14 modules and esophageal cancer into 10 modules.

On the basis of the module already divided, the module with strong specificity is selected for the key analysis. For each module, the first principal component method is used to calculate the module's eigenvector gene (ME). The eigenvector genes of each module are linearly combined by the molecular expression value vectors in the module, and their values are highly correlated with the molecular expression in the whole module. The correlation coefficients of all known cancer markers and module feature vector genes were calculated and calculated. The maximum and minimum method was used to calculate the difference between the maximum and minimum absolute correlation coefficients of each module. The greater the difference between absolute values, the more difference the module has in suppressing cancer markers. The authors select the top three modules in each network and rank the candidate cancer markers for the molecules in the modules. Candidate cancer markers refer to non-known cancer markers in specific modules, which measure the similarity between the module's eigenvector genes and the expression vectors of all candidate cancer markers. This paper assumes that the greater the correlation between a molecule and the eigenvector gene of the module, the more likely it is that the molecule is the central molecule of the module. In each specific module, each similarity measurement method produces a sort, and each top 10 molecule is selected for analysis and verification. Similarity assessment methods can be divided into two categories: distance-based measurement and correlation-based measurement. The metric based on distance includes Euclidean distance, Chebyshev distance and Chebyshev distance. Correlation based measurement methods include the correlation coefficient and the correlation coefficient.

3   Result

Constructing a gastric cancer-specific gene regulatory network by narrowing the range of regulatory and phosphorylation relationships to the molecular range of abnormal methylation. The characteristics of specific modules, the number of molecules in modules and the number of known cancer markers in these molecules selected by the maximum and minimum method are shown in Table 1. Six modules of the two diseases were enriched and analyzed, including Kyoto Encyclopedia of Genes and Genomes (KEGG), Gene Oncology (GO), Canonical Pathways and Reactome Pathway. The result can be shown as Fig. 1 below:

Fig. 1. The figure of gene regulatory network in this paper

From Fig. 1, we knows that The correlation coefficient and the correlation coefficient method were verified by 4 molecules, of which 3 molecules coincided. The TPR molecule predicted by the correlation coefficient method was used to analyze gastric cancer patients by using TPR probe. TPR has been found to inhibit gastric cancer. The only molecule in the correlation coefficient method was UBQLN2, and the expression level of UBQLN2 in the cancer cells was higher than that in the adjacent tissues. Distance-based method and correlation-based method have distinct differences in results. Euclidean distance and Chebyshev distance have only one identical molecule to be verified, while the SOCS1 distance is based on the first two. Studies on the anti-tumor mechanism of SOCS1 have shown that SOCS1 can inhibit the proliferation of 80% of gastric cancer cell lines, which means that SOCS1 can be used as a new treatment for gastric cancer. The correlation coefficient is outstanding, and 6 molecules are fully verified. The correlation coefficient is also verified by 5 molecules. Compared with the correlation coefficient method, only TIMM17B is missing.

The experimental results show that the distance-based measurement method is not as good as the correlation-based method in the five modules. In the three distance based measurement methods, the Manhattan method does not lose to the other two in each module. In the two correlation based measurement methods, the correlation coefficient is superior to the correlation coefficient in the 5 modules.

4   Conclusion

The specific networks of gastric cancer and esophageal cancer were constructed based on a variety of data. The interaction between transcription factors and microRNAs, the regulation of transcription factors to genes and the regulation of microRNAs to genes, and the phosphorylation modification of proteins are considered. These relationships are used as the links of specific networks, and the abnormal methylation molecules in gastric cancer and esophageal cancer are used as nodes of networks. Such operations can simplify the network scale. Methylation is very important in epigenetics and is of great help to the study of cancer markers [14]. Five similarity measures are used to calculate the ranking of candidate molecule by predicting candidate cancer markers within the specific module and combining the characteristic genes of the module. Five methods of similarity assessment were compared and the accuracy of the five methods was evaluated through literature verification. Experiments show that in distance-based and correlation-based methods, correlation-based methods perform better. Further, in the two correlation based methods, Pearson correlation coefficient method performs better than Spielman correlation coefficient method in this experimental data. In the future, similar methods can be applied to other complex diseases to explore the molecular mechanism of complex diseases.

Acknowledgments. This work was supported by grants from The National Natural Science Foundation of China (No. 61862056), the Guangxi Natural Science Foundation (No. 2017-GXNSFAA198148) foundation of Wuzhou University (No. 2017B001), Guangxi Colleges and Universities Key Laboratory of Professional Software Technology, Wuzhou University.

References

1. Kamalnath, M., Rao, M.S., Umamaheswari, R.: Rhizophere engineering with beneficial microbes for growth enhancement and nematode disease complex management in Gherkin (Cucumis anguria L.). Sci. Hortic. **257** (2019)
2. Engelward, B., Ngo, L.P., Swartz, C., et al.: High throughput microarray technology for genotoxicity and cytotoxicity. Environ. Mol. Mutagen. **60**, 37 (2019)
3. Shukla, A.K., Singh, P., Vardhan, M.: A new hybrid wrapper TLBO and SA with SVM approach for gene expression data. Inf. Sci. **503**, 238–254 (2019)
4. Yasuda, H., Kobayashi, S., Costa, D.B.: EGFR exon 20 insertion mutations in non-small-cell lung cancer: preclinical data and clinical implications. Lancet Oncol. **13**(1), E23–E31 (2012)
5. von Mering, C., Krause, R., Snel, B., et al.: Comparative assessment of large-scale data sets of protein-protein interactions. Nature **417**(6887), 399–403 (2002)
6. Li, J., Qin, S.K., Xu, J.M., et al.: Randomized, double-blind, placebo-controlled phase III trial of apatinib in patients with chemotherapy-refractory advanced or metastatic adenocarcinoma of the stomach or gastroesophageal junction. J. Clin. Oncol. **34**(13), 1448–1454 (2016)
7. Enzinger, P.C., Mayer, R.J.: Medical progress - esophageal cancer. N. Engl. J. Med. **349**(23), 2241–2252 (2003)
8. Edgar, R., Domrachev, M., Lash, A.E.: Gene Expression Omnibus: NCBI gene expression and hybridization array data repository. Nucleic Acids Res. **30**(1), 207–210 (2002)
9. Chamberlain, J.R., Deyle, D.R., Schwarze, U., et al.: Gene targeting of mutant COL1A2 alleles in mesenchymal stem cells from individuals with osteogenesis imperfecta. Mol. Ther. **16**(1), 187–193 (2008)
10. Shinozaki, K., Yamaguchi-Shinozaki, K., Seki, M.: Regulatory network of gene expression in the drought and cold stress responses. Curr. Opin. Plant Biol. **6**(5), 410–417 (2003)

Mobile Communication De-noising Method Based on Wavelet Transform

Yixin Tang[(⊠)]

Southwest Jiaotong University (Xipu Campus),
Pidu District, Chengdu, Sichuan 611756, People's Republic of China
ell6y3t@leeds.ac.uk

Abstract. Signal and information processing is one of the most rapidly developing subjects in information science in the past 20 years. Signal processing mainly includes signal noise elimination, feature extraction and edge extraction. Signal noise elimination is the most extensively method in signal processing. Traditional signal noise elimination methods such as pure time domain method, pure frequency domain method, Fourier transform and windowed Fourier transform have their own application defects. The wavelet transform is a new temporal frequency joint analyze method developed in 1982. This method has good time domain and frequency localization characteristics, It is widely used in wavelet transform. With the continuous development of wavelet theory, the idea of wavelet packet transform came into being. Wavelet packet transform is a more sophisticated time-frequency joint analysis method than wavelet transform. Wavelet analysis is a new signal processing method, which decomposes various frequency components into non-overlapping frequency bands, which provides an effective method for signal filtering, signal-to-noise separation and feature extraction. In this paper, the wavelet transform is used to eliminate noise in the channel and image.

Keywords: Wavelet transform · Mobile communications · Noise elimination method · Signal de-noising

1 Introduction

It is inevitably affected by noise in the process of transmission and acquisition of mobile communication information. Because signals with less noise interference, the content of the channel transmission can be identified based on people's experience. For signals with large noise interference, it is often necessary to use professional noise cancellation technology for processing, and the transmission signal can be re-identified after removing the interference [1]. How to effectively eliminate the noise of the transmitted signal is an important issue in the field of mobile communication signal transmission processing.

Conventional signal noise elimination methods include Fourier transform based noise elimination methods. Since the Fourier transform starts, local characteristics of the transmitted signal [2] cannot be processed. The result removes the detail features in the noise signal. In recent years, median filtering, mean filtering noise elimination

© Springer Nature Singapore Pte Ltd. 2020
M. Atiquzzaman et al. (Eds.): BDCPS 2019, AISC 1117, pp. 351–358, 2020.
https://doi.org/10.1007/978-981-15-2568-1_49

processing methods, although noise elimination methods can retain some details than the traditional Fourier transform-based features, the simulation results show that the median filtering noise elimination method is used to treat Including the type of impulse noise, it also leads to excessive detail loss, average filtering noise elimination and smoothing methods, image edge information is destroyed as [3]. Low-pass filtering and Wiener filtering also have large differences in the noise elimination effect when dealing with different noise types. These methods rarely consider the local color of this transmitted signal, it is difficult to retain the detailed information of some signals after noise elimination [4].

These problems can be solved with wavelet analysis. The point of wavelet theory is derived from the extension and translation of signal analysis [5]. There was a time when time-frequency localization analysis project with fixed window size (IE window area), variable window shape, time window and variable frequency window. It is the low frequency part that has higher frequency resolution and lower time resolution, However, the high frequency part has a high time resolution and a low frequency resolution, which is very suitable for detecting transient abnormalities in normal signals and displaying their components [6]. The noise elimination method of wavelet analysis transform compensates for the shortcomings of the previous methods in terms of local detail loss. Starting from the details of the signal transmitted during the mobile communication process, the wavelet transform can eliminate noise while preserving the details of signal [7]. The noise elimination means based on wavelet transform can successfully separate the image coefficients and noise coefficients of the signal to be processed after wavelet transform, and obtain the noise elimination signal after separation and wavelet reconstruction. Noise elimination methods based on wavelet transform usually include modulus maxima method, space domain method and threshold method [8]. Among them, the threshold noise elimination method based on wavelet transform is a commonly used noise elimination method, which is mainly divide soft threshold method, hard threshold method and improved adaptive threshold method [9]. Through simulation experiments, hard threshold value and existing threshold value function noise elimination images have the disadvantages of noise residual and signal detail blur.

2   Research of the Methods

2.1   Overview of Wavelet Analysis

The fundamental difference is that the localized properties of wavelet and sine wave are different. From a macro perspective, Fourier analysis is the entire domain analysis, Using a single time domain or frequency domain signal characteristics [10]. The wavelet analysis is the local time-frequency analysis way combining frequency domain and time domain. As a time-frequency analysis method, wavelet analysis has many essential advances compared with Fourier analysis. It can extract a lot of useful information from the signal. It is a unified processing framework for various time-frequency analysis and other signal processing methods, multi-scale analysis, and sub-band coding. The fast algorithm for analyzing and solving practical problems brings great convenience, sound and image. It has a good application in the fields of graphics, communication, biomedicine, mechanical vibration, and computer vision. Wavelet analysis is a high-tech of signal information acquisition and processing, and has been widely recognized internationally.

2.2 Selection Method of Threshold

Wavelet noise elimination is divide three steps: wavelet decomposition, high-frequency coefficient threshold quantization and signal reconstruction. The noise elimination effect lie on the choose of wavelet basis, the determination of the number of wavelet decomposition layers, the selection of the threshold function and the threshold estimation way. One of the most important parts is how to choose the threshold function and how to quantization the threshold. Because the noise is a random signal, the variance is unknown. In nowadays noise elimination process, the threshold must be estimated first. According to the sample selection of the threshold, the estimation is estimated. The principle of the signal determines a uniform threshold and then remains above the threshold coefficient and the coefficient is removed below the threshold. At present, there are four popularly used threshold estimation methods. It includes: fixed threshold method, adaptive threshold selection method based on Stein's unbiased likelihood estimation principle, heuristic threshold method and maximal minimum threshold method.

In the noise elimination experiment of Gaussian white noise as a signal, it is found that the minimum nuisance and the threshold rule are used for noise elimination, and only the partial coefficient is set to 0, and about 3% is reserved. De-noising is performed using the Sqtwolog and heusure rules, all wavelet coefficients are changed to 0, and noise elimination is completed.

It can be seen that the Minimaxi and SURE threshold rules are relatively conservative. These two thresholds are useful for extracting weak signals when only a small fraction of the high frequency information containing the noise signal is in the noise range. The Sqtwolog and heusure rules have relatively complete noise elimination effects and are more effective in noise elimination, but it is easy to de-conscribe useful high frequency signals as noise and noise.

According to the relationship between the whole and noise elimination part, it can be divide global threshold noise elimination and hierarchical threshold noise elimination. The threshold global processing is to filter the high frequency coefficients obtained by each order wavelet decomposition with the same threshold, and the threshold grading processing filters each order wavelet decomposition based on the threshold. In theory, the layered threshold is selected according to the characteristics of each layer coefficient, and the processing of noise signals is more flexible.

2.3 Signal Wavelet Transform

Set for an exact definition of the wavelet function: for any function bits (t), the Fourier transform to bits (omega), if meet the conditions:

$$C_{\Psi} = \int_R \frac{|\psi(\omega)|^2}{|\omega|} d\omega < \infty \tag{1}$$

Says $\Psi(t)$ as a fundamental wavelet or Parent wavelet functions. The wavelet generating function was scaled and evaluated, and its scaling silver is set as a scaling silver is set as b. In addition, its scaling function is:

$$\Psi_{a,b}(t) = |a|^{-1/2}\Psi[(t - b)/a] \tag{2}$$

Call it a wavelet basis function.

For a basic wavelet, the persistent wavelet conversion of any affect $f(t) \in L^2(R)$ is defined as:

$$W_{f(a,b)} = <f(t), \Psi_{a,b}(t)> = \int_R f(t)\overline{\Psi}_{a,b}(t)dt = |a|^{-1/2} \int_R f(t)\overline{\Psi}(\frac{t-b}{a})dt \tag{3}$$

Its inverse transformation is:

$$f(t) = C_{\Psi}^{-1} \int_{-\infty}^{\infty} \int_{-\infty}^{\infty} \Psi_{a,b}(t)W_f(a,b)\frac{da}{a^2}db \tag{4}$$

2.4   Determination Method of De-noising Affect

The basis for judging the noise elimination effect are Signal-to-noise ratio and minimum mean square error. For the sake of getting the best noise elimination effect, not only to choose the appropriate wavelet function, but also to determine the optimal number of decomposition layers, choose the appropriate threshold. For the noise signal with a certain signal-to-noise ratio, the wavelet function, the decomposition layer number and the threshold method are improved respectively. Through a large number of comparative simulation experiments, the best noise elimination method is found. Then the noise elimination method is improved to make the noise elimination method universal.

3   Experiment

Wavelet packet analysis is a new way based on wavelet analysis. With wavelet analysis, Wavelet packet analysis will offer a better way for signal analysis.

In the processing of the signal, it can analyze according to the characteristics of the signal. The band is adaptively selected to match the signal spectrum, which is more suitable for analyzing non-stationary signals. This chapter mainly discusses the threshold-based wavelet transform noise elimination method. Through design experiments, the wavelet threshold is set to process the noise of the transmitted signal, also the simulation consequence are used to verify the effect of wavelet transform processing in the transmission process of mobile communication signals.

To verify the effectiveness of the algorithm, standard median filtering (SMF) (3 × 3 window), limit median filtering (5 × 5 window), adaptive median filtering and algorithm were compared. This paper combines the objective and subjective criteria of the channel noise elimination quality assessment method to evaluate the noise cancellation effect of the transmitted signal. When we talk about the objective standard, the peak SNR (Signal to Noise Ratio) parameter is used to analyze the effects of various noise cancellation methods, namely 0PSNR (standard/adaptive/extreme median).

3.1 Noise Processing by Wavelet Transform

The signal de-noisingprocess of wavelet transform can be divided into the following four processes:

First of all, The noise signal wavelet decomposition, choice of wavelet, and then for noise signal with n level wavelet decomposition.

Secondly, determine the optimal wavelet base for wavelet signal decomposition: for a given entropy criterion, compute the best tree and determine best wavelet base.

Thirdly, the quality of wavelet decomposition coefficients: for each wavelet disintegrate coefficients (especially low-frequency decomposition coefficients), select the appropriate threshold to quantify the coefficients.

Finally, the signal is reconstructed by wavelet transform: according to n-layer wavelet packet decomposition coefficients and threshold quality coefficients, the signal is reconstructed by wavelet packet.

4 Discuss

4.1 Selection and Analysis of Threshold

Wavelet noise elimination is divide three steps: wavelet decomposition, high-frequency coefficient threshold quantization and signal reconstruction. The effect of noise elimination lie on the choose of wavelet basis and the determination of the decomposition level of wavelet, the selection of the threshold function and the threshold estimation method. One of the most important parts is how to choose the critical value function and how to quantify the critical value. Because the noise is a random signal, the variance is unknown. In fact noise elimination course, the extremity must be estimated first. According to the sample selection of the threshold, the estimation is estimated. The principle of the signal determines a uniform threshold and then remains above the threshold coefficient and the coefficient is removed below the threshold. In nowadays, there have four Currently used threshold estimation ways: fixed threshold means, adaptive threshold selection method, heuristic threshold method and maximal minimum threshold method.

In the noise elimination experiment of Gaussian white noise as a signal, it is found that the minimum nuisance and the threshold rule are used for noise elimination, and only the partial coefficient is set to 0, and about 3% is reserved. De-noising is performed using the Sqtwolog and rules, all wavelet coefficients are changed to 0, and noise elimination is completed.

It can be seen that the Minimaxi and SURE threshold rules are relatively conservative. These two thresholds are useful for extracting weak signals when only a small fraction of the high frequency information containing the noise signal is in the noise range. The Sqtwolog and rules have relatively complete noise elimination effects and are more effective in noise elimination, but it is easy to de-conscribe useful high frequency signals as noise and noise.

According to the relationship between the total and the part, it can be divide global threshold noise elimination and hierarchical threshold noise elimination. The threshold global processing is to filter the high frequency coefficients obtained by each order wavelet decomposition with the same threshold, and threshold grading processing filters each order wavelet decomposition based on the threshold. In theory, the layered threshold is selected according to the characteristics of each layer coefficient, and the processing of the noise signal is more flexible.

4.2   Determination of De-noising Effects

Wavelet noise elimination is divide three steps: wavelet decomposition, high-frequency coefficient threshold quantization and signal reconstruction. The effect of noise removal rely on the selection of wavelet bases, the determination of wavelet decomposition levels, the selection of threshold functions and the method of threshold estimation. The key part is the way to choose the threshold function and how to quantify the threshold. Since the noise is a random signal, the variance is unknown. In the common noise elimination process, the threshold must be estimated first. The estimation is based on the sample selection of the threshold. The key of the signal is to first determine a uniform threshold and then remain above the threshold coefficient to move the coefficient below the threshold. At present, there are four common estimation methods.

In the noise elimination experiment of the Gaussian white noise signal, it was found that the minimum pollution method and the threshold method were used to de-noise, and only the bias coefficient was set to 0, and about 3% was reserved. De-noising is performed using the Sqtwolog and rules, and all wavelet coefficients are changed to 0 to complete the noise elimination.

It can be seen that the minimum and threshold rules are relatively conservative. These two thresholds are useful for extracting weak signals when only a small fraction of the high frequency information containing the noise signal is in the noise range. The Sqtwolog rule and the rule have relatively complete noise elimination effects and better noise elimination effects, but it is easy to eliminate noise useful high frequency signals into noise and noise.

According to the relationship between the whole and the part, it can be divide global threshold de-noising and hierarchical threshold de-noising. The threshold global processing is to filter the high frequency coefficients obtained by each order wavelet decomposition by the same threshold, and the threshold grading processing filters each order wavelet decomposition according to the threshold. The layered threshold is selected according to the characteristics of each layer coefficient, and the processing of the noise signal is more flexible. Wavelet noise elimination and wavelet packet de-noising are carried out below and compared with the noise elimination effect based on Fourier transform. The comparison consequence are exhibition in the following Table 1:

Table 1. Signal-to-noise ratio after de-noising of each method

Original signal to noise ratio	5	10	15
Fourier transform de-noised signal-to-noise ratio	8.9	16.3	25.6
Signal to noise ratio (SNR) after de-noising by wavelet transform	12.3	21.6	30.9
Wavelet packet transform de-noising signal to noise ratio	12.8	22.3	31.8

By using the threshold function to de-noising the transmitted signal, the improved threshold function can change the detail processing of transmitted signal better from subjective analysis than the noise image. Compared with other signals, the improved threshold function in this paper is superior to other function methods in terms of distortion and ambiguity of transmission signals. In order to further evaluate the de-noising effect of the transmission signal with noise, the evaluation index is used as shown in Fig. 1.

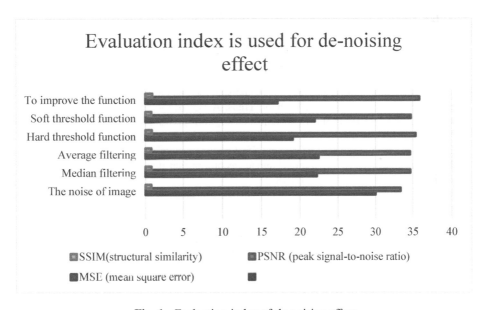

Fig. 1. Evaluation index of de-noising effect

5   Conclusion

After the de-noising effect comparison experiment found: First, the use of wavelet method to remove various signals (especially the non-stationary signal) contains white noise is the key, select the appropriate wavelet decomposition layer according to the specific signal, and choose a suitable threshold rule The signal can be de-noised best. Secondly, as the signal-to-noise rate of the signal to be analyzed decreases, the SNR (Signal to Noise Ratio) rate of the de-noised signal will be improved, but the noise elimination effect is not very obvious, which needs further study.

The key theories of wavelet noise elimination are discussed, and the advantages and disadvantages of various noise reduction methods are compared. At the same time, some understanding and explanation of wavelet noise elimination have been carried out.

It is not difficult to see from this study that in the research and development of mobile communication, the noise elimination processing of transmission signals is the key to research and development. It is feasible to use the existing wavelet transform denoising method.

References

1. Costa, F.B., Monti, A., Paiva, S.C.: Over current protection in distribution systems with distributed generation based on the real-time boundary wavelet transform. IEEE Trans. Power Delivery **99**, 1 (2017)
2. Upadhyay, S.K., Singh, R.: Abelian theorems for the Bessel wavelet transform. J. Anal. **23**, 1–12 (2017)
3. Tsai, M.J., Hsieh, C.Y.: The visual color QR code algorithm (DWT-QR) based on wavelet transform and human vision system. Multimedia Tools Appl. **78**(15), 21423–21454 (2019)
4. Amri, B.: Toeplitz operators for wavelet transform related to the spherical mean operator. Bull. Braz. Math. Soc. News **1**, 1–24 (2018)
5. Kumar, S., Sarfaraz, M., Ahmad, M.K.: Denoising method based on wavelet coefficients via diffusion equation. Iran. J. Sci. Technol. Trans. A Sci. **6**, 1–6 (2017)
6. Li, J., Tong, Y., Li, G., et al.: A UV-visible absorption spectrum de-noising method based on EEMD and an improved universal threshold filter. RSC Adv. **8**(16), 8558–8568 (2018)
7. Abdelkader, R., Kaddour, A., Derouiche, Z.: Enhancement of rolling bearing fault diagnosis based on improvement of empirical mode decomposition de-noisingmethod. Int. J. Adv. Manuf. Technol. **97**(13), 1–19 (2018)
8. Castillo, J., Mocquet, A., Saracco, G.: Wavelet transform: a tool for the interpretation of upper mantle converted phases at high frequency. Geophys. Res. Lett. **28**(22), 4327–4330 (2018)
9. Genakos, C., Valletti, T.M., Verboven, F.: Evaluating market consolidation in mobile communications. Soc. Sci. Electron. Publishing **33**(93), 45–100 (2017)
10. Pachón, Á., Palomares, U.M.G.: Mid-term frequency domain scheduler for resource allocation in wireless mobile communications systems. Comput. Commun. **97**, 96–110 (2017)

Dynamic Assignment Model of Terminal Distribution Task Based on Improved Quantum Evolution Algorithm

Jingjing Jiang[1(✉)], Sheng Guan[1], and Xiangwei Mu[2]

[1] School of Digital Technology, Dalian University of Science and Technology,
Dalian 116052, Liaoning, China
70612101@qq.com

[2] School of Maritime Economics and Management, Dalian Maritime University,
Dalian 116026, Liaoning, China

Abstract. Most logistics and express delivery enterprises will allocate distribution tasks according to fixed distribution service areas at the end of their distribution business. It is impossible to set up reasonable distribution resources for the dynamic distribution demand that changes frequently and is unevenly distributed, which causes the unbalanced workload of each terminal distribution node and further leads to the distribution scheduling management chaos and other problems. A distribution task assignment model considering distribution cost, resource utilization and workload ratio difference is established to solve the problem of terminal distribution task assignment. The stability of quantum group is adopted as the exit criterion to avoid the problem of early withdrawal and invalid iteration of the algorithm, and the mechanism of quantum variation and elimination is introduced to enhance the search ability of the algorithm for feasible solutions. The experimental results show the scheme presented by the algorithm can effectively alleviate the uneven distribution of distribution tasks and effectively reduce the overall distribution cost that compared with the distribution scheme according to distribution area.

Keywords: Terminal distribution · Quantum evolutionary algorithm · Quantum stability · Quantum mutation

1 Introduction

The task assignment of the terminal distribution network mainly solves how to allocate a number of distribution tasks to a number of distribution networks. The end of the distribution network task allocation is the beginning of the "last mile" distribution business process, in the related research, long-distance distribution problem has been well solved, and at the end of "the last kilometer" distribution became the urgent problems at present stage [1]. The relevant empirical research also proves that the service quality of terminal distribution is very important to the development of e-commerce [2]. The current distribution task allocation method is faced with problems such as inability to adjust flexibly according to dynamic demands, insufficient utilization of distribution resources and unbalanced workload. In this paper, considering

© Springer Nature Singapore Pte Ltd. 2020
M. Atiquzzaman et al. (Eds.): BDCPS 2019, AISC 1117, pp. 359–367, 2020.
https://doi.org/10.1007/978-981-15-2568-1_50

distribution node ratio of utilization, distribution costs and capacity building distribution in three aspects: the task allocation model, and the traditional quantum evolutionary algorithm was improved, enable it to dynamically generate to adapt to the demand for real-time changes the distribution of task allocation scheme.

2 Problem Analysis

Compared with the middle and front links of express business, terminal delivery is faced with more complex business scenarios, such as large scale of customers, variable locations, complex goods types and uncertain quantity. In the condition of mass distribution tasks, reasonably allocating each end node distribution tasks will help distribution path optimization and scientific scheduling. It will be one of the key issues that express logistics enterprises needs to solve in enhancing the user experience and optimizing logistics resources configuration.

2.1 Problem Hypothesis

Assumption need to make task allocation for $K(k = 1, 2, 3....K)$ terminal distribution nodes and $I(i = 1, 2, ..., I)$ delivery requirements in terms of distribution, known each distribution capacity of distribution network b_k, distribution capabilities for a distribution node all the sum of the load distribution resources. For each delivery demand d_i, l_{ki} is set to node location and distribution location, all distribution of distance matrix L of the nodes and cargo demand, matching x_{ki} is defined as the distribution network and distribution requirements

$$x_{ki} = \begin{cases} 1, & \text{Distribution task i is assigned to distribution network k.} \\ 0, & \text{Don't assign.} \end{cases}$$

All matching x_{ki} matrices X is a task assignment scheme for all distribution points.

$$X = \begin{pmatrix} x_{11} & \cdots & x_{1I} \\ \vdots & \ddots & \vdots \\ x_{K1} & \cdots & x_{KI} \end{pmatrix}$$

2.2 Problem Objective

(1) End-node utilization rate R

Maximization of utilization, distribution terminal node R is to make full use of the single node distribution ability, as much as possible for each distribution outlets assigned tasks within the scope of its distribution capabilities, a scheme of cubed out as follows:

$$R = \frac{\sum_{i=1}^{I} \sum_{k=1}^{K} x_{ki}(d_i/b_k)}{K} \tag{1}$$

(2) Total cost of delivery scheme C

The cost function C in this paper uses the actual travel distance l_{ik} between the location of the terminal distribution node and the location of distribution demand to describe. The total cost of an allocation scheme is:

$$C = \sum_{i=1}^{I} \sum_{k=1}^{K} l_{ki} x_{ki} \tag{2}$$

(3) Differences in working proportion of terminal nodes σ

A distribution node at the end of the work load ratio is assigned to the node distribution task demand combined with the ratio of the nodes distribution capacity. μ is set to represent the average of the workload distribution ratios of all terminal distribution nodes. Then, the workload difference degree of the distribution scheme of a distribution task outlet is:

$$\sigma = \sqrt{\frac{1}{K} \sum_{k=1}^{K} \left(\frac{\sum_{i=1}^{I} d_i x_{ki}}{b_k} - \mu \right)^2} \tag{3}$$

2.3 Model Construction

Based on the above model parameters, hypothesis conditions and target function analysis, the distribution task distribution model is as follows:

$$\max \ Z = \begin{pmatrix} w_1 & w_2 & w_3 \end{pmatrix} \begin{pmatrix} R \\ (C+1)^{-1} \\ (\sigma+1)^{-1} \end{pmatrix} \tag{4}$$

$$s.t. \quad \sum_{k=1}^{K} x_{ki} = 1, \forall i \tag{5}$$

$$0 \le \sum_{i=1}^{I} d_i x_{ki} \le g b_k, \quad \forall k, g \ge 1 \tag{6}$$

$$d_i x_{ki} \le b_k, \ \forall i, \ \forall k \tag{7}$$

$$i = 1, 2, \ldots, I \tag{8}$$

$$k = 1, 2, \ldots, K \tag{9}$$

3 Algorithm Design and Improvement

Compared with the traditional evolutionary computation method, Quantum Evolutionary Algorithm is characterized by large space search with small population size and strong global search capability [3, 4]. Because of the parallelism of quantum computation, the complexity of the algorithm can be greatly reduced [5, 6], which is widely used to solve combinatorial optimization problems that require a large amount of computational space [7–9].

3.1 Improvement of Algorithm Mechanism

3.1.1 Algorithm Exit Mechanism Improvement

The stability of quantum population is used as one of the criteria for algorithm exit. The more uncertain the quantum state, the higher the searching ability of quantum groups [10]. The stability of quantum individuals is defined as

$$S(Q) = \frac{\sum_{n=1}^{K \times I} |\alpha_n^2 - \beta_n^2|}{K \times I} \tag{10}$$

3.1.2 Algorithm Elimination Mechanism

In order to ensure that the feasible quantum conforms to the constraint formula (5), the variation method of the quantum of the feasible solution is mainly aimed at the quantum-encoded bits corresponding to any two columns in the scheme, and the exchange probability is related to the stability of the quantum individual S(Q).

$$P^{ij} = \begin{cases} 0, & th_s^{ij} > S(Q) \\ 1, & th_s^{ij} \leq S(Q) \end{cases} \tag{11}$$

3.2 Algorithm Process

The process of the algorithm is divided into six stages: creating quantum swarm, calculating fitness, individual optimization, exit judgment, algorithm evolution and decoding

Step 1. Quantum group of initialization: when making quantum population initialization, task allocation should be given at the end of first distribution node related parameters of the model, given the number of individual quantum, the length of each individual quantum bits, random the initial measurement threshold and individual quantum bit, quantum states. $K, I, L, \ b_k, d_i, \ w_1, w_2, \ w_3, g \ P(K \times I) th_n$ $\alpha = \beta = \sqrt{0.5}$

Step 2. Individual fitness calculation of quantum: determine the initial measurement value of each individual quantum. The fitness of each individual quantum is obtained.

Step 3. Select the optimal quantum individuals: if the present feasible solution quantum group, will be feasible solution of quantum individual and history of the optimal individual fitness, save the quantum states of quantum individual fitness maximum feasible solution. If no feasible solution exists, the quantum state of the optimal non-feasible solution individual is preserved.

Step 4. Algorithm exit judgment: the judgment condition for algorithm exit is that the quantum group entropy is greater than a given threshold value, and the successful algorithm exit judgment will enter step 6, otherwise, the algorithm will turn to step 5.

Step 5. Quantum swarm evolution: the angle increment of a given quantum revolving door, set the rotation of the strategy and each individual quantum bits of quantum state was improved, generating a new generation of quantum groups, according to the formula (11) for possible quantum individual mutation, generate new quantum groups, while eliminating the same size of the worst quantum individuals. The algorithm turns to step 2.

Step 6. Optimal quantum individual decoding: the measurement value of the quantum state of the historically optimal quantum individual is analyzed into the task assignment scheme X and the output results.

4   Experimental Analysis

4.1   Experimental Data and Algorithm Parameter Setting

The distance matrix data is shown in Table 1, in which b_k represents the distribution capacity of each terminal distribution node and d_i is listed as the transportation demand of each distribution task.

Table 1. Experimental data of customer points and distribution nodes

	n1	n2	n3	n4	n5	d_i
c1	0.39	0.67	0.23	0.47	0.24	0.03
c2	0.31	0.02	0.56	1.14	0.85	0.04
c3	0.26	0.55	0.40	0.64	0.31	0.03
c4	0.80	1.05	0.50	0.30	0.45	0.05
c5	0.73	1.01	0.54	0.12	0.23	0.07
c6	0.54	0.80	0.74	0.72	0.43	0.01
c7	0.60	0.89	0.69	0.54	0.29	0.04
c8	0.65	0.87	0.31	0.46	0.46	0.01
c9	0.08	0.22	0.41	0.92	0.62	0.09
c10	0.20	0.46	0.50	0.80	0.47	0.02
c11	0.74	0.93	0.37	0.54	0.57	0.01
c12	0.55	0.85	0.50	0.34	0.04	0.04
c13	0.84	1.12	0.62	0.07	0.36	0.10
c14	0.45	0.58	0.12	0.74	0.59	0.08
c15	0.48	0.66	0.11	0.62	0.50	0.02

(*continued*)

Table 1. (*continued*)

	n1	n2	n3	n4	n5	d_i
c16	0.37	0.62	0.16	0.52	0.31	0.03
c17	0.85	1.13	0.63	0.06	0.36	0.04
c18	0.21	0.37	0.57	0.93	0.61	0.02
c19	0.07	0.30	0.31	0.83	0.54	0.10
c20	0.13	0.28	0.29	0.86	0.58	0.10
c21	0.30	0.54	0.11	0.61	0.38	0.03
c22	0.15	0.19	0.37	0.94	0.66	0.05
c23	0.19	0.28	0.28	0.88	0.62	0.10
c24	0.25	0.47	0.13	0.68	0.44	0.05
c25	0.33	0.63	0.32	0.52	0.21	0.01
c26	0.93	1.21	0.70	0.12	0.44	0.04
c27	0.10	0.20	0.42	0.94	0.64	0.06
c28	0.47	0.77	0.51	0.47	0.15	0.04
c29	0.63	0.91	0.73	0.58	0.33	0.03
c30	0.49	0.77	0.34	0.36	0.14	0.02
b_k	0.76	0.3	0.88	0.6	0.48	

4.2 Analysis of Experimental Results

The evolutionary process of the algorithm is shown in Fig. 1, which shows the evolution process of the optimal fitness of quantum groups, the average fitness of modern times and the stability of quantum groups.

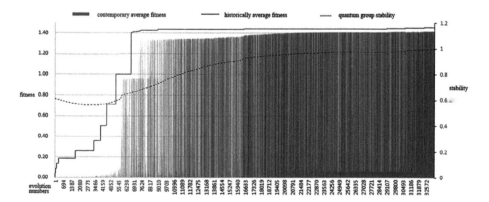

Fig. 1. Schematic diagram of algorithm evolution process

Each distribution requirements according to the distribution area of task allocation scheme as shown in Fig. 2(a), including distribution node at the end of the distribution capabilities through the size of the rectangle, said five PeiSongDian distribution in different distribution, service areas, terminal distribution node distribution demand point is only responsible for the service area distribution, distribution of nodes distribution ability is proportional to the size and distribution. As shown in Fig. 2(b), the improved quantum evolutionary algorithm generates the distribution task distribution scheme according to the distribution task demand. The distribution task in the dotted line box will be assigned to the distribution node in the box for distribution.

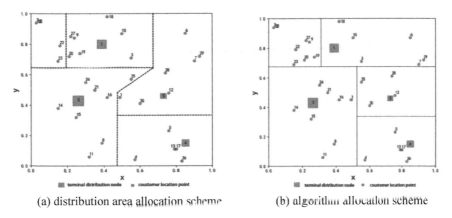

(a) distribution area allocation scheme   (b) algorithm allocation scheme

Fig. 2. Distribution point and distribution task distribution diagram

The workload allocation and delivery cost of each terminal distribution node corresponding to the two schemes are shown in Table 2.

Table 2. Workload allocation and distribution cost table of terminal distribution node

Distribution node	Workload mix		Delivery cost (distance)	
	Distribution area distribution	Algorithm distribution	Distribution area distribution	Algorithm distribution
1	0.2	0.55	2.44	1.05
2	1.8	0.63	1.49	0.5
3	0.3	0.27	1.54	1.64
4	0.5	0.5	0.67	0.67
5	0.23	0.44	0.54	1.61
Total	3.03	2.39	6.68	5.47

4.3   Improvement Effect Analysis

The improved quantum evolutionary algorithm is better in accuracy and stability than before the improvement, with the mean of the optimal fitness increased by 10%, and the optimal fitness obtained after 10 runs is more centralized and stable, with the standard deviation of the fitness reduced by 94% compared with before the improvement.

However, due to the introduction of the mutation elimination mechanism, the improved algorithm needs a longer computational process to achieve better quantum convergence speed. The evolution process is divided into three stages with the first feasible solution () that satisfies all constraints and the two time nodes that find the optimal quantum as the marker bits: $F(Q_{best})^{pun} > 1$

(1) The initial stage, At this stage, the main task is to find the feasible solution, for the next phase of the variable elimination provide feasible solution of quantum operation.
(2) The optimization stage. At this stage, more feasible solutions are available for the quantum group, which makes the average fitness of the quantum group increase by leaps and bounds.
(3) The degradation stage. At this stage, as quantum group stability gradually increased, the average fitness of each generation of quantum group is more and more intense, in the picture also shows that, in the late of quantum evolutionary quantum group differences between each generation is not big, quantum groups become more stable, algorithm search ability degradation.

Before algorithm improvement, the average ratio of initial stage, optimization stage and degradation stage is 33%, 10% and 57%, and the average ratio after improvement is 21%, 53% and 26%. Although the improved algorithm need more calculation process, but a larger proportion of the optimization stage, evolution is mostly used to search a better feasible solution, thus increase the computing time of the improved algorithms can effectively improve the ability of algorithm to search the feasible solution, entered the stage of degradation in advance to avoid premature convergence.

5 Conclusions

An end-distribution task assignment model based on quantum evolutionary algorithm is established to solve the end-distribution task assignment problem. The exit condition of quantum evolutionary algorithm is improved to avoid the problem of early exit and invalid iteration. The quantum mutation and elimination mechanism is introduced to enhance the search ability of the algorithm.

Although algorithms and models obtained good effect in the experiment, but in practice still need according to the actual business needs to adjust the related parameters in the model and algorithm, such as the delivery cost in the distribution of distance computing needs best path analysis by using geographic information system, can also be computed according to the actual business due to the uneven distribution of the task to task the heavier marki extra subsidies, and so on and so forth. In addition, the distribution capability of the terminal distribution node also needs to be analyzed and quantitatively evaluated according to the distribution vehicles attached to the node, the performance of the deliverer and the historical distribution service quality.

References

1. Simon, H., Hans, D.H.: Last mile concepts delivery in e-commerce an empirical approach. Softw. Knowl. Inf. Manage. Appl. **8**, 1–6 (2014)
2. Quanwu, Z., Junping, Z., Ya, L.: Research on optimization of urban distribution network of large retail enterprises based on O2O. China Manage. Sci. **25**(09), 159–167 (2017). (in Chinese)
3. Yupeng, L., Junmei, W., Zhao, W.: Research on the "last kilometer" rapid distribution method of cold chain logistics. Ind. Technol. Econ. **36**(01), 51–60 (2017)
4. Clan-Bin, Z.: Based on the decision-making mechanism of centralized distribution center suppliers' collaborative distribution of master and subordinate. J. Syst. Manage. **26**(03), 577–582 (2017)
5. Mengke, Y., Xiaoguang, Z.: Construction of co-delivery mode of express terminal under the background of "Internet +". J. Beijing Univ. Posts Telecommun. Soc. Sci. Edn. 17(06), 45–50 (2015)
6. Qiang, G., Yuowu, S.: Application of improved two-chain quantum genetic algorithm in image denoising. J. Harbin Inst. Technol. **5**, 140–147 (2016)
7. Fu, F., Ruijin, W.: Three-valued quantum genetic algorithm and its application. J. Univ. Electron. Sci. Technol. **1**, 123–128 (2016). (in Chinese)
8. Ozawa, T., Asaka, T.: Analysis of a dynamic assignment queueing model with poisson cluster arrival processes. J. Oper. Res. Soc. Jpn. **6**, 189–193 (2017)
9. Zhiyong, L., Liang, M., Huizhen, Z.: Function optimization quantum bat algorithm. J. Syst. Manage. **23**(05), 717–722 (2014). (in Chinese)
10. Tirumala, S.S.: A quantum-inspired evolutionary algorithm using gaussian distribution-based quantization. Arab. J. Sci. Eng. **13**(5), 214–218 (2017)

Some Applications of Big Data Mining Technology on Education System in Big Data Era

Xinkai Ge$^{(\boxtimes)}$

Telecommunications Engineering with Management,
Beijing University of Posts and Telecommunications, No 10, Xitucheng Road,
Haidian District, Beijing 100089, China
gexinkai2017@bupt.edu.cn

Abstract. With the development of network technology, the world has stepped into the big data era. The application of big data mining technology in the field of education has provided a new development path for education, realized the informatization of education and promoted the better development of education. To enhance the usage of big data mining technology as well as help to improve the current education system, this paper gives 4 specific aspects of education that can apply data mining technology which are daily management of schools, teaching and teaching reform, enhancing self-directed learning and providing new ideas respectively to illustrate how applying big data mining technology and exploiting valuable data for rational use can improve the quality of education.

Keywords: Big data · Data mining · Education · Teaching

1 Introduction

With the continuous development of technology, the era of big data has come. Through a variety of big data analysis techniques, potential value can be obtained from massive data. Diversified values make big data technology widely used in various fields of society. The traditional education concept in the era of big data has also changed, and education big data mining research has also become a hot spot of social concern [1]. Traditional teaching has long been in the predicament of backward teaching concept and single teaching mode. This not only leads to low enthusiasm for students, but also hinders students' understanding and mastery of knowledge. In order to effectively play the role of educational information, it is necessary to apply big data technology to mine valuable educational information, better serve education, and improve the quality of education.

The application of Data Mining in Education field in China started relatively late, and the search was conducted in China Knowledge Network with the key words of "education data mining". The earliest EDM related academic papers were published in 2002, and the number of papers in this field increased slowly in the following years. It is divided into three stages: budding stage (2002–2012), rising stage (2013–2014),

M. Atiquzzaman et al. (Eds.): BDCPS 2019, AISC 1117, pp. 368–374, 2020.
https://doi.org/10.1007/978-981-15-2568-1_51

rapid development stage (2015 – now). During these 3 phases, the volume of literature in the EDM field is beginning to explode as well as a great of mobile educational platform such as MOOC and Coursera have appeared in China. However, the role of big data mining technology in education is still very limited, and its advantages are not fully realized in many aspects.

This paper illustrates some applications that will help to enhance the usage of big data mining technology in education field and hope to provide some enlightenments to the further implementation of big data technology.

2 Concept and Characteristics of Big Data Mining and Educational Data Mining

2.1 Big Data Mining Concept

Data mining (DM) is an emerging discipline. It was born in the 1980s and is mainly used in the field of artificial intelligence research for business applications. From a technical point of view, data mining is the process of obtaining implicit, potentially valuable information and knowledge that is not previously discovered from a large number of complex, irregular, random, fuzzy data. From a business perspective, data mining is to extract, transform, analyze some potential laws and values from a large database, and obtain key information and useful knowledge to assist business decisions [2].

Big Data Mining (BDM) refers to Data Mining technology that is used under big data environment. It's one of the most important technology to excavate big data and transform massive amounts of data into useful information.

2.2 Educational Data Mining Concept

Educational Data Mining (EDM) refers to the use of relevant big data mining methods and techniques to process, analyse and model the massive data generated in the educational process, identify and solve problems in the educational process, and improve the quality of education [3–5].

The purpose of implementing educational data mining activities is to find, discover and extract value information that can improve the quality of teaching from various educational data information, so as to provide supportive opinions for educators and help analyse the problems encountered in each learning session. Problems, timely change plans, improve efficiency, help educators understand the various states of students from multiple perspectives, change plans appropriately, and improve the quality of teaching [6]. The education data mining process is mainly divided into the following steps: data acquisition, data pre-processing, data mining model establishment, and data interpretation. The process is shown in Fig. 1.

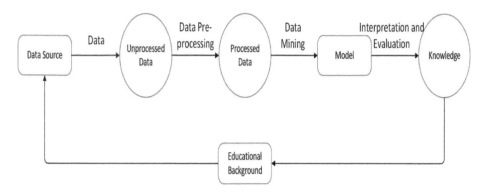

Fig. 1. The process of Educational data mining

2.3 Characteristics of Educational Data

Educational data mining mainly involves 5 technical support whose technical principles are complex. They are prediction, clustering, relationship mining, distillation for human judgment and discovery with models. Educational data possesses many characteristics which can be summarized as:

(1) *Hierarchical:* Educational data is hierarchical. That is, the keystroke level, the question layer, the conversation layer, the student layer, the classroom layer, the teacher layer, and the school layer are nested in layers. For existing online education, the hierarchical classification may be slightly different, but the structure of the data is still hierarchical.

(2) *Possessing time and sequence attributes with contextual background:* The time attribute includes the timestamp of learning, the length of learning, etc.; the sequence indicates how the concepts are related to each other in learning, and how to sort the exercises and teaching; the context is very important for interpreting the results and judging the model.

3 Problems in Current Education System

3.1 Students' Learning Ability Stays Unchanged

For now, most educators in basic education field still stick to the traditional ways of teaching. Traditional teaching methods is a kind of teacher-centred method using books as carriers. Its main form is that teacher explain and students listen with some practice after class. In this process, students receive knowledge from books and teachers, which is, sometimes, very quick and efficient. But students' ability to study and learn by themselves can hardly be improved in this way.

3.2 Low Enthusiasm to Learn

The traditional teaching methods' center lies mainly on teachers and students take knowledge passively. Thus, the atmosphere of the class is closely related to the teaching level of the teacher. If the teacher is not able to make the class active and interesting, which is often the case as it's difficult to make every class interesting just by speaking in the teacher podium, it's very likely that students get board lose enthusiasm to learn by themselves.

3.3 Difficult to Manage

In the era when big data has not yet arrived, the data collection and analysis of student information is so costly that the information we collect is often limited to academic achievement. But with the development of the Internet, data information has exploded, making it easier for us to access student information. As the overall number of students continues to rise, the sharp increase in educational data has increased the difficulty of teaching management. How to deal with a large amount of educational data and use it to strengthen teaching management has become a major problem.

4 Applications of Big Data Mining Technology in Education

4.1 Applied to Daily Management of Schools

The mining of educational data is of great significance to management departments, teachers in big data era, students and technology researchers [7]. The school's teaching management database records the work and study, research activities, social practice, punishment and rewards of all teachers and students. The leaders use data mining technology to conduct in-depth analysis of teaching resources and management data to find out the teachers and students. An inherently hidden connection between common behaviours or activities. In the management, process monitoring, risk warning, classification management, trend prediction and other measures have been adopted to change the unscientific phenomenon of the previous qualitative and fuzzy analysis and evaluation of teachers' classroom teaching quality, which can improve the current assessment management mode of the school and achieve intelligent and accurate Efficiently manage the work of the school and provide an effective scientific decision-making basis for teaching applications and school development.

4.2 Applied to Teaching and Teaching Reform

Generally speaking, teachers' teaching will adopt a variety of teaching methods according to actual needs, the main purpose is to better complete the teaching tasks. Applying big data mining technology is to mine association rules and regression linear analysis data, and further analyse the mined data.

In the obtained educational information, data information that is conducive to the absorption of students' knowledge will be collected, organized and analysed, the teaching methods will be optimized, the quality of teaching will be improved, and the efficiency of students' lectures will be greatly improved.

The application of big data mining technology is to study the students' individualized learning methods through the intelligent campus cross-platform learning excellent teaching mode, participate in the network learning, but also pay attention to cultivating students' hobbies and interests, help students to correct learning behaviours, let Students better master the learning rules and provide students with learning resources suitable for him. The student's learning method is optimized, and the learning mode is improved under the guidance of the materials, and the quality of the students' classroom teaching will be improved.

Rain Classroom is a very good example of applying data mining technology in teaching and teaching reform. Through the combination of PPT and WeChat, Rain Classroom allows students to establish a bridge of communication with teachers before, during and after class, so that classroom interaction remains online. It flexibly integrates the online and offline learning methods, subtly arranges the teaching activities in the classroom, and rationally arranges some teaching activities outside the classroom, making the teaching design more open and manoeuvrable. Through the pre-class preparation-real-time classroom-class data collection of the whole course teaching activities, the teacher's teaching is transformed from experience teaching to data-driven. The whole-cycle and full-time quantitative data assists the teacher in judging and analysing the student's learning situation, so as to adjust the teaching progress and the teaching rhythm makes the teaching process visible and controllable [8].

4.3   Applied to Student Learning and Enhance Self-directed Learning

By mining educational big data, students can analyse, research and process related information such as academic achievements, hobbies, consumption trajectories, behaviour records and reward punishment databases, and quickly obtain student identification results, so as to promptly prohibit and correct students' bad learning. behaviour. This can not only reduce the workload of teachers, but also avoid the teacher's lack of objective and fair subjective arbitrary evaluation of students' preconceptions. Using the results of educational big data to scientifically assess students' learning behaviours, the advantages are reasonable feedback of student information, stimulating students' interest in learning, discovering students' individual needs and teaching students in accordance with their aptitude. According to the student's personal information, academic performance, online learning track, personality characteristics, knowledge structure and other related information, the basic characteristics of the students are excavated, and the macro guidance and micro help students to constantly correct their learning behaviour. Students are no longer limited to the classroom teaching of a teacher at the school. They can choose the excellent courseware of famous scholars from relevant courses around the world, and learn according to their own time and needs. This not only improves the learning efficiency, but also cultivates Learning interest [9]. It is beneficial for teachers to guide students to improve their personality, correct their own learning behaviours and improve their learning ability by comparing pre-established standards of student behaviour and actual analysis of students' personality characteristics through data mining techniques, which is conducive to the overall development of students' comprehensive quality.

There are already some related apps in the app market, among which the most well-known is probably MOOC. Short for Massive Open Online Courses, MOOC adheres to the "life"-based teaching philosophy, flexible teaching model, ingenious teaching design, rich curriculum resources, efficient learning methods, effective teaching management, and scientific teaching evaluation, which are well received by users. Under big data era, its unique teaching philosophy and innovative teaching model extend knowledge to different areas of knowledge, which not only enriches the learning content, but also adds to the fun of learning [10].

4.4   Providing New Ideas for Construction, Application and Sharing of Educational Resources

Teacher's teaching and student's learning activities can be realized based on teaching resources. In the past, teachers' independent research and development and education authorities allotted to build teaching resources, while teachers' independent research and development are prone to low resource technology and poor usability. The drawback of high repetition rate, the resources allocated by the government can only meet the needs of most teachers and students, and cannot meet the individual needs. Excavating educational big data provides new ideas for the construction of educational resources, provides technical support for the construction of teaching resources, and allows the judgment of quality resources to be based on evidence. It enables teachers and students to easily use and share educational resource data stored in the cloud. It can also analyze a large number of unstructured data resources, mine hidden useful information, and enjoy data resources that meet their individual needs. Duplicate construction of teaching resources and waste of quality resources.

5   Conclusions

The role of big data mining technology is to be able to excavate more valuable information in massive amounts of information and data. Applying it into the field of education can mine valuable education information and improve the utilization of information so as to serve education better. This paper introduced the concept of data mining as well as its applications in education, it's obvious that with the popularity of the Internet, the application of big data mining technology to analyze students' interest in learning and provide students with educational information of interest can obtain good educational results.

Acknowledgements. Thanks for Beijing University of Posts and Telecommunications as it provided me a lot of resources and help to finish this paper.

References

1. Li, H.: Teaching exploration of data mining in big data environment. Comput. Era (02), 54–55 (2017)
2. Li, P.: Data mining technology and its applications in big data era. J. Chongqing Three-Gorges Univ. (3), 45–47 (2016)
3. Zhou, Q., Mou, C., Yang, D.: Research progress on educational data mining: a survey. J. Softw. **26**(11), 3026–3042 (2017)
4. Baker, R.S., Inventado, P.S.: Educational Data Mining and Learning Analytics, pp. 61–75. Springer, New York (2017)
5. Shu, Z., Xu, X.: Data mining and analysis of college students' satisfaction education from the perspective of learning analysis. E-Educ. Res. (5), 39–44 (2018)
6. Jiang, Q., Zhao, W., Wang, P., et al.: Realization of individual adaptive online learning analysis model based big data. China Educ. Technol. (01), 85–92 (2016)
7. Wei, S.: Learning analytics: mining the value of education data under the big data era. Mod. Educ. Technol. (2), 5–11 (2013)
8. Luo, J.: The application of rain classroom in education. Course Educ. Res. (36), 147 (2019)
9. Chen, J.: MOOCs and foreign language teaching studies in the big data era—challenges and opportunities. Technol. Enhanc. Foreign Lang. Educ. (01), 3–8 (2016)
10. Fu, M.: Reflections on the use of MOOC in the big data era to develop foreign literature teaching in colleges and universities. Guangxi J. Light. Ind. (27), 113–115 (2016)

Analysis on the Reform Path of College English Teaching in Local Engineering Colleges Under the Background of Big Data

Xiangying Cao$^{(\boxtimes)}$ and Juan Huang

Department of Foreign Languages, Nanchang Institute of Technology,
Jiangxi, China
104379295@qq.com

Abstract. The information revolution in the era of big data has had a huge impact on education, which has led to changes in teaching methods. This paper analyzes the problems in college English teaching in local engineering colleges from the aspects of college English tutorials, teaching resources, methods and teaching objects. The transformation of big data technology based on mobile Internet will have a full impact on college English teaching in China, which will promote the new environment of English teaching. On the basis of analyzing the new teaching environment generated by big data, explore how to further utilize big data information technology and resource reform local colleges and universities college English teaching, it mainly includes: constructing a three-dimensional curriculum system, creating a mixed teaching mode of "Mooc-class, micro-class + flip classroom", and integrating college English teaching resources.

Keywords: Big data era · Information transformation · Local engineering colleges · College English teaching

1 Introduction

The newly developed "College English Teaching Guide" emphasizes the deep integration of modern educational technology and foreign language teaching. It emphasizes that college English teachers integrate information technology elements in the process of classroom teaching design and implementation, making full use of digital resources to expand teaching content and exploring classroom-based teaching. The online teaching mode of the online course helps students develop towards active learning, independent learning and personalized learning [1].

The fast trend of IT and the Internet have brought people in the 21st century into a new time. The popularity of devices such as tablets, mobile phones, mobile smart terminals, and cloud computing has made today's networks full of all kinds of data, called big data. It affects everyone's life, work and study, and it also has a great impact on today's society. In the era of big data, the educational environment has undergone earth-shaking changes, which has greatly affected college English teaching. Relying on big data, the arrival of the "Internet +" era has changed our way of life and thinking, the

M. Atiquzzaman et al. (Eds.): BDCPS 2019, AISC 1117, pp. 375–381, 2020.
https://doi.org/10.1007/978-981-15-2568-1_52

education industry has been deeply affected by it, the birth and prevalence of educational software, the fun and targeted improvement of mobile education apps for people making full use of the fragmentation time provides great convenience, enriches the means of learning, and improves the efficiency and quality of learning. Taking the English education in the era of big data as the blueprint for analysis, we can look at the direction and path of education improvement in the era of big data.

Local engineering colleges have emerged in the field of English teaching, such as the disconnection between curriculum and practice, the disconnection between content and demand, the imbalance between classroom and environment, and the imbalance between learning and application. It is urgent to explore the path of teaching reform and innovative methods. In combination with related issues, the reform should be strengthened from the three-dimensional curriculum system construction, the integration of teaching resources, the creation of teaching environment, and the combination of production, study and research. Through the implementation of three-dimensional teaching, development of teaching platforms, establishment of flipping classrooms and project teaching to achieve teaching innovation. The clear path of reform can point out the direction of college English reform in applied undergraduate colleges, and propose innovative methods to realize the in-depth discussion of English teaching problems. Therefore, we can strengthen the analysis of applied English talents from a new perspective.

2 Important Concepts

2.1 The Concept and Characteristics of Big Data

The US Internet data center defines "big data" as new technology architecture that captures value from high-volume data through high-speed capture, discovery/analysis. It can be summarized as four English letters V, that is, a larger volume, a higher diversity (Variety), a faster generation speed (Velocity) and a fourth driven by the combination of the first three "V" factor - Value [2].

2.2 Engineering Colleges

Engineering colleges are specialized in engineering teaching schools that are developed based on application-oriented disciplines that address the needs of production and other industries, combined with the technical experience accumulated in production practices [3].

3 Big Data Promotes a New Environment for English Teaching

The transformation of big data technology based on mobile Internet will have a comprehensive impact on college English teaching in China, which will lead to a new environment for English teaching.

3.1 Massive Teaching Resources Will Weaken the Status of Traditional Textbooks

At present, most of the textbooks have problems such as too old corpus and not resonating with English learners. Although the current English textbooks are constantly being revised, the limitations of factors such as the overall framework of the textbooks, the concept of writing, and the length of writing will lead to the embarrassing situation of "one publication is out of date". In the era of big data, the network and cloud storage platform can provide learners with a large number of real-time, vivid English learning resources, reducing the absolute importance of teaching materials in teaching, such as: the US President's weekly TV speech, *Economists'* English texts and audio materials, articles on *National Geographic*, etc. These novel, cover-free, and easily accepted materials cannot be used because of the specific principles of textbook writing. The textbooks are compiled in a hurry, but they can be quickly applied to college English teaching through connected mobile platforms and mobile terminals [4].

3.2 Changes in the Role of Teachers and Students

In the traditional college English teaching process, teachers are in a dominant position. The teacher analyzes, summarizes and supplements the content of the textbook to form a lesson plan and handouts, so as to impart the book knowledge to the students. The classroom is basically a teacher's own voice, and there is very little interaction between teachers and students. In the long run, this will inevitably lead to teacher burnout and ultimately affect the teaching effect. At the same time, students are in a passively accepted position, and it is difficult to truly convert language input into language output. Therefore, the phenomenon of "dumb English" is very serious. At present, the number of English classrooms in domestic universities is more than 20–50, and some even nearly 100. This kind of "big water flooding" teaching method obviously contradicts the teaching rules of English. Big data technology can help solve the above problems. For example, teachers can use the relevant software to create micro-courses, decompose the knowledge, and push them to students through network platforms or mobile terminals. Students can complete learning tasks independently. This method can help students eliminate anxiety in learning and improve their interest in learning. The role of the teacher is also changed from the leader to the instructor. The student becomes the main body of the teaching activity and truly realizes the "student-centered" teaching philosophy. In addition, with the powerful screening ability of big data, the teacher can summarize the problems that occur in the students' learning, classify the students according to the level, and then carry out targeted teaching to realize the "teaching according to their aptitude" in the true sense [5].

3.3 Changes in Academic Assessment Methods

Traditional college English teaching adopts a summative assessment method. The basic method is mostly a test paper for the final exam. However, this assessment method is too utilitarian and cannot objectively and dynamically reflect the students throughout the teaching process.

The comprehensive performance of the students in the teaching process is neglected. In addition, another drawback of this type of assessment is that teachers' academic feedback to students is always lagging behind. This can be partly explained by the fact that the backwardness of technology has resulted in a single evaluation method. In the context of big data, online platforms and mobile terminals have made it possible for teachers to evaluate students in multiple ways. Teachers' evaluation of students' academics is more extensive, including learning evaluation and growth evaluation. Using big data technology to test, evaluate, screen and identify students, so as to analyze each student's English application ability, and adjust the teaching plan and teaching content accordingly, so that students can maintain their interest in learning English [6].

4   The Reform Path of College English Teaching in Local Engineering Colleges Under the Background of Big Data

4.1   Building a Three-Dimensional Curriculum System

The content of college English teaching is composed of three parts: general English, special purpose English and intercultural communication. General English is to develop students' language skills in listening, speaking, reading, writing and translating, including vocabulary, grammar, discourse and pragmatic knowledge. Specialized English is divided into academic English and professional English. It is to train students to use English for industry and academic communication and the ability to work; intercultural communication courses are to develop students' cross-cultural awareness to achieve more effective communication to prevent and avoid misunderstandings arising from different cultural expectations. Colleges and universities should integrate these three parts into the school-based curriculum, and combine them to provide more learning content for students [7].

4.2   Creating a Mixed Teaching Mode of "Mooc Class, Micro Class + Flip Classroom"

Flipping the classroom is an emerging teaching model that emerged under the situation of the "Internet +" era. Learners watch the teaching videos before class to complete the knowledge transfer, and complete the internalization of knowledge through various forms of teaching in the classroom [8], which is opposite to the two stages of knowledge transfer and knowledge internalization in the traditional learning process [9]. Their research results show that flipping classroom teaching meets the individualized needs of learners' English learning, and enhances learners' independent learning ability and English comprehensive application ability. In order to realize the integration of emerging teaching modes and high-quality teaching resources, it is necessary for local engineering colleges to construct and implement a mixed teaching mode of college English flip classroom combining MOOC and micro-course.

(1) "Mooc class + flip classroom" teaching mode

Compared to traditional courses, it has its own differences in input and output that are conducive to the learner's language. First, use the Internet platform to combine modern information technology with curriculum teaching. Secondly, there are essential changes in the time and space of the classroom, the number of lectures, the motivation of learning, the subject of learning, and the way of interaction. Finally, platforms, teachers, learners, and learning resources can be based on deep interactions. The reform of college English teaching in local engineering colleges should make use of the development of Moocs to create a personalized teaching model suitable for college English teaching, which is based on the flipping classroom teaching mode of MOOC, to achieve student-centered and cultivate students' English application ability and self-learning ability.

This kind of teaching mode has different tasks before, during and after the class. Before the class, the teacher needs to carefully select the video of the MOOC that is suitable for the language learning of the students. The students repeatedly watch, listen and complete the language input during the self-learning stage. At the same time, teachers should also design student self-learning project activities and results reporting requirements to prepare for classroom teaching. As a classroom instruction in the language output stage, students present the results of group cooperative learning in the form of oral reports or written reports, and teachers make periodic evaluations. After class, students use the online learning platform to present problems in learning, and show the outstanding learning outcomes and learning experiences of each group. Teachers participate in it, supervising and motivating students to sum up better learning methods for independent learning.

(2) "Micro-class + flip classroom" teaching mode

The micro-course is recorded by the instructor for a certain knowledge point or teaching session, and can be used to share short videos with students through the network. Hu (2011) pointed out that micro-classes have the following characteristics compared with MOOCs: prominent topics, short and precise, easy to use, and realistic [10]. Micro-curriculum is an important prerequisite for flipping classroom realization, and it is an extension and expansion of classroom teaching. Under the context of 'Internet +", the continuous development of Internet and mobile smart devices has promoted the online and mobile learning based on micro-courses to become the trend of reform and development of college English teaching mode.

4.3   Integrating College English Teaching Resources

Applied undergraduate colleges should also fully leverage the development trend of the human data era to integrate college English teaching resources and gain more advantages in the field of English education. In the process of resource integration, we should also take the training of students' English application ability as the leading factor, and combine the professional needs of students to complete valuable teaching resources from various network resources such as audio and video. Teachers should give more play to their autonomy and use various teaching techniques to freely integrate various teaching resources. Teachers should use textbooks as a support point in their teaching. Under the premise of clarifying the content of teaching materials, the main body should be refined with the help of big data. The teaching content radiates outward, screening the latest audio and video corpus suitable for students to learn for classroom teaching, and mobilizing students' classroom learning enthusiasm with the latest corpus and topic. After class, teachers can use the Internet mobile terminal to push, check, and answer questions about related learning content.

5   Conclusion

The arrival of the times of big data has spawned a new teaching model motivated by big data in college English teaching. Through analysis, it can be found that local engineering undergraduate colleges should also recognize the impact of the era of big data on college English, and use the advantages of big data to improve the reform and development of English teaching. To this end, institutions should also combine the current curriculum of college English teaching with the actual disjointed, disconnected content and needs, classroom and environmental imbalances, from the three-dimensional curriculum system construction, teaching resources integration, teaching environment creation, production, study and research, etc. The direction of English teaching reform is clear, and through the implementation of three-dimensional teaching, development of teaching platform, establishment of flipping classroom and project teaching to achieve teaching innovation, and then combined with the actual needs of school English teaching to better complete the training of local engineering applied English talents.

Acknowledgments. This work is supported by the project of education reform in Jiangxi Province: *An Empirical Study of Teaching Mode of Interpretation on Eco-translatology* (JXJG-18-25-6).

References

1. Wang, S.: Interpretation of the essentials of college english teaching guide. Foreign Lang. (3), 2–10 (2016). (in Chinese)
2. Wang, X.: Research on the mixed teaching mode of language courses under mobile terminals. Henan Soc. Sci. (11), 96–100 (2015). (in Chinese)
3. Xie, Y., Zhu, Z.: Analysis of factors influencing the quality of mixed teaching in colleges and universities. China Distance Educ. (10), 9–14 (2012). (in Chinese)
4. Xiang, R.: Research on the innovation of college english teaching model in the era of big data——taking the "Rotating Classroom" of Bohai University as an example. J. Lanzhou Coll. Educ. (05) (2016). (in Chinese)
5. Tao, X.: Analysis of the challenges of MOOCs to college english teaching in the age of big data. J. Xihua Univ. (Soc. Sci. Ed.) (05) (2015). (in Chinese)
6. Yang, Y.: Innovation and informationalization of college English teaching mode in the era of big data. Foreign Lang. E-Learn. (04), 56–59 (2017). (in Chinese)

7. Wang, L., Yang, F.: Research on the reform and development of college English teaching under the background of "Internet +" era. Heilongjiang High. Educ. Res. (8), 159–162 (2015). (in Chinese)
8. Zhang, J.: Analysis of the key factors of the "Flip Classroom" teaching model. China Distance Educ. (10), 59–64 (2013). (in Chinese)
9. Wang, S., Zhang, L.: Research on college English learners' acceptance of flipping classroom. Mod. Educ. Technol. (3), 75–76 (2014). (in Chinese)
10. Hu, T.: Micro-course: a new trend of regional education information resources development. Res. Electro-Educ. (10), 61–65 (2011). (in Chinese)

Design of Educational Robot Platform Based on Graphic Programming

Aiping Ju[1(✉)], Deng Chen[1], and Jianying Tang[2]

[1] School of Computer Science and Engineering, Wuhan Institute of Technology
East Lake New Technology Development Zone, No. 206 Optics Valley First
Road, Wuhan, China
1154697126@qq.com, chendeng8899@163.com
[2] School of Artificial Intelligence, Hubei Business College, 632 Xiongchu
Street, Hongshan District, Wuhan, China
3121697302@qq.com

Abstract. This paper designs an educational robot platform according to the needs of robot education for primary and middle school students. The mechanical mechanism design of the robot platform is carried out, and the environment sensing module based on multi-sensor information fusion is designed. The robot platform control is realized based on the ADRC control algorithm. The chassis control board uses the STM32 control board with the STM32F405RG microcontroller as the core. The board integrates MicroPython firmware, which enables access and control of the underlying hardware through the Python scripting language. The sensor modules such as the inertial navigation module and the vision module are connected with the Raspberry Pi development board. The raspberry PI development board conducts path planning according to multi-sensor information fusion, and sends control instructions to the chassis control board through the serial port. On this basis, the Python plug-in management system is designed to realize the graphical programming control robot.

Keywords: Graphical programming · Scratch3 · Python · ROS

1 Introduction

With the rapid development and popularization of artificial intelligence and robot technology, the research and application of robot education for primary and middle school students have been paid more and more attention by more and more research institutions and educational service enterprises [1].

The robot education for primary and middle school students takes educational robot as the carrier to guide students to learn the basic principles, programming operation, hardware operation and structure design of robot, cultivate team spirit and logical thinking ability, and guide teenagers to create science and technology. Different from college students and researchers, primary and secondary school students generally have little professional knowledge of computer programming. The corresponding robot programming environment requires simple operation and friendly interaction, which

© Springer Nature Singapore Pte Ltd. 2020
M. Atiquzzaman et al. (Eds.): BDCPS 2019, AISC 1117, pp. 382–391, 2020.
https://doi.org/10.1007/978-981-15-2568-1_53

can provide users with intelligent auxiliary functions, with certain interestingness and visualization, and can stimulate students' enthusiasm for learning and creation.

The design and development of Python graphical programming software tools for robot education, combined with the hardware resources of educational robots, provide a friendly interface, reduce the threshold of programming learning, stimulate students' enthusiasm and interest in learning programming, is of great significance to promote the development of robot education [2, 3].

Aiming at the needs of robot education for primary and middle school students, this paper applies the basic principles of Python in hardware programming to realize the function of graphical programming control robot, reduce the difficulty of educational robot graphical programming, and improve the versatility of educational robot graphical programming software system.

2 Related Work

2.1 Robot Graphical Programming

With the rapid development of computer technology, non-computer professional programming user groups want programming tools to be simple and powerful, while professional programmers expect to use intuitive, easy-to-use and readable programming languages. Unlike text-based programming languages, graphical programming languages do not require or require a large amount of written text code. Instead, users drag and drop different programming components to match each other with a grammatically appropriate computer program that performs a specific function. Therefore, in the field of computer programming, graphical programming is the only way to make human-computer interaction more smooth.

Stasinos has designed a graphical programming scheme in 2014, which can carry out graphical programming for robots. Robo Flow, a flow-based graphical programming language, was also introduced in 2015, allowing users to control tasks by writing robots on a graphical programming platform. Graphical programming is also being used in education, the Internet of things and smart homes [4–6].

Scratch is a graphical programming tool developed by MIT's Lifelong Kindergarten Group, which is open to young people. There are currently the original version (1.4 version), version 2.0 (adding clone blocks, Lego and Makey makey to expand the building blocks), version 3.0 (adding music, brushes, video detection, text reading, translation and other optional download expansion blocks). Scratch makes the expression of the program realized by the user dragging the mosaic of different shapes of graphic elements, and has a wide range of applications in the field of education [7, 8].

With robots gradually infiltrating into many aspects of human life, robot programming is a major difficulty, so robot graphical programming is one of the future trends. The rise of maker education has promoted the development of educational robots, and prompted some companies to pay more attention to the robot graphical development platform, and a certain amount of robot graphical programming software has emerged.

The LEGO Mindstorm series of graphical programming systems are currently well-suited and mature graphical programming software for robot control. The system can realize many logically complex and versatile robot control, but does not support multi-thread synchronization in programming. In 1999, the European cooperative research program AMIRA proposed the concept of graphical programming of robots. In the robot graphical programming, the operation of the robot is designed as a specific icon, and the function code is written through the process schema, so that the user does not need to think too much about the strict requirements of the computer program syntax, and then put more energy on how to effectively achieve the task goal. At the same time, the parameter setting and operation of the robot hardware are more concise and clear [9, 10].

In the current graphical programming research, the main shortcomings are as follows: (1) The lack of uniformity of various graphical programming systems. For different system implementations, there are different grammar definitions. At present, each manufacturer has its own graphical programming language, but it has not formed a relatively mature specification, and many graphical programming systems are very similar. The system is not innovative enough, graphical programming system interaction is not beautiful enough, the system ability is not strong enough. (2) The existing graphical programming system has high coupling between subsystems, which is not conducive to system modification and improvement. This leads to the limited types of existing graphical programming systems, the lack of cross-platform, and the lack of support for the expansion of programming components. (3) Due to the design or implementation defects of the current graphical programming system, the execution efficiency and compilation efficiency of the graphical programming system are not high. (4) With the development of parallelization technology, high-level programming languages can better support multithreaded programming. However, the existing graphical programming system, in terms of model selection, only supports thread branching, but does not support synchronization between threads.

2.2 Python and ROS Programming

There were Python2 and Python3 versions during the development of Python. Python3 had a lot of changes compared to Python2 and was not compatible downward. But many platforms developed by Python2 have been ported to Python3, and the Python3 development community is very active.

Common embedded systems that can use Python include Zerynth, PyMite, MicroPython, etc. The most common application is MicroPython. MicroPython is a language variant of Python that is refactored using ANSI C at the bottom of the software. It contains a complete parser, compiler, virtual machine, real-time system, garbage collector and can run on the single chip support library, so that the software code can be directly compiled and run on the single chip, and can be compatible with Python 2 and Python3. In product development, it can run offline, which brings great convenience to debugging [11].

ROS is an open source operating system based on robots. It can provide many features similar to traditional operating systems, such as underlying device control, hardware abstraction, interprocess messaging, and feature pack management. In addition, ROS is a distributed network system in which communication between different nodes is realized by topic, service or parameter server. For example, in the navigation and positioning of mobile robot, TF node controls coordinate transformation; Environmental data are collected by lidar and odometer, and transmitted data are received through Laser Scan node. AMCL algorithm receives sensor data and processes it, and displays the processing results through AMCL node. One node subscribes to different topics published by other nodes, and the topics published by itself are also subscribed by other nodes. The message is the carrier of the topic, and the communication between each other is completed through the message. This system structure ensures real-time update of location data and synchronous transmission. Node is the main calculation execution process in the ROS calculation chart. They are independent of each other and decouple the code and function, improving the fault tolerance and maintainability of the system. ROS integrates multiple mainstream function libraries and has multiple sensor drivers [12, 13].

3   Educational Robot Platform Design

3.1   Robot Platform Mechanical Structure Design

The mechanical mechanism design of the robot is the basis of hardware design and software design, which will directly determine the moving speed of the robot, the minimum space size that the robot can pass, and the maximum range of motion that the robot can reach.

The robot designed in this paper adopts the two-wheel differential drive mode, in order to ensure the balance of the robot in the process of moving, a steering wheel with the function of even landing is added in front of the robot. As shown in Fig. 1, the robot platform includes the frame, steering wheel, steering electric drive mechanism, driving wheel, driving module, battery module and control module.

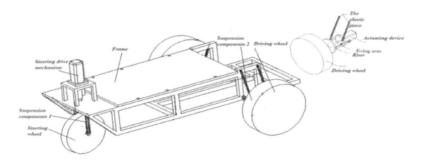

Fig. 1. Mechanical structure design of robot platform.

The frame is horizontally arranged, the steering wheel is vertically disposed at the front end of the frame through the shock absorbing assembly 1, and the shock absorbing assembly 1 is rotatably connected with the frame, the axle of the steering wheel is rotatably connected with the shock absorbing assembly 1, and the steering drive mechanism is mounted on the vehicle. The front end of the frame is coupled to the shock absorbing assembly 1 for driving the steering wheel to turn left and right. The driving wheel sets are respectively vertically disposed on the two sides of the rear end of the frame through the damper assembly 2, and the driving mechanisms are respectively mounted on the two damper components in one-to-one correspondence, and are connected with the corresponding axles of the driving wheels for drive the drive wheel to rotate. The second shock absorbing assembly 2 includes a vertical plate, a swing arm, an elastic member, and a connecting frame. The vertical plates are vertically arranged, and the two swing arms are horizontally spaced and parallel to each other, and the two ends are flush. The same end of the two swing arms is fixedly connected to the vertical plate, and the other ends of the two swing arms are respectively rotatably connected to the lower end of one side of the connecting frame. The two elastic members are in one-to-one correspondence with the two swing arms. One end of the elastic member is respectively rotatably connected with a middle portion corresponding to the longitudinal direction of the upper end of the corresponding swing arm, and the other ends of the elastic member respectively extend obliquely upward to be connected to the upper end of the connecting frame. The driving wheel is disposed on a side of the corresponding vertical plate facing away from the connecting frame, and the axle of the wheel is vertically rotatably connected with the vertical plate. The power drive system is mounted on the vertical plate or the swing arm, and its drive end is connected to the axle of the drive wheel. Two connecting brackets are respectively mounted on the two sides of the rear end of the frame, and the two driving wheels are located on the side where the two connecting brackets are away from each other.

3.2   Environmental Awareness Based on Multi-sensor Information Fusion

It is the primary problem to be solved and the most basic function to be possessed by the mobile robot to perceive the surrounding environment in real time and determine its own position. Whether the positioning accuracy is good or not and whether the positioning can be completed will directly determine whether the mobile robot can perform the following tasks well.

The education robot platform designed in this paper needs to perceive various environmental information of the site, including longitude and latitude, obstacles, target distribution of manipulator grasping, posture, speed, direction and other information, which requires a variety of sensors to provide different sensing information.

The inertial navigation module is used to provide the robot with GPS absolute coordinates (longitude and latitude signals) and relative position coordinates (magnetic deflection signals to calculate the robot's posture). The visual module is used to acquire surrounding environmental information for the robot, and the robot can update and grab the data of the target quantity to complete the path planning algorithm. The photo-electric sensor module is used to provide the robot with information about the number of targets to be captured, so that the robot can be optimized in calculating the path planning algorithm. Motor drive and coding module are used to realize robot motion control and speed measurement.

Environmental perception based on multi-sensor information fusion studies how to fuse sensor data, process and integrate them to obtain the internal relations and rules of all kinds of information, so as to eliminate useless and wrong information, retain correct and useful components, and finally achieve the optimization of information, real-time and effective acquisition of environmental information. In the process of running education robot platform, the use and collection of the robot location information, at the same time using the visual module to collect image information to visual processing board, the robot using these images to create or modify the environment information, through position alignment, path planning and other related algorithms, produce control command to the motion control system.

3.3   Autonomous Positioning and Navigation

The autonomous positioning and navigation in this paper includes the positioning and navigation of the robot itself, grasping targets and obstacles, as shown in Fig. 2. Visual positioning is a new positioning technology which appears with the improvement of computer technology and image engineering level. The panoramic vision sensor, whose field of vision covers 360° horizontal perspective, can provide a large amount of image information with one image, and with its broad field of vision, it can efficiently find landmark and other feature information in the environment, and there is no problem of image Mosaic. The above features make panoramic vision sensor very suitable for the positioning task of mobile robot.

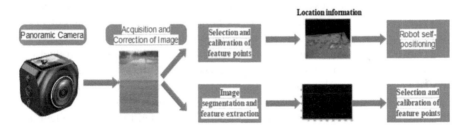

Fig. 2. Autonomous localization of robot.

Self-positioning of the robot means that the robot is operating in an environment where the road signs we have set up are known to exist, and their positions in the coordinate system are known. The robot can obtain the image information of the road sign pattern through the panoramic camera mounted on it, and the absolute position of the camera in the three-dimensional world can be obtained through the selection of feature points and computer calibration technology. The actual position can be corrected for the position of the robot compared to our preset target position.

The positioning of grasping objects and obstacles is to acquire panoramic images of the surrounding environment through panoramic camera acquisition. The grasping objects, obstacles and identification are separated from the images with grasping objects and surrounding environment. The goal of image segmentation is to divide a complete image into areas of interest according to the requirements. Feature extraction is the main purpose of image segmentation. In this paper, the purpose of image segmentation is to take captured objects and obstacles as targets and other environmental information as background, so that captured objects and obstacles can be extracted from the image and the corresponding image can be segmented into multiple grids. In practical application, the position of the captured target can be determined according to the color, and then the number of captured targets in each grid can be calculated. The center point of the grid with the largest number of captured targets can be selected as the next moving target.

3.4   Robot Platform Control Based on ADRC Algorithm

This paper design the drive controller Control driving wheel speed and the rotation Angle, its applying active disturbance rejection control (ADRC) Control algorithm to Control driving wheel speed and Angle. Since the robot platform is equipped with various manipulator, the manipulator will produce a large reaction force to the robot platform in a short time when it moves. At the same time, because the frame is a three-wheel structure, the steering wheel needs to bear part of the steering force of the frame. The traditional PID controller cannot guarantee the smooth operation of the robot platform in this case, and it is easy to fail to respond timely or over-respond. Compared to PID controllers, ADRC controllers are model-independent, responsive, easy to use, and flexible [14, 15].

As shown in Fig. 3, the core of ADRC is mainly composed of three modules: tracking differential units, extended state observer (ESO), and nonlinear state feedback. For motor speed control, the target speed is input into the tracking differentiator and output the tracking speed and tracking acceleration (tracking speed equals target speed and tracking acceleration is target acceleration). The ESO input is the product of the actual velocity, the output voltage U and the coefficient b0, respectively. Where the observed velocity is equal to the actual velocity, the observed acceleration is equal to the actual acceleration, and the observed disturbance is the total disturbance inside and outside the system. After the observed disturbance is divided by b0, the voltage U_0 of the nonlinear state feedback is subtracted, which is the voltage U of the output to the motor. The nonlinear state feedback outputs voltage U_0 according to velocity error (the difference between the tracking speed and the observation speed) and acceleration error (the difference between the tracking speed and the observation speed). If both errors are zero, then U_0 is zero. The coefficient b0 represents the corresponding relationship between voltage and rotational speed RPM [16]. The ADRC controls the motor speed according to the set b0. If the actual speed is different from the set speed, the corresponding deviation is considered to be the observation disturbance, whether caused by the change of external resistance or the inaccurate estimation of internal parameters of the motor. The ADRC controller directly compensates to U output after dividing the observed disturbance by b0.

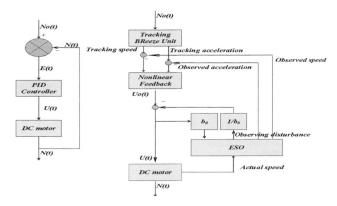

Fig. 3. PID controller and ADRC controller structure block diagram.

4   Experiment and Test

According to the previous mechanical structure design, the prototype of education robot platform is shown in Fig. 4 (a). Fig. 4(b) and (c) are respectively the speed diagrams of the prototype motor using PID and ADRC controller. In the test process, the motor should keep accelerating and decelerating. The yellow line is the target speed and the red line is the actual speed. The two lines of the ADRC controller's target speed and actual speed almost coincide, with almost no overshoot, no vibration, no static error, and the acceleration and deceleration increase by more than 30%, noise reduction by 2db, and energy efficiency increase by 5%. The control performance of ADRC controller is obviously better than that of traditional PID controller.

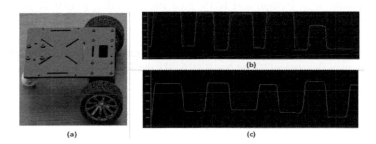

Fig. 4. Robot prototype and controller algorithm test.

Figure 5 is a graph of Scratch3 support for Python and ROS test results. In Fig. 6 (b), the Scratch graphical programming statement communicates with the ROS via a message mechanism to control the forward and backward movements of the real robot.

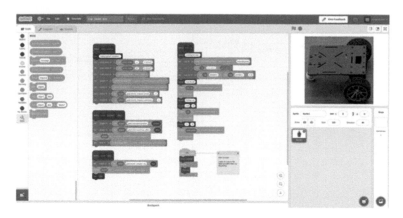

Fig. 5. Scratch3 supports Python and ROS testing.

5   Conclusion

This paper designs the educational robot platform based on the graphical programming language Scratch for the needs of robot education for primary and secondary school students. The mechanical mechanism design of the robot platform is carried out, and the environment sensing module based on multi-sensor information fusion is designed. The robot platform control is realized based on the ADRC control algorithm. On this basis, the Python plug-in management system is designed to implement Scratch3 support for Python and ROS. The experimental results show that the control performance of the ADRC controller of the educational robot platform prototype is significantly better than the traditional PID controller, and based on the platform, Scratch3 can communicate with Python and ROS, thus implementing graphical programming control robot.

Acknowledgments. This paper was supported by the Science and Technology Research Project of Hubei Education Department (Grant No. B2016482), the Scientific Research Project of Hubei Business College (Grand No. KY201806).

References

1. Wang, X., Zhang, S.: Research on robot teaching in primary and secondary schools for maker education. Modern Educ. Technol. **26**(8), 116–121 (2016)
2. Vitousek, M.M., Kent, A.M., Siek, J.G., et al.: Design and evaluation of gradual typing for python. ACM Sigplan Not. **50**(2), 45–56 (2014)
3. Rossum, G.V.: Python 2.7.10 Language Reference (2015)
4. Konstantopoulos, S., Lydakis, A., Gkikakis, A.E.: Embodied visual programming for robot control. In: ACM/IEEE International Conference on Human-Robot Interaction, pp. 216–217. ACM (2014)

5. Alexandrova, S., Tatlock, Z., Cakmak, M.: RoboFlow: a flow-based visual programming language for mobile manipulation tasks. In: IEEE International Conference on Robotics and Automation, pp. 5537–5544. IEEE (2015)
6. Trower, J., Gray, J.: Blockly language creation and applications: visual programming for media computation and bluetooth robotics control. In: ACM Technical Symposium on Computer Science Education, p. 5. ACM (2015)
7. Sáez-López, J.M., Vázquez-Cano, E.: Visual programming languages integrated across the curriculum in elementary school: a two year case study using "Scratch" in five schools. Comput. Educ. **97**, 129–141 (2016)
8. Sengupta, P., Dickes, A., Farris, A.V., et al.: Programming in K-12 science classrooms. Commun. ACM **58**(11), 33–35 (2015)
9. Kiss, G.: Using the lego-mindstorm kit in german computer science education. In: IEEE International Symposium on Applied Machine Intelligence & Informatics (2010)
10. Rahul, R., Whitchurch, A., Rao, M.: An open source graphical robot programming in introductory programming curriculum for undergraduates. In: IEEE International Conference on Mooc, Innovation and Technology in Education, pp. 96–100. IEEE (2015)
11. Khamphroo, M., Kwankeo, N., Kaemarungsi, K., et al.: MicroPython-based educational mobile robot for computer coding learning. In: Information & Communication Technology for Embedded Systems (2017)
12. Martinez, A., FernANdez, E., et al.: ROS Robot Programming. Mechanical Industry Press (2014)
13. Zhang, R., Liu, J., Chou, Y.: Programming and implementation of educational robot based on scratch and ROS. J. Artif. Intell. Robot. **07**(04), 178–183 (2018). https://doi.org/10. 12677/AIRR.2018.74021
14. Dong, Q., Li, Q.: Current control of BLDCM based on fuzzy adaptive ADRC. Micromotors **3**, 355–358 (2010)
15. Zhao, C., Huang, Y.: ADRC based input disturbance rejection for minimum-phase plants with unknown orders and/or uncertain relative degrees. J. Syst. Sci. Complex. **25**(4), 625–640 (2012)
16. Estrada, N., Astudillo, H.: Comparing scalability of message queue system: ZeroMQ vs RabbitMQ. In: Computing Conference (2015)

Influencing Factors of Employee's Environmental Scanning Behavior in Big Data

Yang Yang[⊠], Zhenghua Zhao, Chunya Li, and Yan Li

Business School, Nantong Institute of Technology,
Nantong 226000, Jiangsu, China
295031129@qq.com

Abstract. This paper combined cognitive theory, motivation theory, competitive intelligence management and control theory with theory of planned behavior to optimize the model of employees' environmental scanning in big data, and proposed the research hypothesizes. And then using the questionnaire survey to collect data, we tested the research hypothesizes by using AMOS17.0. Finally, we proposed the competitive intelligence management strategy for the enterprises to promote their employees' environmental scanning in big data, according to the result of research.

Keywords: Big data · Environmental scanning · Theory of planned behavior · Theory of management and control of competitive intelligence

1 Introduction

With the development of big data, the innovation and upgrading of environmental scanning system welcome the opportunities. Big data environment not only improves the ability, consciousness and opportunity of individual environmental scanning behavior, but also promotes the formation of group wisdom with its participatory and bottom-up cultural characteristics, and gives birth to a new intelligence model, the full-staff intelligence model.

Under the circumstance of Chinese characteristics, the concept of environmental scanning or competitive intelligence is usually avoided. Especially "intelligence" is always easy to associate with spies. Therefore, intelligence work in China usually belongs to part-time work with a high degree of business embeddedness, which is usually undertaken by personnel of some specific departments (marketing department, after-sales service department) [1]. Although employees play a key role in the collection and transmission of environmental information in big data. Previous studies mainly focused on the environmental scanning behavior of individuals (especially executives) in traditional environment, or the impact of big data on employees' information acquisition and business [2]. Few studies have explained the special and key tribute of employees.

This paper focuses on the influencing factors of employee's environmental scanning contribution behavior in big data environment, aiming at identifying which factors can explain the changes of employee's environmental scanning behavior.

M. Atiquzzaman et al. (Eds.): BDCPS 2019, AISC 1117, pp. 392–398, 2020.
https://doi.org/10.1007/978-981-15-2568-1_54

2 Model Development and Research Hypotheses

2.1 Theory of Planned Behavior

The theory of planned behavior (TPB) is evolved on the basis of the theory of rational behavior. The key of TPB is to study behavior with human as the core. As shown in Fig. 1, TPB holds that individual behavior is determined by intention, which is influenced by attitude, subjective norms and control of perception [3].

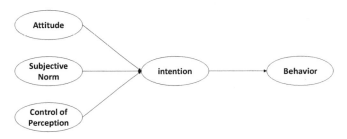

Fig. 1. TPB

In view of the complexity of environmental scanning in big data, TPB is helpful to grasp the psychological and behavioral characteristics of individual environmental scanning behavior from the micro level.

2.2 Research Hypotheses

The research model proposed in this paper (Fig. 2) is an extension of TPB. Considering that TPB holds that the generation of behavior is directly dependent on the inadequacy of intention to act, individual behavior is also affected by other factors such as organization and task. Therefore, we introduce the organizational support from the competitive intelligence management control theory in order to fully understand the influencing factors of employees' environmental scanning behavior in big data.

Environmental scanning intention is a necessary condition for employees to carry out behavior of environmental scanning. In previous studies, Intention has been proved to be a reliable variable for explaining behavior, which can predict behavior more accurately. According to the three-tier structure of TPB, intention has a direct impact on behavior [3]. Therefore, the following assumption is put forward:

H1: Intention of environmental scanning has a significant positive impact on behavior.

2.2.1 Cognition of Environmental Scanning

Cognition of environmental scanning consists of basic usability cognition. Vuori pointed out that the enhancement of employees' basic knowledge of environmental scanning had a positive impact on their attitude towards environmental scanning [4].

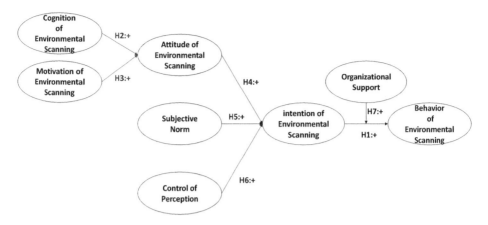

Fig. 2. The conceptual model

Jiao and Pei pointed out that using big data tools can effectively improve the performance of employees, therefore attitudes are significantly enhanced [5]. Thus, the following assumption is put forward:

H2: Cognition of environmental scanning has a significant positive impact on attitudes.

2.2.2 Motivation of Environmental Scanning

Motivation of environmental scanning consists of internal and external motivations. Pérez-González confirms that when employees think that the application of big data tools in work is interesting and challenging, it will generate a strong internal drive, hoping to obtain satisfaction and pleasure in the application of big data technology, then employees will not be resistant to the implementation of big data tools [6]. Le Bon and Merunka pointed out that external motivations such as promotion and other people's recognition had a positive impact on their attitudes towards intelligence contribution when they studied the intelligence contribution of marketers [7]. Thus, the following assumption is put forward:

H3: Environmental scanning motivation has a significant positive impact on environmental scanning attitude.

2.2.3 Attitude of Environmental Scanning

Attitude of environmental scanning consists of instrumental and emotional attitude. Du pointed out that employee's instrumental attitude had a significant impact on their intention, when he studied the application of big data tools using in environmental scanning [8]. The existing research shows that the application of big data not only improves the efficiency and quality of environmental scanning, and also brings pleasure and satisfaction [6]. Therefore, users have a strong emotional attitude towards them. Thus, the following assumption is put forward:

H4: Attitude of environmental scanning has a significant positive impact on intention.

2.2.4 Subjective Norms

Subjective norms consists of directive and demonstration norms. Previous studies shows that employees are usually accustomed to accept the definition of functions and roles stipulated by the organizational system, which can enhance employees' intention to scan the environment. Therefore, the following hypothesis is put forward:

H5: Subjective norms have a significant positive impact on the intention of environmental scanning.

2.2.5 Control of Perception

Control of perception is composed of self-efficacy and control. Du pointed out that big data bring huge amounts of data for environmental scanning, so it needs a lot of time and energy to scanning, and the results of scanning are usually uncertain [8]. In this scenario, if employees have a higher ability to scan, they will prefer to scan. Le Bon and Merunka have shown that when employees perceive that their environmental scanning behavior is supported, empowered and feedback, they will feel lower psychological pressure to scan [7]. Accordingly, the following assumption is put forward:

H6: Control of perception has a positive effect on intention of environmental scanning.

2.2.6 Organizational Support

Organizational support consists of executive support, tool support of formal control and emotional support of informal control. Lei and Shen, Le Bon and Merunka pointed out that the construction of environmental scanning system is a first-hand project, which can not be separated from the support of executive and the leading role of practice [7, 9]. More empowerment and emotional support for employees' environmental scanning will be helpful to enhance employees' environmental scanning intention and promote their behavior [10, 11]. Accordingly, the following assumption is put forward:

H7: Organizational support has a positive regulatory effect on the transformation of intention to behavior of environmental scanning.

3 Data Analysis and Research Results

3.1 Reliability and Validity Analysis

After data collection, the valid questionnaires are analyzed as follows. Cronbach's alpha coefficient is used to test the reliability of the scale. When the coefficient is greater than 0.7, it shows that the reliability of the questionnaires is high and can it be further analyzed. As shown in Table 1, all dimensions of Cronbach's alpha were above 0.7.

Table 1. Reliability and aggregation validity test results

Latent variable	Number	Cronbach's alpha	AVE	CR
Cognition	3	0.942	0.898	0.963
Motivation	6	0.964	0.853	0.972
Attitude	3	0.964	0.857	0.947
Control of perception	4	0.899	0.852	0.959
Subjective norm	4	0.922	0.737	0.918
Intention	3	0.927	0.826	0.934
Organizational support	4	0.943	0.795	0.939
Behavior	5	0.936	0.777	0.946

For aggregation validity, Average Variance Extracted (AVE) and Composite Reliability (CR) were used to test and analyze. The results show that CR of each potential variable are greater than 0.8, which indicates that the reliability of each potential variable is better; AVE of each potential variable is greater than 0.5, which indicates that the aggregation validity of each potential variable is better. For the determination of discriminatory validity, this study compares AVE square root with factor correlation coefficient. The results show that AVE square root of each potential variable is larger than the correlation coefficient of each potential variable, indicating that each potential variable has better discriminatory validity. In conclusion, the scale used in this paper has good validity.

3.2   Hypothesis Testing

This paper uses AMOS 17.0 to validate the model. According to the revised suggestion, only one path is modified at a time, and the rationality of the relationship between variables is considered comprehensively. The path between environmental scanning cognition and environmental scanning motivation, and between subjective norms and control of perception are added. Finally, the modified model analysis results are shown in Fig. 3.

The final fitting results of the revised structural equation show that all indicators except GFI meet the recommended criteria, but considering the complexity of the model and the GFI > 0.85, according to previous expert research recommendations, the revised model in this study is acceptable. At the same time, the table shows that all paths reach significant level, and the results show that the model is acceptable. H1, H2, H3, H4, H5, H6 are all valid.

This paper explores the moderating role of organizational support in the relationship between environmental scanning intention and environmental scanning behavior through regression analysis. As shown in Table 2, the R^2 of model 2 is larger than that of model 1, and the value of Sig. is less than 0.05. Therefore, H7 is valid.

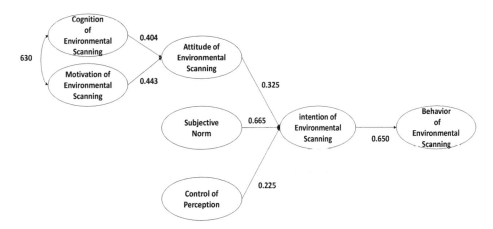

Fig. 3. Modified structural equation model and its results

Table 2. Testing the regulatory effect of organizational support

Model	Independent variable	β	Sig.	R^2	F
1	Environmental Scanning Intention (ESI)	0.631	0.000	0.867	591.691
	Organizational Support (OS)	0.386	0.000		
2	Environmental Scanning Intention	0.592	0.000	0.869	403.421
	Organizational Support	0.394	0.000		
	ESI × OS	0.059	0.037		

4   Discussion

Due to the problems of sample and depth of investigation, this paper is only a preliminary study. But, according to the research, we can learn that TPB can better explain the environmental scanning behavior of employees in big data environment. And subjective norms have the most significant impact on the intention. In view of this, enterprises should optimize and even reengineer their own management and control system and process, and bring environmental scanning into the management and control of employees' roles, functions and positions. Secondly, organizational support plays a positive moderating role between intention and behavior. In this regard, enterprise leaders should pay more attention to environmental scanning, actively promote the construction of big data infrastructure for environmental scanning.

Acknowledgements. This work was supported by the 13th Five-Year Plan of Jiangsu Province "Key Construction Discipline Project of Business Administration Level 1" (Grant No. SJY201609) and the Philosophy and Social Science Fund of Education Department of Jiangsu Province (Grant No. 2018SJA1248).

References

1. Bao, C., Xie, X., Zhang, Y., Li, N.: Competitive intelligence system of enterprise. China Inf. Rev. (8), 33–37 (2001)
2. Deiser, R., Newton, S.: Six social-media skills every leader needs. McKinsey Q. **1**, 62–67 (2013)
3. Ajzen, I.: Perceived behavioral control, self-efficacy, locus of control, and the theory of planned behavior. J. Appl. Soc. Psychol. **32**(4), 665–683 (2002)
4. Vuori, V.: Social media changing the competitive intelligence process: elicitation of employees' competitive knowledge. Rev. Met. **56**, 1001 (2011)
5. Jiao, W., Pei, L.: Studies in social media value identification and implementation under the background of competitive intelligence. Inf. Doc. Serv. **2**, 37–42 (2015). (in Chinese)
6. Pérez-González, D.: Social media technologies' use for the competitive information and knowledge sharing, and its effects on industrial SMEs' innovation. Inf. Syst. Manag. **34**(3), 291–301 (2017)
7. Le Bon, J., Merunka, D.: The impact of individual and managerial factors on salespeople's contribution to marketing intelligence activities. Int. J. Res. Mark. **23**(4), 395–408 (2006)
8. Du, P.: Enterprise competitive intelligence work under the background of big data: opportunities, challenges and promotion strategies. J. Lib. Inf. Sci. **3**(09), 62–66 (2018). (in Chinese)
9. Lei, H., Shen, X.: Research in Web2.0 application tools for knowledge management processes—a case study in Taiwan. Mod. Inf. **34**(3), 87–89 (2014). (in Chinese)
10. Alqahtani, F.H.: The adoption of Web 2.0 within enterprises: employees' perspectives (2013)
11. Saxena, D., Lamest, M.: Information overload and coping strategies in the big data context: evidence from the hospitality sector. J. Inf. Sci. **44**(3), 287–297 (2018)

Rural E-commerce Development Under the Background of Big Data

Zhitan Feng[✉]

School of Commercial, Nantong Institute of Technology, Nantong 226002,
Jiangsu, China
Fengzhtan@126.com

Abstract. With the upgrading of consumption, the demand for consumer quality has been continuously improved, which has brought strong demand guidance and huge market for e-commerce. The development of rural e-commerce provides a huge demand for consumption upgrades in rural areas, and also provides a broad market for e-commerce companies, thereby promoting the sustainable development of the entire industry. This paper starts with the development of rural e-commerce and builds a rural e-commerce data service platform, which is explained from the construction of infrastructure services and big data analysis platform. On this basis, the further development of rural e-commerce benefits from big data analysis and application. Finally, from the improvement of rural e-commerce infrastructure, network-based facilities construction, building agricultural products features and brands, strengthening the introduction and training of electric merchants, etc., to improve the development space of rural e-commerce, to achieve farmers' income, electronic. The business industry is growing rapidly.

Keywords: Big data · Rural e-commerce · Development research

1 Introduction

With the upgrading of consumption, the demand for consumer quality has been continuously improved, which has brought strong demand guidance and huge market for e-commerce. In China, more than 28 million farmers have been employed in rural network sales to transfer to the tertiary industry, which has promoted industrial integration. In 2018, 3,202 Taobao villages were discovered nationwide, an increase of 51% year-on-year. The annual sales of these Taobao Villages on the Alibaba retail platform exceeded RMB 220 billion. Correspondingly, the number of Taobao Towns has grown rapidly. In 2018, the number of Taobao Towns nationwide reached 363, an increase of nearly 50% year-on-year. The development of e-commerce demonstration villages has brought new forms of business and new industries to rural e-commerce, highlighting regional characteristics. In addition, Alibaba, Taobao, Suning Tesco and other platforms to promote the rural market, but also provide a broader platform and requirements for rural products to go out and farmers to re-employment. Nowadays, no matter whether it is the consumer or sales side of rural e-commerce, relying on big data technology to achieve timely updating of data, the quality of rural products and sales are tracked throughout the whole process, and rural consumption can be analyzed in time to capture the consumption needs of farmers. These all achieve the productivity of data, improve the targeting of product sales, and upgrade the consumption of farmers [3, 4, 8].

2   Status of Rural E-commerce Development

With the in-depth development of national e-commerce into rural comprehensive demonstration, the genes of the Internet are gradually infiltrating into all corners of China's rural areas, allowing more poor people to use the Internet, and letting agricultural products out of the countryside through the Internet has gradually become a reality. The development of e-commerce in rural areas has brought important development opportunities for rural poverty alleviation. Farmers' employability has improved and entrepreneurial enthusiasm has increased [1, 2, 6, 7].

2.1   Grassroots Entrepreneurial Demonstration to the Surrounding Area and Gradually Formed an Industrial Cluster

In the process of Internet development, young people in rural areas took the lead in realizing sales of products through e-commerce platforms, realizing grassroots entrepreneurship, increasing income, and promoting the development of rural economy. This demonstration effect, due to the rural geo-structure, the effect spreads faster, and from imitation to cooperation, e-commerce demonstration village, demonstration town are formed. In rural areas, the development of e-commerce has led to the integration of downstream industries, logistics distribution systems, payment systems, financial accounting systems and other industries, and gradually formed industrial clusters. The product supply chain system has been reshaped and the agricultural industry structure has been upgraded.

2.2   "Government Support + Platform Empowerment" Promotes the Rapid Development of Rural E-commerce Development

The e-commerce platform represented by Alibaba, Suning, and Pinduoduo has gradually tapped the rural market to realize the growth of traffic and transactions. The massive online market environment created by enterprises has provided grassroots entrepreneurs with a low threshold for entrepreneurial channels. E-commerce poverty alleviation is one of the important ways to promote poverty alleviation. In the process of rural e-commerce development, governments at all levels play an administrative advantage, provide policy formulation and service supply from different levels and in different ways, and inclusive and support rural grassroots entrepreneurial behavior.

2.3   Product Upgrade Needs to be Improved to Avoid Homogenization Competition

The development of rural e-commerce promotes the upgrading of agricultural industrial structure. However, the rural e-commerce represented by "Taobao Village" has fierce competition for homogenized products, with less product differentiation and extensive products. Many stores have invested in R&D of products. Less, resulting in slow product upgrades, price competition has become the norm, and farmers' income and stability have been affected. The rural logistics system, especially the cold chain logistics system, is in urgent need of gradual improvement, especially in poor areas. The perfect road transportation and logistics system provide important support for the uplink and downlink of products.

3   Building a Rural E-commerce Analysis Platform Based on Big Data

Based on big data, build a rural e-commerce service platform and form a data platform integrating farmers, consumers and big data service centers to realize data sharing and integration, improve the targeting of rural e-commerce development, and provide farmers' income and Convenience of consumption [10]. This kind of construction needs to establish two aspects of the corresponding infrastructure and big data analysis platform, which involves cloud resource configuration, data, security, authentication, and service.

3.1   Construction of Infrastructure Services

Relying on big data to build a rural e-commerce platform, it is necessary to use cloud computing and other technologies to establish a basic platform and support environment for the informatization construction of rural e-commerce platforms [5].

(1)   Realize hardware and software asset management

The establishment of rural e-commerce platform is designed to hardware assets and software assets, in which hardware assets are designed to the necessary hardware environment such as building and computer room. Software assets include middleware, operating system, database, security software, platform management software, and business application system, other software, etc. To achieve effective management of hardware and software assets requires rural e-commerce to provide a technology platform. At the same time, the speed of e-commerce enterprises is closely related to the flow rate.

(2)   Implement platform service management

First of all, service governance, the rural e-commerce platform will appear inconsistent data definition standards, through the service management to achieve data standardization, unification, thus facilitating sharing and statistical analysis, and ultimately achieve the national rural e-commerce users and consumption big data precipitation.

Secondly, with the development of social e-commerce, the sources of information are diversified, which requires the integration of multiple information acquisition platforms, storage and management on the basis of data uniformity, and the timely and effective information transmission through the broadening of information acquisition channels. The limitation of space, so as to achieve the aggregation and analysis of rural e-commerce data.

3.2 Construction of Big Data Analysis Platform

Through the platform to provide support for the analysis of changes in the rural market environment, analyze the changes and trends in rural purchase demand, and provide support and technical judgment for e-commerce corporate strategy and business activities. Provide the most effective data support for rural entrepreneurs by analyzing the situation of rural products. Through the intelligent equipment management of various community e-commerce points and self-reporting cabinets, the first-hand information of rural e-commerce is recorded. Realize regional and multi-faceted collection of rural e-commerce transaction data, realize data visualization, intelligent words, and improve data processing efficiency. According to the statistical data, the refined analysis is carried out. This analysis needs to complete the rural e-commerce data sorting through cluster analysis and deep learning, machine learning and other algorithms to achieve all-round analysis, and the variability and difference between the statistics. This will better mine the value of data, realize value added, and provide efficient and accurate support services for rural economic development. These data analysis is also conducive to the government and enterprises to grasp the current development of the rural areas, to grasp the direction of rural industrial upgrading and agricultural development. In the end, resource optimization is realized, farmers' incomes increase, and sales of e-commerce products increase, achieving multi-win.

4 Big Data Analysis Applied to the Development of Rural E-commerce

4.1 Improve Rural E-commerce Infrastructure Construction

The construction of rural e-commerce infrastructure includes transportation, network development, logistics system and financial service measures. Through big data analysis, it can be seen that the development of rural e-commerce is based on rural network construction, mobile signal coverage and traffic fluency. Internet coverage is the basis for development, and the development of transportation provides reliable guarantee for product access. The solution to the "last mile" problem of rural e-commerce provides important support for its development, and this last mile includes the establishment of logistics transit centers, distribution points and personnel, through the analysis of big data, the perfection of logistics facilities and rural e-commerce the level of development has a strong positive correlation. The development of rural e-commerce entrepreneurship is inseparable from the support of funds, providing strong financial support services, which will have a greater impact on rural industrial upgrading and brand building. In the field of rural financial development, there is still a large blue ocean space, and the rural financial gap is About 300 million yuan, how to make up for the funding gap and realize the development of the rural financial blue ocean field has become a problem that China needs to solve at present.

4.2 Featured Products and Brand Creation

How to make specialty agricultural products go out and increase farmers' income is an important channel for rural development. At this stage, farmers cannot effectively master network-related technologies, product sales are difficult to carry out quickly, and e-commerce promotion has not been played. Based on the above problems, big data platform is needed as a support, combined with rural production, network platform use and buyers to establish channels to share and integrate big data application technology. According to the sales of big data, grasp the customer's consumption habits and characteristics, improve the category and quality of featured agricultural products, and thus comprehensively improve the level of rural development. Attracting traditional rural business entities, rural brokers, cooperatives and e-commerce to integrate, guide and encourage brand registration and construction, form multi-party cooperation with e-commerce as a platform, establish quality monitoring mechanism, product R&D center and brand operation concept.

4.3 Electric Businessman Training

In rural areas, cultivating talents and retaining talents is a difficult problem in the development of e-commerce. More professionals with considerable professional knowledge and abilities are needed to better mine useful data. In cultivating talents, we must first solve the problem of farmers' understanding, raise the peasants' Internet thinking, and guide farmers to pay attention to data analysis. In addition, it is necessary to make full use of the resources and abilities of the university students and village cadres as the representative, and encourage the university students to return to their hometowns [9]. The government provides necessary policy support, thus forming a relatively perfect rural electric merchant training system.

5 Conclusions

At present, it is necessary to follow the objective laws of informatization development, based on the national conditions and the local conditions, promote the upgrading of rural e-commerce, and promote and improve the modernization of agriculture and rural areas. The development of rural e-commerce must also be adapted to local conditions and take advantage of the trend. This paper starts with the development of rural e-commerce and builds a rural e-commerce data service platform, which is explained from the construction of infrastructure services and big data analysis platform. On this basis, the further development of rural e-commerce benefits from big data analysis and application. Finally, from the improvement of rural e-commerce infrastructure, network-based facilities construction, building agricultural products features and brands, strengthening the introduction and training of electric merchants, etc., to improve the development space of rural e-commerce, to achieve farmers' income, electronic. The business industry is growing rapidly.

Acknowledgements. Jiangsu University Philosophy and Social Science Research Fund Project "Research on Innovation Mechanism of Rural E-Commerce Development Model Based on Synergistic Effect". Project number: 2019SJA1475; First Level Key Built Discipline Projects of the Business Administration under Jiangsu Provincial "the 13th Five-Year Plan". Project number: SJY201609; Nantong Institute of Research professor and doctoral research project (201823).

References

1. Yu, Y., Lu, S., Gao, Y.: Research on rural e-commerce precision poverty alleviation in the background of big data. China New Commun. **21**(16), 157 (2019)
2. Gao, X., Yang, Z., Zhang, Y., Gao, Y., Jiang, S.: In the context of big data, the new ideas of Linyi City e-commerce to help Yimeng rural get rid of poverty and get rich. China Mark. **12**, 37–38 (2019)
3. Yang, H.: Research on the development model of rural e-commerce under the background of "Internet+". China Mark. **11**, 184–192 (2019)
4. Ren, H.: Research on the influencing factors and improvement path of rural e-commerce development from the perspective of big data. Inf. Commun. **03**, 270–271 (2019)
5. Li, S.: Research on the construction of rural e-commerce logistics model based on cloud computing. Natl. Circ. Econ. **04**, 20–21 (2019)
6. Ma, Q., Chen, Z.: Research on rural e-commerce precision poverty alleviation innovation model under the background of big data-taking Luxi County of Guizhou Province as an example. Modern Mark. (later) **01**, 181–184 (2019)
7. Cao, X.: Research on rural e-commerce development strategy from the perspective of precision poverty alleviation. Shanxi Agric. J. **21**, 1–5 (2018)
8. Hu, X.: Research on the development strategy of rural e-commerce in Southern Anhui under the background of big data. Wirel. Interconnect Technol. **15**(17), 126–127 (2018)
9. Feng, Z.: Analysis of the development path of rural e-commerce in Jiangsu under the background of big data. Enterp. Technol. Dev. **08**, 44–45 (2018)
10. Guo, J.: Research on the technical service management of rural e-commerce platform based on the underlying integration of G company. Beijing University of Chemical Technology (2018)

Regional Energy Internet Optimization Configuration Method Exploration

Wen Li[1(✉)], Kai Ma[2], Jiye Wang[1], Hao Li[1], Chao Liu[1], Bin Li[1], and Chang Liu[1]

[1] China Electric Power Research Institute, Beijing, China
149070802@qq.com
[2] State Grid Beijing Electric Power Company, Beijing, China

Abstract. Optimizing the configuration problem is the first problem that needs to be solved for regional energy Internet construction. Unreasonable configuration schemes will not only make the economic benefits of regional energy Internet unable to meet expectations, but also may cause large grid-side energy volatility and affect the safe and stable operation of grid-connected devices. Therefore, based on the particle swarm optimization algorithm, this paper proposes a regional energy Internet optimization allocation method based on the grid-side energy volatility, maximizing the economic benefits of regional energy Internet and promoting the efficient use of energy.

Keywords: Regional energy internet · Optimization configuration · Particle swarm optimization

1 Introduction

With the large number of new components such as distributed power, distributed energy storage, and electric vehicles connected to the power grid, the structure and layout of the power distribution system will undergo major changes in the future. In order to promote energy utilization and energy Internet construction, governments at all levels have issued a The series of measures encourages qualified enterprises and scientific research units to manage their own source network reserves and build a regional energy Internet [1, 2].

As one of the evolution trends of energy development and utilization, regional energy Internet can reasonably integrate various elements of source network and storage, and effectively promote the rational consumption of energy and the optimal utilization of consumption, thus promoting the efficient use of energy [3, 4]. Based on graph theory and energy distance concept, the literature [5] proposed an optimized layout method for energy stations and pipeline networks in regional energy Internet. Literature [6] proposes a two-stage regional energy Internet collaborative optimization strategy based on adaptive model predictive control.

Although there have been some research results on regional energy Internet at home and abroad, the problem of optimal allocation is the first problem to be solved in the construction of regional energy Internet. The current research has not been involved. In order to solve the above problems, this paper proposes a regional energy Internet

© Springer Nature Singapore Pte Ltd. 2020
M. Atiquzzaman et al. (Eds.): BDCPS 2019, AISC 1117, pp. 405–411, 2020.
https://doi.org/10.1007/978-981-15-2568-1_56

optimization allocation method based on Particle Swarm Optimization (PSO) and constrains the network side energy volatility to maximize the economic benefits of regional energy Internet and promote energy efficient use.

2   Regional Energy Internet Power Mathematical Model

In order to propose a reasonable regional energy Internet optimization configuration method, it is first necessary to establish a power mathematical model of each element in the regional energy Internet. This paper establishes a typical regional energy Internet case, as shown in Fig. 1.

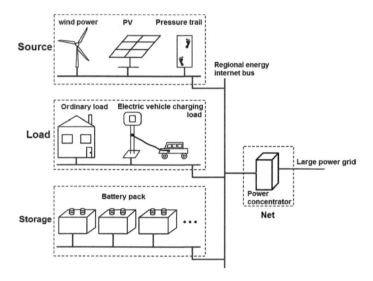

Fig. 1.   Schematic diagram of regional energy Internet case

2.1   Source Class Element Power Mathematical Model

The source elements in the regional energy Internet case include distributed wind power generation, distributed photovoltaic power generation, and power generation units such as pressure walkways in the field of new power generation technologies. The power mathematical model for such elements can be expressed as:

$$P_{Si}(t) = -N_{Si} \times \eta_i(t) \times P_{\max(Si)} \tag{1}$$

Among them, N_{Si} is the number of configuration of element i, $P_{Si}(t)$ is the power output of t-time, $P_{\max}(S_i)$ is the maximum power output, $\eta_i(t)$ the ratio of the actual power output to the maximum power output is estimated.

2.2 Mathematical Model of Charge Class Element Power

The charge elements in the regional energy Internet case include electric vehicle charging load and normal load. The normal load is expressed as $P_L(t)$, and the electric vehicle charging load can be expressed as:

$$P_C(t) = N_C \times L_C(t) \tag{2}$$

Among them, N_C is the number of charging piles, $P_C(t)$ is the charging load at time t, and $L_C(t)$ is the estimated charging pile load value at time t.

2.3 Mathematical Model of Storage Element Power

The storage element in the regional energy Internet case is a battery pack, and its power mathematical model can be expressed as:

$$P_E(t) = -[P_C(t) + P_L(t)] + \sum_{i=0}^{n} P_{Si}(t) \tag{3}$$

Among them, if:

$$\left[N_E \times E_0 + \int_t^\cdot P_E(t)\,\mathrm{dt} \right] = 0 \tag{4}$$

And the battery pack is in a discharged state, then $P_E(t) = 0$.
If:

$$\left[N_E \times E_0 + \int_t^\cdot P_E(t)\,\mathrm{dt} \right] \geq N_E \times E_{\mathrm{bat}} \times D_{\mathrm{OD}} \tag{5}$$

And when the battery pack is in the charging state, $P_E(t) = 0$.
Where N_E is the number of configurations, E_0 is the initial charge, $P_E(t)$ is the power value at time t, E_{bat} is the rated capacity, and D_{OD} is the discharge depth.

2.4 Mathematical Model of Net Class Element Power

The network element in the regional energy Internet case is a power concentrator connected to a large power grid. The mathematical model of the transmission power can be expressed as:

$$P_R(t) = -P_C(t) - P_L(t) - P_E(t) + \sum_{i=0}^{n} P_{Si}(t) \tag{6}$$

Where $P_R(t)$ is the transmission power value at time t of the power concentrator, and the value of the positive value indicates that the regional energy Internet feeds back power to the large power grid, and the negative value indicates that the large power grid supplies power to the regional energy Internet.

3 Regional Energy Internet Optimization Configuration Method

3.1 Restrictions

In order to propose an optimal configuration method for regional energy Internet, it is first necessary to establish constraints on the energy volatility of the network side. The constraints established in this paper include the extreme value constraint and the variance constraint of the network side energy volatility.

The extreme value constraint of the grid side energy volatility is:

$$\begin{cases} \left[-P_c(t) - P_L(t) - P_E(t) + \sum_{i=0}^{n} P_{Si}(t)\right]_{max} \leq \Delta P_{max} \\ \left[-P_c(t) - P_L(t) - P_E(t) + \sum_{i=0}^{n} P_{Si}(t)\right]_{min} \leq \Delta P_{min} \end{cases} \tag{7}$$

Where ΔP_{max} is the maximum power that the regional energy Internet can feed back to the large grid; ΔP_{min} is the maximum power that the large grid can transmit to the regional energy Internet.

The variance of the grid side energy volatility is:

$$D\left[-P_C(t) - P_L(t) - P_E(t) + \sum_{i=0}^{n} P_{Si}(t)\right] \leq D_{max} \tag{8}$$

Where Dmax is the set variance limit value.

3.2 Objective Function

The optimized configuration method proposed in this paper is to maximize the economic benefits of regional energy Internet. Therefore, the objective function of the optimized configuration method is selected as the annual economic benefit expression of the regional energy Internet:

$$f = \max \mathrm{EP} = -A + B - C \tag{9}$$

A is the annual investment cost of the regional energy Internet, which includes the annual average initial investment and annual operation cost of the equipment:

$$A = \left[\sum_{i=1}^{n} \left(N_{Si} \times \frac{C_{ini-Si}}{T_{Si}}\right) + N_C \times \frac{C_{ini-C}}{T_C} + N_E \times \frac{C_{ini-E}}{T_E}\right] \\ + \left[\sum_{i=1}^{n} \left(N_{Si} \times C_{ope-Si}\right) + N_C \times C_{ope-C} + N_E \times C_{ope-E}\right] \tag{10}$$

Among them, C_{ini-Si}, C_{ini-C} and C_{ini-E} are the initial investment of source and storage elements respectively, and T_{Si}, T_C and T_E are the service life of source and storage elements respectively, C_{ope-Si}, C_{ope-C} and C_{ope-E} respectively The annual operation and maintenance cost of the source storage element.

B is the income that can be obtained during the regional energy internet operation within one year:

$$B = M_C \int_t P_C(t)\,\mathrm{dt} + M_L \int_t P_L(t)\mathrm{dt} \tag{11}$$

Among them, M_C and M_L are respectively the electric vehicle charging price and the ordinary load price of the regional energy Internet.

C is the amount of electricity transaction generated by the interaction between regional energy Internet and large power grid:

$$C = V \int_t [P_C(t) + P_L(t) + P_E(t) - \sum_{i=0}^{n} P_{Si}(t)]\mathrm{dt} \tag{12}$$

Among them, V is the exchange price of regional energy Internet and large power grid.

3.3   Solution Flow

As a solution algorithm for optimization problems, PSO algorithm has been widely used in different fields [7–9]. The fastness and accuracy of the solution are in line with the implementation requirements of the regional energy Internet optimization configuration problem.

Any position in the PSO search space is expressed as:

$$\vec{s} = [N_{S1}, N_{S2}, \cdots, N_{Sn}, N_C, N_E] \tag{13}$$

Assuming that a population with m particles searches in the (n + 2) dimensional space, the starting position and velocity of particle i at the tth iteration are:

$$\vec{s}_i(t) = [s_{i1}(t), s_{i2}(t), \cdots, s_{i(n+2)}(t)] \tag{14}$$

$$\vec{v}_i(t) = [v_{i1}(t), v_{i2}(t), \cdots, v_{i(n+2)}(t)] \tag{15}$$

Let p(t) be the optimal position of the individual, which is defined as the position where the target f is the largest in the historical trajectory of particle i after t iterations; let q(t) be the overall optimal position, which is defined as t iterations The position of the target f in the historical trajectory of all particles in the particle group.

The velocity update formula and position update formula of particle i at the t + 1th iteration are:

$$\vec{v}_i(t+1) = \omega\vec{v}_i(t) + c_1 a(p(t) - \vec{s}_i(t)) + c_2 b(q(t) - \vec{s}_i(t)) \tag{16}$$

$$\vec{s}_i(t+1) = \vec{s}_i(t) + r\vec{v}_i(t+1) \tag{17}$$

Where ω is the inertia weight; c1 is the weight of the tracking p(t), called "self-cognition"; c2 is the weight of the tracking q(t), called "social cognition"; a and b are 0 to 1 Random number between; r is the position update constraint factor.

According to the experience [10], the selection of each parameter in the solution formula is shown in Table 1.

Table 1. Selection of specific parameters of PSO algorithm

Name	Value
Number of particles (m)	32
Inertia weight (ω)	1
Self-awareness (c1)	2
Social cognition (c2)	2
random number (a)	Randomly generated by the rand function
random number (b)	Randomly generated by the rand function
Location update constraint factor (r)	1

Based on the above content, the solution process of regional energy Internet optimization configuration based on PSO algorithm proposed in this paper is shown in Fig. 2.

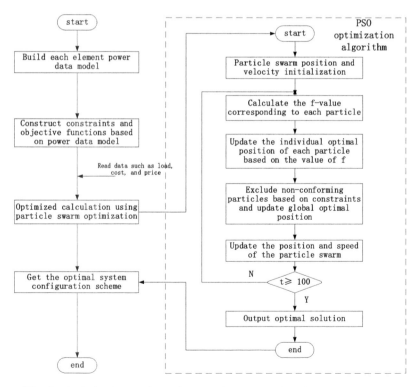

Fig. 2. Optimization configuration solution process based on PSO algorithm

4 Conclusion

Based on the PSO algorithm, this paper proposes a regional energy Internet optimization configuration method, which can obtain the optimal configuration scheme under the premise of grid-side energy volatility constraints, and maximize the economic benefits of regional energy Internet. A case study verifies the effectiveness of the method. In the future, it is necessary to simulate the randomness more closely with the weather forecast data.

References

1. National Development and Reform Commission, National Energy Administration, Ministry of Industry and Information Technology. Guidance on promoting the development of "Internet +" smart energy (2016). (in Chinese)
2. National Energy Administration. Notice on the publication of the first "Internet +" Smart Energy (Energy Internet) demonstration project (2017). (in Chinese)
3. Chen, J., Huang, Y., Lu, B.: Regional energy internet information physics modeling and control strategy. Electric Power Autom. Equipment **36**(12), 1–10 (2016)
4. Cao, L., Wang, M., Shen, X., Zhang, Q., Yan, L.: Research on regional energy internet in Beijing, Tianjin and Hebei. Power Syst. Clean Energy **33**(3), 125–130 (2017)
5. Chen, J., Huang, Y., Lu, B.: Research on the optimization of "station-net" layout of regional energy internet. Proc. CSEE **38**(3), 675–684 (2018)
6. Yin, H., Liu, C., Zhao, J., Geng, H., Li, Y.: Two-stage collaborative scheduling of regional energy internet based on adaptive model predictive control. Mod. Electr. Power **35**(4) (2018). (in Chinese)
7. Cheng, Z., Dong, M., Yang, T., Han, L.: Parameter identification of photovoltaic cell model based on adaptive chaotic particle swarm optimization algorithm. Trans. China Electrotechnical Soc. **29**(9), 245–252 (2014)
8. Lu, J., Miao, Y., Zhang, C., Ren, H.: Optimal scheduling of power system with wind farm based on improved multi-objective particle swarm optimization. Power Syst. Prot. Control **41**(17), 25–31 (2013)
9. Liu, X., Wen, J., Pan, Y., Wu, P., Li, J.: Optimal power flow control of DC grid using improved particle swarm optimization algorithm. Power Syst. Technol. **41**(3), 715–720 (2017)
10. Chen, Y., Xu, J., Xu, G., Pan, P., Ma, G.: Research on maximum access capacity of grid-connected wind farm based on particle swarm optimization. Renew. Energy **35**(9), 1347–1351 (2017)

Design of Personalized Service System for Home-Based Elderly Care Based on Data Fusion

Rongqing Zhuo$^{(\boxtimes)}$ and Xin Sun

Network Data Center, Communication University of Zhejiang, Hangzhou, Zhejiang, China
zhuorq@cuz.edu.cn

Abstract. At present, China is in the rapid development stage of population aging, home care service model in the care system in the basic position, but this model is fraught with problems. The transformation and upgrading of the traditional home-based care service model can not only improve the quality of life of the elderly in their later years, but also is of great significance to the healthy and sustainable development of the service industry. Starting from the connotation and demand of endowment personalized service that occupy the home, in the data mining technology for technical support, analysis of data mining application in home endowment personalized service demand and the demand of the platform, is proposed based on data fusion of the wisdom of the endowment the overall architecture of the platform, wisdom endowment platform can be divided into infrastructure layer, data layer and application layer. The infrastructure layer includes hardware resources such as intelligent pension terminal and network deployment, and the data support layer includes data and data fusion. The research of this paper also shows that the personalized service system for home-based elderly care based on data fusion can meet the actual needs and has strong practicability.

Keywords: Data fusion · Home-based care · Personalized service · Elderly service industry

1 Introduction

Population aging has become an unstoppable trend in the 21st century and a problem faced by the whole world. However, this problem is more serious in China, and population aging has always been a problem that cannot be ignored [1]. According to the public data, by the end of 2018, China's elderly population aged 60 and above has reached 270 million, accounting for 17.5% of the total population, and 220 million aged 65 and above, accounting for 12.1% of the total population. In the next 20 years, China's elderly population will continue to grow rapidly [2]. As the aging degree of China's social population continues to increase, China's "421" family structure has gradually taken shape, leading to a large number of empty nest families in China, which brings greater challenges for the country's pension [3]. The old man became weaker as he got older. Most elderly people suffer from 2–3 chronic diseases with

© Springer Nature Singapore Pte Ltd. 2020
M. Atiquzzaman et al. (Eds.): BDCPS 2019, AISC 1117, pp. 412–419, 2020.
https://doi.org/10.1007/978-981-15-2568-1_57

multiple complications, long course of disease and recessive diseases. They need continuous comprehensive care services in medical and life, which brings great challenges to the existing pension service system and medical and health care system [4]. How to improve the ability and efficiency of home-based care service, provide comprehensive services and care for the elderly, and build an intelligent home-based care service system for the majority of the elderly has become one of the most hot and difficult issues in China [5].

In the field of home-based care, the existing achievements are mainly concentrated in the fields of care management system, call system and health monitoring system [6]. Pension management system includes two functions, one is telemedicine, in many hospitals and oral diagnosis has been laid telemedicine network, so many elderly people through the pension management system at home; The second is remote monitoring, which can monitor the physical condition, behavior and movement of the elderly at any time, and judge whether the elderly have potential disease threats and fall behaviors through some sign processing algorithms [7]. The call system monitors the physical signs of the elderly through various sensors, such as the infrared sensor, which can monitor the temperature of the elderly, and these monitoring results will be transmitted to the mobile phone of the elderly doctor [8]. The calling system can also receive doctors' orders and communicate with the elderly, so as to facilitate the elderly to consult and seek help from doctors, and convey the thoughts and requirements of the elderly to doctors, thus becoming a bridge of communication between the elderly and doctors [9, 10].

Based on data fusion, the home-based old-age care platform proposed in this paper realizes accurate acquisition of real-time information of the elderly through the use of existing technologies, conducts data fusion processing of real-time information, and provides personalized services for the elderly according to the results of processing. Such systems process and analyze collected data through data fusion and display the data to users using server technology, Web technology and Android technology. Traditional pension platforms only provide video monitoring for the elderly, without analyzing their vital signs and activities, let alone providing various services for the elderly through their real-time information. This system realizes the design of home care system for the elderly in all aspects, so that the elderly can truly enjoy the wisdom of home care.

2 Method

2.1 Data Integration and Calculation

Data fusion technology is a kind of automatic information comprehensive processing technology, which means to use computer technology to analyze and synthesize multi-sensor or multi-source information to complete decision-making and estimation tasks. Data generation includes information fusion, multi-sensor fusion and data fusion. The basic principle of data fusion includes five steps: data collection, feature extraction, pattern recognition, data association and data synthesis. Data fusion technology can effectively improve data transmission speed and information accuracy, and save system resources.

Data fusion is a kind of comprehensive information processing technology, which USES the theories and technologies of many traditional and new subjects, including communication, signal processing, pattern recognition, decision theory, optimization technology in traditional subjects, as well as artificial intelligence and neural network and other fields emerging in recent years. Many data fusion algorithms come from these fields, such as bayesian algorithm and SVM algorithm. These algorithms are described in detail below.

As a common algorithm in data fusion, bayesian method represents the measurement uncertainty with conditional probability and calculates the posterior probability through bayesian inference. Its logical expression can be expressed as:

$$P(X = x) = \sum_{i=1}^{n} P(Y = y_i)P(X = x|Y = y_i) \tag{1}$$

The computational complexity of bayesian algorithm is moderate, but it requires that all assumptions are independent, and prior probability and conditional probability should be given. It takes a lot of time to calculate all the probabilities to ensure the accuracy and consistency. Based on the confidence distance theory, a data fusion method based on bayesian estimation is proposed to process the measurement data of multiple sensors of the same type for the same parameter.

SVM algorithm is a machine learning and pattern classification method based on statistics and structural risk minimization. It USES the nonlinear transformation defined by kernel function:

$$\gamma = y(w^T x + b) = y.f(x) \tag{2}$$

The original feature space is dimensionally augmented to find the optimal linear classification interface in the high dimensional space. SVM algorithm can solve the problems of small sample, non-linear and high dimension, but the training sample is also limited.

2.2   Demand Analysis

General home old man because of the age, memory drops, physical disability, physiological function decline, daily operation capacity reduction factors, such as some accident may happen at home, so the hidden trouble in security according to the old people who live alone and living demand puts forward applying data fusion to the old man's life that occupy the home, remote health monitoring service for the old man, home security alarm service, business operation by life, request to answer call service, and the old man indoor and outdoor location services, etc.

(1) Remote health monitoring service
   The class service with advanced and sophisticated wearable devices for the elderly medical measurement indicators, such as breathing rate, pulse, blood pressure, blood sugar, continuously monitoring ecg, body temperature, etc., and transmit the data to the pension service platform, platform according to the physiological parameter data for medical and behavioral data fusion processing, comprehensive analysis it is concluded that the old man's health.

(2) Home security alarm service
   This kind of service mainly provides the prevention and alarm service for the elderly. Once the danger occurs, the alarm device will timely alarm to remind the elderly, while the remote service center can also get the alarm information in the first time, the staff can arrive in time, reduce the loss, so that the elderly's personal safety and property get strong protection.

(3) Life business acceptance service
   This kind of service is mainly aimed at solving the problem that the children of the elderly are not at home for a long time due to work, so they cannot accompany the elderly all the time, and the elderly need help and care in life due to their old age and physical inconvenience. The home-based care service platform will provide all kinds of door-to-door services according to the real-time needs of the elderly, so that the elderly can feel the care and give them the warmth of family.

(4) Request response call service
   This kind of service is mainly considered from the poor physical condition of the elderly living alone and the risk of emergencies. Through this service, when the elderly encounter emergency, major things or general help such as sudden illness, home fire, accidental fall can request rescue, there is an emergency button at the intelligent terminal, the elderly can press according to the need.

(5) Indoor and outdoor positioning services for the elderly
   The main purpose of these services is to facilitate the elderly children and service staff to quickly find the specific location of the elderly. When the elderly need to be rescued in an accident, the platform can quickly and accurately determine the specific position of the elderly through indoor positioning technology to shorten the time to rescue the elderly. When the elderly lost or emergency, family members and the center service staff can also locate the specific location of the elderly in the first time, so that the elderly get timely assistance, to ensure the safety of the elderly.

3   Model Building

3.1   Data Fusion Model

The data fusion module is used in the processing of the data support layer of the intelligent pension platform to process the massive data of the elderly in the intelligent pension platform, including the basic information of the elderly, health records, business data, video data, location information and other heterogeneous data. The data fusion model architecture is shown in Fig. 1 below.

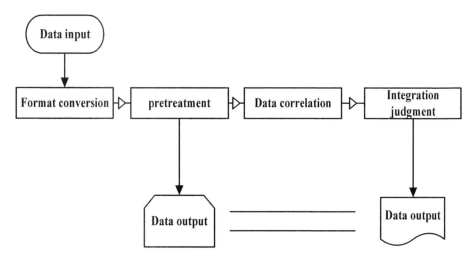

Fig. 1. Architecture diagram of data fusion model

Based on the traditional data fusion model, this model is improved to make it more suitable for the intelligent pension platform. On the basis of the data fusion model proposed by the intelligent pension platform, medical data fusion is applied to the model, so as to conduct the fusion diagnosis of physical signs for the elderly.

3.2 Prediction Model of System Reasoning

Every data fusion processing of the intelligent pension platform can be regarded as a data fusion task. Every data fusion task needs to pass data input, format conversion, data preprocessing, data correlation, target identification and fusion judgment, and finally output the results in the target identification and fusion judgment.

In the data preprocessing stage, time and space fusion is usually carried out on sensor data, that is, a series of data in this period are firstly fused, and then multi-sensor data are fused. Data association according to different data fusion processing, select the related data that can coordinate the target identification, and conduct certain correlation processing on these data to make the final target identification result more accurate. The main function of fusion judgment is to make a comprehensive judgment on the result of multi-object recognition so as to obtain a final processing result and output the result to the data for the application layer to use. The data output object is the result of each step of data fusion, that is, the information processed according to the requirements of data fusion.

4 Discuss

4.1 Realization of Cloud Service Platform

There are two types of users in the whole home-based care personalized service system, and two different operating software are provided respectively, which are the operating terminal used by the service personnel and the client terminal used by the elderly. In the useful port home page of service personnel, the upper area is information prompt bar, the left area is navigation menu, and the right area is various sub-pages. The left navigation menu diagram is divided into five categories, namely, information management, medical health, active care, services and activities, and security rescue. The sub-modules of medical health system are emergency rescue, chronic disease management, health information management, free physical examination management, appointment registration management. Information management interface will add, modify, delete the elderly information and other functions as well as the elderly family information tracking. In the service and activity interface, information such as recent medical care activities and physical examination activities for the elderly will be displayed.

4.2 Implementation of Mobile Client

In the mobile client login interface, the elderly can choose to remember the password and automatic login, click login to enter the main interface. If you are not registered, you can enter the registration interface through the registration button and complete the registration by entering the user name, password and your phone number in the registration interface. After logging in successfully, you can enter the main interface of the system. On the top of the main interface is some elderly activities or consulting recommendations, which can allow children to help the elderly in the home to sign up or recommend. The middle part is the real-time status of the elderly, including location information and physiological status. Below the real-time status is the latest notification, including all kinds of activity notification, abnormal alarm notification, reminder care notification and diagnosis report of the elderly. There are five columns at the bottom, respectively the home page, medical health, service activities, security rescue and system Settings. Click the medical health in the five columns to enter the interface of the medical health subsystem. There are six functions including remote monitoring, article browsing, health video viewing, free physical examination, diagnostic process and complaining.

4.3 System Test

Evaluating the performance of a system of gold standard is always the user experience, in order to verify the performance of this system, the research of this paper considers the degree of the old man's health, self-care ability status, cultural background, such factors as the quality of system factors to establish different types of old people to the health of the different grades (EHC), including the self-care, mobility, interface level, a disabled old man, and their experience in 40%, 30%, 20%, 10% of the grade weight to the merits of the judge system, test results are shown in Table 1 and Fig. 2 below.

Table 1. Test results of the system from different indicators

Elderly health	EHC value	Mark	Weighted score	Score
Self-service level	1	95	40%	38
Mobility level	5	91	30%	27.3
Intermediate level	10	84	20%	16.8
Disabled elderly	15	74	10%	7.4
Final score				**89.5**

The test results show that the elderly are generally satisfied with the evaluation of the system, which verifies the feasibility of the system.

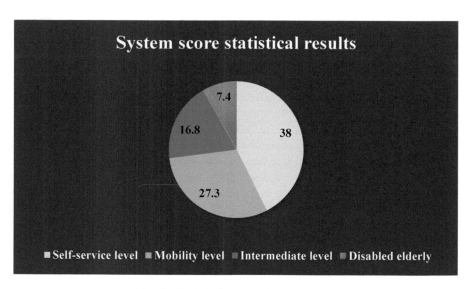

Fig. 2. Statistical results of system score

5   Conclusion

In this paper, data fusion technology is introduced into the intelligent pension platform to design a personalized home-based pension system. This paper studies and analyzes the relevant technologies used in the pension platform based on data fusion. Firstly, the concept and algorithm of data fusion are introduced. According to the actual requirements of the system in this paper, the overall scheme design of the intelligent pension platform based on data fusion was elaborated, and the overall design of the system was completed. In the test of the system, the system got a good test result.

Old-age care is a long process. In this process, data fusion technology not only brings better care to the elderly, but also enables the elderly. They can acquire knowledge and information through simple and convenient operation, participate in social life more deeply, and enhance the close relationship with different groups. The elderly are no longer passive service recipients. In the near future, what we hope to see is that the elderly group pays more attention to the experience of life process, the dusk is warmer and more loving, and the elderly can spend the last sunset of life gracefully.

Acknowledgement. This work was supported by Zhejiang Basic Public Welfare Research Project (Research and Implementation of Personalized Service System for Home-based Old-age Care Based on Data Fusion: LGF18F020004).

References

1. Long, X., Yang, P., Guo, H.: A CBA-KELM-based recognition method for fault diagnosis of wind turbines with time-domain analysis and multisensor data fusion. Shock. Vib. **319**(11), 1–14 (2019)
2. Rodriguez-Garcia, M., Batet, M., Sánchez, D.: Utility-preserving privacy protection of nominal data sets via semantic rank swapping. Inf. Fusion **45**(43), 282–295 (2019)
3. Jeske, D.R., Xie, M.G.: Special issue on data fusion. Appl. Stoch. Model. Bus. Ind. **45**(2), 332 (2018)
4. Pu, W., Liu, Y.F., Yan, J.: Optimal estimation of sensor biases for asynchronous multi-sensor data fusion. Math. Program. **170**(1), 357–386 (2017)
5. Yin, C.B., Sun, S.Q., Ren, P.: Research and application of internet of things in the field of home-based care for the aged. Adv. Mater. Res. **926**(930), 2582–2585 (2017)
6. Chen, M.C., Kao, C.W., Chiu, Y.L.: Effects of home-based long-term care services on caregiver health according to age. Health Qual. Life Outcomes **15**(1), 208–219 (2017)
7. Gautun, H., Bratt, C.: Caring too much Lack of public services to older people reduces attendance at work among their children. Eur. J. Ageing **14**(2), 155–166 (2017)
8. Ang, L., Tabu, M.: Conceptualising home-based child care: a study of home-based settings and practices in Japan and England. Int. J. Early Child. **50**(1), 1–16 (2018)
9. Hamirudin, A.H., Walton, K., Charlton, K.: Feasibility of home-based dietetic intervention to improve the nutritional status of older adults post-hospital discharge. Nutr. Diet. J. Dietit. Assoc. Aust. **74**(3), 217–219 (2017)
10. Wood, E.M., Zani, B., Esterhuizen, T.M.: Nurse led home-based care for people with HIV/AIDS. BMC Health Serv. Res. **18**(1), 219–221 (2018)

Image Tampering Location and Restoration Watermarking Based on Blockchain Technology

Wangsheng Fang, Yuyao Wang$^{(\boxtimes)}$, and Xiangyu Wang

Jiangxi University of Science and Technology, Ganzhou 341000, Jiangxi, China
1109424767@qq.com

Abstract. With the development of multimedia technology and network communication, digital watermarking technology has become one of the research hotspots in the field of signal processing and information security in recent years. In this paper, an effective fragile block watermarking method for image tampering localization is proposed. Compared with the existing algorithms, it has better localization effect. With the idea of blockchain, the algorithm first forms a linear cyclic chain from all blocks in the image on the basis of the key, which is called blockchain. Then the watermarking of the image block is hidden in the subsequent blocks, which ensures the content of the image block and the security of the embedded watermarking. Theoretical analysis and experimental results show that this method can resist VQ and collage attacks well, and has high positioning accuracy.

Keywords: Image watermarking · Blockchain technology · Image restoration

1 Introduction

With the development of multimedia technology and network communication, digital products are becoming more and more popular, and their scope of transmission is becoming wider and wider. At the same time, digital media resources are becoming easier to copy, maliciously tamper spreads without the permission of copyright owners. Digital watermarking technology hides special symbols in digital products plays a huge role. Therefore, digital watermarking [1–3] technology has become one of the research hotspots in the field of signal processing and information security in recent years.

A self-embedding watermarking algorithm for color image detection and recovery after tampering is proposed by Gerring et al. The authentication bits and recovery bits are generated by image information and embedded into the original image as watermarks. By comparing the hash values extracted from the image blocks with the authentication bits embedded in the watermarks, we can judge whether the image is tampered. Song Peifei proposed a high-quality recoverable semi-fragile watermarking algorithm, which can reduce the embedding capacity and improve the quality of the watermarked image while ensuring the restoration effect, while taking into account the embedding capacity and security of the watermarking. Li Shuzhi et al. proposed a watermarking algorithm for tamper location and restoration based on adaptive

© Springer Nature Singapore Pte Ltd. 2020
M. Atiquzzaman et al. (Eds.): BDCPS 2019, AISC 1117, pp. 420–428, 2020.
https://doi.org/10.1007/978-981-15-2568-1_58

classification of image blocks. The image blocks are divided into simple texture blocks and complex texture blocks. The generated eigenvalues are adaptively generated according to the texture complexity as the recovery information. It can effectively improve the quality of tampered image restoration and enhance the concealment of image watermarking [4, 5].

In this paper, an effective fragile block watermarking method for image tampering localization is proposed. Compared with the existing algorithms, it has better localization effect. With the idea of blockchain, the algorithm first forms a linear cyclic chain from all blocks in the image on the basis of the blockchain. Then the watermarking of the image block is hidden in the subsequent blocks, which ensures the content of the image block and the security of the embedded watermarking. In tamper detection, considering the consistency of blocks and their adjacent blocks and subsequent blocks, the legitimacy of test blocks is determined by comparing the number of adjacent blocks and subsequent blocks that are inconsistent with the blocks to be tested. By detecting the relationship between the security strength of the watermarking and the size of the block, the localization accuracy of the proposed method is compared with existing fragile block-level watermarking algorithm. Theoretical analysis and experimental results show that this method can resist VQ and collage attacks well, and has high positioning accuracy.

2 Image Watermarking Protection Method Based on Blockchain Strategy

2.1 Blockchain

Blockchain, as a new information technology, uses time stamp and digital cryptography technology. Transaction records are recorded in data blocks composed of time series, and data are stored in distributed databases by consensus mechanism to form a permanent and unalterable unique data record, and achieve the purpose of trusted transactions without relying on any central organization. Blockchain distributed storage is to store all records distributed on multiple accounting nodes in the whole network. Damage or loss of a single node will not affect other nodes, and data errors or tampering of a single node will not have any destructive impact on the overall data. According to the idea of blockchain technology, each block divided by the image and the embedded watermarks are regarded as blocks in the blockchain. Blocks are the basic units of blockchain structure, which are composed of block heads containing metadata and block bodies containing transaction data. The block head contains the time stamp, random number, ID of the previous block and ID of the block. The watermarking information in each image block contains the information of the previous image block. The watermarking is connected in a certain order, and the embedded watermarking information in the image forms a blockchain. Using the characteristics of blockchain distribution and high redundancy, the results of digital watermarking are changed from traditional single storage center to multi-node storage, and the consistency of digital watermarking is achieved.

2.2   Digital Watermark

Digital Watermark is an application of computer algorithm to embed protection information of carrier files. Digital watermarking technology is a means of information hiding [6, 7]. Its principle is to embed digital watermarking directly into protected digital products, such as multimedia files, text documents, digital software, etc. Another way to realize digital watermarking is through indirect representation, which can effectively protect copyright [8]. If the original image is I, the watermarking is W and the key is K, the embedding process can be described as follows:

$$I_w = F(I, W, K) \tag{1}$$

where F is a watermarking embedding algorithm.

The following expressions represent two commonly used image watermarking embedding methods.

$$\begin{aligned} v_i^w &= v_i + \alpha w_i \\ v_i^w &= v_i(1 + \alpha w_i) \end{aligned} \tag{2}$$

Among them, the original image v_i and the embedded watermarking image v_i^w are respectively represented as the component of the watermarking signal w_i and α is the strength of the watermarking, which are used to adjust the detectability of the watermarking [9, 10].

2.3   Image Watermarking Based on the Blockchain Technology

This section introduces the image watermarking method based on blockchain. In order to ensure the invisibility of the watermarking, the algorithm divides the Y-component image into 4 * 4 blocks, then calculates the information entropy and edge entropy of each block according to the HVS characteristics, and takes the area in which the two values are in the middle as the watermarking embedding area. Let k1, K2 represent the key, X is the original image, W stand for the watermarking, Y is the watermarked image; Xi, Wi and Yi are the corresponding image blocks, and the image is regarded as a set of image blocks, satisfying $X_i = \{x_i\}, i = 1, 2, \ldots, N$, $W_i = \{w_i\}, i = 1, 2, \ldots, N$, $Y_i = \{y_i\}, i = 1, 2, \ldots, N$ $X_i = \{x_i\}, i = 1, 2, \cdots N$. Xi and Yi are the contents of image blocks, N is the number of image blocks, MN is the number of pixels of image blocks, A8(Xi) is the set of eight other image blocks in the neighborhood of image blocks. In this paper, the image is divided into image block structures, and the watermarking corresponding to different image blocks is regarded as blockchains and connected in series. For the sake of security, the method proposed in this paper adopts the strategy of watermarking generation: randomly select an image block from the image at a time, generate the corresponding image watermarking according to the content of the image, and randomly select a watermarking information from the other image blocks and store it in the corresponding image block.

Information entropy and edge entropy in digital image contain important information of image. Generally, information entropy is regarded as a global measure, which only depends on the probability distribution of the intensity values of all pixels. The regions with larger entropy usually contain important feature information of the image, so embedding watermarking in them will make it easy for human eyes to find the embedded watermarking information, which leads to poor invisibility of the watermarking algorithm. If the regions with smaller entropy value are selected to embed watermarking, the robustness of the watermarking will be poor. Therefore, this paper chooses the entropy value. The middle region is the watermarking region to be embedded. The information entropy H1 and edge entropy H2 of a digital image f are calculated as follows:

$$H_1 = \sum_{i=1}^{255} P(f_i) \log_2 \frac{1}{P(f_i)} \tag{3}$$

$$H_2 = \sum_{i=1}^{255} P(f_i) \exp(u(f_i)) \tag{4}$$

where P(fi) denotes the probability of an event i, $P(f_i) \in [0, 1]$ and $\sum_{i=0}^{255} P(f_i) = 1, u(f_i) = 1 - P(f_i)$. $P(f_i) \in [0, 1]$ and $\sum_{i=0}^{255} P(f_i) = 1, u(f_i) = 1 - P(f_i)$. The process of watermarking embedding is shown in Fig 1. Firstly, a random image R of the same size as the original image is generated according to the input key k2, and the partition of the image block is consistent with the original image. Next, the hash value S (m, n, Xi) of the current image block Xiis calculated by SHA1 algorithm, and the corresponding watermarking image block Yi is generated.

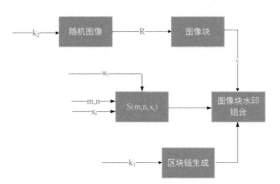

Fig. 1. Principle of blockchain Image watermarking

3   Experimental Results

The test images input in the experiment are all gray-scale images. The test results are measured by tampering ratio (TR), false acceptance rate (FAR) and false rejection rate (FRR). The calculation methods are as follows:

$$TR: \ R_t = \frac{N_t}{N} \times 100$$

$$FAR: \ R_{fa} = \left(1 - \frac{N_{td}}{N_t}\right) \times 100\% \tag{5}$$

$$FRR: \ R_{fr} = \frac{N_{vd}}{N - N_t} \times 100\%$$

Among them, Nt is the number of tampered image blocks detected, Ntd is the number of tampered image blocks correctly detected, Nvd is the number of false detected image blocks. Ideally, FAR and FRR should be 0. The experimental image size should be 512×512. The size of the watermarking image is 64×80. Peak Signal to Noise Ratio (PSNR) is used to evaluate the quality of watermarked images.

$$PSNR = 10 \times \log_{10}(\frac{(2^n - 1)^2}{MSE}) \tag{6}$$

MSR is the mean square deviation between the original image and the current image. The PSNR value between 30 and 40 indicates that the image quality is slightly distorted but acceptable. If the PSNR value is less than 30, the image quality after processing is poor. The result of the watermarking embedding area is shown in Fig. 2.

(1) **Visibility Testing of Watermarking Algorithms**

The invisibility can be divided into subjective and objective judgment criteria. Subjective judgment refers to the difference between the images before and after embedding watermarks by comparing the human eyes. If the human eyes can not recognize the difference, it shows that the invisibility of watermarking algorithm is better. The objective judgment criteria is measured from data by calculating the performance index values. Watermark invisibility is measured by PSNR value in this algorithm. The original image before watermarking is shown in Fig. 3(a), the watermarked image is shown in Fig. 3(b), and the watermarked image processed by this method is shown in Fig. 3(c). The PSNR value of Fig. 4(c) is 39.64 dB, which shows that the image quality remains acceptable after watermarking.

(2) **Robustness Testing of Watermarking Algorithms**

Robustness is to measure the difference between the original watermarking and the extracted watermarking by calculating some parameter values, and to further judge the anti-attack ability of the watermarking algorithm. It is also an essential condition to measure the watermarking scheme. This algorithm measures the robustness of the

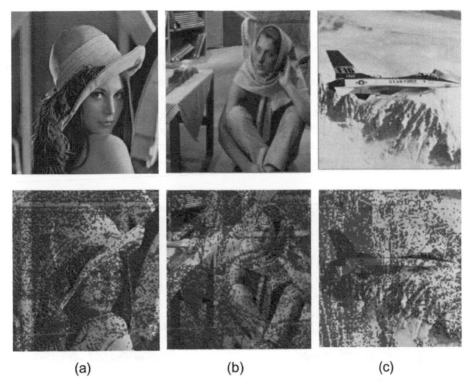

Fig. 2. The original image and the selected area of image watermarking

Fig. 3. Examples of original image, watermarking image and processed image

watermarking scheme by embedding the watermarking image into other images. Shearing attacks are divided into horizontal shearing, intermediate shearing and vertical shearing. The shearing experiments are done and compared with the traditional Makboll and Omar algorithms. The experimental results show that the proposed algorithm has better anti-shearing effect than horizontal and vertical shearing in the middle, and the extracted watermarking information is more complete. When the color image embedded with watermarking is scaled, if it is scaled down, it will lose part of its image information with the image scaling. On the contrary, the extracted watermarking information is relatively complete when the watermarking image is enlarged four times. Using this algorithm, salt and pepper noise and Gauss noise are added to the watermarking image respectively. Through the extracted watermarking information, we can know that when salt and pepper noise is added to the color watermarking image obtained by the algorithm in this chapter. The algorithm adds salt and pepper noise with noise density of 0.002 and 0.01 to the watermarking image. The higher the noise density is, the more dense the added noise, and the more incomplete the extracted watermarking information. Therefore, the effect of salt and pepper noise is better when the density is 0.002 than when the density is 0.01.

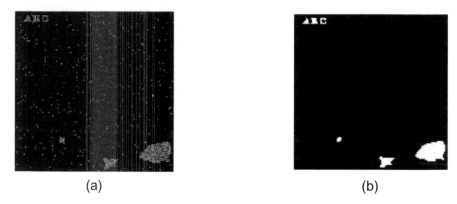

(a)   (b)

Fig. 4. Watermarking extraction and tampering detection results

Figure 4(a) is a differential image of the watermarking information extracted from the given key information. Figure 4(b) shows tamper detection results, in which FAR is 0.66% and FRR is 0.7%. When the proportion of image modified is less than 50%, the FAR and FRR values of the image are close to 0. From the results, it can be seen that after watermarking protection, the tampered image can be detected completely, and the image can be protected.

After tampering detection, all image blocks can be divided into tampered image blocks and untouched image blocks. Specific recovery can be divided into the following four steps:

(1) Tampered image blocks are selected. Find out the mapping block of each block and determine whether the mapping block has been modified. If it does not go to the next step, if it has been tampered with, jump to step 3.
(2) Recovery bits are extracted from mapping blocks, and tampered image blocks are replaced by recovery bits.
(3) Choose the adjacent 3 * 3 neighborhood blocks of the tampered block, and use the average value to replace the tampered image block.

Fig. 5. Tamper recovery results

(4) The image is restored to the spatial domain by IDCT transformation, and finally the YCbCr space is changed back to the original color space of the image and the original image is restored. If the original graph itself is YCbCr space, the recovery graph can be obtained directly without changing (Fig. 5).

4 Conclusions

In this paper, an image watermarking protection method based on blockchain strategy is proposed, which solves the problems of security and location accuracy of watermarking. The experimental results show that the proposed method can effectively resist vector quantization (VQ) and collision attacks. The next research direction is mainly focused on the watermarking protection of degraded or distorted images, and the performance of various image watermarking protection methods is compared.

References

1. Botta, M., Cavagnino, D., Pomponiu, V.: A successful attack and revision of a chaotic system based fragile watermarking scheme for image tamper detection. AEU Int. J. Electron. Commun. **69**(1), 242–245 (2015)
2. Wong, P.: A public key watermark for image verification and authentication. In: Proceedings of IEEE International Conference on Image Processing, Chicago, IL, pp. 425–429 (1998)
3. Holliman, H., Memon, N.: Counterfeiting attacks on oblivious block-wise independent invisible watermarking schemes. IEEE Trans. Image Process. **9**(3), 432–441 (2000)
4. Fridrich, J., Goljan, M., Memon, N.: Cryptanalysis of the Yeung-Mintzer fragile watermarking technique. J. Electron. Imaging **11**, 262–274 (2002)
5. Wong, P., Memon, N.: Secret and public key image watermarking schemes for image authentication and ownership verification. IEEE Trans. Image Process. **10**, 1593–1601 (2001)

6. Fridrich, J.: Security of fragile authentication watermarks with localization. In: Proceedings of SPIE. Security and Watermarking of Multimedia Contents, San Jose, CA, vol. 4675, pp. 691–700 (2002)
7. Bravo-Solorio, S., Calderon, F., Li, C.T., et al.: Fast fragile watermark embedding and iterative mechanism with high self-restoration performance. Digit. Signal Process. **73**, 83–92 (2018)
8. Benrhouma, O., Hermassi, H., El-Latif, A.A.A., et al.: Chaotic watermark for blind forgery detection in images. Multimedia Tools Appl. **75**(14), 8695–8718 (2016)
9. Botta, M., Cavagnino, D., Pomponiu, V.: Fragile watermarking using Karhunen-Loève transform: the KLT-F approach. Soft. Comput. **19**(7), 1905–1919 (2015)
10. Rajput, V., Ansari, I.A.: Image tamper detection and self-recovery using multiple median watermarking. Multimedia Tools Appl. 1–17 (2019)

Innovation of Self-service System of College Mental Health in the Age of Big Data

Tiantian Zhang[1,2(✉)] and Xu Chen[1,2]

[1] Jiangxi Vocational College of Mechanical & Electrical Technology,
Nangchang, China
1012924152@qq.com
[2] Nanchang University College of Science and Technology, Nanchang, China

Abstract. College students must not only master the professional knowledge they have learned, but also learn to interact with others. Due to various pressures, the psychological burden of students is also increasing. The service to students' mental health is imminent. In the era of big data, information is diversified, so it is necessary to mine the information suitable for students'characteristics from the massive information. In view of the above problems, in the era of big data, this paper proposes data mining technology and mental health self-service system in Colleges and universities to achieve innovative service effect. By applying K-means clustering algorithm to data mining, it is found that 58.6% of the students choose self-resolution or let it go naturally when they encounter psychological problems. This paper further gives suggestions on the object-oriented, content and technology of mental health self-service in order to realize the innovation of mental health self-service system in Universities in the era of big data and provide reference for mental health service.

Keywords: Big data · Universities and colleges · Mental health self-service system · Data mining

1 Introduction

Nowadays, with the gradual expansion of enrollment in Colleges and universities, more and more college students are pouring into the campus. The age of college students is generally around 20 years old. They do not touch society. Their daily necessities are all arranged in advance by parents and teachers. These college students' thoughts and values are also vulnerable to interference from external affairs [1, 2]. College is also a small society. College students need to grow up gradually on campus. In this critical period of cultivating ideas and values, students' mental health problems can't be ignored [3]. Therefore, special attention should be paid to the mental health problems of College students, and innovative ways and means should be constantly innovated to cope with mental health problems.

In dealing with mental health problems, many scholars have carried out relevant research. In [4], the author points out that some people think that mental health is a stigma, and the relevant personnel should focus on the attitude of patients with mental

M. Atiquzzaman et al. (Eds.): BDCPS 2019, AISC 1117, pp. 429–435, 2020.
https://doi.org/10.1007/978-981-15-2568-1_59

illness, rather than the intervention measures to reduce stigma. According to the world mental health survey, psychological disorders are common among college students. Attacks usually occur before college admission. Early detection and effective treatment of these diseases at the early stage of college life may reduce brain drain and improve educational and psychological functions [5]. In sports circles, people often pay attention to the achievements of athletes, but neglect the great psychological pressure that athletes endure in training and competition [6]. In [7], the author focused on the mental health of elite athletes. After searching the relevant records and screening, we found that elite athletes experienced higher incidence of mental disorders than the general population. Drug abuse may also be one of the causes. Now is the era of big data, from the massive data application of related technologies can excavate valuable information. In [8], the author captures abundant data about consumer phenomena through technology to better understand the impact of big data on various marketing activities, so that companies can make better use of their advantages. In [9], the author uses big data to solve policy problems, and makes medical treatment and fire protection more productive by data-driven, which effectively avoids the waste of resources. In [10], because the traditional theories and methods used in the integrated environmental performance evaluation are not good in speed and accuracy, it is proposed to use large data for environmental performance evaluation. Large amount, fast speed and high diversity of large data make environmental assessment more reasonable. In the era of data explosion, the emergence of big data technology can better solve complex problems and provide satisfactory solutions.

Although the self-service system of mental health in Colleges and universities has achieved some results in mental health service, with the arrival of the era of big data, the current mental health education work is obviously unable to meet the needs of the management and education of college students. Faced with the problems of work, the integration of big data technology and mental health self-service system is the direction and goal of future development.

2   Method

2.1   Characteristics of the Big Data Age

Big data is an extreme information management and processing problem in one or more dimensions beyond the processing capacity of traditional information technology. Large data is a large-scale, diversified, complex and long-term distributed data set generated by a variety of data sources. Faced with numerous and complicated information, the technology and method related to data are attached importance to. Therefore, the era of big data is coming. In the era of big data, the control of information data is the control of resources. If the relevant information from different channels can be effectively distinguished and sorted out, we can gain a more comprehensive and profound understanding of relevant events, thereby improving the efficiency of work and allocating resources scientifically.

2.2   Large Data Information Storage Technology

In the business intelligence industry, the scale of data sets collected and analyzed is growing rapidly, which makes the cost of traditional data warehouse solutions too high to meet the task in this field. Hadoop is an open source framework based on research results published by Google under Apache. It is used to store and process large-scale data sets on commercial hardware. It replaces traditional data warehouse solutions with its advantages of low cost, high fault tolerance and scalability.

There are two main components of Hadoop, one is storage, the other is processing. Storage refers to Hadoop Distributed File System HDFS, which provides a stable file system for high reliable mass data storage. MapReduce is MapReduce. It provides a distributed programming model for high performance computing and processing of mass data.

2.3   Data Mining Method

Data mining refers to the extraction of valuable and regular data from massive data by means of machine learning, statistical learning and other technical means. Clustering analysis is one of the main tasks of data mining. K-means clustering algorithm is used to mine mental health data. The process of the algorithm is as follows:

(1) K samples are selected from sample set S as initial clustering centers, where $S = \{x_1, x_2, \cdots, x_n\}$ and $K = \{c_1, c_2, \cdots, c_k\}$.

(2) According to the mean value of each clustering sample, the distance d between each sample and the central object is calculated, and then the corresponding objects are re-divided according to the minimum distance. If two p-dimensional sample data points are $x_i = (x_{i1}, x_{i2}, \cdots, x_{ip})$ and $x_j = (x_{j1}, x_{j2}, \cdots, x_{jp})$, the distance between the sample and the central object is as follows:

$$d(x_i, x_j) = \sqrt{(x_{i1} - x_{j1})^2 + (x_{i2} - x_{j2})^2 + \cdots + (x_{ip} - x_{jp})^2} \tag{1}$$

The average distance of all sample data points is:

$$Meandist(S) = \frac{2}{n(n-1)} \times \sum_{i \neq j, i,j=1}^{n} d(x_i, x_j) \tag{2}$$

(3) Recalculate the mean of each sample (central object).

(4) Cycle the above steps (2) and (3) until the value of the objective function remains unchanged or less than the specified threshold. The objective function is the square error criterion function. The formula is as follows:

$$\sigma_i = \sqrt{\frac{\sum_{i=1}^{n_i} (x_i - c_i)^2}{|C_i| - 1}} \qquad (3)$$

Where ci is the centroid of the same class of data objects.

(5) At the end of the algorithm, K clusters are obtained.

3   Experiment

This paper extracts and analyses the data left in the mental health self-service system of colleges and universities. Because there is no uniform standard for the construction of mental health system, the data attribute fields in the system database may be different. There may be data duplication, data missing and even data errors in the retained data. Therefore, when extracting data, some abnormal data are discarded and most commonly used psychological related indicators are taken as the criteria.

After data cleaning, K-means clustering algorithm in data mining method is used for clustering analysis. The combination of big data technology and mental health self-service system makes mental health service more scientific.

4   Results

Result 1: Students' help-seeking on mental health.

The result of big data mining is that most students tend to deal with psychological problems, and these results are representative. Using big data and data mining technology for analysis. When students have mental health problems, they have the choice to ask for help actively, to choose to solve themselves or to let go. The specific proportion is shown in Fig. 1.

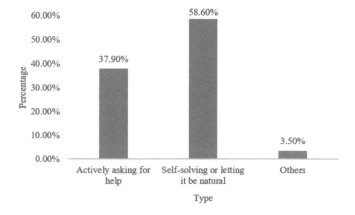

Fig. 1. Students' help-seeking on mental health

From Fig. 1, we can see that more than half of the students choose to self-expulsion or let it go naturally when they encounter psychological problems. 37.9% of students choose to ask for help when they have mental health problems. Further analysis found that college students are in adolescence, if there are psychological problems, they will not easily listen to outside advice. Students with psychological problems tend to move alone. It is difficult for students and teachers around them to find out the psychological changes of students in the first time.

Result 2: Number of Internet users among students.

In the era of big data, the network is very developed, no matter how old it is, it will touch the network more or less. At the end of December 2018, the total number of netizens in China reached 8.292, of which 311.9 million were students, accounting for 37.6% of the total number of netizens, as shown in Fig. 2.

As can be seen from Fig. 1, the network penetration rate of college students is very high. Nowadays, mobile phones are not only a necessity of College Students' daily life, but also an indispensable tool for college students to carry out network activities. Therefore, the psychological health self-service system combined with adolescence and big data era will be the best choice for the psychological service needs of contemporary students.

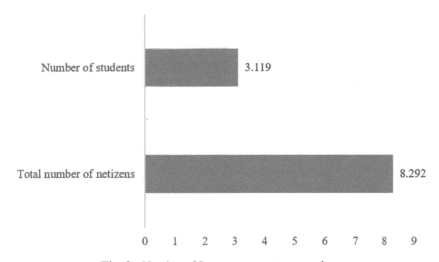

Fig. 2. Number of Internet users among students

With the continuous development of society and the accumulation of psychological problems of College students, a large number of psychological counseling services are needed. However, the national psychological service industry started late, and the corresponding consultants and service platforms can't keep up with the psychological needs of contemporary college students. So, it is urgent to build a platform to satisfy more college students to obtain psychological services and let more students get services at the first time. Now there are many kinds of software and forums, the information is messy, and there is no system. Therefore, it is necessary to integrate the advantages of online psychological services. In the era of big data, it is necessary to reform and innovate the mental health self-service system in Colleges and universities according to the requirements of practical development. Psychological health self-service system includes developmental demand group, crisis early warning intervention group and psychological popular science knowledge demand group for all students on the service object; includes all aspects of students' psychology in content, such as academic pressure, interpersonal distress, emotional confusion, personality development, identification and prevention of mental illness; achieves psychological service in technology, and provides psychological services in the following aspects: learning pressure, interpersonal distress, emotional confusion, personality development, identification and prevention of mental illness, etc. Online psychological testing, automatic collection and operation of psychological data and self-psychological report, online psychological counseling, psychological pleasure and psychological training services.

5   Conclusion

As a group with higher social and cultural level, the arrival of big data era brings opportunities and challenges to the innovation of mental health education for Contemporary College students. Digital mental health self-service system can provide scientific and dynamic data for colleges and universities in time, which can solve the mental health problems of College students. It is not only a barometer of psychological work, but also a digital management, training and service for the teaching staff to help the innovation of campus management.

References

1. Kecui, C.: On the humanization, individualization and institutionalization of college students management. New Campus (early) (2), 155 (2017)
2. Huan, D., Zheng, Yu., Zhengdi, Y.: Research on the guiding mechanism of college students' ideological and moral education. Ind. Technol. Forum **17**(22), 198–199 (2018)
3. Bo, Z.: Research on the influence of network culture on the values of post-95 college students. Beijing Educ. (Med. Educ.) **838**(11), 26–31 (2018)
4. Thornicroft, G., Mehta, N., Clement, S., Evans-Lacko, S., Doherty, M., Rose, D., Henderson, C.: Evidence for effective interventions to reduce mental-health-related stigma and discrimination. Lancet **387**(10023), 1123–1132 (2016)
5. Auerbach, R.P., Alonso, J., Axinn, W.G., Cuijpers, P., Ebert, D.D., Green, J.G., Nock, M.K.: Mental disorders among college students in the world health organization world mental health surveys. Psychol. Med. **46**(14), 2955–2970 (2016)
6. Jingru, W., Yuan, L., Guannan, Z., Lin, C.: The influence of tactical and psychological factors on the results of kayak athletes. Contemp. Sports Sci. Technol. **7**(19), 194–195 (2017)

7. Rice, S.M., Purcell, R., De Silva, S., Mawren, D., McGorry, P.D., Parker, A.G.: The mental health of elite athletes: a narrative systematic review. Sports Med. **46**(9), 1333–1353 (2016)
8. Erevelles, S., Fukawa, N., Swayne, L.: Big data consumer analytics and the transformation of marketing. J. Bus. Res. **69**(2), 897–904 (2016)
9. Athey, S.: Beyond prediction: Using big data for policy problems. Science **355**(6324), 483–485 (2017)
10. Song, M.L., Fisher, R., Wang, J.L., Cui, L.B.: Environmental performance evaluation with big data: theories and methods. Ann. Oper. Res. **270**(1–2), 459–472 (2018)

Comparative Analysis of ARIMA Model and Neural Network in Predicting Stock Price

Enping Yu$^{(\boxtimes)}$

Shanghai University, Shanghai 201800, China
yepepm@163.com

Abstract. ARIMA model is a time series model, which is used to analyze and predict the mean value of the sequence. Neural network can be used to predict in various fields. In this essay, the neural network algorithm is used for the financial time series to predict the trend of stock price change, and the results are compared with the traditional ARIMA. It is found that the neural network algorithm can better predict the change of stock price. In the empirical analysis, Python is used to establish three models for analysis and prediction, which can provide a more appropriate model reference for investors to judge the short-term trend of stock prices and portfolio decision-making.

Keywords: ARIMA model · The neural network · Prediction

1 Introduction

With the development of China's economy and society and the continuous reform of financial and securities markets, stocks have become an integral part of the national economic development, and stock speculation has become a topic of great concern to the people. Stock price trends are closely related to every investor. How to judge the future trend of stock price more accurately and provide more reliable investment decision for investors has become an important issue that puzzles experts and scholars. Stock price data is time series data, so we use statistical methods to study in the financial field. Although previous studies have found that the stock price changes are not completely random and predictable, there are still large uncertainties and volatility in the price changes. Therefore, we still need to explore ways to predict the future trend of stocks.

At present, there is basic analysis and technical analysis in the analysis of the stock market. There are also many people who build models to predict future tendency of stock. In recent years, more and more scholars have applied machine learning to stock forecasting. Lots of researches use Neural network algorithm, and prediction effects are more accurate. Compared with ARIMA model, (Schumann Lohrbach 1993) [1] finds neural network algorithm performs better. A lot of research results confirm that the prediction effect of neural network is better than that of ARIMA model. Neural network [2] has a certain reference value in the prediction of financial time series because of its strong non-linear mapping ability. However, it is a hotspot for financial researchers to make models of network become more accurate.

© Springer Nature Singapore Pte Ltd. 2020
M. Atiquzzaman et al. (Eds.): BDCPS 2019, AISC 1117, pp. 436–445, 2020.
https://doi.org/10.1007/978-981-15-2568-1_60

This paper mainly studies the forecasting of the closing price of Ping An stock in China. The methods adopted are ARIMA model and neural network. The forecasting effect of different methods on the closing price of Ping An stock is compared by two models [3].The full text is divided into five parts. The first part is the introduction of research background, the second part is the theoretical knowledge of ARIMA model, the third part is the introduction of ANN and LSTM model, the fourth part is the empirical research, and the fifth part is the summary.

2 ARIMA Model

2.1 ARMA Model Theory

ARMA model is a combination of AR model and MA model. It is mainly used to describe stationary stochastic processes. The model can be expressed as follows:

$$y_t = \varphi_1 y_{t-1} + \varphi_2 y_{t-2} + \ldots + \varphi_p y_{t-p} + \varepsilon_t - \theta_1 \varepsilon_{t-1} - \theta_2 \varepsilon_{t-2} - \ldots - \theta_q \varepsilon_{t-q} \quad (1)$$

ARIMA model is actually an extension of ARMA model. Most of the time series data in daily life do not satisfy the requirement of stationarity and cannot directly use ARMA model. Generally, As soon as the data be used is not stable, we use different methods to adjust it. Then the ARMA (p, q) model is used to model the stationary sequence, so the ARIMA model is generated. ARIMA (p, d, q) model is called differential autoregressive moving average model. D is the number of difference needed for stationarity of data.

2.2 Stationarity Test and White Noise Test

In order to establish ARMA model, we should first test the characteristics of the data, draw the data diagram of closing price of Ping An stock in China. Then, this essay use method-ADF to get further judgement. The ADF test is a regression of formula (2) to construct ADF test statistics, formula (3).

$$\Delta u_t = c + \delta u_{t-1} + \sum_{i=1}^{p-1} \beta_i \Delta u_{t-i} + \varepsilon_t \quad (2)$$

$$ADF = \frac{\hat{\delta}}{S(\hat{\delta})} \quad (3)$$

Among them, $S(\hat{\delta})$ is the standard deviation of the sample with parameter $\hat{\delta}$. After calculating the value of ADF, we compare it with the critical value to see if we can reject the original hypothesis $\delta=0$, and if we refuse, the original sequence will be stable. Usually we think that stock prices are not random, that is, there is a certain correlation between time series. If the time series data is completely random, then we call this series as random, in this time it is insignificance for us continue modeling. In this essay, the method of LB test is used to test white noise.

2.3   ARMA Model Order Determination Method

According to the time series data of stock closing price, coefficients of correlation are obtained. According to the coefficient, the appropriate model is selected. We select AR (p) model when the autocorrelation coefficient is tailed and the partial autocorrelation coefficient is p-order truncated, we choose MA(q) model. Otherwise, we construct ARMA (p, q) model. the minimum p, q can be judged by combining AIC and BIC criteria. AIC and BIC formulas are as follows (4):

$$AIC = 2a - 2\log_e(S)$$
$$BIC = a\ln(m) - 2\log_e(S)$$

$$(4)$$

The amount of model parameters is a, the likelihood function is S. Choosing the model with minimum AIC and BIC as the best model.

3   Artificial Neural Network and LSTM Model

3.1   Artificial Neural Network

Artificial neural networks are very important operational models in the field of artificial intelligence. They are also called collections of link units of artificial neurons. They are many other frameworks based on machine learning algorithms. A basic ANN structure including three elements: the input layer, the hidden layer, and the output layer. Generally, a fully connected neural network, that is, each neuron in the current layer is connected to all neurons in the previous layer, and each connection has a weight. But there are no connections between neurons on the same layer. Its structure is shown in Fig. 1 below.

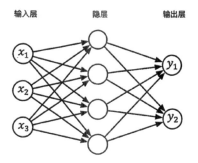

Fig. 1. Neural network structure

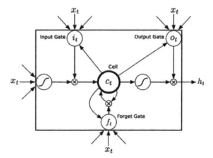

Fig. 2. LSTM model structure

3.2 LSTM Model

LSTM model is a deformed structure of cyclic neural network [4]. In view of the ordinary RNN, the memory unit is added to every unit of neural in the hidden layer, In this condition, what the time series including - memory information is controllable, and every moment passing through each unit of the hidden layer, numbers of controllable gates are passed (forgotten gates, input gates), candidate gates, output gates), the function of them are collecting previous information and current information and then control the memory and forgetting degree [5], As for this, the RNN networks have function of l remember in the Long term. The LSTM model structure is shown in Fig. 2.

4 Empirical Analysis

4.1 Data

In order to compare the fitting effect between traditional model and neural network and choose the appropriate algorithm to better predict the fluctuation of future stock closing price, the sample data selected in this paper is from January 2, 2008 to September 12, 2019. The daily closing price is 2799 data. The data source is Netease Finance.

When using the neural network, we first normalize the closing price data [6], and then set the hidden layer and the number of neurons to get the optimal neural network through constant adjustment. Finally, the prediction results of the three models are compared.

4.2 Establishment of ARIMA Model

4.2.1 Stationarity Test

Draw the untreated timing chart of the closing price of the stock, as shown in Fig. 3 below.

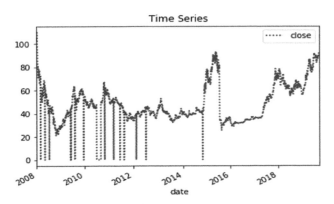

Fig. 3. Sequence diagram

From Fig. 3, we can see that the original sequence of the closing price of Ping An Stock in China from 2008 to 2019 is generally stable, and the closing price of Ping An Stock in individual time varies greatly. Therefore, it cannot be accurately judged whether it is a stationary sequence or not. Furthermore, ADF test was used, and the sequence was differentiated and then the sequence diagram and ADF test results were observed. As shown in the table below.

Table 1. Close ADF results

T-Value	−3.795
P-value	0.003
Lags	6
Critical Values: −3.43 (1%), −2.86 (5%), −2.57 (10%)	
Null Hypothesis: The process contains a unit root	

What the result showing in the Table 1 is the ADF test result of the original sequence is −3.795, and the significance level is less than the corresponding critical value at 0.01, 0.05, and 0.1, indicating that the original sequence is stationary, and the ARMA model can be directly established. However, since there are some large fluctuations in the data from the timing diagram, the first-order difference is also made to the data and the model is built, which is not shown here.

4.2.2 White Noise Test
Before building the ARMA model, white noise test is done on the sequence. The LB test in Table 2 shows that there is a significant correlation between the closing prices of the stock, so the model can be continued construct.

4.2.3 Model Order Determination
When the ARMA model is fixed, the autocorrelation and partial autocorrelation coefficients are obtained for the sequence. Figure 4 shows the autocorrelation and partial autocorrelation coefficients of the closing price. It can be seen that the autocorrelation coefficient of the closing price sequence is smeared. The partial autocorrelation coefficient is a fourth-order truncation. Empirically, AR (4) model should be selected. In order to get the best fitting model, ARMA models with different orders should be established and the AIC and BIC values of models with different orders should be compared. The model corresponding to the smallest values should be selected for fitting.

Table 2. Lyung_BoX Test

	Lyung_BoX	P-value
1	2572.422745	0
2	5080.809455	0
3	7542.476891	0
4	9942.988788	0
5	12280.44682	0

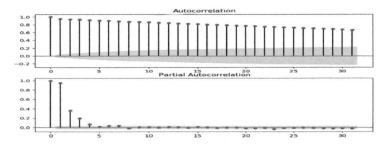

Fig. 4. Autocorrelation and partial autocorrelation

Table 3 lists AIC and BIC of ARMA model in different order. It is found that ARIMA (1, 1, 1) model has the best fitting effect, and the RMSE of model fitting is 14.2413.

Table 3. Index of ARMA (p,d,q)

Model	AIC	BIC	RMSE
AR(1)	17358.68	17376.49	14.11414
AR(2)	16759.31	16783.05	14.19033
AR(3)	16507.03	16536.72	14.21818
AR(4)	16456.45	16492.07	14.22118
ARMA(1,1)	16409.52	16433.27	14.19851
ARMA(1,2)	16411.42	16441.11	14.19731
ARMA(2,1)	16411.43	16441.12	14.19975
ARMA(2,2)	16411.92	16447.54	14.19731
ARIMA(1,1,1)	16402.77	16426.52	14.24132
ARIMA(1,1,2)	16403.16	16432.84	14.23824

4.2.4 Model Fitting and Forecasting

After setting up ARIMA (1, 1, 1) model, the fitting effect, coefficient and significance of the model are explored, and the model is used to predict the future closing price of Ping An stock in China.

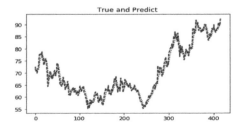

Fig. 5. Fitting chart of real and predicted

ARIMA Model Results

Dep. Variable:	D.close	No. Observations:	2798
Model:	ARIMA(1, 1, 1)	Log Likelihood	-8197.387
Method:	css-mle	S.D. of innovations	4.530
Date:	Thu, 10 Oct 2019	AIC	16402.773
Time:	20:55:08	BIC	16426.520
Sample:	1	HQIC	16411.345

	coef	std err	z	P>|z|	[0.025	0.975]
const	-0.0046	0.029	-0.158	0.875	-0.062	0.053
ar.L1.D.close	-0.0065	0.028	-0.228	0.820	-0.062	0.049
ma.L1.D.close	-0.6570	0.021	-30.845	0.000	-0.699	-0.615

Fig. 6. ARIMA (1, 1, 1) model results

In Fig. 5, blue and red lines describe true value and predicted value respectively and reasonable fitting is obtained. The significance of the model and coefficients is further studied. Figure 6 shows that the coefficient constant terms of ARIMA (1,1,1) model and AR model are not significant, while the coefficient terms of MA model are significant. According to the autocorrelation graph of the residual error, the fitting effect of the model is better and the information extracted is sufficienting.

4.3   ANN and LSTM Model

4.3.1   Data Preparation
In this paper, 2799 samples are selected. When using artificial neural network training model, the first 2000 samples are chosen to be training set, and rest of samples are chosen to be test set.

4.3.2   Model Prediction Effect
In the training artificial neural network and LSTM model, the hidden layer [7] is selected as 12 and the activation function relu is selected. Table 4 shows R2 fitted by two different models. The fitting effect of LSTM model is better than that of ANN. The R2 of training set is 0.854, and that of test set is 0.982, which is higher than that of ANN. In addition, from the fitting chart, it can be seen that the prediction value of LSTM model is nearly to authentic value than that of artificial neural network, obviously, this prediction effect is more accurate Fig. 7.

4.4   Comparison of Traditional Model and Neural Network Forecasting Effect

It can be seen from Table 5, the model of neural network [8] is more precise. The RMSE of ARIMA (1, 1, 1) is 14.24132. The RMSE of artificial neural network test set is 0.002743, and that of LSTM [9] model test set is 0.001558. The LSTM model has the smallest prediction error when the prediction error of the neural network is small. It is concluded that among the three models which are compared in this paper, the prediction effect is LSTM model > ANN > traditional ARIMA (1,1,1) model.

Table 4. R^2 of Train and Test

Model	Train R^2	Test R^2
ANN	0.850	0.975
LSTM	0.854	0.982

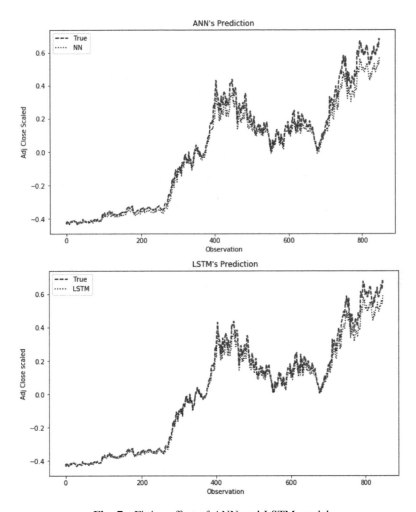

Fig. 7. Fitting effect of ANN and LSTM model

Table 5. RMSE of Models

Model	RMSE
ARIMA(1,1,1)	14.24132
ANN_train	0.012772
ANN_test	0.002743
LSTM_train	0.012432
LSTM_test	0.001558

5   Conclusion

Stock market has become an inseparable part of national life. The analysis and prediction of financial time series is an important way to research on volatility of stock markets. This paper chooses the closing price of Ping An stock in China from January 2, 2008 to September 12, 2019 as sample data, and makes an empirical analysis of the time series. ARIMA (1, 1, 1) model and ANN and LSTM models were established to predict and compare.

It is found that ARIMA (1,1,1) model can predict future stock closing price fluctuations to a certain extent, and it has certain use value. However, compared with the neural network algorithm, traditional model effects not very ideal [10].

Actually, stock price can be easily affected by a number of elements especially by the fluctuation of economy. This paper predicts the financial time series by establishing a single model. Although it has been fitted with good results, it has not considered whether the combined model will have a more significant impact on the prediction effect. Therefore, it is necessary to continuously explore innovations and try to use different methods to choose more effective models which can reduce the error of forecasting.

References

1. Schumann, M., Lohrbach, T.: Comparing artificial neural networks with statistical methods within the field of stock market prediction. In: Proceeding of the Twenty-Sixth Hawaii International Conference on System Sciences, vol. 4, pp. 597–606 (1993)
2. Le Cun, Y., Bengio, Y., Hinton, G.: Deep learning. Nature 521(7553), 436–444 (2015)
3. Xiong, R., Nichols, E.P., Shen, Y.: Deep learning stock volatility with google domestic trends (2015). arXiv preprint, arXiv:1512.04916
4. Hochreiter, S., Schmidhuber, J.: Long Short-term Memory. Neural Comput. 8, 1735–1780 (1997)
5. Heaton, J.B., Polson, N.G., Witte, J.H.: Deep Learning in Finance (2016). arXiv preprint, arXiv:1602.06561
6. Specht, D.F.: A general regression neural network. Revue De Physique Appliquée. 2(6), 1321–1325 (1991)

7. Hu, X.T., Wang, H., Xu, L., et al.: Predicting stock index increments by neural networks: The role of trading volume under different horizons. Expert Syst. Appl. **34**(04), 3043–3054 (2008)
8. Di Persio, L., Honchar, O.: Artificial Neural Networks Architectures for Stock Price Prediction: Comparisons and Applications. Int. J. Circuits Syst. Sign. Proces. **10**, 403–413 (2016)
9. Kim, H.Y., Won, C.H.: Forecasting the volatility of stock price index: a hybrid model integrating LSTM with multiple GARCH-type models. Expert Syst. Appl. **103**, 25–37 (2018)
10. Hinton, G.E., Salakhutdinov, R.R.: Reducing the dimensionality of data with neural networks. Science **313**, 504 (2006)

Traditional Supermarkets Using Internet of Things

Qiang Li, Dong Xie$^{(\boxtimes)}$, and Jiayue Wang

Information School, Hunan University of Humanities, Science and Technology,
Loudi 417000, Hunan, China
287566288@qq.com

Abstract. At present, since several reasons such as the development of economy, the change of technology, the improvement of people's living standards and the change of shopping from the pursuit of low prices, buyers pay more attention to experience and service. Traditional supermarkets are facing great challenges. Based on Internet of Things technology, this paper proposes solutions for traditional supermarkets to save costs, change business models and improve profitability according to the predicament and current situation of traditional supermarkets.

Keywords: Traditional supermarket · Internet of Things · RFID

1 Current Difficulties of Traditional Supermarkets

The traditional supermarket is still under the management mode of customers' self-selecting and queuing. With the rapid development of society, people's living standards have been constantly improved. Compared with decades ago, the material life of modern urban white-collar workers is much richer [1, 2]. People's wages have gone up, that's what supermarket operators want to see, and it's a fact that the economy is driving retail growth. However, in recent years, large supermarkets in China have been confronted with the problems of poor management and difficult profits.

1.1 Impact of E-Commerce

With the change of e-commerce, taobao, jd, Tmall and other online supermarkets have a huge impact on physical supermarkets.

1.2 Shopping Experience

With the continuous improvement of people's living standards, shopping is no longer limited to the attraction of low price promotion. They focus on convenient and fast shopping experience [3–5]. Now when people go into supermarkets, they often find that although supermarket promotions are increasing, the checkout desk is overcrowded and the long waiting time overdraws consumers' preference for supermarkets, which is not conducive to long-term development.

© Springer Nature Singapore Pte Ltd. 2020
M. Atiquzzaman et al. (Eds.): BDCPS 2019, AISC 1117, pp. 446–452, 2020.
https://doi.org/10.1007/978-981-15-2568-1_61

1.3   Profit Difficulties

The main way for supermarkets to make a profit is to make a small profit but sell more goods. However, large supermarkets have large scale, large floor space and large number of employees. Whether it is the increasing rent cost, the increasing salary treatment of employees, and the decrease of profit caused by excessive promotion, all affect the final profit of supermarkets.

1.4   Tedious Price Tag Replacement in Supermarkets

Traditional supermarkets mostly use paper labels to display the information of goods. People can see the existence of these small labels on each shelf of the supermarket. On the one hand, it provides basic information for customers and also brings some convenience to the management of the supermarket. However, according to incomplete statistics, a large supermarket has an average of 25,000 paper price tags, and 39,000 price tags need to be replaced every week [6].

Some promotional products even need to be replaced four times a week, a total of nine people are required to work hard for a week, and the workload is heavy. Moreover, traditional labels are usually paper labels, which can cause a lot of waste. At the same time, they are easy to make mistakes and lose. They are not suitable for frequent revision. Since the price labels do not match the actual prices of commodities, this causes customers to complain or even compensate, which has a negative impact on supermarket management and customers' shopping experience.

1.5   Complex Commodity Inventory Checking

Commodity inventory checking is an important means of checking the existing commodities of enterprises, ensuring the consistency of accounts and strengthening commodity management. Through checking, we can find out the real situation of commodity inventory quantity, whether the varieties and specifications are consistent, whether the system inventory is consistent with the actual situation, etc. After checking and verifying, it is determined that the system and the actual inconsistency should be adjusted in order to achieve the consistency of the system quantity and the actual quantity.

At present, this work is still carried out by combining manual and hand-held devices. However, due to the errors in the number of goods counted by the staff or the omission of some goods, the accuracy of the work is often not ideal, and the operation status can not be controlled 100%. As a result, a great deal of labor, equipment and time are wasted, which aggravates the ever-increasing cost.

2   Concept and Prospect of Internet of Things

2.1   The Concept of Internet of Things

Through radio frequency identification (RFID) device, infrared sensor, global positioning system, laser scanner and other information sensor equipment, according to the agreed protocol, any item and the Internet connection, information exchange and communication, in order to achieve intelligent identification, location, tracking, monitoring and management of a network.

2.2 Supermarkets Using Internet of Things

As early as 2003, metro opened the first "supermarket of the future" in Germany with the sales mode and applied technology that were realized by the concept and technology of RFID supermarket. A company provides related services for supermarkets in China, said that at present it could add such technological elements as color LCD screen, RFID technology, media broadcasting technology and wireless network Wi-Fi technology to supermarket shopping carts to realize the collection of advertisement rotation, shelf goods and corresponding advertisement intelligent automatic on demand, commodity information retrieval, store navigation, promotion information inquiry and convenient information around the store.

Inquiry, calculator and other interactive functions are one of the interactive intelligent terminals of commodity information. Intelligent shopping cart provides customers with timeliness and convenience. What is important is that it also brings the interaction between businesses and customers, which also brings higher sales to supermarkets [7].

In 2003, Wal-Mart first proposed to use RFID technology to replace the traditional bar code and become the information carrier of supermarket goods. Other retail giants, such as Tesco and medlon, followed Wal-Mart's lead. The report shows that this technology has saved Wal-Mart nearly $8.5 billion in annual costs in inventory management, sales management and after-sales service.

Today's customers pay more attention to the experience and efficiency of shopping, and RFID is the technology that can be innovated in the supermarket. But at present, no matter in foreign or domestic, no company has fully applied RFID technology to supermarket management, and its potential value has not been fully developed.

3 Reforming Traditional Supermarkets

This paper proposes a plan for the reform of traditional supermarkets from the following three aspects:

3.1 Electronic Price System

Adopting a kind of wisdom supermarket electronic tag system based on Internet of things technology, through wireless ZigBee technology to realize electronic tag information modification, only need a coordinator gateway can control all the tags in the same ZigBee network nodes, complete information update, convenient supermarket management information such as the price of real time maintenance.

Theoretically, the maximum number of ZigBee connections is 65,000, which basically meets the needs of the goods in the supermarket. Electronic price tag system can save a lot of paper labels and printing supplies, but also can save a lot of manpower, thus saving the cost of the supermarket [9].

3.2   RFID Management of Commodities

The supplier is required to complete the embedding of electronic tags synchronously during the commodity production, and the RFID reader is installed at the entrance of the supermarket warehouse and on the smart shelf. When goods enter the warehouse from the entrance of the supermarket warehouse, the label is scanned by the reader at the entrance of the supermarket, and the information in the label is automatically entered into the supermarket database. [8] Intelligent shelves use RFID reader to count goods in real time. When the number of certain goods on the shelf is reduced to a certain value by the detection of RFID reader, and a commodity is approaching or has exceeded the shelf life, the intelligent shelf will promptly remind supermarket staff to deal with it.

It not only solves the complex inventory problem, but also greatly strengthens the supermarket managers 'control over commodities, especially in dealing with the shortage of commodities or the problem of expired commodities. Because RFID tags are provided by suppliers, the cost output of supermarkets on RFID tags is greatly reduced.

3.3   Smart Shopping, a Combination of Various Payment Methods

Customers who want to buy conveniently can find shelves in supermarkets where they need to buy goods through mobile phone applications, scan barcode payment, and reduce queuing at cash desks. For those who want to experience shopping or those who are not educated enough to use smart devices skillfully, they can still use the traditional way to pay for shopping [10].

4   Function Realization

4.1   Electronic Price Tag System

Electronic price tag system is divided into coordinator gateway, electronic price tag node and supermarket database shown as Fig. 1.

Adjust the commodity price information of the server database through supermarket workstation or handheld terminal, and then transfer the changed price to the electronic price tag node on the supermarket workstation or handheld terminal, so that the commodity price information of the POS machine of the cash register is consistent with the server and electronic price tag.

4.2   Commodity RFID Management

The structure of RFID commodity management process is shown in Fig. 2.

When goods are produced, the embedded RFID is completed synchronously. When entering the supermarket warehouse, the imported RFID reader scans the goods information and transmits it to the server database through wireless network. The intelligent shelf counts the goods in real time by using the RFID reader. When the RFID reader detects that the number of certain goods on the shelf decreases to a certain amount, and a product is approaching the shelf life or has exceeded the shelf life. Intelligent shelves transmit information to servers. Workers can find out the lack and expiration of goods through workstations or handheld terminals and process them in time.

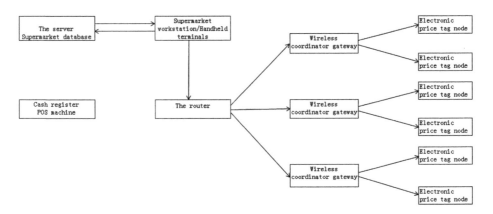

Fig. 1. Electronic price tag system

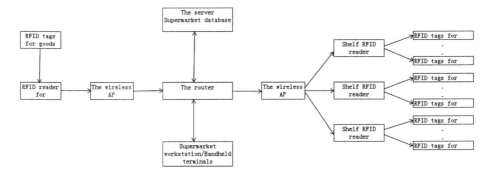

Fig. 2. RFID commodity management process structure

4.3   Smart Shopping, a Combination of Multiple Payment Methods

The schematic diagram of intelligent shopping process is shown in Fig. 3.

In fact, many large supermarkets, such as Wal-Mart, yonghui and bu bu gao, have launched convenient shopping, which is to scan the barcode of goods for payment by mobile phone App, and there is no need to queue up at the checkout desk for settlement. However, there is still no function of product positioning and navigation. In the navigation mode, a layout map of the supermarket will appear which is like a map navigation to guide customers to the shelf of goods they need to buy.

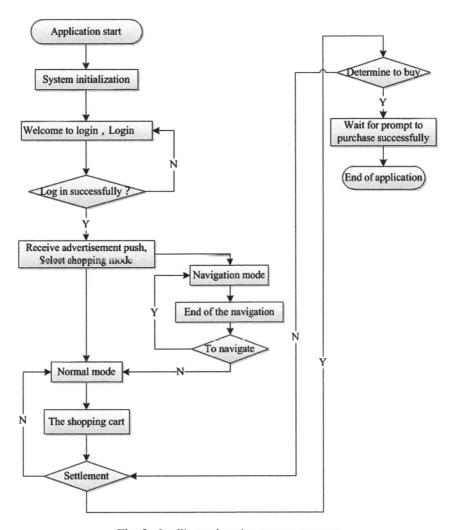

Fig. 3. Intelligent shopping process structure

5 Conclusions

In view of the current situation of traditional supermarkets, this paper puts forward three reform schemes. Although there will be some investment in the early stage, there are mature products on the market of electronic price tags. The price of an electronic price tag can be directly purchased and used for up to five years. The cost of paper labels and human resources saved is huge. It can also further save costs for traditional supermarkets, so as to enhance profitability. Moreover, the use of RFID tags can accelerate the transformation of traditional supermarkets into smart supermarkets, radically change the traditional queuing mode, improve customer shopping experience, increase commodity conversion rate, and create more profits.

References

1. Zhang, P., Xu, H., Zhang, Z.: Application of Internet of Things in retail industry. Fujian Comput. (1), 18–19 (2011)
2. Li, C.: Framework scheme of intelligent supermarket based on RFID technology. Softw. Eng. (2), 14–116 (2010)
3. Pu, L., Jiang, Z.: Design of smart electronic price tag system for supermarkets based on Internet of Things technology. Lab Res. Explor. (12), 147–151 (2017)
4. Sun, S., Liu, Y.: Research on RFID tag collision prevention algorithm of smart supermarket based on Internet of Things. Telev. Technol. **42**(9), 142–146 (2018)
5. Chen, P., Gao, H., Chen, J., et al.: Intelligence supermarket based on Internet of Things. Telev. Technol. **37**(1), 21–26 (2013)
6. Wang, H., Zheng, Y.: Application of Internet of Things technology in fresh food e-commerce. E-commerce (09), 54–55 (2019)
7. Deng, Z.: RFID technology and its application. Chin. Foreign Entrepreneurs (25), 131 (2019)
8. Brandl, H.B., Griffith, S.C., Farine, D.R., Schuett, W.: Wild zebra finches that nest synchronously have long-term stable social ties. J. Anim. Ecol. (2019)
9. Wang, L., Shao, M., Gu, S., Sun, Y.: Application of RFID technology in smart supermarket mobile phone client. Comput. Knowl. Technol. **12**(03), 102–103 (2016)
10. Li, T.: Intelligent supermarket system design based on Internet of Things. Sci. Technol. Vis. (05), 179–180+142 (2018)

Adaptability of Professional Setting and Local Industrial Structure of Applied Undergraduate Colleges in Dongguan Under the Background of Big Data

Yan Wang[✉]

Guangdong University of Science and Technology, Dongguan 523083, China
48916761@qq.com

Abstract. Recently, new technologies such as the Internet of Things, cloud computing, and big data have penetrated into various fields of urban life, which have brought great changes to people's lives. "Dongguan Big Data Development Plan (2016–2020)", the document clearly points out that the combination of Dongguan "Smart City" and "Dongguan Manufacturing 2025" will promote the rapid development of big data industry, and the Internet of Things and cloud computing will grow into the pillars of Dongguan Industry. This paper analyzes the data of the four major applied undergraduate colleges in Dongguan, such as the professional setting, the number of enrolled students, and the scale of professional setting, and the connection between the pillar industries and emerging industries in Dongguan. The professional settings of the four local undergraduate colleges in Dongguan are inferior to the pillar industries in Dongguan. This provides a reference for better service and local economy in local undergraduate colleges in Dongguan.

Keywords: Big data background · Dongguan · Applied undergraduate college · Regional industrial structure · Adaptability

1 Status Quo of Professional Structure of Applied Undergraduate Colleges in Dongguan

There are 4 applied undergraduate colleges in Dongguan, including Dongguan University of Technology, Guangdong University of Science and Technology, Dongguan Polytechnic City College, and Sun Yat-sen University Xinhua College. Dongguan Institute of Technology was established in 2002. It is the only public undergraduate college in Dongguan. The other three institutions are full-time applied private undergraduate colleges (Table 1).

According to 2012, the Ministry of Education issued the "General Undergraduate Professional Catalogue (2012)" [1] (hereinafter referred to as "Catalog") documents. The data of this "Catalogue" is as of 2012, with a total of 12 disciplines, professional 92 classes, 506 professional. According to the professional situation of the enrollment

© Springer Nature Singapore Pte Ltd. 2020
M. Atiquzzaman et al. (Eds.): BDCPS 2019, AISC 1117, pp. 453–460, 2020.
https://doi.org/10.1007/978-981-15-2568-1_62

Table 1. Enrollment plans of various applied undergraduate colleges in Dongguan in 2019

Applied undergraduate college	Plan enrollment	Number of students	Number of majors
Dongguan Institute of Technology	4490	20338	48
Guangdong Institute of Science and Technology	5406	17605	31
Dongguan Polytechnic City College	5235	23002	38
Sun Yat-sen University, Xinhua College	5036	23560	46
Total	20138	84505	163

(Thinking that the data comes from the school admissions brochure)

of 4 applied undergraduate colleges in Dongguan in 2019, it is found that there are 95 majors in the four applied undergraduate colleges in Dongguan, involving 9 categories (see Table 2). The engineering, management, and economics majors are more abundant, and they can better reflect the needs of Dongguan's industrial development.

Table 2. The application of undergraduate professional structure in Dongguan in 2019

Serial number	Subject category	Professional title	Number	Number of people
1	Education	Primary education, music education, Chinese language and literature, social sports guidance and management	4	782
2	Science	Medical Imaging Technology Human Geography and Urban and Rural Planning Physical Geography and Resource Environment	3	642
3	Engineering	Applied Chemistry Food Quality and Safety Chemical Engineering and Technology Energy Chemical Engineering Energy and Power Engineering	5	783
4		Environmental Engineering Civil Engineering Engineering Management Architecture Safety Engineering	5	950
5		Industrial Design Machinery Design and Manufacturing and Automation Intelligent Manufacturing Engineering Material Forming and Control Engineering Machinery Electronic Engineering Industrial Design Industrial Engineering Financial Materials Engineering Materials Science and Engineering Polymer Materials and Engineering Automotive Services Engineering Robot Engineering Electronic Information Engineering Biomedical Engineering Building Electrical and Intelligent Printing engineering	16	1756
6		Electronic Information Engineering Technology Communication Engineering Software Engineering Network Engineering Cyberspace Security Optoelectronic Information Engineering IoT Engineering Information Management and Information Systems Data Science and Big Data Technology Electronic Information Science and Technology Information Resource Management	11	2686

(continued)

Table 2. (*continued*)

Serial number	Subject category	Professional title	Number	Number of people
7	Literature	Chinese Language and Literature Radio and Television Studies English Japanese Business English Portuguese Spanish Communication Network and New Media Secretary	10	2389
8	Management	Business Administration Accounting Human Resource Management Logistics Engineering Investment Culture Industry Management Administrative Management E-commerce Marketing Tourism Management Logistics Management Public Relations	12	3917
9	Economics	International Economics and Trade Financial Management Investment Science Auditing Economic Statistics Taxation Financial Engineering Insurance Internet Finance	9	3186
10	Art	Broadcasting and Television Directoring Fashion Design and Engineering Visual Communication Design Environmental Design Product Design Digital Media Art Art Design Musicology	8	1574
11	Law	Legal work	2	684
12	Medicine	Health Service and Management Hearing and Rehabilitation Nursing Kang Kang Therapeutics Pharmacy	5	786

Note: The above data is compiled according to the college admissions website.

2 The Characteristics of Dongguan's Economy and Industrial Structure

2.1 Economic Characteristics of Dongguan City

In 2018, the economic operation of Dongguan was generally stable and stable, and the annual production value exceeded 800 billion yuan. Calculated at comparable prices, the growth rate was 7.4% over the previous year. The growth rate was higher than the national average (6.6%) and the province (6.8%). The total amount ranked 4th in the 21 provinces in the province. Ranked 5th. In terms of industries, the added value of the primary industry was 2.504 billion yuan, up 7.4%; the added value of the secondary industry was 402.721 billion yuan, up 6.9%; the added value of the tertiary industry was 422.634 billion yuan, up 7.9%. In 2018, the proportion of the three industries in Dongguan was adjusted to 0.3:48.6:51.1 [2], and the proportion of service industry remained high. The proportion of high-end manufacturing has increased. The added value of advanced manufacturing and high-tech manufacturing industries accounted for 52.5% and 39.1% of the added value of industrial enterprises above designated size, up 2.0 and 0.1% points respectively. The private economy accounted for 49.4%, an increase of 0.3% points; the general trade proportion was 51%, an increase of 5.2% points.

In general, the primary industry has risen steadily, the output of major agricultural products has grown rapidly, the production of the secondary industry has been generally stable, the structure of domestic and foreign investment has been further optimized, the development of the service industry in the tertiary industry has continued to accelerate, and the emerging service industry has become the main driving force (see Table 3).

Table 3. Proportion of the three major industrial structures of Dongguan's national economy in 2016–2018 [3]

Years	Primary industry	Secondary industry	Tertiary industry
2016	0.3	46.5	53.2
2017	0.3	47.2	52.5
2018	0.3	48.6	51.1

Note: The above data is quoted from "Dongguan City Statistical Report on National Economic and Social Development (2016–2018)", Dongguan Statistics Bureau

2.2   Characteristics of Industrial Structure in Dongguan

Dongguan City, known as the world factory, is a famous city, and the manufacturing industry naturally supports the economic development. In recent years, the added value of the five pillar industries in Dongguan has continued to grow. Dongguan relies on electronic information, equipment manufacturing, textile and garment, food and beverage, furniture manufacturing and other "five pillars, four characteristics" industries continue to optimize and upgrade, a new generation of electronic information, robots, smart terminals, new energy vehicles and other emerging industries continue to grow and develop, And the forward-looking layout of technological innovation carriers such as spallation neutron source science equipment (see Table 4) [4].

Table 4. Characteristics of industrial structure in Dongguan

Five pillar industries	Four characteristic industries	Emerging industry
Electronic information manufacturing Electrical machinery and equipment manufacturing Textile clothing, footwear and hat manufacturing Paper and paper products industry Food and beverage processing industry	Toys and stationery industry manufacturing Furniture manufacturing Chemical manufacturing Packaging and printing industry	A new generation of information technology High-end equipment manufacturing New material field New energy field Life sciences and biotechnology

3   Adaptability Analysis of Professional Structure and Industrial Structure of Applied Undergraduate Colleges in Dongguan

According to the new "Three Industry Classification Regulations" [5], the professional settings of Dongguan applied undergraduate colleges are divided according to the production scope of each industry, and the majors corresponding to the primary industry are not; the main industries corresponding to the secondary industry are environmental engineering, 37 majors in civil engineering, electrical automation technology, applied electronics technology, industrial design; major industries related to the tertiary industry include business administration, international economics and trade, financial engineering, English, and law.

3.1 Matching and Matching of the Applied Undergraduate Professional Structure and the Three Major Industries in Dongguan

It can be seen from Table 5 that the application-based undergraduate professional structure of Dongguan City does not match the industrial structure, mainly as follows: the specialty of the first industry agriculture and forestry is not available. The number of professions that are connected to the secondary industry is relatively small, accounting for only 39% of the secondary industry, and does not match the 49.3% ratio of the secondary industry in Dongguan. However, there are more professions in connection with the tertiary industry. In 2019, the number of professions in the tertiary industry accounted for 61% of the total number of enrollment majors, and the number of enrollment students accounted for 75%. This kind of docking phenomenon will cause imbalance in talent supply and demand, and structural imbalances hinder the undergraduate professional structure and the balanced development of the three major industrial structures.

Table 5. Matching of the professional structure of applied undergraduate colleges in Dongguan and the three major industries

	The three industries account for the percentage of GDP/%	Application-oriented undergraduate colleges	Application-oriented undergraduate colleges/%	Application-oriented undergraduate colleges	Percentage of students enrolled in applied undergraduate colleges/%
Primary industry	0.3	0	0%	0	0%
Secondary industry	48.6	37	39%	2176	19%
Tertiary industry	51.1	58	61%	9138	80%

Note: The proportion of the three major industries is the average of 2016–2018.

3.2 Matching and Matching of Applied Undergraduate Professional Structure and Pillar Industry

The five pillar industries in Dongguan, the demand for engineering professionals in the four characteristic industries and emerging industries Exuberant, but there are only a few applied undergraduate colleges in Dongguan that have related majors that train the above types of talents, and the distribution is uneven [6]. At present, the professional

Table 6. Professional status of the largest number of enrolled students in applied-type undergraduate colleges in Dongguan in 2019

Professional title	Enrollment	Professional title	Enrollment
Accounting	1338	English language	610
Software engineering	715	Business English	585
International economy and trading	679	Mechanical design and manufacturing and automation	520
Financial management	665	Electronic information engineering	508

setting of applied undergraduate colleges in Dongguan is mainly set around the tertiary industry and service industry, and the number of students enrolled is also more specialized, such as international economics and trade, accounting and other majors. 10% (see Table 6), showing a professional surplus and insufficient employment. And the major industries that Dongguan needs to develop, such as toys and cultural and sports goods manufacturing, packaging and printing, chemical manufacturing, new generation information technology, high-end equipment manufacturing, life sciences and biotechnology, are in professional settings. The number of enrolled students is seriously lagging behind, and it is obviously behind the development of local industries.

4 Put Forward the Strategy of Adapting the Professional Setting and Industrial Structure of Dongguan Applied Undergraduate Colleges

4.1 Cultivate the Concept of High-Level and Distinctive Professions and Establish Professional Groups

The Ministry of Education, the National Development and Reform Commission, and the Ministry of Finance's "Guiding Opinions on Guiding the Transformation of Some Local Undergraduate Universities to Applied Types" proposes "establishing a platform for cooperation and development of industrial enterprises. The professional clusters of school-enterprise cooperation realize full coverage" and "establishing a close-knit industrial chain, the professional system of the innovation chain... adjusts the professional setting around the industrial chain and the innovation chain to form a special professional cluster" [7]. Combining the advanced manufacturing industries such as new energy, 3D printing and robots that Dongguan is vigorously developing, especially intelligent manufacturing engineering, robot engineering, Mechanical design and manufacturing and its automation pillar industry. And Dongguan will focus on supporting high-end electronic information industry, biotechnology, new generation Internet and other strategic emerging industries.

The four applied undergraduate colleges in Dongguan should follow the general rules and principles of professional setting, and combine the advantages and characteristics of disciplines to better serve the new profession of regional cultural industry development. At the same time of docking industrial clusters, it highlights the characteristics and advantages of the school's disciplines, and better reflects the idea of building clusters according to the integration of disciplines and disciplines. Focus on the cultivation of professional characteristics that closely related to the development of local pillar industries and strategic emerging industries, promote the coordinated development of other professions in professional clusters, form professional clusters with superior characteristics, guide high-quality resources to gather in professional clusters, and support the continuous deepening of professional connotation construction [8].

4.2 Strengthening the Top-Level Design and Establishing a Sound Professional Setting and Adjustment System

The professional setting should take into account factors such as the development of the whole school, the concept of running a school and the orientation of running a school. The school leaders deeply understand the importance of professionalism to the development of school education. Local applied high-level universities must always serve local economic and social development. Accurately connect with the local industrial structure, actively serve the major strategic needs of the industrial structure adjustment of Guangdong Province and the industrial demonstration zone of Dongguan City, and continuously optimize the professional structure and layout. In order to accurately connect local strategic emerging industries or major pillar industries, form scientific and reasonable professional structure and professional group advantages, ensure the forward-looking of professional settings, and promote the sustainable development of undergraduate majors.

Local undergraduate colleges should consider the issue of professional setting from a strategic perspective, pay attention to the top-level design of the system, and strengthen the organization. Strengthen the establishment of professional leaders, professional leaders to grasp the overall position of the school, industry talent demand, teacher team construction, subject professional support, practical training conditions, professional development ideas, professional layout and student resources, and regularly report relevant information to the leadership [9]. Finally, at the school level, give full play to the role of academic committees, hold professional adjustments and set up meetings, and strengthen the discussion and communication of academic committees.

In particular, we must fully mobilize the enthusiasm of the secondary colleges and listen to the opinions of professors, academic leaders, professional leaders, and key teachers during the establishment and adjustment of the profession.

4.3 Establishing an External Argumentation Mechanism for Third-Party Professional Settings

In the professional construction, do not seek more but refinement, and seek to match the industrial structure of Dongguan. First of all, fully understand the relationship between Dongguan undergraduate education and Dongguan's industrial development. In the professional setting, actively seek the help of the government, industry associations and third-party organizations, and give full play to their role in connecting market demand and professional assessment [10]. Secondly, government departments should effectively play an administrative role in coordinating and optimizing the professional settings of undergraduate colleges in this city. Taking Dongguan Institute of Technology as the leader, based on the industrial structure of Dongguan, based on the needs of employment, the undergraduate majors are divided into categories. Finally, establish a multi-party consultation mechanism for the establishment of colleges and universities, and establish specialized institutions to carry out all aspects of the docking of enterprises in colleges and universities, such as the establishment of a professional consultation committee for colleges and universities, jointly negotiate policies and methods for professional setting and reform, and regularly release changes in industrial structure and The development report provides a reference and basis for the professional setting and adjustment of colleges and universities.

References

1. Ministry of Education of the People's Republic of China. Undergraduate professional catalogue (including professional catalogue) [EB/OL], 15 May 2012. http://www.zyyxzy.cn/gpw/shtml/bulletin/36.shtml. (in Chinese)
2. Dongguan Municipal Bureau of Statistics. 2018 Statistical Report on National Economic and Social Development of Dongguan [EB/OL], 17 April 2019. http://tjj.dg.gov.cn/website/flaArticle/art_show.Html?code=nj2014&fcount=2. (in Chinese)
3. Dongguan Municipal Bureau of Statistics. 2016 Statistical Communique of National Economic and Social Development of Dongguan [EB/OL], 17 September 2017. (in Chinese)
4. Dongguan Municipal People's Government. Dongguan Industrial Structure Adjustment Plan (2008–2017) [EB/OL], 2 February 2010. http://www.dgkp.gov.cn/showart_4384.htm. (in Chinese)
5. Statistics Bureau of the People's Republic of China. Provisions for the division of three industries [EB/OL], 14 January 2013. http://www.stats.gov.cn/tjsj/tjbz/201301/t20130114_8675.html. (in Chinese)
6. Dongguan Municipal People's Government. 2015 Dongguan City Government Work Report [EB/OL], 18 January 2015. http://xxgk.dg.gov.cn/publicfiles/business/htmlfiles/0101/15/201507/922953.htm. (in Chinese)
7. Yin, X.: Thoughts on optimizing the structure of higher vocational education based on regional industrial structure adjustment. Vocational Education Forum, no. 33 (2018). (in Chinese)
8. Sun, W., Liu, X.: Research on the coordinated development of local higher vocational education and regional industrial structure. Liaoning Vocational College, no. 5 (2016). (in Chinese)
9. Samantha, A., Ristan, G., Jennifer, R., Petra, V.-B., Phillip, G.: The impact of provision of professional language interpretation on length of stay and readmission rates in an acute care hospital setting. J. Immigr. Minor. Health **21**(5), 965–970 (2019)
10. Sellier, A.-L., Scopelliti, I., Morewedge, C.K.: Debiasing training improves decision making in the field. Psychol. Sci. **30**(9), 1371–1379 (2019)

Port Logistics Efficiency Evaluation Based on DEA Model

Wendi Lu[(✉)]

Department of Economics and Management,
Dalian University, Dalian, Liaoning, China
Lwd1187401676@163.com

Abstract. First, this paper uses data envelopment analysis (DEA) to analyze the logistics efficiency of four major ports in Liaoning Province. Then based on the results of data envelopment analysis (DEA) analysis, the relationship between logistics efficiency and its influencing factors is studied. The results show that the overall pure technical efficiency of port logistics in Liaoning Province has been maintained at a high level and the state is stable. The change in overall technical efficiency is mainly due to changes in scale efficiency and needs to be further strengthened. The economies of scale need to be further improved, and the port logistics efficiency in Liaoning Province still has great potential for development. The Tobit model is used to empirically study the influencing factors of port logistics efficiency on the results of data envelopment analysis (DEA). The results show that the economic development level of the port city and the development level of the secondary industry have a positive impact on the development of port logistics, and the effect is remarkable.

Keywords: Port logistics · Efficiency evaluation · Influencing factors

1 Introduction

Port logistics is a special form of the logistics industry. The development of port logistics has always been the top priority of coastal cities. Port logistics efficiency evaluation is one of the important indicators for evaluating port logistics development over the years. In recent years, despite the rapid development of China's logistics industry and its considerable progress, the development of the port logistics system has not yet been perfected, directly affecting and restricting the development of the entire regional economy. This paper evaluates the efficiency of port logistics efficiency in major ports of Liaoning Province, analyzes its main influencing factors, and puts forward specific implementation strategies, which is of great significance to the development of port logistics.

Roll and Hayuth [1] fi first applied the dea method to evaluate the efficiency of 20 ports and proved the effectiveness of the method. Tongzong [2] studied the advantages of several major international ports and compared them. Then evaluate the efficiency of four international ports in Asia and several other international ports. It shows that the relationship between port efficiency and port size is not clear from the final results.

© Springer Nature Singapore Pte Ltd. 2020
M. Atiquzzaman et al. (Eds.): BDCPS 2019, AISC 1117, pp. 461–467, 2020.
https://doi.org/10.1007/978-981-15-2568-1_63

Wu and Goth [3] use the DEA method to evaluate the number of containers entering and leaving the port per unit time, which provides an effective means of discovering problems and improvements in container transportation. Lai [4] proposed to study the DEA method selection evaluation indicators from two aspects of finance and service industry. Studies have shown that port logistics assessment should select indicators that have a significant impact on the port and the mode of operation of each port based on characteristics and characteristics. Barros and Athanassioul [5] proposed a method to improve the efficiency of port logistics based on the evaluation results of the data envelopment model. Yun [6] introduced the two-stage relative evaluation of DEA, and used the DEA model to empirically analyze 13 ports in China, verified the applicability of the model, introduced a two-stage DEA model and conducted an empirical analysis. The applicability of the model was verified in 13 ports in China. Huang, Peng et al. [7] used the Data Envelopment Analysis Model (DEA) to calculate the efficiency between Zhejiang port traffic and economic growth, and proposed the development strategy of the logistics industry from the perspective of ocean economy. Although most scholars have analyzed and evaluated the efficiency of port logistics, they lack an analysis of the relationship between influencing factors and logistics efficiency. Data Envelopment Analysis (DEA) was first developed in 1978 by Charness and Cooper [8]. The main idea of data envelopment analysis (DEA) is to make comparable types of the same type. Each unit has multiple input units and output units. The linear planning method is used to determine whether each decision unit DEA is valid. Since there are no estimated parameters in the DEA model, the influence of subjective factors can be avoided and the error can be reduced. It is a relatively effective evaluation and is very effective and objective in the multi-input and multi-output mode. Data Envelopment Analysis (DEA) results are not affected by any human factors, and are objective and superior to the above methods.

On this basis, the logistics efficiency of main ports in Liaoning Province is studied from two aspects. On the one hand, data envelopment analysis (DEA) can directly perform data envelopment analysis (DEA) when evaluating each decision unit, and it does not need to undertake input and output functions. This paper refers to a large number of literatures, through the selection of appropriate evaluation indicators, using the data envelopment analysis model (DEA), the logistics efficiency of the four major ports in Liaoning Province in the past six years. On the other hand, according to the analysis results of logistics efficiency of major ports in Liaoning Province, Tobit is adopted. Regression model is used to analyze the correlation between each influencing factor and logistics efficiency, and some suggestions are put forward.

2   Data Envelopment Analysis (DEA)

Data Envelopment Analysis (DEA) is an objective method for assessing the relative efficiency of multiple input-output units, where the CCR and BCC models are representative models. Therefore, this paper selects the most representative CCR model and BCC model in the DEA model. The former is used to assess the overall effectiveness of decision-making departments in terms of effectiveness and technical effectiveness, while the latter is used to assess the effectiveness of pure technology.

In the past studies of port efficiency, cargo throughput or container throughput is the most commonly used output indicator [5–7]. This paper still chooses cargo throughput and container throughput as output indicators. By referring to the indicators selected by the scholars in the past literature, the input indicators selected in this paper are mainly the number of berths, the length of the terminal, the total number of container loading and unloading bridges, the number of yard machinery stations and the annual throughput of container terminals. This paper considers that the quantity and length of berths are important indicators for measuring the size of a port or terminal, which directly determines the production capacity of the port. In view of the availability of sample data, the number and length of berths are selected as input indicators. Several major large ports in Liaoning Province include Dalian Port, Yingkou Port, Jinzhou Port, Dandong Port, Huludao Port and Panjin Port. The development scale of Huludao Port and Panjin Port differs greatly from the other four ports. It is also impossible to fully satisfy the selection of the evaluation index. Therefore, we only conduct DEA evaluation and influencing factors analysis on the four port areas of Dalian Port, Yingkou Port, Dandong Port and Jinzhou Port. China's port planning and investment recovery period is generally 5 years, so objectively choose a year with an interval greater than 1 cycle and no more than 2 cycles (5–10 years). In addition, the DEA method requires that the number of decision units should be more than twice the sum of the input and output indicators, so the evaluation results are more accurate. Therefore, this paper chooses every port year from 2010 to 2017 as a decision-making unit (DMU) to evaluate the port logistics efficiency of Liaoning Province vertically. The data mainly comes from the China Port Yearbook, Liaoning Statistical Yearbook and various journal articles and Internet portals.

3   Tobit Regression Model

The effective value interval calculated by the dea model is 0–1. The general least squares method can not accurately analyze the DEA model. The regression coefficient analysis of the DEA model is very likely to have deviations in parameter estimates and different situations [9]. To this end, Tobin proposed the Censored Regression Model using the maximum likelihood method in 1958 [10].

In this paper, the logistics efficiency value derived from the port DEA model is used as the dependent variable, which will affect the internal and external factors of port logistics efficiency such as foreign trade level and secondary industry development level as independent variables. In view of the above research results, we selects the factors related to port logistics efficiency for Tobit analysis. The factors select in this paper include foreign trade level, port scale, economic development level, secondary industry level and tertiary industry level.

4   DEA-Based Port Efficiency Analysis

This paper uses the deap2.1 software to calculate the variable income ccr and bcc models. The results are shown in Table 1.

Table 1. Results of logistics efficiency evaluation of 4 ports in Liaoning Province

Port	Years	crste	vrste	Scale	Scale remuneration
Dalian Port	2010	0.761	1.000	0.761	irs
	2011	0.779	0.927	0.840	irs
	2012	0.829	0.941	0.881	irs
	2013	0.856	0.913	0.937	irs
	2014	1.000	1.000	1.000	–
	2015	1.000	1.000	1.000	–
	2016	0.879	0.882	0.997	irs
	2017	0.911	1.000	0.911	drs
	Mean	0.877	0.958	0.916	
Yingkou Port	2010	0.781	1.000	0.781	irs
	2011	0.853	0.975	0.875	irs
	2012	0.862	0.929	0.928	irs
	2013	0.911	0.930	0.979	Irs
	2014	1.000	1.000	1.000	–
	2015	1.000	1.000	1.000	–
	2016	0.963	0.997	0.966	drs
	2017	0.953	1.000	0.953	drs
	Mean	0.915	0.979	0.935	
Dandong Port	2010	0.538	1.000	0.538	irs
	2011	0.536	0.867	0.618	irs
	2012	0.692	0.915	0.757	irs
	2013	0.775	0.946	0.820	irs
	2014	0.935	1.000	0.935	irs
	2015	0.860	0.901	0.955	irs
	2016	0.953	0.981	0.972	irs
	2017	1.000	1.000	1.000	–
	Mean	0.786	0.951	0.824	
Jinzhou Port	2010	1.000	1.000	1.000	–
	2011	0.869	0.899	0.966	irs
	2012	0.974	0.999	0.975	irs
	2013	1.000	1.000	1.000	–
	2014	1.000	1.000	1.000	–
	2015	0.942	0.958	0.983	irs
	2016	0.942	0.958	0.983	irs
	2017	0.917	0.945	0.970	irs
	Mean	0.963	0.975	0.987	

This paper analyzes the data of four ports in Liaoning Province for nearly 8 years. First of all, Dalian Port will come into effect in 2014 and 2015, and it is technically effective. Other years are not valid for DEA. From the point of view of pure technical efficiency, pure technical efficiency of Dalian Port in 2011–2013 and 2016 did not reach 1. Only from the perspective of scale efficiency, from the table, Dalian Port has the same scale efficiency value in other years except 2014–2015. The results show that the main factors leading to the low level of logistics in Dalian Port from 2010 to 2017 are low scale efficiency and large profit margin.

Yingkou Port and Dalian Port are comparatively similar. In 2014 and 2015, DEA is effective on average, while technology and scale are effective. In other years, non-DEA is effective. In terms of pure technical efficiency, Yingkou Port's pure technical efficiency did not reach 1 in 2010–2013 and 2016. In terms of economies of scale, except for 2014–2015, the net technical benefits of Yingkou Port reached 1. Explain that the scale problem is also the main reason for the low overall efficiency of Yingkou Port.

Dandong Port has always been on the low side in terms of comprehensive efficiency. Before 2014, the comprehensive efficiency value has been below 0.8. After 2015, the comprehensive efficiency value has been rising and will reach DEA validity by 2017. In 2010, 2014 and 2017, the net technical efficiency of Dandong Port reached 1, while scale efficiency was low, especially in 2010–2011, which did not reach 0.7. It can be seen that the main reason for the low overall efficiency of Dandong Port DEA is the inefficiency of scale. However, Dandong Port has been increasing investment and expanding its scale in recent years, so its scale efficiency value has been rising. From the perspective of scale compensation, Dandong Port has entered the stage of increasing scale, which can increase the speed of investment and continue to expand the scale. This shows that Dandong Port has developed better in recent years, can increase the investment speed, continue to expand the scale, and improve the functional quality of the port.

Finally, the comprehensive efficiency of Jinzhou Port is the highest among the four ports. The validity period of Jinzhou Port DEA is 2010, 2013–2014, which is technically effective and effective, while other years are invalid. From the perspective of pure technical efficiency, the pure technical efficiency values of Jinzhou Port in 2011–2012 and 2015–2017 have not yet reached 1. In terms of scale efficiency, the scale efficiency values of Jinzhou Port in 2011–2012 and 2015–2017 reached 1, respectively. It can be seen from the table that the values of pure technical efficiency and scale efficiency of Jinzhou Port have not changed much over the years, and the value of scale efficiency is slightly higher than the value of pure technical efficiency. Overall, the average logistics efficiency of Jinzhou Port in 2010 was 0.963 in 2017, the highest among the four ports. The average pure technical efficiency of Jinzhou Port is 0.975, second only to Yingkou Port, and the average scale efficiency is 0.970. As the number of countries decreases, the overall development situation will be better. The efficiency of Jinzhou Port is very high in all aspects, mainly because of its complete functions, large scale, intensive, high degree of information and dense routes.

5   Analysis of Influencing Factors

On this basis, regression analysis was performed on the established tobit model using eviews 8.0 software. The results of the regression model are shown in Table 2.

It can be seen from the regression results in Table 2 that the coefficient of determination R-squared is 0.822474, indicating that the goodness of fit is good, and it is

Table 2. Tobit model regression results

Variable	Coefficient	Std. Error	t-Statistic	Prob.
Constant (C)	0.072188	0.122572	0.588942	0.5610
Foreign trade level	4.99E-08	1.59E-08	3.135335	0.0042
Port size	0.001674	0.000188	8.921625	0.0000
Port city economic development level	0.532603	0.244732	2.176276	0.0388
Secondary industry level	0.043406	0.019326	2.246011	0.0334
Tertiary industry level	0.024286	0.021633	1.122635	0.2719
R-squared	0.822474			

reasonable to use the model for DEA regression analysis. The regression results show that the level of significance of foreign trade is $P < 0.05$, the coefficient is only 4.99E-08 > 0, and the influence coefficient is small, indicating statistically significant. Among them, the impact of foreign trade level on the efficiency of port logistics is positive, but the degree of influence is limited, and the role is not particularly large. The significance level of the port scale is $P < 0.05$, and the coefficient is 0.001674 > 0, but it is not very high, indicating that the efficiency of port logistics changes positively with the change of port size, but the speed of change is not very fast. The significance level of economic development level is $P < 0.05$, the coefficient is 0.532603 > 0, and the coefficient value is higher, indicating that it is statistically significant, and the economic development level of port city is highly positively correlated with the improvement of port logistics efficiency, and between the two. The impact is greater. The significance level of the development level of the secondary industry is $P < 0.05$, the coefficient is 0.0334, and the coefficient value is higher, indicating that the statistical significance is more significant. The impact of the development level of the secondary industry on the port logistics efficiency is similar to that of the port city development. The development of the secondary industry has great potential for improving the efficiency of port logistics. The significance level of the development level of the tertiary industry is $P > 0.05$, and the coefficient is 0.024286 > 0, indicating that the development of the tertiary industry and the efficiency of port logistics are proportional, but the impact between them is not significant.

6 Summary and Recommendations

In general, the province's port efficiency development is uneven, and regional differentiation is obvious. From 2010 to 2017, the overall technical efficiency of major ports in Liaoning Province showed an upward trend. In the efficiency results of various ports in Liaoning Province, pure technical efficiency has been at a relatively high level and is relatively stable. By establishing the Tobit model, the influencing factors of port logistics efficiency in Liaoning Province from 2010 to 2017 were analyzed. The study found that the economic development level of port cities has a more positive impact on the port logistics efficiency of Liaoning Province than the internal factors such as total foreign trade import and export and port size.

In order to improve the service level of the port function, improve the service efficiency of the port and increase the throughput of the port. At the same time, we will deepen the optimal allocation of port resources and rationally integrate port resources to further improve port logistics efficiency. In view of the above analysis, several suggestions were made:

(1) First of all, efforts should be made to improve the scale efficiency of the port, match the scale with the input and output, make full use of the scale of the port under construction, and reduce the port idle rate. Then, avoid blind construction and port construction investment, promote the construction of various port infrastructure, and constantly improve the port function services.

(2) In addition to scale efficiency, we also need to improve the pure technical efficiency of the port. We focus on improving the technical level and management level. First of all, the production and management of the port should be strengthened to reduce the failure rate of mechanical equipment. Then, improve the utilization rate, loading and unloading efficiency and the quality of mechanical equipment, and improve the professional level of port berths.

(3) In recent years, the development speed and scale of each port are different. First of all, the development of each port should be positioned correctly. For different functions of each wharf, ports should be reasonably divided and resources allocated.

Acknowledgement. The author is thankful to the reviewer and the editor for their constructive suggestions and comments to improve the initial version of our manuscript.

References

1. Hayuth, Y.R.Y.: Port performance comparison applying Data Envelopment Analysis (DEA). Maxitime Policy Manag. **20**(2), 153–161 (1993)
2. Tongzong, J.: Efficiency measurement of selected Australian and other international ports using data envelopment analysis. Transp. Res. Part A Policy Pract. **35**(2), 107–122 (2001)
3. Wu, Y.-C.J.: Fuzzy linear programming based on statistical confidence interval and interval-valued fuzzy set. Eur. J. Oper. Res. (2010)
4. Lai, M.-C., Huang, H.-C., Wang, W.-K.: Designing a knowledge based system for bench marking: a DEA approach. Knowl. Based Syst. **24**, 662–671 (2011)
5. Barros, C.P., Athanassion, M.: Efficiency in European seaports with DEA: evidence from Greece and Portugal. Marit. Econ. Logist. **6**(6), 122–140 (2004)
6. Yun, L., Ji, C., Li, Q., et al.: Comprehensive efficiency measurement of port logistics - study based on DEA two-stage relative evaluation. In: Liss - International Conference on Logistics. DBLP (2011)
7. Huang, Y., Peng, J., Huang, Y., et al.: Efficiency evaluation between port logistics and economic growth by DEA: a case study of Zhejiang Province. J. Appl. Sci. **14**(20), 2594–2600 (2014)
8. Charnes, A., Cooper, W.W., Rhodes, E.: Measuring the efficiency of decision making units. Eur. J. Oper. Res. **2**(6), 429–444 (1978)
9. Greene, W.H.: On the asymptotic bias of the ordinary least squares estimator of the tobit model. Econometrica **49**, 505–513 (1981)
10. Tobin, J.: Estimation of relationships for limited dependent variable. Econometrica **26**(1), 24–36 (1958)

Business English Word Frequency Analysis Method Based on Big Data

Yichen Xing[⊠]

Shandong Women's University, Jinan 250000, Shandong, China
25870439@qq.com

Abstract. With the rapid development of information, big data has increasingly become an important strategic resource and reliable action basis for various activities. At the same time, in the trend of economic globalization, the economic links between countries are increasingly close. As a global language, English is widely used in various business activities. In this case, business English is getting more and more attention. In business English for key word, this study search China hownet, ten thousand, d PuSanDa database, 1585 pieces of the sifting effective literature as the analysis sample, using statistical analysis software SATI3.2 literature citations information, social network analysis software Ucinet6.0 and .net the Draw, using word frequency analysis and spectrum analysis method of combining the knowledge, comprehensive and integrated analysis of the business English category of highlight the word frequency and large data for subsequent business English research to provide the reference. The research results of this paper show that the research of big data is increasing, and the research content of business English is gaining increasing attention.

Keywords: Big data · Business English · Word frequency analysis method · Social network analysis

1 Introduction

With the rapid development of informatization, big data has increasingly become an important resource and reliable basis for countries around the world to make decisions and engage in various activities [1]. In the wave of global big data, the scale and application ability of various reliable data owned by a country or region has become an important symbol and symbol to measure the comprehensive strength of a country or region [2]. However, business English plays a role of bridge of communication in the exchanges of economic activities between countries. The use of business English is playing an immeasurable role in promoting the competitive development of a country or region, and is becoming a strong driving force and support for promoting economic and social development [3]. Especially since the beginning of the 21st century, the application of business English worldwide has reached an unprecedented level [4]. Therefore, under the background of the frequency of the use of business English in the era of big data to make research more help to the economic development of the thermal field and space, every country and region to adapt to today's social integration into the world, to go global must, at the same time also is the realistic requirement of

M. Atiquzzaman et al. (Eds.): BDCPS 2019, AISC 1117, pp. 468–475, 2020.
https://doi.org/10.1007/978-981-15-2568-1_64

accelerating to promote regional economic and social development and effective way, have important practical significance and the practical application value [5].

The rise of big data technology promoted the development of linguistics, based on the statistical analysis of large-scale corpus, foreign researchers found that 90% of the natural speech is made up of chunk, vocabulary and syntax differences do not exist between category, vocabulary and structures is a completely free to completely fixed combinations continuum, there are many between both lexical and syntactic characteristics of semi-fixed structure [6]. Some foreign scholars refer to word blocks as "prefabricated phrases", which can be stored as a whole unit in people's mental word bank. With only minor processing, they can form sentences that conform to syntax and become fluent language to fill some conceptual gaps [7]. Domestic scholars believe that business English is a linguistic phenomenon and a social functional variant of English [8]. As a kind of special-purpose English in international economic activities, business English is increasingly showing its importance. Some domestic researchers have made relevant studies on business English lexical chunks [9]. They discussed the lexical system and translation of business English and concluded that lexical acquisition of business English can accelerate language processing and output, which is of great significance for promoting second language acquisition and improving the fluency and accuracy of language output. Therefore, in the research and analysis of business English, the correct use of word blocks directly affects the application quality of business texts [10].

At present, big data is widely used in the field of business in China, and its economic benefits are increasingly obvious, which provides a broad application basis for exploring the word frequency of business English. In this study, visual software such as SATI3.2, Ucinet6.0 and Net Draw were used, word frequency analysis and knowledge graph analysis were combined to objectively and truly show the hot topics and trends of business word frequency based on big data, so as to explore effective ways and innovative ways for accelerating the global economic development.

2 Methods

2.1 Core Concepts

(1) Big data

With the rapid development of big data, there are more and more expressions of the concept of big data, but it is difficult to make a unified statement. In fact, big data is not only about the large amount of data, but also includes some features that are different from "massive data" and "very large data". The concept of big data is relatively vague. The large capacity of data alone cannot distinguish it from the concepts of "massive data" and "super-large data". Big data is not only about the large volume of data, but also about the semi-structured and unstructured forms of data. The generation and processing of data is closely related to the ever-changing environment, and the tools for processing are diverse. After consulting a large number of literatures, it can be seen that the concept of big data is not completely unified. In general, it refers to the full access to a large number of data resources available for information, including text, image, digital, audio, behavior track and other recordable data.

(2) Word frequency analysis method

The generation of word frequency analysis method has a profound historical background. The generation and development of bibliometrics provide it with theoretical basis and application examples, while the wide application of big data, especially the large number of machine-readable databases, provides it with technical background and application place. If bibliometrics gave birth to it, then big data analysis is the midwife who prompted it. Words pour statistics is not easy, because it requires that the statistical sample, the extent of statistics is generally manual statistics are difficult to do, which directly affect the precision of the reliability and the conclusion, the data is only as big data technology, especially after the creation of a large database for intelligence detection element, word frequency statistics has the means of rapid and enough samples. Information retrieval system with millions of documents, data and facts can directly provide enough statistical samples, and big data technology also makes such a huge number of statistics not a problem.

2.2   Research Methods

Literature measurement analysis method widely used by current academic circle, Web of Science (WOS), and other foreign database platform development of application software, such as social network analysis software Ucinet, embedded open-source software Pajek, Netdraw and Mage, scientific metrology research visual information analysis software Citespace software Bibexcel, literature, etc., in general, although these methods developed function, analysis, high sensitivity, but need special data input format, is not suitable for domestic commonly used database data analysis.

In order to better and faster to master the frequency of a business application, this study using SATI3.2, Ucinet6.0 and .net the Draw visualization software, such as by using word frequency analysis, knowledge map analysis, objectively and truly show under the background of big data business English application hotspot and trend, to accelerate the economic and social development of our country road to explore effective ways and innovation.

Many statistical data show that the distribution of business English index words in the document database obeys normal distribution. For example, a study of guides, indexes, catalogues and bibliographies shows that in each case, the distribution of quantities is normally distributed in terms of the number of words in the bibliography and the number of entries in the index at the back of the book:

$$l(x) = \frac{1}{xa\sqrt{2\pi}} \exp\left\{ -\frac{1}{2}\left[\frac{\log x - m}{g} \right] \right\} \tag{1}$$

Where m is the average, "is the standard deviation of the distribution. Let m and s both go to infinity, and then this will approach a parameter, and then the logarithm distribution will be converted to ziff's law. When the length of the text increases, the relationship between the number of different words D and the length of the text N satisfies the following equation:

$$\log D = \beta \log N + \log K \qquad (2)$$

Here K and beta are text-dependent constants.

3 Experimental

The data sources adopted in this study are the three most frequently used databases in China at present: CNKI China academic journal network publishing general database CNKI, wanfang China academic dissertation database and weipu Chinese science and technology journal full text database. The retrieval period is August 30, 2019. With "business English" as the key word, 1672 relevant literatures can be retrieved from the three databases mentioned above, and 87 articles with the same publication and high similarity of literatures can be screened out. Finally, 1585 articles are selected as the research samples.

In the concrete experiment link, this study adopts the combination of word frequency analysis and knowledge map analysis, first of all, with the help of literature citations information Statistical analysis tools SATI3.2 (Statistical Analysis Toolkit for Infor SATI) software research focus "business English" word frequency analysis, then using social network analysis software Ucinet6.0 and Net - the draw into A visual image, objective truth to show its current situation, And further in-depth analysis of "business English" application research and its development trend. Among them, the word frequency analysis method can be used to find out the research hotspots of "business English", systematically and objectively reveal the research trends and application trends of business English application under the background of big data, and use the social network knowledge mapping method to objectively and truly show the hot words in the thermal points of business English found by word frequency analysis.

4 Discuss

4.1 Statistical Analysis

Specifically, this paper USES the literature distribution word frequency analysis method to sort out the application hotspots, development trends and research trends of business English. As mentioned above, 1585 articles from 1672 literatures were selected as effective research samples by using the three databases mentioned above (CNKI, wanfang and weipu) as data sources and "Business English" and "business foreign language" as search keywords. The annual distribution of these literatures is shown in Table 1 below.

Further research from the data of CNKI, through advanced retrieval set retrieval conditions, respectively is given priority to with business English inscription, with devaluation, settlement, cost, allowance for keywords, retrieval conditions for communication etiquette, economic management, business activities, to set the search terms

Table 1. Distribution of "Business English" literature since 2009

Year	2009	2010	2011	2012	2013	2014
Article number	18	23	33	65	87	98
Year	2015	2016	2017	2018	2019	Until August 30, 2019
Article number	115	158	189	256	309	234

for accurate model at the same time, set the source category to all journals. The search results show that a total of 18,921 articles about big data have been published since 2009, and the trend of the number of papers published each year is shown in Fig. 1. The data in the figure is derived from the published data of CNKI data center.

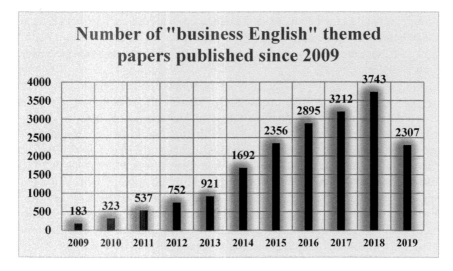

Fig. 1. The number of articles published on the topic "Business English" since 2009

4.2   Word Frequency Analysis of "Business English"

In order to study the relationship between different keywords and business English, the collinear coefficient calculation formula of Salton, McGill and Leydesdorff was adopted in this paper to compare the co-occurrence matrix of two-stage keywords. The co-occurrence coefficient between words with higher word frequency and business English was calculated as shown in Table 2 below.

As can be seen from Table 2, in the first stage, the co-occurrence coefficient of each keyword and business English is significantly higher. Except that the co-occurrence coefficient between data processing and data mining is higher than 0.2, all the others are lower than this value, and most of them are 0. The collinear coefficient of big data is relatively close to that of data mining. The analysis shows that this stage is the initial stage of business English, mainly focusing on distributed data entry, and there is still a long way to go before it can be widely used in business English.

Table 2. Co-occurrence coefficient statistics of Business English keywords

Phase number	First stage		Second stage	
	Keyword	Frequency	Keywords	Frequency
1	Cost	45	Devaluation	211
2	Settlement	69	Security	109
3	Business activities	113	Cost	76
4	Economic management	54	Social etiquette	265
5	Business activities	196	Business activities	342

*Data came from cnki statistics center

In the second stage, the co-occurrence coefficient of each key word frequency and business English is significantly higher than that in the first stage, and the co-occurrence coefficient of all words with high frequency is greater than 0, while the co-occurrence coefficient of most words with data mining is not 0. The emergence of new business English terms such as "security fund", "depreciation" and "settlement" indicates that the application of business English has been in a stage of rapid development.

Based on the existing research results in China, the paper was cited from high to low in order of citation frequency, and the first 500 articles were selected from two stages for research. The literature metrology software SATI3.2 was used for keyword analysis. By summarizing and sorting the similar word frequency of keywords, the top ten keywords in each stage were counted, as shown in Figs. 2 and 3. The data in the figure came from the author's statistical arrangement. By comparing the frequency distribution and proportion of keywords in the two stages, it can be seen that the mining and application of business English in the context of big data is becoming more and more extensive.

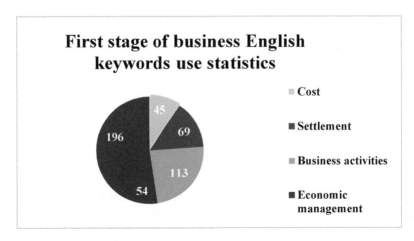

Fig. 2. Key word frequency statistics of business English in the first stage

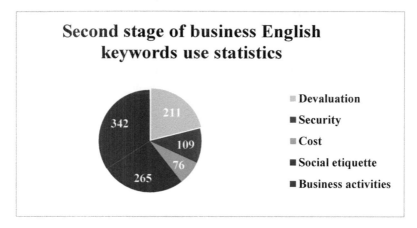

Fig. 3. Key word frequency statistics of business English in the second stage

5 Conclusion

The research in this paper shows that in today's world, the speed of informatization is rapidly increasing, and the role of big data technology is increasingly prominent. It has become an important symbol to measure whether a country is integrated into the development trend of modern economy and society and possesses important strategic resources. With the deepening of international communication, the important role of business English in the economic and social development of various countries and the value of strategic resources are more and more obvious. The more developed countries are, the more thorough the analysis of business English will be, and the more abundant the benefits will be. At the same time, the study of this paper also shows that the high-frequency vocabulary of business English is moving towards internationalization, and the style of business English contains a large number of professional terms, multi-word combination, idiomatic expression word blocks and new word blocks with the characteristics of The Times, and continuous in-depth research is the best way to maintain a dominant position in the fierce competition.

References

1. Monaghan, P., Chang, Y.N., Welbourne, S.: Exploring the relations between word frequency, language exposure, and bilingualism in a computational model of reading. J. Mem. Lang. **93**(13), 1–21 (2017)
2. Pina, J., Massoudi, B.L., Chester, K.: Synonym-based word frequency analysis to support the development and presentation of public health quality improvement taxonomy. J. Public Health Manag. Pract. **25**(8), 1–7 (2019)
3. Chen, C., Hong, J., Zhou, W.: The method and application to construct experience recommendation platform of acupuncture ancient books based on data mining technology. Chin. Acupunct. Moxibustion **37**(7), 768 (2017)

4. Yang, C., Zhang, J., Wang, Z.: Research on time-frequency analysis method of active earth pressure of rigid retaining wall subjected to earthquake. Environ. Earth Sci. **77**(6), 232–237 (2018)
5. Zheng, J., Pan, H., Yang, S.: Adaptive parameterless empirical wavelet transform based time-frequency analysis method and its application to rotor rubbing fault diagnosis. Signal Process. **130**(31), 305–314 (2017)
6. Margusinoframiñán, L., Cidsilva, P., Menadecea, Á.: Intelligent MONitoring system for antiviral pharmacotherapy in patients with chronic hepatitis C (SiMON-VC). Farm Hosp. **41**(31), 68–88 (2017)
7. Boettger, R.K., Ishizaki, S.: Introduction to the special issue: data-driven approaches to research and teaching in professional and technical communication. IEEE Trans. Prof. Commun. **61**(4), 352–355 (2018)
8. Grover, V., Chiang, R.H.L., Liang, T.P.: Creating strategic business value from big data analytics: a research framework. J. Manag. Inf. Syst. **35**(2), 388–423 (2018)
9. Festa, G., Safraou, I., Cuomo, M.T.: Big data for big pharma. Bus. Process Manag. J. **24**(1), 1110–1123 (2018)
10. Brock, T.R.: Performance analytics: the missing big data link between learning analytics and business analytics. Perform. Improv. **56**(7), 6–16 (2017)

Considerations on the Relationship Between Accounting Conservatism and Firm Investment Efficiency Based on Large Data Analysis

Xuanjun Chen[✉]

Department of Construction Management and Real Estate,
Chongqing Jianzhu College, Chongqing 400072, China
535023045@qq.com

Abstract. In recent years, large data analysis methods have been widely used in many fields. For example, it can be applied to environmental art design, computer technology application and so on. More and more companies are investing, so the importance of accounting conservatism and investment interest is increasing gradually. Based on big data, this paper explores the relationship between accounting conservatism and firm investment interest. The endings indicate that accounting conservatism has the following effects on investment profits. First, it can restrain excessive investment and second, it can aggravate underinvestment. If accounting conservatism restrains over-investment, the regression coefficient of accounting conservatism is −0.0265, and there is a remarkable change at the level of 4%. If this conservatism aggravates under-investment, the regression coefficient of accounting conservatism and underinvestment is 0.0130, which is remarkable at the level of 0.9%.

Keywords: Big data analysis · Accounting conservatism · Investment efficiency

1 Introduction

Big data refers to the data set that cannot be acquired, managed and processed by conventional software tools in a certain period of time. Big data has five characteristics: large capacity, fast speed, diversity, low density and authenticity. We often use big data as predictive analysis, user behavior analysis or other advanced data analysis methods.

With the continuous development of financial condition and the continuous improvement of people's living standards, more and more companies will make various investments. The principal-agent problem and information asymmetry will lead to insufficient investment or over-investment. This problem can be alleviated by high-quality accounting information, and the quality of accounting information can be improved by robust accounting information. In article [1], the author uses the staggered formulation of the state-level M & a law as the exogenous growth of the threat of corporate acquisition, and examines the impact of a standardized corporate control market on accounting soundness. In article [2], the author provides empirical evidence

© Springer Nature Singapore Pte Ltd. 2020
M. Atiquzzaman et al. (Eds.): BDCPS 2019, AISC 1117, pp. 476–482, 2020.
https://doi.org/10.1007/978-981-15-2568-1_65

of the impact of management risk incentives on financial reporting robustness. The results show that firms with higher risk incentives can gain economic benefits from more conservative accounting methods. In article [3], the author uses a set of robust data collected in recent years to study the conservatism, to prove comprehend how the conservatism affects the relationship between stock returns and accounting variables. The results show that when the variables related to accounting conservatism are included in the analysis, the explanatory power of the income level and the estimated value of income change coefficient is improved. In article [4], the impact of audit quality on Turkish accounting conservatism is discussed. The results show that accounting conservatism is a complement to companies in Turkish business environment. In article [5], the author examines the influence of introducing competition on conditional accounting conservatism. The results show that there is a significant positive correlation between enterprise-level import competition and conditional conservatism. In article [6], the author examines whether aggregate, conditional and unconditional conservatism is related to economic growth. The results show that this paper facilitates the current debate on accounting conservatism and expands the literature on macroeconomic effects of the attributes of overall financial reporting. In article [7], the author explores the correlation between the number of suppliers and customers and accounting conservatism. The results show that the higher the industry competition density, the more suppliers and customers, the more significant the earnings conservatism. In article [8], the purpose of the study is to investigate whether these two types of accounting conservatism (conditional and unconditional) can mitigate the risk of Jordanian company operating cash flow decline. The endings indicate that the two types of accounting conservatism have a remarkable positive influence on cash holdings. In article [9], the author analyses the impact of foreign investment tendency on accounting conservatism, and the results indicate that foreign investment is positively correlated with accounting conservatism. In article [10], the author aims to determine the impact of corporate complexity on accounting conservatism. The results show that accounting conservatism varies with information asymmetry caused by complex environment.

Based on the analysis of large data, this paper systematically analyses the relation between accounting conservatism and firm investment interests. The endings show that accounting conservatism can restrain over-investment and aggravate underinvestment.

2 Method

2.1 Large Data Analysis Method

Text big data analysis technology refers to the representation, understanding and extraction of words, grammar, semantics and other information contained in unstructured text strings, mining and analyzing the existing facts, and quantifying the feature words extracted from text to express text information. Since each work of art has corresponding text description information, first of all, it is necessary to consider the storage of data. The data storage scheme uses HDFS distributed file system to store large files in streaming data access mode. The text retrieval technology is implemented on Elastic Search. EELASTIC Search provides distributed full-text retrieval through clusters. An Elastic Search cluster can be composed of multiple nodes and can be dynamically increased.

2.2 Calculating Method of Accounting Conservatism

When calculating the annual accounting conservatism index of a company corresponding to an observation value, the main measurement method applicable to Chinese listed companies can choose the accounting conservatism index method or the negative accrual method. This paper chooses KW model, which has been widely used in recent domestic literatures, as an improved model near BASU model.

$$\frac{EPS_{a,b}}{P_{a,b-1}} = \beta_{0,b} + \beta_{1,a,b}DR_{a,b} + \beta_{2,a,b}R_{a,b} + \beta_{3,a,b}R_{a,b} * DR_{a,b} + \varepsilon_{a,b} \tag{1}$$

The model (2) and model (3) are substituted into model (1), and then regression is carried out. The values of accounting robustness can be calculated by estimating the values of lambda (0, b), lambda (1, b), lambda (2, b), lambda (3, b), and then back to model (2).

Among them: EPS represents earnings per share; P represents the closing price of the corporation's stock last year; RET is the stock return rate; D is the fictitious variable.

$$\text{G-Score} = \beta_{3,a,b} = \mu_{0,b} + \mu_{1,b}Lev_{a,b} + \mu_{2,b}Size_{a,b} + \mu_{3,b}Mb_{a,b} \tag{2}$$

$$\text{C-Score} = \beta_{3,a,b} = \lambda_{0,b} + \lambda_{1,b}Lev_{a,b} + \lambda_{2,b}Size_{a,b} + \lambda_{3,b}Mb_{a,b} \tag{3}$$

The calculation of accounting conservatism is as follows (Table 1):

Table 1. Accounting conservatism

Variable	Coef	Std. err	t	$P > |t|$
D	0.0532	0.0229	2.33	0.020
Ret	0.0036	0.0253	0.14	0.886
Size	0.0001	0.0012	0.05	0.962
Mb	−0.0026	0.0004	−5.83	0.000
Ret*Lev	0.02236	0.0069	3.24	0.001
Ret*D	−0.1560	0.0790	−2.03	0.043
Ret*D*Size	0.0087	0.0036	2.41	0.016
Ret*D*Mb	0.0010	0.0016	0.60	0.547
Ret*D*Lev	−0.0315	0.0222	−1.42	0.156
Size	0.0116	0.0006	20.05	0.000
Mb	−0.0002	0.0001	−1.67	0.096

(*continued*)

Table 1. (*continued*)

Variable	Coef	Std. err	t	P > |t|
Lev	−0.0213	0.0038	−5.64	0.000
D*Size	−0.0026	0.0010	−2.51	0.012
D*Mb	−0.0012	0.0005	−2.43	0.015
D*Lev	0.0124	0.0067	1.86	0.063

2.3 Calculation Method of Investment Efficiency

This paper regards the investment efficiency as the explanatory variable. This paper uses the residual measurement pattern to gauge the investment profit of enterprises. The measuring method model of investment profit in this paper is as follows:

$$INVEST_{a,b} = \beta_1 + \beta_2 INVEST_{a,b-1} + \beta_3 CASH_{a,b-1} + \beta_4 SIZE_{a,b-1} + \beta_5 LEV_{a,b-1}$$
$$+ \beta_6 MB_{a,b-1} + \beta_7 RET_{a,b-1} + \beta_8 AGE_{a,b-1} + YEAR \qquad (4)$$
$$+ INDUSTRY + \varepsilon_{a,b}$$

The relevant variables are explained as follows (Table 2):

Table 2. Richardson model related variables

$INVEST_{a,b}$	New investments in the current period	Company a b − 1 year (expenditure on acquisition and build of changeless investments, intangible investments and other standing investments + net cash paid by affiliated companies and other business company net cash recovered from disposal of standing investments, intangible assets and other standing investments)/total assets at the beginning of the year
$INVEST_{a,b-1}$	New invest in the previous period	The calculation method is the same as the novel investment at present
$CASH_{a,b-1}$	Shanghai Futures Currency Fund	Company a b − 1 monetary fund/initial total assets
$SIZE_{a,b-1}$	Enterprise scale	a company b − 1 logarithm of general investments at the early stage
$LEV_{a,b-1}$	Asset-liability ratio	Company a b − 1 total liabilities divided by initial total assets
$MB_{a,b-1}$	Market share ratio	b − 1 share price/net asset per share of company a
$RET_{a,b-1}$	Return on equity	a company b − 1 year return on individual shares considering reinvestment of cash dividend
$AGE_{a,b-1}$	Listing date	Natural logarithm of listing years
$YEAR$	Annual dumb variable	Set 6 dummy variables for 7 years in total
$INDUSTRY$	Industry dumb variable	According to the industry classification of the CSRC, there are eleven industries and ten industry virtual variables are set up

3 Data

In order to make this study go on normally, we deleted ST, financial and insurance categories, missing data and listed companies after 2006, and finally left 8693 research samples. The reason for deleting ST companies is that these companies cannot guarantee normal production and operation activities, which will affect the normal conduct of research. The reason for deleting financial and insurance listed companies is that their business activities and management control are different from those of general listed companies.

4 Result

4.1 Accounting Conservatism Can Restrain Overinvestment

From the regression results of the whole sample, we can see that the symbols and significance of β_0 are in line with expectations. The regression coefficient of accounting conservatism (CSCORE) is −0.0265, and there is a significant change at the level of 4%. It can be seen that accounting conservatism can restrain excessive investment. The specific outcomes are displayed in Chart 3.

Chart 3. Empirical return endings of accounting conservatism and overinvestment

Variable	Overinv		
	Full sample	High cash holdings	Low cash holdings
CSCORE	0.0282** (−2.01)	0.0239 (−1.18)	−0.0393* (−1.94)
TOBINQ	0.0032*** (4.81)	0.0019** (2.40)	0.0058*** (5.08)
BD	0.0564*** (8.54)	0.0473*** (5.02)	0.0619*** (6.43)
NDTS	0.0153 (0.24)	0.0764 (0.87)	−0.0801 (−0.88)
Age	−0.0002 (−0.96)	−0.0000 (−0.07)	0.0003 (−1.20)
Cycle	−0.0118*** (−6.13)	−0.0136*** (−5.49)	−0.0097*** (−3.19)
Tunnel	−0.0142 (−1.63)	−0.0227* (−1.86)	−0.0060 (−0.49)
SOE	−0.0011 (−0.62)	−0.0024 (−1.03)	0.0008 (0.28)
Cons	0.0392*** (5.50)	0.0320*** (3.02)	0.0407*** (4.09)

4.2 Accounting Conservatism Aggravates Underinvestment

From the regression results of the whole sample, we can see that the symbols and significance of β_0 are in line with expectations. The regression coefficient of accounting conservatism (CSCORE) and underinvestment (UNDERINV) is 0.0130, and there is a significant change at the level of 0.9%. This shows that accounting conservatism will aggravate underinvestment. The specific results are shown in Fig. 1.

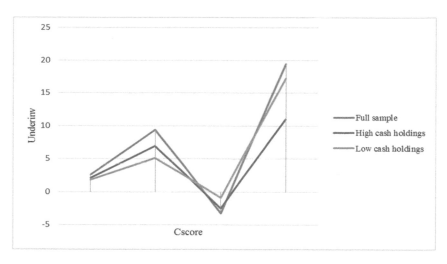

Fig. 1. Empirical regression endings of accounting conservatism and underinvestment

5   Conclusion

Based on a large number of data, this paper analyzes the relationship between accounting conservatism and investment efficiency protection, and draws the following conclusions. Firstly, accounting conservatism can restrain over-investment, but at the same time it will aggravate underinvestment. In addition, there are still some short-comings and shortcomings in this paper, which will be slowly corrected in future research.

Acknowledgements. This paper was supported by a grant from The Youth Research Foundation of Chongqing Jianzhu College (Grant No. QN2015003).

References

1. Khurana, I.K., Wang, W.: International mergers and acquisitions laws, the market for corporate control, and accounting conservatism. J. Acc. Res. **57**(1), 241–290 (2019)
2. Hu, C., Jiang, W.: Managerial risk incentives and accounting conservatism. Rev. Quant. Financ. Acc. **52**(3), 781–813 (2019)
3. Xia, B.S., Liitiainen, E., De Beelde, I.: Accounting conservatism, financial reporting and stock returns. Account. Manag. Inf. Syst. **18**(1), 5–24 (2019)
4. Mohammed, N.H., Ismail, K.N.I.K., Amran, N.A.: Audit quality and accounting conservatism. J. Account. Financ. Audit. Stud. **5**(2), 1–23 (2019)
5. Burke, Q.L., Eaton, T.V., Wang, M.: Trade liberalization and conditional accounting conservatism: evidence from import competition. Rev. Quant. Financ. Acc. **53**(3), 811–844 (2019)

6. Do, C., Nabar, S.: Macroeconomic effects of aggregate accounting conservatism: a cross-country analysis. J. Int. Financ. Manag. Account. **30**(1), 83–107 (2019)
7. Kao, H.S., Tsai, M.J.: Implications of suppliers and customers and industry competition density on accounting conservatism. Global J. Bus. Res. **13**(1), 1–14 (2019)
8. Hamad, A., Mohammad, A.M., Al-Mawali, H.: Does accounting conservatism mitigate the operating cash flows downside risk? J. Soc. Sci. Res. **5**(2), 472–483 (2019)
9. Ji, S.H., Ryu, Y.R.: Investment tendency of foreign investor and accounting conservatism. J. Digit. Converg. **17**(3), 153–160 (2019)
10. Silva, A.D., Ganz, A.S., Rohenkohl, L.B., Klann, R.C.: Accounting conservatism in complex companies. Rev. Contab. Financas **30**(79), 42–57 (2019)

Design of Higher Vocational Education Auxiliary Platform Based on Cloud Computing Technique

Miao Yu$^{(\boxtimes)}$

Chongqing Three Gorges Vocational College, Chongqing, China
87571946@qq.com

Abstract. Along with the popularity of cloud computing technique today, it has been realized to help people get rid of pen and paper writing and cumbersome report calculation, get rid of the limitation of time and space, as long as you have a network, you can use cloud computing technology to accomplish a variety of tasks quickly and efficiently. Higher professional education is an significant link in the higher education institution. How to effectively run the country's higher vocational education is related to the overall development of China's education cause an important topic. This paper starts from the major problems and development direction in the construction of teaching resources in China's higher vocational colleges, explore the use of modern cloud computing and mesh resources, USES the information technology of cloud computing service mode, service architecture of high quality teaching resource sharing platform in higher vocational colleges is proposed, implementation maximize the use of teaching resources in higher vocational colleges, crack education teaching resources problem such as "gaps between the rich and the poor", promote the level of education modernization, for the cloud platform repository construction goals provide beneficial theoretical support and technical guarantee. The research results of this paper show that in process of design teaching at the higher vocational schools, the advantages of cloud computing can be fully utilized to create a good cloud computing education environment and give full play to teaching assistance function of cloud computing, which is conducive to more effective learning for learners.

Keywords: Cloud computing technology · Education support platform · Higher vocational education · Construction of teaching resources

1 Introduction

Teaching information of technical construction is a very important sign of modernization to education in higher vocational school, and modernization of transformation of education and teaching system to engineering are closely related, and the construction of higher vocational school education auxiliary platform is essential part of the construction of digital campus information, thus establishing auxiliary platform dedicated to the education of higher vocational education is very important and necessary [1]. The education assistance platform of higher vocational school is an

M. Atiquzzaman et al. (Eds.): BDCPS 2019, AISC 1117, pp. 483–491, 2020.
https://doi.org/10.1007/978-981-15-2568-1_66

integrated digital teaching resource based on certain norms and standards that collects and integrates various media materials [2]. But in actual construction process, attaches great vital to the create and ignore the real operation of practical value, the collection resources and ignore the importance of the depth of integration, results in a practical low data information resources in the platform, fuzzy retrieval, affect the sustainable development strategy, and easy to form a "data island" data information resources sharing is poor [3, 4].

Education information public service platform is the core part of education information support system and basic platform environment for education information services and applications. In recent years, it has become the key content of education information construction at national and local levels. National medium and long-term plan for education reform and development (2020–2030) proposed in the major project of "the application of IT in education" that "Public service platform construction that effectively shares and covers all levels of education", which has achieved initial results in the construction of various key universities in China. What is more noteworthy is that along with the rapid evolution and continuous promotion of cloud computing science, the education cloud service model has gradually become a new idea for the construction of education information public service site in higher vocational colleges [5, 6]. The 10-year development plan for education informatization (2019–2035) issued by the ministry of education minister in March 2018 figured out that the existing resources should be fully integrated and cloud computing technology ought to form an comprehensive development path of resource allocation and services, so as to build a stable, reliable and low-cost cloud service fine example for higher vocational education. At present, some regions are trying to public cloud computing technology and the higher vocational education information service platform to integrate, in some places has launched the early education cloud construction projects, and implementation of the "sea of clouds plan" of education cloud construction projects, etc., all of these are of higher vocational education basis on cloud computing information public service platform construction actively explore [7].

In recent years, on account of the rapid growth of information technology and the vigorous promotion of national education director, virtual simulation teaching and training platform with cloud computing as the core has been applied and promoted in various higher vocational colleges [8]. The use of this platform is for vocational education to construct, the integration of multimedia technology, virtual simulation technology and network communication technology, etc., creating real learning for students, practice training environment, and to present the strong interactivity, situational, perceptual, etc., to motivate the students' thirst for knowledge, can significantly improve the students' professional technology level, for the current practice, practice and professional education are facing difficult employment problem is a good pattern to dilution [9]. The concept of cloud calculate applied in higher vocational college education is a newness pattern, which can provide great help for the construction of resource base of educational auxiliary platform in higher vocational college. It can awareness the unified allocation and manipulation of resources, and has an important value and function for the construction of the education auxiliary platform of higher vocational colleges [10]. Based the analyze of the superiority and characteristics of cloud computing, this paper hammer at the construction planning and functional structure of education message MTPS on account of cloud computing.

2 Method

2.1 Core Concepts

(1) Cloud computing technology

The so-called CloudComputing is a new computing mode that provides various resources such as software and hardware to end users in the form of services. In this model, users to install the application and the software is not operation on the user's personal laptop or terminal devices such as mobile phone, but to install and run on the Internet in large-scale cloud computing provides server colony, user must deal with the data is not stored locally, but stored in the Internet cloud computing services provided by the center database.

The company that provides cloud computing services is take charge of maintaining and managing the equipment in these DC (data centre), providing sufficient computing power and memory space for the user to access these services at any time, anywhere, with any terminal device that can be connected to the Internet. In contrast to the traditional pc-centric computing approach, in the cloud computing model, hardware and software resources are acquired on the "cloud" side. Users do not need to purchase high-end computer configuration and install a lot of software, as long as they can access the Internet, they can enjoy the software and hardware resources provided by the "cloud" at any time and place. As a result, users no longer have to worry about losing data due to damaged hard drives, being busy with virus attacks on their computers, or whether their personal computers don't have enough storage space, hardware needs to be upgraded, or software needs to be updated.

(2) Supporting platform for higher vocational education

The auxiliary teaching platform is to provide teaching assistance through the Internet, and the auxiliary teaching platform in higher vocational school is to provide scholastic with an auxiliary teaching form based on the Internet. This learning mode can reduce the dependence of higher vocational learners on classroom teaching, and the learning time, place and content can be determined by the learners themselves, which greatly improves the degree of learning freedom and enables the learners to obtain massive learning resources. Traditional classroom teaching and learning materials, knowledge points and teaching progress, in the teaching mode based on network resources, students' subjective initiative will get into full play, learning to selectivity according to their understanding and knowledge level, ability to learn, so that they can make good use of time, greatly improve learning efficiency.

Based on multimedia and CNT (computer networking technology), the educational auxiliary platform makes the learning content realize the network. Cloud computing is a neo-technology with the development of network technology. Cloud computing can provide mass data services According to the characteristics of higher vocational education and needs, and it can also take full advantage of share the resources of each network node. The effective and reasonable combination of cloud computing technology and higher vocational education is an vital way to realize informationization of education. The application of cloud calculate technology in higher vocational education in China. It has profound effects on the cause of higher vocational education and may bring about changes in the education industry.

2.2 Combination Method of Cloud Computing and Educational Auxiliary Platform Design

Cloud teaching, on account of the technology of "cloud calculate" is a variety of teaching services, such as server, bandwidth, software system, data resources and technical resources such as migration to the cloud, learners can log on to the cloud anywhere, anytime access to learning resources, learning plan, keep the cloud services, such as the private data for personalized autonomous mobile learning. Cloud computing technology is not limited to the hardware and software bottlenecks of the computer itself, most applications and services are stored on the server, and educational recipients can access a large number of resources only through the Internet. Network education platform based on cloud computing technology, economic effectively for teachers and students can not only provide the high quality of network education services, in the meantime brings to the teaching of convenient for collaboration and resource sharing and storage space is infinite convenience condition, in addition, the technical threshold low convenient building personalized teaching informationization environment, support the student through any terminal equipment at any time and place into the personalized learning, training students' development of advanced thinking ability and wisdom of crowds, truly make learning is everywhere, it's fast and convenient to the interflow between the teachers and the scholastic, upgrade educational quality has very fatal practical significance and forward-looking theoretical guiding significance.

The design architecture of cloud computing combined with educational assistance platform is shown in Fig. 1 below. Students can conduct education-related actions through the education cloud, which includes the infrastructure layer, application development layer and service sharing layer.

3 System Implementation Scheme

3.1 System Structure Design

The platform is mainly composed of operating system, user interface, server and various resources. The architecture of education assistance platform based on cloud computing technology is mainly composed of client layer, client tier, application layer, management layer, data layer and infrastructure layer. The application tier is the service provider of all kinds of the whole teaching platform, the management provides all kinds of maintenance for the online platform, the data layer is the guarantee of all kinds of

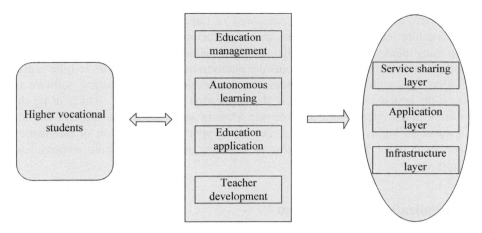

Fig. 1. Design architecture diagram of educational assistance platform

resources of the platform, and the infrastructure layer is the hardware foundation of the online platform.

Applications can use cloud platform, cloud and cloud storage tertiary structure, based on the mobile B/S architecture, to provide Internet users online platform system based on web browser, interface protocols based on Internet, to provide users with platform online browsing, online learning and data uploads and downloads and online communication, and other functions, system of all kinds of data in sync. The system platform as a whole adopts C/S framework, and individual functions can adopt B/S framework. The server runs on Cent OS 7.4, Windows server 2016. The system chooses open source software to undertake Web service software development function. The relevant development tools of the teaching platform system are shown in Table 1 below.

Table 1. System platform development tools

Nomenclature	Software tools used
Operating system platform	Cent OS 7.4, Windows server 2016
Web server	Apache Tomcat\IIS
Java runtime environment	Sun JDK7.0
Database	My SQL data management software
Development tools	Visual Studio 2017\Eclipse J2EE
Development of language	Java\JSP\Objective - C and C++

3.2 System Implementation Planning

The specific steps of the system are mainly divided into three steps, the first line is to build the development environment, the second is to establish the system directory structure, the third step is the design and establishment of each task function module.

The system of the total cloud is composed of three levels: the lowest level is the infrastructure level, which supports the infrastructure as a service and is the basic supporting platform for the public service of education information. The middle application development layer adopts the platform-as-a-service model to offer a professional education application engine for the development of teaching software or education system. The top layer is the application service layer, which using the software-as-a-service model to cover the learning needs of all the teachers and the scholastic in higher vocational colleges and support the application of information technology in all aspects of education including teaching, scientific research and administer.

4   Realization of the System

4.1   Platform Construction

The advantage of adopting layered design for the education auxiliary platform of higher vocational colleges proposed in this paper is that when the business logic, data processing and display effect change, it can realize the upgrading requirements of partial modules without destroying the overall structure. It is based on the high quality teaching resources sharing platform in higher vocational school in content should be colorful, inclusive; In terms of column setting, it should have a powerful functional operation system. Therefore, this paper determines the content and column setting of the platform on the basis of full survey on the current situation of teaching resource information sharing application and service demand in higher vocational school.

(1)   Platform content

Content of auxiliary education terrace in higher vocational colleges is mainly composed of three core databases: public curriculum resource database, professional teaching resource database and in-post practice resource database. In addition, also including education theory, the excellent lesson plans and courseware, teaching video, high-quality goods curriculum, teaching material construction, teaching resources, database, simulation training, simulation class, typical cases, on display, virtual experiment, internships, work-integrated learning, professional work, comprehensive evaluation, market research report, the employment situation analysis and so on education teaching resources and information.

(2)   Column setting

On the basis of the requirements of teaching in higher vocational school, respectively set up with the corresponding column, specific include: resource distribution, resource download, resource retrieval, platform message, online communication, news, user messages, online examination, job submission, network classroom, network storage, network mailbox, teachers, field information, test registration and enrollment, scores query etc.

In addition, this article utilizes Linux and WebServices technologies on top of the current cloud services infrastructure and USES the open-source Eucalyptus cloud platform software architecture, which requires very little platform hardware and can be deployed in both hybrid and private clouds. Eucalyptus allows customers to quickly and easily create service clouds tailored to their specific needs. In addition, Eucalyptus has secure virtualized servers, networks, and storage capabilities to reduce costs, improve maintenance, and provide user self-service.

Cloud service platform ultimately relies on the hardware resources of the underlying client and server, so building an appropriate underlying hardware environment can provide strong hardware support for resource management and application, enabling users to access the underlying resources through the application layer and platform layer. The construction process is as follows:

First, build a small local area network and configure the network environment. This LAN selects several ordinary PCS as server resources and storage units of resource pool layer, including one server as back-end virtual node machine, one server as user and group configuration management, and one server for cloud resources management.

Second, install the operating system. Select the Linux operating system as the operating environment for the cloud service platform and install it on each server.

Finally, install third-party software: select JavaJDK, apache-ant, MySQL, apache-tomcat as the supporting tools, and install the Xen environment with the en kernel configured.

4.2 System Operation Test

The key to the stability test of cloud computing technology architecture is the characteristics of the terminal in the actual running process. At present, there is no optimal construction standard for cloud computing software framework. Relevant organizations and enterprises put forward the terminal Internet of things framework. The experiment took the Internet of things software framework as the verification target and conducted the design and research of the stability test system.

The basic principle of the stability test of the architecture of the education assistance platform of higher vocational colleges supported by cloud computing technology is similar to the continuous test. It is based on the automated test method and tests the running stability of the software through the driver module of the mobile terminal. In the case that the test quantity is uncertain, operation stability is taken as the test index, and the results are present in Table 2 and Fig. 2 below.

The laboratory finding reveal that the stabilization test results of the cloud computing based higher vocational college education assistance platform architecture designed in this paper have less error and better stability performance.

Table 2. System stability test results

Test module	Test amount/thousand	Output/ms	Expected result/ms
Release resources	100	54	52
Download resources	200	71	69
Online communication	300	89	71
User message	400	93	90
Online exam	500	101	104
Online class	600	112	114
Test registration	700	117	121
Results the query	800	134	133

*Data in the table is derived from the author's test results.

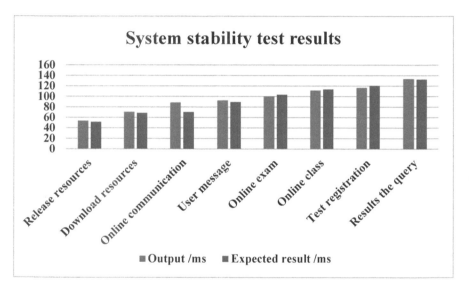

Fig. 2. System stability test results

5 Conclusion

With the evolution of cloud computing technique, online learning has turn into a new mode for all kinds of students to learn. Especially for higher vocational colleges with relatively weak educational resources, the necessity of developing an auxiliary education platform with high stability, large storage space and low cost is self-evident. The cloud platform has a powerful data storage capability, and users can download the data they need from the cloud at any time. With the support of the powerful hardware resources of the cloud platform, users' various requests will be quickly responded. Therefore, for the majority of vocational college students, the education assistance platform based on cloud computing will be a learning platform with huge data resources, convenient operation and use, and powerful network functions.

Acknowledgements. Teaching reform of "big data technology" under the mechanism of deep integration of school and enterprise and skill competition Item number 193435.

References

1. Kolluru, S., Varughese, J.T.: Structured academic discussions through an online education-specific platform to improve Pharm.D. students learning outcomes. Curr. Pharm. Teach. Learn. **9**(2), 230–236 (2017)
2. Steele, J.R., Jones, A.K., Clarke, R.K., et al.: Use of an online education platform to enhance patients' knowledge about radiation in diagnostic imaging. J. Am. Coll. Radiol. **14**(3), 386–392 (2017)
3. Ning, W., Chen, X., Lan, Q., et al.: A novel wiki-based remote laboratory platform for engineering education. IEEE Trans. Learn. Technol. **131**(99), 1 (2017)
4. Pratte, G., Hurtubise, K., Rivard, L., et al.: Developing a web platform to support a community of practice: a mixed methods study in pediatric physiotherapy. J. Contin. Educ. Health Prof. **38**(1), 1–3 (2018)
5. Jones, O.A.H., Spichkova, M., Spencer, M.J.S.: Chirality-2: development of a multilevel mobile gaming app to support the teaching of introductory undergraduate-level organic chemistry. J. Chem. Educ. **95**(7), 1216–1220 (2018)
6. Lindgren, R., Morphew, J., Kang, J., et al.: An embodied cyberlearning platform for gestural interaction with cross-cutting science concepts. Mind Brain Educ. **13**(1), 53–61 (2019)
7. Edwards, G., Hellen, K., Brownie, S.: Developing a work/study programme for midwifery education in East Africa. Midwifery **59**, 74–77 (2018)
8. Moyer, A., Shinners, J.: Interprofessional continuing education in the ambulatory care setting. J. Contin. Educ. Nurs. **48**(9), 390–391 (2017)
9. Nguyen, U.N.T., Pham, L.T.H., Dang, T.D.: An automatic water detection approach using Landsat 8 OLI and Google Earth Engine cloud computing to map lakes and reservoirs in New Zealand. Environ. Monit. Assess. **191**(4), 13–14 (2019)
10. Krishnaswamy, V., Sundarraj, R.P.: Impatience characteristics in cloud-computing-services procurement: effects of delay horizon and situational involvement. Group Decis. Negot. **45**(11), 1–30 (2019)

Constructing Mobile Network Platform for Sports Resources Sharing of Universities-Enterprise Alliance Under Big Data

Wensheng Huang[✉]

Sports Teaching Department,
Dalian Polytechnic University, Dalian 116034, China
hws02@163.com

Abstract. Under the environment of big data internet, aiming at the phenomenon of periodic idleness of university sports resources and lack of enterprise sports resources, this paper builds a bridge between University and enterprise, studies and designs a mobile network platform for university sports resources sharing in universities-enterprise alliance, enhances the degree of socialized sharing of university sports resources, and promotes the healthy development of enterprise sports culture. According to the large data of Internet, the construction and application of the mobile sharing platform of universities-enterprise sports resources network alleviates the contradiction between supply and demand of universities-enterprise sports resources, enhances the exchange and cooperation between universities and enterprises, and effectively promotes the realization of the national strategic goal of national fitness for all.

Keywords: Big data · Internet · Universities sports resources · Universities-enterprise alliance · Mobile network platform

According to relevant investigations, universities have abundant high-quality sports resources, not only modern sports venues and facilities, but also professional sports teachers [1]. The social sharing of sports resources in universities can not only solve the problem of lack of sports facilities and scientific sports knowledge for the general public, but also give full play to the guiding role of university teachers in national fitness [2]. With the popularity of Internet mobile network equipment, mobile sports resources network sharing platform has become an urgent need to disseminate and develop sports culture. In this context, based on the Internet mobile micro-messaging network, Tencent small program is used to build a mobile network platform for sports resources alliance between universities and enterprises. Taking the mobile network platform as an example, the network resources of universities-enterprise alliance can be shared through mobile micro-messaging client. This paper focuses on this mobile network platform to discuss and study the social sharing of sports resources in universities.

© Springer Nature Singapore Pte Ltd. 2020
M. Atiquzzaman et al. (Eds.): BDCPS 2019, AISC 1117, pp. 492–498, 2020.
https://doi.org/10.1007/978-981-15-2568-1_67

1 Research Object and Method

1.1 Research Object

Universities-enterprise Alliance on sports resources mobile network sharing platform.

1.2 Research Method

The methods of documentation, logical analysis, comparative study and questionnaire survey were used. Through reading a large number of documents and relying on the national strategy of national fitness for all, this paper studies the effectiveness of the construction of mobile network sharing platform for sports resources of University-Enterprise alliance. By means of questionnaires and on-the-spot interviews, it concludes the differences of socialized sharing of sports resources in Universities before using the mobile network platform, carries out comparative analysis, and explores the characteristics of the network sharing platform for sports resources in Universities. Function and existing problems.

2 Research Results and Analysis

2.1 Analysis on the Necessity of Building a Mobile Network Platform for Sports Resources Sharing of Universities Enterprise Alliance

2.1.1 It is the General Trend for University Sports Resources to Serve the Society

As early as 2006, the State General Administration of Sports published "the Eleventh Five-Year Plan for Sports", emphasizing that the relevant departments of education should speed up the opening of sports resources in Universities to the society [3]. In 2009, the State Council formally put forward "the National Fitness Program" and issued the Regulations on National Fitness to encourage public and private universities to open sports resources to the society [4]. In 2016, the State Council put forward the National Fitness Program (2016–2020) and the Outline of "Healthy China 2030" again. The socialization and marketization of sports resources in Universities has become the current trend of sports development in Universities. The opening of sports facilities in Universities to the outside world and charging a certain fee can not only alleviate the economic burden of the country and Universities themselves, but also promote the continuous development of sports in Universities [5]. In order to promote the smooth development of national fitness activities, Universities should change their thinking, dare to innovate, and actively explore ways of social sharing of sports resources in Universities [6]. Universities should give full play to their advantages in sports facilities, teachers, information resources and scientific research achievements, and make more contributions to "healthy China" [7].

2.1.2   Reasons for the Low Openness of Sports Resources to Society in Universities

(1) The blockade of sports resources in Universities. Sports facilities are located in Universities, there is no means of external propaganda, only on the campus network to publish open venues and open time, can not meet the needs of external propaganda, making most people unable to understand the university sports resources.

(2) There are no specific policies and regulations for the opening of sports facilities in Universities. There are no specific policies and regulations restricting the exercise of foreign personnel in Universities. The ambiguity of the open area and the exercise area of teachers and students affects the normal life of teachers and students and extracurricular sports activities.

(3) There are potential safety hazards in the opening of sports venues in Universities. While sharing sports facilities with the society, increasing the mobility of social personnel on campus increases the possibility of security risks to a certain extent, students personal safety and the safety of University property [8], as well as sports injuries caused by the exercise of foreign sports personnel in the venue, resulting in disputes inside and outside the University also become the reasons why Universities do not support the opening of sports venues to the outside world.

2.1.3   Feasibility of Opening University Sports Resources to Enterprises

(1) The supply and demand of university sports resources and enterprise sports culture. The resources of university sports facilities are open to the society in response to the call of the national policy. With the development of economy, the salary level of university teachers is limited, which causes the need of teachers resources export [9]. In recent years, with the continuous development of enterprise sports culture, the demand for enterprise sports competitions and exercises is also rising. The large number of university sports facilities, in line with the competition standards, strong professionalism of teachers, just in line with the needs of enterprise sports competition and exercise.

(2) Enterprise competition, exercise organization, purpose is strong, easy to manage. Enterprise competition and exercise are highly organized, which correspondingly reduces the random mobility of social personnel on campus, reduces the conflicts between teachers and students and social personnel. Enterprise competition and exercise have strong purpose, and reduces the chaos of wearing high-heeled shoes into plastic fields.

(3) Increase the income of universities, and push back the opening of university sports resources to the society. Enterprise competition needs a large number of venues, organizers, planners, referees, etc. Enterprise exercise has the characteristics of large number of people, large demand for venues, and professional guidance of coaches. The utilization of university sports resources and the growth of economic income push back the further opening of university sports resources to the society [10].

2.2 Architecture Analysis of Mobile Network Platform for Sports Resources Sharing in University-Enterprise Alliance

2.2.1 Characteristics of Mobile Network Platform for Sports Resources Sharing in University-Enterprise Alliance

(1) Publicity. Through the integration of network platform, university sports resources have realized information publicity. Enterprises can inquire through the network platform, understand the use of university sports facilities, according to the needs of free choice.
(2) Convenience. Enterprises enter the mobile Wechat client for simple operation, to achieve sports knowledge browsing, fitness technical consultation, venue booking and other functions.
(3) High efficiency. Through the mobile Wechat client network platform, users can consult related questions online, and university physical education teachers can give real-time answers to the questions, giving full play to the efficiency of mobile phone client.

2.2.2 The Running Content of Sports Resources Sharing Mobile Network Platform of University Enterprise Alliance

(1) Sports facilities module. Enterprises can learn about the specific situation of university sports facilities, charging standards, opening time and insurance purchases. According to their actual needs, choose the most appropriate venues.
(2) Competition organization module. Universities can organize various competitions in conjunction with enterprises. Through the organization and guidance of professional physical education teachers, it can effectively promote exchanges among enterprises while competing.
(3) Sports professional guidance module. (1) Sports skill module; (2) Fitness method module; (3) Health knowledge module

2.3 Effectiveness Analysis of Mobile Network Platform for Sports Resources Sharing in University-Enterprise Alliance

2.3.1 University Sports Resources Sharing Before Using "Mobile Network Platform for Sports Resources Sharing of University-Enterprise Alliance"

(1) Taking some universities in Liaoning Province as an example, they have abundant resources of physical education teachers. Through questionnaires, the number of full-time physical education teachers who use their spare time to participate in part-time sports-related work in enterprises and society is relatively small, accounting for only 12.2% of the survey population.
(2) In the use of university stadiums and gymnasiums, the rate of shared use of stadiums and gymnasiums with enterprises and society is low, accounting for 19.6% of the survey venues. Before using the Internet mobile network platform,

the supply and demand information of sports resources in Universities and enterprises is not smooth, there is a lack of effective management platform, less funds for stadium maintenance, and there are security risks in the opening of stadiums and gymnasiums, which lead to the low utilization rate of stadiums and enterprises and society.

2.3.2   University Sports Resources Sharing After Using "Mobile Network Platform for Sports Resources Sharing of University-Enterprise Alliance"

(1)   After the network sharing platform of university sports resources was put into use, the utilization rate of venues increased, and the number of sports instructors increased, which provided certain remuneration for PE teachers and enhanced their enthusiasm for participation. Its participation in the field also expanded from the original referee, aerobics to badminton, basketball and other fields (Table 1).

Table 1. Sports culture of part-time enterprises after using network platform in some universities in Liaoning Province.

Name of University	Number of full-time physical education teachers	Number of part-time enterprise sports culture instructors	Part time rate
Dalian Polytechnic University	24	8	33.3%
Dalian University of Foreign Languages	16	3	18.7%
Dalian Jiaotong University	30	12	40%
Dalian Ocean University	26	7	26.9%
Shenyang Jianzhu University	29	10	34.4%
Liaoning University of Traditional Chinese Medicine	15	2	13.3%
Shenyang Agricultural University	32	9	28.1%

(2)   After using the Internet mobile network platform, the opening degree of resources sharing in some university stadiums and Gymnasiums in Liaoning Province has been significantly improved, and the opening time has also been greatly increased. Enterprises and individuals who rent campus sports facilities through the platform have increased substantially, effectively increasing the utilization rate of the venues.

From Table 2, we can see that the utilization rate of venues has reached 74.2%. Only some cold door sports venues such as rugby are vacant. The opening of University venues and the improvement of utilization rate mainly include the following aspects: (1) The supply and demand information is more unimpeded, so that enterprises can choose suitable time to rent venues. (2) Reduce the potential safety hazards and sports injury disputes such as personnel conflicts inside and outside schools. (3) The utilization rate of stadiums and teachers has been improved, and the further opening of university stadiums and gymnasiums has been pushed back for the increased economic benefits of the university.

Table 2. Sharing of stadium resources with social enterprises after using network platform in some universities of Liaoning Province

Name of University	Name and quantity of stadiums and gymnasiums	Amount used	Utilization rate
Dalian Polytechnic University	Basketball, volleyball, badminton, etc. totaled 9 pieces	9	100%
Dalian University of Foreign Languages	Basketball, volleyball and rhythmic gymnastics totaled 12 pieces	8	66.7%
Dalian Jiaotong University	Badminton, table tennis, gym totaled 15 pieces	11	73.3%
Dalian Ocean University	Basketball and Volleyball 5 pieces	5	100%
Shenyang Jianzhu University	Basketball, volleyball, badminton and swimming pool totaled 24 pieces	17	70.8%
Liaoning University of Traditional Chinese Medicine	Aerobics and rhythmic gymnastics totaled 4 pieces	2	50%
Shenyang Agricultural University	Football, basketball, volleyball, badminton and rugby totaled 28 pieces	20	71.4%

3   Conclusion

Through the construction and use of Internet mobile network platform, the supply-demand relationship between universities and enterprises is formed, which makes up for their shortcomings, plays a positive role in promoting the spread of sports spirit and culture, promoting the development of healthy sports, and can also create certain social and economic benefits. At the same time, we should also see the low popularization rate, insufficient propaganda, slow development and function of the network platform for sports cultural exchange between universities and enterprises, constantly overcome the limitations, and scientifically expand the effectiveness of the network platform, so that more universities and enterprises can form sports alliances, better play its strong application value and achieve self-promotion. A win-win situation between mental health and economic efficiency.

References

1. Han, Z.: Overview of Management Thought, pp. 130–138. Economic Management Publishing House, Beijing (2006)
2. Tian, X.: Reasons, problems and development path of marathon boom. J. Nanjing Inst. Phys. Educ. (Nat. Sci. Ed.) **15**(3), 88–92 (2016)
3. Qian, X., Tian, Y., Luo, B.: School sports stadiums and stadiums opening to the society: from the perspective of functional release of sports social organizations. J. Shenyang Inst. Phys. Educ. **38**(1), 39–51 (2019)
4. Lu, Y., Yu, Y.: Sports reform and national fitness in China: review, prospect and care for social facts. J. Shanghai Inst. Phys. Educ. **43**(1), 1–6 (2019)
5. Zhao, H.: Research on the construction of university sports network information service platform from the perspective of national fitness. J. Liaoning Norm. Univ. (Nat. Sci. Ed.) (6), 283–288 (2016)
6. Li, J.: The construction of harmonious sports in the new era - research on the sharing of sports resources between schools and communities. J. Shandong Inst. Phys. Educ. **26**(7), 93–96 (2010)
7. Li, Y., Zhang, W., Zhang, X., et al.: On the path choice of university sports resources sharing in China. J. Nanjing Inst. Phys. Educ. **24**(5), 104–107 (2010)
8. Notice of the State Administration of Sports on the Issue of the "Twelfth Five-Year Plan for the Development of Sports Undertakings" - Law Library - 110 Net, 28 March 2011. http://www.110.com/fagui/law_378406.html
9. Jing, L., Tang, L., Zhu, Y.: Research on the present situation and development strategy of community physical exercise in universities. Sports Cult. Guide (1), 136–140 (2015)
10. Town, L.: On the national fitness culture. Sports Cult. Guide (3), 35–40 (2015)

Micro-level Innovation Governance Ability Improvement Based on Internal Drive

Lingni Wan[1(✉)] and Fang Yang[2]

[1] Department of Public Administration, Wuhan Donghu University,
Wuhan, Hubei, China
317135275@qq.com
[2] Department of Business Administration, Wuhan Business University,
Wuhan, Hubei, China

Abstract. The purpose of this paper is to study the effective way to improve the ability of science and Technology Governance from the perspective of enhancing internal drive. Based on the report data of 31 Listed Companies in the chemical raw materials and chemical products manufacturing industry of Jiangsu Province from 2013 to 2017, this study uses Pearson analysis method to quantify the effect of three-level internal driving factors in scientific and technological innovation activities. The results show that the overall factors, cost factors and organizational factors are the weak links of internal drive. Conclusion: in order to improve the ability of science and technology governance, we need to build innovative organizational model, optimize the structure of senior management team, and deal with three groups of relationships in innovation activities. Gradually form the internal power of scientific and technological innovation of enterprises.

Keywords: Internal drive · Innovation governance ability · Promotion path

1 Introduction

1.1 Research Background

In today's complex and ever-changing economic environment, facing increasingly fierce market competition, enterprises must enhance their core competitiveness by improving their technological innovation ability if they want to maintain sustained and healthy development for a long time. This is because technological innovation ability is the soft power of enterprises, which can bring value to enterprises and promote their rapid growth.

1.2 Literature Review

Through combing the existing literature, it is found that most scholars currently focus on the concept of science and technology governance [1, 2]. Composition of science and technology management system [3, 4]; Scientific and technological governance ability [5, 6]; Evaluation index of scientific and technological governance ability [7, 8]; The path of scientific and technological governance [9, 10] and other aspects discussed

M. Atiquzzaman et al. (Eds.): BDCPS 2019, AISC 1117, pp. 499–505, 2020.
https://doi.org/10.1007/978-981-15-2568-1_68

the scientific and technological governance ability of enterprises. There are two deficiencies in the research: first, the research breadth is not enough. The current literature mainly studies the technological innovation driving and governance of enterprises from the external and macro levels. Second, the research content is not deep enough, lack of analysis on the mechanism of the internal driving factors of science and technology governance system, which affects the application of the research.

1.3 Research Program

Based on the microscopic governance ability of science and technology as the object, starting from the enterprise technology innovation drive, analyzes the role of internal factors in enterprise technology innovation mechanism, report data by PEARSON analysis method is used to calculate all kinds of drive factors on enterprise innovation activities within the force strength, find the weak link, targeted enterprise management science and technology innovation ability the ascension path.

2 The Function Mechanism of Internal Driving Factors in Micro-science and Technology System

Firstly, this section clarifies the internal factors that can affect the activities of scientific and technological innovation, then combs its mechanism, and finds its reflection in the financial data, which lays the foundation for the follow-up correlation research.

2.1 Three-Level Inside-Driving Factors of Scientific and Technological Innovation

According to strategic management theory, the internal driving factors of scientific and technological innovation can be divided into overall driving factors, functional driving factors and business driving factors. The overall driving factor is the highest driving level of an enterprise, which is embodied in the drive of entrepreneurship and the drive of incentive mechanism to stimulate innovation spirit. Functional driving factors arise from the middle management level of enterprises, including sales department, R&D department, production department, purchasing department, human resource department, financial department, etc. Business drivers are generated by the execution process of the grass-roots innovation department. For manufacturing enterprises, R&D department is the main execution department of scientific and technological innovation.

2.2 Progressive and Evolutionary Relations of Internal Driving Factors

The internal driving factors of enterprise's three-level scientific and technological innovation interact with each other and influence each other. These relationships form the driving force of enterprise's scientific and technological innovation and ultimately achieve the goal of scientific and technological innovation. The process of interaction and interaction first shows progressive relationship, secondly shows evolutionary relationship. See Fig. 1 below specifically:

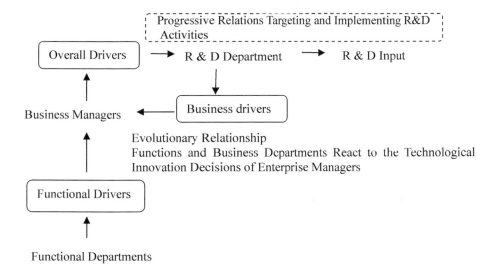

Fig. 1. The mechanism of internal driving factors in science and technology system

2.3 Explanation of Financial Parameters of Internal Driving Force

In order to reasonably calculate the relationship between enterprise internal driving factors and the intensity of scientific and technological innovation activities, this study uses financial reporting data of listed companies as parameters to quantify each factor. The overall driving factors are entrepreneurial spirit, and the financial parameters are: age, gender and educational background of senior managers. Functional driving factors include sales factors, production and procurement factors, financial factors, business factors and so on. Business drivers, namely R&D factors, are the number of R&D personnel and the proportion of R&D personnel. The larger the two parameters, the greater the role of R&D factors.

3 Analysis of the Driving Force of Technological Innovation in Chinese Enterprises

In this section, the financial report data of 31 listed companies were used to quantify the internal driving factors and the intensity of technological innovation activities of enterprises. SPSS software was used to study the correlation between the two, and to obtain the driving force intensity of various internal driving factors in the technological innovation activities of enterprises, so as to find the weak links driven by technological innovation of Chinese enterprises.

3.1 Sample Selection

In this paper, the research sample selection of 31 in Jiangsu province in 2013–2017 chemical raw materials and chemical products manufacturing industry listed company financial report data as sample, the sample data from the east wealth network by the data released in March 2018, the total number of samples is 131, 42, the effective sample of different sample form listed in the following Table 1:

Table 1. Sample composition list

Number	Year	Sample size	Percentage (%)
1	2013	1	2.38
2	2014	8	19.05
3	2015	13	30.95
4	2016	19	45.24
5	2017	1	2.38
Total		42	100

3.2   Analysis Method and Indicator Description

The research used SPPS software to analyze the correlation of PEARSON. Six types and ten financial parameters were selected to describe the strength of each factor. The specific indicators and the performance of each factor are listed in Table 2 below.

Table 2. A list of financial reporting indicators and their factor performance

Number	Indicator name and number	Financial parameters and numbers	Performance factors
1	Executives Indicators (G)	Average Age of Executives (G1), Proportion of Male Executives (G2), Proportion of Senior Executive Graduate Students and Professors (G3)	Overall drivers
2	R&D personnel Indicators (Y)	Number of R&D personnel (Y1), R&D personnel proportion (Y2)	Business drivers
3	Cost-income ratio Index (C)	Cost-income ratio (C1)	Production and procurement drivers
4	Sales Indicators (X)	Advance to Operating Revenue Ratio (X1), Total Operating Revenue (X2)	Sales Department Driven
5	Financing structure Index (R)	Asset liability ratio (R1)	Financial sector driving
6	Organizational Efficiency Index (E)	Inventory Turnover Rate (days) (E1)	Organizational Behavior Driven
7	R&D Input Index (K)	R&D Input Amount (K1), R&D Input Percentage of Operating Revenue (K2)	Science and Technology Innovation Intensity

3.3   Relevance Analysis Results

The PEARSON analysis of internal driving factors and technological innovation activities of enterprises is calculated by the bivariate correlation analysis method in SPSS software. The financial parameters have been tested by Shapiro-Wilk test. The results of the analysis are listed in Table 3 as follows:

Table 3. A list of relevance analysis results of inner driving factors of innovation

Number	Financial parameters	R&D input amount (K1)			R&D input percentage of operating revenue (K2)		
		Pearson correlation	Sig. (2-tailed)	N	Pearson correlation	Sig. (2-tailed)	N
G1	Average Age of Executives	.221	.160	42	−.248	.113	42
G2	Proportion of Male Executives	.161	.310	42	−.285	.067	42
G3	Proportion of Senior Executive Graduate Students and Professors	−.007	.963	42	.488**	.001	42
Y1	Number of R&D personnel	.723**	.000	42	−.180	.254	42
Y2	R&D personnel proportion	.067	.674	42	.514**	.001	42
C1	Cost-income ratio (%)	–	–	–	−.647**	.000	42
X1	Advance to Operating Revenue Ratio	–	–	–	−.441**	.003	42
X2	Total Operating Revenue (Billion Yuan)	.600**	.000	42	–	–	–
R1	Asset liability ratio (%)	.153	.334	42	−.441**	.003	42
E1	Inventory Turnover Rate (days)	.154	.331	42	.213	.175	42

**Correlation is significant at the 0.01 level (1-tailed).
*Correlation is significant at the 0.05 level (1-tailed).

As can be seen from the table above, the number of R&D personnel (Y1) and total operating income (X2) are related to the amount of R&D investment (K1). The proportion of highly educated personnel (G3), the proportion of R&D personnel (Y2), the cost-income ratio (C1), the ratio of advance receipts to income (X2), the ratio of assets to liabilities (R1) and the ratio of R&D investment to operating income (K2) are related. However, the 2 parameters of executive average age (G1) and executive sex ratio (G2) did not correlate with R&D input indicators (K1 or K2). The analysis results show that the cost-income ratio has a negative correlation with K2 parameters. According to the discussion in the second part, cost and procurement factors should have a positive impact on innovation input. In addition, organizational efficiency indicators have no correlation with R&D input indicators (K1 or K2). In addition, in the process of data collection and collation, it is found that the degree of capitalization of sample R&D costs is very low. Only 5 of the 31 enterprises have the process of R&D investment capitalization. Except the Hongda Xingye Co., Ltd., the capitalization ratio of R&D expenditure of other enterprises in each year has not exceeded 60%.

3.4 Weak Links in the Driving Force of Scientific and Technological Innovation of Chinese Enterprises

Firstly, the age of senior executives has no impact on the investment of enterprises in scientific and technological innovation, and the decision-making team of Chinese enterprises is relatively aging. The average age of the executive team in the sample ranged from 40 to 53. Secondly, there is no correlation between the gender structure parameters of the senior management team and the R&D investment of the enterprise, and the gender imbalance exists in the decision-making team of the enterprise in China. In the sample, more than 70% of the executive team members were male. Finally, cost factors have negative driving force on innovation input. Finally, the financial parameters of organizational efficiency index have no correlation with R&D investment, that is to say, the operation mode of Chinese enterprises centering on supply, sales and inventory activities fails to effectively drive scientific and technological innovation.

4 Conclusions

In view of the above problems, this paper proposes to lay two foundations, balance the three groups of relations, and form a way to improve the scientific and technological governance ability of business promoting innovation and innovation promoting a virtuous cycle of business.

(1) Laying a good organizational foundation and building an organizational structure with scientific and technological innovation as the core
(2) Lay a good foundation for power and diversify the senior management team structure in terms of age and gender
(3) Balance the relationship between innovation and norms and effectively guide the direction of scientific and technological innovation
(4) Balancing the relationship of resource allocation in innovation activities, R&D investment should be classified and managed.
(5) Balancing the relationship between innovation and cost to improve the effect of scientific and technological innovation.

References

1. Chen, T., Feng, F.: Dynamic evaluation and improvement path of governance capacity of China's regional science and technology innovation system. J. Dalian Univ. Technol. (Soc. Sci. Ed.) 37(1), 44–50 (2016). (in Chinese)
2. Sun, F.: A comparative study of science and technology governance at home and abroad. Sci. Dev. 103(6), 34–44 (2017). (in Chinese)
3. Wu, J., Sun, R., Ma, L.: Modernization of science and technology governance system: concept, characteristics and challenges. Sci. Technol. Manag. 36(8), 3–9 (2015). (in Chinese)

4. Xibao, Yu, C., Jiang, Z.: Framework and mechanism of technology collaborative governance – based on the perspective of value structure process relationship. Sci. Res. **34**(11), 1615–1624, 1735 (2016). (in Chinese)
5. Zhong, C., Chen, Y., Zhang, C., Huang, X.: Research on the influence of technological innovation governance capability on the performance of technological innovation governance. Sci. Technol. Prog. Countermeas. **18**(8), 1–8 (2019). (in Chinese)
6. Yin, Z.: From science and technology management to science and technology governance. People's BBS (7), 56–57 (2019). (in Chinese)
7. Li, Y.: Evaluation of scientific and technological innovation ability of internal colleges in universities from the perspective of innovation drive. J. Fuzhou Univ. (Philos. Soc. Sci. Ed.) **31**(01), 34–41 (2017). (in Chinese)
8. Wu, N., Ma, Z., Gu, G.: Empirical research on the performance evaluation of cooperative R&D of small and micro technological enterprises: based on the perspective of resource integration. Sci. Technol. Prog. Countermeas. **33**(24), 109–115 (2016). (in Chinese)
9. Ding, Z., Jinhua: The way to improve the governance ability of scientific and technological innovation in Qingdao. J. Party School CPC Qingdao Munic. Comm. Qingdao Adm. Inst. **256**(4), 46–49 (2019). (in Chinese)
10. Jie, K., Yong, Y., Hu, H.: Research on evaluation framework system of science and technology innovation policy based on the whole process. Sci. Technol. Manag. Res. **18**(02), 25–30 (2019). (in Chinese)

WSN Coverage Optimization Based on Artificial Fish Swarm Algorithm

Changlin He[✉]

Information Technology Service Center, Hexi University,
Zhangye 734000, Gansu, China
40632622@qq.com

Abstract. To solve the problem of unreasonable random distribution of sensor nodes and low network coverage in sensor networks, a coverage optimization method based on artificial fish swarm algorithm is proposed. Firstly, the current research status of WSN coverage was analyzed, the node coverage and regional coverage in WSN on the basis were analyzed, the corresponding mathematical model was established, the WSN coverage optimization program in view of the artificial fish swarm was obtained. Finally, MATLAB was used for the simulation experiment, and the simulation results showed that the introduction of this method improved the node coverage in WSN effectively, the coverage area was huger at the same amount of nodes. Moreover, the algorithm can get the optimal solution in the global scope, and reach the better network coverage optimization effect with less sensor nodes, and the number of iterations was decreased significantly.

Keywords: Wireless sensor networks (WSN) · Artificial fish swarm algorithm · Coverage optimization

1 Introduction

As the multidisciplinary leading research field, WSN was applied in the military field at the early time. Along with the constant technological development, the WSN has been widely used in the civilian areas currently. The research and application has becoming a vital topic increasingly at the current information science, while the coverage optimization problem of WSN is an important research direction for WSN. However, it is difficult and even impossible to replace and maintain the sensor or supplement the electricity of sensor for the WSN is deployed in the battlefield, field, underwater and other complicated environment generally, therefore, how to maximize the coverage of WSN with the minimal number of nodes and strive to extend the life cycle of the network becomes one of the key problems for the research of sensor network [1]. The network coverage is the primary problem for the research of WSN and an important index to measure the service quality of WSN [2]. Solving the coverage optimization problem of WSN well can have a profound impact on the development of WSN technology, therefore, the research on the coverage optimization of WSN has the vital scientific research value.

© Springer Nature Singapore Pte Ltd. 2020
M. Atiquzzaman et al. (Eds.): BDCPS 2019, AISC 1117, pp. 506–513, 2020.
https://doi.org/10.1007/978-981-15-2568-1_69

The early coverage optimization algorithm was developed on the basis of graph theory and test generally [3, 4] with some deficiencies. The graph theory algorithm assumes that the sensor node can be found in any monitoring area and it is inconsistent with the actual situation. The detection algorithm cannot ensure the complete coverage of network, and it is only suitable for scaled WSN. During the past years many researchers take advantage of the high search speed, powerful processing ability and other advantages of these algorithms, introduced into the solution of WSN coverage problem and achieved the great results along with the development of intelligent optimization algorithm. The introduction of these algorithms improves the network coverage, saves the network energy consumption and improves the overall performance of WSN to a certain extent, while they also have some deficiencies which cannot be overcome. Literature [5] applied the standard genetic algorithm into the coverage optimization of WSN, and the research result showed that although there was certain validity for the method, it was easy to fall into the premature convergence. Literature [6] used the method to settle the mathematical model with established WSN to get the optimal coverage set of node in the network. The algorithm can improve the network coverage, while it was easy to fall into the local optimal value sometimes, the global optimal value cannot be obtained, thus influencing the coverage effect. Literature [7] mixed the chaotic motion and particle swarm algorithm. It overcame the slow solving speed, while it was no block to go down in the local extreme value. Meanwhile, the complexity of the algorithm was higher, and the solving process was very complex. Literature [8] proposed a hybrid genetic algorithm, the exact solution can be solved rapidly, while the energy of network node was not considered, and the calculation was complex relatively, and it was not conducive to sensor node with limited energy.

2   Mathematical Model of WSN Coverage

The detection range of each node in WSN is limited, and the insufficient number of node can lead to the perception shadow and blind spot in the network due to the unreasonable node distribution under the random distribution mode. The huge redundancy perception nodes must be invested to get the better node density in order to realize the overall coverage of target region, thus improving the cost of network greatly and bring the difficulty for the later maintenance. Therefore, the research of WSN coverage optimization problem can rationalize the layout of node, fully play the role of each node, improve the perception ability, information collection ability and survival ability of the network effectively and enhance the QoS of the network.

Controlling the distribution of mobile nodes via the coverage optimization of network can eliminate the shadow and blind spot in the detection region effectively. However, the large-scale node adjustment means the huge network maintenance cost for the sensor network containing a large number of nodes. Therefore, how to complete the distribution optimization of sensor network node at smaller cost has become the key problem for the later management.

The optimal coverage of WSN refers to optimize the configuration of various resources in WSN via the rational distribution of network sensor node and other means under the condition of limited sensor network node energy, wireless network communication bandwidth, network computing capacity and other resources, thus improving various QoS.

The task of node distribution optimization is to conduct the dynamic adjustment on the location of node via the node mobility, and use the minimum sensor nodes to get the larger coverage under the premise of maintaining the network connectivity. Meanwhile, the network connectivity can be guaranteed at full coverage when the communication radius among the nodes is maintained at twice of the perceived radius. Therefore, WSN coverage problem can be defined as follows: supposing the sensor node collection $C = \{c_1, c_2, c_3, \cdots, c_N\}$ is existed, a subset of $|C'|$ is sought to maximize the coverage of $R_{cov}(C')$, and the number of node of $|C'|$ is the minimum.

2.1 Node Coverage

Supposing the monitoring region A is two-dimensional plane, N Sensors in the area is covered by sensors with the same parameters. The coordinate of each node is known and the effective monitoring radius is r, the sensor node collection can be expressed as $C = \{c_1, c_2, c_3, \cdots, c_N\}$, wherein $c_i = \{x_i, y_i, r\}$ is used to represent the circular region of node c_i with the center of (x_i, y_i) and the monitoring radius of r. Assuming the monitoring region A is composed by $m \times n$ pixel regions, the covered event of pixel point (x, y) by the sensor node i is defined as r_i, and the probability $P\{r_i\}$ for the occurrence of the event is that $P_{cov}(x, y, c_i)$ of pixel point (x, y) covered by the sensor node i.

$$P\{r_i\} = P_{cov}(x, y, c_i) = \begin{cases} 1, & f(x - x_i)^2 + (y - y_i)^2 \leq r^2 \\ 0, & otherwise \end{cases} \tag{1}$$

Equation (1) indicates that when the distance between the pixel point (x, y) to the sensor node i is smaller than the sensing range r, the pixel point (x, y) is regarded as covered by the sensor i.

$$P\{\bar{r}_1\} = 1 - P\{r_i\} = 1 - P_{cov}(x, y, c_i) \tag{2}$$

In Eq. (2), $\bar{r}_i$ is the supplement of r_i, representing the event that the Sensor node i does not contain the pixel point (x, y). If they are unrelated, the following relationship exists:

$$P\{r_i \cup r_j\} = 1 - P\{\bar{r}_i \cap \bar{r}_j\} = 1 - P\{\bar{r}_i\} \bullet P\{\bar{r}_j\} \tag{3}$$

As long as the pixel point (x, y) is covered by one node from the node set, then we can assume that the pixel point (x, y) are covered by the nodes. Therefore, the probability of pixel point (x, y) covered by the node set is the union set of r_i. Assuming all random events r_i are independent, the coverage of node set C can be calculated according to the following formula:

$$P_{cov}(x, y, c) = P\{_{i=1}^{N} Y r_i\} = 1 - P\{_{i=1}^{N} \bar{r}_i\} = 1 - \prod_{i=1}^{N} [1 - P_{cov}(x, y, c_i)] \tag{4}$$

Equation (4) indicates that if the pixel point (x, y) is not covered by all nodes, the pixel point (x, y) is the uncovered point, otherwise, it shall be regarded that the pixel point (x, y) is covered by the node set.

2.2 Regional Coverage

There are $m \times n$ pixels in the monitoring region A, and the area at each pixel can be expressed as $\Delta x \times \Delta y$, whether each pixel is covered can be measured via the node set coverage $P_{cov}(x, y, c_i)$, the regional coverage $R_{area}(C)$ of node set C is defined as the ratio between the coverage area $A_{area}(C)$ of node set C and the total area A of monitoring region A, namely:

$$R_{area}(C) = \frac{A_{area}(C)}{A_s} = \frac{\sum_x^m \sum_y^n P_{cov}(x, y, c)}{m \times n} \tag{5}$$

3 WSN Coverage Based on Fish Swarm Algorithm

3.1 Artificial Fish Swarm Algorithm

Artificial fish swarm algorithm is an embodiment of bionic optimization algorithm raised by Dr. Li and others in the academic paper of A Kind of Optimizing Mode based on Animal Autonomy published in 2002 [9]. Several typical behaviors applicable for the fish swarm algorithm were summarized and extracted in the paper, the article held that there were following typical behaviors of fish school during predation: foraging behavior, gathering behavior, following behavior and random behavior as well. The fish swarm algorithm was one type of effective optimizing algorithm featured with the concurrency, simplicity, quick away from the local extremum and fast optimizing speed. The algorithm was improved on the basis by many scholars subsequently [10].

3.2 WSN Coverage Based on Fish Swarm Algorithm

WSN coverage was optimized with artificial fish swarm algorithm. The algorithm was as follows:

Initialization: the sensor was replaced by the artificial fish, each artificial fish represented a sensor, and the coordinate of artificial fish was the location for the placed sensor.

Foraging: the position of fish school was set as $A(x, y)$ at present, the distance with the nearest artificial fish was d_{min}, one point A of artificial fish was determined randomly in the field of vision. If the minimum distance from B point to other artificial fish except A point was more than d_{min}, it shall be regarded that the food concentration at B point was better than that in A point, and the current artificial fish shall forward towards B point. If the condition was not met, one step shall be moved randomly. It tried to reduce the overlapping among the sensors and maximize the covering region via the foraging behavior.

Following behavior: the distance between the current artificial fish AF_A to the nearest artificial fish AF_B was set as d_{min}. If $d_{min} > 2 \times$ detection distance, one step shall be forward towards the direction of AF_B. If $d_{min} < \left(\sqrt{3}/_2 \right) \times$ detection distance, one step shall be forward towards the direction opposite to AF_B. If the condition was not met, one step shall be moved randomly. The distance between the sensor can be judged via the following behavior. If there was any blind point between the adjacent sensor, let the approach; if the distance between the adjacent sensor was too close, pull them away. The blind region or overlapping of the sensor coverage can be solved effectively via the following behavior.

Random behavior: artificial fish's moving one step randomly at the random step length within the field of vision was the default behavior for the following behavior and foraging behavior.

Group bulletin board: the location information of each artificial fish was written into the bulletin board and saved after the following behavior and foraging behavior were implemented in turn. In this way, all artificial fishes' location information were recorded in the bulletin board after an iteration was completed.

Judgement on the condition for algorithm ending: the current target region was set as divided into $m \times n$ pixel points, and the distance d from each pixel point to the artificial fish shall be calculated in turn. Once d was smaller than the detection distance, it shall be regarded as covered, and it was set as 1. Finally, 1 was accumulated, the total amount of covered pixel points were obtained, and the current coverage was calculated with Eq. (5). If the coverage requirement (threshold) was met, the iteration was terminated. Otherwise, the next iteration would be continued in (2) until the coverage requirement was met.

4 Experimental Simulation

4.1 Initial Condition Setting and Environment for Simulation

Assuming WSN monitoring region was 100×100, there were 60 sensor distribution nodes, the perceptive radius of sensor was 10. The method to improve WSN coverage in the monitoring area is to use artificial fish swarm algorithm. The selected parameters were as the following below: total number of artificial fish N = 60, Visual = 10, Step = 3, Try_number = 2, T-value = 0.9. 50 times were operated in a row under MATLAB R2016a environment, and the average was obtained for the result.

4.2 Experimental Result and Analysis

The experimental results were shown in Figs. 1 and 2. The center of each circle in the diagram represented the location of sensor node, and the circular region represented the perceptive range of a sensor node. The artificial fish swarm algorithm was used to figure out the sensor node coverage program in Fig. 1, and the obtained WSN node coverage result was shown in Fig. 2. Seen from Fig. 2, WSN node distribution was uniform relatively, and the redundancy between nodes was less relatively. The experimental result showed that when the coverage reached over 90%, the average iteration number of 59 was needed for the artificial fish swarm algorithm. The artificial fish swarm algorithm realized the WSN node coverage optimization goal well, it was a kind of effective WSN node coverage optimizing method, and the WSN coverage optimization rate can be improved.

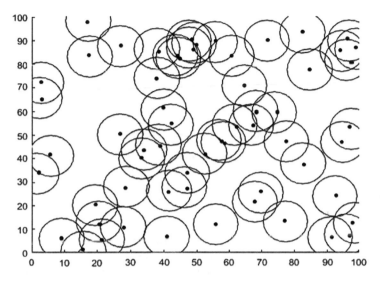

Fig. 1. Randomly deployed sensor node distribution

GA and PSO algorithms were taken for the comparison test under the same simulation environment in order to note the superiority of this method to settle the WSN coverage optimization problem, and the test results were shown in Fig. 3.

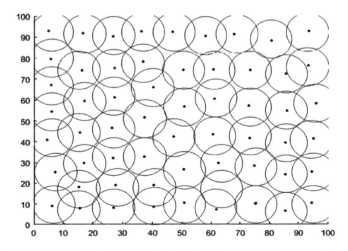

Fig. 2. Distribution of sensor nodes based on artificial fish swarm algorithm

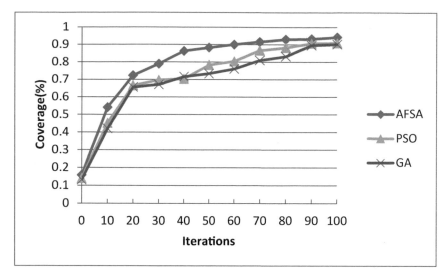

Fig. 3. Coverage comparison

Seen from Fig. 3, GA and PSO covered the sensor node, and the coverage can reach the threshold required in the experiment within the limited number of iterations. However, 90% of coverage can be reached at about 60 times of average iteration of ASFA algorithm, while 90 times were needed for PSO and GA algorithm. In order to realize the overall optimization of WSN node distribution more effectively. It is a good choice to use artificial fish swarm algorithm. The maximized coverage can be realized within the shorter time, and it was featured with broader using scope.

5   Conclusions and Future Work

The sensor node in WSN is featured with the fixed sensing radius. If any point in the monitoring region is located in the sensing radius of the sensor, the point is considered to be covered by the sensor. There is the huge application prospect of the main technology of WSN in each field, and the network coverage problem is one of key problems of WSN. In order to resolve the network coverage problem, a WSN coverage optimization algorithm based on artificial fish swarm algorithm is proposed. ASFA algorithm was applied in the coverage optimization of WSN, and it was compared with the coverage optimization based on GA and PSO. The experimental results showed that WSN coverage optimization design based on ASFA can realize the distribution optimization of sensor node with smaller cost, thus effectively improving the network coverage and enhancing the network effectiveness and survival ability. It was featured with the better optimization effect and faster speed when compared with GA and PSO algorithm.

References

1. Akyildiz, I.F., Su, W., Sankarasubramaniam, Y., et al.: A survey on sensor networks. IEEE Commun. Mag. **40**(8), 102–114 (2002)
2. Dimple, B.: Maximum coverage heuristics (MCH) for target coverage problem in wireless sensor network. In: 2014 IEEE International Advance Computing Conference (IACC), pp. 300–305. IEEE (2014)
3. Ren, Y., Zhang, S.D., Zhang, H.K.: Theories and algorithms of coverage control for wireless sensor networks. J. Softw. **17**(3), 422–433 (2006)
4. Wang, X.Q., Yang, Y.T., Sun, T., et al.: Research on the gird based coverage problem in wireless sensor networks. Comput. Sci. **33**(11), 38–39, 78 (2006)
5. Zeng, Y.L., Chen, J., Zheng, J.H.: Genetic algorithm-based sensor network coverage-enhancing approach. Comput. Eng. Appl. **45**(11), 89–91 (2009)
6. Huang, Y.Y., Li, K.Q.: Coverage optimization of wireless sensor networks based on artificial fish swarm algorithm. Appl. Res. Comput. **30**(2), 554–556 (2013)
7. Liu, W.T., Fan, Z.Y.: Coverage optimization of wireless sensor networks based on chaos particle swarm algorithm. J. Comput. Appl. **31**(2), 338–340, 361 (2011)
8. Lin, M.J., Su, C.H., Wang, Y.: Study on optimization algorithms in wireless sensor networks. Comput. Simul. **28**(3), 178–181 (2011)
9. Li, X.L., Shao, Z.J., Qian, J.X.: An optimizing method based on autonomous animates: fish-swarm algorithm. Syst. Eng. Theory Pract. **11**, 32–38 (2002)
10. Jinag, M.Y., Yuan, D.F.: Artificial fish school algorithm and application, pp. 261–269. Science Press, Beijing (2012)

Organizational Architecture of Smart Urban Planning Simulation Model in Big Data Environment

Minghui Long[✉]

Wuhan Donghu University, Wuhan, Hubei, China
642604051@qq.com

Abstract. With the deepening of the pilot work of smart cities, the traditional two-dimensional vector data lacks intuition in expression, and can no longer meet the needs of scientific and refined work. At present, most top-level planning of smart cities tend to pay more attention to information service construction, neglecting the ultimate demand of urban development centered on the economical layout of urban space and the rational allocation of factor resources. In view of this, based on the design of urban planning and simulation model under big data environment, this paper explores the inherent law of urban operation through big data analysis of the problems existing in urban operation, and shapes the large data space of smart city according to data dimension, business system set and data exchange catalogue to effectively improve the scientific level of urban planning and construction of smart cities, provide scientific prediction for sustainable urban development, put forward reasonable intervention measures and public policies through planning means, and alleviate or improve the urban living environment, traffic congestion, pollution and other urban problems.

Keywords: Smart City · Top-level design · Planning and simulation model · Big data

Based on the application of new generation information technology such as Internet of Things and Cloud Computing, as well as the tools and methods such as Wikipedia, Social Network, Fab Lab, Living Lab and Integrated Method, Smart City creates an ecology conducive to innovation emergence, realizes comprehensive and thorough perception, ubiquitous interconnection, intelligent integration and sustainable innovation characterized by user innovation, open innovation, mass innovation and collaborative innovation. At present, there are two driving forces for the development of smart cities. One force is represented by Internet of Things technology, cloud computing technology and mobile Internet ecosystem. The other force is an open urban innovative ecosystem that has matured slowly under the humane conditions of modern high-knowledge society. One of them is as the influence factor of modern innovative high-tech, the other is as the influence factor of modern innovative new economy. Innovation is an important driving force for the development of smart cities. The combination of VR virtual reality technology, AR augmented reality technology and three-dimensional simulation technology with the smart city can truly integrate the operational carrier of smart city.

© Springer Nature Singapore Pte Ltd. 2020
M. Atiquzzaman et al. (Eds.): BDCPS 2019, AISC 1117, pp. 514–521, 2020.
https://doi.org/10.1007/978-981-15-2568-1_70

1 Foundation of the Building Smart City

1.1 Establish the Management System of Smart City Construction

The construction of an innovative the smart city requires the cooperation of a new generation of information technology and innovative technology, including VR virtual reality technology, AR augmented reality technology and three-dimensional simulation technology. Now the integration and progress of information and efficient communication technology has gradually broken down the barriers of data sharing at the level of information and knowledge, and the boundaries of innovation have become more infinite. It is based on this pattern that the new forms of innovation promote the disappearance of boundaries at all levels of social organizations. Innovative forms of expression are transformed from different modes of production and manufacturing to service-oriented categories, which promotes the transformation of industrial performance, the management and operation mode of national public organs and the urban presentation form in the same direction [1].

Intelligent processing is not the end of the process of information use. Smart cities should also have the open application ability of information. It can send all kinds of information processed to the information demanders through the network or operate the control terminals directly, so as to complete the Value-added Utilization of information. The application of information in smart cities should be characterized by openness, not only by the unified control and distribution of information by the government or urban management departments, but also by the establishment of an open information application platform so that individuals, enterprises and the other individuals can contribute information to the system and enable individuals to exchange the information through the system of smart cities. This will make full use of the existing capabilities of the system, and enrich the information resources of smart cities and is conducive to promoting the birth of new business models.

1.2 Infrastructure Construction of Smart City Construction

It needs a broad foundation of the Internet of Things and the interconnection of all the things. Information is more open and sharable to a greater extent, rather than in an isolated state. It is also necessary to use artificial intelligence to analyze all kinds of data collected in cloud computing centers, to conduct in-depth mining and effective integration and utilization. Smart cities with information adaptability need to have information adaptability that fits well with social knowledge and information demands. Through the establishment of real-time dynamic response information interactive service at present, the new form of the smart city is constructed in an open form with users as the core and people-oriented source. The relevant facilities, target points and landmark areas built in the city can be labeled and explained in detail in the form of AR augmented reality. The information can be linked to the dynamic change information update of government departments at any time. By building such an information adaptability form, we can provide real-time, efficient and accurate information services for urban management, life and services. The formation of the smart city is mainly through pioneering innovative technology and manifestations (including virtual reality technology and three-dimensional simulation technology), and collaborative innovative information adaptation technology (augmented reality technology), to achieve the degree of convergence of intellectual attributes.

Wide coverage of information perception network is the basis of the smart city. Any city has a huge amount of information resources. In order to get the city information more timely and comprehensively and to judge the city situation more accurately, the central system of a smart city needs to have the ability to exchange all kinds of elements of the city. The information perception network of smart city should cover all dimensions of time, space and object, and can collect information of different attributes, forms and densities. The development of Internet of Things technology has provided more powerful help for information collection in smart cities. Of course, "extensive coverage" does not mean to collect all-round information in every corner of the city, which is neither possible nor necessary. The information collection system of the smart cities should be guided by moderate demand of the system, and excessive pursuit of comprehensive coverage will increase costs and affect efficiency [2].

2 Organizational Framework of the Smart City Simulation Model

Infrastructure support system layer includes the integration of basic hardware and software. The basic hardware includes computer room, wired and wireless network, servers, terminals, handheld and peripherals, instruments and sensors, and the basic software includes operating system, database software, middleware software, backup system, etc. The Smart cities should make unified planning and intensive construction of traditional and decentralized urban networks, computer rooms, computing and storage resources, information security and basic data centers.

Material Link Perception Layer. It provides the ability of intelligent perception and execution of the environment. Through perception equipment, execution equipment and transmission network, it can identify, collect, monitor and control the elements of infrastructure, environment, equipment and personnel in the city.

Network Communication Layer. It provides network communication infrastructure consisting of large capacity, high bandwidth, high reliability optical network and city-wide coverage wireless broadband access network for the smart cities. It includes core transmission network with Internet, telecommunication network, radio and television network as the main body, cellular wireless network providing wireless access services, and some special networks such as cluster private network.

Computing and Storage Layer. It includes software resources, computing resources and storage resources, which can help to provide data storage and computing and related software environment resources for the smart cities and to protect the upper level of the relevant data needs [3].

Smart Application Layer. Intelligence application and Application Integration based on industry or field, such as the smart government, smart transportation, smart public service, smart medical treatment, smart park, smart community, smart tourism, etc., are established on the perception layer, network communication layer, computing and storage layer, data and service integration layer, providing information applications and services for the public, enterprise users, urban management decision-making users, etc. [4].

Safety Guarantee System. Safeguarding data and services are the important pre-requisite for the construction of "the smart city". While providing intelligent services, the smart cities inevitably need to acquire and store all kinds of information about urban management and service objects, which contains a large number of information with strong privacy, and need to be protected. In the practical application, the technical means should ensure the requirement of efficient and stable operation of "the smart city" system, strengthen the implementation of security provisions and measures, and improve the infrastructure of the "the smart city" information system. Build a unified security platform for the smart city, realize unified entrance, authentication, authorization and log recording, involving all horizontal levels.

Operation and Maintenance Management System. Based on various urban resources and collected data, the suppliers and service objects of urban services should be organically linked through the technical means of "smart city", which provides an integrated operation and maintenance management mechanism for smart city, involving all horizontal levels [5].

Construction Management System. The construction of the smart city should break the traditional service mode divided by administration and industry, implement the integration of government administration and one-stop intelligent application of service, actively promote the improvement of management and service level, improve the satisfaction of urban residents, the competitiveness of enterprises and the level of government administration.

3 Application Scenario of the Smart City Simulation Model

The Smart city's information perception is based on a variety of information networks, and "deep interconnection" requires multiple networks to form effective connections, realize the mutual access of information and the scheduling operation of access equipment, and realize the integration and three-dimensional information resources. In smart cities, we can also see that connecting multiple separate and independent networks into interconnected networks can greatly increase the degree of information interaction and enhance the value of the network to all members, thus significantly enhancing the overall value of the network, and forming a stronger driving force, attracting more elements to join the network, forming positive feedback of network node expansion and information increment in smart city. When natural disasters such as earthquakes, floods and typhoons occur, or when man-made disasters such as trampling and terrorist attacks occur, urban emergency simulation can simulate the best emergency plan in the first time, give rescue guidance in the shortest time, and strive for more favorable space for rescue and improve rescue efficiency. With the help of the new generation of information technology such as Internet of Things and Cloud Computing, physical infrastructure, information infrastructure, social infrastructure and commercial infrastructure in cities can be connected by means of perception, interconnection of things and intellectualization. On the basis of policy support and complete infrastructure, the application scenarios of the smart cities are becoming more and more abundant [6].

3.1   Urban Construction Planning Simulation

The purpose of "collaborative sharing" of the cities is to break these barriers and form a unified urban resource system so that "resource isolated island" and "application isolated island" will no longer appear in cities. In the collaborative sharing the smart city, any application link can start the related application after authorization and operate its application link, so that all kinds of resources can give full play to their maximum value according to the needs of the system. So that, it can simulate the plan and effect more quickly and truly, and check whether the plan meets the planning requirements and building standards according to the evaluation criteria of construction projects.

Based on the architecture of SOA and Web Service technology, open standard interfaces such as network processing service, network element service, network map service, sensor observation service and sensor planning service of Open Geographic Information Alliance are used to complete the collaboration of subsystems through directory service. Integration system integrated application ability: the smart city integrated decision-making platform contains many application subsystems. In system function design, it fully integrates the functions, modes, designs, exhibitions, resources of each subsystem to achieve the unity, consistency and integrity of each subsystem. Strengthen the independent ability of each subsystem: By adopting the service-oriented architecture and using the intelligent decision center to provide event-oriented, model-oriented, data and decision-making service management interface. The interaction channel of each subsystem based on the intelligent decision center is established, which greatly reduces the coupling of each subsystem [7].

3.2   Urban Real-Time Simulation

The Smart city has a huge and complex information system, which is the basis of its decision-making and control. In order to truly realize "smart", the city also needs to show the ability to process the vast amount of information intelligently. This requires the system to analyze the data according to the constantly triggered needs, extract the required knowledge, make judgments and predictions independently, so as to achieve "smart". In order to realize intelligent decision-making and give control instructions to the corresponding executing equipment, the ability of self-learning is also needed in this process. Intelligent processing is shown in the macroscopic value-added of information extraction, that is, after information is processed and transformed within the system, its form should be transformed, become more comprehensive, more specific and easier to use, and the value of information has been enhanced. In technology, the new information technology application mode represented by cloud computing is a powerful support for intelligent processing. Establish a digital virtual city synchronized with the current city, collect dynamic data from all aspects of the city with real-time monitoring, and establish a city management framework based on the Internet of Things and services. Virtual city can experience the existing road traffic system in real time from any angle and observation scene. It can integrate simulation in virtual the city, carry out re-planning and upgrading, optimize the level of urban governance, and make the city "smarter" [8].

3.3   The Smart Community

The Smart community is based on community residents as the core of service, starting from people's well-being. It aims to provide safe, efficient and convenient intelligent management and services for owners by creating the smart community to provide convenient services for people, and to fully meet the owners' survival, development and living needs. The Smart community applications have been reflected in all aspects of people's lives. The Smart buildings and the smart homes further extradite intelligent applications from the community to the family. For example, the face recognition entrance guard in smart community is just to solve the problem of the convenient access to residential industry, which has greatly made the wisdom of residents' life and improved the quality of life.

3.4   Smart Transport

On the basis of smart transportation, smart transportation makes full use of the Internet of things, cloud computing, Internet and other high-tech to gather traffic information in the field of transportation, and provides management and control the support for all aspects of the field of transportation and the whole process of traffic construction management, so that the transportation system has the ability of perception, analysis, and control in the regional, urban and even the larger space-time range. For example, traffic monitoring and operation system can use Internet of Things technology to complete traffic control and command of multiple intersections and key areas under the unified portal, and solve the information fusion and coordination problems of urban traffic control, command and congestion management; the traffic light power monitoring system monitors the working status of traffic light power supply or emergency power supply system at intersections, and reports the operation status in real time. When a failure occurs, it reports to the cloud management platform in time to deal with it quickly, which improves the image of the government [9].

3.5   Smart Sanitation

Smart sanitation relies on the Internet of Things and Internet technology to manage people, vehicles, things involved in sanitation management in the real time, which is an important guarantee to improve the quality of sanitation. For example, smart toilets, through sensors to achieve toilet detection, environmental monitoring, flow statistics and other functions, to solve the current difficulties in finding public toilets, poor environment, and stacking, management problems; The smart sanitation monitoring platform can locate the location of sanitation workers, and it has SOS alarm function and monitoring equipment, which can collect real-time acquisition of video information of sanitation facilities. It can also realize the functions of electronic scheduling, trajectory playback and event reporting, in order to improve the quality of sanitation operation and reduce the cost of sanitation operation [10].

4   Conclusions

To some extent, the level of urban development represents the comprehensive strength. It is imperative to build a long-term sustainable development goal of a city. The continuous iteration of urban model provides quantitative methodological support for the study of geography, planning, the management and other fields. It is mainly reflected in the deep excavation of the operation mechanism of urban factors through the calculation model, and the refinement of the analysis results through rich visualization means.

From the perspective of urban thematic activities, this paper expounds the construction idea and organizational structure of the urban simulation model based on element interaction activities, and takes urban problems as the guidance, and integrates the elements system of various industries from the perspective of system theory. On this basis, through the acquisition, integration and realization of the system elements information, simulation and dynamic collaboration of industrial, transportation, community, culture, ecology and other elements of urban resources, focusing on the excavation and analysis of large data of residents' activities, simulation modeling and practical scenario analysis of the quality of urban spatial development are carried out. Supported by the urban data perception, storage and management platform, a smart urban planning simulation application platform is constructed, which integrates urban spatial elements simulation system, the urban resources optimization allocation simulation system, urban spatial development policy evaluation system, and urban operation dynamic simulation system, so as to make urban planning and construction more scientific, systematic and provide a reliable basis for the efficient operation of urban planning.

Acknowledgements. This work was supported by the grants from Wuhan Donghu University School level educational research project (2019) No. 25 Document, 190008.

References

1. Cao, Y., Zhen, F.: Organizational framework of smart city simulation model. Technol. Guide (9), (2018). (in Chinese)
2. Zhuang, X.: Simulation of smart city planning model under big data. Modern Electron. Technol. (6), (2018). (in Chinese)
3. Wan, B.: Urban simulation helps to build a new smart city. Chin. Constr. Informatiz. (11), (2018). (in Chinese)
4. Fan, Q.: The practical significance of urban simulation. Chin. Constr. Informatiz. (7), (2018). (in Chinese)
5. Cao, Y., Zhang, S., Zhen, F.: Research on organizational structure and standard system of smart city planning simulation model. Constr. Technol. (7), (2017). (in Chinese)
6. Wu, P.: Research on the evaluation of the development potential of information service in smart city. Xiang Tan Univ. (12), 1–3 (2017). (in Chinese)

7. Yu, H.: Application research of intelligent city information service based on big data theory: a case study of Qiqihar City in Heilongjiang Province. Electron. Commer. (05), (2019). (in Chinese)
8. Guo, Y.: Application of big data and smart city technology in urban and rural planning. Build. Mater. Decor. (05), (2019). (in Chinese)
9. Zhang, Q.: Research on the practice of 3D aided design in complex urban planning. Intell. Build. Intell. City (08), (2018). (in Chinese)
10. Xiao, Y., Deng, Q.: Application of virtual reality technology in urban planning. Value Eng. (06), (2015). (in Chinese)

On Electric Power Consumed and Economic Development in Fujian Province Under the Background of Big Data

Zhenglong Leng[1], Yujing Zhang[2(✉)], and Changyong Lin[3]

[1] State Grid Fujian Electric Power Company, Fuzhou 350001, China
[2] Beijing SGITG-Accenture Information Technology Co., Ltd.,
Beijing 100032, China
yujing19830@163.com
[3] Power Economic Research Institute of State Grid Fujian Electric
Power Company, Fuzhou 350001, China

Abstract. With the continuous improvement of the intelligent management level of the power system, it will be possible to use the power big data to conduct "real-time monitoring" of the current state of economic development. In recent years, with the optimization of the industrial structure of Fujian province in China, the economic and power demand of various industries in Fujian province has also changed greatly. Therefore, studying the relations between power consumption and economy in Fujian's sub-sectors has important guiding significance for the Fujian provincial government and power grid companies. A Granger causal nexus exam method was adopted to probe the relations between power consumption and economy in Fujian province. The results show that bidirectional causal relationships existed between electric power consumed as well as economic development at aggregate level and sectors of the primary industry; the industry has a one-way causal nexus from electric power consumption to economic development, while construction industry, transportation, warehousing and postal services have a one-way causal coupling nexus from economic development to electric power consumed.

Keywords: Big data · Electric power consumed · Economic development · Granger test method

1 Introduction

The coupling relation between energy and economy has always been the focus of research [1, 2], but there is a lag in energy statistics [3]. Therefore, in recent years, a large number of scholars have turned their research perspectives to examine the relations between electricity and economy [4].

In recent years, Fujian has continuously optimized its industrial structure and actively integrated into the construction of "one belt and one road". The economic development and electric power consumed at industrial level in Fujian province have changed greatly. So the study of the causal nexus between power consumed and economic development in different industries in Fujian province has a guiding role for

© Springer Nature Singapore Pte Ltd. 2020
M. Atiquzzaman et al. (Eds.): BDCPS 2019, AISC 1117, pp. 522–529, 2020.
https://doi.org/10.1007/978-981-15-2568-1_71

Fujian provincial government and Fujian electric power company. First, predict the future economic trend of Fujian province through electric power consumed, and provide some ideas for the government to formulate economic policies of Fujian province under the new situation. Second, predict the future power demand of Fujian province through the economic development trend, and provide some guiding suggestions for the power companies to formulate power planning and marketing strategies.

There is a wide range of area that considered the relations of electricity and economy. Gurgul and Lach [5] discussed the relations between total power consumption, economic development, industrial power consumption and employment in Poland; Shahbaz [6] explored the relations between industrialization, power consumption and CO2 in Bangladesh; Shahbaz [7] analyzed the impact of electric power consumed and economic development in Portugal; Abosedra [8] studied the causal relations between electricity and economy in Lebanon etc. [9–12].

For China, Lai [13] analyzed the causal relations between Macau's power consumption and economy; Liu [3] analyzed the nonlinear causality between power consumption and GDP in Beijing's three major industries; Du [14] studied the relations between power consumption and economy in Fujian province, but he did not study the power economic relations of the industry (industry) from the perspective of industry segmentation. In recent years, Fujian has been continuously optimizing its industrial structure. Therefore, here we makes effort to exam the causal relations between electric power consumed and the economy at industrial level in Fujian, so as to provide directions for the Fujian provincial government to formulate industrial development policies and Fujian power grid to formulate power demand services in various industries.

2 Fujian Province's Economy and Power Situation

2.1 Economic Development of Fujian

Looking at the economic development of Fujian Province in a longitudinal way, since 2008, the economy of Fujian Province has been developing at a steady and rapid pace, and the total value of production has crossed the two trillion yuan. In 2008, Fujian's total GDP increased by more than 1 trillion US dollars, and in 2018 it exceeded 30,000 megabytes, of which about 3.3 times in 2008, the incremental increase reached 240 million US dollars. The economic growth rate of Fujian Province has been on an upward trend. The growth rate in 2010 and 2011 exceeded 20%. Except for a slight decline in 2015, the growth rate has remained in the range of 10%–15% since then, as shown in Fig. 1.

The industrial structure can show the economic and technological links between industries and the level of industrial development. The essence of this is the distribution of resources among various industries, which is one of the basic factors affecting economic development. As shown in Fig. 2, the GDP of the first industry in Fujian Province is low, and its economic growth fluctuates greatly. In 2017, agriculture showed negative growth. The growth of the secondary and tertiary industries in Fujian Province is rapid and relatively stable. The growth rate of the secondary industry has

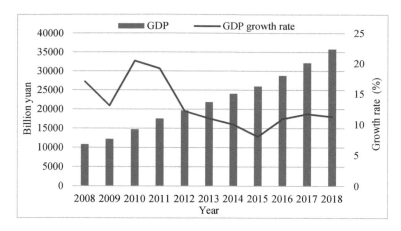

Fig. 1. Overall economic growth rate of Fujian Province in 2008–2018

remained between 5% and 15% in recent years, and the growth rate of the tertiary industry has remained between 10% and 17.5%. The GDP of the secondary industry increased from more than 500 billion yuan in 2008 to the first breakthrough in the trillion-dollar mark in 2012, reaching more than 1 trillion yuan, continuing to grow at a high speed, reaching more than 1.7 trillion yuan in 2018; the tertiary industry In 2008, more than 400 billion yuan to more than 1.6 trillion yuan in 2018, the total increase of more than 1.1 trillion yuan, about 3.7 times in 2008.

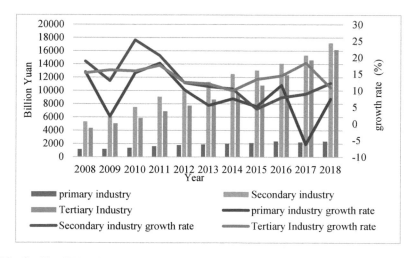

Fig. 2. The GDP of the three major industries in Fujian Province from 2008 to 2018

2.2 Power Consumption of Fujian

At present, China's economic development has entered a new normal, the economic structure has been continuously upgraded, and the economy has shifted from a high-speed growth stage to a high-quality development stage. Vigorously promoting the construction of ecological civilization and actively promoting the energy production and consumption revolution have gradually become the consensus of the whole society. Similarly, the power industry is facing a new situation in which the country's overall industrial structure is improving to high quality, energy efficiency and energy efficiency.

In recent years, driven by the two factors of sustained macroeconomic stability, new business conditions and booming emerging industries, Fujian's power consumption demand has generally improved, and the consumption structure optimization effect is obvious. From 2009 to 2018, the whole society's electricity consumption in Fujian Province showed a steady growth trend, and the year-on-year growth rate showed the overall trend of decreasing first and then increasing, as shown in Fig. 3. Among them, in 2018, the total electricity consumption of Fujian Province reached 114.238 billion kWh, a year-on-year increase of 7.08%, and the growth rate increased by 2.51% year-on-year. The power consumption situation has improved from the previous year.

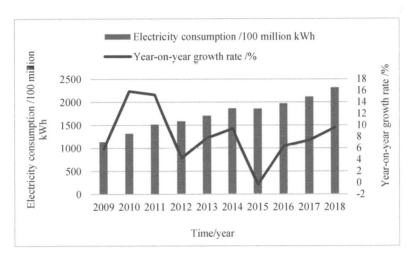

Fig. 3. Fujian Province 2009–2018 power consumption and growth rate

3 Data and Methodology

3.1 The Data

Here an empirical research applies the historical annual GDP and power consumed from the year 1995 to 2017 in Fujian. The time series of power consumption is denoted as EC, and economic development is denoted as EG. Therefore, the data of power consumption and economic development of the total amount, agriculture, forestry, fishery and animal husbandry, manufacture, construction, transportation, warehousing and postal services, as well as wholesale and retail, were denoted as EC, EG, EC1, EG1, EC2, EG2, EC3, EG3, EC4, EC5 and EG5 respectively. The electric power consumed information was from the Fujian Provincial Bureau of Statistics, and the economic development information was from the National Bureau of Statistics of China.

3.2 The Methodology

(1) Unit root test (URT)

The URT is mostly employed to exam stationarity of sequence data. If there is a unit root process (URP), it is regarded as a non-stationary sequence, otherwise it is considered as a stationary series. The Augmented dickey-fuller (ADF) test was utilized in this paper to verify stability of the sequence.

(2) Cointegration test

Cointegration refers to several variables produced by URP. If such a linear combination nexus exists, the residual of this combination can be created by a stable process, and this combination becomes cointegration between variables. The Johansen-Juselius co-integration (JJ) test was used in this study.

(3) Granger causal test

Granger causal nexus means that under the condition that the historical sequence of A and B is included, if the prediction result of B is better than that the predict without adding the past information of variables A and B, if A is helpful to predict the coming fluctuation of B, and then A is considered to be the Granger cause to B.

4 Empirical Results

4.1 URT

The outcomes of ADF URT on historical data are shown in Table 1.

Table 1. Outcomes of unit root tests

Variable	Levels	First differences
EC	0.039	–
EG	0.0024	–
EC1	0.2443	0.0018
EG1	0.8437	0.0306
EC2	0.7692	0.0039
EG2	0.9864	0.0811
EC3	0.0913	–
EG3	0.0031	–
EC4	0.9906	0.0024
EG4	0.6821	0.0158
EC5	0.1379	0.0377
EG5	0.9442	0.0474

All of the results are presented in Table 1, after the first order difference Stable at 5% significance level, co-integration test for first-order stationary data.

4.2 Testing for Co-integration

The above the first-order stationary sequence is tested to examine whether or not there exists a long-term equilibrium relations. The outcomes are presented in Table 2.

Table 2. JJ Co-integration test results

Variable	Trace statistic	5% critical value	Prob.**	No. of CE(s)
EC1-EG1	27.73666	15.49471	0.0005	None *
	11.13389	3.841466	0.0008	At most 1 *
EC2-EG2	28.0319	18.39771	0.0017	None *
	8.207744	3.841466	0.0042	At most 1 *
EC4-EG4	39.67836	15.49471	0	None *
	6.195447	3.841466	0.0128	At most 1 *
EC5-EG5	18.78247	15.49471	0.0154	None *
	5.952637	3.841466	0.0147	At most 1 *

The long-term stable coupling relations between power consumption and economy of various industries are available from Table 2.

4.3 Causality Results

Co-integration analysis reveals that there exists a long-term coupling relations between power consumed and economy. To further probe the coupling relations, a Granger causal method is exploited, and lag order is chosen 1–4. The outcomes are listed in Table 3.

Table 3. Granger causal outcomes

Null Hypothesis:	Prob.			
	Lag = 1	Lag = 2	Lag = 3	Lag = 4
LEG does not Granger Cause LEC	0.0517	0.106	0.0484	0.6602
	*Reject	Accept	** Reject	Accept
LEC does not Granger Cause LEG	0.0011	0.0403	0.0042	0.0328
	*** Reject	** Reject	*** Reject	** Reject
LEG1 does not Granger Cause LEC1	0.0438	0.0641	0.0105	0.304
	** Reject	* Reject	**Reject	Accept
LEC1 does not Granger Cause LEG1	0.1124	0.0106	0.0259	0.0608
	Accept	**Reject	**Reject	*Reject

(continued)

Table 3. (*continued*)

Null Hypothesis:	Prob.			
	Lag = 1	Lag = 2	Lag = 3	Lag = 4
LEG2 does not Granger Cause LEC2	0.9248	0.9966	0.9412	0.8911
	Accept	Accept	Accept	Accept
LEC2 does not Granger Cause LEG2	0.0021	0.0197	0.0531	0.1823
	***Reject	**Reject	*Reject	Accept
LEG3 does not Granger Cause LEC3	0.0028	0.0199	0.0754	0.14
	***Reject	**Reject	*Reject	Accept
LEC3 does not Granger Cause LEG3	0.1108	0.2888	0.1477	0.1377
	Accept	Accept	Accept	Accept
LEG4 does not Granger Cause LEC4	0.0215	0.0511	0.118	0.0023
	**Reject	*Reject	Accept	***Reject
LEC4 does not Granger Cause LEG4	0.4548	0.3489	0.5797	0.3079
	Accept	Accept	Accept	Accept
LEG5 does not Granger Cause LEC5	0.1465	0.0082	0.0147	0.0676
	Accept	***Reject	**Reject	*Reject
LEC5 does not Granger Cause LEG5	0.9237	0.0786	0.0407	0.0548
	Accept	*Reject	**Reject	*Reject

***, **, * refer to 1%, 5%, and 10% significance level.

The results in table 2–8 show that in the total amount analysis, there exists a bi-directional causal nexus between power consumed and economy, but the degree and duration of the mutual influence between the two are not the same. For different industries, the causal nexus between power consumed and economy vary, so does the significance level. For sectors in primary industry, and wholesale and retail there is a bi-directional causality as well. And manufacture industry has a one-way causal relation from power consumption to economic development, while the construction Industry, transportation, warehousing and postal services have unidirectional causal relations from economic development to power consumed.

5   Empirical Results

From the causal exam result, it is known that power consumption is the Granger causality of Fujian's economic development. Therefore, the Fujian provincial government and the grid company should pay more attention to the planning and construction of electricity. This will ensure the reliability of power supply and ensure that Fujian's economic development can be stable and sustainable.

From the perspective of industry, for sectors in primary industry, and wholesale and retail industry there exists a bi-directional causal nexus. Therefore, the economic development of these two industries indicates that the power demand of Fujian province will increase in the future, and power grid companies should make power planning and power marketing strategies in advance. Fujian provincial government should promote the modernization of agricultural electrical appliances, which can stimulate economic booming for the above sectors. The industry has a unidirectional causal relation from power consumption to economy. Therefore, implementing energy conservation and emission reduction in such sectors will reduce economic development level. Hence, the Fujian province needs to optimize its industrial structure as well as to reduce the economic damage caused by aggressive environmental policies. The construction, transportation industry, warehousing and postal services are one-way Granger causality from economic development to power consumed. Therefore, when the economy of both industries are improving, it will indicate that the future electric power demand of these two industries will increase, and power grid companies should make power planning and power marketing strategies in advance.

References

1. Bakirtas, T., Akpolat, A.G.: The relationship between energy consumption, urbanization, and economic growth in new emerging-market countries. Energy **147**, 110–121 (2018)
2. García-Gusano, D., Suárez-Botero, J., Dufour, J.: Long-term modelling and assessment of the energy-economy decoupling in Spain. Energy **151**, 455–466 (2018)
3. Liu, D., Ruan, L., Liu, J.C., Huan, H., Zhang, G.W., Feng, Y., et al.: Electricity consumption and economic growth nexus in Beijing: a causal analysis of quarterly sectoral data. Renew. Sust. Energy Rev. **82**, 2498–2503 (2018)
4. Zhang, C., Zhou, K., Yang, S., Shao, Z.: On electricity consumption and economic growth in China. Renew. Sustain. Energy Rev. **76**, 353–368 (2017)
5. Gurgul, H., Lach, L.: The electricity consumption versus economic growth of the Polish economy. Energy Econ. **34**, 500–510 (2012)
6. Shahbaz, M., Uddin, G.S., Rehman, I.U., Imran, K.: Industrialization, electricity consumption and CO2 emissions in Bangladesh. Renew. Sust. Energy Rev. **31**, 575–586 (2014)
7. Shahbaz, M., Benkraiem, R., Miloudi, A., Lahiani, A.: Production function with electricity consumption and policy implications in Portugal. Energy Policy **110**, 588–599 (2017)
8. Abosedra, S., Dah, A., Ghosh, S.: Electricity consumption and economic growth, the case of Lebanon. Appl. Energy **86**, 429–432 (2009)
9. Acaravci, A., Ozturk, I.: On the relationship between energy consumption, CO2 emissions and economic growth in Europe. Energy **35**, 5412–5420 (2010)
10. Alam, M.J., Begum, I.A., Buysse, J., Rahman, S., et al.: Dynamic modeling of causal relationship between energy consumption, CO2 emissions and economic growth in India. Renew. Sust. Energy Rev. 15:3243–3251 (2011)
11. Amri, F.: The relationship amongst energy consumption (renewable and non-renewable) and GDP in Algeria. Renew. Sust. Energy Rev. **76**, 62–71 (2017)
12. Bildirici, M.: Defense, economic growth and energy consumption in China ☆. Procedia Econ. Finan. **38**, 257–263 (2016)
13. Lai, T.M., To, W.M., Lo, W.C., Choy, Y.S., Lam, K.H.: The causal relationship between electricity consumption and economic growth in a gaming and tourism center: the case of Macao SAR, the People's Republic of China. Energy **36**, 1134–1142 (2011)
14. Du, Y., Hu, P.F., Lin, H.Y., Xiang, K.L., Zheng, H., Deng, J.P., et al.: Cointegration analysis between electricity consumption and economic growth in Fujian Province. In: 2017 IEEE Conference on Energy Internet and Energy System Integration (Ei2), pp. 148–52 (2017)

Seismic Anomaly Information Mining Based on Time Series

Hongwei Li and Ruifang Zhang[✉]

Shanxi Earthquake Agency, Taiyuan 030021, Shanxi, China
lhw_one@163.com

Abstract. The trend turning of seismic observation series contains important seismic anomaly information. The turning points are extracted, and the potential change patterns and laws are found according to the wave characteristics of the data. On the basis of summarizing the analysis trend turning characteristics of time series, this paper proposes a method of extracting trend turning points based on the combination of feature points and vector corners by optimizing the extraction algorithm of trend turning points of time series. This method can quickly and accurately extract the turning point of time series. Taking the short-level BM_1–BM_4 line of Linfen station as experimental data, the reliability of the algorithm is verified, and the seismic prediction rule is combined to make the seismic predictable efficiency R higher as the optimization target of anomaly information mining. Iterative calculated all the turning points of time series, and quantify the change range of the trend before and after the turning point, and obtain the effectiveness evaluation result of the line for earthquake prediction.

Keywords: Time series · Feature points · Vector corners · Information mining · Effectiveness evaluation

1 Introduction

Time series information mining is one of the hot research topics in the field of data mining in recent years, and is widely used in astronomy, meteorology and finance [1–3]. A one-dimensional time series is a quantitative description of the evolution of a thing, which contains the inherent information of the law of the development of things. The trend turning of seismic monitoring data series is an important information of seismic anomaly changes, and it is also a kind of seismic precursor anomaly with high frequency. At present, the identification and extraction of such anomaly mainly rely on morphological analysis and empirical judgment, lacking mathematical quantitative methods and having subjective influence. Since the time series data recorded by different observation instruments have obvious differences in data characteristics, the time series observed by the same observation instrument exhibit different data characteristics in different time periods, which are represented by the fluctuation characteristics, time series trend turning point and the representation of how time series changing over time, of the data. There are many ways to represent time series patterns [4], and the piecewise linear representation method is most commonly used [5, 6]. Because the Piecewise Linear Representation (PLR) has the characteristics of intuitive form, flexible distance

© Springer Nature Singapore Pte Ltd. 2020
M. Atiquzzaman et al. (Eds.): BDCPS 2019, AISC 1117, pp. 530–539, 2020.
https://doi.org/10.1007/978-981-15-2568-1_72

measurement and simple calculation, it is often applied to time series information mining. The core idea of time series PLR is to replace the original time series with a number of straight line segments adjacent to each other. The difficulty lies in accurately determining the position to set the piecewise point [7]. Many scholars have done a lot of research on this core problem. For example, Pratt et al. [8] proposed a straight line piecewise method based on significant points; Yi et al. [9], Keogh et al. [10] used the average value of a given window length instead of the original time series PAA piecewise representation algorithm; Park et al. [11] proposed a time series piecewise method based on feature points. After finding the piecewise point, the least squares algorithm is used to fit the straight line segments on both sides of the piecewise point [12]. The current PLR algorithm uses a single fitting error as the threshold for the original data piecewise, and the piecewise effect is not so ideal. The objective of the piecewise-approximation is to minimize the residuals sum of squares between the original time series and its linear approximation. Under the limitation of this objective function, the turning point may not be accurately obtained. Through the above analysis and research, while using segmental linear represent the original time series data, some feature points in the original series play a major role in controlling the variation characteristics of the time series, including extreme points reaching a certain range and the point at which the rate of change increases in a monotonic interval.

Aiming at the above problems, in order to reduce the deviation caused by subjective judgment, and scientifically, rapidly and timely extract this type of seismic anomaly information, this paper proposes a method based on the combination of feature points and vector angles to study the trend turning point of quantitative extraction time series. The anomaly amplitude is defined by the size of the vector angles, and the seismic anomaly information contained in the observation data is deeply mined in combination with the corresponding rules of the seismic case, and the prediction results are evaluated.

2 Methods and Models

2.1 Data Preprocessing

First, by observing the observation log table, remove various human and environmental interferences of the original observation data, and data interpolation is performed to ensure the integrity and continuity of the data. Secondly, The linear interpolation method is used to complete the missing processing and ensure that there are no missing time series included. The first-order differential calculation is performed on the data, and the differential time series is scanned point by point to determine whether it exceed a given threshold (3 times mean square error), and if it is exceeded, the data step is determined and there will be a step correction; finally, according to the research problem, wavelet decomposition method is used to filter out high frequency data and retain the band data relatively low frequency.

Since the time series comes from different observation instruments and research periods, and in order to analyze and compare the differences among the trend changes of different observation points in the same area, the time and amplitude for each

observation curve are normalized to make each observation curve comparable on the
same time scale and amplitude of change, so the time series of the observation data and
the series of observations are uniformly normalized to 0–1, and the normalization
formula is as follows (1):

$$norm(x_i) = \frac{x_i - \min(x)}{\max(x) - \min(x)} \tag{1}$$

2.2   Quantification of the Turning Amplitude

In order to better compare the amplitude of the trend before and after the different
turning points, the vector angle is used to quantitatively analyze the anomaly amplitude
(Fig. 1). If the observed curve changes to AB before the turning point and the observed
data is BC after the turning, then the formula α for representing the anomaly amplitude
can be obtained by calculating the Eq. 2 of the vector angle. In addition, the magnitude
of the anomaly amplitude is independent with whether the observation curve is
clockwise or counterclockwise.

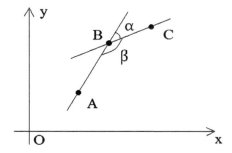

Fig. 1. Vector angle schematic diagram

$$\alpha = a \cos \frac{\overline{AB}.\overline{BC}}{|\overline{AB}|.|\overline{BC}|} \tag{2}$$

The vector AB and the vector BC are obtained by linear fitting according to a given
minimum period window. If the fitting error is too big, the amplitude of the trend
turning cannot be truly reflected, so the fitting error needs to be judged. If the vector
AB is linearly fitted by time series $x = [x_1, x_2 \ldots x_n]$, the derived fitting time series
obtained by filling vector AB with linear interpolation method is recorded as
$x^c = [x_1^c, x_2^c \ldots x_n^c]$, and the fitting error of the fitting series and the original series is
calculated by Eq. 3:

$$E = \sqrt{\sum_{i=1}^{n} (x_i - x_i^c)^2} \tag{3}$$

The fitting error is an important index to measure the difference between the fitting time series and the original time series. Under the same minimum period window, the smaller the fitting error, the more the trend turning amplitude can be truly reflected.

2.3 Extraction of Feature Points

The time series data has the characteristics of large scale and high dimension. If extracted the feature points directly on the original observation curve, the computational complexity will be high and will be influenced by noise. Therefore, it is necessary to extract preliminary observation curve of the original time series. Most of the existing PLR methods have problems in that the screening accuracy of the turning points and the extraction of points. According to the variation characteristics of time series, the observation curve is scanned point by vector angle method, and the time series of Fig. 2(b) is obtained. It is found that the maximum value of the vector angle time series corresponds to the extreme point of the original observation curve and the position of the point with large change rate, so this paper proposes a new turning point extraction algorithm, which can obtain the position of the feature point of the original observation series by finding the maximum value of the time series based on the vector angle time series. It can be seen from Fig. 2 that the feature points of the time series contain points with large change rate in the extreme points and monotonic intervals. At this time, the time series change trend before and after the feature points is completely different. Therefore, this method can effectively extract the natural piecewise point of the time series trend turning.

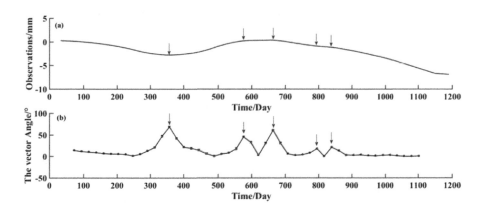

(a) Original observation curve   (b) Result of vector turning's point-by-point scan

Fig. 2. Feature point extraction result process based on vector angle

3   Experiments and Results Analysis

3.1   Experimental Data

In this paper, the short-level BM_1–BM_4 line of Linfen station in the south of the Shanxi fault zone is selected as the experimental data. The line is in the west boundary of the Linfen Basin, monitoring the Luoyunshan fault activity. The slope of the fracture section is about 70°, and the trend is N30°E. The SE is a normal fault with the east plate descending and the west plate rising, and its fault throw is about 150 m. Due to the influence of the main fault, the small faults in the sub-level of the area are relatively developed and consistent with the main structure. The leveling line has 3 measuring sections, 4 leveling points (Fig. 3). The lines BM_1–BM_4 have been observed since 1990, nearly 30 years' observation provides continuous, reliable and accurate data. and accumulated rich seismic case information for the study of the Luoyunshan fault activity. The seismic case data of *M4* or higher-level earthquakes within 200 km of the station are selected.

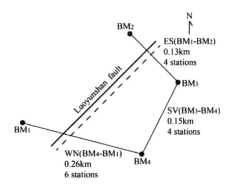

Fig. 3. Linfen short-level survey line distribution map

3.2   Anomaly Point Extraction

In this paper, the data is observed by the daily sampling rate. After the data interpretation, normalized, and removing the linear trend, the preprocessing time series is shown in the black line of Fig. 4, because the fault activity are mainly low-frequency activities, so the time series requires further low-pass filtering to filter out high-frequency interference signals as much as possible. According to the previous research case, the fluctuation information less than 180 days is filtered out, and the trend change series of the fault activity is obtained (Fig. 5 red line); then based on this time series, the sliding parameter with a window length of 180 days and a step size of 10 days are selected, obtain the time series of vector angle by point-by-point scan (Fig. 5). It can be seen from Fig. 5 that the maximum value point of the vector angle time series corresponds to the trend turning point of the preprocessed observation curve, and the time series change trend before and after the turning point represents the active state of the fault. In the previous research and related theoretical models, it is confirmed that when the stress and strain are accumulated to a certain extent, the time series of seismic observation may have a trend turning phenomenon before the earthquake. These trend turning points may have certain indication significance for earthquake prediction. We tentatively set such a turning point for the trend as an anomaly point that may be related to the earthquake.

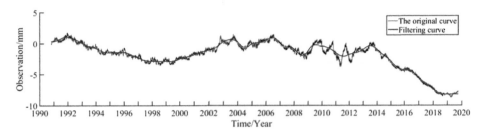

Fig. 4. Preprocess curve and low-pass filtering result

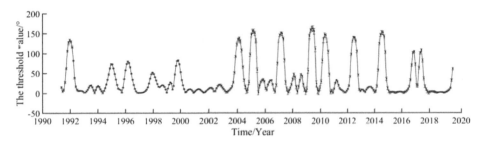

Fig. 5. Time series of vector angle

3.3 Effectiveness Evaluation of Earthquake Prediction

The R value scoring method was proposed by Xu Shaozhen in 1973, and the improved R value scoring method was given in 1983. The R value scoring method currently used mainly considers the earthquake prediction rate $\frac{N_R}{N_A}$ and the forecast time occupancy rate of the actual earthquake prediction $\frac{T_R}{T_A}$.

Take $0°$–$180°$ as the search range of the trend turning threshold, 0–720 days as the search period of the forecast period, and the length of the straight line fitting window before and after the trend turning point is 180 days, and the optimal R is iteratively calculated according to the loop of the corresponding case of the earthquake, and the threshold under the optimal R. The formula is:

$$R = \frac{N_R}{N_A} - \frac{T_R}{T_A} \tag{4}$$

In the formula, N_R is the number of reported earthquakes, N_A is the total number of earthquakes to be forecast, T_R is the forecast occupation time, T_A is the total time of the forecast study.

Table 1 is the statistical table of the trend turning anomaly and effectiveness evaluation of the short-level BM$_1$–BM$_4$ line in Linfen station. The number of occurrences of this type of anomaly since the observation is given in detail, and the accurate number of earthquakes reported and the missing number of earthquakes reported under the given rules of the earthquake case. The optimal forecast period and optimal threshold according to the optimal R value. Table 2 shows the detailed anomaly occurrence time of each test item and the abnormalities corresponding to the earthquake.

Table 1. Trend turning anomaly and effectiveness evaluation of Linfen level BM$_1$–BM$_4$ line

Anomaly	Accurate report	Missing report	R	Optimal forecast (day)	Optimal threshold (°)	Lower limit magnitude (M)
10	4	3	0.33	259	85	4

Table 2. Anomaly time and earthquake corresponding anomaly of Linfen level BM$_1$–BM$_4$

Turning starting time	Whether accurate	From the time to the earthquake (day)	Earthquake time	Longitude of earthquake	Latitude of earthquake	Magnitude
1991-11-21	N					
2003-07-11	Y	137	2003-11-25	111.62	36.17	4.6
2004-07-05	N					
2006-06-15	N					
2008-08-03	N					
2009-06-29	Y	209	2010-01-24	110.77	35.57	4.8
2011-07-19	N					
2013-07-09	N					
2015-09-27	Y	167	2016-03-12	110.83	34.93	4.4
2016-04-04	Y	258	2016-12-18	112.45	37.59	4.3

Figure 6 is the corresponding earthquake under the maximum value of the point-by-point scanning vector angle time series of the short-level BM_1–BM_4 line of Linfen station in Shanxi. The red dotted line is the optimal threshold line (turning angle), and the solid line of blue line is the optimal forecast period. The arrow indicates the moment of the earthquake, red represents the earthquake that occurred during the optimal forecast period, and black represents the earthquake that occurred outside the optimal forecast period. The results show that the data trend turns more frequently after 2002, and the amplitude is larger. There are 4 earthquakes stronger than *M4* in this period, and all of them occur in the optimal forecast period. Before 2002, the data changed slowly. During this period, there was total of 3 earthquakes stronger than *M4*, which did not occur during the optimal forecast period. The analysis suggests that there may be a certain relationship between the enhanced regional stress fields after 2002.

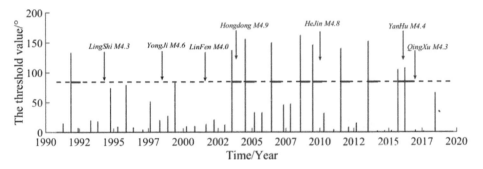

Fig. 6. Vector angle time series maximum value and earthquake corresponding situation under optimal threshold

Figure 7 shows the spatial distribution of earthquakes under the corresponding rules of the earthquake case. The red is the accurately reported earthquake and the blue is the inaccurately reported earthquake. From the spatial distribution, the Linyi level BM_1–BM_4 line has certain indications for the earthquakes in Linfen Basin, Yuncheng Basin and Taiyuan Basin.

Figure 8 is the spatial distribution of the forecasting efficiency R map in the forecast period-threshold coordinate system. The position of black pentagram is the optimal threshold and the R value under the optimal predictable period. The results show that the optimal threshold of the Linfen level BM_1–BM_4 line is 85°, and the optimal predictable period of the Linfen level BM_1–BM_4 line is 259 days. The result map shows the predictable efficiency evaluation result of the line more intuitively. According to the actual retrospective results, the 4 earthquakes occurred during the forecast period have a trend turning phenomenon in the original observation series before the earthquake. The trend turning anomaly of Hongdong *M4.9* and Hejin *M4.8* before the earthquake have been recognized by relevant experts so that he reliability of the algorithm is further demonstrated.

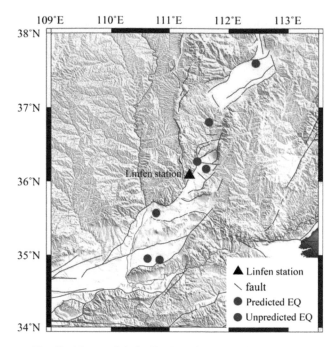

Fig. 7. The spatial distribution of predicted earthquakes

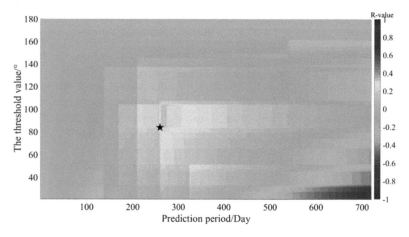

Fig. 8. Distribution of predictable efficiency R map

4   Conclusion

On the basis of summarizing the piecewise linear representation, this paper proposes an extraction algorithm based on the combination of feature points and vector angles to obtain the time series of vector angles by using the time window sliding method. On the basis of that extract the maximum point of the series, so that the trend turning point of the original observation series can be accurately obtained. Compared with the previous work in this aspect, the accuracy and objectivity have been improved. By using this method, the initial time of anomaly can be accurately determined, and objective basic information for quantitatively extracting seismic anomaly information is provided. In this paper, the short-level BM_1–BM_4 line of Linfen station is used as the experimental data, and the predicted efficiency turning point is evaluated. It has certain comparability with the actual manual interpretation, which further verifies the reliability of the algorithm and can be used for the mining of seismic anomaly information of time series.

Acknowledgements. This work was supported by The Earthquake Tracking Task of CEA, No. 2018010214; The Earthquake Tracking Task of CEA, No. 2019010218; Research Project of Shanxi Earthquake Agency, No. SBK-1923.

References

1. Ziegel, E.R.: Analysis of financial time series. Technometrics **44**(4), 408 (2010)
2. Jaeger, H.: Observable operator models for discrete stochastic time series. Neural Comput. **12**(6), 1371–1398 (2014)
3. Sadeghzadeh, N., Afshar, A,, Menhaj, M.B.: An MLP neural network for time delay prediction in networked control systems. In: Chinese Control and Decision Conference, pp. 5314–5318 (2008)
4. Lin, J., Keogh, E., Lonardi, S., et al.: A symbolic representation of time series, with implications for streaming algorithms. In: ACM SIGMOD Workshop on Research Issues in Data Mining and Knowledge Discovery, pp. 2–11. ACM (2003)
5. Maciej, K., Grazyna, S.: An approach to dimensionality reduction in time series. Inf. Sci. **26**(6), 15–36 (2014)
6. Pavlidis, T., Horwitz, S.: Segmentation of plane curves. IEEE Trans. Comput. **23**(8), 860–870 (1974)
7. Li, L., Su, X., Zhang, Y., et al.: Trend modeling for traffic time series analysis: an integrated study. IEEE Trans. Intell. Transp. Syst. **16**(6), 3430–3439 (2015)
8. Pratt, K.B., Fink, E.: Search for patterns in compressed time series. Int. J. Image Graph. **2**(1), 89–106 (2002)
9. Yi, B.K., Faloustsos, C.: Fast time series indexing for arbitrary Lp norms. In: Proceedings of the 26th International Conference on Very Large Data Bases, pp. 385–394. Morgan Kaufmann Publishers Inc., San Francisco (2000)
10. Keogh, E., Chakrabarti, K., Pazzant, M.: Dimensionality reduction for fast similarity search in large time series databases. J. Knowl. Inf. Syst. **3**(3), 263–286 (2000)
11. Park, S., Kim, S.W., Chu, W.W.: Segment-based approach for subseries searches in series databases. In: Proceedings of the 16th ACM Symposium on Applied Computing, pp. 248–252. ACM Press, New York (2001)
12. Fuentes, J., Poncela, P., Rodriguez, J.: Sparse partial least squares in time series for macroeconomic forecasting. J. Appl. Econom. **30**(4), 576–595 (2015)

Inheritance and Innovation of Traditional Architectural Interior Design Techniques

Xianqiong Liao[(✉)]

Guangzhou Nanyang Polytechnic College,
Guangzhou 510925, Guangdong, China
290463763@qq.com

Abstract. The Chinese traditional architectural decoration art has been brilliant and splendid. In ancient time, it possessed advanced technology, which gave birth to Chinese traditional Craftsman's spirit. Combining with the author's design practice, this paper studies the innovative application of traditional architectural decoration Craftsman's spirit in interior design so as to enhance the awareness of the traditional craftsman's spirit, excavate its essence, and create a modern comfortable interior space by combining with modern design techniques.

Keywords: Traditional architectural decoration · Craftsman's spirit · Interior design · Innovation

1 Introduction

The concept of redesign is mentioned in the book Design in Design written by Kenya Hara, a Japanese designer and redesign is innovation. This process of changing the known into the unknown is creative behavior, such as the application of dome arch and screen in interior design, which reflects the reproduction of art, function and technology in traditional wood architecture. At the same time, new technology and new materials are used to recreate these traditional cultural elements, to produce new products that meet the requirements of contemporary aesthetics and function, and to realize the recreation of culture. It can be seen that design is a kind of goal-oriented solving activity, is a process of recognition and creation of cultural forms through the collection and processing of existing technology, knowledge and information [8].

The results of our design are presented in various forms that people can perceive. The most common one is the specific visual form that we are familiar with, such as the appearance of buildings and products we see. Dwelling in the cave of primitive society, dry-rail architecture in Hemudu period and the caves in the northwest loess plateau, are the civil structure of the early buildings, which embody the practical principles of architecture and is also one of the three elements of architecture. As a world-famous cultural heritage, Dunhuang murals have become a treasure in the treasure house of world culture and art with their huge scale and exquisite techniques [2]. Their rich and colorful contents depict the image of God and the relationship between God and man, and convey people's good wishes, which reflect the religious and cultural thoughts of that time.

© Springer Nature Singapore Pte Ltd. 2020
M. Atiquzzaman et al. (Eds.): BDCPS 2019, AISC 1117, pp. 540–549, 2020.
https://doi.org/10.1007/978-981-15-2568-1_73

The interior design space without spiritual connotation can only be said to be an empty shell, and the connotation expression is reflected in the interior spatial layout, decorative structure and so on. The traditional architectural decoration contains the connotation of the Craftsman's spirit, and it is needed for us to understand, grasp, inherit the Craftsman's spirit in the interior design and promote the development of interior design.

2 The Connotation of Craftsman's Spirit

Craftsman's spirit is a kind of professional spirit. It is the embodiment of professional ethics, professional ability and professional quality, is a kind of professional value orientation and behavior expression of the practitioners. The basic connotation of "craftsman's spirit" includes dedication, leanness, focus and innovation [9].

3 Craftsman's Spirit Embodied in Chinese Traditional Architectural Decoration

Throughout China's historical development, the form of Chinese traditional culture is various, including the cultural and artistic thoughts in various historical periods, which have had a far-reaching impact on the development of ancient architecture. Relevant works on design such as the Book of Diverse Crafts, Art of Garden-building, Exploitation of the Works of Nature, and Ying-tsao-fa-shih have been published accordingly, as well as the formed construction technology, these are the embodiment of the craftsman's spirit of traditional architectural decoration. Understanding these traditional cultures is an important practice of inheriting Chinese traditional culture in modern design. It is also of great significance to carry out design innovation.

In particular, the wisdom embodied in Chinese classical architecture, such as the wooden structure technology of Chinese classical architecture, the creation of arches, the plane of "interval" of the unit, the structure of lifting beams, tenon through and shaft fence. The roof forms of the hip roof, gable and hip roof, overhanging gable roof, double-sloping roof, and the vertical-ridge roof, as well as the colorful paintings and carvings of various components, and decoration of doors and windows with algae wells, etc., all these technologies and art contain the craftsman's spirit, reflecting innovation.

4 Inheritance of Craftsman's Spirit of Chinese Traditional Architectural Decoration in Interior Design

The literal meaning of "inheritance" is transmission, continuation and inheritance, while the word innovation originates from Latin, which contains three levels of meaning, one is renewal, the other is creation of new things, and the third one is change [6].

In my understanding, inheritance is the reappearance of culture, is the protection and transmission of the original culture, such as the legend of dragon and phoenix, which represent the auspicious patterns in Chinese traditional culture, and have been expressing auspicious wishes since ancient times. In the design, we will not change the forms of dragon and phoenix. If a dragon is changed into a non-dragon and a phoenix is changed into a non-phoenix, then how can we trace back to the origin? And then the real image of dragon and phoenix will disappear. It can be seen that the correct inheritance in the design is very important, to pass on, to continue, abandon the bad, so that good culture can be passed on from generation to generation.

However, it is not enough for design to rely solely on inheritance. Innovation, especially the innovation in interior design is badly needed. We can't distort the correct cultural form, especially in soft loading design, but it can't be applied mechanically, otherwise it will become the reproduction of antiques, so it has become especially important for the redesign of known cultural forms.

It can be seen that innovation is not only a process of abandoning old things, old ideas, and pursuing fresh and strange things and ideas, but also a process of cognition, absorption and re-creation of known cultural forms. In the process of interior design, we cannot be separated from the consumer's understanding of the known residential space environment, or the perception of the comfort degree of space use, the concept of design style. And these are the cultural characteristics (design style, ergonomics, decoration material characteristics, etc.) of interior design. The cultural part is the cognitive psychology model that consumers have built in the process of long-term use. In the process of interior design, we should tap into this potential demand, understand the aesthetic orientation, lifestyle, values of contemporary people, create more valuable design, lead new design and create a new design culture through innovation.

5   Innovation of Craftsman's Spirit Chinese of Traditional Architectural Decoration in Interior Design

5.1   Embodiment of Traditional Aesthetic Thoughts

5.1.1   Confucianism

During the Warring States Period, Confucius initiated Confucian aesthetics, advocated "rites and music" and "benevolence and righteousness", advocated "loyalty and forgiveness" and "the golden mean" and emphasized the beauty of neutrality, to an appropriate extent, going beyond the limit is as bad as falling short. His artistic expression showed tranquility, gentleness and honesty and sincerity, and implicit beauty. In the space design, the space should be decorated without being bothered or reduction, just be perfect. "The Analects of Confucius Yongye" says: "Too straightforward temperament is rude, too respectful etiquette is superficial, the appropriate temperament and etiquette is what mature people should look like." "If simplicity surpasses ornamentation, it will be rude, if ornamentation surpasses simplicity, it will be superficial, if the proportion of simplicity and ornamentation is appropriate, then it can become a gentleman." That is, the unity of knowledge and intrinsic quality is the gentleman. In design, text is understood as texture, quality refers to grain, which also reflects the relationship between content and form, which reflects the unity of function and form. Design is not only a simple list of patterns, textures, colors, but to be refined and trade-offs, it should not be limited to the surface decoration, should be combined with the function (Fig. 1 author's design of his house). When Confucius talked about the music and dance of Shao, he said, "The art form is wonderful and the content is very good." When talking about the music and dance of Wu, he said, "The art form is very beautiful, but the content is worse." It embodies that beauty is an artistic form, goodness is content, and beauty should be united with goodness [6]. Figure 2 is a new Chinese-style living room of a villa designed by the author, in which, speed dial is applied into the ceiling. The author drew lessons from the large-square checkboard of ancient building, and then simplified, dissolved the excess beams in the middle so as to achieve the unity of function and form.

Fig. 1. Author's design of his house

Fig. 2. Design of the living room Source: Designed by the author

5.1.2   Taoism

The eleventh chapter of Laozi's Tao Te Ching, "Thirty spokes converge into a hole in a hub. Only when the hub is hollow can the car function. Knead and clay make pottery into utensils. Only when the utensils are hollow can the utensils function. Building a house by opening doors and windows, with the emptiness of the walls and walls, the role of the house can only be achieved. Therefore, something is convenient for people and "nothing" plays its role." [11] This is a classic philosophy of interior design. It has strong artistic value for architecture and interior design. It is of the understanding of the spatial relationship of space concept. In interior design, attention should be paid to the rational use of space, the use of white space, staggering, intersection, the use of borrowed scenery, the combination of virtual and real, etc. In interior design, such as the circular floor cover in the new Chinese design, moon gate, the dining room space designed by the author in Fig. 3, the leisure space and dining room space are organically integrated through the moon gate partition, so that the space is smooth, open, with good ventilation and lighting effects.

"Thirty-spoke and One Hub" is about space, we can further understand that "creating space, retaining space", and the interior design is not a houseful of piles, but to properly use the space, embodying the design philosophy of virtual reality, the application of this philosophy is the most obvious in Chinese painting, such as Qi Baishi's "frog voice ten miles out of the mountain spring", "ancient temple hides in deep mountain", and "the horse's hoofs are covered with the fragrance of flowers after visiting the spring." In the practice of interior design, the use of white space can leave more imagination and use white space for owners. "Emptiness is something, something is empty", space design is to create an artistic conception of "silence speaking". In the Song Dynasty, Fan Yiwen's "Preface to Four Needs" says that "does not take the virtual as the virtual, but the reality as the virtual". As shown in Fig. 4, the author's new Chinese-style living room fully embodies the philosophy of leaving white space.

Fig. 3. A new Chinese dining room Source: Designed by the author

Fig. 4. A new Chinese living room Source: Designed by the author

Another classic of Laozi's Tao Te Ching is "the better the music, the farther it is, the lower the image, the better the image and the more magnificent." It reflects the relationship between the virtual and the reality. No matter how beautiful the appearance of the building is, if there is no white space for internal room, then the building cannot be used. So, there should be corresponding image to it. As the saying goes, the architecture is solidified music [4]. Like music, it has melody, rhythm and sense of rhythm. The white space in the architecture is equivalent to the rest and pause in the music.

Another example is the philosophy of "less is more, more is confused", which is well combined in Japanese design. For example, Muji design gives people a kind of philosophical charm of "major principles (basic principles, methods and laws) being extremely simple", the design pursues nature, conciseness, but without losing modern beauty. Mies's principle of "less is more", and the design of his Barcelona German Pavilion is the perfect embodiment of this philosophical classic. Gropius, the Bauhaus Principal, said, "If the product only whose appearance is decorated and beautified cannot better give the full play to its effectiveness, then this beautification may also lead to the destruction of the product form." Therefore, in interior design, we should make the best use of everything. If it is not necessary, do not increase entities.

5.1.3   Harmony Between Man and Nature

Harmony between heaven and man originated from Confucius in the Spring and Autumn Period and the Warring States Period, developed from Dong Zhongshu's interaction between heaven and man; in the Han Dynasty. It is an important manifestation of the integration of Confucianism, Taoism and Buddhism. The philosophical essence of "harmony between man and nature" contains the harmonious coexistence between man and nature. Traditional gardens and architecture all embody the spiritual core of "harmony between man and nature". The green, ecological and humanistic design advocated by the contemporary era is the interpretation of the idea of harmony between man and nature. Beijing Xiangshan Hotel designed by I. M. Pei uses the decorative elements, temperament and style of Suzhou Gardens and the combination of Beijing Quadrangle and modern design concepts to integrate naturally and organically into the human environment. Open space layout, in accordance with the situation, the high and low yard with multi-story buildings is created.

Japanese architect Sakamoto, winner of the Pritzker Prize in 2014, has focused on paper architecture, he hated waste, and even proposed that he was not interested in green, ecological and environmental protection. He said that the goal of using paper was not to "build", but to "demolish this building" so as to recycle and reuse paper. The network shell structure made of paper tubes in the Japanese Pavilion of the World Expo in Hanover, Germany in 2000, was removed and transported back to Japan to make schoolbooks for pupils after the closing of the pavilion. His paper buildings played an important role in the disaster with the features of simple, fast, waterproof, fire-proof and earthquake-proof.

5.2   The Application of Traditional Cultural Symbols

Chinese traditional cultural symbols are diverse in content and rich in subject matter. They are abstract and concrete, including natural, philosophical, mythological, ethnical, regional and folk auspicious symbols. They embody the traditional aesthetic taste and cultural connotation of the Chinese people [1]. The symbol "卐" character expressing abstract philosophical concept, like the spiral shape of Taiji, implies that there is no beginning and no ending, and life is never ending. The "卐" pattern is most used in Chinese window decoration. There are auspicious symbols to express the ancient people yearning for a better life, such as luck, longevity, happiness and wealth, dragon and phoenix, mandarin ducks, bats, etc. There are also architectural features of different ethnic regions, including color, roof, decoration, doors and windows, partitions, railings [7]. All reflect the characteristics of different regions and different architectures, such as the splendid style of royal architecture, the restrained and low-key civil buildings. Figure 5 are the porch cases designed by the author, who used the lattice and traditional auspicious symbols.

In the design practice, the traditional cultural symbols can be directly quoted or used for reference. And flexible design innovations can be achieved by means of deconstruction, transformation, reorganization or symbolic metaphor.

Fig. 5. New Chinese porch Source: Designed by the author

5.3   Reference of Traditional Craft

Chinese traditional craftsmanship has made brilliant achievements in ancient times, especially in the field of ancient architecture, such as arch-fighting structure and mortise-tenon structure of architectural art [3]. There are also traditional arts and crafts (architectural paintings, sculpture, ceramic art, etc.). bucket-arch structure has the beauty of structural decoration in ancient wooden structure architecture. It can also be used for reference in modern interior design, such as the combination of modern building materials and the form of bucket-arch, as shown in Fig. 6.

The traditional mortise and tenon joint structure is widely used in ancient buildings and furniture. It is famous all over the world for its good mechanical properties, no nails and glue, and easy disassembly [10]. Mahogany furniture is widely used in modern interior furniture especially. There are also the use of ceramic art, such as the Grain Concert Hall in Seoul, Korea, which makes use of the ceramic "sound like the chime stone" and the reflective characteristics of ceramics. The concert hall is a warm-colored, high-temperature color glaze cuboid of different sizes constructed through the elaborate arrangement of pottery artists, being quiet, gentle and elegant. The combination of ceramic materials and building materials is a technological breakthrough, the integration of science and art, thereby achieving the perfect cooperation of artists, musicians, architects, as shown in Fig. 7.

Fig. 6. Application of bucket-arch structure Source: a network picture

Fig. 7. Korean Grain Concert Hall designed by Zhu Legeng Source: a network picture

Fig. 8. A square stool Source: a network picture

In short, traditional crafts adopt traditional and modern techniques to achieve the combination of traditional and modern techniques [5]. Figure 8 is a set of furniture design, a tea table assembled with acrylic materials and logs, a square stool created by vacuum sealing technology.

6 Conclusion

First of all, we should show respect for the traditional architectural decoration craftsman' spirit, make good use of traditional culture, tap its essence, the cultural connotation behind things, and the underlying "essence" and "phenomenon". Secondly, for the traditional architectural decoration cultural symbols, we should examine the time culture they represent and understand the underlying cultural value and significance. Interior design cannot copy rigid traditional things, nor simply imitate antiquity, imitate the shape of ancient utensils and the original decoration. The design is not the reproduction of antiques, but the use of new materials, new technology to reproduce the essence of traditional culture by means of deconstruction, deformation, synthesis and other design techniques. Finally, we should learn to be flexible. In interior design, we should not use the same way of thinking, we should learn to be flexible and keep up with the public aesthetic thinking. The combination of contemporary design semantics with keeping up with modern people's life needs makes design culturally rich and stylish.

Acknowledgements. This paper is financially supported by the project: "Innovative Research Team of Environmental Art Design" of Guangzhou Nanyang Polytechnic College in 2018: (Project No. NY-2018CQ2TD-01), the Provincial Project of Guangdong Provincial Education Department in 2017, "Application Research on Lingnan Traditional Culture in Modern Architectural Design" (Project No. 2017GkQNCX125) and the 13th Five-Year Plan for the Development of Philosophy and Social Science in Guangzhou in 2019, "The Creative Transformation and Innovative Development of Lingnan Traditional Architectural Culture in the Context of Cultural Confidence——Taking Liwan District of Guangzhou as an Example" (Project No. 2019GZGJ228).

References

1. Zhao, Y.: Analysis of the inheritance of Chinese traditional culture in modern interior design. Home Drama (9), 135–137 (2018)
2. Liu, L.: Research on the application of traditional elements in architecture and decoration design. Constr. Mater. Decor. (8) (2019)
3. Guan, T.: Application of traditional techniques in modern architectural decoration design. Constr. Mater. Decor. (6) (2019)
4. Nie, Y.: Inheritance and development of Chinese traditional auspicious patterns in modern Chinese interior decoration design. Shaanxi Normal University (2018)
5. Zhu, S.: Inheritance and innovation of Chinese traditional elements in new Chinese interior design. Wuhan Textile University (2016)
6. Sun, D.: Traditional Chinese Culture and Contemporary Design, 2015th edn. Social Sciences Literature Publishing House, Beijing (2015)

7. Zhou, W.: Oriental Culture and Design Philosophy, 2017th edn. Shanghai Jiaotong University Press, Shanghai (2017)
8. Yuan, Y.: Design in Design, 2017th edn. Guangxi Normal University Press, Guilin (2017)
9. On the spirit of craftsman (5) (2017). http://www.wenming.cn/
10. Luo, Y.: On the traditional mortise and tenon structure in China. Shanxi Archit. (24) (2009)
11. https://www.daodejing.org/

Analysis of Digital Twins and Application Value of Power Engineering Based on BIM

Yingfu Sai[1], Tianjia Zhang[1], Xin Huang[1], and Chong Ding[2(✉)]

[1] Inner Mongolia Electric Power Information and Communication Center, Hohhot 010000, China
[2] China Electric Power Enterprise Association Electric Power Development Research Institute, Beijing 100000, China
525871545@qq.com

Abstract. With the continuous advancement of computer and Internet technologies, the application of BIM technology in power engineering construction has also been widely developed. The paper proposes the integration of BIM technology and digital twinning technology to realize the informationization and intellectual construction of power engineering. Based on the BIM power engineering project management platform, the paper introduces digital twinning technology, and applies advanced methods such as Internet of Things and big data to give real-time dynamic information to the BIM model, and establishes power engineering digital twins approach from the perspective of the whole process and all participants. In addition, the application value of BIM technology and digital twin technology in power engineering is analyzed. Through the integration of BIM technology and digital twin technology, digital and intelligent management of power engineering construction can be realized, and real-time perception, resource sharing and application capabilities of smart building can be promoted.

Keywords: Digital twins · Project management · Comprehensive informationization · Smart building

1 Introduction

In recent years, with the advancement of computer and communication technology, China's power industry has gradually developed in the direction of informationization and intelligence [1]. The development of building information model (BIM) provides favorable support for the realization of power engineering digital twin [2, 3]. At present, digital twin technology has been applied in aerospace, smart city, intelligent manufacturing, smart transportation, health care and other fields and has achieved good application results [4].

BIM is a digital representation of the physical and functional characteristics in the facility project. The information in the BIM is inserted and updated and modified to support and reflect collaborative work of different parties at different stages of the project [5]. However, most information in building information models will not be

M. Atiquzzaman et al. (Eds.): BDCPS 2019, AISC 1117, pp. 550–558, 2020.
https://doi.org/10.1007/978-981-15-2568-1_74

passed to the next stage after the creation of the 3D information model in the design stage. It has led to the true value of the information model not being fully tapped.

Therefore, based on the traditional power information model, the concept of digital twin is introduced into power engineering. Digital twin refers to the simultaneous production and updating of the building products of the physical world and the digital building information model in the virtual space during the construction process to result the completely consistent delivery [6]. Through the real reflection and real-time interaction between physical engineering and virtual engineering, the full integration can be realized, including full element, full process and full business data. The data assets including the digital twin projects, digital archives and the other data asset are formed in order to lay a solid foundation for the realization of multi-project, enterprise-level, group-level data assets. By studying the information given to the model at each stage, a large amount of information data can be mapped from the physical world to the information world, realizing the comprehensive informationization and digital construction of power engineering.

2   The Approach to BIM Digital Twinning

Digital twin in the power industry mainly includes three concepts: the real product in the physical world, the virtual product in the virtual world, and the data and information for the accurate mapping between the physical world and the virtual world [7]. The digital twin of the construction process requires a digital model, real-time management information, and a comprehensive intelligent network. So the traditional construction mode can no longer meet the requirements, and the highly integrated information platform need be utilized to provide data support and guidance for project management decision-making.

The article studies the digital twin of power engineering based on BIM. The engineering information data is received from the whole life process of engineering construction, including planning, design, bidding, construction, inspection and the operation and maintenance management, and the data is given to the information model to enrich the model content. Through the transmission of the digital model, the information loss between the participants and the various stages of the project construction can be reduced to improve the construction efficiency of the project. The information model can provide management information for the various stages of project construction by continuously expanding various information. Therefore, this paper studies the information content given to the model at each stage of construction.

2.1   Engineering Planning Stage

The planning phase is the most critical phase in the power engineering project life cycle. At this stage, through the economic and technical analysis of different investment schemes, the best scheme is selected. The statistical results show that the impact of the planning stage on the project cost is high, 85%–95%. Therefore, investment estimation plays an important role in the program selection [8].

In the planning stage, terrain information, geological structure, existing buildings information, climatic conditions of the construction site and so on are collected Through the combination of BIM and digital twin, and a large number of physical information data is mapped to the information model by using the GIS and real-world modeling techniques. At the same time, through BIM technology and digital twin technology, each program can be compared and selected. The various program plans can be fully demonstrated by the powerful twin data service platform. Finally, the appropriate scheme is chosen to ensure the quality of the power transmission and transformation project and the interests of the builder. The digital twin information of the model in the planning stage is shown in Fig. 1.

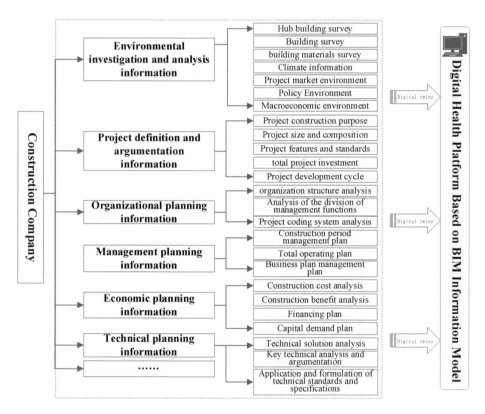

Fig. 1. The digital twin information of the model in the planning stage

2.2   Engineering Design Stage

Engineering design is the key stage of project cost management. It plays a decisive role in the construction period, construction cost, project quality and whether it can exert better economic benefits after construction. According to relevant statistics, the impact of the design stage on project cost is above 75%, so it is necessary to improve the design quality and optimize the design scheme.

In the design stage, based on the BIM digital twinning model established in the planning stage, the design unit establishes a three-dimensional information model of the project. The corresponding information, including project information, model component size, scale, angle, color and other geometric information, need to be given to the model. The information of the original engineering project model is very limited. It only contains the project geometry information for the project cost, in the absence of attribute information and material information. Therefore, on the basis of the original information model, with the continuous deepening of the design, the design unit gradually inputs the attribute information such as the parameters and materials of the equipment. The digital twin information of the model in the design stage is shown in Fig. 2.

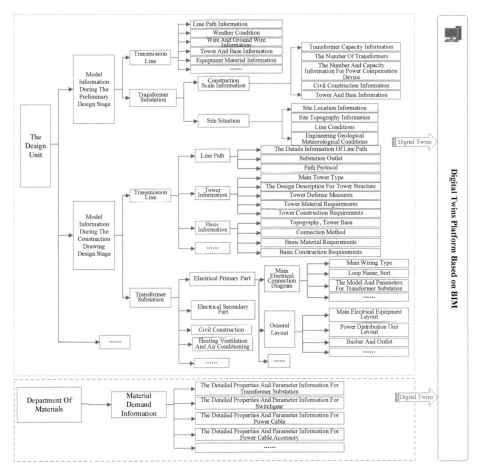

Fig. 2. The digital twin information of the model in the design stage

2.3 Engineering Construction Stage

In the power engineering construction process, the information model is basically abandoned after the design unit has completed the design task, and a large number of model design work done in the previous period cannot continue to exert its value [9]. It believes that digital twinning technology can be combined with BIM technology in the construction phase, and the information model of the design phase can be reused by giving new information.

Before the engineering construction, the physical environment information around the construction site is obtained by using GIS or total station or sensing technology, and the actual environmental information is converted into digital model information through real-world modeling technology or three-dimensional modeling technology. The site planning in the construction phase is different from that in the planning phase and the design phase. From the perspective of the construction unit, it involves the information on the arrangement of water, electricity, temporary buildings, temporary road and so on. The digital real-time information is generated by digital twinning technology, and the information model can dynamically change in real time according to the sensing technology in the physical environment.

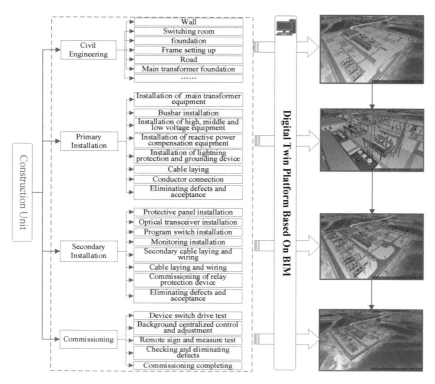

Fig. 3. The digital twin information of the model for the construction unit in the construction stage

In the process of engineering construction, the attribute information is continuously increased, such as the transformer equipment model, the installation location, the installation information of the tower. With the construction of actual power engineering, the corresponding digital information model is also enriched. Digital twin allows managers to dynamically adjust in the digital system.

In addition, the information model can provide more detailed data information for all aspects of the entire project through the collection of construction project progress, cost, resources, management, physical performance and other information. Each participant involved in the project can use the building information model to obtain the engineering information that they want, and can make appropriate adjustments based on construction conditions.

In order to realize the digital twin of the model, in the construction phase, each participant is required to assign a technical staff to enrich the information contained in the model. The technical staff of the construction unit are responsible for simultaneously updating the construction status and related information of the virtual model in the digital twin platform. The supervisory unit needs to update the supervisory status synchronously in the digital twin platform. Figures 3 and 4 show the digital twin information of the model for the construction unit, owner unit and supervision unit during the construction phase.

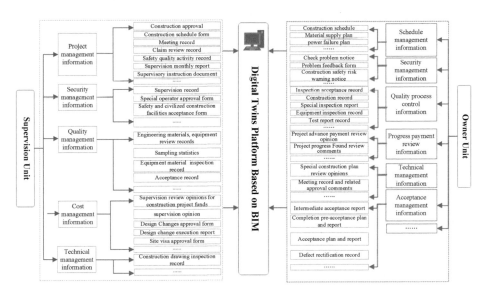

Fig. 4. The digital twin information of the model for the owner unit and supervision unit in the construction stage

2.4 Engineering Operation and Maintenance Stage

The information model still needs to play an important role after the project is delivered. The BIM and digital twin technology can help the operation and maintenance unit to carry out project management, so that the project can be managed more effectively during a reasonable service life. The site information can be obtained by using BIM digital twinning technology, such as the operating status of the transformer substation and the real-time status of the transmission line. The BIM digital twinning technology can perfectly simulate the operation status of the project, including the operation cycle and operating environment, and it is easy to evaluate the structural performance and resource consumption of the device.

3   The Application Value of BIM Digital Twin in Power Engineering

3.1   The Target Management Value During the Life Cycle

Through BIM digital twinning technology, it can dynamically manage project construction goals in real time. In schedule management, by correlating the task schedule with the project digital twinning model, it is possible to visualize the construction schedule and simulate the design and construction process. In terms of cost management, digital twinning model can promote the communication and coordination of project cost management in the whole process, improve the calculation accuracy and calculation speed of engineering quantity, and ensure the efficiency and accuracy of multi-dimensional cost analysis [8]. In terms of safety management, BIM twin model can dynamically monitor the construction site in real time, accurately locate the construction personnel and be able to alert the safety range to prevent safety risks. In quality management, digital twins on the construction site can monitor concealed projects in real time, prevent quality risks. Also, the BIM digital twinning model has certain application value in engineering object management such as material equipment management and file management.

3.2   All Participant Application Value

Project participants can make rational use of BIM-based digital twinning models.

For the project builder, the BIM digital twin model can obtain the current status of the base, three-dimensional landform and other information in real time, which is convenient for the construction unit to carry out visual comparison and selection, three-dimensional measurement, substation site selection, transmission line selection, initial cost and construction period estimation, and environmental assessment, etc.

For the design company, it can use the BIM digital twinning model generated in the previous planning stage, and carry out 3D design based on it, which can be combined with the accurate information of the site for modeling. Through the BIM digital twinning platform, the collaborative design between various professions can be realized, and the communication cost can be reduced and can make the design coordination and improve the design efficiency.

The construction enterprise reuses the digital twinning model in the design phase and continuously increases the real-time engineering information during the construction phase. The construction unit management personnel can use the virtual digital twin model to monitor the construction of the project and update the construction progress in real time to ensure that the original plan matches the actual progress. In addition, the platform can carry out detailed management work such as site layout, 3D technical disclosure, schedule control, site safety management, finding progress risk points, real-time statistical engineering quantity and price, and progress payment [10].

The supervision company can manage and control the remote security and real-time environment of the project by using the real-time dynamic BIM model. The BIM digital twins platform is used to realize digital management of engineering supervision, including digitalization of engineering supervision information, paperless engineering supervision files, information transmission network, and intelligent information retrieval [11, 12].

4 Conclusion

Based on the concept of digital twinning, taking BIM model as the carrier and taking the data as the core, it sorts out the realization path of digital engineering and the information content of the digital twinning model in the whole process of engineering construction, and the application value of BIM-based power engineering digital twinning is obtained. The application of BIM technology and digital twin technology in power engineering can effectively promote the informationization and digital construction of China's power industry.

References

1. Wang, L., He, Y., Han, Y.: Research on power engineering construction management based on BIM technology. In: IEEE 2018 International Conference on Robots & Intelligent System (ICRIS), Changsha, China, 26–27 May 2018, pp. 107–109 (2018)
2. Zhang, H., Gao, Y., Ding, C.: Research of BIM theoretical system based on big data thinking. Project Manag. Technol. (5) (2019)
3. Smith, P., Sbspro, J., Overruns, P.C.: BIM & the 5D project cost manager. Procedia Soc. Behav. Sci. **119**, 475–484 (2014)
4. Qi, Q., Tao, F.: Digital twin and big data towards smart manufacturing and Industry 4.0: 360 degree comparison. IEEE Access **6**, 3585–3593 (2018)
5. Liang, J., Zhang, M., Yuan, Y.: Construction project collaboration management based on the BIM. In: International Conference on E-business & E-government (2011)
6. Zhuang, C., Liu, J., Xiong, H., et al.: Connotation, architecture and trends of product digital twin. Jisuanji Jicheng Zhizao Xitong/Comput. Integr. Manuf. Syst. CIMS **23**(4), 753–768 (2017)
7. Liu, C., Zhou, Q., Xu, L., Yin, Y., Su, Q.: Research and application of key technologies in "intelligent, transparent and green" digital twin construction site. Constr. Technol. **48**(01), 9–13 (2019)
8. Lili, Z.: Control and management of construction project cost. J. Landsc. Res. **8**(4), 7 (2016)
9. Sun, K., Liu, R.: Inheritance and innovation of engineering management informatization. Front. Eng. Manag. **1**(1), 76 (2014)
10. Smith, P.: Project cost management – global issues and challenges. Procedia Soc. Behav. Sci. **119**, 485–494 (2014)

11. Jian, W.: Study on management information system for construction engineering supervision based on ubiquitous environment. In: IEEE International Conference on Ubi-Media Computing. IEEE (2010)
12. Yongling, B., Shiliang, L., Xing, Z.: Digitalization management mode of engineering supervision qualification and its application. Technology Supervision in Petroleum Industry (2018)

Western Culture MOOC Teaching in the Age of Big Data

Wei Yao[1(✉)] and Yanping Huang[2]

[1] School of Foreign Language, Wuhan University of Technology, Wuhan, China
yaoweivivian@qq.com
[2] International College of Chinese Studies,
Fujian Normal University, Fuzhou, China

Abstract. In this era of information technology, big data has had a huge impact on education. As a new online product of big data, MOOC offers a good platform for students to learn efficiently, effectively and individually. Western Culture course is a selective English course for college students in many universities of China. For a long time, Western Culture courses in China's universities have been taught by the traditional teaching mode: teachers are the instructors in teaching activities; students are the recipients of knowledge. In terms of teaching materials, textbooks in the traditional Western Culture courses emphasize western cultural history, containing many cultural concepts, which are difficult for students to learn and memorize. It is not conducive to stimulate students to learn. Students are easily fed up with the teaching and learning. This study researches the application of MOOC in the Western Culture course in a key university of China. It analyzes the advantages of applying MOOC in the Western Culture course, and proposes the corresponding curriculum design.

Keywords: Big data · Massive Open Online Course · Western Culture

1 Introduction

The rapid development of Internet technology has opened up a new era—the era of big data, which has been regarded as one of the top 10 revolutions in the coming decade [8]. Big data does not just mean "a lot of data". It has been defined with 5V's: Volume, Velocity, Variety, Variability and Value [6, 10]. Big data contains huge amount of data, the size of which is beyond our imagination. It is even difficult for some software to process and analyze. The big data not only generates economic value, but also has potential scientific and social value. The advent of the big data era has enabled humans to have the opportunity to use these tremendous, extensive, various, comprehensive and systematic data in many fields to explore the world and to acquire knowledge that was impossible in the past. Big data has penetrated into all aspects of the world, affecting people's value systems, knowledge systems and lifestyles. The field of education is also impacted by its rapid development. The typical application of big data in education is MOOC. Undoubtedly, MOOCs are an extremely useful and valuable emerging trend in online learning. It could provide a flexible and cost-effective method for large numbers of students from all over the world to enjoy high-quality education [5].

© Springer Nature Singapore Pte Ltd. 2020
M. Atiquzzaman et al. (Eds.): BDCPS 2019, AISC 1117, pp. 559–564, 2020.
https://doi.org/10.1007/978-981-15-2568-1_75

Western Culture Course is a selective English course for college students in many universities of China. With the globalization process, there are more and more cross-cultural communication among different countries. Countries all over the world depend on each other in many fields, such as economy, politics, science and technology. The teaching objective of Western Culture course is to help students be familiar with the customs, social conventions, history, literature and important achievements in the Western countries and acquire necessary communication skills to cope with possible culture conflicts in future workplace. For a long time, Western Culture courses in China's universities have been taught by the traditional teaching mode: teachers are the instructors in teaching activities, and students are the recipients of knowledge. In terms of teaching materials, textbooks in the traditional Western Culture courses emphasize western history, containing many abstract cultural concepts, which are difficult for students to learn and memorize. It is not conducive to stimulate students to learn. Students would be easily fed up with the teaching and learning.

MOOCs in the era of big data have brought great challenges on traditional education. In the context of this era, it's necessary for English teachers to seek changes and reforms in Western Culture teaching. The emergence of big data can facilitate the teaching of Western Culture. Some teaching content or cultural connotations that are difficult to express under traditional means can be visually displayed by some online video resources. Proper selection of certain useful resources can deepen students' understanding of some concepts, stimulate their interest in western culture learning, and fundamentally reform traditional education and learning models. To arouse effectively students' interest in the course, an efficient way is to build MOOC course with the aids of rich, vivid and meaningful resources in the big data. This study attempts to introduce western culture by means of a large number of video resources to stimulate students to learn, to think and discuss critically.

2   The Development of MOOC in the Era of Big Data

MOOC is the abbreviation of Massive Open Online Course. Big data provides information support for its popularization. The concept of MOOCs in digital education appeared firstly in 2008 [9]. It caused widespread concern in 2011 [3]. The year after that, some top universities in the United States started to establish online learning platforms and offered students free courses online. The Coursera and edX set up by Stanford, together with Udacity jointly established by Massachusetts and Harvard University, became the leading online MOOC institutions in the world. In December of the same year, 12 top universities in the UK followed the trend of MOOC in the era of big data and launched the Future Learn platform. Currently, the above platforms serve millions of enrollments.

MOOC integrates a variety of social networking tools and various forms of digital resources to form a learning tool rich in diverse curriculum resources. The characteristics of the MOOC is that the learning object, teaching methods, course content and educational concept are completely open. MOOCs, compared to traditional classroom courses, offer the chance for students to learn at any place, any time. Besides, MOOCs have no limitation for the numbers of students who can register and participate in the courses [4]. MOOCs comprise a complete teaching mode which includes all the important elements of teaching, such as, participation, feedback, homework, discussion, evaluation and exams. Moreover, the MOOCs platform can collect automatically a large amount of learning data during the teaching process, for instance, the study time required for each student, students' participation in question answering and discussion. By analyzing the systematic massive data, the teacher can learn and predict students' learning habits and problems.

The MOOCs institutions mentioned in the above are only typical representatives and leaders in this new area. More and more online learning websites for primary education, secondary education, lifelong education, vocational education and other fields are also rapidly emerging. MOOC has become an important way of education internationalization and popularization. By the end of 2017, it's estimated that there were more than 800 higher education institutions offering about 94,000 MOOCs for 81 million students all over the world [7].

3   The MOOC Design for Western Culture Course

3.1   The Teaching Content Design

The main contents of the MOOC for Western Culture course include teaching objectives, syllabus, teaching videos, culture videos, teaching instructions, exercise and practice, communication and discussion.

One of the teaching objectives of Western Culture course is to develop students' communication skills in order to meet their needs for effective cross-cultural communication in their future work and social interactions. The syllabus should be formulated to ensure that the course content is student-oriented, and stimulate the initiatives of both teachers and students.

In order to keep up with the pace of the western society, updated online resources should be adopted as the reading and listening materials. Traditional Western culture courses take ancient Greek and Roman culture, medieval Christian culture, Western modern culture, and history as teaching materials. Students generally complain about the tiresome and cumbersome curriculum. Teaching resources related to western culture in big data are plentiful and of different kinds. This study selects some primary themes from traditional Western Culture curriculum and reorganizes them according to students' demands and interests.

The teaching content includes 9 sections: culture introduction, the overview of the United Kingdom, the overview of the United States, Greek Mythology, Religions, Bible stories, Superstitions, Hall of Fame and Superhero Culture. According to the theme of each section, teachers collect related video and text materials, design teaching videos to explain cultural knowledge embedded in the materials. The online MOOC consists of cultural videos, teaching videos, teaching instructions, extended reading materials, exercise and practice, as well as an online interactive platform. Teaching videos are the core of the MOOC curriculum, and teaching instructions are summary of teaching content.

3.2   The Teaching Methodology Design

The shortcomings of MOOCs were found in many researches, such as lacking of instructor involvement, inability to interact with students, absence of physical and synchronous experience of campus life [1, 2]. The concept of blended teaching based on MOOC has already led hot debate. To achieve more efficient and effective learning, teachers should integrate online and offline teaching reasonably.

The teaching methods of Western Culture course include online MOOC and classroom teaching (Seen in Fig. 1). Online, with the guidance of the teaching instructions, students are required to watch relevant videos, read articles chosen from big data resources, and then watch teaching videos produced by teachers according to the themes of Western Culture curriculum. Finally, to make use of the information they have learnt, they need to interact in the online chatroom. In the classroom, with the facility of multimedia technology, group collaboration, task report, discussion, debate and other forms of teaching activities are conducted to explore more culture information concerning the theme. This teaching mode frees teachers from the cumbersome teaching of definitions and explanations of some abstract concepts. Teachers will have more time to answer questions and cultivate individualized teaching to foster students' capability in dealing with cross-cultural difficulties.

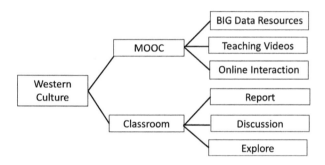

Fig. 1. Teaching design for Western Culture

3.3   The Assessment System Design

Since the teaching content and the teaching model of the course are reformed, it is also necessary to pay attention to the reform of the course Assessment system. With the blended teaching mode, students have a high degree of participation in both MOOC and classroom teaching. A combination of formative assessment and summative assessment is needed to replace the traditional assessment model. In order to better integrate the assessment and teaching, the assessment system of this course was designed under the following considerations. Firstly, in order to stimulate students to do self-study, the scope of assessment should be expanded. It should take into consideration online study time and online study assessment, the participation of classroom study as well as theme report presentation. Moreover, the theme report presentation of each group should be jointly evaluated according to assessment from peers and teachers. Quiz for each unit and final examination for the course is the assessment of the learning outcome, which are also an essential portion of a balanced assessment system.

4 Conclusions

The teaching of Western Culture course in the era of big data features information-ization and digitization. It requires teachers to have the ability to integrate, analyze data from different data sources in different periods. It also requires teachers to have the ability to effectively explore and research the value behind the data, and to extract more valuable information and data, so as to improve the teaching system and realize the all-round connection of data and teaching content. The big data has infinitely expanded educational resources. The role of teachers and students in the teaching process are changed in this context. Furthermore, it promotes reforms in teaching and learning model, which brought immeasurable impact on education. In the era of big data, English teachers should seize the opportunity and make active respond to the trend in order to achieve more efficient teaching effect.

Acknowledgements. This work was supported by No. 201702031022, Industry University Cooperation Collaborative Project of China's Ministry of Education.

References

1. Xiao, C., Qiu, H., Cheng, S.M.: Challenges and opportunities for effective assessments within a quality assurance framework for MOOCs. J. Hosp. Leis. Sport Tour. Educ. **24**, 1–16 (2019)
2. Chiu, K.F., Hew, K.F.: Factors influencing peer learning and performance in MOOC asynchronous online discussion forum. Australas. J. Educ. Technol. **34**(4), 16–28 (2018)
3. Hew, K.F., Hu, X., Qiao, C., Tang, Y.: What predicts student satisfaction with MOOCs: a gradient boosting trees supervised machine learning and sentiment analysis approach. Comput. Educ. **145**, 103724 (2019)
4. Chan, M.M., Barchino, R., Medina-Merodio, J.-A., de la Roca, M., Sagastume, F.: MOOCs, an innovative alternative to teach first aid and emergency treatment: a practical study. Nurse Educ. Today **79**, 92–97 (2019)
5. Morales, M., Barchino Plata, R., Medina Merodio, J.A., Alario-Hoyos, C., Hernández Rizzardini, R., De la Roca, M.: Analysis of behavioral intention to use cloud-based tools in a MOOC: a technology acceptance model approach. J. Univ. Comput. Sci. **8**, 1072–1089 (2018)
6. Lozada, N., Arias-Perez, J.: Big data analytics capability and co-innovation: an empirical study. Heliyon **5**, e02541 (2019)

7. Shah, D.: By the numbers: MOOCs in 2017. https://www.class-central.com/report/mooc-stats-2017/. The Hong Kong Polytechnic University (2016). Handbook on the PolyU's quality assurance framework, mechanism and process for academic departments. The Hong Kong Polytechnic University 45, Hong Kong (2018)

8. Shaikh, A.R., Butte, A.J., Schully, S.D., et al.: Collaborative biomedicine in the age of big data: the case of cancer. J. Med. Internet Res. **4**, e101 (2014). https://doi.org/10.2196/jmir.2496

9. Siemens, G.: Massive open online courses: innovation in education. Open Educ. Resour. Innov. Res. Pract. **5**, 5–15 (2013)

10. Willems, S.M., et al.: The potential use of big data in oncology. Oral Oncol. **98**, 8–12 (2019)

Landscape Sand Table System Based on Deep Learning and Augmented Reality Technology

Pan Ou$^{(\boxtimes)}$, Moran Chen, and Qingfeng Yu

School of Instrument Science and Optoelectronic Engineering,
Beihang University, Beijing 100191, China
chenmoran0928@163.com

Abstract. Landscape sand table has been widely used in many fields, and higher requirements have been put forward for landscape sand table system. Therefore, the traditional landscape sand table system cannot meet the needs of current use. In addition, the traditional research has great limitations and cannot meet the requirements of high restoration and precision of the landscape sand table. In this paper, on the basis of deep study and augmented reality, combine the two, the in-depth study in augmented reality, and with landscape as the research object, combined with image processing and network technology knowledge, based on extensive analysis of the data source, this paper proposes a method to build a landscape sand table system with higher efficiency and restoration degree. This method is a deep learning algorithm constructed through parameter optimization and experimental verification, which can accurately identify scene information and realize augmented reality effect. In this paper, the stochastic gradient descent algorithm was used to conduct an experiment to improve the accuracy of the target detection in the landscape sand table. The experimental results also showed that the augmented reality technology could make the landscape sand table with high reproducibility and stereoscopic effect, and the use of the augmented reality effect could make the landscape sand table show better development.

Keywords: Deep learning · Augmented reality · Landscape sand table ·
Stochastic gradient descent

1 Introduction

As a model based on topographic maps, sand table has a long history in China and was mainly used in the military field at the earliest time [1]. With the development of society, sand table has been widely and deeply applied in other fields. There are many kinds of sand table, such as terrain sand table, military sand table, construction sand table and so on. In recent years, the sand table also began to be applied to the landscape, thus the landscape sand table appeared. As a landscape sand table, it is required to have high reducibility and visualization of the original appearance of things [2, 3]. However, the traditional sand table system simply carries out a basic restoration of the landscape with the help of image processing, geographic positioning and other

© Springer Nature Singapore Pte Ltd. 2020
M. Atiquzzaman et al. (Eds.): BDCPS 2019, AISC 1117, pp. 565–570, 2020.
https://doi.org/10.1007/978-981-15-2568-1_76

knowledge. Therefore, the efficiency and accuracy are not high, and the most authentic features of the landscape cannot be restored, affecting the overall sense of use [4].

In recent years, the research on the landscape sand table system has become a hot topic, but it is generally limited to the research on traditional landscape sand table which do not combine with deep learning and augment reality. And the research on the landscape sand table system based on deep learning alone cannot guarantee the intuitive of the sand table, so a broader method must be explored [5, 6]. With the development of science and technology, augmented reality technology has gradually entered people's vision. Due to its characteristics of three-dimensional registration, real-time interaction and the combination of virtual and real, it can restore things accurately to the greatest extent, so it is gradually applied in many fields, such as military field and tourism industry. The high reducibility requirements of the landscape sand table system coincide with the function of augmented reality technology, which can further promote the accuracy and high reducibility development of the landscape sand table system [7, 8]. The research on sand table system at home and abroad is still focused on the field of a lot of manual production, no automatic detection and augmented reality effect, so the relevant research is not deep enough, and the establishment of sand table system that cannot meet higher requirements has many limitations.

In order to make up for the research gap, based on the landscape sand table as the research object, combined with image processing and network technology knowledge, to create a library of landscape database from different data sources, research the influence of data on network model generalization ability, this paper higher degree of precision methods of landscape sand table detection system, and proposes a higher precision of system landscape layout algorithm, the optimized parameters and experiment is constructed based on the technology of deep learning and enhance landscape automatic identification network framework [9, 10]. The establishment of the algorithm framework of this kind of landscape sand table system is of great guiding significance to the establishment of the future high-precision and high-restoration sand table system, and lays a certain foundation for the later research on relevant aspects [11].

2   Method

2.1   Core Concepts

The meaning of augmented reality technology is a new high-tech technology that generates three-dimensional information synchronously with the help of computer technology and network technology. For example, AR technology is one of them. Because this technology is developed on the basis of virtual reality, virtual reality is to create a new virtual world, while augmented reality is more focused on the combination of virtual and real. The advantage of augmented reality technology is that it breaks through the virtual world that is only limited to the computer-generated virtual world presented by virtual reality and cannot truly see and perceive the real world that exists. The essential difference between augmented reality technology and virtual reality technology is that augmented reality technology can present the real world to users while allowing users to see the virtual illusion based on the real world at the same time.

The real world and the virtual world can be bridged by augmented reality technology, which is a supplement to the virtual world and an extension of the real world. Augmented reality technology has three main characteristics, namely, three-dimensional registration, real-time interaction and the combination of virtual and real, which use additional information to enhance the user's sensory perception of the real world, instead of the real world will be replaced by the virtual world.

2.2 Stochastic Gradient Descent Algorithm

After selecting the network model for the landscape sand table system, some optimization methods can be used to study the network training parameters. The algorithm usually used to train the parameters of the CNN model is stochastic gradient descent, which USES back propagation to calculate the gradient to update the weight, and automatically learns the parameters by continuously reducing the output value of the loss function. Back propagation is the most commonly used algorithm in multi-layer networks. The core of this algorithm is to apply the chain rule to calculate the influence of each weight on the loss function in the network, so as to find the global optimal solution of the network. The goal of SGD algorithm is to minimize the loss function so as to find the optimal parameter with good generalization ability in the network model. In order to quantify the role of SGD algorithm, the form of loss function is defined as:

$$C(w, b) = \frac{1}{2n} \sum_{x} \prod y(x) - a \prod^2 \tag{1}$$

Where said the network weights in the collection, a collection of b is the bias, n is the number of samples of each batch, a for output when the input is x, y (x) said the expectations in the corresponding output or samples for each category value, the classification of the label in advance sum is losing λ aspirant λ x in the general training. SGD algorithm the basic idea is through the random input sample small batch training, to continuously make the loss calculation function C (w, b) the output value reduced gradient, update the network parameters at the same time, to find global minimum value when the loss function of the optimal solution, the network weights and bias of the specific update methods you can use the following formula, said the vector for, is a small positive number:

$$w \rightarrow w' = w - \frac{\eta}{m} \sum \frac{\partial C(w, b)}{\partial w} \tag{2}$$

$$b \rightarrow b' = b - \frac{\eta}{m} \sum \frac{\partial C(w, b)}{\partial b} \tag{3}$$

3 Experiment

Step1: collect sample data and set up the database. As the research object of this paper and the core component of the training deep learning identification network, the authenticity of the data source and the clarity of the picture quality of the landscape sand table are two factors that should be considered when selecting the data. In this paper, the digital landscape sand table with the widely used landscape data sets.

Step2: set up the network architecture structure. In this paper, the research applies CNN network which is chosen to train landscape database, and the identification accuracy by comprehensive consideration of time consumption, existing equipment performance and the amount of landscape factors, such as classifying landscape image CaffeNet model, can not only ensure higher recognition accuracy, also can save a lot of time cost, at the same time avoid the deeper network in fitting the emergence of the phenomenon.

Step3: data integration based on the random gradient descent algorithm. The collected data sources and the built network system were built into the stochastic gradient descent algorithm formula through deep learning, and the final results were integrated and analyzed. On this basis, the conclusion of constructing landscape sand table system is drawn.

4 Discuss

4.1 Analysis of Test Results

In order to verify the feasibility of the sand table landscape automatic recognition method proposed in this paper based on deep learning and augmented reality technology, we mainly conducted some tests from the following three aspects. Firstly, the experimental landscape data set is fully trained. The data includes Building, Vehicle, Mountain, Road, River and Forest 6 types of landscape. Batch size and training epochs are adopted to make the trained model not only have high accuracy, but also have good generalization ability. Secondly, compared with the traditional GD (Gradient Descent Algorithm) Algorithm, the detection results of each category are more accurate, which indicates that the use of random Gradient Descent is superior to the traditional Gradient Descent. Finally, the result also shows that the recognition of multi-scale targets also has a high recognition ability. The training set and test set are only used in the training process of the network model, and do not participate in the verification of the network model's generalization ability and other performance after the training. The final specific results are shown in Table 1 below. All the data in the table are sorted out by the author after calculation.

Table 1. Test results

Sample	Building	Vehicle	Mountain	Road	River	Forest
Num of training samples	4243	3390	3022	3988	3932	3043
Num of testing samples	521	502	592	492	490	321
Num of training epoch	400	400	400	400	400	400
Batch size	16	16	16	16	16	16
Precision of GD algorithm	92.7%	94.6%	94.1%	93.9%	91.6%	93.2%
Precision of SGD algorithm	96.3%	97.2%	98.1%	96.9%	95.4%	97.3%

4.2   Design and Implementation of AR Sand Table System

Through above calculation and model test it is easy to find, based on the basis of depth study to optimize the new algorithm, with a collection of augmented reality could be further strengthens the reality environment of people awareness, to feel and what to simulate a whole new world and enhance our awareness of the real world, let the real world and simulated by computer technology has become increasingly blurred the line between the virtual world. So on this basis, through a large number of calculations designed a new system. The system meets the arithmetic requirement of CNN and stochastic gradient descent method and passes the objective test. And for all kinds of elements in the three dimensional system for editing, not only keep the basic analysis functions of geographical information system (measurement functions, generating contours, generate terrain profile, horizon, threat analysis, etc.), 3D virtual technology characteristic function (in the scene observed freely according to the requirements in the scene, can realize Angle of translation, rotation, near and far scene function such as scaling, fixed-point scaling), more according to the characteristics of the different application requirements in the field of design, querying, editing features, its special annotations. According to the author's statistics, the scholars have high expectations for the landscape sand table technology based on deep learning and augmented reality technology. The specific industry situation that they hope to widely promote is shown in Fig. 1. The data in the figure is the result of the author's analysis after investigation.

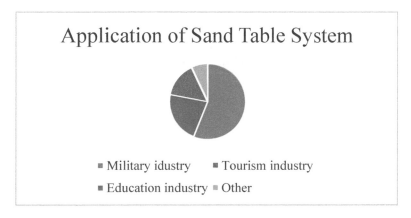

Fig. 1. The expected application industry of landscape sand table system in the future

5   Conclusion

Augmented reality (ar) landscape sand table system innovatively combines the substantiality of the traditional sand table with the virtuality of the new electronic sand table to combine virtual information on the real world, so as to ensure accurate acquisition of information while still having a good three-dimensional effect. It has the advantages of high reappearance accuracy, stereoscopic visualization, short production period, convenient handling and high stability. On the basis of deep learning, with the development of augmented reality technology and the continuous improvement of geographic information data, the augmented reality landscape sand table system will be widely applied in various other industries to provide users with more intuitive, convenient and intelligent services.

References

1. Bo, W., Xia, W., Fei, C.: Pavement crack recognition based on aerial image. Acta Optica Sinica **37**(8), 81–104 (2017)
2. Xie, Y.: Design and implementation of 3D electronic sand table in intelligent scenic area. Surv. Mapp. Spat. Geogr. Inf. **40**(10), 89–91 (2017)
3. Yang, H.W., Liu, Y., Xu, M.: Design and practice of landscape sand table model. Geospatial Inf. **08**(6), 95–97 (2018)
4. Zhang, X.T., Duan, J.M., Yu, L.J.: Sand table model making of garden landscape. Shandong Industr. Technol. **25**(10), 1403–1412 (2017)
5. Liu, N., Liang, M.X., Chen, L.L.: Reconstruction of 3D virtual landscape sand table in Haishi art park. J. Hebei North Univ. (Nat. Sci. Ed.) **33**(3), 472–481 (2017)
6. Chesnokova, O., Purves, R.S.: From image descriptions to perceived sounds and sources in landscape: analyzing aural experience through text. Appl. Geogr. **93**, 103–111 (2018)
7. Wang, Y.Y., Xie, H.: Research on the monitor terminal of container yard control system sand table model GIS-based. Chin. J. Environ. Eng. **7**(11), 4513–4518 (2017)
8. Bernhardt, S., Nicolau, S.A., Soler, L.: The status of augmented reality in laparoscopic surgery as of 2016. Med. Image Anal. **37**(2), 66–90 (2017)
9. Grubert, J., Langlotz, T., Zollmann, S.: Towards pervasive augmented reality: context-awareness in augmented reality. IEEE Trans. Vis. Comput. Graph **23**(6), 1706–1724 (2017)
10. Rehman, U., Shi, C.: Augmented-reality-based indoor navigation: a comparative analysis of handheld devices versus google glass. IEEE Trans. Hum. Mach. Syst. **47**(1), 140–151 (2017)
11. Alhaija, H.A., Mustikovela, S.K., Mescheder, L.: Augmented reality meets computer vision: efficient data generation for urban driving scenes. Int. J. Comput. Vis. **2**, 1–12 (2017)

Analysis of Student Financial Aid Based on Big Data Analysis

Jingang Bai[✉]

Chifeng University, Chifeng 024000, China
1461942076@qq.com

Abstract. Since the reform and opening up, China's financial aid for poor students has gone through three different stages of development, and each stage has its own characteristics. In order to effectively carry out the aid work for poor students, it is necessary to analyze and evaluate the effect of the aid, so as to achieve better aid effect. The ultimate starting point of student aid is to prevent any student from going to school because of poverty. However, in the actual funding work, there are many problems, such as difficulty in identifying funding, ineffective funding, and confusion in the implementation and management of funding. In order to solve this problem, this paper proposes a method based on big data analysis to analyze the effect of student aid. Through the analysis and guidance of this method, we can make better use of the state's financial aid, so that students who really need financial assistance can get effective help and help them complete their studies smoothly.

Keywords: Big data analysis · Poverty assistance · Funding analysis · Funding confirmation

1 Introduction

Over the past 40 years of reform and opening up, China's economic level has been developing rapidly, but there are still many students whose families are poor and unable to complete their studies due to various reasons. In order to solve this major problem, China has gradually introduced the implementation of nine-year compulsory education, but also formulated a series of financial assistance system for poor students, to ensure that poor students have difficulties in school, to help them successfully complete their studies. However, there are many problems in the implementation and implementation of student aid. Only by fully analyzing the effect of the aid can the work of the aid be better and the students who really need the aid get the aid.

At present, there are many different policies at home and abroad for the methods of student aid measures. In document [1], the author introduces a new performance-based funding (PBF) model developed by Texas for poor students in community colleges. In the literature [2], the author studies the changes of academic performance in low-income school districts caused by the reform, and the results show that school resources have a great influence on educational achievement. In the literature [3], the author focuses on the comparative analysis of the employment results of the graduates from 2015 to 2016 and the graduates funded by the British Higher Education Grants

© Springer Nature Singapore Pte Ltd. 2020
M. Atiquzzaman et al. (Eds.): BDCPS 2019, AISC 1117, pp. 571–577, 2020.
https://doi.org/10.1007/978-981-15-2568-1_77

Committee. Through the study, the author analyses whether the poor students have also been properly improved in employment after receiving the grants. In the literature [4], the author introduces that many states in the United States have formulated performance fund subsidies in order to encourage students to complete degree courses in public universities. There are many ways of subsidizing policy, but how to ensure that the existing subsidizing policy can maximize the use of special subsidizing funds and effectively subsidize students in need is an urgent problem to be solved. With the advent of the information age, our information data storage is becoming larger and larger, and information processing technology has been developed rapidly. Big data analysis, as a popular data analysis method, plays an important role in many aspects. In the literature [5], in order to meet the growing challenges of agricultural production, we need to better understand the complex agricultural ecosystem. In this document, the author objectively analyzed the whole complex agro-ecosystem by using large data analysis. In the literature [6], the author uses data collected by sensors everywhere in the city to analyze and study the operation of smart cities using big data analysis technology. In document [7], the author applies large data analysis to the management of security cluster, which can optimize the implementation system of control plane. In the literature [8], the author applies large data to medical data analysis to help doctors correctly judge patients' condition and analyze their condition. In reference [9], the author proposes a storage plan scheme supporting big data based on wireless big data computing, and optimizes it with hybrid method, including genetic algorithm for storage plan and game theory for daily energy scheduling. In the literature [10], the author has made some contributions to international news stream literature by using big data analysis method in the global media agenda.

In view of the current results of student aid, this paper uses big data analysis method to fully understand the direction and implementation effect of the financial aid. Through data analysis, we can know which students really need to be subsidized and how much the scope of the financial aid provides a great reference value.

2 Method

The principle of big data analysis is to collect a large amount of data and then use the research method of big data to analyze the data. This big data analysis method can effectively solve the analysis requirements that the current database technology cannot meet, and improve the timeliness of data processing operations and the speed of problem-solving response. The big data analysis method used in this paper is based on Hadoop architecture.

2.1   Design of 2.1 Map/Reduce Computing Architecture

Under the condition of constructing Hadoop cluster name node and data node, we use Map and Reduce to develop user programming, so that we can use the partitioning program to calculate the requirement by subsidizing the flow data. This program is divided into many sub-tasks. The system uses data it points to process and summarize the sub-tasks. When deleting the data, first clean up the data, count the unusual data in the data, analyze the relationship between the unusual data and the normal data, and get the proportion of the abnormal data in all the data. Here we use Map/Reduce computing architecture to implement student-funded data cleaning and data slicing process, as shown in Fig. 1 below.

Fig. 1. Data cleaning and data slicing process flow

2.2 Task Configuration and Distribution in Map Phase

The system distributes data cleaning and slicing implemented by our program to a specific Map task. Slaver roles are all handed over to some data nodes of Map, and then the task of dispatching name nodes is also handed over to the data nodes of Map for operation.

2.3 Task Merging and Data Storage in Reduce Phase

Here, the job input data stream in Reduce phase is equivalent to the output data stream in Map phase described earlier.

After each student subsidizes the data key pair according to the relation of two rows arrangement, we process the data with Reduce function, which is the necessary precondition step for the unification of the output data processed by Map function in Reduce stage. Because the operation time of data in Map/Reduce phase is different and uncertain, the time to save these data in the database is also uncertain.

3   Experiment

The innovation of the funding mode needs to consider comprehensively how to scientifically identify the poor students, deeply analyze the students' funding needs, closely track the funding process, and scientifically and uniformly evaluate the funding results, so as to better improve the existing funding mode.

3.1   Scientific Recognition of Poverty-Stricken Students

At present, the way to identify poor students is generally through the method of proof material, observation and certification, and through visits to classmates and democratic evaluation. The proof material method is a complicated procedure and relies too much on written materials. It is sometimes difficult to effectively guarantee the authenticity and reliability of the proof material, and there are a lot of fraudulent acts. The observation and certification rules are mostly based on the observation of the students' consumption behavior by teachers in peacetime, and the students' real poverty situation can be understood by visiting their families; and the methods of democratic evaluation. Similar to the observation method, the object of observation has changed from a teacher to a classmate of the observee, but this method is easily restricted by the personality and relationship of the observee, and sometimes it is difficult to draw scientific conclusions.

3.2   Developing Individualized Approaches to Funding

The traditional way of financial aid often adopts the one-size-fits-all form, which emphasizes the single material form of financial aid and is difficult to fully meet the needs of students with financial difficulties in their study, life and growth. Through large data analysis, we can effectively allocate and individualize funding resources, which is also an important means to maximize the benefits of limited funding funds. On the basis of taking care of the basic needs of poor students, we use quantitative indicators to determine the factors causing poverty in students' families, adopt different funding standards for different poverty levels, and ensure that the limited funding is used on the cutting edge.

3.3   Implementing Effective Management Measures

Developing a real-time tracking and dynamic adjustment mechanism is an important part of our efficient funding. The large data integrated information management platform is used to establish the early warning and elimination mechanism for each student to realize the dynamic management of the subsidized objects. Establishing an early warning mechanism should give timely attention to the students who are subsidized, and other students who are not subsidized should also pay timely and effective attention to their lives, studies, psychology and so on. If there are difficulties in these areas, early warning and help measures should be formulated in time. By using the big data integrated management platform and prediction function, the management system of information and support platform such as campus card, bank card and campus network for the assisted students will be established. Through the analysis of students' consumption data, academic data and network data, more scientific screening and early warning and eliminating measures will be formulated.

4   Result

Result 1: More scientific funding results

Through the effective analysis of large data, we have come up with a more scientific way to manage financial aid. For students who suffer from different reasons, we have adopted different proportion of reward and assistance methods to effectively use the limited funding to the maximum extent. On the target of subsidized students, we also formulate relevant elimination mechanism through data analysis. For subsidized students, we analyze their consumption behavior by accessing their bank card and campus card. If the student's consumption behavior for a long time has been basically close to the level of mass consumption, it will stop or reduce the degree of subsidization of the student. Learn to use the results of big data analysis to improve existing funding methods.

Result 2: Teacher and student satisfaction is higher

Since the implementation of student aid management method based on big data analysis, it has been unanimously recognized by teachers and students. To some extent, the traditional one-size-fits-all financial aid method cannot solve the problem of students' difficulties very well, but the financial aid method after big data analysis can better distribute the financial aid to students in different situations with different proportions of the amount of financial aid, and strive to make sure that every student will not be unable to complete their studies because of poverty. The results of a survey on the satisfaction of 600 students in a school are shown in Fig. 2. From the Fig. 2, we can see that more than 80% of the students are satisfied with the funding method, more than 20% of the students are very satisfied, while only 0.2% of the students feel very unsatisfactory.

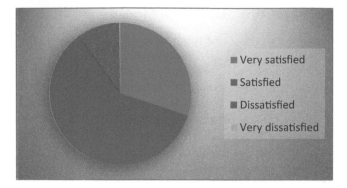

Fig. 2. Satisfaction survey on funding scheme implementation

5   Conclusion

Through the analysis of the effect of student aid based on big data analysis, the financial aid program is more scientific, and the purpose of making rational use of the financial aid is really achieved. It also does not omit any students who need financial aid, let alone let students who do not meet the criteria of poverty aid receive the funds that should not have been received. Only by scientific tracking system and perfect financial aid management methods can every poor student not be deprived of school because of poverty, so that every student can study with ease and grow happily.

References

1. Zhang, K., Zuo, W., Chen, Y., et al.: Beyond a Gaussian denoiser: residual learning of deep CNN for image denoising. IEEE Trans. Image Process. **26**(7), 3142–3155 (2017)
2. Gondara, L.: Medical image denoising using convolutional denoising autoencoders. In: 2016 IEEE 16th International Conference on Data Mining Workshops (ICDMW), pp. 241–246. IEEE (2016)
3. Lefkimmiatis, S.: Non-local color image denoising with convolutional neural networks. In: Proceedings of the IEEE Conference on Computer Vision and Pattern Recognition, pp. 3587–3596 (2017)
4. Zhang, K., Zuo, W., Zhang, L.: FFDNet: toward a fast and flexible solution for CNN-based image denoising. IEEE Trans. Image Process. **27**(9), 4608–4622 (2018)
5. Xie, Y., Gu, S., Liu, Y., et al.: Weighted Schatten p-norm minimization for image denoising and background subtraction. IEEE Trans. Image Process. **25**(10), 4842–4857 (2016)
6. Jin, K.H., McCann, M.T., Froustey, E., et al.: Deep convolutional neural network for inverse problems in imaging. IEEE Trans. Image Process. **26**(9), 4509–4522 (2017)
7. Choi, K., Fazekas, G., Sandler, M., et al.: Convolutional recurrent neural networks for music classification. In: 2017 IEEE International Conference on Acoustics, Speech and Signal Processing (ICASSP), pp. 2392–2396. IEEE (2017)

8. Nah, S., Hyun Kim, T., Mu Lee, K.: Deep multi-scale convolutional neural network for dynamic scene deblurring. In: Proceedings of the IEEE Conference on Computer Vision and Pattern Recognition, pp. 3883–3891 (2017)
9. Lakhani, P., Sundaram, B.: Deep learning at chest radiography: automated classification of pulmonary tuberculosis by using convolutional neural networks. Radiology **284**(2), 574–582 (2017)
10. Kruthiventi, S.S.S., Ayush, K., Babu, R.V.: DeepFix: a fully convolutional neural network for predicting human eye fixations. IEEE Trans. Image Process. **26**(9), 4446–4456 (2017)

Tracking Method of Agricultural Products Logistics Based on RFID Technology

Jianbo Wang and Jiayue Wang[✉]

Information School, Hunan University of Humanities, Science and Technology,
Loudi 417000, Hunan, China
984790270@qq.com

Abstract. With more and broader concerns paid to the quality safety of farm products, RFID technology is also gradually known by the public with the development of the Internet of things, so how to use RFID technology to build better agricultural products logistics information management system to improve the quality safety of agricultural products has become a hot topic. The establishment of the agricultural products logistics system based on radio frequency identification technology (RFID) can effectively ameliorate the efficiency of agricultural products logistics, reduce losses in the process of agricultural products logistics, realize the effective tracking and tracing of agricultural products, prevent the harm of fake and inferior products to authentic products, and ensure the safe consumption of consumers. This article simply introduces the present situation of China's agricultural products logistics and information management, analysis of the agricultural products logistics information management system based on radio frequency identification technology (RFID) application in the practical significance, and puts forward the agricultural products logistics information management system based on radio frequency identification technology (RFID) in the specific design ideas and implementation process, to further improve the quality and safety of agricultural products in our country, reduce logistics cost, speed up the information flow at the same time realize the traceability information in different stages of the agricultural products, to enhance the food safety, ensure the safety of consumer spending.

Keywords: Agricultural products logistics · Food safety · Quality traceability · Radio frequency identification technology (RFID)

1 Introduction

RFID, commonly known as electronic tag, is a non-contact automatic identification technology, which can identify multiple tags at the same time when identifying high-speed moving objects. The identification work does not want manual intervention, and it can be applied to various harsh environments [1, 2]. It automatically identifies the target object through the radio frequency signal, and carries on the mark, the registration, the storage and the management to its information, has the efficiency high, the anti-counterfeiting ability is strong, the operation is quick and convenient and so on characteristic. The disadvantage is that user promotion and update costs are very high [3]. In the field of material flow, RFID technology can be put into use to product

© Springer Nature Singapore Pte Ltd. 2020
M. Atiquzzaman et al. (Eds.): BDCPS 2019, AISC 1117, pp. 578–583, 2020.
https://doi.org/10.1007/978-981-15-2568-1_78

assembly and production management, automatic storage and inventory management, product material flow tracking, self-motion supply chain management, product anti-counterfeiting and other sides. A large number of RFID label can improve the level of logistics operation management [4, 5].

In recent years, RFID technology has developed rapidly. From a global perspective, the United States is the leader in the apply of RFID technology, leading world in terms of RFID standard, software and hardware development and application [6]. The European standard follows the us-led EPC global standard, and in closed system applications, Europe and the us are basically at the same stage. Japan has put forward a set of UID standards relying on some domestic manufacturers, and it still has a long way to go in internationalization [7]. For emerging countries like South Korea, the importance of RFID has been highly valued by the government, but RFID standards are still vague [8]. RFID technology in China developed late, but developed rapidly, and has been widely applied in catering, logistics, retail, manufacturing, medical treatment, identity identification, payment and other fields, enhancing public safety, improving people's quality of life, and improving the economic benefits of enterprises [9]. According to the prediction of ABI, in the next few years, China's RFID industry will maintain a sustained growth trend of 30%–50% per year [10].

The application of RFID science in agricultural products logistics, as a special branch of the material flow industry, is an important part of agricultural logistics and rural logistics, including a series of relation such as products, collection, processing, packaging, store, transportation and sales of agricultural products [11]. In the process of agricultural products logistics, we should ensure the quality and quality safety of agricultural products to the greatest extent, reduce losses and prevent pollution, so as to maintain and augment the worth of agricultural product. The development goal of agricultural products logistics is to increase the added value of agricultural products, save circulation costs, speed up circulation, cut away unnecessary losses and avoid market risks to some extent. RFID technology is more and more widely used in agricultural products logistics, mainly in agricultural products transportation and distribution management, warehousing management, circulation processing management and traceability management.

2 Method

2.1 Traceability Method

The system can trace the path of information management system from the origin of agricultural and sideline products to the whole process of selling agricultural products from labeling and initializing the labels. Dijkstra algorithm is used to settle the shortest routing problem of a singly source spot, and its formula is as follows:

$$M \quad in \quad .z = \sum_{i=1}^{m} \sum_{i=1}^{n} c_i x_i \tag{1}$$

If S is the set of the end points of the shortest path, and if the user of the subshortest path is j, it is conceivable that this path is either (I, j), the weight of the path whose length is from I to j, or the sum of the weights. Therefore, in general, the length of the next shortest path of this segment must be:

$$D[k] = \min\{D[k]\}, j \in (J - S) \tag{2}$$

2.2 Data Collection Method

Data collection data collection is the most important part of a logistics message administration system. Whether the information system is effective depends largely on whether the data collection is accurate. Therefore, the authenticity, accuracy and timeliness of the data must be guaranteed in the data collection stage.

In this system, it is necessary to install data acquisition devices in all links of the trade in agricultural and sideline products. Starting from source of supply chain, raw data should be collected at the place where agricultural products are produced, and corresponding electronic labels should be made for agricultural products to record their original production information. RFID technology can be used in the processing of agricultural products in the crude material and made-up articles for the rapid collection of information and input into the database to achieve information.

2.3 Information Processing

Agricultural products because of its own characteristics caused its information management system database information is huge. Therefore, how to deal with huge data in real time and carry out real-time feedback has become the primary problem to be solved. Therefore, the construction of this information management system also requires the construction of a cloud computing platform, which USES the computing resources of cloud computing to gather all information and data resources in the cloud for processing. At the same time, the introduction of the cloud platform can also make it very convenient for consumers to check the quality and safety of agricultural products in real time through computers, mobile phones and other Internet tools.

3 Experiment

The main idea of the study on the tracking means of agricultural products material flow account of RFID technology is to conduct processing research through Dijkstra algorithm. In the distribution operation module of the logistics system, enterprises need to establish a transportation-oriented model in the database. The model generally takes the minimum transportation cost as the optimization target, and considers the factors related to the transportation volume of the transportation route. After the system collects the information of transportation route and transportation volume through GPS, GIS and RFID, the data is input into the path optimization algorithm library to calculate the optimal path.

The following is the algorithm steps of agricultural product logistics tracking method research based on RFID technology:

Step 1. Determine the route weight. Firstly, an supplementary complexor D is introduced, and each element D[I] on behalf of the length of the shortest path currently found from distribution center I to each intermediate user. Its initial state is: if there is a feasible route from I to each intermediate user, D[r] is the weight of the route.

Step 2. Calculate the shortest path. If S is the set of endpoints of the obtained shortest path, it can be proved that the next shortest path (let its endpoint user be x) or the route (I, x) can be proved by contradiction. If there is a vertex on this path that is not in S, it means that there is a path whose endpoint is not in S and whose length is shorter than this path. However, this is impossible, so their endpoint must be in S, that is, the hypothesis is not true.

Step 3. Repeat the operation. A total of n − 1. Thus, the shortest path from I to the rest of the users on the graph is a sequence increasing in length according to the path length.

Step 4. Calculate and analyze according to the above steps, the shortest way can be used to realize the logistics tracking of agricultural products.

4 Discuss

4.1 Follow-up Management Analysis

Agricultural product circulation information tracking, in fact, is the product in the sales network circulation process and data management. Agricultural products sales tracking system to product delivery information as the source of information tracking. The manufacturer is the information source of the traceability system of agricultural products. At this node, the label is written into the product identification code of the corresponding product and attached to the agricultural products. As the unique number of the product in the circulation process, it is written into the label by the product manufacturer. It cannot be modified in the circulation process and can only be read by the node reader. As an agricultural product manufacturer, the operations to be completed at this node include: product release, product expiration monitoring, product theft, etc.

4.2 Information Detection and Analysis

The system can be divided into the internal testing management platform and the testing platform for ordinary consumers and regulatory departments when carrying out the information query of agricultural products circulation. The internal inspection of the enterprise refers to the circulation nodal enterprise of agricultural products according to the problems of product quality fed back by consumers or quality supervision departments, and the circulation process of the product is reverse inquired, so as to find the problems in time and improve. Testing for ordinary consumers and quality supervision departments means that ordinary consumers can quickly and conveniently understand the information of the products they buy through this platform, and can report and inquire about the problems in product quality through this platform. Qualitative inspect branch through the platform can manage the information in meat production enterprises, agricultural products safety information, and in the event of a product safety accidents timely query get detailed product information and quickly locate the source of the problem, rapid positioning and problem products to products of the same batch of suspected problems and recall in a timely manner, will be the scope of damage control in as small as possible. Meat enterprises can combine the system with decision-making management and other functions to improve the utilization rate and competitiveness of resources. The tables designed by the manufacturer event database include: product information table, product outbound table, product in-store table, product theft table, product expiration table, manufacturer information table and reader management table. The product information table is used to record each commodity code and its corresponding product information. It is the source of agricultural products traceability system information. The laboratory finding of product message are shown in Table 1 and Fig. 1.

Table 1. Product information test result table

Property	Detection method	Accuracy
EPC	Date set 1	70%
Product_Number	Date set Spark	80%
Guarantee_period	Date set 2	99%
Product_Date	Date set 2 graphs	98%

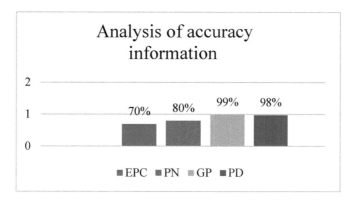

Fig. 1. Product information test results

On the basis of following the requirements of the traceability system of agriculture products cycle, combining ours research results of rf-based product circulation system and data processing based on complex event processing technology, the prototype system of traceability of agricultural products circulation process suitable for China's national conditions was designed and realized.

5 Conclusion

At present, the system based on technology is just in the initial stage in China, and the research on complex event processing technology is one of the few challenging subjects. This kind of system is very close to the research needs of the current agricultural product tracking and traceability system based on, so it has great practical significance. RFID technology tracking and traceability system, to achieve automatic management of product information, has become a global industry to improve management, improve quality, enhance product competitiveness of a major means. Agricultural industry is a comprehensive, complex, and is closely related to people's health industry, with the intensification of market competition, the need to constantly improve the quality of their products, agricultural products industry use based on the technology of RFID system to improve product reputation, improvement of agricultural product industry products flow management is the so-called hindering.

Acknowledgements. This work is supported by the Application and Development Project of Loudi, Hunan Province, China (2019YJ001).

References

1. Bibi, F., Guillaume, C., Gontard, N.: A review: RFID technology having sensing aptitudes for food industry and their contribution to tracking and monitoring of food products. Trends Food Sci. Technol. **62**(432), 91–103 (2017)
2. Tao, L., Pan, J., Xi, M.: Improving agricultural straw preparation logistics stream in bio-methane production: experimental studies and application analysis. Biotech **7**(5), 283 (2017)
3. Huang, F., Yang, B.Z., Jiang, Y.: Research on SOA application in logistics tracking system of tropical agricultural products. J. Northeast. Agric. Univ. **49**(17), 3411–3417 (2018)
4. Qian, J., Fan, B., Zhang, X.: Temperature monitoring in cold chain chamber based on temperature sensing RFID labels. Trans. Chin. Soc. Agric. Eng. **33**(21), 282–288 (2017)
5. Turri, A.M., Smith, R.J., Kopp, S.W.: Privacy and RFID technology: a review of regulatory efforts. J. Consum. Aff. **51**(2), 329–354 (2017)
6. Seco, F., Jiménez, A.R.: Smartphone-Based cooperative indoor localization with RFID technology. Sensors **18**(1), 266–273 (2018)
7. Łopucki, R., Klich, D., Gielarek, S.: Do terrestrial animals avoid areas close to turbines in functioning wind farms in agricultural landscapes? Environ. Monit. Assess. **189**(7), 343–456 (2017)
8. Jayatilaka, A., Ranasinghe, D.C.: Real-time fluid intake gesture recognition based on batteryless UHF RFID technology. Pervasive Mob. Comput. **34**, 146–156 (2017)
9. Hao, Z., Qiu, B., Zhang, K.: A new risk assessment model for agricultural products cold chain logistics. Ind. Manag. Data Syst. **117**(9), 1800–1816 (2017)
10. Yang, B.: Machine learning-based evolution model and the simulation of a profit model of agricultural products logistics financing. Neural Comput. Appl. **54**(128), 261–277 (2019)
11. Ali, I., Nagalingam, S., Gurd, B.: A resilience model for cold chain logistics of perishable products. Int. J. Logist. Manag. **29**(1), 147–162 (2018)

Art Design Methods
Based on Big Data Analysis

Dong Shao$^{(\boxtimes)}$

Dalian Neusoft University of Information, Dalian, China
Shaodong@neusoft.edu.cn

Abstract. With the progress of science and technology and the continuous development of social economy in the new era, art design has become an indispensable factor to promote economic development. The development of Internet technology provides favorable conditions for the expansion of the field of big data, and the collaboration of big data cloud computing enables the Internet to achieve efficient operation. Under the background of data, the combination of Internet and cloud computing technology can satisfy more functions, especially provide more powerful conditions for promoting the development of art design. In recent years, the research on big data analysis and art design methods has been deepening, which makes it easy for us to find that in the art design research based on big data analysis, we should pay attention to the characteristics of current information development, and combine digital information with network technology to improve work efficiency. By studying art design based on big data analysis, this paper analyzes the existing problems of art design under the background of big data, and puts forward solutions according to the existing problems, which provides better design ideas and methods for art design and facilitates the rapid development of art design.

Keywords: Big data analysis · Art design · Method study · Genetic coding algorithm

1 Introduction

We are now in an era of explosive data growth. With the progress of science and technology, the emergence of the Internet, mobile phones and so on, make the data in the speed of rapid growth beyond people's imagination. According to the estimation of the research institute, the amount of data has been increasing rapidly, which includes not only the growth of data stream, but also the growth of entirely new types of data [1]. For the massive and rapidly changing big data, storage is no longer the ultimate goal. How to obtain valuable information from the data and apply it to the industry is the significance of big data analysis. At present, with the improvement of people's living standards and spiritual pursuit, people pay more attention to artistic design [2, 3]. Art design is a very complicated process, art design collection, design, production process contains a large amount of data information, such as the kinds of information about design and art design is relatively abundant, level is relatively complex, only rely on traditional processing methods for data analysis of the efficiency is too low, can not

© Springer Nature Singapore Pte Ltd. 2020
M. Atiquzzaman et al. (Eds.): BDCPS 2019, AISC 1117, pp. 584–589, 2020.
https://doi.org/10.1007/978-981-15-2568-1_79

update the design idea and method, which creates a lag of art design, hindered the further development of art design. Therefore, in order to meet the needs of the era of art design, it is necessary to combine it with big data and conduct research on art design methods based on big data analysis [4, 5].

The revolutionary development mode of art design has been widely recognized by art designers, and the rational application of big data analysis can significantly reduce the amount of labor of designers and provide good support for their creative design. In general, under the influence of big data analysis technology, art design methods have been significantly improved [6]. There are also many researches on art design by domestic and foreign scholars, mainly including the establishment of art design model, analysis of art design problems and strategies, and digital art design, etc. However, due to the limitation of data content, the results obtained to some extent lack scientific nature and accuracy. In addition, the existing analysis and research on art design fail to meet the requirements of the development of The Times and effectively combine it with big data analysis [7, 8]. Therefore, in general, there is still a big gap in this area of research.

In order to make up for in art research field is blank, this article through the study of art and design, based on the analysis of large data is analyzed based on the background of big data problems in art design, and through the genetic encoding algorithm put forward effective countermeasures to solve the problem of art and design, for art design provides a better design ideas and methods, promote the rapid development of art design. It not only fills the gap in the previous study of art design, but also, to some extent, provides help for the later research on relevant aspects [9, 10].

2   Method

2.1   Core Concepts

With the continuous and extensive application of big data analysis technology, a variety of big data computing methods emerge in an endless stream. Among them, the evolutionary algorithm originated in the 1950s is still in use today. Evolutionary algorithm is a kind of bionics algorithm which simulates the law of genetic evolution in nature. With the deepening of the research on evolutionary algorithm, product innovation design through genetic algorithm has become an important auxiliary design method, which is also the focus of recent years. Application of standard genetic algorithm to build product modeling refinement design system, so that the original design scheme is constantly improved. The design of a genetic algorithm initial solution construction method with Pareto front smoothness and uniformity in mind can make the computation-intensive problem of complex industrial product optimization design well solved, which is conducive to the improvement of artistic design quality. Through the above studies, it can be seen that genetic algorithm has a good guiding significance for the assistance of artistic design. Therefore, this paper adopts genetic algorithm to solve the problems of pattern deformation (mutation operation) and combination (crossover operation) in the process of art design. Different from the standard genetic algorithm, this paper USES real coded genetic algorithm to describe the contour curve of artistic design patterns. In addition, the fitness value of iterative evolution is obtained by assigning user satisfaction scores with different weights, in which users are divided into ordinary users and professional users.

2.2 Real Coding Genetic Algorithm

Genetic algorithm breaks through the traditional thinking, through the analysis and generalization of art design, abstracts the patterns needed for technical design, and generates rich gene bank of patterns by the crossover and mutation operation of real coded genetic algorithm. After image material collection, it abstracts the patterns needed for artistic design and USES 2-position plane to generate contour curve of patterns, I = 0, 1, 2, 3... N represents the contour curve. In this paper, real number coding is used to represent the curve:

$$Ocur = \langle (\rho_0, \gamma_0, \theta_0), (\rho_1, \gamma_1, \theta_1), \ldots, (\rho_n, \gamma_n, \theta_n) \rangle \qquad (1)$$

Where rho I represents the radius of the curve of the upper half of the polar Angle theta I; Theta I represents the radius of the lower half of the curve of theta I. By the principle of genetic algorithm, the initialization of random population, according to the probability PC pick 2 curves Ocur and Ocur2 Settings for the parent individuals, and on the two lower and curve exchange in polar Angle (ϕ, bits) between the point, can produce new 2 is the individual:

$$Ocur1' = \langle (\rho_{10}, \gamma_{10}, \theta_{10}), \ldots (\rho_{2m}, \gamma_{2m}, \theta_{2m}), \ldots, (\rho_{2k}, \gamma_{2k}, \theta_{2k}), \ldots, (\rho_{1n}, \gamma_{1n}, \theta_{1n}) \rangle \qquad (2)$$

The above operation is a crossover operator. In addition, in order to satisfy the psychological needs of consumers to the greatest extent, the fitness function was determined with user satisfaction as the demand and the artistic pattern assembly program flow was designed. Divide users into ordinary user a and professional user b, and set different weights 0.4 and 0.6 for them respectively. Then, the modeling features of art design patterns can be scored as follows:

$$X_j = 0.4a + 0.4b \qquad (3)$$

3 Test

Step 1: analyze the art design process and the production process of art design; The process of art design is relatively complex. Through the above analysis, it can be concluded that the whole process can be divided into three parts: art collection process, art production process and art output process. Only after the comprehensive analysis of each process can the relative accurate data be obtained.

Step 2: through the analysis and generalization of art design patterns, abstract the patterns required for art design, and use the crossover and mutation operation of real coding genetic algorithm to generate a rich gene pool of patterns; The related data obtained in the previous step are substituted into the formula of genetic coding, and the data are sorted out and classified. In the process of substitution, we must pay attention to the type of art design data.

Step 3: determine the fitness function with user satisfaction as the demand and design the assembly program flow of artistic patterns. The score of art design is based on the satisfaction of art demand, and the final score is an important criterion to judge the achievement of art design. On the basis of customer satisfaction, the design of artistic patterns is more reliable.

4 Experiment

4.1 Test Results and Analysis

The parameters of the genetic algorithm are set as follows: population size 100, termination algebra 150, crossover probability 0.4, mutation probability 0.002. The art design result with the highest user satisfaction is obtained by applying the art pattern assembly method based on real coding genetic algorithm. It can be seen that the proposed method can not only enrich the designer's design ideas, but also make the design process intelligent and automatic. In addition, the average user satisfaction of art design can reach about 90%. Specific indicators are shown in Table 1, and the data in the table are calculated by the author.

Table 1. Artistic design results under genetic algorithm

Sample	Pi	Yi	a	b
Ocur1	102	154	0.4	0.6
Ocur2	132	124	0.4	0.6
Xj	89.9%	91.2%		

*Data came from sorting of algorithm results

4.2 Art Design Methods and Strategies

(1) Improve the use and management of big data analysis technology

In order to improve the integration of big data analysis technology in environmental art design, it is also necessary to strengthen the use and management of big data analysis technology in environmental design, strictly follow the principle of use, better apply big data analysis technology into art design, and build a more complete technical art design of big data analysis. First of all, it is necessary to be clear about the purpose of art design, the content of design and the final effect to be presented. Secondly, a new form of art design should be adopted, and bold and individual innovation should be carried out on the basis of big data analysis, so as to endow the design with new soul.

In addition, it is necessary to adhere to the dual requirements of scientifica and authenticity, which are based on big data analysis. Finally, and most importantly, we should learn to make use of big data to analyze technology in a reasonable way, avoid abuse, and keep the spiritual symbol and core content of the design itself.

(2) Enhance the market integration of environmental art design

To improve the integration of big data analysis technology in art design, it is necessary to enhance the integration between art design and market, deepen its close relationship with other industries in society, and provide more solid knowledge and technical support for the digital technology of environmental art design. As is known to all, in today's digital media era, the integration of disciplinary knowledge is very frequent and close, and nothing can exist and develop independently. Although the subject knowledge involved in art design is diverse, its independence is quite obvious. And with the continuous development of society, more and more new industries, or is different from traditional forms of building materials and building new technology, such as these require art design has great inclusive, always with the development of the society from all walks of life pace and development needs, and constantly enrich their knowledge and cultural conservation. Further improve the market integration force, in order to change in the development of always have a foothold, not to be eliminated by the society. In addition, we should take a long-term view, not only limited to local design development, to achieve the international development of environmental art design industry as soon as possible.

Based on the results of big data analysis, the future development trend of art design is obtained, as shown in Fig. 1 below.

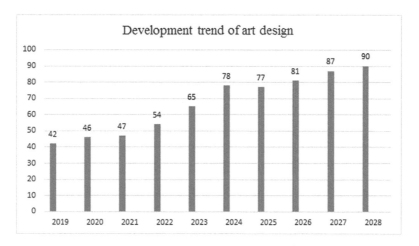

Fig. 1. Future development trend of art design

5   Conclusions

To sum up, with the progress of The Times and the development of science and technology, the application of big data analysis to the process of art design is the development of the art industry itself, social development and people's life needs. In such a situation, the designers should be its own previous design concept to fully shift, in the big change of art and design on the basis of data analysis methods and strategies of the large data analysis in the application of art design effect into full play, thus promote the effective development of art design, effectively improve the quality of art design and quality.

References

1. Lo'ai, A.T., Mehmood, R., Benkhlifa, E.: Mobile cloud computing model and big data analysis for healthcare applications. IEEE Access 4(99), 6171–6180 (2017)
2. Bo, T., Zhen, C., Hefferman, G.: Incorporating intelligence in fog computing for big data analysis in smart cities. IEEE Trans. Ind. Inform. 13(5), 2140–2150 (2017)
3. Kamilaris, A., Kartakoullis, A., Prenafeta-Boldú, F.X.: A review on the practice of big data analysis in agriculture. Comput. Electron. Agric. 143(5), 23–37 (2017)
4. Sun, J., Jeliazkova, N., Chupakhin, V.: Erratum to: ExCAPE-DB: an integrated large scale dataset facilitating big data analysis in chemogenomics. J. Cheminformatics 9(1), 17 (2017)
5. Kung, S.Y.: Discriminant component analysis for privacy protection and visualization of big data. Multimed. Tools Appl. 76(3), 3999–4034 (2017)
6. Drovandi, C.C., Holmes, C., Mcgree, J.M.: Principles of experimental design for big data analysis. Stat. Sci. Rev. J. Inst. Math. Stat. 32(3), 385 (2017)
7. Balint, T.S., Pangaro, P.: Design space for space design: dialogs through boundary objects at the intersections of art, design, science, and engineering. Acta Astronaut. 134(3), 41–53 (2017)
8. Sugiura, S., Ishihara, T., Nakao, M.: State-of-the-art design of index modulation in the space, time, and frequency domains: benefits and fundamental limitations. IEEE Access 5(99), 21774–21790 (2017)
9. Sun, J.: Studio teaching mode of art design subject of vocational college: a case study of "art workshop" teaching mode of Nanjing vocational institute of industry technology. J. Landsc. Res. 3, 118–120 (2017)
10. Bates, V.: 'Humanizing' healthcare environments: architecture, art and design in modern hospitals. Des. Health 2(1), 5–19 (2018)

Influence of Artificial Intelligence Development on Employment in Beijing

Jialin Cui[1(✉)], Aiqiao Qian[1], and Fan Yu[2]

[1] Department of Digital Media Technology,
North China University of Technology, Beijing, China
434480041@qq.com
[2] School of Geomatics and Urban Spatial Informatics,
Beijing University of Civil Engineering and Architecture, Beijing, China

Abstract. With the rapid development of artificial intelligence (AI), there has been a structural change in employment in the labor market in Beijing, which makes the relationship between AI and Beijing's labor market structure change from one-way influence and passive adaption to two-way interaction and active adaption, and finally to the ideal state of dynamic matching, which will be of strategic significance to promote the high-quality development of Beijing's economic society. Therefore, the government should systematically build a policy system for coordinating industry, talents, employment and entrepreneurship, social security, education and training, and regional linkage, and starts from four paths of industrial development, labor market structural adjustment, psychological counseling, and regional linkage to solve the employment structure in Beijing, reduce the cost of labor employment conversion, and improve the ability of labor employment, so as to achieve the dual objectives of scientific management and smooth transition of the labor market. In addition to the advance response of the government, it is also necessary to give full play to the self-organization, self-adaptation and self-regulation mechanisms of the market, and clarify the boundary between the government and the market regulation.

Keywords: Artificial intelligence · Employment · Influence

1 Introduction

In 2017, the State Council of China issued the "Development Plan for the New Generation of AI". To promote the development of AI has become a national strategy, and AI will become the core driving force of industrial transformation. In the context of Beijing's efforts to build a national science and technology innovation center and develop top ten industries such as AI, AI industry has also become a new important economic growth point in Beijing. At present, the development of AI in Beijing is in a period of policy dividend. Policies and measures to promote the development of AI in Beijing have been issued successively, such as the "Guiding Opinions on Accelerating Scientific and Technological Innovation and Cultivating AI Industry in Beijing", and the "Action Plan for Cultivating AI Industry in Zhongguancun National Independent Innovation Demonstration Zone (2017–2020)". Beijing's AI industry will develop

© Springer Nature Singapore Pte Ltd. 2020
M. Atiquzzaman et al. (Eds.): BDCPS 2019, AISC 1117, pp. 590–597, 2020.
https://doi.org/10.1007/978-981-15-2568-1_80

rapidly under the promotion of policy dividend for a long time in the future. Relevant reports show that by 2020, the number of AI enterprises in Beijing will reach more than 500, the industrial scale will reach 50 billion yuan, and the driving scale for related industries will reach more than 500 billion yuan. With the rapid development of AI, there has been a structural change in employment in the labor market in Beijing, which has triggered people's concerns about "machine substitution" and technology unemployment. The progress of AI technology not only has positive creation effects on employment, but also has negative destruction effects on employment, such as the unemployment of medium skilled labor force, the psychological panic caused by technological progress, the uneven allocation of labor resources of high and low skilled labor force, etc. If these negative effects are not paid attention to, they will certainly lead to social problems, which is not conducive to the harmony and stability of Beijing. Meanwhile, the negative effects will also cause the "unbalanced and inadequate development" of the labor market, which is in contradiction with the "people's growing needs for a better life", and needs to be solved. What industries, occupations, jobs and people are involved in the impact of AI on employment in Beijing? What is the current impact? What is the future trend? How to deal with the negative impact? How to jointly promote the progress of AI technology and the improvement of Beijing's economic growth and employment quality? The following will discuss these issues.

In a narrow sense, AI refers to the simulation and application of human brain. In a broad sense, AI refers to the simulation and application of all intelligence, including computer vision, natural language understanding and communication, voice recognition and generation, robotics, game and ethics, deep learning, etc. In terms of the number of AI enterprises in Beijing, 2015 and 2016 are the breakout periods, with 226 and 237 enterprises established, respectively. The total number of companies established in two years is the same as that in the previous years. Horizontally, the AI industry is mainly distributed in Haidian District and Chaoyang District of Beijing, with enterprises accounting for 62% and 29% respectively.

From the perspective of application fields, the key application fields of AI in Beijing include unmanned driving, new retail, smart medical care, smart home, smart city, etc. Unmanned driving has a leading position in China, with many firsts, such as the first issuance of the "Document to Standardize the Promotion of the Actual Road Test of Self-Driving Vehicles", the construction of the first unmanned driving pilot road in Yizhuang of Daxing District, and the planning of establishing the first unmanned driving test operation base in Shunyi in Beijing. New retails, such as Freshhema, have been arranged, and Jingdong's unmanned logistic vehicle distribution has started trial operation. The smart home industry is also developing rapidly, with 414 enterprises, ranking the second in China. AI medical enterprises are favored by capital, and 33 representative enterprises have obtained more than A rounds of financing.

The rapid development of AI technology and industry will have a huge impact on Beijing's manufacturing and service industries. The influence of AI on manufacturing industry is different from "automation" and "industrial Internet". Automation refers to the large-scale production by machines without any need of people, and its core is automatic production by machines. "Internet + manufacturing" is the application of Internet tools to the manufacturing industry, emphasizing the supply and demand docking. The "intelligent" manufacturing is the application of AI technology to the manufacturing industry. On the basis of digitalization and networking, machines independently cooperate with the changes of elements and human work to realize independent and flexible production. In addition to improving the production efficiency, optimizing the industrial structure and reconstructing the international division of labor, the application of AI technology in the manufacturing industry has profoundly changed the employment market in Beijing, resulting in structural unemployment. 50% of the existing jobs may be replaced, and the employment population in the manufacturing industry in Beijing will be reduced, for example, the use of intelligent technologies such as 3D five axis CNC machine tools of QM and digital drilling may replace about 1000 people. American scholars Acemoglu et al. paid attention to the impact of industrial robots on the labor market of the United States. They believed that the large-scale application of industrial robots had a significant negative correlation with employment and wages. At present, the substitution effect of industrial robots on the labor market is greater than the creation effect. For every 1000 workers, if there is one more robot, the proportion of employed population will be reduced by 0.18–0.34%, and the wage will be reduced by 0.25–0.5% [1].

Similarly, the application of AI technology in the service industry has brought new influence to Beijing's service industry. For example, the China Construction Bank announced the official opening of the first unmanned bank in China, without security guards, lobby managers or tellers, which are replaced by face recognition, service robots and intelligent teller machines. Deloitte's financial robot, JD's UAV logistics, Beijing sanitation's unmanned cleaning car, Haidilao's intelligent restaurant and other AI are widely used in Beijing's service industry, and the substitution of machines for human will become more and more common. The "2019 World Development Report" released by the World Bank predicts that 51% of China's work content has intelligent potential, which is equivalent to the impact of 394 million full-time human working hours.

2   The Influence of AI on Posts

AI has a different impact on employment than previous technical progress. Most of the previous technical progress provides tools for labor and greatly improves labor productivity, while the technical progress of AI brings about subversive changes in the nature, means and progress of work, making the labor forms in the workplace more diverse, including human labor, human-machine cooperation, interactive labor, and robot labor. However, the progress of AI technology is still in the stage of man-machine cooperation. The influence of AI on posts includes both positive and negative effects. The negative effects include post polarization and post substitution. For post polarization, with the development of AI technology, the problem of employment "polarization" brought by technical progress is increasingly prominent [2], which is reflected in the relative increase in the number of low-skilled and high-skilled labor, and the relative decrease in the number of medium-skilled labor. In addition, the wage level and the employment quality of high-skilled labor is improved, the wage level of medium-skilled labor is decreased and the employment quality is difficult to guarantee, and the relative wage of low-skilled labor is increased and the income gap is widened [3]. According to the "2017 World Development Report" issued by the World Band, the polarization of China's labor market has emerged, and the polarization of Beijing's labor market is more prominent. The second is the post substitution effect. Cao and Zhou think that the substitution effect of AI is more obvious than any technical progress in the past [4], which is reflected in the reduction of the number of posts and the change of the post structure. Taking the manufacturing industry as an example, the number of front-line jobs in production has decreased, the number of jobs in R&D design, machine development, and management and maintenance has increased, and the skill requirements of employees have changed, requiring employees to have creative thinking ability, digital ability and social ability. The nature of work task is often the man-machine cooperation. Machines and people form a new balance of mutual cooperation, supplement work and collaborative work. For example, a printing equipment manufacturing enterprise located in Daxing District of Beijing adopts the intelligent transformation scheme, with 220 employees replaced from January 2017 to June 2018, and a petrochemical product manufacturing enterprise located in Fangshan District upgrades its DCS intelligent system, which has replaced about 8000 employees since 2014 [5]. The research results of Manyika et al. show that 57% of posts in over 50 countries are affected by AI technology [6], 47% of posts in the United States are at high risk of being replaced [7], and 55% of posts in Japan are at risk [8]. Through the analysis of crawling data from Internet recruitment websites, in 2017, in Beijing's Internet industry, the demand for general technical posts such as IT testing, and Internet operation and maintenance, which are highly procedural and can be replaced by artificial intelligence technology, decreased by more than 20%, while the demand for tellers and general bookkeepers decreased by more than 30%. Many workers are facing job transformation or moving to work in the underdeveloped areas of AI industry. In addition, the technology unemployment caused by AI will cause workers' anxiety and psychological panic.

In addition to the above negative effects of AI technology progress on employment, there are also some positive effects such as post creation, post cooperation and employment quality improvement. There are two aspects in the creation of posts by the progress of AI technology. One is the increase of the labor demand caused by the increase of AI application business volume. The other is the generation of new types of posts by AI, such as algorithms, AI trainers, intelligent equipment maintenance, and AI system interpreters [9, 10]. As far as Beijing is concerned, the judgment of the first trend still needs to wait for the deepening and overall layout of AI technology. The second trend is clear. According to the job recruitment data crawler results of Internet recruitment websites such as Zhaopin, the demand for talents of AI in 2017 increased by 179% compared with a quarter of 2016, nearly three times of the demand for talents in 2016. Not only that, the salary level of Beijing AI engineers is high, with an average annual salary of 120000–150000 yuan, which is 40000–70000 yuan higher than the average annual salary of other industries. The average annual salary of senior AI experts is more than 500000 yuan. At the same time, the flow rate of AI engineers is very high, and the per capita job flow rate accounts for more than 50% in 1–3 years. The competition between companies for talents is the main reason for talent flow [11].

The biggest difference from the previous technical progress is that the combination of AI and work is deeper and more complex. AI changes the nature of work tasks. At present, China's AI technology is in the stage of incomplete maturity, most of which is in the stage of man-machine cooperation. There are more man-machine cooperation and man-machine interaction than complete replacement. Machines and people form a new man-machine symbiosis of mutual cooperation, supplement work and collaborative work. The progress of AI technology has also greatly improved the quality of employment, freed people from repetitive and dangerous tasks, and enabled them to do more valuable, interesting and meaningful work.

3 Coping with the Negative Effect of AI on Beijing's Employment in Advance

From the perspective of industry and people, starting from two aspects of labor market impact and psychological impact response, considering the constraints of Beijing's population relief policy, the government should formulate policy objectives at the short-term tactical level, the medium-term strategic level and the long-term vision level, systematically construct coordinate policies such as regional linkage, industrial development, entrepreneurship and employment, education, vocational skills training, talent introduction and training, and social security foundation, and design the operable realization path and countermeasures. In addition to the advance response of the government, it is also necessary to give full play to the self-organization, self-adaptation and self-regulation mechanisms of the market, and clarify the boundary between the government and the market regulation.

3.1 Carrying Out Policy Objective Analysis

According to the benchmarking city, combined with Beijing's new urban planning, urban functional positioning, AI industry development objectives and the characteristics of AI enterprises, the paper outlines the ideal state of the total labor market and the structure of Beijing's AI industry in stages. The short-term tactical goal is to study the adaptability by 2020, that is, how the labor market adapts to the adjustment brought by the progress of AI technology. The medium-term strategic goal is to study the matching problem by 2025, that is, how to precisely match the labor market and the industrial structure of AI. The long-term strategic goal is to study the dynamic matching problem by 2030, that is, how the labor market actively matches the adjustment brought by technical progress.

3.2 Building a Systematic and Coordinated Policy System

The government should design policies that support and coordinate each other, such as AI industry development, labor market adjustment, human capital investment such as education and training, talent introduction and social security, and regional linkage. There are four paths, namely regional linkage, industry promotion, entrepreneurship and employment, and psychological counseling, to formulate countermeasures to actively adapt to and dynamically match the progress of AI technology and industrial development. The industry promotion is to actively develop new industries and new types of AI, and enlarge the effect of AI on employment creation. The employment and entrepreneurship is to change and adjust the education and training system in the labor supply side to dynamically match the labor market effect caused by the trend of AI technology progress. Specifically, it includes the reform and adjustment of the education system, the opening of new AI + X related majors in colleges and vocational colleges, the opening of AI related courses in existing majors, the opening of related courses in primary and secondary schools, the increase of the supply of high-end AI R&D personnel and the high-skilled labor force, and the improvement of labor force to adapt to the skill level in the context of AI through AI + training. At the demand side of labor force, the implementation of policies in coordination with the trend of AI should be promoted from the macro level, such as the introduction of high-level AI talents, the training of AI R&D personnel, the work conversion of the impacted personnel, the skill training of low and medium skilled personnel, the income buffer and redistribution, the social security, and the labor relation adjustment. At the micro level, the government should make precise efforts to formulate intervention policies and measures for key industries, occupations and groups in the area of new technology destruction such as AI. At the same time, the government should expand the entrepreneurial driving effect of AI technology progress, build a double helix system that AI technology progress drives employment and entrepreneurship and promotes AI technology breakthrough and application, and construct a multi-functional system that entrepreneurship promotes employment. The psychological counseling is to formulate psychological intervention measures for unemployment panic according to the degree of psychological impact of various industries, occupations and people, focusing on the psychological intervention of middle skilled people. The regional linkage is to design policies and measures for industrial, labor market and talent linkage development with Xiongan New Area under the policy constraints of population dispersal and relieving non-capital function.

4   Conclusions

The impact of AI technology progress on employment in Beijing is not only beginning to appear in the manufacturing industry, but also in the financial industry, logistics express industry, catering industry and other service industries. The impact of AI technology progress on Beijing's employment is not only a challenge, but also an important opportunity for the adjustment of Beijing's employment structure and the transformation and upgrading of labor skill structure. The government should seize the opportunity and make the relationship between AI technology progress and Beijing's employment structure change from one-way impact and passive adaptation to two-way interaction and active adaption, and finally to the ideal state of dynamic matching under the joint action of the external technology impact and the internal industry and employment structure adjustment, which will be of strategic significance to promote the high-quality development of Beijing's economic society. Different from the technological impact of the three industrial revolutions of steam, electricity, and computer and

Internet in history, the progress of AI technology is the structural problem of labor market caused by the current situation of Beijing's economic society, such as the upgrading of industrial structure, the population dispersal, and the integration of Beijing, Tianjin and Hebei, which may lead to the structural shortage of labor supply and the shortage of high skilled and innovative talents. Therefore, the government should build a systematic policy system of AI industry development, labor market adjustment, human capital investment such as education and training, talent introduction and social security, and regional linkage, design an operational realization path and countermeasures to reduce the risk of technical unemployment and solve the employment structural contradiction that laborers' skills do not match the industry, effectively reduce the employment conversion cost of labor force, and rebuild laborers' knowledge and skills. In addition, the government should coordinate the relevant strategies and policies in the fields of entrepreneurship and employment, talent introduction and training, scientific and technological development, industrial promotion, education and training, and social security in Beijing, actively respond to and take comprehensive measures to "suppress the destructive effect of AI, and focus on three aspects: cultivating new industries of AI to expand employment and entrepreneurship opportunities, improving the quality of labor force to meet the needs of high skilled jobs, and adjusting the panic caused by psychological impact of labor force. The government should also formulate differentiated employment promotion and social security support policies for different industries, posts and groups, improve the quantity and quality of employment, enable workers to share the "dividend" brought by the progress of AI technology, and realize the harmonious development of people, technology and economy.

References

1. Acemoglu, D., Restrepo, P.: Robots jobs: evidence from US labor markets. Econ. Rev. **12**, 123–1554 (2017)
2. Lai, D., Li, C., Meng, D.: China's Labor Market Development Report, vol. 2, pp. 15–22. Beijing Normal University Publishing Group, Beijing (2018). (in Chinese)
3. Lv, S., Zhang, S.: Employment "polarization" in China: an empirical research. China Economic Quarterly **12**(2), 56–62 (2015). (in Chinese)
4. Cao, J., Zhou, Y.: The research progress on the influence of artificial intelligence on economy. Econ. Perspect. **2**, 16–26 (2018). (in Chinese)
5. Zhang, Y.: The effect of machine replacement on labor employment in Beijing manufacturing-based on observations from six companies. Hum. Resour. Dev. China **10**, 136–146 (2018). (in Chinese)
6. Manyika, J., Chui, M., Miremadi, M.: A Future that works: automation, employment, and productivity. Mckinsey Co. **3**, 224–236 (2017)
7. Frey, C., Osborne, M.: The future of employment: how susceptible are jobs to computerization? Technol. Forecast. Chang. **12**(5), 16–23 (2013)
8. David, B.: Computer technology and probable job destructions in Japan: an evaluation. J. Jpn. Int. Econ. **43**(01), 77–87 (2017)
9. Ji, W.: The dilemma and solution of labor relations in start-up enterprises. Hum. Resour. Dev. China **20**, 69–76 (2016). (in Chinese)

10. Li, H., Lai, D.: Internet and income distribution, management science and engineering in the process of china's development. Hum. Resour. Dev. China **1**, 749–752 (2008). (in Chinese)
11. Shen, W.: On the technical characteristics of contemporary artificial intelligence and its influence on workers. Contemp. Econ. Res. **4**, 15–27 (2018). (in Chinese)

Strategies on Traditional Chinese Medicine Translation Teaching in the Age of Big Data

Yan Gong and Yawei Yu$^{(\boxtimes)}$

School of Humanities, Jiangxi University of Traditional Chinese Medicine,
Nanchang, China
252457910@qq.com

Abstract. The era of big data brings both challenges and opportunities for Traditional Chinese Medicine translation teaching. Firstly, it analyzes the characteristics of the era of big data and the cognitive features and learning habits of students. Then the major problems in current TCM Translation teaching are analyzed such as insufficient practice of TCM translation, too much emphasis on the translation knowledge and skills while less on the cultivation of thinking; less class hours, and insufficient feedback from teachers. Then combined with the characteristics of the era of big data, the strategies of TCM translation teaching reform are put forward: make full use of resources on network and technology, use TCM translation corpus, use network platform and fully explore student big data in order to provide references for Traditional Chinese medicine translation teaching.

Keywords: Big data · Traditional Chinese medicine translation teaching · Strategies

1 Introduction

With the increasing self-confidence of Traditional Chinese culture and the wide spread of Traditional Chinese Medicine, higher requirements are put forward for TCM translation and its teaching. The cultivation of TCM translators and the improvement of TCM translation ability have attracted much attention. The increasing popularity of the Internet and various types of mobile terminals has brought human into a new era of Big Data [9]. The arrival of the era of big data has brought about profound changes in the field of education [10]. The field of foreign language teaching may be fundamentally changed. How to use the big data, the network to cultivate students' self-learning ability and habits, as well as to improve the participation and teaching effect of TCM translation classrooms is really worth exploring.

2 Big Data and Learner Characteristics

With the advent of the Internet+ era, the popularity of computers and networks, the amount of data that people can access is growing. In 2011, McCarthy Consulting first proposed the concept of big data [1]. It refers to a large collection of data, with large

© Springer Nature Singapore Pte Ltd. 2020
M. Atiquzzaman et al. (Eds.): BDCPS 2019, AISC 1117, pp. 598–603, 2020.
https://doi.org/10.1007/978-981-15-2568-1_81

capacity, high speed, diversification, low value density, with authenticity as its universal feature [2]. The massive data generated by the era of information explosion, as a new medium, is infiltrating into all areas of social life, and it also promotes reform and innovation in the field of education [8], it provides a new perspective for our cognition, ideas and methods, and becomes a powerful force for reform in the field of higher education [3]. The huge amount of data makes quantification possible [4]. A large number of learner data is collected for teaching decision-making and also supports personalized learning [5]. Many software and programs related to English learning have emerged, providing resource support for autonomous learning and facilitating online communication [6]. Besides, various shared courses, micro-courses, Mooc, etc. emerged.

At present, most of the college students are growing up with digital technology and Internet technology. Compared with previous students, their study habits and requirements have distinct characteristics of the times [7]. Books, chalk and simple ppt teaching are no longer sufficient for their needs. In the classroom, they are more inclined to ask questions, discuss and demonstrate rather than just being the recipient of knowledge. After class, they are also more accustomed to learning and communicating through the Internet and social media.

3 Problems in Traditional Chinese Medicine Translation Teaching

3.1 Emphasis on the Teaching of Translation Knowledge and Skills, Lack of Practice

In The traditional teaching mode the teaching of translation knowledge and skills on teachers' side is emphasized, while translation practice on student's part is far from sufficient. The ultimate goal of TCM translation course is to cultivate students' practical ability in TCM translation. It is necessary to have enough practice to internalize translation knowledge and skills to form translation capabilities. Statistics show that the majority of class time were spent on lecturing TCM translation skills. The common teaching procedure of TCM translation is first lecturing about translation skills followed by giving examples, then giving assignment and check homework by providing so-called standard answers. Due to the limited time in class, not enough time is left for translation practice which is crucial to the internalization of translation skills. What's more, because of the great number of students, teachers hardly have enough time to go over homework of every student. Therefore, the lack of necessary feedback and individual instruction to student prevent the improvement of translation skills.

3.2 Emphasis on the Translation Knowledge and Skills, Lack of Concern on Translation Thinking

Unlike the focus on mastering TCM translation skills, translation thinking which plays an important role in actual translation work, is far more underestimated in traditional translation class. Translation thinking is referred to the thoughts and consideration behind translation. Without it, a translator may not be able to use translation skills flexibly. The translation work of traditional Chinese medicine is complicated since it is closely related to ancient Chinese culture, which is difficult to understand, for instance, words are often polysemy and contains literary rhetoric as well. In today's vigorous promotion of Chinese medicine culture, with Chinese medicine going to the world today, translators must be more flexible. That is to say, the cultivation of Translation thinking, not just mechanical mastery of translation skills, should be the aim of teaching. Students should know not only how to translate, but also to know why this translation technique is used. The traditional translation course pays too much attention to the translation skills of Chinese medicine, neglecting the cultivation of translation thinking. According to some statistics, in a translation class, 80% of the time is used for teachers to teach translation knowledge and skills. Students have little time and opportunity to discuss and think.

3.3 Lack of Class Hours, and Feedback to Individual Student

Since translation course takes relatively few hours compared with other English courses, an average of 32 periods in a term, it's not likely that both lecturing and practice can be fully carried out. The review and reflection of student translation works is the key to student improvement. Traditional classrooms are limited by the number of students and class hours. Teachers usually assign post-class translation exercises, but they can hardly give feedback to each student's homework every time. Students' misunderstandings, learning questions and expressing errors cannot be promptly dealt with, which seriously affect the learning effect.

4 TCM Translation Teaching Strategies in the Era of Big Data

4.1 Make the Most of Various Resources Online and New Technology

Course books on TCM translation often have problems of being out of date which usually greatly weakens the learning interest, while the Internet and cloud storage platforms have a large number of fresh, instant, and vivid information resources that can be effectively supplemented as course resources. Teachers can use teaching technology tools to integrate various resources, using the textbook as the center to radiate, and screening the appropriate corpus and topics suitable for students to learn. After class, the mobile terminal can be used to push, check and answer relevant learning content.

In the era of big data, with the development of information technology, various online courses, such as micro-courses, moocs, flipped classrooms, etc. which are considered the major trend of future curriculum development, have come into existence and soon become the focus of attention in recent years. In terms of the translation course of TCM, the innovative use of flipped classroom model by the use of micro courses and moocs can provide better solutions to the problems in traditional teaching. In the traditional mode of teaching, as the three stages, pre-class, in-class, after-class are separated to each other, communication and internalization are limited due to the limit of time and space. Whereas in the flipped classroom mode, the three links can be seamlessly connected. With the reversed teaching procedure, the cognition cycle can be completed. Moreover, it makes it possible for teachers to deal with the difficulties, focus on tackling learning problems rather than just pouring out knowledge. Traditional teaching is teacher-centered, where teachers do most of the talking, instead, the mixed teaching model such as the micro-course and the moocs, are student-centered, where the students are given full play to the learning autonomy. Through the application of TCM translation micro-courses, teachers can optimize TCM translation teaching. To be specific, Students learn the language and skills through micro-course before class and come to the classroom well prepared with questions. Teachers don't have to explain translation skills all the time any longer, but focus on the problems, hence sufficient time can be left for students to translate and practice. Classroom activities like forming groups, discussing, consulting and reporting on learning outcomes will help students internalize the translation knowledge they have learned.

Besides, advanced technology can also be integrated into TCM translation teaching. For instance, in the aspect of TCM interpreting, computer simulation technology or virtual reality technology can be used to design a virtual oral communication scene, so that students wearing VR headset feel as if placed in the real situation of TCM international exchange, such as at an international conference or being an accompanying interpreter, to improve their skill response ability.

4.2   Make Use of TCM Translation Corpus

Corpus, the product of the Big Data Era and network computers, is a large-scale electronic text library that is scientifically sampled and processed. It is a specialized and real language material, usually a collection of professionally collected, useful objects and useful resources. The corpus has the characteristics of large capacity, large scale, multi-languages, continuous development of related software, and expanding application scope. At present, corpus-based translation teaching has received extensive attention (Wang Kefei, 2006). For TCM translation, corpus is an important auxiliary tool for TCM translation and teaching. Teachers should first be familiar with the use of TCM translation corpus, and then teach students how to use the corpus. Secondly, use the corpus according to different teaching stages and different teaching objectives. The biggest problem in TCM translation is the translation of technical terms since they are closely related to Chinese culture and lack of unified standards. For example, teachers guide students in the use of online Chinese medicine dictionary or TCM terminology corpus (online) to query TCM professional vocabulary, technical terms, related grammar and sentence structure in order to avoid common errors in grammar and word, or continue to use corpus to do modifications based on the teacher's opinion after the teacher reviews the translation. In the translation classroom teaching, the teacher can instruct the students to retrieve a variety of translated text corpora of a certain term from the Chinese medicinal parallel corpus. The Chinese and English contrasts are presented to the students, and the students are grouped and analyzed to discuss the accuracy and rationality of the evaluation. This can not only help students understand a variety of translation skills, but also help them to think about translation strategies, by way of which help them form their own translation ideas, and finally establish a foundation for the study and exploration of translation theory.

4.3 The Use of Online Communication Platform

Besides introducing TCM translation network resources, we should also make full use of online platform such as Weibo, WeChat, QQ, Superstar Learning pass (Chaoxing Xuesaitong) and other platforms to establish a bridge between teachers and students, between students and students, between discipline experts and teachers. In such an environment, communication can happen 24 h in a day, either inside or outside the classroom, so that the learning tasks can be uploaded in real time, the learning content communicated with each other, the questions promptly given feedback, and the problem quickly answered by experts.

4.4 Fully Explore Big Data on Learners

Learner big data refers to data on students' reading, attendance, class performance, grades, etc. it can also provide support for teaching, especially in the teaching process. For example, through the mining of these teaching data teachers can have a better knowledge of the state of teaching. And they can dig deeper into the big data of learning. For instance, with the screening effect of big data, the problems that occur in the process of students' learning can be summarized, and the students can be classified according to the level, so as to carry out targeted teaching or learning activities for students. What's more, it is also important to conduct analysis to understand student learning habits and needs, identify regular characteristics, and provide customized learning resources. Only in this way catering to students' individual needs can be achieved in the real sense.

5 Conclusions

In the context of Big Data Era, TCM translation teaching faces enormous challenges and new opportunities as well. In the teaching of TCM translation, teachers should conform to the development of the times, with good awareness of reforming the current teaching by the full and rational use of technology and big data. Besides, teacher need to adopt advanced teaching methods, be a good organizer of teaching activities, providers of teaching resources, in order to stimulate students' interest in learning and to cultivate independent learning ability.

Acknowledgements. This work was supported by 2014RW010.

References

1. Li, Y., Zhai, X.: Review and prospect of modern education using big data. Procedia Comput. Sci. **129**, 341–347 (2018)
2. Klašnja-Milićević, A., Ivanović, M., Budimac, Z.: Data science in education: big data and learning analytics. Comput. Appl. Eng. Educ. **25**(6), 1066–1078 (2017)
3. Kim, Y.H., Ahn, J.-H.: A study on the application of big data to the Korean college education system. Procedia Comput. Sci. **91**, 855–861 (2016)
4. Eynon, R.: The rise of big data: what does it mean for education, technology, and media research? Learn. Media Technol. **38**(3), 237–240 (2013)
5. Nav, S.N.: Teaching Writing for Academic Purposes to Multilingual Students: Instructional Approaches. John Bitchener, Neomy Storch, Rosemary Wette, xiii + 220pp. Routledge, New York (2017). ISBN 978-1-138-28421-0. International Journal of Applied Linguistics 29(1), 149–151 (2019)
6. Kuşçu, S., Ünlü, S.: Teaching translation: a suggested lesson plan on translation of advertising through the use of authentic materials. Procedia Soc. Behav. Sci. **199**, 407–414 (2015)
7. Zhang, M.: Teaching translation with a model of multimodality. Asia Pac. Transl. Intercult. Stud. **2**(1), 30–45 (2015)
8. Zhou, J., Hu, J., Chen, Y.: Big data-driven foreign language teaching and scientific research innovation - a seminar on foreign language teaching and research innovation under the background of big data and a review of the first "Love the Future" foreign language education technology science and technology innovation workshop. Foreign Lang. Electrification Teach. (187), 118–122 (2019)
9. Wang, H.: College English writing teaching reform in the age of big data. Modern Distance Education Research (3) (2014)
10. Wang, K.: Corpus translation studies——new research paradigm. Chin. Foreign Lang. (3), 8–9 (2006)

Relationship Between Rice Growing Environment and Diseases-Insect Pests Based on Big Data Analysis

Yichen Chen and Wanxiong Wang[✉]

College of Resources and Environmental Science,
Gansu Agricultural University, Lanzhou 730070, China
wangwx@gsau.edu.cn

Abstract. Rice output directly affects China's food security. Diseases - insect pests are one of the main agricultural disasters that affect the stable and high yield of crops. With the change of the environment, the variety and scale of diseases and insect pests are increasing. Every year, rice in many areas suffers from diseases and insect pests, which causes serious loss of national food production. Therefore, the research on the relationship between rice growing environment and DIP is increasingly important. At the same time, the emergence of big data analysis enables the growth data of rice and the data of DIP to be analyzed and processed quickly, which provides convenience for this research to a large extent. On the basis of big data analysis, this paper takes rice cultivation as the research object, and through the follow-up investigation of rice DIP in various regions, it can understand the occurrence of rice DIP in different regions in successive years. The relationship between the occurrence of rice diseases and pests and the growth environment factors was discussed, and the climate background and coupling mechanism of the occurrence and prevalence of rice diseases and pests were revealed, so as to predict the occurrence trend of major rice diseases and pests in the region, and to provide basic data for the establishment of disease and pest prediction model.

Keywords: Big data analysis · Growth environment · Diseases and insect pests (DIP) · PCA algorithm

1 Introduction

The planting area of rice accounts for about one third of the total planting area of China's food crops, and its output accounts for more than 40% of the total production of China's food crops. The production of rice directly affects the daily food problems of Chinese citizens and the national food security [1]. In the process of rice cultivation, problems of rice pests and diseases often have the most direct impact on the quality and yield of the crops. Rice DIP occur in many areas of China every year, which has a great negative impact on rice production [2, 3]. And various growth environment factors have a profound impact on the reproduction of bacteria, insect activities, such as light, soil, temperature and other factors. By means of monitoring and forecasting the growth environment of DIP, it is necessary to make a comprehensive analysis of the growing

© Springer Nature Singapore Pte Ltd. 2020
M. Atiquzzaman et al. (Eds.): BDCPS 2019, AISC 1117, pp. 604–609, 2020.
https://doi.org/10.1007/978-981-15-2568-1_82

environment of rice in its growing period, so as to control DIP to the greatest extent and reduce losses. Therefore, it is of great significance to study the relationship between rice growing environment and DIP [4, 5]. At present, big data analysis technology provides convenience for this research, especially in data analysis. The research on the relationship between rice growing environment and DIP based on big data has gradually become the focus of attention.

Under the above background, many researchers analyzed and determined many environmental factors affecting diseases and pests from multiple aspects. They investigated all the factors of environmental factors, most of which were general and the research on rice was not typical, so the research results were often inaccurate [6, 7]. In addition, due to the large amount of data, it is often difficult to deal with, and it is easy to cause the omission and error of the research data, which will more or less affect the scientific nature of the research on the relationship between rice growing environment and DIP. So there are a lot of gaps in this area.

Blank in order to make up for this study, in this paper, based on large data analysis, with principal component analysis (PCA) algorithm as the theoretical basis, the mathematical model for rice growth environment factors as the research object, using the algorithm formula, finally it is concluded that the growth of environmental factors most closely associated with DIP in the four factors, determine the type of the main factors influencing the plant DIP, which makes the accuracy of the results are obtained by the, and improve the operation efficiency of data [8, 9]. This study played a positive role in the prevention and control of rice diseases - pests in China, and provided a valuable theoretical basis for further research on the relationship between growing environment and rice diseases and pests [10].

2   Method

2.1   PCA Algorithm Application Conditions

Diseases - insect pests of rice are affected by many environmental factors. Many researchers have determined many environmental factors that affect DIP from multiple aspects. Most of the environmental factors are general and the research on rice is not typical, so the research results are often inaccurate. Visible, in order to study the plant DIP and the relationship between the growth environment of rice, the first thing to find out the different environmental factors, need to study to determine the main factors influencing the rice plant DIP, to determine the quantitative relationship between environmental factors and plant DIP of rice to create favorable conditions. Therefore, in this paper, principal component analysis, or PCA algorithm, will be used to extract and analyze the main characteristic components of the growth environment impact factors and determine the main factors affecting DIP, which ensures the accuracy of the results and improves the operation efficiency of the data.

2.2   PAC Algorithm

PCA algorithm. It extracts the main feature components of multidimensional vector data to achieve the purpose of compression of vector dimensions, so as to reveal the internal laws of things and make reasonable explanations for various indicators. The mathematical model is as follows. With n samples, each sample has p items (influence factors), namely X1, X2, X3..., Xp, and get the original variable database matrix:

$$X = (X_{11}, X_{12}, \ldots X_{1P}) = (X_1, X_2, \ldots X_P) \tag{1}$$

Where, Xi = (x1i, x2i... Xni)T, I = 1, 2..., p; The linear combination of the data in the p-item index of the data matrix x is as follows:

$$F_1 = a_{11}x_1 + a_{12}x_{12} + \ldots + a_{1p}x_p \tag{2}$$

Where, aip is the coefficient, which satisfies a2i1 + a2i2 + a2i3 + ... + a2ip = 1, I = 1, 2, 3..., p; X1, x2..., xp is the corresponding value of each index; According to the above formula, F1, F2..., Fp is the new p irrelevant variables; F1 is x1, x2..., the linear combination of xp has the largest difference in China, which is the first principal component of the original data, and F1, F2... The proportion of Fp in the total variance decreases successively.

Principal component analysis is solved. First, the original data matrix is standardized to eliminate the influence of variables on the order of magnitude or dimension. Each element in the matrix (1) will be processed by the formula to get a new element. Then, the covariance or correlation coefficient matrix is calculated according to the normalized matrix. Then according to the correlation coefficient matrix R, the corresponding eigenvalue and eigenvector are calculated. Finally, the principal components are determined according to the corresponding eigenvalues and eigenvectors. When the general eigenvalue ratio is 1 and the cumulative contribution rate reaches 85%, the first k indexes corresponding to it are taken as the principal components, indicating that these principal components contain the main information of the original variable.

3   Experiment

Step 1: conduct an investigation on the establishment of standard sites for serious hazards and major pests from the large database, determine the location of each standard site with GPS positioning, set markers, and indicate the host area represented. According to the biological learning ability of rice pests and the classification of different rice. SPSS statistical software was used to analyze the original data.

Step 2: obtain the variance contribution rate and cumulative contribution rate of each component according to the correlation coefficient matrix, and then compare which three components' characteristic roots (i.e. characteristic values) are greater than 1. Finally, extract the three principal components with SPSS statistical software.

Step 3: conduct the principal component analysis of the three principal components in the previous step, reduce the dimension of the principal components, and then obtain the corresponding correlation coefficients of each index variable in the principal components according to the formula, and record them.

Step 4: divide the data obtained in the previous step by the square root of the eigenvalues corresponding to the principal components, and then obtain the coefficients corresponding to each index in the three principal components. Then the obtained coefficients are verified to check whether they meet the requirements. By using the principal component analysis method, a large number of influential factors were successfully reduced in dimension, and the newly generated principal component formula was used to recalculate the parameters of each factor in the growing environment of rice.

4   Discuss

4.1   Analysis and Discussion of Test Results

In this paper, based on large data analysis, with principal component analysis (PCA) algorithm as the theoretical basis, the mathematical model for rice growth environment factors as the research object, using the algorithm formula, the main factors that influence the successful rice growth environment of the dimension reduction to four principal components, including terrain, temperature, precipitation, illumination of the four major components, and generate the four principal component equations. It can be seen that PCA algorithm is an effective tool in dealing with a large number of complex variables. It also provides research ideas and reference value for further determining the quantitative relationship between impact factors and building energy consumption. The variance contribution rate and cumulative contribution rate of the four principal components are shown in Table 1. The data in the table are obtained by SPSS system analysis and collation.

Table 1. Table of values of main components

Composition	Starting			Extract		
	Eigenvalue	Variance contribution rate	Cumulative contribution rate	Eigenvalue	Variance contribution rate	Cumulative contribution rate
Soil	4.108	45.643%	45.643%	4.108	45.643%	45.643
Temperature	1.921	21.345%	66.988%	1.921	21.345%	66.988%
Precipitation	1.64	18.227%	85.219%	1.64	18.227%	85.215%
Light	1.59	17.567%	83.236%	1.59	17.567%	83.231%

*Data came from the SPSS analysis

4.2   Relationship Between Diseases and Pests of Rice and Growing Environment

The occurrence and development of rice DIP are affected by climate, biological factors, soil factors and external environmental conditions of human activities. It is the result of the joint action of various biological and non-biological factors. The change of rice growth environment conditions is closely related to the life activities of DIP.

However, as the growing environment of rice is complicated in many aspects, only the following major environmental factors obtained by big data are selected to illustrate the relationship between DIP. Specific DIP in recent years are shown in Fig. 1, the data source is agricultural data network.

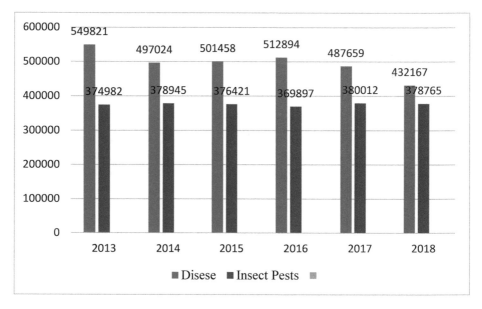

Fig. 1. Rice DIP in China from 2013 to 2019

The first is the impact of the terrain where rice grows on DIP. According to the above calculation, it can be concluded that the areas with poor growth, less vegetation, frequent human and animal activities, high temperature and dry, and few natural enemies are conducive to the occurrence of DIP. The second is the effect of temperature. Can be seen from the historical data, the nearer to as climate change, especially the effects of global warming, different parts of the plant DIP of rice CeBaoDian trend of annual average temperature is rising steadily, under this conditions conducive to plant DIP of rice plant DIP of rice variety on the increase, invasive species more and more, potentially dangerous invasive plant DIP of rice species are also growing. The occurrence area of rice diseases and pests is positively correlated with the annual average temperature, that is, the annual average temperature is one of the main factors affecting the annual occurrence of rice diseases and pests. With the increase of the annual average temperature, the occurrence of rice diseases and pests is on the rise. Then there is the effect of precipitation. Rainfall can change the weather and soil moisture, thereby affecting the water content of host plants and the development of pests themselves. Through calculation, it can be found that the lower the annual precipitation is, the larger the area of DIP. Finally, the light factor. The effects of light on pests are mainly manifested in three aspects: light wave, light intensity and photoperiod. Light waves are closely related to phototaxis of pests. Light intensity mainly affects the diurnal rhythm of pests such as feeding, roosting, mating and spawning, and is related to the color and concentration of pests. Photoperiod is an important factor causing diapause and dormancy.

5 Conclusion

In the process of rice planting, DIP are the most easily encountered problems, which will directly affect the yield and quality of rice. Therefore, this problem must be paid high attention to by rice growers and researchers of DIP. Grasp the relationship between growth environment and the occurrence and development of plant DIP of rice, rice plant DIP forecasting, combined the technology of prevention and control of agricultural, biological, chemical control technology and management technology, take measures in advance, can provide favorable environment for the healthy growth of rice, promote the sustainable development of rice planting industry of our country, and ensure the food security of our country.

References

1. Han, Y.-Q., Wen, J.-H., Peng, Z.-P.: Effects of silicon amendment on the occurrence of rice insect pests and diseases in a field test. J. Integr. Agric. **17**(10), 40–49 (2018)
2. Zhang, S., Zhang, C., Ding, J.: Disease and insect pest forecasting model of greenhouse winter jujube based on modified deep belief network. Trans. Chin. Soc. Agric. Eng. **33**(19), 202–208 (2017)
3. Nyamwasa, I., Li, K., Rutikanga, A.: Soil insect crop pests and their integrated management in East Africa: a review. Crop Prot. **106**(10), 163–176 (2018)
4. Xu, H.-X., Yang, Y.J., Lu, Y.-H.: Sustainable management of rice insect pests by non-chemical-insecticide technologies in China. Rice Sci. **24**(2), 61–72 (2017)
5. Sattler, C., Schrader, J., Farkas, V.M.: Pesticide diversity in rice growing areas of Northern Vietnam. Paddy Water Environ, **16**(2), 1–14 (2018)
6. Li, C.M., Lei, C.X., Liang, Y.T.: As contamination alters rhizosphere microbial community composition with soil type dependency during the rice growing season. Paddy Water Environ, **15**(3), 1–12 (2017)
7. Jabran, K., Riaz, M., Hussain, M.: Water-saving technologies affect the grain characteristics and recovery of fine-grain rice cultivars in semi-arid environment. Environ. Sci. Pollut. Res. **24**(14), 12971–12981 (2017)
8. Prasad, R.: Knowledge-based decision tree approach for mapping spatial distribution of rice crop using C-band synthetic aperture radar-derived information. J. Appl. Remote Sens. **11**(4), 1 (2017)
9. Fan, X., Fang, K., Ma, S.: Assisted graphical model for gene expression data analysis. Bioinformatics **38**(1), 2364–2380 (2019)
10. Lo'ai, A.T., Mehmood, R., Benkhlifa, E.: Mobile cloud computing model and big data analysis for healthcare applications. IEEE Access **4**(99), 6171–6180 (2017)

Analysis of Operation and Maintenance of Power Distribution Network Management Technology Under the Background of Big Data Era

Yiyuan Ding[✉]

State Grid State Grid International Development Co., Ltd.,
Beijing 100031, China
xily@enacs.com.cn

Abstract. The emergence of the Internet has brought tremendous changes to the global economy and society. The emergence of e-commerce has pushed the world to enter the era of network business faster. The great changes in big data to enterprises require power companies to reform traditional financial management methods through corresponding measures. Accelerate the implementation of digital management. Power distribution network management is an important part of strengthening user electricity consumption experience and promoting the development of power enterprises. Power companies should pay attention to the problems existing in power distribution network management, improve the technical level of power distribution network management, and ensure the power supply stability of power systems. This paper explores the problems existing in the domestic power distribution network management under the background of the big data era, and explores the operation and maintenance measures of the power distribution network management technology.

Keywords: Big data era · Power distribution network · Management technology · Operation · Maintenance

Foreword

The era of big data refers to the era in which various industries quickly grasp and analyze the information they need. Today's world is based on network information. People's communication and people's lives are inseparable from data information. Therefore, for a company, mastering data information is an advantage in competition. The accelerated development of the domestic industrialization process has promoted the improvement of the social and economic level, and also improved people's quality of life, making the power facilities more widely used in people's daily life and production activities [1]. It is extremely important to construct a scientific power distribution network management system and strengthen the management level of power distribution network management technology to ensure the power supply security of the power system and create a good power environment for users [2].

© Springer Nature Singapore Pte Ltd. 2020
M. Atiquzzaman et al. (Eds.): BDCPS 2019, AISC 1117, pp. 610–615, 2020.
https://doi.org/10.1007/978-981-15-2568-1_83

1 Power Distribution Network Management Technology Under the Background of Big Data Era

With the support of emerging technologies such as big data, companies can quickly obtain valuable information from massive data and seek new opportunities for the development of enterprises. In the process of operating the power distribution network system, the management technologies used are collectively referred to as power distribution network management technologies. In the process of safe production, it is possible to take targeted measures for the operating state, so that the power distribution network management technology has played a huge role. The distribution network is not only related to the quality of life of urban residents, but also closely related to the economic development of the city and has a positive impact on the progress of society [3]. The operation, maintenance and management of the distribution network mainly lies in the following aspects: equipment operation and operation capability level, maintenance quality and test level, level of live operation and ability to handle power failure, communication communication plan, power outage arrangement rationality, personnel quality level And training work, etc.

2 Problems in the Process of Power Distribution Network Management Under the Background of Big Data Era

2.1 Distribution Network Structure Problem

The scientific power distribution network structure can ensure that the power system can normally perform its performance and can effectively reduce power loss. At present, the distribution line loss rate of the power system is still high, which is a difficult and important problem to be solved in the domestic power distribution network management work. The reason for the difficulty in controlling the power distribution network line loss rate is mainly due to the lack of attention to the construction of power distribution network, the unreasonable design of transmission lines, and the unqualified structure of the grid.

2.2 Distribution Network Problem

Unreasonable distribution network will increase the loss of electric energy and power grid facilities, and interfere with the normal power consumption behavior of power users. At present, the power supply forms of common power distribution network systems are mostly single power supply, single circuit and tree type. However, these distribution networks have some problems in terms of security, and it is difficult to ensure stability during operation [4] degree. In the event of a fault in the power distribution network system, it may directly cause the main line switch to trip, causing a large-scale blackout. When a certain line of the power distribution network system is overhauled, it may also cause a power outage.

2.3   Distribution Network Facilities Management Issues

Most of the supporting facilities of the power distribution network system are directly exposed to the outdoor environment, and the corresponding protective measures are lacking [5]. This relatively harsh operating environment may cause the facility to be prone to various failures and shorten the use time of the equipment. The cost of the power distribution network system is high, and the maintenance has high requirements for the professional skills of the staff. As a result, after the installation of the power facility, if there is no serious fault, it is difficult to obtain maintenance and maintenance. Accelerate the aging speed of power distribution network facilities.

2.4   Distribution Network Management Specialization Needs to Be Improved

The low level of specialization in distribution network management is a major problem in the current management of power distribution networks [6]. In terms of management methods, some staff members also adopt traditional management models and backward management methods, failing to pay attention to the application of information technology and automation technology, and lacking awareness of target management. In terms of human resources, the number of managers with higher professional skills is less, and some staff members lack a clear understanding of their responsibilities, which makes it difficult to improve their management efficiency. In the process of power distribution network management, there is a lack of perfect specifications for the operation and maintenance process of power distribution network. The management of maintenance and operation operations is not strict, and the operation and maintenance information is not tracked, resulting in frequent accidents.

3   Operational Measures of Power Distribution Network Management Technology Under the Background of Big Data Era

3.1   Distribution Network Planning

In the modern social production and construction, the overall planning of the power distribution network can directly affect the stable operation of urban construction and greatly promote people's production and life [7]. Effective power distribution network planning is conducive to promoting the improvement of power system operation quality and meeting the needs of current urban construction. And promote the core competitiveness of power companies to improve resource efficiency. Power distribution network technology is very important for urbanization and can directly affect the stability of social production. As far as the current situation is concerned, there are mainly the following problems in the management of power distribution network: First, the structure of power distribution network is too single, the structure levels are complex, and the primary and secondary relationships cannot be correctly distinguished. Secondly, the stability of the power distribution network is very poor. In terms of safety production, it has a great destructive effect; finally, the efficiency of intelligent power production is very low, and the power system optimization construction cannot be carried out according to the actual production needs. From the above several perspectives, power companies want to enhance their core competitiveness [8]. They must consider the use of new technologies and equipment, improve the planning and construction of safety production as a whole, optimize the structure of power resources, and make reasonable innovations. In addition, the power distribution network system should be optimized to promote the improvement of management methods, minimize the cost investment, and promote the development of the core competitiveness of power companies.

3.2   Distribution Network Operation

The safe and stable operation of the power distribution network is very important for power supply and has a great improvement on the economic benefits of enterprises [9]. Conducting safe production in accordance with the prescribed production standards of enterprises can enable better management of enterprise resources, promote the enhancement of core competitiveness, and ensure the safe and stable operation of distribution networks. In combination with the existing problems in the power distribution network and replacing the old equipment, the power company should adopt a reasonable planning scheme to achieve reliable power distribution network management. In combination with an effective distribution network construction plan, effective coordination of power consumption is carried out, and under the premise of safe production, the operation efficiency of the power system is promoted to meet the needs of social production. And should be based on the actual situation, the operation and maintenance of power production equipment, reduce the occurrence of security risks.

During the operation of the distribution network, there are often some faults due to long-term high-load operation. These faults come from multiple regional lines or different types of power distribution equipment. Different operational management measures are needed to deal with different faults and problems. To ensure the safe operation of the distribution network. The power supply enterprise shall formulate scientific management systems and strategies according to the actual operation environment and conditions of the distribution network, change the previous management work concept, continuously improve the management level, and pay attention to the supervision of the critical lines and the operational status of the equipment. By taking reasonable measures to solve it, it is necessary to improve the maintenance and maintenance system of equipment, increase the capital and manpower investment in distribution management, and maintain the stable operation of the distribution network.

3.3   Maintenance Measures for Power Distribution Network Management Technology

For the current social production, the reform of the power system has gradually deepened, bringing an increasingly fierce market competition environment to power companies. Therefore, in terms of the safe and stable operation of the power system, promoting the improvement of the power distribution network management solution and optimizing the quality of the management service are very important for management and maintenance. After the safety accident occurs, the fault can be further reduced through good maintenance work. The problem arises and coordination arrangements are made to reduce accidents caused by unstable power distribution networks. And after the accident, the power company has a well-organized coordinated dispatch, which is very important for the power distribution network, and can well maintain the smooth flow of the power system. For some common power system problems, the relevant units should do timely protection work, reduce the occurrence of safety hazards, rationally use the surrounding environment, reduce the impact of other factors on the power system, and adopt new types under the requirements of capital and technical conditions. The technical means can carry out more effective power distribution network management work, and comprehensively drive the efficiency of distribution network. Combine safety production and operation and maintenance to reduce the occurrence of power distribution network failures. And with the development of the times, the safe and stable operation of the power system should be combined with the needs of the current market, optimize the innovation and construction of power distribution network management technology, and promote the overall development of the economic benefits of enterprises.

The construction and improvement of the distribution network is an inevitable requirement for the normal operation of the power system network. Once the distribution network fails during operation or causes hidden dangers and risks due to untimely maintenance, it will have extremely serious consequences for the stable power supply and transmission of the power system. Bad effects. According to the previous maintenance work experience, the staff must master the relevant professional knowledge and operational skills of various lines and equipment for maintenance, pay sufficient attention to the occurrence of faults, and take active and effective action measures to deal with various faults and hidden dangers [10]. At the same time, improve the relevant systems and operational mechanisms for maintenance management, make up for institutional defects, and improve the level of maintenance technology.

3.4   Strengthen Data Management of Power Distribution Network

Power distribution network operation data is the reference for power distribution network maintenance work. Strengthening the management level of distribution network data and properly storing data information is an important task to improve the efficiency of power distribution network troubleshooting and ensure the stability of distribution network operation.

Power companies should do a good job in archiving power distribution network data. The distribution network protection data protection scope is wide, and the staff should timely sort and sort the operation data to effectively improve the efficiency of data search, which is the basis for ensuring the smooth follow-up data investigation. The archive management work can divide the distribution network operation data according to the relevant national management standards, and make appropriate adjustments according to the actual situation.

The data of the power distribution network needs to be constantly updated and adjusted. If there are data recording errors or data omissions, it may cause serious power accidents. The power distribution network data management personnel need to improve their data management capabilities to ensure that the data update of the distribution network power facilities can be consistent with the data in the data.

4   Conclusion

The management of power distribution network is the basis for ensuring the stability and safety of the power system. Power companies should strengthen their emphasis on distribution network management, continuously analyze the problems and their causes in distribution network management, and adopt corresponding methods to adjust and improve. Power companies should actively innovate management methods, improve automation, strengthen the scientific layout of distribution network structure, strengthen the maintenance and inspection of distribution networks, pay attention to the management of distribution network facilities, ensure that power distribution facilities can operate safely, and avoid large-scale power outages. Occurs to improve the safety of electricity users and promote the long-term development of power companies.

References

1. Yu, X., Qi, X., Yu, Z., et al.: Analysis and prediction: a research on the operation and maintenance cost of power distribution network. In: International Conference on Systems & Informatics. IEEE (2018)
2. Yang, Y.: Analysis of operation and maintenance of power distribution network management technology under new situation. Sci. Technol. Innov. Appl. 62(1), 198 (2017)
3. Lin, H.T.: Analysis of operation and maintenance of power distribution network management technology under new situation. Scientist 31(12), 57–62 (2017)
4. Li, Y.: Analysis of operation and maintenance of power distribution network management technology under the new situation. East China Sci. Technol. Acad. Ed. 29(1), 88–92 (2017)
5. Li, X.D.: Operation and maintenance analysis of power distribution network management technology. Enterp. Technol. Dev. 32(16), 84 (2013)
6. Li, K.R.: Data processing technology for power operation monitoring in big data era. Glob. Market 28(21), 32–36 (2017)
7. Lu, G.Y., Zhao, C.X., Zhao, Y.: Analysis of effective operation and maintenance of computer database. Appl. Mech. Mater. 539(2), 345–348 (2014)
8. Siyuan, F.: Overview of application of big data technology in power distribution system. Proc. CSEE 52(6), 158–163 (2018)
9. Suwnansri, T.: Asset management of power transformer: optimization of operation and maintenance costs. In: 2014 International Electrical Engineering Congress (iEECON). IEEE (2014)
10. Arimoto, K., Hattori, T., Takata, H., et al.: Continuous design efforts for ubiquitous network era under the physical limitation of advanced CMOS. IEICE Trans. Electron. 90(4), 657–665 (2007)

Analysis of Enterprise Employee Performance and Corporate Benefit Correlation Based on Big Data Analysis

He Ma[(⊠)] and Hui Wang

Department of Human Resource Management,
Dalian Neusoft University of Information, Dalian, China
mahe@neusoft.edu.cn

Abstract. With the development of the economy, enterprises have experienced explosive growth, and the business management data generated by enterprises is also massive. How to extract valuable information from massive data to help companies achieve better development prospects has become one of the hot issues of concern. Based on this, this paper proposes a big data analysis method from the perspective of enterprise employee performance, and explores the correlation between enterprise employee performance and company benefit. By preprocessing the financial data of several companies and performing feature analysis, the association rule apriori mining algorithm is used to analyze the correlation between employee performance and company benefit. The results show that corporate benefits are positively correlated with corporate employee performance. The higher the performance level of the employees, the better the company's economic benefits.

Keywords: Performance management · Big data · Corporate benefit

1 Introduction

The value measurement of human resources is an important part of human resource management, and it is also the focus of the company's exploration [1]. Good human resource management can help the company form a good culture and help reduce the company's operating costs and operational risks. The key to human resource management lies in the evaluation of employee performance [2]. The company's existence is mainly for profit. If the company's efficiency is at a low level all the year round, then an in-depth investigation is needed to reverse the unfavorable situation. What is the relationship between employee performance and corporate effectiveness? This has always been a hot issue that top management has always been concerned about.

Performance evaluation is related to many aspects, such as salary, education and training, rewards and punishments [3]. Performance evaluation as a management tool, through the evaluation of the performance of the company's employees, can reward the employees who work hard, punish the employees who are passive and unsatisfied, and then make the employees work harder and devote themselves to the work as much as possible [4]. Performance appraisal seems to be one of the most common practices for

© Springer Nature Singapore Pte Ltd. 2020
M. Atiquzzaman et al. (Eds.): BDCPS 2019, AISC 1117, pp. 616–621, 2020.
https://doi.org/10.1007/978-981-15-2568-1_84

global companies because it helps companies operate more orderly [5]. As the saying goes, "No rules can't do things." Through this kind of restraint and encouragement mechanism, the company continues to operate. There are many factors influencing the company's benefits. For example, the external business environment and business policies will have more or less impact, but there is also an influential factor that is often overlooked, that is, the performance of employees within the company. If the company's operations are likened to a war, then the performance of the company's internal staff is the embodiment of the soldier's personal ability and morale. The external business environment and business policy are like guns in the hands of soldiers. Whether or not to win a war depends on the soldier's will to fight, combat capability and gun performance. Similarly, the company's ability to achieve good results depends on the performance of the company's internal staff and the external business environment, business policies.

Big data analysis is the processing of massive data [6]. Now that we have entered the era of big data, mining information that is useful to ourselves from massive data often has unexpected effects. In [7], the author uses big data technology to analyze public interests and then promote innovation. In [8], the author applies big data analysis technology to the study of insurance data, in order to explore potential customers and improve the accuracy of product marketing. Big data analysis in the medical field can provide clinical decision support, disease surveillance, etc. for related health care [9, 10]. Big data analytics technology is uniquely positioned to handle massive amounts of data. There are hundreds of companies now, and their data is huge. Through the mining of enterprise data, it is possible to ascertain the correlation between employee performance and company benefits. The amount of data in a company has grown exponentially, and to explore the relationship between employee performance and corporate effectiveness, information needs to be mined from massive data. Based on this, this paper proposes the use of big data analysis technology to analyze the correlation between employee performance and company efficiency, in order to provide reference for the operation management of related enterprises, so that enterprises can operate smoothly in the long run.

2 Method

2.1 Data Preprocessing

The data needed to analyze corporate employee performance and company benefits is massive. These data may be missing or even abnormal. Therefore, the data needs to be pre-processed. The data pre-processing operations need to solve various types of problems, including the processing of abnormal data, the processing of missing data, changing data attributes, and discretizing data. When processing missing data, if the distance of the sample attribute is measurable, the average value of the attribute's effective value is used to interpolate the missing value. If the distance is not measurable, use the mode of the attribute's valid value to interpolate the missing value.

In addition, you need to standardize the data, that is, scale the properties of the data samples to a reasonable range. Data normalization operations are designed to avoid the effects of different magnitudes of different attributes of the sample. Because the magnitude difference will result in a larger magnitude attribute dominated, algorithms that rely on sample distance are very sensitive to the magnitude of the data.

2.2 Feature Analysis

After the data is preprocessed, the amount of data is still huge, and feature analysis is necessary. The feature analysis mentioned here is a way to replace the overall content by extracting some individual features from the performance of many employees and the company's benefit data. Although this approach may not be comprehensive, it can quickly find valuable information. In this study, the individual characteristics selected for the performance of enterprise employees include work quality, cooperation spirit, mastery of knowledge required for work, initiative, diligence, workload, attendance rate, etc. For the company's benefits, the selected characteristics include sales net profit margin, asset turnover rate, equity multiplier, and return on equity. By analyzing the characteristics of data such as employee performance and company benefits, it is convenient to analyze the correlation between the two.

2.3 Association Rules Mining Algorithm

Association rules can explore the interactions between things from disorganized data, and help to adjust and optimize related decisions. For example, the correlation analysis between employee performance and company benefit studied in this paper can get the answer from the association rule mining algorithm. The expression for an association rule is:

$$X \Rightarrow Y \tag{1}$$

In the above formula, X is the former term and Y is the latter term. Let data set D be a collection of all things, each thing T is a collection of data items and T is non-empty. Suppose that T has two itemsets, X and Y. These two items are not empty and have no intersection. Then $X \Rightarrow Y$ is called an association rule in the set D.

Here, the Apriori algorithm commonly used in the association rule algorithm is used to analyze the correlation between employee and company benefits. Proceed as follows:

(1) Scan all transaction data sets in order to get frequent 1-item sets.
(2) According to the frequent $(k - 1)$-item set from the previous iteration, transform into the candidate k-term set C_k by function, and filter the support degree of all elements in C_k, and then perform the frequent k-item set L_k.
(3) Step (2) is repeated, and the iterative process is stopped when there is no new frequent item set generated by the process.

3 Data

This article uses employee performance data and company benefit data from 1,500 companies. In order to ensure the authenticity of the analysis method in this paper, the acquired data are designed into various types of companies, such as joint ventures, sole proprietorships, state-owned companies, private companies, joint-stock companies, and so on. Among these companies, the types of jobs they can be engaged in can be divided into production-oriented enterprises, trade-oriented enterprises, and online-commerce enterprises, all of which are 500. Enterprise employee performance data includes work quality, cooperation spirit, mastery of knowledge required for work, initiative, diligence, workload, attendance rate, etc. The company's benefit data includes sales net profit margin, asset turnover rate, equity multiplier, and return on equity. Through the use of the company's massive operational data, it is easy to find out the correlation between the performance of the company's employees and the company's benefits.

4 Results

Result 1: Analysis of the performance evaluation of enterprise employees.

According to the evaluation method of employee performance, the quality of work (X1), the spirit of cooperation (X2), the knowledge required for work (X3), initiative (X4), degree of diligence (X5), workload (X6), attendance rate (X7) and so on. The commonality and characteristics of employees in performance evaluation are summarized, and the performance of employees is further identified. Take A, B, and C as examples. The performance levels of these three companies are shown in Fig. 1.

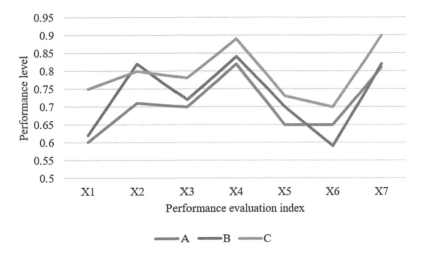

Fig. 1. Performance levels of employees in three companies

As can be seen from Fig. 1, the overall performance level of Company C is the highest among the three companies, and the gap between the two companies is significant. Although company C's level of cooperation is slightly lower than company B, the gap is almost negligible. The overall performance level of Company B is slightly higher than that of Company A, but the workload of Company B is significantly lower than that of Company A. The employee performance level is measured as a maximum of 1, the average performance level of company A is 0.68, the average performance level of company B is 0.70, and the average performance level of company B is 0.81.

Result 2: Analysis of the correlation between employee performance and company benefit.

Based on the analysis results of enterprise employee performance evaluation, the association rule Apriori mining algorithm is used to mine the company benefit data and enterprise employee performance. Considering that there may be large differences in the profitability of companies of different sizes, in order to reduce the interference of other factors, the selected companies A, B, and C are basically the same size and normalized to make the final result more reasonable. After applying the association rule algorithm, the employee performance level and company benefit of the three companies are placed under the same metric, and the specific situation is shown in Fig. 2.

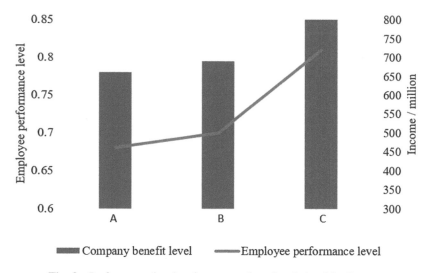

Fig. 2. Performance level and company benefit relationship diagram

As can be seen from Fig. 2, the company's benefits are positively related to the results of the company's employee performance evaluation. If the performance evaluation level of the enterprise employees is higher, the corresponding benefit level of the company will also show a higher advantage. After obtaining the data mining results, further investigations into the three companies revealed that the company's product quality has improved if employees actively work hard at work. After the quality is improved, the product's reputation is steadily improved, and the product sales are better, which will undoubtedly help the company increase its economic benefits. In turn, looking at companies with poor employee performance levels, the company's performance is not optimistic. In the long run, the company's efficiency has declined.

5   Conclusions

In the era of big data, there is a wealth of information in the vast amount of data. By using big data analysis technology, this paper studies the relationship between employee performance and company benefit, and finds that if the performance evaluation level of enterprise employees is high, then the company's economic benefits also show corresponding advantages. In the actual business operation, the management of the company should pay attention to the process of employee performance evaluation, and enhance the enthusiasm of employees for the company through performance evaluation, thereby increasing the economic benefits of the company. The company's economic benefits are good, and the welfare of employees is naturally improved. This is a two-way mutually beneficial situation. I hope that the research in this paper can provide reference for related companies and lay the foundation for enterprise development.

Acknowledgements. This work was supported by Science and Technology Innovation Think Tank Project of Liaoning Science and Technology Association (project number: LNKX2018-2019C37), and Project of Dalian Academy of Social Sciences (project number: 2018dlskyb223).

References

1. Yang, Y.: Talking about the role of corporate values in the management of human resources in state-owned enterprises. Taxation **12**(30), 246 (2018)
2. Wan, X.: A brief account of performance appraisal in human resource management. China Bus. Theory **761**(22), 80–81 (2018)
3. Lidinska, L., Jablonsky, J.: AHP model for performance evaluation of employees in a Czech management consulting company. Cent. Eur. J. Oper. Res. **26**(1), 239–258 (2018)
4. Peng, H., Si, Q., Cao, Y.: The role of performance evaluation in nursing management. Baotou Med. **40**(2), 91 (2016)
5. Yu, L., Liu, Y.: Research on enterprise employee performance evaluation system based on AHP method. J. Jilin Eng. Technol. Teach. Coll. **33**(6), 105–108 (2017)
6. Wang, H., Wu, X.: "Internet +" era coal mine big data application analysis. Coal Sci. Technol. **44**(02), 139–143 (2016)
7. Hardy, K., Maurushat, A.: Opening up government data for Big Data analysis and public benefit. Comput. Law Secur. Rev. **33**(1), 30–37 (2017)
8. Lin, W., Wu, Z., Lin, L., Wen, A., Li, J.: An ensemble random forest algorithm for insurance big data analysis. IEEE Access **5**, 16568–16575 (2017)
9. Lee, C.H., Yoon, H.J.: Medical big data: promise and challenges. Kidney Res. Clin. Pract. **36**(1), 3 (2017)
10. Dimitrov, D.V.: Medical internet of things and big data in healthcare. Healthc. Inform. Res. **22**(3), 156–163 (2016)

Revealing Land Subsidence in Beijing by Sentinel-1 Time Series InSAR

Jisong Gou[1], Xianlin Shi[1(✉)], Keren Dai[1,2], Leyin Hu[3],
and Peilian Ran[1]

[1] College of Earth Sciences, Chengdu University of Technology,
Chengdu 610059, China
shixianlin06@cdut.cn
[2] State Key Laboratory of Geohazard Prevention
and Geoenvironment Protection, Chengdu University of Technology,
Chengdu 610059, Sichuan, China
[3] Beijing Earthquake Agency, Beijing 100080, China

Abstract. Beijing, one of the biggest cities in China, has suffered land subsidence for a long time. It has been reported from the study in 2005–2017 the maximum subsidence rate reaches up to more than 10 cm/year. In this paper, the subsidence in Beijing will be revealed by time series InSAR with Sentinel-1 TOPS data during 2017–2018 for the first time. The time series analysis was performed by SBAS-InSAR method. The annual subsidence rate and time series subsidence of Beijing were acquired. The results revealed that eastern Chaoyang district and northwestern of Tongzhou district were the severe subsidence areas in Beijing with more than 10 cm/year subsidence rate, indicating that the subsidence was continuous in recent years and relative measures should be taken by the government.

Keywords: Land subsidence · InSAR · Beijing · Sentinel-1

1 Introduction

Beijing, one of the biggest cities in China, has suffered land subsidence for a long time. According to a large number of reports, newspapers, studies, etc., the land subsidence in Beijing has last a long time with wide coverage and rapid subsidence rates. Land subsidence in the city would bring potential threats to the urban infrastructure. The subsidence distribution map is crucial for supporting relative disaster prevention and reduction.

As a space-based geodetic technique, interferometric synthetic aperture radar (InSAR) has exhibited great potential in many fields [1–3], such as monitoring city subsidence [4–6], with the advantages of wide coverage and high accuracy. The subsidence in Beijing from 2005–2017 has been studied with many researchers. For example, Chen et al. [7] investigated land subsidence in Beijing region with

M. Atiquzzaman et al. (Eds.): BDCPS 2019, AISC 1117, pp. 622–628, 2020.
https://doi.org/10.1007/978-981-15-2568-1_85

Envisat ASAR data and TerraSAR-X data from 2003–2011, finding that the severe subsidence was seen in the eastern Beijing with a rate of more than 10 cm/year. Deng et al. [8] monitored the subsidence in Beijing plain during 2003–2014 based on ASAR images and RadarSat-2 data, revealing that the maximum deformation rate reach up to-124 mm/year. Hu et al. [9] applied time series InSAR based on 22 Sentinel-1 (2015–2017) SAR data to analyze the relationship between geological faults and land subsidence in Beijing Municipality. The land subsidence in Beijing after 2017 was not revealed by InSAR yet.

In this paper, the land subsidence in Beijing during 2017–2018 was revealed by Sentinel-1 time series InSAR. The mean velocity subsidence map and time series results were acquired. The subsidence details on the area with severe subsidence were analyzed.

2   Study Area and Datasets

To analyze subsidence in Beijing from 2017–2018, we used 22 SAR images from Sentinel-1 satellite, which were acquired from 12th August 2017 to 26th July 2018 with C-bands. The main parameters of these Sentinel-1 data were shown in Table 1.

Table 1. Key parameters of Sentinel-1 data

Parameters	Value
Incidence angle	38.85 (degrees)
Pixel spacing in range	2.33 (m)
Pixel spacing in azimuth	13.93 (m)
Wavelength	5.55 (cm)
Center range	902000 (m)

As shown in Fig. 1, 73 interferograms were generated according to the given threshold values, i.e. 125 m as the perpendicular baseline threshold and 48 days as the temporal baseline threshold. The acquisition date of each image was labelled and their interferometric pairs were connected with dashed lines. The spatial and temporal baselines of the datasets were presented. A robust network can be seen with these plenty of interferometric pairs, which can ensure the accuracy of the results.

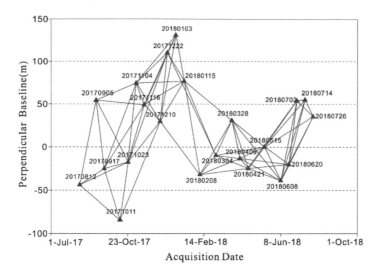

Fig. 1. The spatial and temporal baselines of the Sentinel-1 datasets

3   Methodology

Assuming N+1 SLCs covering the study area were acquired, one master image was selected and all the rest of the images would be coregistered to it. According to the given perpendicular baseline threshold and the temporal baseline threshold, interferometric pairs were formed as shown in Fig. 1. Then interferograms were generated in the interferometric process by removing the flattening effect and topographic effect with external digital elevation model (DEM). The interferograms would be filtered by the Goldstein method and then be unwrapped by the minimum cost flow method.

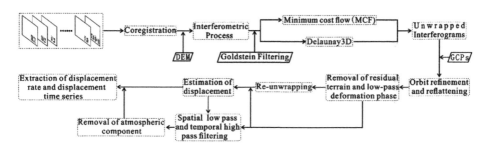

Fig. 2. Flowchart of SBAS-InSAR method

The unwrapped phase in an arbitrary coherent pixel (x, y) can be modeled as:

$$\Delta\phi_i(x,y) = \Delta\phi_{disp} + \Delta\phi_{topo} + \Delta\phi_{atm} + \phi_{res} \qquad (1)$$

Where $\Delta\phi_{disp}$, $\Delta\phi_{topo}$, $\Delta\phi_{atm}$ and $\Delta\phi_{res}$ denote the phase components from displacement along the radar line of sight (LOS) direction, topographic error, atmospheric artifacts and decorrelation/thermal noise, respectively. The $\Delta\phi_{disp}$ and $\Delta\phi_{topo}$ can be expressed as:

$$\Delta\phi_{disp} = \frac{4\pi}{\lambda}[d_{t_2}(x,y) - d_{t_1}(x,y)] \qquad (2)$$

$$\Delta\phi_{topo} = \frac{4\pi}{\lambda}\frac{B_{\perp i}}{r\sin\theta}\Delta Z(x,y) \qquad (3)$$

λ and θ are wavelength and incidence angle, respectively. $dt_2(x,y)$ and $dt_1(x,y)$ are the cumulative displacements at time t_2 and t_1, respectively. $\Delta Z(x,y)$ is the topographic error between the actual elevation and the DEM. Refinement and reflattening will be performed after selecting GCPs (Ground Control Points). The components of residual height and low-pass deformation phase were removed from the re-flattened interferograms before the Re-unwrapping process. After the phase re-unwrapping [10], the displacement velocity was firstly estimated based on the linear model. Before extracting the time series displacements, atmospheric correction is performed by spatial high-pass filtering and temporal low-pass filtering. Finally, the mean velocity map and time series displacements maps were obtained and geocoded. The flowchart was shown in Fig. 2 and the details of this method can be referred to [11].

4   Result and Discussion

Following the procedures as described in Sect. 3, the time series InSAR analysis was performed. The mean velocity subsidence map was shown in Fig. 3. The boundary of Beijing was presented with black line and the locations of each district were briefly shown with red solid dot with white edge. The mean subsidence rate reached up to near 12 cm/year in the line-of-sight direction (positive value means far away from the sensors). There were three main subsidence zones with blue color can be seen.

As shown in Fig. 3, a slight subsidence zone was located in the south of Changping District, with a rate of 3–8 cm/year. Two severe subsidence zones were located in eastern Chaoyang district and northwestern of Tongzhou district with a maximum rate of 12 cm/year. The other subsidence zone was located in the Langfang city with a rate of 4–7 cm/year, which was near the south boundary of Beijing.

Assuming that the image acquired at 12th August 2017 was with relative zero displacement as the reference image, the time-series result was shown in Fig. 4. It can be noted that the subsidence in Beijing was accumulated with time all over the whole year. The spatial pattern of the subsidence zones formed gradually.

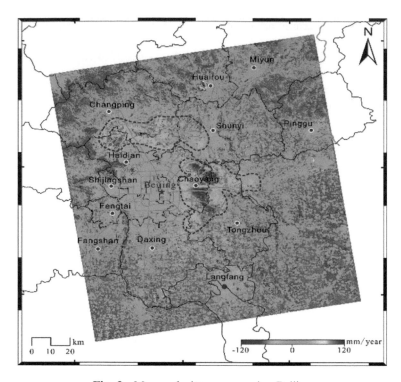

Fig. 3. Mean velocity map covering Beijing

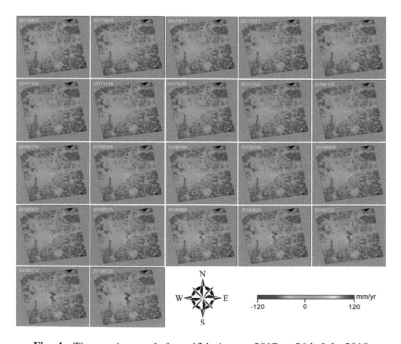

Fig. 4. Time series result from 12th August 2017 to 26th July 2018

The severe subsidence zones were shown in Fig. 5 with the enlarged map, revealing that eastern Chaoyang district and northwestern Tongzhou district were the severe subsidence zones in Beijing with more than 10 cm/year subsidence rate. These two areas between the Beijing 5th ring road and 6th ring road shown in Fig. 5 were the most severe subsidence areas in Beijing, where Jinzhan area Shawo area, Guanzhuang area, Heizhuanghu area and Taihu area were located.

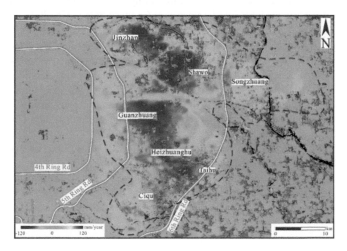

Fig. 5. Enlarged displacement map over subsidence zone in Tongzhou-Chanyang district

5 Conclusion

This paper revealed the subsidence with Sentinel-1 TOPS time series InSAR in Beijing during 2017–2018. The results showed eastern of Chaoyang district and northwestern Tongzhou district were the severe subsidence zones in Beijing with more than 10 cm/year subsidence rate, indicating that the subsidence was continuous in recent years and relative measures should be taken by the government.

Acknowledgements. This work was funded by the science for earthquake resilience under the project No. XH18001Y, the National Natural Science Foundation of China (No. 41801391, No. 41771444), Sichuan Provincial Key Technology Support Program (No. 2013FZ0021).

References

1. Delouis, B., Nocquet, J.M., Vallée, M.: Slip distribution of the February 27, 2010 Mw = 8.8 Maule Earthquake, central Chile, from static and high-rate GPS, InSAR, and broadband teleseismic data. Geophys. Res. Lett. **37**(17) (2010)
2. Henderson, S.T., Pritchard, M.E.: Decadal volcanic deformation in the Central Andes Volcanic Zone revealed by InSAR time series. Geochem. Geophys. Geosyst. **14**(5), 1358–1374 (2013)

3. Liu, L., Millar, C.I., Westfall, R.D., Zebker, H.A.: Surface motion of active rock glaciers in the Sierra Nevada, California, USA: inventory and a case study using InSAR. Cryosphere **7**, 1109–1119 (2013)
4. Dai, K., Liu, G., Li, Z., Li, T.: Extracting vertical displacement rates in Shanghai (China) with multi-platform SAR images. Remote Sens. **7**, 9542–9562 (2015)
5. Liu, P., Li, Q., Li, Z., Hoey, T.B.: Anatomy of subsidence in Tianjin from time series InSAR. Remote Sens. **8**, 266 (2016)
6. Ng, A.H.-M., Ge, L., Li, X., Zhang, K.: Monitoring ground deformation in Beijing, China with persistent scatterer SAR interferometry. J. Geodesy **86**(6), 375–392 (2011)
7. Chen, M., Tomás, R., Li, Z., Motagh, M.: Imaging land subsidence induced by groundwater extraction in Beijing (China) using satellite radar interferometry. Remote Sens. **8**, 468 (2016)
8. Deng, Z., Ke, Y., Gong, H., Li, X., Li, Z.: Land subsidence prediction in Beijing based on PS-InSAR technique and improved Grey-Markov model. GIScience Remote Sens. **54**, 797–818 (2017)
9. Hu, L., Dai, K., Xing, C.: Land subsidence in Beijing and its relationship with geological faults revealed by Sentinel-1 InSAR observations. Int. J. Appl. Earth Obs. Geoinf. **82**, 101886 (2019)
10. Xu, W., Ian, C.: A region-growing algorithm for InSAR phase unwrapping. IEEE Trans. Geosci. Remote Sens. **37**(1), 124–134 (1999)
11. Guo, J., Hu, J., Li, B.: Land subsidence in Tianjin for 2015 to 2016 revealed by the analysis of Sentinel-1A with SBAS-InSAR. J. Appl. Remote Sens. **11**(2), 026024 (2017)

Scale Adaptation of Text Sentiment Analysis Algorithm in Big Data Environment: Twitter as Data Source

Yaoxue Yue[(⊠)]

Shandong Polytechnic College, Jinan, Shandong, China
330612305@qq.com

Abstract. The research of text sentiment analysis under big data is of great significance to the researchers of information science. Through this kind of text sentiment analysis based on big data, it can greatly optimize the efficiency and cost. The data set used in this paper is sentiment140 produced by Stanford University. In this paper, based on the traditional emotional analysis, a text emotional analysis based on big data environment is proposed, and the accuracy and generalization ability of the algorithm are compared. The experimental results show that the efficiency advantage of text emotion analysis based on big data is more obvious. In the aspect of resource utilization of system environment, with the increase of the number of system cores and nodes, the operation speed of the whole system is also accelerated.

Keywords: Big data · Sentiment analysis · Text analysis · Machine learning

1 Introduction

With the continuous development of science and technology, great changes have taken place in people's clothing, food, housing and transportation. Nowadays, people can travel, shop, chat and make friends through a mobile phone. During this period, huge data information will be generated. Therefore, how to make good use of these huge data information is very important for enterprises and even national governments, especially the importance of emotional analysis is self-evident.

Emotion analysis can judge a person's mood change and political attitude well, so the research of emotion analysis based on machine learning becomes very hot. In literature [1], the author analyzes the data set in tweets by using unsupervised and supervised machine learning methods. The results show that the proposed unsupervised method can achieve 80.68% accuracy compared with the dictionary based method (the latter has an accuracy of 75.20%). In document [2], the author uses the multi tag corpus in the field of film to analyze the emotion in conversational texts. In literature [3], the author studies the current situation of text sentiment analysis from viewpoint to sentiment mining. In reference [4], the author studies the text sentiment analysis in code transformation using the joint factor graph model. In reference [5], the author published a unique method, in which physiological disorders can be analyzed to determine users' emotional status. The basis of loading physiological data and generating emotional

© Springer Nature Singapore Pte Ltd. 2020
M. Atiquzzaman et al. (Eds.): BDCPS 2019, AISC 1117, pp. 629–634, 2020.
https://doi.org/10.1007/978-981-15-2568-1_86

analysis comes from a semiotic analysis framework, in which all symbols, signified symbols and symbols are recognized and utilized to complete the transformation. This method uses time-based slope clustering algorithm to provide real-time human emotion evaluation report based on clustering frequency. In document [6], the author puts forward a multi-disciplinary work in order to alleviate this problem by standardizing the transformation between emotion models. Specific contributions are: semantic representation of emotional transformation; API proposal for services that perform automatic transformation; reference implementation of such services; verification of proposals by integrating different emotional models and use cases of service providers. In [7], the author applied the research of emotion analysis to the analysis of emotion and attention of autistic children. In literature [8], the author studied the influence of emotion on the best brand analysis emotion on gender brand recall, and the research showed that there was a positive correlation between advertising effect and emotion; brand recall was different according to gender. In literature [9], the author used emotional analysis technology to enhance the ability of positive emotional experience in comedy performance in social occasions. In literature [10], the author studied the importance of emotional expression in social communication and the role of regulating emotional expression in social dysfunction and interpersonal problems.

In this paper, under the big data platform spark, by trying a variety of different emotion analysis algorithms, and applying these algorithms to different data sets for training, through experimental comparative analysis, let researchers reasonably choose the optimal emotion analysis algorithm and configuration platform to achieve the optimal cost efficiency.

2   Method

This paper mainly studies the scale adaptation problem of emotional analysis of text based on big data environment. In other words, in the same field and language environment, by comparing the traditional emotional analysis algorithm and the emotional analysis algorithm under big data that we now put forward, we can test the model adaptation effect under different sizes and different data sets. Specifically, the problems studied in this paper are: the comparison of operation efficiency of emotion analysis algorithms in different operating environments under big data; the operation efficiency of various emotion analysis algorithms in different operating environments; how to optimize the operation efficiency of emotion analysis algorithm with different number of nodes and CPU cores configured in the system; the different size of data set, the different situation The training effect of sensory analysis algorithm. In this paper, the big data sentiment analysis is based on Twitter as the data source.

2.1   Platform Construction of Emotion Analysis Model

At present, the important platforms for big data emotion analysis algorithms are Hadoop, Mars, spark, twister, haloop, phonenix, etc. The big data emotion analysis platform used in this paper is spark, which has better memory iteration mechanism, faster calculation efficiency, greater fault tolerance and better scalability than other emotion analysis platforms spark. Moreover, the library of spark platform already contains a large number of commonly used machine learning libraries and can be programmed with Python. Compared with the old-fashioned sklearn, the library call of spark is more convenient.

The algorithm analysis framework adopted in this paper is shown in Fig. 1:

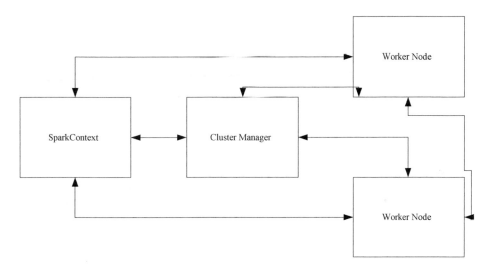

Fig. 1. Spark emotion analysis framework

2.2 Emotion Analysis Algorithm in Big Data

The algorithm in this paper is based on machine learning algorithms such as naive Bayes, random forest, decision tree, logical regression and support vector machine, which are obtained by parallel operation. The reasons for choosing these machine learning algorithms are as follows: usability, these algorithms have been widely used in traditional text sentiment analysis; strong scalability, these machine learning algorithms can be applied to larger data sets; comparability, these algorithms have been very mature in model evaluation indicators. It needs to be pointed out that in addition to the classification algorithms used in this paper, there are many classification algorithms, such as gradient lifting tree, multi-layer perceptron and convolutional neural network, etc., but because the parameter debugging of these algorithms is very complicated, so this paper does not test these algorithms.

3 Experiment

During the experiment, we took 80% of twitter data as training set and 20% as test set, and trained at least three times for all training set data. In terms of parameter setting, we follow the following principles: first, availability, that is to say, the selected parameters should make the model available, and the classification result of operation is relatively accurate and reliable; second, equivalence, that is, when designing parameters, we should try to make the algorithm run in different environments with the same parameters as far as possible, so as to make more use of the model. Ratio analysis. In the decision tree algorithm, we set the feature tree as 3000 and the maximum depth of the tree as 10; in the random algorithm, we set the feature tree as 3000 and the maximum depth of the tree as 10; in the logical regression algorithm, we set the iteration step size as 1000; in the support vector machine algorithm, we use the iteration step size as 1000.

In the experiment, the evaluation indexes we used are mainly from the classification effect and operation speed. The indicators we use for classification effect are accuracy, accuracy and recall. The indicators we use for classification efficiency include running time and acceleration ratio.

4   Result

Result 1: Compare operating efficiency.

In order to study the running efficiency of different algorithms, we use 150 m medium-sized datasets to run tests on traditional, single node spark and cluster spark respectively. During the test, we use four core 4G operating environment for traditional and single node spark, and three four core 4G operating environments for cluster spark (Fig. 2).

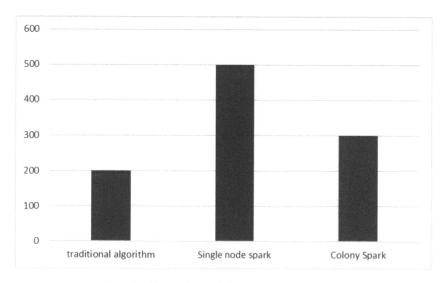

Fig. 2. Comparison of algorithm operation rate

Result 2: Comparison of results from different algorithms.

In This experiment compares the accuracy of different algorithms in the classification results. Similarly, the data set used in Experiment 2 is also a medium-sized 150 m data set, and the evaluation indexes are accuracy, accuracy and recall rate. The experimental results are shown in Fig. 3.

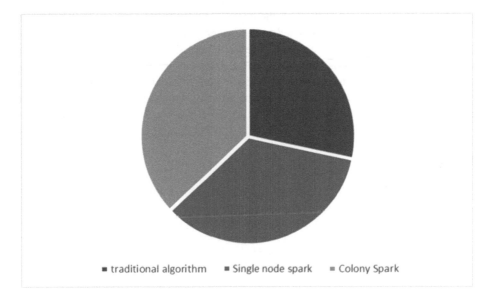

Fig. 3. Comparison of accuracy

From Fig. 3, we can see that the algorithm runs with the highest accuracy on the cluster spark. In addition, the accuracy of spark single node is very close to that of cluster spark, which fully shows that for spark, expanding the number of nodes from single node to multiple nodes does not affect the accuracy of classification results. In general, when the number of CPU cores in the system increases, it means that more resources can be used for task execution, so it will show better scalability. In addition, for different algorithms, their running time will be different, and with the increase of the number of CPU cores, the running speed does not increase linearly with the number of CPU cores. Experiments show that the algorithm utilization rate in the cluster environment is higher when the number of four cores is higher than when the number of eight cores is higher. From the experimental data comparison of various algorithms, we find that Nb algorithm has the shortest running time and the best classification effect.

In the process of experiment, the running speed of emotion analysis algorithm in a single node has a certain relationship with the size of data set. When the size of data becomes larger, the running time of the algorithm also becomes longer. Experiments show that when the data set size is less than 80 m, the running time of the algorithm used in the experiment increases slowly in the classification task, while when the data set size is more than 80 m, the running time of the algorithm increases rapidly.

5 Conclusion

This paper is based on the text sentiment analysis under the big data environment, taking twitter data as the data source, in the process of sentiment analysis, we use five machine learning classification algorithms, and test the performance of various algorithms in different operating environments and different size datasets through experiments, mainly considering the accuracy of algorithm classification and algorithm running time. Two aspects. Through the experimental results, we can see that the text emotion analysis algorithm proposed in this paper has very good scalability and high accuracy. At the same time, we can also see the superiority of spark platform in dealing with text emotion analysis. However, the experiment in this paper also has some shortcomings, such as the data set is relatively single, and it can not provide a high-quality operating environment on the test equipment, which are all the areas that need to be improved in the future.

References

1. Rout, J.K., Choo, K.K.R., Dash, A.K., et al.: A model for sentiment and emotion analysis of unstructured social media text. Electron. Commer. Res. **18**(1), 181–199 (2018)
2. Duc-Anh, P., Matsumoto, Y.: EMTC: multilabel corpus in movie domain for emotion analysis in conversational text. In: Proceedings of the Eleventh International Conference on Language Resources and Evaluation (LREC 2018) (2018)
3. Yadollahi, A., Shahraki, A.G., Zaiane, O.R.: Current state of text sentiment analysis from opinion to emotion mining. ACM Comput. Surv. (CSUR) **50**(2), 25 (2017)
4. Wang, Z., Lee, S.Y.M., Li, S., et al.: Emotion analysis in code-switching text with joint factor graph model. IEEE/ACM Trans. Audio Speech Lang. Process. (TASLP) **25**(3), 469–480 (2017)
5. Rajkumar, A., Rajkumar, A.: Human Emotion Assessment Based on Physiological Data Using Semiotic Analysis. U.S. Patent Application 14/825,181[P], 16 February 2017
6. Sánchez-Rada, J.F., Iglesias. C.A., Sagha, H., et al.: Multimodal multimodel emotion analysis as linked data. In: 2017 Seventh International Conference on Affective Computing and Intelligent Interaction Workshops and Demos (ACIIW), pp. 111–116. IEEE (2017)
7. Egger, H.L., Dawson, G., Hashemi, J., et al.: Automatic emotion and attention analysis of young children at home: a ResearchKit autism feasibility study. npj Digital Med. **1**(1), 20 (2018)
8. Wang, W.C., Pestana, M.H., Moutinho, L.: The effect of emotions on brand recall by gender using voice emotion response with optimal data analysis. In: Moutinho, L., Sokele, M. (eds.) Innovative Research Methodologies in Management, pp. 103–133. Palgrave Macmillan, Cham (2018)
9. Sayette, M.A., Creswell, K.G., Fairbairn, C.E., et al.: The effects of alcohol on positive emotion during a comedy routine: a facial coding analysis. Emotion **19**(3), 480 (2019)
10. Chervonsky, E., Hunt, C.: Suppression and expression of emotion in social and interpersonal outcomes: a meta-analysis. Emotion **17**(4), 669 (2017)

Regional Value-Added Trade Network: Topological Features and Its Evolution

Shan Ju[✉]

School of International Economics and Trade, Shandong University of Finance
and Economics, Jinan 250014, Shandong, China
jushantopic@163.com

Abstract. Using the OECD-ICIO (2018) Database, this paper constructs an
ASEAN+3 value-added trade network and calculates its topological features,
including the network as a whole and degree centralities of nodes from 2005 to
2015. Among aggregate topological features, the biggest fluctuation is index of
average degree with downward trend while the other four indicators of average
geodesic distance, reciprocity correlation, average clustering coefficient and
degree assortativity are stable. In addition, the values of degree assortativity
have been negative. From the perspective of nodes centralities, Japan and China
have been top two according to the results of out-degree centrality and out-
closeness centrality of nodes while Malaysia, Singapore, Vietnam, and Thailand
rank high on the indices of in-degree centrality and in-closeness centrality of
nodes. Brunei Darussalam has been less involved in ASEAN+3 value-added
trade network during the sample period. The hub countries such as Japan, China,
Malaysia and Singapore need to advance the regional economic cooperation
based on the advantages of countries and existing production network.

Keywords: Regional value-added trade · Complex network · Topological
features

1 Introduction

During the past decades, the countries in East Asia and Southeast Asia have grown
rapidly to be one of the most important regions in the global production network. The
interdependence among them has been strengthened. To be more specific, China and
some Southeast Asian countries import key components from Japan or Korea, then
export final products to the United States or countries in Europe after processing
assembly [1, 2]. Based on the complexity features, this paper analyses value-added
trade relations among ASEAN+3 (Association of Southeast Asian Nations Plus Three)
countries from a complex network perspective.

In recent years, the complex network method has been applied to analyze value-
added trade with the big increase of the value-added trade in world economy. By
constructing directed or undirected, weighted or unweighted value-added trade network
based on different world input-output databases and computing the topological features
of them, the value-added trade relationships among countries in the world are presented. It helps us to understand the real situation of global vertical specialization.

© Springer Nature Singapore Pte Ltd. 2020
M. Atiquzzaman et al. (Eds.): BDCPS 2019, AISC 1117, pp. 635–641, 2020.
https://doi.org/10.1007/978-981-15-2568-1_87

[3–6] However, there has no research on the value-added trade network relations among ASEAN+3 countries until now. Therefore, considering the importance of ASEAN+3 countries as a whole in the global production network, this paper constructs an ASEAN+3 value-added trade network which is directed and unweighted. And, the topological features of it are calculated as well.

The rest of this paper is organized as follows. In part two, the method for constructing an ASEAN+3 value-added trade network and the required data are introduced. In part three, the topological features of the ASEAN+3 value-added trade network are computed and analyzed. In part four, the results and policy implications are discussed.

2 Method and Data

2.1 Method for Constructing ASEAN+3 Value-Added Trade Network

The ASEAN+3 value-added trade network in this study is directed and unweighted. The orientation of the edge in the ASEAN+3 value-added trade network is set as follows [4]:

$$\overrightarrow{a_{ij}} = \begin{cases} 1, & \text{if } \dfrac{FVA^{ij}}{EX^j} > 0.01 \\ 0, & \text{otherwise} \end{cases} \tag{1}$$

In Eq. (1), $i \neq j = 1, 2, \cdots N$, FVA^{ij} is the value-added created by economy i in the total value of export of economy j, EX^j is the total value of export of economy j, the $N \times N$ adjacency matrix $AM = [a_{ij}]$ is the value-added connections between economies of ASEAN+3.

2.2 Data for Constructing ASEAN+3 Value-Added Trade Network

The inter-countries input-output data are available now from WIOD (World Input-Output Database), OECD (Organization for Economic Co-operation and Development), ADB (Asian Development Bank) and other databases. These databases have different characteristics in terms of the number of countries, the number of industrial sectors and the time span. Considering no database covers the overall 13 countries in ASEAN+3, OECD-ICIO (2018) is the best choice because it contains 10 countries except Cambodia, Laos and Myanmar. In fact, the participation of the three countries in the global and regional production network is quite low. In this paper, we use the 10 economies as the nodes of the ASEAN+3 value-added trade network (see Table 1).

To avoid iterative calculations, GVC Index database are available from research institute for global value chains in University of International Business and Economics. In this paper, in order to calculate FVA^{ij} and EX^j in Eq. (1), we obtained 880 files from the UIBE GVC Index.

Table 1. The 10 ASEAN+3 value-added trade network economies

Acronym	BRN	CHN	IDN	JPN	KOR
Name	Brunei Darussalam	China	Indonesia	Japan	Korea
Acronym	MYS	PHL	SGP	THA	VNM
Name	Malaysia	Philippines	Singapore	Thailand	Vietnam

3 Topological Features of ASEAN+3 Value-Added Trade Network

3.1 Mapping ASEAN+3 Value-Added Trade Network

After constructing ASEAN+3 value-added trade network from 2005 to 2015 based on the computed results of Eq. (1), we draw network graphs using the Cytoscape 3.7.1 tools. Due to limited space, only the network graphs for 2005 and 2015 are shown here (see Fig. 1). The circles in Fig. 1 represent node countries in the ASEAN+3 value-added trade network. The size of each node is proportional to its total degree.

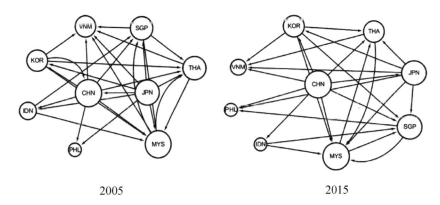

2005   2015

Fig. 1. Network graphs of total foreign value added in exports in 2005 and 2015.

On the one hand, from Fig. 1 we could find that the degrees of Malaysia and Thailand are also relatively high in addition to Japan, China, Singapore and Korea. And compared to 2005, the degree of Vietnam in 2015 is significantly smaller. On the other hand, Japan has been the main value-added supplier in the regional network from the direction of it. China's position as a value-added supplier is increasing with more arrows pointing outward in 2015 than in 2005. Singapore, Thailand, Malaysia and Vietnam are major importers of added value in the network both in 2005 and in 2015.

3.2 Aggregate Topological Features of ASEAN+3 Value-Added Trade Network

1. Aggregate Network Metrics

The main topological features of one directed complex network are average degree (k), average geodesic distance (D), reciprocity correlation (R_B), degree assortativity (r), average clustering coefficient (C) and so forth. The formulas for calculating the above aggregate network topological features are as follows [7, 8]:

$$\langle k \rangle = \frac{1}{N} \sum_{i=1}^{N} k_i \tag{2}$$

$$D = \frac{1}{K} \sum_{i \neq j} d_{ij} \tag{3}$$

$$R_B = E_B / E \tag{4}$$

$$r = \frac{E^{-1} \sum_i j_i k_i - [E^{-1} \sum_i \frac{1}{2}(j_i + k_i)]^2}{E^{-1} \sum_i \frac{1}{2}(j_i^2 + k_i^2) - [E^{-1} \sum_i \frac{1}{2}(j_i + k_i)]^2} \tag{5}$$

$$C = \frac{1}{N} \sum_{i=1}^{N} C_i \tag{6}$$

In Eqs. (2) to (6),

k_i is the number of neighbors of the node v_i in the network which is called the degree of the node;
K is the total number of node pairs with paths;
d_{ij} is the number of edges of a path with the fewest edges between two nodes v_i and v_j in the network;
E_B is the number of bidirectional edges in the directed network;
E is the number of all edges of the network;
$C_i = E_i / [k_i(k_i - 1)]$, E_i is the actual number of edges.

2. Results of Aggregate Topological Features of ASEAN+3 Value-Added Trade Network

We calculate the five aggregate topological features of ASEAN+3 value-added trade network from 2005 to 2015 according to Eqs. (2) to (6), the results of them are shown in Fig. 2.

Firstly, during the period of 2005 to 2015, among aggregate topological features of ASEAN+3 value-added trade network, the biggest fluctuation is index of average degree with downward trend, which indicates that the network is getting looser. One of the reasons is that some countries like China, Vietnam and Indonesia participated in value-added cooperation outside the ASEAN+3 region. Secondly, the three indicators of average geodesic distance, reciprocity correlation and average clustering coefficient are stable. Therefore, degree of integration, bidirectional supply-demand relations and

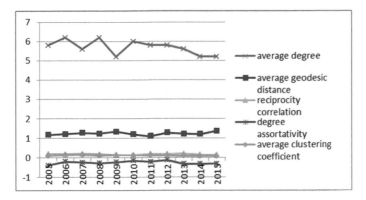

Fig. 2. Aggregate topological features of ASEAN+3 value-added trade network from 2005 to 2015.

the degree of clustering among the 10 countries are stable. Thirdly, the values of degree assortativity have been negative and it reflects that the ASEAN+3 value-added trade network is disassortative, that is, core-periphery relationships among countries in the network has been existing.

3.3   Centralities of Nodes of ASEAN+3 Value-Added Trade Network

1. Centralities Metrics

Four indicators are used in this study to measure centralities of nodes in a directed complex network: out-degree and in-degree centralities $(C_{D_{out}}$ and $C_{D_{in}})$, out-closeness and in-closeness centralities $(C_{C_{out}}$ and $C_{C_{in}})$. The formulas for calculating them are as follows [9, 10]:

$$C_{D_{out}}(v_i) = k_i^{out}/(N-1) \tag{7}$$

$$C_{D_{in}}(v_i) = k_i^{in}/(N-1) \tag{8}$$

$$C_{C_{out}}(v_i) = (N-1)/\left[\sum_{j=1,j\neq i}^{N} d_{ij}\right] \tag{9}$$

$$C_{C_{in}}(v_i) = (N-1)/\left[\sum_{j=1,j\neq i}^{N} d_{ji}\right] \tag{10}$$

All parameters in the Eqs. (7) to (10) have the same meanings as above.

2. Results of Centralities of Nodes

Due to limited space, all results cannot be shown here in the form of tables or figures. Instead, we represent the typical characteristics of them.

Out-degree centrality reflects the situation of one node's supply of the added value for the total exports of other countries in the same network. Regarding the values of out-degree centrality during the sample period, the top two are Japan and China from

2005 to 2012 while China and Japan from 2013 to 2015 with the maximum of 8 and the minimum of 6. Country in the third place is Korea except in the years of 2011 and 2012 when Indonesia is in the third place. The last three countries have been Brunei Darussalam, Philippines and Vietnam respectively with values of out-degree centrality for all years are 0.

In-degree centrality reflects the situation of one node's using added value from other countries in the same network in its total export production. On the values of in-degree centrality during the sample period, Malaysia ranks first in the remaining 8 years, except for the years of 2006 to 2008 when it ranks the second. The in-degree centrality values of Malaysia in all years are 6 or 5. Countries with several years in the top three are Singapore (9 years), Thailand (7 years) and Vietnam (6 years) while countries with several years in the bottom two are Brunei Darussalam (11 years) and Indonesia (6 years).

The larger the values of out-closeness or in-closeness centralities of one node are, the easier it establishes a value-added trade relationship outward or inward, that is, the node is at the center of the network. As for the values of out-closeness centrality during the sample period, the top two and the last three countries have been completely same as those of out-degree centrality. While on the values of in-closeness centrality, countries with several years in the top three are Malaysia (11 years), Vietnam (9 years), Singapore (8 years) and Thailand (5 years) while countries with several years in the bottom two are Brunei Darussalam (11 years) and Indonesia (6 years).

4   Conclusions

Based on complex network theory, this paper uses the OECD-ICIO (2018) database to construct an ASEAN+3 value-added trade network and measures the main topological features of the network from 2005 to 2015. Relevant conclusions are as follows:

As an important economic cooperation region, the ASEAN+3 value-added trade network has evolved from 2005 to 2015. In general, this network is getting looser. It reflects the structural change that countries including China and Vietnam in this network began to participate more in the global value-added trade cooperation with countries outside the region. With core-periphery relationships among countries in the network been existing, whether the hub countries cooperate or not does matter.

Based on the results of measuring out-degree centrality and out-closeness centrality of each node, Japan and China are the main suppliers of the added value for the total exports of other countries in this network, at the same time, it is easier for them to establish a value-added trade relationship outward with other nodes in the network. On the other hand, based on the results of measuring in-degree centrality and in-closeness centrality of each node, Malaysia, Singapore, Vietnam, and Thailand are the main added value users in its total export production, also, it is easier for them to establish a value-added trade relationship inward with other nodes in the network. All results of node centralities indicate that Brunei Darussalam has been less involved in ASEAN+3 value-added trade network during the sample period.

Therefore, the original advantages of ASEAN+3 value-added trade network still exist. The core-node countries need to strengthen regional economic cooperation by reducing tariffs and non-tariff barriers, expanding the scope of open industries and so forth.

Acknowledgements. This work was supported by the National Social Science Fund of China, "On the Status and Influences of China in the Evolution of Global Value Chains with the Perspective of Complex Network" (No. 17BJL114).

References

1. Chan, S.: The belt and road initiative: implications for China and East Asian economies. Copenhagen J. Asian Stud. **35**(2), 52–78 (2018)
2. Obashi, A., Kimura, F.: Deepening and widening of production networks in ASEAN. Asian Econ. Pap. **16**(1), 1–27 (2017)
3. Cepeda-López, F., et al.: The evolution of world trade from 1995 to 2014: a network approach. J. Int. Trade Econ. Devel. **28**(4), 452–485 (2019)
4. Amador, J., Cabral, S.: Networks of value-added trade. World Econ. **40**(7), 1291–1313 (2017)
5. Amador, J., et al.: Who's who in global value chains? A weighted network approach. Open Econ. Rev. **29**(5), 1039–1059 (2018)
6. Long, T., et al.: Exploring the competitive evolution of global wood forest product trade based on complex network analysis. Physica A Stat. Mech. Appl. **525**, 1224–1232 (2019)
7. Hossain, Md.M, Sameer, A.: A complex network approach towards modeling and analysis of the Australian airport network. J. Air Transp. Manag. **60**, 1–9 (2017)
8. Attar, N., Aliakbary, S.: Classification of complex networks based on similarity of topological network features. Chaos Interdiscip. J. Nonlinear Sci. **27**(9), 091102 (2017)
9. Des Marais, D.L., et al.: Topological features of a gene co-expression network predict patterns of natural diversity in environmental response. Proc. R. Soc. B Biol. Sci. **284**(1856), 20170914 (2017)
10. Zhou, C., et al.: Topological mapping and assessment of multiple settlement time series in deep excavation: a complex network perspective. Adv. Eng. Inform. **36**, 1–19 (2018)

Analysis on the Impact of FDI on Regional Innovation in China from the Perspective of Big Data

Sitong Lin[✉]

School of Economics, Shanghai University,
Chengzhong Road 20, Shanghai, China
Hdlstfq2013@163.com

Abstract. Based on the balanced panel data of 31 provinces in China from 2003 to 2016, this paper studies the impact of FDI on the innovation ability of China's provinces through an econometric model. The results show that FDI has a significant positive U-shaped effect on the quantity of provincial innovation, and there are localization phenomena. However, according to the results of characteristic facts, FDI basically has a positive U-shaped effect on innovation quality of provinces, but in the central provinces represented by Heilongjiang, FDI has an inverted U-shape effect on innovation quality. At the same time, it is found that the relationship between FDI and innovation depends on the R&D expenditure, the level of economic development and the degree of knowledge protection in each province. More attention should be paid to the quality of FDI introduction and corresponding measures should be taken to promote innovation.

Keywords: FDI · Innovation quality of city · Innovation quantity of city

1 Introduction

Since the 1980s, international capital flows have become increasingly active, and direct investment has become the main mode of capital flows and the main channel for developing countries to obtain external resources. By the end of the 1980s, cumulative foreign direct investment had reached $1.5 trillion. Since the 1990s, the scale of international direct investment has been expanding and growing rapidly. At the same time, the influence of FDI on the development of host countries, especially developing countries, is also expanding.

The Chinese government has been vigorously encouraging innovation. According to the Global Entrepreneurship Monitor report, China's innovation index is 15.53, higher than that of the United States, the United Kingdom and Japan. A large number of studies have confirmed that encouraging innovation can promote economic growth and help solve employment problem. According to the SAIC, the number of newly registered enterprises and the total registered capital in China reached a record high in 2015.

While China's overall innovation capacity continues to improve, the amount of FDI China absorbs is also increasing. However, the distribution of FDI in China is not

© Springer Nature Singapore Pte Ltd. 2020
M. Atiquzzaman et al. (Eds.): BDCPS 2019, AISC 1117, pp. 642–649, 2020.
https://doi.org/10.1007/978-981-15-2568-1_88

equal. Obviously, we can know that the total amount of FDI absorbed by the eastern provinces are much higher than those in the central and western regions. In addition, according to the GEM 2016 report released by Tsinghua university, there are highly active, active and inactive areas of innovation in China. After Grossman first put innovation in an open economic environment to analyze the impact of international trade and FDI on domestic innovation, scholars began to pay attention to the relationship between FDI and innovation.

However, what role does FDI play in the innovation capacity of Chinese provinces? Considering that innovation should include both the quantity aspect and quality aspect, how should we investigate the impact of FDI on the quantity and quality of innovation? In order to answer the above questions, this paper uses the balance panel data of 31 provinces in China from 2003 to 2016 to reveal the impact of FDI on the innovation capacity of each province through the fixed-effect model, and analyzes the heterogeneity of regional differences. This paper has certain practical significance for each province on how to make better use of foreign capital.

2 Literature

Innovation is not only the creation and output of one's own knowledge, but also the exploration of the external environment and the discovery and utilization of innovation opportunities. Moreover, the probability of individual innovation is a function of expected innovation rewards. FDI influences the expected income of potential innovators, which in turn influences the innovation behavior. However, the research on the spillover effect of FDI on innovation has not reached a consistent conclusion. There are mainly two viewpoints: crowding in effect and crowding out effect.

First, some scholars believe that FDI will promote domestic innovation. There are certain paths and mechanisms for FDI to generate technology spillovers for local enterprises in the host country. Blomstrom and Kokko summarized the paths of FDI spillovers into imitation learning and competition [1]. Kinoshita [2], Zhang and Ouyang [3] further classified FDI spillover effect into demonstration effect, competition effect, correlation effect and personnel training effect. Li classified FDI spillover effects into three types: direct learning effect, correlation effect and R&D localization, among which, direct learning effect includes demonstration effect, human capital flow effect and technology or R&D cooperation [4]. Some scholars (Chen) used the theoretical model of spillover mechanism in FDI industry constructed by Wang & Blomstrom to construct two models with labor productivity of domestic enterprises and foreign enterprises as each other's explained variable and explanatory variable, and verified Wang & Blomstrom's viewpoint that competition promotes technological progress [5]. Fan found that the net effect of FDI on China's domestic R&D investment was negative [10]. Wang honglin collected the panel data of China's science and technology openness and FDI industry level and investigated the influence of FDI on the independent innovation ability of China's national enterprises through regression analysis. The research supports the viewpoint of "facilitation theory" [11]. Some other scholars made empirical studies on the relationship between FDI and technology transfer in China since 1981 through causal relationship test and co-integration relationship test based on China's data, and believed that FDI was an important reason for China's technological progress (Chen et al.) [12].

However, some scholars believe that FDI will inhibit innovation. When transnational corporations occupy a large market share in a certain country and the technological level gap is too large, there is no evidence that FDI can play a role in the technological progress of the host country (Kokko) [6]. Haddad and Harrison analyzed the data of Moroccan companies and proved that FDI with higher technology level would not necessarily bring about the improvement of domestic R&D capacity [7]. Aitken and Harrison analyzed panel data of Venezuelan enterprises and found that FDI actually had a negative impact on R&D of domestic enterprises. From the perspective of human capital allocation structure, other scholars concluded that FDI would have negative effects on the development of science and technology in the host country [8]. Romer points out that the opportunity cost of human capital investment in the R&D sector is in fact the wage income of employees employed in the final product sector. Given the total amount of human capital, high wages in the final product sector will attract the transfer of human capital from the R&D sector to the final product sector, resulting in a decline in human capital investment in the R&D sector and thus inhibiting the growth rate of new products. Therefore, in the long run, the entry of foreign capital will have a negative impact on the long-term economic growth of the host country by changing the structure of human capital allocation [9].

3   Model and Data

3.1   Data

The data is composed of panel data of 31 provinces from 2003 to 2016. The data of invention patent is from the database of the national intellectual property office (2003–2016), the data of foreign direct investment is from the wind database, the GDP is from the statistical yearbook, and the R&D data is from the statistical yearbook of China's science and technology.

The explained variable is innovation capacity. The number of authorized invention patents is used to measure the quantity of regional innovation, and the regional innovation sensitivity is used to measure the quality of regional innovation. Through the comprehensive consideration of the quantity and quality of regional innovation, the spill-over effect of FDI can be more accurately examined. The data were obtained from SIPO (Chinese State Intellectual Property Office).

In this paper, the lnFDI and $\ln^2$FDI which is the square of the logarithm of foreign direct investment are used to represent the intensity of foreign direct investment. The data comes from wind database (Table 1).

Table 1. Description of data variables

Variable	Calculation method	Data source
lnpat	The logarithm of the quantity of invention patents granted	SIPO
lnlingm	The logarithm of patent originality	Calculated by the author
lnFDI	The logarithm of patent originality	Wind database
Ln^2FDI	The square of the lnFDI	Calculated by the author
lnGDP	The logarithm of GDP	China statistical yearbook
lnRD	The logarithm of R&D expenditure	Science and technology yearbook
lnjan	The logarithm of patent enforcement cases	SIPO

3.2 Model Setting

In the following model setting, we will try to build a nonlinear econometric model to test the relationship between foreign direct investment and the innovation ability of various administrative regions in China.

$$\ln pat = \alpha_0 + \beta_1 \bullet \ln FDI_{it} + \beta_2 \bullet \ln^2 FDI_{it} + \gamma \bullet X_{it} + N_i + T_t + \varepsilon_{it} \qquad (1)$$

$$\ln lingm = \alpha_1 + \delta_1 \bullet \ln FDI_{it} + \delta_2 \bullet \ln^2 FDI_{it} + \gamma \bullet X_{it} + N_i + T_t + \eta_{it} \qquad (2)$$

In the model, i represents the province and t represents the year. Lnpat represents the quantity of regional innovations, which is represented by the logarithm of the number of provincial invention patent applications. Lnlingm represents the quality of regional innovation, expressed by regional innovation sensitivity. In terms of regional innovation sensitivity, it measures how quickly a region can accept newly generated creative genes. At present, it is an index calculated according to patents. LnFDI represents regional direct investment, which is represented by the logarithm of FDI. Ln^2FDI represents the square of the logarithm of FDI of the province and is used to show the non-linear effect.

In this paper, logarithmic lnFDI and logarithmic square ln^2FDI are used to represent the intensity of foreign direct investment. If the coefficient of the linear term is significantly negative and the coefficient of the quadric term is significantly positive, it indicates that FDI in the province has a positive u-shaped curve effect on innovation.

In addition, N_i is a fixed effect of provinces to capture the influence of unobservable regional differences. T_t is a time-fixed effect that captures the impact of a common shock. ε_{it} and η_{it} represent the disturbance term. X represents the control variable, including (1) the level of economic development, which is measured by taking the logarithm of the annual gross national product of a province. (2) scientific and technological research investment (ln RD), which is measured by taking the logarithm of the R&D expenditure of the province. (3) strength of intellectual property protection (lnjan), which is represented by the logarithm of the number of patent enforcement cases in the province in this paper.

4 The Empirical Results

4.1 Basic Regression Results

In this paper, the double fixed effect model and the least square method are used for regression. The heteroscedasticity problem can be solved to some extent by setting the robust standard deviation or clustering robust standard error. And this paper added control variables to reduce the possibility of omission of variables. Table 2 shows the regression result, respectively using lnpat which represents innovation quantity and lnlingm which represents innovation quality. The results compared between lpat and lnlingm are as follows:

Table 2. Panel model regression results

	1 lnpat	2 lnpat	3 lnpat	4 lnpat	5 Lnlingm	6 Lnlingm	7 Lnlingm	8 Lnlingm
lnFDI	−0.039	−0.052	−0.855***		−0.561*	0.632***	−0.596***	
	−0.16	−0.42	−5.9		−1.89	−3.1	−3.16	
ln²FDI	0.037***	0.001	0.034***		0.048***	−0.026***	0.022***	
	−3.34	−0.21	−5.01		−3.85	−2.93	−2.83	
LlnFDI				−0.433***				−0.433***
				−2.83				−2.83
Lln²FDI				0.018**				0.018**
				−2.48				−2.48
lnGDP		−8.749***	2.498**	1.72		−3.760***	0.387	1.72
		−9.36	−2.32	−1.5		−3.56	−0.44	−1.5
lnRD		0.998***	0.515***	0.533***		0.651***	0.323***	0.533***
		−33.6	−5.4	−5.22		−19.26	−4.85	−5.22
lnjan		0.034*	0.042**	0.036**		0.033	0.046***	0.036**
		−1.92	−2.51	−2.12		−1.59	−3.72	−2.12
_cons	1.809	34.675***	−6.794	−5.752	6.657***	12.016**	5.259	−5.752
	lnpat	lnpat	lnpat	lnpat	Lnlingm	Lnlingm	Lnlingm	Lnlingm
	−1.27	−7.87	−1.48	−1.21	−3.85	−2.42	−1.46	−1.21
Year	No	No	Yes	Yes	No	No	Yes	Yes
Id	No	No	Yes	Yes	No	No	Yes	Yes
N	433	388	388	359	350	322	322	359

Note: (1) t values in parentheses (2) *, **, and *** indicate significant statistical levels at 10%, 5%, and 1% respectively.

In Table 2, this paper controlled the regional and time fixed effects in column (3), added control variables at the same time, and conducted regression with robust standard error. The regression results showed that lnFDI was significantly negative and ln²FDI was significantly positive. Table 3 also shows that the protection of intellectual property rights, the level of economic development, and the level of R&D investment are significantly positive, indicating that these control variables have a significant promotion effect on the quantity of innovation. At the same time, considering that there is a certain time lag between the implementation of policies and the emergence of innovative effects, in column (4), the core explanatory variable is delayed for one period. LlnFDI and Lln^{2}FDI are variables that lag behind lnFDI and Lln^{2}FDI for one period respectively. As can be seen from the results, they basically show a significant positive u-shape, and the control variables are all significantly positive.

In Table 2, the regression results showed that lnFDI was significantly negative and ln^{2}FDI was significantly positive in column (7). In the column (7) and (8), control variables have a significant promotion effect on innovation quality. At the same time, considering that there is a certain time lag between the implementation of policies and the display of innovation effects, in column (8), the core explanatory variables in this paper lag for a period, LlnFDI and ln^{2}FDI are respectively variables of lnFDI and ln^{2}FDI lag for a period, and it can be seen from the results that they basically show a significant positive U shape.

4.2 Heterogeneity Analysis

As we have seen before, the influence of FDI on innovation in eastern regions such as Beijing and Shandong, as well as western regions represented by Tibet, seems to be positive u-shape. For the central region represented by Heilongjiang, there is a significant inverted U shape. So, we try to prove it by empirical results.

Table 3. Regional regression results: influence on the quantity of innovation

	lnpat			lnlingm		
	East	West	Center	East	West	Center
lnFDI	−0.451***	−0.943***	−1.423***	−1.025***	−0.320**	1.792***
	−2.78	−5.3	−3.17	−3.54	−2.49	−4.37
ln^{2}FDI	0.005	0.041***	0.075***	0.036***	0.018***	−0.070***
	−0.65	−5.11	−3.71	−2.93	−2.94	−4.14
lnGDP	−1.868	4.920***	4.628***	−0.516	3.366***	5.193***
	−1.51	−3.11	−2.82	−0.52	−5.35	−3.27
lnRD	0.799***	0.038	0.277*	0.595***	−0.117	0.301***
	−5.39	−0.25	−1.79	−5.25	−1.6	−2.97
lnjan	0.076***	0.032	0.008	0.065***	−0.004	−0.001
	−2.9	−1.3	−0.27	−3.49	−0.38	−0.05
	lnpat			lnlingm		
	East	West	Center	East	West	Center
_cons	9.291*	−13.333*	−12.995	8.835**	−7.717**	−32.762***
	−1.68	−1.71	−1.51	−2.14	−2.33	−4.03
Year	Yes	Yes	Yes	Yes	Yes	Yes
Id	Yes	Yes	Yes	Yes	Yes	Yes
N	163	120	105	160	68	94

Note: (1) t values in parentheses (2) *, **, and *** indicate significant statistical levels at 10%, 5%, and 1% respectively.

According to the non-linear regression, we can see that the influence of FDI on innovation quantity is almost positive u-shaped in all regions. Basically, in terms of the innovation quantity, when it is on the left of the threshold value, with the increase of FDI, the amount of regional innovation keeps decreasing, and when it reaches the threshold value, the amount of innovation keeps increasing with the increase of FDI.

We can see that in the eastern and western regions, the influence of FDI on innovation quality is still positive U shape. It's basically significant at a significance level of 1%. However, for the central region, the influence of FDI on innovation quality presents an inverted U shape.

5 Conclusion

This paper adopts an econometric model to reveal the non-linear effect of FDI on innovation ability. No matter from the overall point of view or from the sub-regional point of view, FDI has a significant positive u-shaped effect on the number of provincial innovations, and there is localization phenomenon. However, according to the results of characteristic facts, the overall and sub-regional impact of FDI on the innovation quality of provinces showed a positive u-shape, while in the central provinces represented by Heilongjiang, FDI had an inverted u-shape effect on the innovation quality.

This study has some practical significance.

(1) Different policies should be formulated according to local conditions. Specifically, for less developed provinces such as Heilongjiang, the scale of FDI is still on the left side of the inverted u-shaped curve, and FDI is significantly positively correlated with entrepreneurship. This means that FDI is an important driving force for innovation, and the governments of these provinces should continue to maintain an open attitude towards foreign investment, reduce the restrictions on foreign investment access, and strive to increase the overall innovation capacity.

However, if the scale of FDI is close to the threshold of the inverted u-shaped curve and its role in promoting innovation gradually weakens, then more FDI may not be a good choice in these provinces. Therefore, these provinces should focus on absorbing FDI from high-tech industries. While promoting the transformation and upgrading of economic structure, they should further improve the overall level of entrepreneurship and pay attention to the maximization of entrepreneurial economic benefits, such as increasing the proportion of high-tech entrepreneurship and international entrepreneurship.

As for some provinces where the relationship of FDI and innovation ability presents positive U shape. When the foreign capital first came into local market, there may be crowding out effect. But based on this paper, there will be crowding in effect. So local government should be more patient, and should take a long-term view of the impact of foreign investment on local enterprises' innovation. (2) Considering that the control variables basically play a significant positive role, the local government should provide more incentives for entrepreneurship from the perspective of improving the institutional environment, encourage independent R&D and innovation, and reduce dependence on foreign capital. The government needs to raise the consciousness of ownership to avoid the dominance of foreign capital. At the same time, we will strengthen protection of intellectual property rights to prevent "free-riding". We also encourage more investment in technology research and development, and constantly improve the level of economic development to provide a solid foundation and a dynamic policy environment for enterprise innovation.

References

1. Kokko, A., Tansini, R., Zejan, M.: Local technological capability and spillovers from FDI in the outsourcing of Ayan manufacturing sector. J. Dev. Stud. **32**(4), 602–611 (1996)
2. Kinoshita, Y.: Technology Spillovers through Foreign Direct Investment (1998). www.cerge.cuni.czPDFwpwp139.pdf
3. Zhang, J., Ouyang, Y.: Foreign direct investment, technology spillover and economic growth: an empirical analysis of Guangdong data. Economics (quarterly) (3) (2003)
4. Li, P.: Impact of international technology diffusion on technological progress in developing countries: mechanism, effect and countermeasures analysis (2007)
5. Chen, T.: Intra-industry spillover effects of foreign direct investment. Economic Science Press (2004)
6. Kokko, A.: Technology, market characteristics, and spillovers. J. Dev. Econ. **43**, 279–293 (1994)
7. Haddad, M., Arrison, A.: Are there spillovers from direct foreign investment? Evidence from panel data for Morocco. J. Dev. Econ. **42**, 51–74 (1993)
8. Aitken, B., Arrison, A.: Do domestic firms benefit from direct foreign investment? Evidence from Venezuela. Am. Econ. Rev. **89**(3), 605–618 (1999)
9. Romer, P.: Endogenous technological change. J. Polit. Econ. **98**, 71–102 (1990)
10. Fan, C., Hu, Y., Zheng, H.: Theoretical and empirical study on the impact of FDI on technological innovation of domestic enterprises. Econ. Res. (01), 89–102 (2008)
11. Wang, H., Li, D., Feng. J.: FDI and independent R&D: empirical research based on industry data. Econ. Res. (02), 44–56 (2006)
12. Chen, G., et al.: Empirical study on the relationship between foreign direct investment and technology transfer. Sci. Res. Manag. (3) (2000)

Solving Small Deformation of Elastic Thin Plate by ULE Method and Computer Simulation Analysis

Lijing Liu[✉]

General Ability Teaching Department,
Beijing Information Technology College, Beijing 100015, China
liulj@bitc.edu.cn

Abstract. In this paper, the small deformation and stress of an elastic thin plate with one end fixed and one end simply supported are studied by United Lagrangian - Eulerian method. The kinematic equation and dynamic equation of fluid-solid contact surfaces are put forward by united Lagrangian-Eulerian method. The differential equations for the small deformation of the elastic thin plate were established in a continuous cross-flow of ideal fluid. The expression of deflection is given by the boundary conditions. The Taylor expansion method is used to solve the deformation and stress of the elastic thin plate. The influence of different parameters on plate deformation and stress was analyzed by computer simulation.

Keywords: United Lagrangian-Eulerian method (ULE) · Elastic plate · Deformation analysis · Computer simulation

1 Introduction

This paper deals with the deformation and stress of elastic thin plate in a transversal cross-flow by a new theoretical method. These phenomena have been studied by many authors from a different angle (theoretical algorithm, numerical example analysis and computer simulation, etc.) over the past few years [1–4]. The analysis of deformation and stress of plate and shell in cross-flow have also been studied by some authors [5–7].

United Lagrangian-Eulerian method is used to study the small deformation and stress problems of an elastic thin plate with one end fixed at one end under the condition of lateral flow of ideal fluid. It is a new method of where fluid and structure equations are given in their preferred reference frames. To the pressure of the shell and the velocity of fluid flow, the effect of the deformations of the thin plate was taken into account.

The elastic thin plate was fixed on one end while other was simple supported in a continuous cross-flow of ideal fluid was presented. The pressure reverse flow of the thin plate is assumed to be constant. The physical parameters of the thin plate are: length b, thickness h.

M. Atiquzzaman et al. (Eds.): BDCPS 2019, AISC 1117, pp. 650–656, 2020.
https://doi.org/10.1007/978-981-15-2568-1_89

2 Fundamental Equations

The equation of state of the fluid is written in the Euler reference coordinate system. The steady state equation of the potential flow satisfies the following equations:

$$\nabla^2\phi = 0, p = p_\infty + \frac{\rho_\infty}{2}\left[V_\infty^2 - (\nabla\phi)^2\right] \tag{1}$$

where φ represents the velocity potential, represents the pressure. p_∞, ρ_∞ and V_∞ represent the pressure, mass density, and velocity of the steady flow at infinity, respectively. And φ satisfies the condition

$$\varphi = V_\infty z, (z \to \infty) \tag{2}$$

The kinematics and dynamic kinetic conditions of the fluid and solid interface can be determined [8]:

$$\frac{\partial\varphi}{\partial z} = \frac{\partial w}{\partial y}\frac{\partial\varphi}{\partial y} - v\frac{\partial^2\varphi}{\partial z\partial y} - w\frac{\partial^2\varphi}{\partial z^2} \quad (z = 0) \tag{3}$$

$$Z_1 = p - p_i + v\frac{\partial\mu}{\partial y} + w\frac{\partial\mu}{\partial z}, Z_2 = 0 \, (z = 0) \tag{4}$$

in which v and w are projections of the displacement vectors of the thin plate in the y and z axes, respectively, Z_1 and Z_2 are projections of the external mass forces in the z and y axes, respectively. Where p and p_i are the pressures along outward and inward normal directions.

The theory of minor deflection of the thin plate is used. We obtain the balanced equations [9, 10]

$$\frac{\partial N_{22}}{\partial y} - \rho h\frac{\partial^2 v}{\partial t^2} = 0 \quad \frac{\partial Q_2}{\partial y} + Z_1 - \rho h\frac{\partial^2 w}{\partial t^2} = 0 \quad \frac{\partial M_{22}}{\partial y} - Q_2 = 0 \tag{5}$$

where N_{22}, Q_2, M_{22} are projections of the pull forces, shearing forces and bending moments of internal force in the y axes, respectively.

The middle surface of the shell can not be flexed, we obtain

$$D\frac{\partial^4 w}{\partial y^4} = p - p_i + v\frac{\partial p}{\partial y} + w\frac{\partial p}{\partial z} \tag{6}$$

where $D = \frac{Eh^3}{12(1-v^2)}$ is the shell rigidity, E indicates Young's modulus, v indicates Poisson's ratio, h indicates the thickness of elastic sheet.

3   Theoretical Solutions of Dynamic Equations

When we study the static and dynamic problems of the fluid-solid coupling part, we assume that the stiffness of the plate $(w_1 \equiv 0)$, the velocity potential φ_1 and the pressure p_1 of the fluid introduced, and the potential function φ_2 and the pressure p_2 that cause the bending of the thin plate, we can get:

$$\phi = \phi_1 + \phi_2, \, p = p_1 + p_2, \, w = w_2 \tag{7}$$

Then substituting Eq. (7) into Eqs. (1)–(4) and (6), the following expressions for φ_1 and p_1 can be written

$$\nabla^2 \phi_1 = 0, p_1 = p_\infty + \frac{\rho_\infty}{2} \left[V_\infty^2 - (\nabla \phi_1)^2 \right] \tag{8}$$

$$\phi_1 = V_\infty z, \, (z \to \infty), \frac{\partial \phi_1}{\partial z} = 0 \tag{9}$$

The following expressions for φ_2 and p_2 can be written

$$\nabla^2 \phi_2 = 0, p_2 = -\rho_\infty \left(\frac{\partial \phi_1}{\partial z} \frac{\partial \phi_2}{\partial z} + \frac{\partial \phi_1}{\partial y} \frac{\partial \phi_2}{\partial y} \right) \tag{10}$$

$$\phi_2 = 0, \frac{\partial \phi_2}{\partial z} = \frac{\partial w}{\partial y} \frac{\partial \phi_1}{\partial y} - w \frac{\partial^2 \phi_1}{\partial z^2} \, (z = 0) \tag{11}$$

Equation (6) can be written as:

$$D \frac{\partial^4 w}{\partial y^4} = p_1 + p_2 - p_i + w \frac{\partial p_1}{\partial z} \, (z = 0) \tag{12}$$

This article studies minor deflection of plate, then displacement v is neglected.

Complicated flow field can be decomposed into superpositions of several simple flow fields. Therefore, the method of current sharing and a set of dipoles can be used to solve the problem of flow around the plate. It can be seen that infinite doublets distribute continually in the plate, and the wrap surface of these doublets can be seen the plate's surface.

Therefore, the velocity potential φ_1 can be determined such that

$$\phi_1 = \phi_{11} + \phi_{12} = V_\infty z - \int_0^b \frac{\pi h z}{z^2 + (y-s)^2} ds$$
$$= V_\infty z - \pi h \left(\arctan \frac{y}{z} - \arctan \frac{y-b}{z} \right) \tag{13}$$

Substituting Eqs. (13) into (8), the pressure p_1 can be written as:

$$p_1 = p_\infty + \pi\rho_\infty V_\infty h \left[\frac{y}{y^2 + z^2} - \frac{y-b}{(y-b)^2 + z^2} \right] \tag{14}$$

Substituting Eqs. (13) and (16) into (11), taking into account the Eq. (14), we obtain the following equation

$$\phi_2 = \left(4Ay^3 - \tfrac{15}{2}bAy^2 + 3b^2Ay \right)(-\pi h)\left[\tfrac{1}{2}\ln(y^2 + z^2) - \tfrac{1}{2}\ln\left((y-b)^2 + z^2 \right) \right]$$
$$- \left(Ay^4 - \tfrac{5}{2}bAy^3 + \tfrac{3}{2}b^2Ay^2 \right)(-\pi h)\left(\tfrac{-y}{y^2+z^2} - \tfrac{-(y-b)}{(y-b)^2+z^2} \right) \tag{15}$$

Substituting Eqs. (13) and (15) into (10), the higher order terms of h are neglected, the pressure p_2 can be determined:

$$p_2 = \rho_\infty V_\infty \pi h \left(4Ay^3 - \frac{15}{2}bAy^2 + 3b^2Ay \right)\left(\frac{z}{z^2 + y^2} - \frac{z}{(y-b)^2 + z^2} \right)$$
$$- \rho_\infty V_\infty \pi h \left(Ay^4 - \frac{15}{2}bAy^3 + \frac{3}{2}b^2Ay^2 \right)\left\{ \frac{2zy}{(y^2 + z^2)^2} - \frac{2z(y-b)}{\left[(y-b)^2 + z^2 \right]^2} \right\} \tag{16}$$

Substituting Eqs. (14), and (16) into (12), the following expression can be found

$$24AD - p_\infty + p_i + \rho_\infty V_\infty \pi h \frac{1}{y-b} = \rho_\infty V_\infty \pi h \frac{1}{y} \tag{17}$$

and using Taylor expansions at $y = \tfrac{b}{2}$, the following expressions can be written

$$A = \frac{p_\infty - p_i + \rho_\infty V_\infty \pi h \tfrac{8}{b}}{24D} \tag{18}$$

Consequently, we can write the displacement, the stress and the velocity of flow as follows

$$w = \frac{p_\infty - p_i + \rho_\infty V_\infty \pi h \tfrac{8}{b}}{24D} y^2 \left(y^2 - \frac{5}{2}by + \frac{3}{2}b^2 \right) \tag{19}$$

$$V_y = \frac{\partial \varphi_1}{\partial y} + \frac{\partial \varphi_2}{\partial y}$$

$$= -\pi h \left[\frac{z}{z^2 + y^2} - \frac{z}{z^2 + (y-b)^2} \right]$$

$$- \pi h A \left(12y^2 - 15by + 3b^2 \right) \left\{ \frac{1}{2} \ln \left(y^2 + z^2 \right) - \frac{1}{2} \ln \left[(y-b)^2 + z^2 \right] \right\}$$

$$- \pi h A \left(4y^3 - \frac{15}{2} by^2 + 3b^2 y \right) \left[\frac{2y}{y^2 + z^2} - \frac{2(y-b)}{(y-b)^2 + z^2} \right]$$

$$+ \pi h A \left(y^4 - \frac{5}{2} by^3 + \frac{3}{2} b^2 y^2 \right) \left\{ \frac{y^2 - z^2}{(y^2 + z^2)^2} - \frac{(y-b)^2 - z^2}{\left[(y-b)^2 + z^2 \right]^2} \right\}$$

(20)

$$V_z = \frac{\partial \phi_1}{\partial z} + \frac{\partial \phi_2}{\partial z}$$

$$= V_\infty - \pi h \left[\frac{-y}{z^2 + y^2} + \frac{y-b}{z^2 + (y-b)^2} \right]$$

$$- \pi h A \left(4y^3 - \frac{15}{2} by^2 + 3b^2 y \right) \left[\frac{z}{y^2 + z^2} - \frac{z}{(y-b)^2 + z^2} \right]$$

(21)

$$+ \pi h A \left(y^4 - \frac{5}{2} by^3 + \frac{3}{2} b^2 y^2 \right) \left\{ \frac{2yz}{(y^2 + z^2)^2} - \frac{2(y-b)z}{\left[(y-b)^2 + z^2 \right]^2} \right\}$$

4   Computer Simulation and Discussion

In this section, numerical examples are presented. The test elastic thin plate and fluid flow have the following characteristics: pressure $p_i = p_\infty = 10^5$ Pa, mass density $\rho_\infty = 1000$ kg/m^3, flow velocity $V_\infty = 0.08$ m/s, plate length $b = 1.0$ m, plate thickness $h = 0.001$ m, the material is low carbon steel, Young's modulus $E = 200 \times 10^9$ N/m^2, Poisson's ratio $v = 0.3$. The results are shown in Figs. 1, 2, 3, 4 and 5.

In Figs. 1, 2 and 3 the displacements w are presented by varying the flow velocity, plate thickness and material. The copper and aluminum have the following characteristics: Poisson's ratio $v = 0.33$, $v = 0.33$, It is easy to see that the displacements w are the maximum value which are not at $y = b/2$ but the migration to simply supported end.

In Fig. 4, the velocity V_z along plate's length at $z = -0.0005$ m is presented. It shows that the minimum value of V_z at the middle of the thin plate. It is increased far away from the middle and return to V_∞ at both ends. The velocity V_z along plate's length at $z = -0.0005$ m

In Fig. 5, the variation of velocity V_z with respect z at $y = 0.05$ m. It shows that the value of V_z closely approximates the value of V_∞ at infinite, when the fluid flow is far away from the plate. The value of V_z is reduced close to the thin plate and return to zero at end.

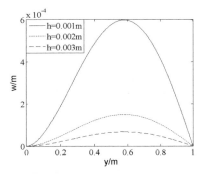

Fig. 1. Deflection curves with different thickness

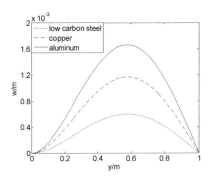

Fig. 2. Deflection curves with different broadside

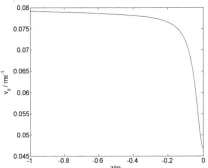

Fig. 3. Deflection curves with low carbon steel, copper and aluminum

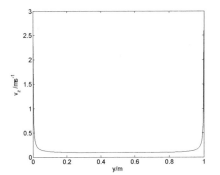

Fig. 4. Velocity along plate's length

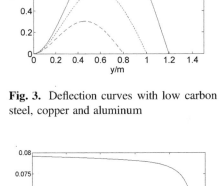

Fig. 5. Variation of velocity V_z with respect z at $y = 0.05$ m

5 Conclusions

With the theory of united Lagrangian-Eulerian method, this paper deals with the deformation and stress of elastic thin plate in a transversal cross-flow by a new theoretical method. The expression of deflection was given and simplified based on the boundary conditions. The theoretical deformation, stress and velocity of the elastic thin plate in a continuous cross-flow of ideal fluid have been derived. The effects of different parameters on the deformation and stress of the thin plate were analyzed by specific examples and computer simulations. The compatible Lagrangian-Eulerian method is used to study the small deformation and stress problems of an elastic thin plate with one end fixed at one end under the condition of lateral flow of ideal fluid.

References

1. Chen, W.Q.: Numerical simulation of flow-induced vibration on two circularcylinders in a cross-flow; Part I: transverse y-motion. Acta Aerodynamica Sinica **23**, 442–448 (2016)
2. Malatip, A., Wansophark, N., Dechaumphai, P.: Fractional four-step finite element method for analysis of thermally coupled fluid-solid interaction problems. Appl. Math. Mech. **33**(1), 99–116 (2017)
3. Kuberry, P.: A decoupling algorithm for fluid-structure interaction problems based on optimization. Comput. Meth. Appl. Mech. Eng. **267**, 594–605 (2016)
4. Schiffer, A., Tagarielli, V.L.: Fluid-structure interaction experiments and simulations. Int. J. Impact Eng. **1**, 34–49 (2018)
5. Bai, X.: Analysis on the deformation and internal force of an elastic thin cylindrical shell in fluid. Eng. Mech. **24**, 47–52 (2017). (in Chinese)
6. Zhou, X., Li, J., Hao, Y.: Model for fluid-solid interaction problem between a thin cylindrical shell and a transversal cross-flow. J. Liaoning Tech. Univ. (Natural Sci.) **32**, 1055–1059 (2016). (in Chinese)
7. Song, X., Bai, X.: Analysis of deformation and stress of the arched shell in potential flow. Chin. J. Appl. Mech. **26**(2), 304–307(2016). (in Chinese)
8. Ильгамов, М.А., Введение, в.: Нелинейную Гидроупругость. Москва: Изд. Наука 15–53 (2017)
9. Xu, Z.L.: Elasticity Mechanics, 4th edn. Higher Eduction Press, Peking (2018)
10. College of Ship and ocean engineering, Shanghai Jiaotong University. Fluid Mechanics, pp. 134–139. Science and Education Press, Peking (2017). (in Chinese)

Evaluation System of Human Settlement Speciality Learning Roadmap Based on Machine Learning

Junzhe Wang$^{(\boxtimes)}$ and Yuan Meng

Urban Construction School, Beijing City University, Beijing 101399, China
wangjunzhe@bcu.edu.cn

Abstract. The application of machine learning is becoming more widespread, especially in the fields of consumption and medical treatment, while its application in field of education is relatively rare. In this paper, based on the diversion requirements of large-scale speciality platform of the human settlement discipline system, artificial method is used to establish evaluation feature items and set up the evaluation system model. Meanwhile, under the condition of the small sample set, the machine learning method is used to train and adjust parameters to form the first edition of speciality potential assessment system combining with the artificial method, which would lay the foundation for the introduction of more courses and samples to upgrade the system in the future, in the hope to construct more accurate and in-depth potential evaluation to guide students with a more effective method.

Keywords: Science of human settlement · Machine learning · Evaluation system of learning roadmap

1 Introduction

The human settlement discipline system was proposed by the academician of Chinese Academy of Engineering, Wu [1, 2]. It consists of the trinity of "Architecture-Landscape-Urban Planning" and integrates through urban design. The integration of the disciplines on Architecture, Landscape Architecture, Urban and Rural Planning requires the practitioners to handle the relationship of "people-construction-urban-nature" correctly and implement the pursuit of a good living environment in material construction to create a comfortable and pleasant human settlement [3].

Currently, under the trend of large-class enrollment nationwide, Beijing City University combines the orientation of school application-based undergraduate education, introduces learning roadmap and relevant evaluation system, and builds a human settlement discipline platform which integrated three specialities of architecture, landscape architecture and urban and rural planning. The learning roadmap is a comprehensive program designed with the aim of improving professional skills [4, 5]. It is a scientific plan for growing process of students with free choice and self-learning. The establishment of the roadmap curriculum system needs to be built with the help of three platforms: the professional foundation course platform for basic skills development in

© Springer Nature Singapore Pte Ltd. 2020
M. Atiquzzaman et al. (Eds.): BDCPS 2019, AISC 1117, pp. 657–662, 2020.
https://doi.org/10.1007/978-981-15-2568-1_90

the first year of undergraduate course, the professional expansion platform of free electives courses and distributional electives courses for speciality and direction development based on interest in addition to their respective specialized courses in the second and third grades. Based on this, the evaluation system of the human settlement professional learning roadmap which can be used to reflect the professional potential of students came into being.

In this process, the information mining of educational data is realized through focusing on the ability assessment and machine learning, and the performance of students in the learning process is evaluated according to indicators, so as to provide recommendations for students' subsequent major and major direction selection which is more in line with their own abilities.

2   Research Methodology

2.1   Guiding Ideology

At this stage, the evaluation system is based on the scores of the freshmen in the fall semester, which can be used to guide students to make major selections. Under the guidance of experienced teachers, the evaluation feature items of the system can be set up manually to establish the evaluation system model. The machine learning method is used under the condition of the small sample set acquired, to obtain the weights of the system evaluation feature items and form the evaluation reports of student professional ability which conform to the actual situation of students.

This study used the K Nearest Neighbor (KNN) classification method to assist in the construction of the learning roadmap evaluation system [6, 7]. The basic idea is that the student samples are described by the scores of a certain number of courses (count the number as N), and each course is expressed as an evaluation index of the student sample. All the samples are classified to different majors through the evaluation index space [8, 9]. Some false information needs to be filtered during the selection of student samples [10]. The data after screening can be used as a sample space. The new freshmen can be use as the resource to be classified, and a certain number of student samples (count the number as K) closest to the new characteristics can be find by searching for the evaluation index space through the KNN classification method, then the new freshmen would be assigned to the to the minimized public classes of the K nearest neighbor samples.

2.2   Evaluation Index

The evaluation system of learning roadmap aims to form a student's professional potential evaluation based on the results of the student's participation in the curriculum and the weight of the curriculum.

The professional courses involved in the professional selection are shown in the following table, in which the impacting factor is divided into weight indicators (Table 1).

Table 1. Names and weights of the participated courses

Number	Courses name	Weight (credit)	Sub items
1	Art foundation	3	Geometry sketch
			Geometry combination sketch
			Still life sketch
			Creative sketch
			Architectural sketch
2	The basis of space form expression	5	Architectural mapping
			Room design and model making
3	Computer aided design	4	Foundation plan (CAD)
			Vertical section (CAD)
			Interior design renderings (PS)
4	History of urban construction in ancient China	2	Close-book examination

Establish the correspondence between the participated courses and the specific requirements in the Architecture, Landscape architecture, and Urban and Rural Planning training programs. Combining the descriptions of quality, knowledge and ability of different majors, the indicators relevant to the curriculum are selected. Take Architecture as an example, the corresponding relation is shown in the following (Table 2).

Table 2. Correspondence Tables for Indicators (Quality, Knowledge, and Ability of Architecture) and the participated courses

Courses name	Indicator 1	Indicator 2	Indicator 3	Indicator 4
Art foundation	✓	–	–	✓
The Basis of Space Form Expression	✓	–	–	✓
Computer Aided Design	–	✓	–	–
History of Urban Construction in Ancient China	–	–	✓	–

Note:

Indicator 1: Mastering the basics of Architecture.

Indicator 2: Mastering the design concepts and technical realizations of Architectural Design Frontier, such as green building design, etc.

Indicator 3: Mastering the basic knowledge related to architecture, such as economy, geography, society, population, ecology, etc.

Indicator 4: Mastering the basic drawing skills, including hand-drawn and computer-aided drafting capabilities.

3 Test Validation

3.1 Test Procedure

The sample data of this study were collected from 167 freshmen majoring in architecture in 2018 to track their academic performance and professional classification during the year. The specific test steps are as follows:

(1) Establish the evaluation features and characteristic weights combing various factors such as importance, representativeness and rationality with the help of experienced teachers;

(2) Process the original scores of students through the features and their weights mentioned above;

(3) Select 17 students randomly as test data after professional classification, and the rest are included in the historical sample data;

(4) Use K nearest neighbor classification method to classify students' major selections, and calculate distance by Euclidean distance and relevant weights;

(5) Obtain different sample sizes and K values according to the voting rule of majority and get the correlation of sample quantities and K value through machine learning.

3.2 Test Results

K-Nearest Neighbour (KNN) classification method is adopted to evaluate the capability of the test samples in this experiment and was verified based on the result of class-division (2 classes of Architecture, 2 classes of Landscape Architecture and 1 class of Urban and Rural Planning). The pseudo-correct data, such as strong intention samples, forced placement samples, were screened out in the process, and the correlation between the number of samples and K value was obtained through machine learning. As shown below (Fig. 1 and Table 3):

Table 3. The values of matching rate

Samples	K value						
	3	5	7	9	11	13	15
100	35%	35%	47%	41%	53%	53%	41%
120	41%	35%	41%	53%	59%	47%	41%
150	41%	41%	59%	65%	53%	47%	47%

Among the 167 experimental data, 17 students are selected as test samples and the remaining students are selected as training samples with clear major selections. The value of the key parameter k is evaluated by adjusting the number of samples. It is observed in the experiment that with the increase of the number of samples, the K value decreases gradually. In the case of existing small samples, it can be concluded that when k is 9 through repeated experiments and comparison, the classification effect is

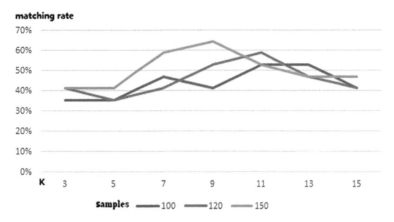

Fig. 1. The values of matching rate

the best, and the matching rate can reach 65%, which can be used to for the decision-making of their major selections for the freshmen.

Since the selection of system evaluation feature items mainly relies on manual completion, which is greatly affected by subjective factors, and the number of samples invested is insufficient, the professional matching rate of students obtained through the evaluation system has not exceeded 70%. The evaluation system will be improved through the expansion of feature items and student sample data in the future work.

4   Summary

The talent cultivation of Architecture, Landscape Architecture, and Urban and Rural Planning under the application-oriented undergraduate training mode, although different in scale and emphasis, has certain similarities. The human Settlement Speciality Learning Roadmap provides more free space for students to choose their major and professional direction. Relying on existing samples for major selection in the process,machine learning makes predictions and calculations through multi-dimensional index evaluation, and forms a preliminary capacity assessment system, which would lay the foundation for the introduction of more courses and samples to upgrade the system in the future, with a view to constructing more accurate and in-depth potential evaluation to guide students with a more effective method.

References

1. Wu, L.: Introduction to Science of Human Settlements. China Architecture & Building Press, Beijing (2001). (in Chinese)
2. Wu, L.: The Research Progress of the Science of Human Settlements (2001–2010). China Architecture & Building Press, Beijing (2011). (in Chinese)

3. Ma, R., Wang, T., Zhang, W., Yu, J., Wang, D., Chen, L., Jiang, Y., Feng, G.: Overview and progress of Chinese geographical human settlement research. J. Geog. Sci. **26**(08), 1159–1175 (2016)

4. Starr, L.: Minchella, D: Learning beyond the science classroom: a roadmap to success. J. STEM Educ. Innovations Res. **17**(1), 52–57 (2016)

5. Oakleaf, M.: A roadmap for assessing student learning using the new framework for information literacy for higher education. J. Acad. Librarianship **40**(5), 510–514 (2014)

6. Denoeux, T.: A k-nearest neighbor classification rule based on Dempster-Shafer theory. IEEE Trans. Syst. Man Cybern. **25**(5), 804–813 (1995)

7. Dasarathy, B.V.: Nearest Neighbor (NN) Norms: NN Pattern Classification Techniques, vol. 13, pp. 21–27. IEEE Computer Society Press, Los Alamitos (1990). Issue No. 100

8. Hattori, K., Takahashi, M.: A new edited k-nearest neighbor rule in the pattern classification problem. Pattern Recogn. **33**(3), 521–528 (2000)

9. An, F., Chen, L., Akazawa, T., Yamasaki, S., Mattausch, H.J.: K nearest neighbor classification coprocessor with weighted clock-mapping-based searching. IEICE Trans. Electron. **99**, 397–403 (2016)

10. Wang, D., Dillon, T.S., Chang, E.J.: A data mining approach for fuzzy classification rule generation. In: IFSA World Congress and 20th NAFIPS International Conference, Joint 9th IEEE (2001)

Smart City Privacy Protection in Big Data Environment

Lirong Sun[(⊠)] and Shidie Wu

School of Economics, Shanghai University, Shanghai, China
slrl996jt@163.com

Abstract. In recent years, smart city has gradually become popular. Smart city of big data plays an active role in urban fine management, easing traffic jams, improving emergency security system, improving urban environment, eliminating information islands, etc. However, smart city management in big data environment is faced with data privacy protection issues, which is necessary in traditional smart city. There are three ways to protect the privacy of smart city in big data environment. One is to enhance the data security and privacy protection awareness of data managers from the moral level; the other is to formulate comprehensive data security and privacy protection laws from the legal level to prevent malicious data theft and abuse; the third is from the security technology level and incorporate the data protection algorithm into the smart city technology framework. In this article, according to the basic characteristics of smart city of big data, on the basis of the basic technical framework of big data smart city, from the three important aspects of big data storage, search and calculation, respectively, big data privacy protection is included in the basic technical framework to improve the privacy protection of smart city.

Keywords: Big data · Smart city · Privacy protection

1 Introduction

With the development of Internet technology, the Internet of things represented by sensor technology began to enter our lives, and various intelligent devices and facilities began to change people's life style. Everything in the society has been transformed into virtualized data. And people live in the world of data. Smart city is a new concept of using the Internet of things, cloud computing, big data, and other new generation of information technology to promote urban planning, management and intelligence [1]. From the concept of smart city, we can find that the establishment of smart city needs the support of big data. That is to say, the establishment of smart city should be based on the collection and processing of urban basic data under big data. The collection and handling of urban data is highly related to the establishment of smart city. It is the key to ensure the healthy and high-quality operation of smart city to obtain appropriate data and output useful information through mining and processing. Although smart city under the big data plays an active role in urban fine management, traffic congestion alleviation, emergency safety system improvement, targeted improvement of urban environment, public data sharing, and elimination of information island, smart city

M. Atiquzzaman et al. (Eds.): BDCPS 2019, AISC 1117, pp. 663–670, 2020.
https://doi.org/10.1007/978-981-15-2568-1_91

management in big data environment [2]. If a large number of data such as personal information, enterprise information and municipal information are leaked, it will seriously affect the social order. So it is urgent to add the protection of data security and privacy into the traditional smart city technology framework.

Big data's collection and output are very different from traditional data. Big data is usually stored in could, which often has data security and privacy problems. In foreign countries, Wu et al. proposed a new model to study the transmission mechanism of data virus, and introduced an incentive mechanism to control the spread of virus and protect data security [3]. Liu Bo et al. Liu Bo constructs an image privacy protection framework for deep learning tools, and proposes two new image privacy measurement methods, which can effectively protect image privacy [4]. In China, Li et al. proposes a privacy protection scheme for urban big data collection in large-scale dynamic environment based on the characteristics of large amount and strong real-time of urban data, which effectively protects data privacy and reduces information loss [5]. Based on the definition and classification of personal information under the background of big data, Wang discusses legal attributes of personal information, the rights and restrictions of the original obligee and controller of personal information [6]. In a word, the above researches on privacy protection are all aimed at a specific harm that big data may cause, but there is no comprehensive solution to the privacy problem under the whole big data environment, and there are few researches on privacy protection of smart cities.

Through previous literature, it can be found that there are not many researches on the privacy protection of smart cities. Most of literature are proposed under the background of big data, and there is no systematic consideration of the possible privacy problems of smart cities under the big data, nor a systematic privacy protection scheme. Therefore, this paper puts forward three privacy protection methods: morality, law and technology from the perspective of privacy issues that may arise in smart city. On the technical level, from the three important aspects of big data storage, search and calculation, privacy protection is included in the technological framework of smart city in big data environment. The above three solutions can solve the privacy protection problem of smart city in some extent.

2   The Impact of Big Data Environment on Smart City

2.1   The Value of Big Data Environment to the Establishment of Smart City

Urban life data has a large scale, diverse structure, fast speed property and so on, which can comprehensively reflect the most real urban life status and provide the most basic information for the good operation of smart city. However, the traditional data processing technology and thinking mode can not dig out enough data. Only by collecting, storing and processing "big data" efficiently and accurately, these basic information can promote the construction of smart city. In the era of big data, the construction of smart city is not a simple informatization and digitalization of various data sources based on experience, but an intelligent "calculation" with effective technical means in the face of large-scale and diverse data, with the purpose of transforming data into useful value more effectively. The construction standard of smart city under big data usually includes functions of benefiting people, precise governance, ecological livability, intelligent facilities, information resources, network security, public security, etc. [7].

According to the construction standards of various massive, heterogeneous, multi-source data and smart city in the city, combined with the current advanced big data processing cutting-edge technology, the basic technical framework of smart city driven by big data is established, which is mainly composed of data layer, storage, calculation and application system layer. The data layer is mainly devoted to the compilation of classification, classification and coding basis of urban planning simulation elements, and the collection and integration of urban planning data. Generally, these data contents mainly include: various track big data, transportation big data, urban residents' consumption data, health care data, social media data, meteorological environment and air quality data, urban map and geographic information etc. The data layer transmits collected data to the big data storage and management layer, which mainly stores and manages the data. In the computing layer, the distributed computing method is mainly used to analyze, mine and intelligently calculate the data, fully extract the value of urban big data, provide sufficient information for data users and support decision-making. The application layer is to build a brand-new platform for urban land allocation, sustainable development and urban development policy based on the urban planning simulation platform, so as to visualize the data and provide services for the citizens.

At present, data collection, storage and processing mainly involve intelligent sensing technology, distributed storage technology and data mining technology etc [8]. These technologies will speed up the establishment process of smart city. The platform based on big data is mainly used in e-commerce, health care, public security, intelligent transportation, telecommunications, finance, environmental management, government management and other fields. Relying on big data can realize the functions of digitalization, refinement, intelligence and socialization of urban management, and constantly improve the detection indicators of urban economy and environment. Through the establishment of smart city management big data platform, urban management resources can be integrated, urban management behavior can be standardized, and urban management efficiency can be improved.

2.2   Data Security Risks of Smart City in Big Data Environment

Through big data mining and analysis technology, we can get more valuable information, but the data security problem is becoming more and more prominent and has become the focus of attention under the smart city construction, which is a major problem in the process of smart city construction. Due to the weak awareness of data protection, hacker attacks, unclear definition of data property rights and other factors, the sensitive data or characteristics represented by data that individual, enterprise, user and other participants are unwilling to be disclosed are disclosed. For example, data enterprises use big data technology to judge users' gender, age, occupation, emotional status, interests etc. according to their data on social platforms. If these data are leaked, it will bring great problems to the social order. Therefore, privacy security must be fully considered in smart city under big data, while legal norms and traditional privacy protection approach cannot meet the privacy protection requirements of smart city under big data. We need to improve the data management and prevention system, encrypt some sensitive data, and ensure the security within the data flow under the building of a smart city.

3   Privacy Protection Path of Smart City in the Environment of Big Data

Since the big data is generally stored on the cloud, although it is convenient for the data owner to store it on the cloud under the management of the cloud storage service provider, the cloud storage service provider is not completely trusted, which leads to the following three problems. First, data owners must verify the data stored on the cloud to prevent data from being destroyed. Second, data may be stored in the form of ciphertext, so data owners need efficient ciphertext search algorithm to search the encrypted data that stored in the cloud. Third, data owners need to securely use the data on the cloud for computing. In the smart city under the big data environment, the data has the characteristics of sharing, transparency, complexity and diversity, and wide-ranging sources. The industry and government have launched the mining of big data in succession. When enjoying the value brought by big data, we should also realize that big data also has threat to our privacy. For management with big data, we need to take different measures according to different situations to prevent our sensitive information from being leaked. This paper argues that there are three ways to solve the problems of privacy protection within the building a smart city: moral, legal and technical.

3.1   From the Moral Level

From the moral level, moral constraints are very important for the privacy protection of smart city within big data environment, which is also a source to solve the privacy protection. Therefore, the society needs to strengthen the data management training for the relevant personnel, enhance the data privacy protection awareness of the data management personnel, and prevent the relevant personnel from divulging the data in charge to the third party for profit, endangering users and other behaviors. At the same time, the society needs to encourage all kinds of participants to actively participate in or know the whole process of privacy protection and big data utilization, which is very important in improving transparency in data utilization and participants' awareness about active privacy protection.

3.2   From the Legal Level

From the perspective of law, the key to privacy protection under big data environment is to rely on the state to formulate perfect laws and regulations according to the existing problems of privacy protection. In the process of legislation, the state needs to fully consider the characteristics of data sharing and transparency of smart cities, formulate comprehensive data privacy protection laws, establish mechanisms such as traceability mechanism, governance mechanism, transaction licensing mechanism and reporting mechanism for data privacy protection, and define the ownership of data property rights. According to the original data, the data set is fully anonymous, the ownership of data owned by enterprises is limited, and relevant punishment regulations are formulated to restrict data managers. From the perspective of law, to a certain extent, to prevent malicious data theft, abuse of data behavior, while improving people's awareness of data rights.

3.3   From the Aspect of Technical Prevention and Control

From the perspective of security technology, traditional data protection methods do not match with it for big data is large scale and unpredictable production speed, so a new data protection technology framework needs to be established. In the whole life cycle of big data from being collected to value, the process of possible data leakage mainly includes the storage, search and calculation of big data. Therefore, according to the basic properties of smart city, from the perspective of technical means, based on the basic technical framework of big data smart city, from the three important aspects of big data storage, search and calculation, respectively, big data privacy protection is included in the basic technical framework, as shown in Fig. 1.

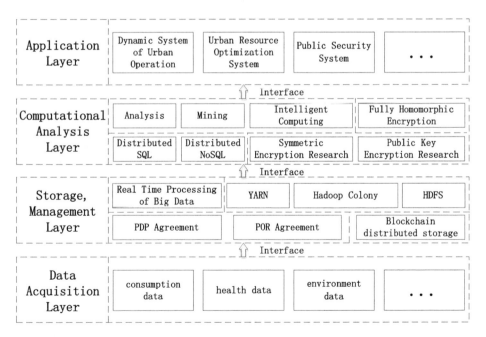

Fig. 1. Smart city technology architecture under big data

As to the big data storage, there are many protocols that can verify the integrity of big data stored in the cloud more efficiently, but these protocols need the personal verification of data owners, and the verification efficiency is not high. Therefore, considering the professional level of data verifiers, it is a good choice to introduce a third-party verification agency. In this way, the main reason that data faces the risk of privacy disclosure in the storage stage is that the integrity verification protocol of big data adopts the third-party audit organization. Therefore, the main privacy protection problem of big data storage is how to design a safe and efficient big data integrity verification protocol that can prevent the data owner's data from leaking to the third-party audit organization. In this paper, adding the third-party verification protocol to the data storage technology architecture, including the PDP protocol to verify data integrity and the POR protocol to allow data recovery [9, 10]. We use the system establishment algorithm, challenge and response algorithm and the verification algorithm to establish the protocol, which will include the data owner, cloud server and third-party auditor. Among them, the core algorithm of the protocol includes: system establishment algorithm, challenge and response algorithm and verification algorithm. In recent years, blockchain has developed rapidly, and its properties such as non tamperability, smart contract and decentralization have been widely used in data privacy protection. This paper believes that the blockchain distributed application can be embedded in the smart city technology framework to solve the privacy protection problem of smart city.

In terms of search, big data may be stored in plaintext or ciphertext, so big data search is divided into two modes: plaintext search and ciphertext search. If big data is stored in plaintext, the corresponding big data search problem is the traditional data query and data publishing problem, otherwise, the big data search is the ciphertext search. Data stored in ciphertext is mainly because this kind of big data is relatively confidential. If the search is not restricted and prevented, the privacy of data will be leaked. Therefore, researchers have designed many plaintext search algorithms to protect privacy. However, many of these search algorithms are not applicable in smart city of big data. For hidden danger of data security in search process, this paper thinks that symmetric encryption algorithm and public key encryption algorithm can be used. For example, in the aspect of symmetric encryption algorithm, Popa et al. Proposed a more secure order preserving encryption algorithm, PLZ13, which plays a huge role in data privacy protection [11].

In the big data environment, data owners or other users usually want to use the big data which has been stored in the cloud, so they may need cloud service providers to calculate specific big data and return the calculation results. However, the big data or calculation results as the input of calculation may be very confidential, so the calculation problem of protecting privacy of big data refers to how to calculate the big data safely without disclosing the confidential data or calculation results to the third party. Obviously, for confidential data, it is unwise to submit it directly to cloud service provider for calculation, so in the calculation of big data, data and calculation results need to be saved in the form of ciphertext. Considering the complexity and diversity of the calculation problems in big data, it is an appropriate choice to use the full homomorphic encryption algorithm in calculation, and use the partial homomorphic encryption scheme based on LWE problem with ideal security.

4 Conclusions

Smart city in big data environment will make people's life more beautiful, but this process is long, and the theoretical research and practical exploration of smart city are only in the initial stage. From the perspective of data collection, we need a variety of basic data collection equipment, such as: video data can be obtained through various cameras, and the passenger data can be obtained through the smart terminal card swiping system used by the citizens. Only by comprehensively collecting urban life data, can we build a firm foundation for smart cities. From the perspective of data processing technology, big data storage, processing, mining and other technologies need to be improved at present, and technical problems are the biggest obstacle in smart city. From perspective of legislation, opening and using of big data must be restricted and guaranteed by law, and the division of data property rights should be clear. Data privacy protection should be considered in building smart city, specifically from three aspects. First, from the moral level, strengthen the data management training for relevant personnel, and enhance data privacy protection awareness of data management personnel. Second, from the legal level, the government should formulate comprehensive data security and privacy protection laws to define the ownership of data property rights and restrict data managers, so that the malicious data theft and abuse can be prevented in some extent. Third path is based on the security technology. Because the first two methods have certain risks, they are only auxiliary means of data security and privacy protection. Therefore, the key to solve the problem is to prevent privacy disclosure from the perspective of technology. Considering the property of big data, the privacy protection of smart city in big data environment needs to improve traditional data privacy protection technology, make full use of data security access control technology, data anonymity technology, data security encryption level, key distribution mechanism and other means to strengthen data security and privacy protection. However, we still need to realize that the current technology for data security and privacy protection is still limited, considering the dual needs of data security and privacy protection, and vigorously develop and deploy privacy protection technology. However, we still need to further study technologies such as anonymity, homomorphic encryption, blockchain, etc., especially strengthen the research on provable formal privacy protection technology, and make up for the lack of short board of privacy protection technology in big data environment. Various methods are implemented together to provide technical and information security and build a technological ecosystem of privacy protection.

References

1. Mohbey, K.K.: The role of big data, cloud computing and IoT to make cities smarter. Int. J. Soc. Syst. Sci. **9**(1), 75 (2017)
2. Zhang, K., Ni, J., Yang, K.: Security and privacy in smart city applications: challenges and solutions. IEEE Commun. Mag. **55**(1), 122–129 (2017)
3. Wu, Y., Huang, H., Wu, N., Wang, Y., et al.: An incentive-based protection and recovery strategy for secure big data in social networks. Inf. Sci. **508**, 79–91 (2020)

4. Bo, L., Ming, D., Tianqing, Z., et al.: Adversaries or allies? Privacy and deep learning in big data era. Concurrency Comput. Pract. Exp. **31**(19), j31 (2019)
5. Li, H., Guo, L., Guo, F., Wang, J., Zhang, W.: Privacy Protection Scheme of Urban Big Data Collection Based on MapReduce Model. J. Commun. **39**(S2), 35–43 (2018). (in Chinese)
6. Wang, Y.: Research on legal issues of personal information development and utilization in big data. Inf. Theory Pract. **39**(09), 19–24 (2016). (in Chinese)
7. Habibzadeh, H., Nussbaum, B.H., Anjomshoa, F., et al.: A survey on cybersecurity, data privacy, and policy issues in cyber-physical system deployments in smart cities. Sustain. Cities Soc. **50** (2019)
8. Offenhuber, D.: The platform and the bricoleur-improvisation and smart city initiatives in Indonesia. Environ. Plan. Urban Anal. City Sci. **46**(8), 1565–1580 (2019)
9. Esposito, C., Pop, F., Huang, J.: Application of soft computing and machine learning in the big data analytics for smart cities and factories. Int. J. Inf. Manag. **49**, 489–490 (2019)
10. Guo, W., Zhang, H., Qin, S., et al.: Outsourced dynamic provable data possession with batch update for secure cloud storage. Future Gener. Comput. Syst. Int. J. Esci. **95**, 309–322 (2019)
11. Paterson, M.B., Stinson, D.R., Upadhyay, J.: Multi-prover proof of retrievability. J. Math. Cryptol. **12**(4), 203–220 (2018)

A Target Tracking Algorithm Based on Correlation Filter and Least Squares Estimation

Yiqiang Lai$^{(\boxtimes)}$

South China Business College, Guangdong University of Foreign Studies,
Guangzhou 510545, China
laiyql982@foxmail.com

Abstract. Target tracking plays a very important role in the civil field and has always been a hot topic for scholars. But there are also problems such as the difficulty of tracking goals. The core part of the algorithm framework proposed in this paper includes target recognition and target tracking models. The target apparent model is solved by using the correlation filter-based tracking algorithm, and the reliability of the block itself is calculated by using each block response graph, and calculate the relationship between multiple blocks in the same frame, and define the contribution of each block to the overall tracking result. In order to solve the problem of difficult target tracking, the core part of the algorithm framework proposed in this paper includes target recognition and target tracking model. In this paper, a tracking algorithm based on correlation filtering is used to solve the apparent model of the target, and the reliability of each block is calculated by the response graph of each block. The experimental results show that the parallel algorithm framework designed in this paper can not only achieve stable and accurate tracking under the interference of masking, deformation and scale change, but also quotes the idea of least squares method. The models and characteristics of the previous least squares method are analyzed and summarized. The target initial value least squares estimation method is improved.

Keywords: Target tracking · Correlation filter · Least squares method · Target partitioning · Tracking algorithm

1 Introduction

Target tracking has become an integral part of various intelligent systems in the field of computer vision, covering applications from civilian to military, from individual to public. In the public domain, in order to prevent various order interference and reduce labor costs, digital monitoring data has great mining value; under the development of Internet and artificial intelligence, people have proposed more effective control methods.

In [1], the author proposes an IMM based target tracking in WSN, called ittwsn, which uses multiple models (velocity and acceleration) to process maneuvering and non maneuvering targets, and uses multiple sensors to detect and identify targets.

© Springer Nature Singapore Pte Ltd. 2020
M. Atiquzzaman et al. (Eds.): BDCPS 2019, AISC 1117, pp. 671–677, 2020.
https://doi.org/10.1007/978-981-15-2568-1_92

In [2], the author proposes a consensus based distributed multi-target tracking algorithm, which is called multi-target information consensus (mtic). The proposed mtic algorithm and its extension to the nonlinear camera model (called extended mtic (emtic)) are robust to erroneous measurements and limited resources. In [3, 4], the author proposes an adaptive strong tracking particle filter algorithm based on particle filter. The forgetting factor and weakening factor are adjusted adaptively according to the residual between the actual measured value and the predicted measured value at each time. The simulation results show that when the state of the target changes suddenly, the algorithm can effectively track the state of the moving target and improve the stability of the system. In [5], the author provides a range dependent beam pattern through the frequency change array (FDA), and proposes a mobile target tracking method to achieve this goal. In addition, a cognitive closed-loop updating scheme is proposed to update the operating parameters in real time to improve the tracking performance of moving targets. The simulation results verify all the proposed methods.

In [6], the author proposes a new peak intensity measure to measure the discrimination ability of the learned correlation filter. The results show that the proposed method effectively enhances the peak value of correlation response, and has higher discrimination performance than the previous methods. In [7], the author proposes a model to improve tracking performance, and deduces efficient dense confidence propagation for the reasoning of the MRF model. A large number of experimental results show that the algorithm has good performance compared with the latest method, and can run in real time. In [8], the author proposes a new compression depth convolution neural network (CNN) function based on correlation filter tracker, and the proposed multi object tracking method has the ability of re identification (Reid). Extensive experiments have been carried out on the Kitti and mot2015 tracking benchmarks. The results show that our method is superior to most of the latest tracking methods. In [9], the author uses a guiding method to find out the mathematical difference between the methods of variation (Rayleigh Ritz), Galerkin and least squares. The advantages and disadvantages of each technology are illustrated. Finally, the convergence rates of the three techniques are obtained. In [10], the author concludes that under a large number of unknown parameters, the one-to-one evaluation of the experiment is a mathematical uncertainty. The non statistical experimental error of thermal analysis hinders the determination of a single parameter set by simultaneous least square method. In this paper, the author discusses several evaluation techniques used to deal with non statistical errors in the process of least square method for evaluating experimental sequences.

In order to solve the problem of difficult target tracking, the core part of the algorithm framework proposed in this paper includes target recognition and target tracking model. In this paper, a tracking algorithm based on correlation filtering is used to solve the apparent model of the target, and the reliability of each block is calculated by the response graph of each block. The relationship between multiple blocks in the same frame is calculated, and the contribution of each block to the overall tracking results is defined.

2 Method

2.1 Design of Bayesian Tracking Framework Based on Correlation Filter

In time t, the i-th block selects AIT from the target view in the model, selects target from the target motion model, selects sit from the state representation type, and selects an observation type to select an OIT to form a parameter set t i t = {AIT, MIT, sit, OIT} to represent the parameters of the i-th block in the multi block Bayesian framework. When applied to the multi block Bayesian framework, the above formula can express the joint posterior estimation of the whole tracking process by combining the weight with the posterior probability of N blocks:

$$P(X_t|Y_{1:t}) = \sum_{i=1}^{N} P(T_t'|Y_{1:t})P(X_t|T_t', Y_{1:t}) \tag{1}$$

Where, P represents the contribution of the i-th block to the whole tracking result. For convenience, it can be redefined as ψ (XT, xit) = P (tit Y 1: T). This chapter defines and understands that P (XT tit, Y1: T) represents the posterior probability of each block.

When the appearance of the target changes, such as shadow, deformation and scale changes, its visual model is vulnerable to damage. In order to solve this problem, this paper defines a variable units to express the obvious influence degree of the i-th block at time t, namely confidence, and defines a function called objective's obvious ability.

$$\varphi(X_t, Y_t, U_t') = P(Y_t|T_t', X_t)\phi(X_t, Y_t, U_t') \tag{2}$$

Where p (YT t i t, XT) represents the likelihood function of the i-th block at time t, and φ (XT, YT, UIT) is used to measure the confidence of the appearance of the block at this time.

2.2 Target Tracking Model Based on Least Square Method

When the measured parameter is linear with the measured parameter, the analytical expression of the measured parameter is obtained by the linear least square method (LLS). Take Zhengdong and Zhengbei as Cartesian coordinate system X-axis Y-axis direction, and measure the target azimuth β (T). Since the mathematical symbol of the target is used under the assumption of uniform linear motion, the measurement equation is obtained:

$$\beta_i = tan^{-1} \frac{D_0 sinb_0 + (t_i - t_0)V_{mx} - \int_0^t V_w sinC_w ds}{D_0 cosb_0 + (t_i - t_0)V_{my} - \int_0^t V_w cosC_w ds} + w_t, \quad i = 1, 2, L, m \tag{3}$$

So the form of iteration:

$$X_{k+1} = X_k - \left(\sum_{i=1}^{m} w_i^2 \nabla r_i(X_k)^T \nabla r_i(X_k) \right)^{-1} \sum_{i=1}^{m} w_i^2 \nabla r_i(X_k)^T \nabla r_i(X_k) \qquad (4)$$

When the appearance of the target changes, such as shadow, deformation and scale changes, its visual model is vulnerable to damage. In order to solve this problem, this paper defines a variable units to represent the obvious influence degree of the i-th block at time t, and it is not difficult to find the observation false point "oto". So, for Δ too ', the sine theorem is applied:

$$\frac{|OT|}{\sin \angle TO'O} = \frac{|OO'|}{\sin \angle OTO'} \qquad (5)$$

However, due to the influence of observation error and sea environment, the observed azimuth information often contains noise. Here, the least square recursive method mostly uses small error noise or Gaussian noise error, which is different from the actual process. If you want a better least square result, you need to do some preprocessing on the observation data, and then perform least square recursion.

3   Experiment

3.1   Experimental Setup

Under the environment of C/C++, the multithreading implementation of block parallel target tracking algorithm based on correlation filter is realized, and the depth optimization is realized by CUDA parallel programming architecture. All experiments were carried out on 24 GB ram and inter-i7-6700k processors. NVIDIA ger force 1070 graphics card and 8 GB video memory were selected as GPU devices. The coefficient in the equation is set to 1, and the learning rate of adaptive updating is set to 0.02.

3.2   Data Sources and Evaluation Indicators

Target tracking is a research hotspot in the field of computer vision. One or more authoritative assessment datasets is one of the contributions. Otb2013 is a data set containing 50 sequences proposed in 2013. It contains 11 tracking difficulties (light change, occlusion, scale change, fast motion, etc.) and introduces 50 more challenging sequences in 2015. These 100 sequences, called otb100, contain 26 gray-scale sequences and 74 color sequences, with a total length of more than 50000 frames.

In the accuracy curve, the Euclidean distance from the tracking target center to the mark position center is defined by the center position error (CLE), and then the average center position error of all frames is used as the overall performance of the sequence.

The success rate curve is represented by the Pascal VOC overlap ratio (VOR) and the real position bounding box of the predicted output, which is defined as: VOR = area (BT BGT)/area (BT BGT), where B is the rectangular box, GT represents the tracking result and the actual target location.

4 Results

4.1 Overall Comparison of Tracking Algorithms

In this paper, multiple partitions are calculated at the same time, and intermediate memory is used to maximize data transmission and reduce the space in the task level in parallel. Therefore, this section compares the overall tracking algorithm of CPU algorithm and GPU algorithm in time series. Finally, the tracking algorithm proposed in this paper is compared with other algorithms, and compared with other mainstream algorithms (Table 1).

Table 1. Comparison of average execution time of tracking algorithm

	cars	sylvester	panda	car24	Human2	liquor	mhyang	Mean
CPU	3.26	3.15	8.22	2.05	2.66	4.09	2.37	3.44
GPU	101.29	100.22	88.19	85.63	122.66	79.99	76.49	91.39
Speedup ratio	50.22	48.08	22.05	61.28	63.57	26.64	39.18	35.16

The above table shows the time-consuming comparison of the whole block tracking algorithm before and after partial sequence parallelization, and adaptively adjusts the forgetting factor and weakening factor according to the residual between the actual measured value and the predicted measured value at each time. The simulation results show that when the state of the target changes suddenly, the algorithm can effectively track the state of the moving target and improve the stability of the system. In the table, the real-time performance of the algorithm and other comparison algorithms are shown in the table. The results show that the block algorithm improves the accuracy of the algorithm and ensures the real-time performance of the algorithm to 100 FPS.

4.2 Analysis of Tracking Model Based on Correlation Filter and Least Squares

Because there is a large peak vibration except for small fluctuation, in order to study the large peak curve and try to filter it, it is necessary to get a large peak curve of system noise close to the actual situation. Therefore, the mean smoothing filtering method is used.

The test simulates the azimuth measurement process, and the results are shown in Fig. 1. Green curve is the azimuth signal $\{Z_m(i)|i = 1, 2, \cdots\}$ obtained by measuring samples, which is the superposition of oscillation noise and large peak value. Because the measurement of azimuth curve is caused by observation noise and system noise. By contrast, the blue curve is the true azimuth.

It is assumed that the small fluctuation of the measurement value is the measurement noise DT (b), which is caused by the fluctuation and measurement error; the system noise f (T) is the human factor (anti tracking) or the observer system. Therefore, you should try to eliminate these interferences, and use the algorithm which is only

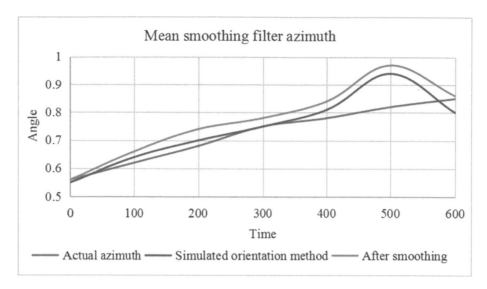

Fig. 1. Noise after smoothing

effective for small noise to track, so as to improve the tracking performance, and derive efficient dense confidence propagation for the reasoning of the model. A large number of experimental results show that the algorithm has good performance compared with the latest method, and can run in real time. Extensive experiments have been carried out on the Kitti and mot2015 tracking benchmarks. The results show that our method is superior to most of the latest tracking methods.

5   Conclusions

The core of the algorithm framework proposed in this paper includes target recognition and target tracking model. The tracking algorithm based on correlation filter is used to solve the visual model of the target, and the reliability of each block is calculated by the response graph of each block. The relationship between multiple blocks in the same frame is calculated, and the contribution of each block to the overall tracking results is defined. Experimental results show that the parallel algorithm framework designed in this paper can not only achieve stable and accurate tracking under the interference of masking, deformation and scale change, but also use the idea of least square method. In addition, the models and characteristics of the previous least square method are analyzed and summarized. The least square estimation method of target initial value is improved.

References

1. Vasuhi, S., Vaidehi, V.: Target tracking using interactive multiple model for wireless sensor network. Inf. Fusion **27**(C), 41–53 (2016)
2. Kamal, A.T., Bappy, J., Farrell, J., Roychowdhury, A.: Distributed multi-target tracking and data association in vision networks. IEEE Trans. Pattern Anal. Mach. Intell. **38**(7), 1397–1410 (2016)
3. Cui, Y., Ren, J., Du, W., Dai, J.: UAV target tracking algorithm based on task allocation consensus. J. Syst. Eng. Electron. **27**(6), 1207–1218 (2016)
4. Li, J.Q., Zhao, R.H., Chen, J.L., Zhao, C.Y., Zhu, Y.P.: Target tracking algorithm based on adaptive strong tracking particle filter. IET Sci. Meas. Technol. **10**(7), 704–710 (2016)
5. Wang, W.Q.: Moving-target tracking by cognitive RF stealth radar using frequency diverse array antenna. IEEE Trans. Geosci. Remote Sens. **54**(7), 1–10 (2016)
6. Sui, Y., Wang, G., Zhang, L.: Correlation filter learning toward peak strength for visual tracking. IEEE Trans. Cybern. **48**(4), 1290–1303 (2018)
7. Rui, Y., Xia, S., Zhen, Z., Zhang, Y.: Real-time correlation filter tracking by efficient dense belief propagation with structure preserving. IEEE Trans. Multimedia **19**(4), 772–784 (2017)
8. Zhao, D., Fu, H., Xiao, L., Wu, T., Dai, B.: Multi-object tracking with correlation filter for autonomous vehicle. Sensors **18**(7), 2004 (2018)
9. Sarkar, T.K.: A note on the variational method (Rayleigh-Ritz), Galerkin's method, and the method of least squares. Radio Sci. **18**(6), 1207–1224 (2016)
10. Várhegyi, G., Szabó, P., Jakab, E., Till, F.: Least squares criteria for the kinetic evaluation of thermoanalytical experiments. Examples from a char reactivity study. Parasitol. Res. **57**(2), 203–222 (2016)

Personality Feature Model of Innovation Motivation and Innovation Ability Based on Big Data Analysis

Weiying Wang[✉]

Jilin Engineering Normal University, Chang Chun 130052, Ji Lin, China
Tougao_010@163.com

Abstract. With the improvement of productivity, the competition between different countries and enterprises shifts from the competition of materials and materials to the competition of talents. One of the characteristics of competitive talents is the ability to innovate. This paper focuses on human innovation motivation and innovation ability, and hopes to establish a personality trait model that can measure innovation ability. This paper mainly adopts the method of big data analysis, and takes the college students as the research object. Through the follow-up survey of 10,000 college students, the results show that the personality model of this paper is used to simulate the innovation ability, and the score and the salary after graduation are synchronized.

Keywords: Innovation motivation · Innovation ability · Personality characteristics · Big data analysis

1 Introduction

With the development of the economy, the country and enterprises are increasingly eager for talents. The ability to innovate is an important capability for senior talents. Therefore, in order to improve the competitiveness of the country and enterprises, more and more countries and enterprises have joined the competition and innovation. Among the talents, there are risks in recruitment. How can we accurately predict the potential innovation ability of candidates? There is no accurate and practical method. If you can design a model, you can give quantitative evaluation indicators according to the applicant's application report, which is a very happy thing for the recruiting unit.

Psychologists' research shows that personality affects people's lives and work. For example, in the literature [1], the author surveyed 766 for the personality characteristics of young/middle-age/old people, and analyzed the personality of different groups of people. Influence, analyze the predictive factors of happiness; literature [2] analyzes the personality from life history, summarizes the life factors that influence personality development, and analyzes the impact of personality on life. The same personality also affects people's creative motives, but the description of personality is multi-faceted. Attributes, people with innovative personality may be embodied in strong creativity, strong independence, strong desire to learn and curiosity Heart, have good study habits,

© Springer Nature Singapore Pte Ltd. 2020
M. Atiquzzaman et al. (Eds.): BDCPS 2019, AISC 1117, pp. 678–683, 2020.
https://doi.org/10.1007/978-981-15-2568-1_93

etc. [3–5], but really want to portray the personality characteristics of innovative talents, there is no good research results.

The research on innovative personality originated from the attention and discussion of creativity and intelligence of Western psychologists. Its original intention and goal is to effectively develop human creative potential by discovering the characteristics that are beneficial to individual innovation, thus producing more abundant innovation results. The college student group is a group with strong motivation for innovation, and college students are a repository of innovative talents for the society. Therefore, many researchers have studied this group [6–8], in order to accurately extract the characteristics of innovative talents, many The team has designed a mature questionnaire, and through the statistical analysis of the questionnaire, the personality characteristics of the innovative talents are finally obtained. In today's big data analysis, by comparing large amounts of data to replace traditional statistical analysis, more accurate results can be obtained, so big data analysis is widely used in statistical field analysis [9–11].

This paper mainly analyzes the personality of innovative ability, and uses the method of survey documents and correlation coefficients to establish a two-level forecasting model and conduct a follow-up survey of 10,000 college students. The results show that the model established in this paper is for the innovative personality of college students. The forecast accuracy rating averaged 89.4%.

2 Methods

2.1 Feature Separation Method

So far, there is no accurate definition to describe personality, but from a psychological point of view, personality does exist, it will affect people's values, human ability, and human personality and work motivation. of. In order to analyze the influence of personality on the ability of innovation, this paper must abstract certain personality attributes. Classic personality traits can include: family inheritance, growth environment, educational experience, interpersonal communication, life performance, self-consistency, and self-control ability; some of these factors are somewhat related, and some are contradictory, in order to Being able to better study the influence of personality, psychology gives the five major personalities, and based on these personality research results, Costa and MCCrae proposed a five-factor personality model, which is neurotic, extroverted, open, and easy-going. Three dimensions of due diligence are evaluated. According to the five dimensions of personality, the researchers built many scales, such as BH, NEO-PI-R, and so on. These tables are designed for 8–10 projects per dimension, depending on the research object.

This paper mainly describes the personality of innovation motivation and innovation ability. Therefore, the main considerations are the growth environment, educational experience, etc., and the research on growth, education and interpersonal relationship, but the relevance of inheritance, family and emotion is not great. The method of establishing personality characteristics in this paper is mainly based on the combination of questionnaires and literatures, and finally selects 120 personality characteristics as the research object. In order to make the analysis results better descriptive, and in order to make the results concise, this paper only selects the sample salary as the index of innovation ability, including monthly salary below 1000 (very low, marked A1), 1000–3000 (low, marked A2), 3000–5000 (medium, mark A3), 5000–10000 (middle upper; mark A4); 10000 or more (high; mark A5), finally establish a feature—revenue information correspondence table, fill out this form according to the questionnaire, Establish an analytical model.

2.2 Synchronization Calculation Method

Synchronization is a method to describe the similarity between similar samples. This paper is mainly used to simplify the model. For the 120 information and the five-level income standard information table, the large sample brings a huge amount of data. The possibility of complete agreement between these data is not great. In order to more prominently characterize the model, this paper first performs similar clustering calculation on the sample. The synchronization calculation method is a method to make the population characteristics of different income standards as uniform as possible.

Synchronization P_{ij} calculation method is as follows:

$$P_{ij} = \frac{\sum_{t=1}^{N} (x_i(t) - \bar{x}_i)(x_j(t) - \bar{x}_j)}{\sqrt{\sum_{t=1}^{N} (x_i(t) - \bar{x}_i)^2} \sqrt{\sum_{t=1}^{N} (x_j(t) - \bar{x}_j)^2}} \tag{1}$$

Where x represents the 120 feature vector of the sample, t represents the tth feature, i represents the ith vector, and N represents the length of the vector.

2.3 Big Data Analysis Method

There are four types of big data analysis methods that can be classified according to their needs: description, diagnosis, prediction, and instructions. This paper is mainly a predictive analysis method. Through big data analysis, a model is built, and then this model is used to predict the innovation ability of people with certain characteristics.

3 Data

The main purpose of the questionnaire "Question Questionnaire on Innovation Motivation and Innovation Ability and Personality Relationship" is to test the relationship between the innovative attributes of the respondents and collect data for the establishment of the model. The survey documents are in the form of scoring. The score of the road is 100 points, and it is scored according to the actual situation. The questionnaire consists of 140 fill-in-the-blank questions, which belong to the five personality characteristics. 20 of the 140 questions are designed to prevent the data distortion design caused by the investigators.

In the same form, the questionnaire was conducted in a survey of 17 universities in a certain province for 6 years. An online questionnaire was issued once a year. Because it cooperated with the research institutes of these universities, it was smoothly carried out in order to facilitate the unification of data statistics. Sexuality, the final questionnaire for the calculation of a total of 100 million, including 53.2% males and 46.8% females; the survey covers four major categories of science and engineering/literature/medicine/art, of which 37% of science and technology, 31% of literature and history Medicine accounts for 22% and art accounts for 10%. From the perspective of students, 10,000 questionnaires accounted for 69% of urban students and 31% of rural students.

4   Results

4.1   Correlation Between Different Dimensions

This paper predicts the subject's innovative ability model based on 120 attributes as shown in Fig. 1. Based on the input 120 subjects, the model calculates the weight values of the five personality attributes using the correlation coefficient, and the five personality attribute weight values summarize the output results.

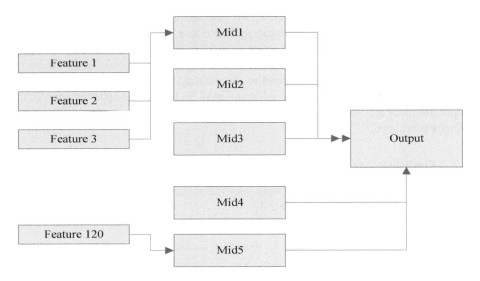

Fig. 1. Innovation capability prediction model based on big data analysis

In this paper, 120 attributes are selected as the input objects of the model, but the contribution of 120 personality attributes to the model is not the same. Therefore, before the prediction, the different personality attributes are trained. The 120 attribute contributions are based on the correlation coefficient values. Weights, this paper selects 8000 samples as training samples and 2000 samples as test samples. The results are shown in Fig. 2.

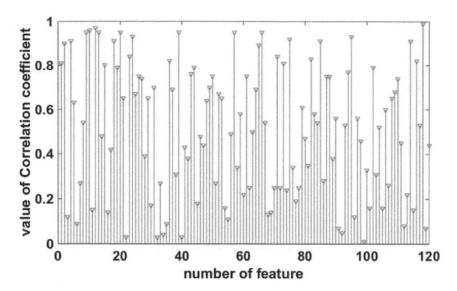

Fig. 2. Weights of different related attribute values

According to the results of Fig. 2, we can see that for different attributes, the contribution is different, and the curiosity/brave/love learning is the highest, all of which are above 0.95, but from the above simulation results, perseverance/judgment is relatively low. Below 0.1, which is inconsistent with the expectations at the beginning of the study, and does not correspond to the existing results based on psychological research. The reason for this non-conformity may be that the parameters used in this paper are more and do not highlight these results, so big data analysis as a result, this weight is reduced.

4.2 Prediction Accuracy

After the simulation training, this paper uses the model to predict 2000 samples. The prediction results are shown in Table 1.

Table 1. Prediction accuracy

	A1	A2	A3	A4	A5	Average
Accuracy	75%	84%	94%	97%	97%	89.4%

From the results in Table 1, it can be seen that from A1–A5, the prediction accuracy shows an upward trend. The lower the income standard, the lower the prediction accuracy. The higher the income, the higher the prediction accuracy, and the average recognition rate is 89.4.%. The main reason for this result is that this paper mainly predicts the ability to innovate. Therefore, the innovative personality model is mainly for modeling people with innovative motives, but the characteristics of people with lower innovation ability are not significant, so the prediction results Lower.

5   Conclusion

Although many studies have studied innovation ability and innovation motivation, there is still no model to calculate and predict innovation ability and innovation motivation. This paper takes the undergraduate students as the research object and establishes two-level prediction based on 120 attributes. From the perspective of forecasting results, the average forecasted result of this paper reaches 89.5%, which can basically meet the requirements of talent evaluation and provide a reliable reference for the evaluation of innovative talents.

References

1. Gomez, V., Krings, F., Bangerter, A., et al.: The influence of personality and life events on subjective well-being from a life span perspective. J. Res. Pers. **43**(3), 345–354 (2009)
2. Csathó, Á., Birkás, B.: Early-life stressors, personality development, and fast life strategies: an evolutionary perspective on malevolent personality features. Front. Psychol. **9**, 305 (2018)
3. Friston, K.J., Lin, M., Frith, C.D., Pezzulo, G., Hobson, J.A., Ondobaka, S.: Active inference, curiosity and insight. Neural Comput. **29**(10), 1–51 (2017)
4. Kahan, D.M., Landrum, A., Carpenter, K., et al.: Science curiosity and political information processing. Polit. Psychol. **38**, 179–199 (2017)
5. Archer, L., Dawson, E., Dewitt, J., et al.: Killing curiosity? An analysis of celebrated identity performances among teachers and students in nine London secondary science classrooms. Sci. Educ. **101**, 741–764 (2017)
6. Andersson, F., Freedman, M., Haltiwanger, J., et al.: Reaching for the stars: who pays for talent in innovative industries? Econ. J. **119**(538), F308–F332 (2010)
7. Wu, Z.-W., Zhu, L.-R.: Cultivating innovative and entrepreneurial talent in the higher vocational automotive major with the "on-board educational factory" model. Eurasia J. Math. Sci. Technol. Educ. **13**(7), 2293–2300 (2017)
8. Wei, W.U., Zhong, Z., Chen, M.: A study on the training mode of discipline-oriented innovative and entrepreneurial talents—taking agricultural, forestry and normal universities as an example. Asian Agric. Res. **10**(12), 80–82 (2018)
9. Tawalbeh, L.A., Mehmood, R., Benkhlifa, E., et al.: Mobile cloud computing model and big data analysis for healthcare applications. IEEE Access **4**(99), 6171–6180 (2017)
10. Bo, T., Zhen, C., Hefferman, G., et al.: Incorporating intelligence in fog computing for big data analysis in smart cities. IEEE Trans. Industr. Inf. **13**(5), 2140–2150 (2017)
11. Sun, J., Jeliazkova, N., Chupakhin, V., et al.: ExCAPE-DB: an integrated large scale dataset facilitating big data analysis in chemogenomics. J. Cheminformatics **9**(1), 17 (2017)

Determination and Check of Basic Parameters of Automobile Clutch Under the Background of Big Data

Fengjuan Wang[1(✉)], Jia Yu[2], Ruiming Ning[3], and Shufeng Jiang[1]

[1] Institute of Mechanical and Electrical Engineering, Qiqihar University, Qiqihar 161001, China
657678158@qq.com
[2] Qiqihar City Vocational Education Center School, Qiqihar 161001, China
[3] Qiqihar Northern Machinery Co., Ltd, Qiqihar 161001, China

Abstract. Clutches play a vital role in the entire automotive transmission system and is also one of the key components. The performance of the clutch will directly affect the braking performance, the cost of use and the service life of the car. The basic parameters of the clutch are critical to the performance, and changes in basic parameters directly affect the structural performance and performance. n today's big data era, many factors should be taken into account when calculating the basic parameters of the clutch. In this paper, a commercial vehicle is taken as an example. In the case of known raw data, the size of the best basic parameters is obtained by a classical algorithm. After verification, the size is qualified and meets the requirements. The check result indicates that this parameter determination method is desirable.

Keywords: Big data · Automobile clutch · Basic parameter size · Parameter check

1 Introduction

The clutch acts as a link in the automotive transmission system for controlling the interruption and connection of the transmission system. The active clutch assembly flywheel of the automobile clutch is connected to the engine of the car, The follower assembly is connected to the first shaft of the gearbox, and the condition that the car runs smoothly is that the main and driven components of the clutch are smoothly combined [1]. When the car in the process of driving wants to achieve shifting, when shifting operation, The clutch inside the car can ensure that there will be no sharp impact between the gears in the gearbox due to shifting. Due to the existence of the clutch, the gearbox can be guaranteed to have a safe working environment under any working conditions [2]. The performance of the clutch mainly depends on the basic parameters and main dimensions of the clutch.

Big data, or huge amounts of data, refers to massive, high-growth, and diverse information assets that require new processing models to have greater decision-making, insight, and process optimization capabilities [3]. The era of big data has arrived, and it

M. Atiquzzaman et al. (Eds.): BDCPS 2019, AISC 1117, pp. 684–690, 2020.
https://doi.org/10.1007/978-981-15-2568-1_94

has set off waves of change in many fields. The determination of the basic parameters of the clutch studied in this paper will also consider the influencing factors from big data [4].

Many articles on the determination of basic parameters of automobile clutches are determined by modern mechanical design methods. The parameter data determined by this method is accurate, but it is still more practical for many beginners, people who do not major in design, and places that need to quickly check parameters. Therefore, the research in this paper reinforces the theory that the traditional classical calculation method can still effectively design the clutch.

2 Design Clutch Conditions

This design of the clutch is used as an example of a commercial vehicle. The basic Parameters required for the design of its clutch, as shown in Table 1

Table 1. The original data of the clutch

Automobile driving form			4 × 2			
Maximum load of the car	1500 kg		The quality of the car	1065 kg		
Maximum parameters of automobile prime mover	Power	38.5 kw	Torque	83 N.m	Rotating speed	5200 r/min
Friction plate maximum outer diameter			160 mm			
Maximum speed per hour			115 km/h			
Engine position			Preposition			
Main reduction ratio			4.479			
First gear ratio			2.8			
Second gear ratio			1.875			
Wheel rolling radius			0.308 m			

3 Determination of Basic Parameters of the Clutch

3.1 Determination of Clutch Torque Capacity T_c

The torque capacity T_c of the clutch should be greater than the maximum torque T_{emax} of the engine. This is to ensure that the clutch reliably transmits the torque from the engine under any circumstances [5]. Its mathematical expression is:

$$T_c = \beta \cdot T_{emax} \tag{1}$$

In the middle β—Clutch backup factor (β > 1)
T_{emax}—Maximum torque (see Table 1 for this design)

The coefficient β can reflect the reliability of the conveying torque and the requirements for the working environment are also high. The design dimension is also closely linked to the coefficient β. This requires that the coefficient β not only meets the reliability in terms of value, but also takes into account that the system will not break when the system is overloaded [6]. According to the impact of big data on the size of the reserve factor, the factors of maneuvering should also be considered in the design [7]. Refer to Table 2 for the range of backup factors for different types of vehicles [5], Select β = 1.5; according to formula (1), the torque capacity of the clutch can be calculated Tc = 124.5 N.m.

Table 2. The selection of reserve coefficient table of different types of cars

Vehicle type	Backup coefficient β
Passenger car or Commercial vehicle with total mass less than 6t	1.2–1.75
Commercial vehicles with a maximum total mass of 6–14t	1.5–2.25

3.2 Determination of the Inner Diameter D of the Friction Plate Outer Diameter D Friction Plate

By empirical formula:

$$D = K_D \cdot \sqrt{T_{emax}} \qquad (2)$$

$$c = \frac{d}{D} \qquad (3)$$

In the middle K_D—Diameter coefficient
c—Inner and outer diameter ratio of friction plate

Table 3. The selection of coefficient of the diameter of the different types of car

Vehicle type	Diameter coefficient K_D
Passenger car	14.6
Commercial vehicle of 1.8 to 14t	16–18.5 (single)
	13.5–15 (double)
Commercial vehicles larger than 14t	22.5–24

Table 4. The selection of friction plate parameters (in part)

Outer diameter D/mm	160	180	200	225	250	280	300	325
Inter diameter d/mm	110	125	140	150	155	165	175	190
Thickness h/mm	3.2	3.5	3.5	3.5	3.5	3.5	3.5	3.5
$1 - c^3$	0.687	0.694	0.700	0.667	0.620	0.589	0.583	0.585
C = d/D	0.676	0.667	0.657	0.703	0.762	0.796	0.802	0.800
Unit area cm^2	106	132	160	221	302	402	466	546

In general, the value of K_D is selected from the range of the diameter coefficient of Table 3, and the range of the ratio c is 0.53 to 0.70 [6]. This design selects $K_D = 17$; c = 0.7. From Eqs. (2) and (3), the value of the initially selected coefficient parameter can be obtained to obtain D = 155 mm, d = 108.5 mm. Refer to Table 4 for part of clutch disc size series and parameters (JB/T9190-1999 standard) [8], Finally, the outer diameter D = 160 mm, the inner diameter d = 110 mm, and the thickness h = 3.2 mm.

3.3 Determination of the Unit Pressure P_0 of the Friction Lining and the Working Pressure F Acting on the Pressure Plate

The formula for calculating the torque capacity of the clutch according to the force of each transmission component in the clutch drive train under working conditions [8]:

$$T_c = \mu z R_c P_0 S_d \tag{4}$$

$$R_c = \frac{1}{3} \cdot \frac{D^3 - d^3}{D^2 - d^2}, \text{ If } c > = 0.6, \text{ then } R_c = \frac{D + d}{4} \tag{5}$$

$$F = P_0 \cdot S_d \tag{6}$$

In the middle z— Working face number of clutch friction plate, z = 2 in this paper

R_c—Average radius of the friction lining (mm), It is assumed that the pressure of the friction plate is uniformly distributed, and its value is calculated according to formula (5) P_0—Unit pressure

S_d—Friction sheet single-sided friction area, the value of this paper is determined according to Table 4. F—Working pressure acting on the platen

Table 5. The scope of friction plate unit pressure

Category		Unit pressure P_0/MPa
Asbestos base	Molding	0.15–0.25
	Weave	0.25–0.35
Powder metallurgy	Copper base	0.35–0.5
	Iron base	
Cermet		0.7–1.5

By introducing the parameters of the known friction plates by the formulas (4) and (5), the unit pressure $p_O = 0.29$ MPa of the friction plate can be solved, and the working pressure acting on the platen F = 3074 N is determined by the formula (6).

4   Checking the Main Parameters of the Friction Plate

(1) The unit pressure p_o of the friction plate is checked. The unit pressure p_o acting on the friction plate is calculated as:

$$p_0 = \frac{4F}{\pi(D^2 - d^2)} \tag{7}$$

Substituting the above calculated working pressure value into (7), the calculation yields $p_o = 0.28$, referring to the unit pressure range of the friction material of Table 5, $[p_o]$ is 0.35–0.50, and $p_o < [p_o]$ is obtained, thereby, the friction plate The unit pressure p_o meets the material requirements.

(2) Friction plate per unit area friction torque T_{CD} check unit area friction torque T_{CD} calculation formula:

$$T_{C0} = \frac{4T_{emax}}{z\pi(D^2 - d^2)} \tag{8}$$

Substituting the known parameters into the above formula, and obtaining $T_{CD} = 0.0039$ N.m/mm2, the allowable unit area friction torque $[T_{CD}] = 0.0060$ N.m/mm2 from the powder metallurgy copper-based material [8], we can get $T_{CD} < [T_{CD}]$, thereby, the friction plate The unit area friction torque T_{CD} meets the material requirements.

(3) The calibration of the total mass m_{CD} of the vehicle per unit area of the friction lining The total mass of the vehicle per unit area m_{CD} Calculation formula:

$$m_{C0} = \frac{4m}{z\pi(D^2 - d^2)} \tag{9}$$

Substituting the known parameters into the above formula, $m_{CD} = 0.0502$ kg/mm2, the total mass of the vehicle $[m_{CD}] = 0.06$ kg/mm2 from the allowable unit area of the powder metallurgy copper-based material, can be obtained $m_{CD} < [m_{CD}]$, thus, the friction plate The total mass m_{CD} of the vehicle per unit area meets the material requirements [9].

(4) The outer diameter D of the friction lining should be selected such that the maximum circumferential speed v_D does not exceed 70 m/s, i.e.

$$v_D = \frac{\pi n_{emax} D \times 10^3}{60} \tag{10}$$

In the middle, v_D—Maximum peripheral speed of friction plate (m/s)
n_{emax}—Maximum engine speed (r/min)
Bring the data into the calculation $v_D = 43.56$, meet the requirements

(5) When the clutch is working, the work done by the sliding friction between the friction plates is called the clutch sliding work w. The greater the slip-grinding work w, the more heat generated by the friction lining during friction, and the higher the temperature associated with it. The temperature continues to increase, and when the temperature reaches a certain level, the friction lining will be burned. In order to prevent this from happening, the frictional work per unit area of the clutch friction working area is lower than the allowable value [10], i.e.

$$\omega = \frac{4W}{\pi z(D^2 - d^2)} \le [\omega] \tag{11}$$

$$W = \frac{\pi^2 n_e^2}{1800} \left(\frac{m_a r_r^2}{i_0^2 \cdot i_g^2} \right) \tag{12}$$

In the middle z—Number of working friction surfaces of the friction disc
$[\omega]$—The allowable friction work of the unit friction area of the clutch friction
plate, the value is shown in Table 6, this paper $[\omega] = 0.4$ J/mm^2
W—The total sliding work generated by the clutch engagement at the start of the
car, calculation formula (12)
ω—Sliding work in unit contact wear area (J/mm^2)
n_e—The motor speed of the normal operation of the car (r/min), the passenger car
takes 2000r/min commercial vehicles to take 1500r/min
i_0—Main reducer ratio
i_g—Transmission gear ratio used when the car starts

Table 6. Friction plate of allowable friction work unit area friction work

Maximum quality of the car (Kg)	≤ 2	2–5	5–12	≥ 15
Unit allowed friction work J/mm^2	0.4	0.6	0.8	1.0

Substituting relevant data into Eqs. (11) and (12), and calculating
$W = 7924.696$ J, $\omega = 0.374$ J/mm^2.

5 Conclusions

Under the background of the rapid development of the times, big data has become the
most important strategic resource of the automobile industry. It is also an ability to
discover the importance of big data and make effective use of it.

In this paper, after considering the factors of automobile clutch parameters from big
data. In this paper, the torque capacity T_C of clutch, the outer diameter D of friction
plate, the inner diameter D of friction plate, the unit pressure Po of friction plate and the
working pressure f acting on the pressure plate are determined and checked. The
purpose of clutch basic parameter design is to ensure that the clutch performance can
meet the requirements, make its structure size as small as possible, and get better design
effect. The basic parameters of the clutch designed in this paper meet the requirements
after checking, and provide guarantee for the best performance of the clutch.

Acknowledgements. 2018 Annual Heilongjiang Provincial Department of Education Basic
Research Business Youth Innovation Talent Project (135309376); 2019 Qiqihar City-level Science and Technology Plan General Project Contract Number (GYGG-201919); 2018 Education
Department's Basic Research Business Achievements Transformation and Cultivation Project
(135309505).

References

1. Zhou, Y.: Cognition and maintenance of automobile clutch. Manag. Technol. SME (11), 159–160 (2018). (in Chinese)
2. Zhu, Y.: Thinking on the design of automobile clutch. New Technol. New Prod. China (08), 58–59 (2019). (in Chinese)
3. Wang, Y., Li, J.: Research on the transformation of automobile marketing model under the background of big data era. China Logist. Purchasing (22), 47–48 (2018). (in Chinese)
4. Wang, H., Wang, Y., Li, Z., et al.: The application of big data in automobile connector mould design. Auto Electr. Parts (12), 66+68 (2018). (in Chinese)
5. Lu, K., Lu, Z., Cheng, Z., et al.: Study on the influence rules of clutch parameters on HMCVT shift performance. Mech. Sci. Technol. Aerosp. Eng. (3), 1–7 (2019). (in Chinese)
6. Wang, M.: An optimum design of the basic parameters of clutches. J. Shiyan Tech. Inst. **13** (02), 46–49 (2000). (in Chinese)
7. Xi, X., Mao, W., Zhang, Z.: Application of industrial big data in improving the quality of new energy vehicles. Auto Time (12), 66–67 (2019). (in Chinese)
8. Sun, J.: Main parameters optimizing of vehicle clutch. For. Sci. Technol. Inf. (04), 52–54 (2005). (in Chinese)
9. Luo, X., Huang, Y., Li, G., et al.: Main parameters and structural machinability of the clutch of Dongfeng-12 walking tractor. Hunan Agric. Sci. (24), 13–15 (2014). (in Chinese)
10. Deng, P.: Optimal design of automobile clutch based on particle swarm optimization algorithm. Wirel. Internet Technol. (02), 64–65 (2017). (in Chinese)

New Features of Editing Role Positioning Under the Background of Artificial Intelligence

Yongxin Liu[(✉)]

Media Academy, Jilin Engineering Normal University, Changchun, Jilin, China
35249827@qq.com

Abstract. As the development and progress of the science and technology, artificial intelligence has been a breakthrough in the field of editing in which editing machine is one of good example. With the continuous development of artificial intelligent on editing industry, human-computer conflict has become one of the hot points in the academic circle, which has brought many discussions at current stage. The purpose of this paper is to explore how to get the positioning for editors and coordinate the relationships between instrumental rationality and humanistic spirit under certain circumstances of artificial intelligence shuffle on the editing industry. The study of this paper is based on the theory of artificial intelligence algorithms to study the evolutionary algorithms for data sources of evidence, which is based on the emergency of artificial intelligence under the background of machine editor and also for the traditional editing on the information production, distribution and management mechanism of the industry as well as editing practitioners professional role with function achieved, connotation of the attributes and skills of editing practitioners to work put forward on new requirements. In this paper, the research results show that the widespread using of artificial intelligence, which does not mean that the editor practitioners will be replaced by editing machine of artificial intelligence because of the inadequacy of the technique on artificial intelligence algorithms to be always unable to break through the symbol manipulation level that is not like human consciousness which is equipped with a self-consciousness that can produce complex concepts, consciousness and meaningful system.

Keywords: Artificial intelligence · Editing industry · Human-computer conflict · Artificial intelligence algorithms · Man-machine conflict · Those in the editing profession · Intelligent editing machine

1 Introduction

The emergence and development process of intelligent editing industry is always under the influence of continuously evolving information communication technology [1]. Since the second half of the 20th century, such as network, mobile, digital, intelligence as the core characteristics of the development of new information communication technology and the rapid popularization, set off a worldwide wave of technology integration and transformation of media change, with newspapers and radio and television as the main body of the public news media form is increasingly tend to integration and intelligent of the impact of new media technology, facing a huge

© Springer Nature Singapore Pte Ltd. 2020
M. Atiquzzaman et al. (Eds.): BDCPS 2019, AISC 1117, pp. 691–698, 2020.
https://doi.org/10.1007/978-981-15-2568-1_95

transformation pressure [2, 3]. In this process, the news editing industry, which is supported by the form of traditional media, is most affected [4]. The introduction of artificial intelligence technology into the editing industry and the continuous improvement of the automation of the production process of the editing industry will inevitably have an impact on the values, production mechanism, consumption pattern, industrial structure and functions of the traditional editing industry, thus causing automation anxiety among the editing practitioners [5]. It is against this background that the so-called artificial intelligent editing machine may gradually challenge or even replace the view of traditional editors.

In the editing industry and academia, in addition to the impact and change of technology and industry, the application of artificial intelligence algorithm technology in the editing industry also causes the consequences of the editing industry in terms of ideology and social politics [6]. Used to play a important role in the transformation of Chinese society rapid exit of the professionalism of the media, business and state industry and characterization of the socialist dominant order quickly occupy the core position in the field of editing, which produced with commercialization and capitalized in the core logic of technology of worship and of social media platforms pan enter-tainment trend of colonization of public discourse [7]. Faced with such a complex situation, people try to analyze the institutional and technical root causes of the dilemma of traditional media from the perspective of operational logic, and actively explore strategies and strategic paths for traditional media to break through. Or from the plight of the traditional editing industry practice, logic analysis to edit the pro-duction mode of industry and technology behind the transformation of the political significance, especially for editing industry outlook of society and the resolution of public construction mission expressed critical concerned, and put forward the recon-struction editor vision industry values in [8]. However, no matter from the perspective of pure technology or from the perspective of social politics, a basic practical problem that we cannot avoid is to what extent the emergence, development and application of artificial intelligence algorithm technology have changed people's concept of editing industry and its technical and social conditions [9].

Undeniably, with the rise of platform media and the development of artificial intelligence technology, the embedded relationship between the performance of new information communication technology and the practice process, social structure and human thinking process of social subjects is further deepened, which indeed poses a challenge to the survival of traditional editorial organizations in the global scope [10]. In the face of the continuous expansion of big data and artificial intelligence tech-nology, these propositions have become urgent issues that editors and researchers have to face. Around this core issue, the purpose of this paper is to use the interdisciplinary technology field of vision, from a special editor industry practice according to the internal logic of human culture, from two dimensions of technology and culture to discuss the relationship between artificial intelligence and edit industry, thus to editor practitioners and industry transformation logic to make critical analysis.

2   Methods

2.1   Evolutionary Algorithm of Artificial Intelligence

Evolutionary algorithms simulate the evolution of organisms in nature. Darwin's theory of evolution pointed out that "natural selection leads to the survival of the fittest". Evolution explains almost everything. "why is this creature like this? That kind of thing. Organisms that are more adapted to their environment are more likely to leave their chromosomes behind. So, in the computer, we simulate the biological selection. Although the researchers who do evolutionary calculation know that evolutionary algorithm is suitable for solving optimization problems without analytic objective function, most evolutionary algorithms tend to assume that the objective function is known when they are designed. The general form of evolutionary algorithm is as follows:

$$D_l(t) = D_l(t-1) + H(l,n) \tag{1}$$

Where, $D_i(t)$ represents a target function in the data cluster.

Relatively speaking, our research and application of data-driven evolutionary optimization are relatively early. Why do artificial intelligence drive optimization? Because many optimization problems in the real world cannot be described by analytic mathematical formulas, their performance can only be verified by simulation or experiment. The general evolutionary optimization algorithm needs to solve the challenge mainly lies in the problem contains a lot of local optimization, large-scale, multi-objective, strong constraints and uncertainty, and artificial intelligence-driven optimization is bound to face the challenge from the data.

2.2   Combined Sort Algorithm

To explore the operation details of artificial intelligence algorithm in text editing, it is not enough to rely only on understanding the evolutionary algorithm of artificial intelligence, or even to grasp its basic technology roughly. To further understand the mystery of artificial intelligence, we must also have a certain understanding of the combined sort algorithm. In the statistical calculation of this algorithm, all the appearing elements search for the most probable parameter location value in the probabilistic model, where the model relies on undiscovered potential variables. The first step is to calculate the expectation and calculate the maximum possible estimate of the hidden variable by using the existing estimate. The second step is to maximize the maximum possible value obtained in the first step to calculate the value of the parameter. The algorithm attempts to find the maximum flow from a traffic network. Its advantage is defined as finding the value of such a flow. The maximum flow problem can be regarded as a specific case of a more complex network flow problem. The sorting of element positions in this algorithm is updated according to the following formula:

$$v_i^d = \omega * v_i^d + c_1 * r_1 * (\rho_i^d - \chi_i^d) + c_2 * r_2 * (\rho_g^d - \chi_i^d) \tag{2}$$

When we solve the machine edit text problem with merge sort algorithm, we can get a series of models, some of which may be the best interpretable, some of which may be the most accurate, and some of which may be overfitting. When you look at all the different models you can pick a few of them. This is one of the advantages of using evolutionary algorithms to solve edit sorting. In general, using evolutionary algorithms to help machine learning to sort words doesn't just mean I can make it learn better, it gives it more possibilities. At the same time, it can be used for parameter optimization, structure optimization, as well as model interpretability and security.

3 Experimental

3.1 Difference Between Artificial Intelligence Editing and Traditional Editing

In essence, the so-called artificial intelligence editing machine or intelligent editing industry is an algorithmic process that converts data into narrative news texts with limited or no human intervention except the initial program setting. It with big data algorithm, artificial intelligence and natural language generation technology as the foundation, the emergence of automated editing profoundly changed the traditional editing industry production and consumption mode, thereby to edit information industry consumer decision-making, political, and social agenda Settings, including objectivity and credibility of the traditional editor value factors, institution of legal and ethical responsibility and the professional status of traditional edit industry practitioners had a profound impact.

Technically, the algorithm is dynamic as a series of data calculation rules that can transform the data input into the specific result output through a specific calculation program for a specific problem. However, algorithm-based news production is a quasi-automatic process realized through natural language generation technology, including three stages: input based on big data, filtering, analyzing and processing data based on relevant features to transform it into semantic structure, and finally presenting it on a specific output platform. Natural language generation and the combination of big data technology makes editing techniques of artificial intelligent hit areas of human creativity and expression, the division of labor in editing activity influence and role assignment, to edit the role of practitioners gradually from direct information filtering, processing, and writing activities to become the application rules of the algorithm and the data set and the indirect role of management. It is inevitable that the editing industry in transition and its practitioners must bring the algorithm and programming ability into the category of the basic professional skills of editing.

Although algorithm editor is still highly dependent on structured data, with highly differentiated or internal conflict data will not be able to effectively deal with complex contexts, but from the perspective of editing industry organizations and professional functions, in addition to the contents related to the core level of thinking of a large part of traditionally held by editing practitioners work will be gradually replaced by artificial intelligence, the basic trend is inevitable.

4 Discuss

4.1 Inherent Limitations of Artificial Intelligence Algorithm Technology

The rapid development of artificial intelligence technology has produced a huge impact on human society, and this impact is even considered to be enough to match the impact of the industrial revolution on human society, so it is called the fourth industrial revolution. However, judging from the current state of technology, although the ultimate goal of scientists since the emergence of the idea and related technologies of artificial intelligence is to realize the ideal of machines imitating or even surpassing the human brain, so far, the process of achieving this goal has not been smooth. This is directly related to the internal technical logic of artificial intelligence.

In terms of the current development state of artificial intelligence technology, whether the editors who have been in the leading position in the intelligent editing industry in modern times are facing the fate of being replaced? Despite the widespread claim that artificial intelligence may replace new creative work subjects such as editors and practitioners in the discourse of popular technicism and commercialism, the author's findings on 55 respondents in this study are quite different, as shown in Table 1 below. At least from the current development of artificial intelligence technology we can't come to the conclusion that in the basic logic, reason involves prediction and rationalization, process the black-box operation, complicated concept learning, simulate the brain from four aspects, such as the feasibility of the first two problems can be summed up in the subconscious and consciousness mechanism, can be summed up in two points after heart feasibility problem of reductionism.

4.2 The Dual Logic of Artificial Intelligence and the Transformation of Journalism

Compared with the practitioners of physical editing, the artificial intelligence editing machine can analyze emotional or soft facts and identify irregular or abnormal data information under the condition of further development, but it still can't understand the subtleties of human expression, and can't form value judgment and sense of moral mission. It is precisely these unquantifiable meanings that affect the personalized writing style of editors and highlight that the editing industry has rich social and political significance in addition to economic effect. The author's survey results of 55 respondents also confirm this conclusion. The statistical table of the survey results is shown in Fig. 1 below. Visible, artificial intelligence can bring new speed to edit industry, to form a new distribution mechanism, but not close to the complex the truth behind the fact, so the future of intelligent editor should consider mainly under the Internet environment redefined, the intelligent editor industry should keep rational expectation, in the edit industry in the application of artificial intelligence technology innovation more to highlight the value of a man.

Table 1. Survey on the future development prospect of artificial intelligence

Category		Number of persons	Proportion
Age of respondents	Under the age of 20	5	9.09%
	20–30	24	43.64%
	30 to 40	11	20.00%
	40–50	8	14.55%
	50 years of age or older	7	12.73%
Record of formal schooling	High school and below	9	16.36%
	University college	19	34.55%
	University degree	21	38.18%
	Postgraduate and above	6	10.91%
Contact frequency	Almost no	2	3.64%
	Very few	9	16.36%
	more	32	58.18%
	often	12	21.82%
View of artificial intelligence will replace human	sure	14	25.45%
	possibly	34	61.82%
	Certainly not	7	12.73%

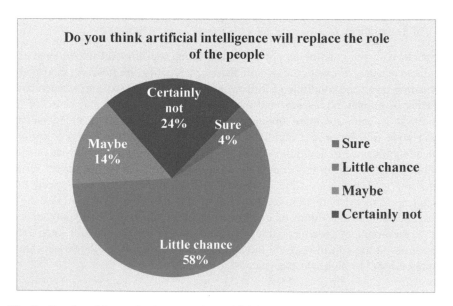

Fig. 1. Results of jump check on whether artificial intelligence will replace human roles

Obviously, artificial intelligence technology always stay in perfecting to analyze information and symbol form and control level, and can't develop independent thinking mind machine, also cannot have human body based on the brain thinking process and on the basis of the unity of body and mind of sexual consciousness of main body formed by the construction ability, the significance of cultural creativity and moral sense.

5 Conclusion

The mission of editorial practitioners is not to compete with artificial intelligence algorithm technology to quantify the efficiency of information production, but to form a benign interactive cooperative relationship with machines, on the one hand, fully grasp the knowledge and skills of artificial intelligence and algorithm technology, so as to better control them; On the other hand, it is necessary to give full play to the creative advantages of human beings in contextualized meaning interpretation and value giving to literal entities. The relationship between them should not be a conflict or a zero-sum game, but a cooperative relationship that gives full play to their respective advantages and coordinates with each other. With the information attribute of text increasingly becoming a function carried by intelligent machine, under the heavy burden of information caused by highly saturated mass information, people's demand for the cultural attribute of text is definitely more urgent, which is exactly the irreplaceable advantage of editors.

Acknowledgements. This paper was supported by the Research and Development Fund Project of JiLin Engineering Normal University (XYB201834).

References

1. Zhao, N., Dong, S.H.: Man-machine cooperated dynamic schedule orienting disturbance in steelmaking. Appl. Mech. Mater. **130–134**, 3396–3400 (2012)
2. Lamichhane, B.R., Persoon, G.A., Leirs, H., et al.: Are conflict causing tigers different? Another perspective for understanding human-tiger conflict in Chitwan National Park, Nepal. Glob. Ecol. Conserv. **11**(3), 177–187 (2017)
3. Neupane, D., Kunwar, S., Bohara, A.K., et al.: Willingness to pay for mitigating human-elephant conflict by residents of Nepal. J. Nat. Conserv. **36**(4), 65–76 (2017)
4. Kennedy, D., Philbin, S.P.: The imperative need to develop guidelines to manage human versus machine intelligence. Front. Eng. Manag. **5**(2), 54–66 (2018)
5. Nibret, B., Yihune, M., Takele, B.: Human-wildlife conflict in Choke Mountains. Ethiopia. Int. J. Biodivers. Conserv. **9**(1), 231–324 (2017)
6. Andreas, G., Andrea, I.: Respecting human rights in conflict regions: how to avoid the u2018Conflict Spiralu 2019. Bus. Hum. Rights J. **2**(01), 109–133 (2017)
7. Brien, M.L., Gienger, C.M., Browne, C.A., et al.: Patterns of human–crocodile conflict in Queensland: a review of historical estuarine crocodile (Crocodylus porosus) management. Wildl. Res. **44**(4), 281–290 (2017)

8. Anand, S., Radhakrishna, S.: Investigating trends in human-wildlife conflict: is conflict escalation real or imagined. J. Asia-Pac. Biodivers. **10**(2), 154–161 (2017)
9. Vanderhaegen, F., Carsten, O.: Can dissonance engineering improve risk analysis of human–machine systems. Cogn. Technol. Work **19**(1), 1–12 (2017)
10. Grande, J.M., Zuluaga, S., Marchini, S.: Casualties of human-wildlife conflict. Science **360** (6395), 1309–1311 (2018)

Examination of Regional Energy Internet Information Management System Based on Source-Grid-Load-Storage

Chao Liu[1(✉)], Xianfu Zhou[2], Jiye Wang[1], Bin Li[1], Chang Liu[1],
Wen Li[1], and Hao Li[1]

[1] China Electric Power Research Institute, Beijing, China
darren770@126.com
[2] State Grid Zhejiang Electric Power Company Lishui Power Supply Company,
Lishui, China

Abstract. In order to promote the efficient use of energy, the rational consumption of energy and the optimal use of consumption, the regional energy Internet model is highly valued by the power and energy sector. Due to the inclusion of new components such as distributed power, distributed energy storage, and electric vehicles, the regional energy Internet presents new features of multi-source convergence and supply interaction, so its requirements for information management systems are high. This paper designs a regional energy Internet information management system based on source network storage, and discusses the regional energy Internet information management system from macro and micro perspectives.

Keywords: Regional energy · Energy interconnection · Information management system

1 Introduction

With the large number of new elements such as distributed power [1], distributed energy storage [2], and electric vehicle [3] connected to the power grid, the structure and pattern of the future power distribution system will undergo major changes. The main features are as follows: Multiple voltage levels constitute a multi-level ring network structure; AC-DC hybrid operation; the large power grid and the micro-grid complement each other and coordinate development; the physical distribution network and the information system are highly integrated; The end of the distribution network presents a trend of multi-source energy integration and supply-demand interaction for users [4–6]. Therefore, the process of energy production, transmission and consumption is increasingly dependent on internet-based information technology. Energy Internet can effectively promote the rational consumption of energy and the optimal use of consumption, thus promoting the efficient use of energy [7].

In order to accelerate the development of the energy Internet, the construction of regional energy interconnection is very important [8], and the information management system is the core part of the regional energy Internet [9]. The construction of regional

© Springer Nature Singapore Pte Ltd. 2020
M. Atiquzzaman et al. (Eds.): BDCPS 2019, AISC 1117, pp. 699–706, 2020.
https://doi.org/10.1007/978-981-15-2568-1_96

energy interconnection information management system is a key solution to realize smart energy use, which can promote the visualization, quantifiability and controllability of energy applications. It collects, correlates and utilizes information of various energy equipment and personnel through "information communication technology" to process and analyze the obtained information, so as to realize effective absorption and application of information [10].

This paper designs a regional energy interconnection information management system based on source-grid-load-storage. Firstly, the main structure of the information management system is analyzed from the perspective of overall architecture; then, the details of the architecture of the information management system are analyzed from a technical perspective; finally, the established regional energy interconnection information management system is displayed. The regional energy interconnection information management system described in this paper is based on the energy Internet model and supported by electric energy. It can integrate a variety of distributed energy sources, realize the interaction of "source, grid, load, and storage", establish a reasonable energy allocation and energy conservation strategy, reduce energy consumption, ensure the terminal energy safety, and achieve regional integrated energy management.

2 The Current Research Status of Domestic and Abroad

In order to promote the efficient use of energy, the rational consumption of energy and the optimal use of consumption, the regional energy Internet model is highly valued by the power and energy sector. In recent years, many researchers around the world have carried out various examination of energy Internet. From the perspective of the current situation, domestic and foreign scholars' research on energy Internet mainly focuses on the characteristics and nature of energy Internet and its architectural pattern.

2.1 Foreign Research Status State

In 2011, Jeremy Rifkin, an American scholar, summarized the four requirements that energy Internet should meet: (1) renewable energy dominates the primary energy; (2) provides large-scale access to distributed power generation systems and distributed energy storage systems, respectively. (3) share energy on a large scale with the support of Internet technology; (4) facilitate the promotion of electric vehicles in the transportation system. On the basis of the smart grid, the German Federal Ministry of Economics and Technology and the Ministry of Environment launched an informational energy plan to meet the needs of the distributed characteristics of the future power supply structure of the power system, and proposed the use of information and communication technologies to complete the comprehensive digital interconnection and computer monitoring and Control to upgrade existing energy supply networks.

In 2011, the concept of energy routers was proposed by American scholar Thilo Krause et al. Energy routers are abstract entities used to model physical entities (including power plants, microgrids, substations, various power terminals, etc.) in network systems that integrate multiple forms of energy (natural gas, electricity, electric vehicles, etc.) in the future. The basic principle is through the energy conversion equipment and energy storage equipment to complete the interaction between the various energy conversion and storage. In 2010, Japan implemented the "Smart Energy Community" program for research in the areas of smart grid and energy. In 2011, Japan borrowed from the idea of Internet interconnection protocols to propose a "digital grid" that transmits energy and information through the IP of different grid devices well.

2.2 Domestic Research Status

In June 2014, the State Grid Corporation's forward-looking project "Energy Internet Technology Architecture Research" was led by the China Electric Power Research Institute and carried out corresponding research work on the future energy Internet architecture. The Chinese Academy of Sciences has set up the "Strategic Research on China's New Generation Energy System" project, focusing on the blueprint for a new generation of energy systems. In February 2016, the Ministry of Industry and Information Technology, the National Development and Reform Commission, and the Energy Bureau officially released the National Energy Internet programmatic document on "Guiding Opinions on Promoting the Development of "Internet+" Smart Energy". The road map for the energy Internet was planned, and the basic principles, guiding ideology, core tasks and organizational implementation of promoting the construction of the energy Internet were pointed out. In March 2016, the national "13th Five-Year Plan" clearly stated: to coordinate infrastructure and energy construction such as communication and transportation; to promote the deep integration of new technologies in the fields of energy and information; to build an energy network with integrated complementary and coordinated development.

3 The Positioning of Regional Energy Interconnection Information Management System

The regional energy interconnection information management system is mainly for comprehensive energy-use areas such as parks, large public buildings, and industrial enterprises, and realizes flexible access to multiple energy sources such as electricity, water, gas, cold and heat. Through comprehensive integration of energy control parameters, energy operation, energy use and other data, the system realizes multi-energy coordination and comprehensive energy efficiency management, and builds an integrated multi-energy complementary regional energy Internet with multi-point access, network sharing, demand perception and ubiquitous iot, so as to improve energy efficiency and reduce energy cost.

4 The Overall Structure of the Regional Energy Interconnection Information Management System

The regional energy interconnection information management system studied in this paper is a way to realize the management of regional-level energy Internet information technology. By dividing the energy in the region into four dimensions: source, grid, load and storage and taking energy data business flow as a clue, the overall structure of regional energy interconnection information management system is proposed for energy monitoring, management, interconnection and interaction.

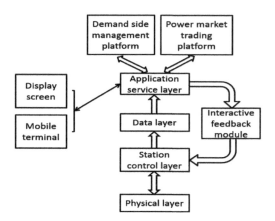

Fig. 1. Overall architecture of the regional energy interconnection information management system

As shown in Fig. 1, the overall architecture of the regional energy interconnection information management system mainly includes the physical layer, the station control layer, the data layer, the application service layer, and the interactive feedback module.

The physical layer associated with the station control layer transmits the real-time information of each terminal device to the station control layer, and the station control layer delivers the real-time policy information of the upper level to the physical layer terminal. The terminal can be classified into a source terminal, a grid terminal, a load terminal, and a storage terminal based on the source-grid-load-storage dimension.

The station control layer is connected to the data layer, and the real-time information of each terminal device is classified and transmitted to the data layer through the server. The station control layer integration includes source-grid-load-storage various station control systems.

The data layer is connected with the application service layer, providing the data information directly to the application service layer and storing it in the database in the application service layer.

The application service layer is first interconnected with the demand side management platform and the power market trading platform of the large power grid. The regional energy Internet can actively participate in demand response and power market through real-time interaction between application service layer and demand-side management platform and power market trading platform of large power grid. Secondly, it is connected with display devices such as display screens and mobile terminals. The user can not only observe the real-time running status of the regional energy Internet through the display screen and the mobile terminal, but also send adjustment instructions to the regional energy Internet through the display screen and the mobile terminal. Finally, it can be associated with the interactive feedback module, which can transmit real-time strategy information to the interactive feedback module, change the real-time operation mode of source, grid, load and storage, and improve the operational economy of the regional energy Internet.

The interactive feedback module is associated with the station control layer, which can classify the policy information into source, grid, class, and storage control strategies and transmit them to the station control layer. The interactive feedback module includes the source class interaction feedback, grid class interaction feedback, load class interaction feedback, and storage class interaction feedback. The interactive feedback module first receives real-time strategy information from the application service layer, and then classifies it into energy supply strategy information of source class, power flow strategy information of grid class, load strategy information of load class and energy storage strategy information of storage class. Finally, it is distributed to the station control layer.

5   The Technical Framework of Regional Energy Interconnection Information Management System

Based on the above overall structure, combined with the technical characteristics of the information system, and based on the "Internet+" energy data, the technical architecture of the regional energy interconnection information management system based on source-grid-load-storage is studied. The architecture is shown in Fig. 2.

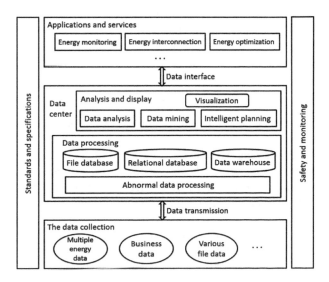

Fig. 2. Technical architecture of the regional energy interconnection information management system

The data acquisition layer collects various types of data based on the energy data transmission standard interface, and the data center layer receives the acquisition layer data and distributes it to the data processing module. The data processing module performs abnormal data cleaning, and at the same time, separates structured and unstructured data for energy data integration and storage to form an energy data center. The analysis and presentation module realizes the visual presentation of energy data based on the energy data center, and realizes data analysis, mining and intelligent planning with advanced Internet technologies such as large data and artificial intelligence. The analysis results are passed to the application and service layer through the data interface. In the overall technical framework, standards and specifications are used as the basis to ensure the integration and sharing of business and data between various links and the entire system. The safety and monitoring are used as the context to ensure the safety and controllability of each link.

The main tasks of each layer in the technical architecture and the key technologies involved are as follows.

5.1   The Data Collection

Data collection is the key to the stable operation of the system platform. In order to ensure the integrity and availability of the data and the driving force of the data, the data collection is required to have the intelligent processing ability from original ecological data to high-quality re-ecological data. Compared with traditional electric energy data, energy sources based on "Internet+" are more diversified, including multiple energy data, business data and various file data. This puts forward higher requirements on data collection ability, interpretation and classification ability and data quality. Therefore, it is necessary to standardize the data format in the process of data collection, and have the preliminary preprocessing ability to facilitate the storage, management and application of subsequent data.

The key technologies involved in the data collection process include entity identification and analysis methods for multi-source data, data cleaning and restoration methods based on edge calculation, accurate positioning of data sources and collection methods of high-quality raw ecological data, traceability management of data evolution, and data loading, stream computing, information transfer technology, etc.

5.2   The Data Center

The data center mainly includes data processing, analysis and presentation.

5.2.1   The Data Processing Module

The data processing module is mainly responsible for the integration and storage of energy data. The integration of energy data refers to the processing of data through high-reliability data integration technology while preserving the original semantics as much as possible to ensure data consistency and relevance. The data storage is the foundation of the data center. It is the place to store energy and information data. It mainly includes structured, semi-structured and unstructured historical data, statistical summary data and data to be shared. The storage technology of energy Internet data not only needs to reduce the cost of storing massive data, but also needs to adapt to diversified unstructured data management requirements and be scalable in data format.

5.2.2   The Analysis and Presentation Module

The analysis and presentation module makes in-depth value mining and diversified display of data in the data processing center. The key technologies involved are learning analysis, data mining and intelligent planning.

Learning analysis and intelligent planning is to use artificial intelligence means to understand and optimize energy environment and energy planning through measuring, collecting, analyzing and summarizing all kinds of data. Learning analysis, data mining and intelligent planning are closely related, and the applied analysis methods are similar. Currently, methods commonly used in the field of learning analysis include content analysis, network analysis, and discourse analysis.

Data mining is an important technical means used by energy producers, consumers, and managers by converting raw energy data of various energy sources into re-ecological data with specific values. Traditional data mining is mostly oriented to small data sets of structured and single objects. The mining focuses on pre-establishing models based on prior knowledge. For the analysis of unstructured, multi-source heterogeneous large data sets, if there is a lack of knowledge reserves, it is difficult to establish an explicit mathematical model. This requires the development of more diversified, holographic, high-dimensional data mining methods and techniques.

In addition to learning analysis, data mining and intelligent planning, it is the goal of data visualization technology to present complex analysis results to users in a more intuitive and easy-to-understand way.

5.3   The Applications and Services

Applications and services leverage the data from data centers to provide intelligent, professional service support to energy users. At present, regional energy Internet application services mainly focus on energy monitoring, energy efficiency management, energy interconnection, energy optimization, intelligent interaction and energy trading.

6   Conclusion

This paper proposes an information management system based on source, grid, load and storage in the regional energy Internet. The research analyzes the overall architecture and technical architecture of the system, and demonstrates the energy interconnection information management system of the China Electric Power Institute based on the content of this exploration. The regional energy interconnection information management system proposed in this paper has a complete and clear structure. In the future, based on this architecture, a reasonable energy allocation and energy conservation strategy can be established to reduce energy consumption, ensure terminal energy security, and achieve regional integrated energy management.

References

1. Shen, X., Cao, M.: Exploration the influence of distributed power grid for distribution network. Trans. Chin. Electrotechnical Soc. **30**(S1), 346–351 (2015)
2. Tao, Q., Sang, B., Ye, J., et al.: Optimal configuration method of distributed energy storage system in distribution network with high penetration of photovoltaic. High Voltage Eng. **42**(7), 2158–2165 (2016)
3. Wang, Y., Cai, C., Xue, H.: Optimized charging strategy of community electric vehicle charging station based on improved NSGA-II. Electr. Power Autom. Equipment **37**(12), 109–115 (2017)
4. Ren, S., Yang, X., Zhang, Y., et al.: A real time optimization strategy for microgrid integrated with schedulable ability and uncertainties. Proc. CSEE **37**(23), 6866–6877 (2017)
5. Ma, T., Cintuglu, M.H., Mohammed, O.A.: Control of hybrid AC/DC microgrid involving storage, renewable energy and pulsed loads. IEEE Trans. Ind. Appl. **53**(1), 567–575 (2017)
6. Ding, M., Chu, M., Pan, H., et al.: Operation optimization modeling and uncertainty analysis for hybrid AC/DC microgrids. Autom. Electr. Power Syst. **41**(5), 1–7 (2017). https://doi.org/10.7500/asps20160531012
7. Zhang, Y., Zhang, T., Meng, F., et al.: Model predictive control based distributed optimazation and scheduling approach for the energy internet. Proc. CSEE **37**(23), 6829–6845 (2017)
8. Hu, S., Liu, L., Zhang, T., et al.: Technology roadmap of energy Internet. Heat. Ventilating Air Conditioning **47**(3), 57–62, 50 (2017)
9. Liu, S., Han, X., Wang, J., et al.: Research on the impact of 'Internet plus' action on the power industry. Electr. Power Inf. Commun. Technol. **14**(4), 27–34 (2016)
10. Zhang, Y., Li, L., Chen, Z., et al.: Fusion between internet and energy system: morphology and technology. Eng. Sci. **20**(2), 79–85 (2018)

Research and Implementation of Contours Automatic Generalization Constrained by Terrain Line

Yan Lan[1,2]([✉]), Yong Cheng[3], and Xian Liu[2]

[1] Faculty of Geosciences and Environmental Engineering,
Southwest Jiaotong University, Chengdu 610031, Sichuan, China
llanyan@qq.com
[2] College of Science, Chengdu University of Technology, Chengdu 610059,
Sichuan, China
[3] Guiyang Engineering Corporation Limited,
China Electric Power Construction Group, Guiyang 550081, Guizhou, China

Abstract. Generalization is the key method of map-scale transformation. Because Contour curve is complex and the topography characteristics are expressed in groups, the result of its generalization directly decides the map quality. For Douglas Algorithms can easily lead to memory overflow when recursion depth is too deep, the two-stack Douglas algorithm is used to extract the feature points from the contours. The reference threshold of the Algorithms key parameter is test and 2 m is selected as the optimal threshold. The terrain lines are automatically recognized and tracked by limiting the angle, which is formed by the three nearest points on the same contour. The contour clusters are automatically generalized under constraint with the terrain lines. The results indicated that the original surface topography was better retained when the terrain feature lines are used to constrain the contour automatic generalization.

Keywords: Contour · Automatic generalization · Two-stack · Douglas algorithm · Terrain line

1 Introduction

Contour automatic generalization is an international problem recognized. In the process of reducing the scale of the map from the original image to the target scale, the same drawing size needs to accommodate more symbols, causing conflict and congestion of the graphic symbols. In order to make the map clear and identifiable, generalization is the most widely used method. The contour is the most important part of the map to express the landform, so research on the automatic generational of the contour has a strong theoretical and practical significance.

The contour lines exist in discrete forms, but the geomorphological features are expressed in groups. In the process of topographic map geomorphology, the topography needs to be taken into account. The single-line synthesis algorithm treats the contour as a single curve, regardless of or less consideration of the spatial relationship between the contours. This type of algorithm is more and more mature. Douglas-Peucker proposed

© Springer Nature Singapore Pte Ltd. 2020
M. Atiquzzaman et al. (Eds.): BDCPS 2019, AISC 1117, pp. 707–713, 2020.
https://doi.org/10.1007/978-981-15-2568-1_97

the famous DP algorithm based on feature point extraction [1], Wang proposed a comprehensive research on automated cartography based on fractal analysis [2]. Since the contour lines express the terrain in groups, the simplification of the contours should take into account the topography. Wu constrained the contour reduction by taking the bottom line [3]. He proposed a three-dimensional DP algorithm to synthesize unformatted discrete points and then re-create contours [4]. The method of automatically extracting the ground line from the contour line or DEM: Chang proposed the section identification and polygon splitting method for extracting the ground line from DEM [5]. Zhang proposed a method for generating terrain feature lines based on contour feature segments to construct a constrained Delaunay triangulation [6].

Based on the previous research results, this paper uses the improved two-stack DP algorithm to achieve ground line extraction and contour reduction to solve the problem of memory overflow caused by Douglas algorithm when the recursion depth is too deep. Then, the extracted geomorphic line is used as the constraint to realize the automatic synthesis of the contour clusters, and the quality of contour automatic generalization is analyzed [7, 8].

2   Two-Stack DP Algorithm Principle

2.1   Principle of DP Algorithm

The principle of the DP algorithm is illustrated by the contour lines shown in Fig. 1. The medium-high line of the figure consists of 22 points of P_1, P_2,..., P_{22}, given a threshold of ε. Connect the first and last points of the curve as the baseline P_1P_{22}, find the distance from the remaining points $P_2 \sim P_{21}$ on the curve to the baseline, find the distance maximum distance value d_{16} and the corresponding point P_{16}, compare d16 with the given threshold ε, if $d_{16} < \varepsilon$, connect P_1 and P_{22} two points, delete the rest of the curve, and the calculation ends; if $d_{16} > \varepsilon$, then keep P_{16}, P_{16} is the split point, divide the curve into two segments P_1P_{16} and $P_{16}P_{22}$, repeat the previous steps for the two curves. Until the distance from all nodes on the curve to the corresponding baseline is less than the reference threshold ε, the algorithm ends and all feature points {P_1, P_2, P_4, P_6, P_9, P_{10}, P_{11}, P_{14}, P_{16}, P_{19}, P_{22}} are recorded. The algorithm flow is shown in Fig. 2.

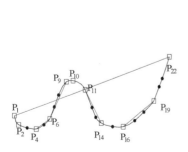

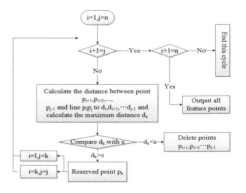

Fig. 1. Diagram of the DP algorithm   **Fig. 2.** DP algorithm flow

It can be seen from Fig. 2 that the DP algorithm continuously divides the curve into two segments for processing. It is a recursive call. Since the number of nodes on the contour line is often large, it is easy to cause memory overflow when the recursion depth is too deep. Based on the above situation, Wang proposed a two-stack algorithm based on DP algorithm using the stack characteristics [9].

2.2 Two-Stack DP Algorithm Principle

The two-stack DP algorithm constructs two stacks A and B on the basis of the DP algorithm, and puts two points on the stack at the beginning and the end of the curve, namely A.push(1), B.push(n). Read the top element of the two stacks i = A.peek(), j = B.peek(). At that time, connect the first and last two points with a straight line, find the maximum distance d_{max} and the corresponding point pk of the remaining points to the baseline. If $d_{max} > \varepsilon$, then A.push(p_k); otherwise, B.push(A.pop()). Reread the top elements of the two stacks and repeat the above until the stack is empty. This shows that all the feature points on the curve have been found.

Taking the curve shown in Fig. 1 as an example, using the two-stack DP algorithm to simplify the contour line, in the curve simplification process, the storage state changes in the A and B stacks are as shown in Fig. 3. The extracted feature points are $\{P_1, P_2, P_4, P_6, P_9, P_{10}, P_{11}, P_{14}, P_{16}, P_{19}, P_{22}\}$.

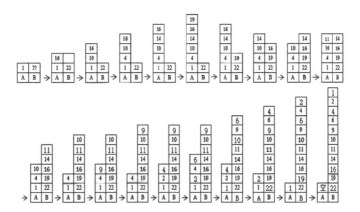

Fig. 3. Node storage state diagram in the B stack

Compared with the DP algorithm, the two-stack algorithm cleverly utilizes the nature of the stack, retains the intermediate feature points of the curve, avoids repeatedly traversing the intermediate feature points, and reduces the data storage and memory space in the calculation process while improve operational efficiency (Fig. 4).

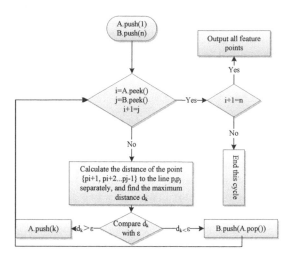

Fig. 4. Two-stack DP algorithm flow chart

3   Contour Automatic Realization

3.1   Contour Pretreatment

The contours need to be pre-processed before the automatic integration of the contours. Contour preprocessing mainly consists of connecting disconnected contour lines and converting contour attributes (converting polylines to 2D polylines) and deleting shorter contours.

3.2   Contour Line Thinning

The contour line is selected to perform the contour selection. The contour height of the integrated contour is 5 m, and the contour height of the original contour is 1 m. Dilution is to retain the curve (the elevation is 5 m integer multiple of the contour), delete other contours.

3.3   Reference Threshold Determination

In the DP algorithm, the size of the threshold is directly related to the effect of the contour integration. The threshold is too large, the number of nodes after synthesis is too small, the edge of the curve is too sharp, and the curve is self-intersecting. If the threshold is too small, the reserved node will be too many, the effect after the integration is not obvious. The results of the contour generalization under different thresholds are analyzed to determine the optimal reference threshold. The results are shown in Fig. 5, the smaller the threshold is, the smaller the contour change, and vice versa [10] (Table 1).

Table 1. Threshold calculation results at 5 m contour

6/10 of the distance	0.0687533348448699 (m)
7/10 of the distance	0.2055143057521 (m)
8/10 of the distance	0.696376664106024 (m)
9/10 of the distance	2.045684798056346 (m)
Maximum distance	85.6182153161175 (m)

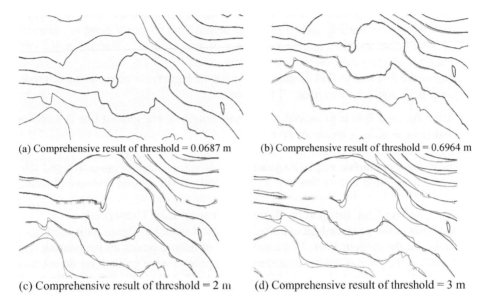

(a) Comprehensive result of threshold = 0.0687 m   (b) Comprehensive result of threshold = 0.6964 m

(c) Comprehensive result of threshold = 2 m   (d) Comprehensive result of threshold = 3 m

Fig. 5. Comprehensive effect under different thresholds

For the number of nodes after synthesis, the results are shown in Table 2. As can be seen from the table, the larger the threshold setting, the fewer the contour nodes after integration, which is the same as the result reflected in the graph in Fig. 5. It shows that when the threshold is 2 m, the newly generated contour is better. In the complex contours, the new contours retain the characteristics of the original curve (Fig. 5(c)).

Table 2. Number of nodes after different thresholds are synthesized

	The threshold is taken at a distance of 6/10	The threshold is taken at a distance of 9/10	The threshold is taken at a distance of 2 m	The threshold is taken at a distance of 3 m
Number of original contour nodes	14563			
Number of nodes after integration	5983	3107	1920	1556
The proportion	41.9%	21.3%	13.2%	10.7%

3.4   Automatic Integration of Contour Clusters

3.4.1   Automatic Extraction of Ground Lines

The authors studied the characteristics of the ground line and proposed the following ground line connection rules:

1. Find the point with the highest elevation from the feature point set as the reference point;
2. When looking for the next point, when the distance between the point to be judged and the reference point are different by an equal distance and the distance is within the tolerance (this article takes 5 times the height), the reference point is connected with the point, and the Point as a reference point, continue to find the point, until this finds all the points on the ground line;
3. Repeat the first step to find another point with the highest elevation as the reference point until you have traversed all the points.

After the local line is generated, the shorter length of the ground line is removed (the experiment takes a length of 10 m).

The threshold is 1 m, and the second stack DP algorithm extracts 4641 feature points in the topographic map. The ground line is automatically connected, and after removing the ground line of length < 10 m, a total of 348 ground lines are obtained.

3.4.2   Contour Line Integration Based on Ground Line Constraints

On the basis of the generated ground line, the second stack DP algorithm is used again to synthesize the contour lines. In the extraction of feature points on the contour line, the factors of the ground line are considered. Any points on the ground line do not need to participate in the threshold comparison and remain directly in the point set. Re-create the contours after all nodes have been selected. The result is shown in Fig. 6, the red curve is the integrated contour line, and the black curve is the ground line. It can be seen from Fig. 6 that the points on the ground line are preserved in the synthesis process, and the comprehensive effect is better.

Fig. 6. Synthesis of the results of the contours based on the ground line

4   Conclusion

In this paper, using AutoCAD platform, using C# to write the program, using the two-stack DP algorithm to extract the ground line from the contour line, and using the extracted ground line as a constraint to simplify the selected contour line, the results show that The second stack DP algorithm performs contour reduction, and the reduced contours have fewer feature points while maintaining the curve features. As the experimental data area is small (0.2 km^2), the landforms is not complex, and the result of the contour automatic generalization remains good. This method is subject to further verification for larger areas and complex landforms.

References

1. Douglas, D.H., Peucker, T.K.: Algorithms for the reduction of the number of points required to represent a digitized line or its caricature. Cartographica: Int. J. Geograph. Inf. Geovisualization **10**(2), 112–122 (1973)
2. Qiao, W., Wei, H.: A study on the automated cartographic generalization by using fractal analysis. Acta Geodaetica Cartogr. Sin. **24**(3), 211–216 (1995). (In Chinese)
3. Hehai, W.: The principle and method of automatic synthesis of landform morphology. J. Wuhan Univ.: Inf. Sci. Edn. **7**(1), 44–51 (1982). (in Chinese)
4. Jin, H., Lifan, F., Lina, H.: Study on the indirect synthesis method of contour line of three-dimensional Douglas-Peucker algorithm. Acta Geodaetica Cartogr. Sin. **42**(3), 467–473 (2013). (In Chinese)
5. Chang, Y.C., Song, G.S., Hsu, S.K.: Automatic extraction of ridge and valley axes using the profile recognition and polygon-breaking algorithm. Comput. Geosci. **24**(1), 83–93 (1998)
6. Yao, Z., Hong, F., Yue, L.: A method for extracting topographic feature lines based on contour lines. Acta Geodaetica Cartogr. Sin. **42**(4), 574–580 (2013). (in Chinese)
7. Guangyi, Y., Xiaopeng, Z., Liwei, W.: Topographic feature point extraction method based on combination of general vision and D-P algorithm. J. Surv. Mapp. Sci. **27**(1), 83–84 (2010). (in Chinese)
8. Ai, T.: The network extraction from contour Lines for contour line generalization. ISPRS J. Photogrammetry Rem. Sens. **62**(2), 93–103 (2007)
9. Wang, Y.: Research and Implementation of Automatic Synthesis Algorithm for Contour Lines. Kunming University of Science and Technology, Kunming (2008). (In Chinese)
10. Bin, J., Junjun, Y.: Ht-index for quantifying the fractal or scaling structure of geographic features. Ann. Assoc. Am. Geograph. **104**(3), 530–540 (2014)

EVA-Based Commercial Bank Value Evaluation Model – Take SPD Bank as an Example

Maoya Huang[✉]

School of Economics, Shanghai University, Shanghai 200072, China
1227147290@qq.com

Abstract. Commercial banks are a significant part of China's banking industry in solving financing problems for local enterprises. However, with the development of economic globalization and the rapid growth of Internet finance, inter-bank competition is increasingly fierce. How to seize market portion in the keen rivalry, achieve more efficient resources distribution, build a more reasonable achievements appraisal system, and optimize the management model of their daily affairs is closely followed by an increasing number of banks and social public. Based on the perspective of corporate financial evaluation, this paper start with analysis of the basic financial condition of Shanghai Pudong Development bank from the four dimensions of capital composition, asset quality, profitability and growth ability. Then starting from the maximization of the internal value of commercial banks, this paper takes the financial statements of SPD bank from 2014 to 2018 as the analysis object, evaluates the internal value of SPD bank by using the EVA achievements appraisal system and compares the internal value with the trading market price. It finds that the market price of SPD bank was underrated to a great extent by the end of 2018.

Keywords: SPD bank · EVA performance evaluation · Value evaluation

1 Introduction

According to the statistics of China Banking Insurance Regulatory Commission, in the first half of 2019, commercial banks achieved a cumulative net profit of 1.13 trillion yuan, with annual growth rate of 6.5%. It can be seen that the overall operation of commercial banks is still relatively stable. Commercial banks are mainly positioned to provide financial services for the development of local industries, so their ability to create value cannot be ignored. In 1982, Stern Steward consulting company redesigned the concept of "residual income" and named it as Economic Value Added in order to better evaluate the worth realization ability of enterprises [1]. In 1990, Stern Steward consulting company published this new performance evaluation index for the first time, and it has been widely recognized and applied worldwide [2]. Economic Value Added (EVA) is different from accounting profit, when measuring the economic profit of an enterprise, it adjusts the accounting profit from the perspective of the owner and comprehensively considers all the capital invested, namely the cost of equity and bond

© Springer Nature Singapore Pte Ltd. 2020
M. Atiquzzaman et al. (Eds.): BDCPS 2019, AISC 1117, pp. 714–722, 2020.
https://doi.org/10.1007/978-981-15-2568-1_98

capital [3]. The adjusted profit should cover the cost of invested capital, that is, if EVA is greater than zero, the business can realize worth for the owner [4].

Financial scholars' research on EVA first originated in foreign countries at the beginning of the 20th century. By the end of the 20th century, EVA evaluation model was accepted by more and more enterprises. After years of development at home and abroad, a relatively mature system has been formed. Most scholars' researches are based on the empirical analysis, performance appraisal and application prospect of EVA evaluation model. For example, Silvia and other scholars applied EVA index to the operating performance evaluation of PT company and found that EVA based performance evaluation was more helpful for enterprises to develop and realize their strategic goals [5]. Eleftherios Aggelopoulos evaluated the value of commercial banks and so on [6].

EVA evaluation model is relatively applicable to the value evaluation of commercial banks [7]. Therefore, this paper takes the financial report information of SPD bank from 2014 to 2018 as this paper's analysis object, and conducts empirical analysis on its intrinsic value based on EVA value evaluation model. It is of practical significance to explore the gap between the evaluation results and the actual situation and the reasons for its formation.

2 EVA Value Evaluation Model

Financial scholars' research on EVA first originated in foreign countries at the beginning of the 20th century [8]. By the end of the 20th century, EVA evaluation model had been accepted by more and more enterprises [9]. After years of development at home and abroad, a relatively mature system has been formed so far.

2.1 General EVA Measurement Model

In the original EVA model proposed by Stern Stewart, there are more than 200 subjects involved in adjustment, but most enterprises only need to select a part for adjustment [10]. According to the provisions on EVA in the "Interim Measures for the Assessment of the Business Performance of Responsible Persons of Central Enterprises", which came into effect on January 1, 2010.

$$EVA = NOPAT - CAP \times WACC \tag{1}$$

NOPAT represents adjusted net operating return after duty, CAP represents adjusted capital and WACC represents weighted average cost of capital. Specific settlement rules are as follows:

NOPAT = net profit + (interest cost + R&D cost adjustment item - non-recurring income adjustment item $\times$ 50%) $\times$ (1 − income tax rate).

CAP = mean of owners' equity + mean of liabilities − mean of interest − free present liabilities − mean of construction in progress.

WACC = ratio of equity capital $\times$ cost ratio of equity capital + ratio of creditor's capital $\times$ cost ratio of creditor's capital $\times$ (1 − income tax rate).

2.2   EVA Measurement Model for Commercial Banks

Commercial banks is different from the industrial enterprises, their main business is associated with all kinds of financial products, and their main income comes from the interest rate difference between deposit and loan. [11] So for the commercial banks, to absorb and make loans do not belong to the investment and financing activities, but belong to the category of main business. In view of the particularity of commercial banks and the adjustment principles of importance and feasibility, EVA measurement models of commercial banks should be different from general EVA models. This paper refers to the basic measurement model EVA = NOPAT − CAP × WACC, based on the current accounting standards of China's commercial banks, and drawing on that past research success and experience of civil and abroad research workers, to adjust the accounting variables in the model, as follows:

① Adjustment of After-tax Net Operating Profit (NOPAT).

For commercial banks, their main operating assets are money, rarely involving inventory, fixed assets and other related items. At the same time, they rarely carry out large-scale research and development, and usually only have a small amount of research and development expenses. Therefore, when calculating the adjusted after-tax operating profit of a bank, it will be different from that of general industrial enterprises. The main calculation method is to adjust based on the account book after-tax net profit according to changes in various impairment provisions, changes in deferred income tax and net non-operating income after duty of the present period, as shown below:

Adjusted after-tax net operating profit (NOPAT) = book after-tax net profit + change in loan impairment provision this year + change in other asset impairment provision this year + present postponed income tax debt − current delayed income tax asset − current non-operating return and expenditure net × (1 − income tax rate)

② Adjustment of Total Capital (CAP).

According to the status of capital investors, capital can be divided into equity capital and debt capital two categories. For commercial banks, the liability formed deposits and made loans to form assets belong to the category of business activity, but not belong to the category of investment and financing activities, so take deposits formed by debt cannot be included in the calculation, the amount of the capital in after deducting operating deposits, the proportion of debt capital will be relatively small, so when calculating the total capital does simplify the processing, only consider the equity capital, without regard to debt capital. On the basis of equity capital, calculation of total capital (CAP) is adjusted according to the balance of impairment provision for various assets, the equilibrium of delayed income tax and after-duty net non-operating income, as shown below:

Adjusted capital (CAP) = shareholders' equity + balance of loan impairment provision + balance of defect provision of other assets − net non-operating income after duty + balance of put-off return tax liabilities at the end of the year − balance of postponed income tax assets this year.

③ Adjustment of Weighted Average Cost of Capital Ratio (WACC).

Due to the relatively small portion of debt capital in the formation of capital sources of commercial banks, debt capital is excluded when counting the cost, that is, WACC is represented by the expense of equity capital. Based on the study methods of previous literatures, this paper adopts the mainstream calculation method CAPM model to determine the equity capital cost ratio of commercial banks.

$$\text{CAPM}: \quad R = Rf + \beta \times (Rm - Rf) \tag{2}$$

That is: WACC = cost of proprietor's capital = $Rf + \beta \times (Rm - Rf)$

The risk-free rate Rf by commercial banks when residents savings and small one-year nominal interest rate as lump sum, beta using daily yields on commercial bank stock market index daily yields unary linear regression, obtained a coefficient is the beta value, using the GDP growing ratio as the market average risk premium rate $(Rm - Rf)$.

Finally, the Economic Value Added (EVA) of commercial Banks in each year can be obtained by combining the above adjusted factors and using the calculation formula of EVA: EVA= NOPAT − CAP × WACC.

3 EVA Based Commercial Bank Value Evaluation Model

3.1 Assumptions of EVA Model

When using EVA to evaluate the overall value of commercial banks, the following three assumptions are made:

The first assumption is that the external macro environment of the enterprise remains stable, such as the level of inflation, the market economy environment and the industry competition environment. Only on this basis, the estimate of the cost of capital and increasing rate of the enterprise in future based on the current conditions is more likely to be true and reliable.

The second assumption is that the ratio of equity capital and debt capital of commercial banks will not change greatly in the future. Since equity capital is used to replace the total capital when solving EVA of commercial banks, under this assumption, the enterprise will not conduct new equity financing, and the debts that have matured in the past will be replaced by equal amount of new debts. The growth of the enterprise mainly depends on the retained earnings of capital.

The third assumption is that commercial banks will continue to operate in the future and the increase or decrease of EVA conforms to certain rules. EVA forecasting method is used to evaluate the value of commercial banks.

3.2 EVA Model Evaluation Method

EVA valuation model can be divided into single stage and two stage. Among them, the single-stage EVA value evaluation model is applicable to enterprises with steadily improved expected business results. The formula is as follows:

$$V_0 = I_0 + \frac{EVA_n}{WACC - g} \tag{3}$$

Where, EVAn represents the economic added value in year t; g stands for sustainable growth rate; I0 represents the total capital at the beginning; V0 represents the current value of the enterprise.

In this paper, the two-stage EVA value evaluation model is mainly adopted to divide the growth of enterprises into two stages. The first stage maintains relatively high growth and the second stage maintains relatively low stable growth. Therefore, the discounted present value of EVA is predicted as follows:

$$V_0 = I_0 + \sum_{k=0}^{n} \frac{EVA_n}{(1 + WACC)^n} + \frac{EVA_{n+1}}{(WACC - g)(1 + WACC)^n} \tag{4}$$

4   Steps to Calculate EVA Value of SPD Bank

Step 1: based on relevant items in SPD bank's income statement, obtain the adjusted after tax net operating profit (NOPAT).

Step 2: obtain the adjusted capital amount (CAP) based on the relevant items of SPD bank's balance sheet.

Step 3: compute the average-weighted cost of capital ratio (WACC) of SPD bank by replacing the cost of owner's capital ratio.

Step 4: on the foundations of the first three steps, the EVA calculation formula is used to calculate the EVA value of SPD bank.

Step 5: evaluate the enterprise value of SPD bank based on the single-stage EVA value evaluation model.

4.1   Step 1

After tax net operating profit (NOPAT) does not consider the capital composition of the enterprise, but only reflects all the after-tax investment revenue produced by the company, which can reflect the real level of profitability of the enterprise. Specific adjustment methods are as follows:

Adjusted NOPAT = book after-tax net return + change of loan impairment provision this year + change of impairment provision of other assets this year + current delayed income tax liability − current postponed income tax asset − current non-operating revenue and expenditure × (1 − income tax rate).

NOPAT value of SPD bank from 2014 to 2018 is exhibited in the following (Table 1):

Table 1. NOPAT value of SPD Bank from 2014 to 2018

Year	2014	2015	2016	2017	2018
NOPAT(RMB million yuan)	59211.75	67955.5	62083.75	51832.25	63938.25

4.2   Step 2

Due to the small proportion of debt capital in the commercial banking system and the deduction of debt interest expense in the calculation of NOPAT, the total capital calculated in this paper is mainly adjusted on the basis of shareholders' equity. Specific adjustments are as follows:

Adjusted capital (CAP) = shareholders' equity + balance of loan impairment provision this year + balance of impairment provision of other assets this year − net income not belonging to business's operating activities + balance of postponed income tax liabilities at the end of the year − balance of delayed income tax assets at the end of the year.

CAP value of SPD bank from 2014 to 2018 are exhibited as follows (Table 2):

Table 2. CAP worth of SPD Bank from 2014 to 2018

Year	2014	2015	2016	2017	2018
CAP(RMB million yuan)	309,805	382,288	445,635	500,752	555,235

4.3   Step 3

Corresponding to the previous adjustment method of total capital, when seeking the WACC of Shanghai Pudong Development bank, only the expenses of shareholder's capital is considered, and the expenses of creditor's capital is not considered. Therefore, the weight-based cost of capital ratio of SPD bank could be replaced by the cost of equity capital ratio. Capital Asset Pricing Model is applied to determine the cost ratio of owner's capital. That is: WACC = cost of shareholder's capital = Rf + beta × market risk premium.

Rf is refers to the rate of return that has no risk, this paper uses Shanghai Pudong Development bank the residents savings and small one-year nominal interest rate as lump sum as the risk-free interest rate, based on the Shanghai composite index volatility, regression analysis was carried out on the commercial bank stock daily yields, seeks for the beta, and USES the average GDP growth between 2014 and 2018 as the market risk premium rate (Rm − Rf).

WACC values of SPD bank from 2014 to 2018 are shown in Table 3:

Table 3. WACC values of SPD Bank from 2014 to 2018

Year	2014	2015	2016	2017	2018
Rate of return without risk	3.30%	3.00%	2%	2%	2%
Beta coefficient	1.319	0.739	0.457	0.753	0.722
Market Risk premium	7.30%	6.90%	6.70%	6.90%	6.60%
WACC	12.93%	8.10%	5.06%	7.20%	6.77%

4.4   Step 4

Finally, based on the above adjusted factors and the calculation formula of economic added value: EVA = NOPAT − CAP × WACC, the EVA of commercial banks in each year can be obtained. EVA values of the SPD bank from 2014 to 2018 are shown in Table 4:

Table 4. EVA value of SPD Bank from 2014 to 2018

Year	2014	2015	2016	2017	2018
EVA(RMB million yuan)	19,154.00	36,993.65	39,526.16	15,799.62	26,375.47

The calculation shows that the economic added worth of the SPD bank in the past five years is positive, which indicates that the operating achievements of the SPD bank is relatively good, the overall EVA value increases first and then decreases. Among them, the EVA value of 2016 is the largest and the most profitable.

4.5   Step 5

This article mainly uses the EVA value evaluation model of two phase, the growth of the SPD bank is spilt into two phases, the first stage to maintain relatively high growth, steady growth of the second stage remain relatively low, predict EVA discounted current worth = present value + current worth of stable growing phase + current worth of high growing stage, formula is as follows:

$$V_0 = I_0 + \sum_{k=0}^{n} \frac{EVA_n}{(1+WACC)^n} + \frac{EVA_{n+1}}{(WACC - g)(1+WACC)^n} \tag{4}$$

Where, EVAt represents the economic added value in year t; g stands for sustainable growth rate; I0 represents the total capital at the beginning; V_0 represents the current value of the enterprise. It is assumed that the increasing rate will achieve the average level of 4% in the next five years. With reference to the GDP growth of developed countries, it is assumed that the increasing rate will stabilize at around 1.5% after the next five years (Table 5).

Table 5. Forecast value of EVA in the next 5 years

Year	2019	2020	2021	2022	2023
EVA (RMB million yuan)	27,430.5	28,527.7	29,668.8	30,855.6	32,089.8
The discount factor	0.9258	0.8572	0.7936	0.7348	0.6803
PV (RMB million yuan)	25,396.3	24,453.4	23,545.5	22,671.4	21,829.7

Therefore, V0 = 26375.47 + 117896.21 + 735481.7 = 879753.38 (million)

According to the data, by the end of 2018, the total issued common shares of SPD bank were 29352080397, and the calculated price per share was 29.97 yuan per share. By December 30, 2018, the latest closing price of SPD bank was 9.80 yuan per share.

5 Conclusions

By using EVA valuation method to estimate the worth of SPD bank, it can be found that the model assessment value of SPD bank share is 29.97 yuan per share, nearly twice higher than its market value of 9.80 yuan per share. Assuming that the empirical results are true and valid, it can be seen that the overall market value of SPD bank is seriously underestimated by the end of 2018.

This paper believes that the reasons for SPD bank's undervaluation in the stock market may include the following aspects: first, the overall downturn of the stock market at the end of 2018, the trading price of most stocks is lower than the intrinsic worth; Second, the industry's overall PE and PB of listed commercial banks are generally low, which leads to certain deviation. Third, there is a serious phenomenon of trading speculation in China's stock market. As a large listed company, SPD bank has relatively open and transparent information and lacks speculation potential. Its stock price fluctuates little in a certain period. Fourthly, the assumptions of the EVA valuation model are relatively strict, that is, the capital structure of commercial banks should remain unchanged. However, in reality, the equity capital and creditor's capital of commercial banks will be dynamically adjusted with the change of their operating conditions.

References

1. Menicucci, E., Paolucci, G.: Factors affecting bank profitability in Europe: an empirical investigation. Afr. J. Bus. Manag. **10**(17), 410–420 (2016)
2. Mendoza, R., Rivera, J.P.R.: The effect of credit risk and capital adequacy on the profitability of rural banks in the Philippines. Ann. Alexandru Loan Cuza Univ.-Econ. **64**(1), 83–96 (2017)
3. Kawshala, H., Panditharathna, K.: The factors effecting on bank profitability. Int. J. Sci. Res. Publ. **7**(2), 212–216 (2017)
4. E-Kassem, R.C.: Determinants of banks' profitability: panel data from Qatar. Open J. Acc. **04**, 103–111 (2017)

5. Silvia, D., Yulianeu, Y., Gagah, E., et al.: Financial performance analysis using economics value added (eva) and market value added (MVA) method in go public telecommunication company. J. Manag. **3**(3), 36–38 (2017)
6. Aggelopoulos, E., Georgopoulos, A.: Bank branch efficiency under environmental change: a bootstrap DEA on monthly profit and loss accounting statements of greek retail branches. Eur. J. Oper. Res. **261**(3), 1170–1188 (2017)
7. Tseng, S.T., Levy, P.E.: A multilevel leadership process framework of performance management. Hum. Resour. Manag. Rev. **10**, 669–726 (2018)
8. Thekdi, S., Aven, T.: An enhanced data-analytic framework for integrating risk management and performance management. Reliab. Eng. Syst. Saf. **12**, 277–287 (2016)
9. Bahovec, V., Barbic, D., Palic, I.: The regression analysis of individual financial performance: evidence from croatia. Bus. Syst. Res. J. **8**(2), 1–13 (2017)
10. Altaf, N.: Economic value added or earning: what explains market value in indian firm? Ecol. Econ. **2**(2), 152–166 (2016)
11. Vig, S., Dumicic, K., Klopotan, I.: The impact of reputation on corporate financial performance: median regression approach. Bus. Syst. Res. J. **23**(4), 40–58 (2017)

Financing Predicament of Small and Medium-Sized Foreign Trade Enterprises in China–Based on Big Data Analysis

Xuewei Li[✉]

School of Economics, Shanghai University, Shanghai 200072, China
327702099@qq.com

Abstract. With the steady expansion of the world economy, China's import and export trade is booming day by day, and foreign trade has become a major factor to promote our country's economic growth. As one of the economic subjects of external trade, small and medium-sized enterprises in foreign trade play an irreplaceable role. However, many small and medium-sized foreign trade enterprises encounter financing difficulties in the process of the development. On the foundation of big data analysis, with the starting point of the financing of small and medium-sized enterprises in foreign trade, firstly, this paper describes the present financing condition of the small and medium-sized enterprises in foreign trade as a whole, and then the reasons are analyzed. The reasons are divided as enterprises themselves, indirect financing, direct financing, private financing, and the government one by one.

Keywords: Big data analysis · Small and medium-sized companies in foreign trade · Financing predicament

1 Introduction

China's cumulative foreign trade import and export amount has 4.62 trillion dollars in 2018, 12.6% larger than the same period last year, and the import volume broke through the US $2 trillion mark for the first time, 15.8% larger than the same period last year; the export volume directly forced 2.5 trillion dollars, 9.9% larger than the same period last year. Among them, small and medium-sized companies in foreign trade are the critical part in promoting our country's economic growth. According to statistical data, the total value of our country's import and export trade reached 24.6 trillion yuan in 2015, and private foreign trade enterprises grew strongly, contributing 1.0295 trillion US dollars, of which about 90% came from SM Enterprises (small and medium-sized enterprises in foreign trade).

Currently, many domestic and foreign scholars study the financing predicament of SM Enterprises in foreign trade and bring forth their own opinions and suggestions, but the research data are relatively old, in the setting of big data era, this research is not comprehensive enough.

This paper uses many years of data for comparative analysis, based on big data analysis of the financing difficulties of SM Enterprises in foreign trade.

© Springer Nature Singapore Pte Ltd. 2020
M. Atiquzzaman et al. (Eds.): BDCPS 2019, AISC 1117, pp. 723–730, 2020.
https://doi.org/10.1007/978-981-15-2568-1_99

Table 1. Import and export trade data for 2014–2018

Unit: Trillions of dollars	Total export and import volume	Total export	Total import	Difference
2014	4.30	2.34	1.96	0.38
2015	3.95	2.27	1.68	0.59
2016	3.68	2.10	1.59	0.51
2017	4.11	2.26	1.84	0.42
2018	4.62	2.48	2.14	0.35

From the Table 1, we are informed that the total amount of imports and exports generally fluctuates and increases, and the total imports and exports also show a fluctuating and rising trend. SM enterprises in foreign trade have a lot of opportunities and space for development.

According to the Statistical Yearbook of China (2018), the contribution to net exports of commodities and services to GDP in 2015 was −1.3%, and the pulling effect to economic growth was weakened to −0.1%. In 2017, the contribution to net exports of commodities and services to GDP was 9.1%, stimulating economic growth by 0.6%. It can be seen that the contribution to net exports of goods and services has been significantly increased. From the weakening of the pulling effect on economic incre-ment to the strengthening of positive economic development, the pulling influence of net exports on China's GDP is becoming more and more significant. However, SM enterprises in foreign trade are realizing the positive direction of export to China's economic increment. Pull plays a vital role. However, with the pace of development of those enterprises, the demand for funds and loans are increasing, and small and medium-sized enterprises in foreign trade have encountered many difficulties in the financing process. The financing dilemma has become one of the biggest constraints to the increment of SM enterprises in foreign trade.

2 The Current Reality of Financing of SM Enterprises in Foreign Trade

2.1 Single Financing Channel

Endogenous financing is the main financing method for those enterprises in foreign trade in the period of establishment and growth, but when enterprises want to expand their scale, relying on self-accumulation and relational lending is far from meeting the capital needs of those enterprises, and external financing is needed [1, 2].

The external channels of financing of SM enterprises in foreign trade are direct financing, indirect financing and private financing. In the aspect of direct financing, because the business scope of those foreign trade enterprises is relatively narrow and the development of enterprises is relatively slow, usually, large companies with capital surplus will not easily finance or borrow directly to those enterprises, so issuing stocks or bonds and other securities has become the main way for SM companies in foreign trade to gain non-banking financing [2]. In the existing stock trade market, it provides a special section for these sized enterprises to facilitate their direct financing.

Regarding the indirect financing, bank credit is the primary form to realize. However, it is still difficult for those foreign trade enterprises with low holdings of fixed assets and current assets to pass the examination of banks.

When it comes to the private financing, owing to the absence of intermediaries, the risk of this lending behavior is increased remarkably, borrowers may fail to recover debts, financiers would encounter the excessive interest rates, and for small-sized companies and medium-sized companies in foreign trade, owing to the absence of the support of corresponding laws and regulations, there exist moral hazards, therefore it is very difficult to finance via the private sector [3–5].

As shown in Table 2 below, firstly, the internal financing of SM enterprises in foreign trade in foreign trade accounts for more than 80% of the financing channels, while the proportion of exogenous financing, such as bank lending, is not more than 10%, and the loans part from non-financial companies is even smaller.

Table 2. Financing structure of SM enterprises in Foreign Trade in China from 2015 to 2017

	Internal financing	Bank advance	Loans from non-financial companies	Other channels
Less than 3 years	87.6%	8.7%	2.5%	1.2%
3–5 years	85.0%	9.0%	4.7%	1.3%
6–10 years	84.5%	9.6%	2.6%	3.3%
More than 10 years	82.8%	9.1%	6.4%	1.7%

2.2 Low Financing Satisfaction

Financing satisfaction refers to whether the amount of funds obtained by SM companies in foreign trade through various financing channels can satisfy the needs of enterprises' funds and ensure their need to expand reproduction. Take Heilongjiang Province as an example, according to the relevant statistics, 1/3 of the SM enterprises in foreign trade in Heilongjiang Province have not received financing or the financing capital satisfaction rate is 30% or less, only nearly 17% of those foreign trade enterprises have met their capital needs, and more than 50% of the enterprises have realized financing, but the financing amount has not been fully met [4]. Therefore, whether through direct or indirect financing, their demand for funds is not available to most small-sized enterprises and medium-sized enterprises in foreign trade.

2.3 High Financing Cost

Because of the large demand for short-term liquidity and the small volume of small-sized enterprises and medium-sized enterprises in foreign trade, the expense of external financing is very high for those enterprises. First, when these enterprises gain financing from banks and other financing institutions, the loan interest rate of these enterprises is generally higher than that of other subjects, and the one-year loan interest rate of these foreign trade enterprises is often higher than the basic interest rate. At the same time, because of the characteristics of small-sized and medium-sized enterprises in foreign trade, such as these enterprises have their own capital, they often have to pay uncertain and unpredictable public relations expenses to acquire loans from financial institutions [6]. Second, when financing is carried out through private lending, the interest rate is often higher than the bank interest rate. For these foreign trade enterprises, there is no doubt that it has increased the cost of financing. The table below shows that the private interest rate is often four times or more higher than the loan interest rate of financial institutions, so for small-sized and medium-sized enterprises in foreign trade, the financing cost of private borrowing is very huge, and this problem also aggravates the financing predicament of these foreign trade enterprises [6, 7] (Table 3).

Table 3. Statement of private interest rates and one-year loan interest rates of financial institutions in Sichuan Province from 2014 to 2018.

Year	2014	2015	2016	2017	2018
Private interest rate	19.6%	19.9%	20.6%	21.52%	22.1%
One-year lending interest rate	5.6%	5.1%	4.35% (Floating)	4.35% (Floating)	4.35% (Floating)

3 The Reasons for the Financing Predicament of SM Enterprises in Foreign Trade

3.1 Unstable Operating Conditions and Low Capacity

According to statistical data, the average life expectancy of these enterprises in China is only 2.9 years, but they are being in a fiercely competitive foreign trade market, owing to the small and medium-sized enterprises in foreign trade have less funds accumulation and poor ability to resist external factors, the survival time in the market is even more short, SM enterprises in foreign trade close down and go bankrupt. Their unstable operating conditions prevent it from providing the relevant "threshold" data needed for financing and unable to meet the conditions required for financing [6]. The life span of those enterprises in foreign trade that do not have access to funds is therefore even shorter, forming a vicious circle.

Simultaneously, there are many small-sized and medium-sized enterprises in foreign trade are newly registered in recent years, registered funds and company own capital are very limited. Many of these companies have small investment scale, low science and technology content, and because of their low entry threshold and relatively fierce market competition, it is difficult for each enterprise to form its own core competitiveness. In this case, those enterprises in foreign trade have poor ability to withstand risks and are more severely affected by the impact of economic fluctuations.

3.2 Financing Innovation Is Low

The main types of international trade finance of small-sized and medium-sized enterprises in foreign trade are goods right or movable property pledge credit business; export tax refund escrow loan; export credit insurance financing; export factoring business; Forfeiting business; supply chain financing. However, because of the limited trade mode supported by domestic banks, the international trade finance chosen by those enterprises referred above are always based on several relatively mature traditional trade financing methods, and the traditional ways can not meet the growing demand for funds of small-sized and medium-sized enterprises in foreign trade.

Table 4. International trade financing statistics of 2016 of China Merchants Bank

Content	Volume (Million)	Amount of money (Million)	Compared with the same period last year	Proportion international amount of trade financing
Inward documentary bills	319.47	12590.40	26.60%	26.74%
Packing loan	89.27	5805.52	−6.25%	12.33%
Export credit insurance	65.03	4430.65	−11.49%	9.41%
Overseas payment	350.23	15909.86	8.99%	33.79%
Bill purchased	63.75	5042.75	2.65%	10.71%
Forfeiting	3.67	480.26	New	1.02%
Others	49.20	2825.07	26.60%	6%
Total	940.62	47084.51	12.31%	100.00%

From the Table 4, we can see that, in international trade financing, the most important varieties are import deposit and overseas payment, and the proportion of financing of other varieties is small, and the loan amount of package loan and export credit is lower than that of in the same period over last year. It can be seen that there are few trade varieties that enterprises in foreign trade can choose according to their own situation when financing international trade through commercial banks.

3.3 Banks Set up Higher Financing Threshold Based on Financing Security Requirements

According to the statistics of the 2017 annual report of Bank of China Co., Ltd., at the end of 2017, the total amount of NPL of the Group increased by 12.466 billion yuan compared with the same period to 158.469 billion yuan, and the NPL rate was 1.45%, decrease 0.01% points from the end of last year. In order to control and reduce their non-performing loan rate at a relatively stable level, and to ensure that the borrower can meet the security of debt service, the bank will be very cautious in the audit of those foreign enterprises, the threshold will be very high, to prevent the enterprise fail to pay back the principal and interest. Industry financing meets the difficulty of financing failure due to the high threshold of banks. According to the relevant investigation, Chinese commercial banks, considering the safety principle of their own operation, will concentrate 80% of the new loan line in the credit rating 3A or 2A enterprises, however, more than 60% of the small-sized and medium-sized enterprises in foreign trade have a credit rating of 3B or less [7, 8].

3.4   High Listing Cost of Direct Finance

When an enterprise is listed in China's equity market, it needs a lot of listing cost, and the cost is divided into explicit cost and hidden cost, which is shown below (Table 5).

Table 5. The main explicit cost and implicit cost of enterprise listing

	General capital (Million)	Less than 200 million	200 million to 400 million	400 million to 600 million	600 million to 800 million	More than 800 million
Motherboard fee	Initial one	30	45	55	60	65
	Annual one	5	8	10	12	15
Small and medium-sized enterprise board fee	Initial one	15	20	25	30	35
	Annual one	5	8	10	12	15
Start up board fee	Initial one	7.5	10	12.5	15	17.5
	Annual listing fee	2.5	4	5	6	7.5

3.5   High Risk of Private Financing

Most of the financing of these foreign trade enterprises have characteristics of less single capital demand and high frequency. Neither direct financing nor indirect financing can satisfy the needs of enterprises to obtain funds quickly. However, the procedures of private financing and borrowing are simple and fast, which can make those foreign trade enterprises obtain the required funds in a very short period of time, so private financing is very popular with those in need of funds [8, 9]. Private financing can help enterprises to solve the problem of an urgent need of funds, promote social investment and stimulate economy's development, but due to the lack of formal supervision and regulation of private financing, there are often some inevitable problems. Business procedures are not complete, there is illegal collection Problems such as the chaos of capital, the serious loss of tax and profit, and so on, may arise [9].

3.6   The Government Support Policy Implementation Is Not in Place

As small-sized and medium-sized foreign trade companies are a momentous part in promoting the development of China's GDP, the Chinese government has also issued relevant preferential and supporting policies for those sized enterprises in foreign trade, such as the implementation of export tax rebates in Bayannur City in 2017 to benefit these foreign trade enterprises; Hunan Province in 2018 to publish a list of export facilitation preferential financing for foreign trade enterprises; and Jiangsu Province in 2019 to promote "Soviet loans" to support small sized and micro-foreign trade enterprises. However, these preferential and supporting policies are not many, and the introduction and updating of policies can not keep up with the continuous development of these foreign trade enterprises to meet their demand for financing, there are still many small-sized and medium-sized enterprises in foreign trade can not rely on these policies. To obtain financing, at the same time, in the process of implementing these preferences and policies at the provincial, municipal and county levels, it is necessary to have supporting detailed policies and personnel to carry out relevant operations [8–10]. However, there is a lag or lack of awareness for the policies and personnel of the relevant units at the provincial, municipal, and county levels, which directly leads to the fact that even if there are preferential policies and support policies, it is hard for those companies in foreign trade to know or obtain practical benefits. All kinds of support and preferential policies cannot be implemented.

3.7   The Government Policy Guarantee Is Insufficient

Because a policy guarantee has the function of transferring financing risk and reducing financing cost, government policy guarantee is one of the critical ways to solve the financing demand of companies. At present, our government has paid attention to the important role of policy guarantee in solving the financing predicament of small and medium-sized enterprises in foreign trade.

Take Beijing as an example, according to the Beijing Credit guarantee Industry Association statistics, the new amount of financing guarantee business of 17 state-owned guarantee institutions in Beijing is 28.035 billion yuan in the first half of 2017, but at present, the policy guarantee of our country is far from enough to meet the increasing financing demand of small-sized and medium-sized enterprises in foreign trade, and the capital satisfaction of those foreign trade enterprises is relatively low [10].

4   Conclusion

Small-sized and Medium-Sized Companies of Foreign Trade in China face many difficulties in contemporary financing activities, which hinders the development of those enterprises. Therefore, it is necessary to realize the causes of financing predicament for small-sized and medium-sized corporations in foreign trade. From this big data analysis, the conclusions are as follows: to tackle these financing problems of those foreign trade enterprises, we need to work together from three sides: enterprises themselves, financial organizations and our government, which can effectively help small and medium-sized corporations in foreign trade to explore new ways of financing.

References

1. Ling, H.: Study on financing dilemma and countermeasures of small and medium - sized foreign trade enterprises. Chin. Bus. Trade **34**, 110–111 (2013). (in Chinese)
2. Berger, A., Udell, G.: The economics of small business finance: the roles of private equity and debt markets in the financial growth cycle. J. Bank. Finance **30**(11), 2945–2966 (2006)
3. Hu, Y., Li, C.: Research on financing of small and medium-sized foreign trade enterprises in china under the background of internet finance. J. Jingdezhen Coll. **33**(04), 109–115 (2018). (in Chinese)
4. Bernanke, B., Gertler, M.: Financial fragility and economic performance. Q. J. Econ. **105**(1), 87–114 (1990)
5. Xu, P.: Research on the problems and countermeasures in the development of gem in china. Soc. Sci. Front. **7**(6), 856–860 (2018). (in Chinese)
6. Jaffee, D.M., Russell, T.: Imperfect information, uncertainty, and credit rationing. Q. J. Econ. **90**(4), 651–666 (1976)
7. Chan, Y., Kanatas, G.: Asymmetric valuation loan agreements. J. Money Credit Bank. **17**(1), 84–95 (1985)
8. Beck, T., Asli Demirguc-Kunt, M.: Financial and legal constraints to firm growth: does size matter. World Bank Work. Pap. (3), 17–19 (2004). (in Chinese)
9. Wu, X.: Research on the present situation and countermeasures of international trade financing of Commercial Banks. Modern Finance **8**, 29–30 (2016). (in Chinese)
10. Whette, H.C.: Collateral in credit rationing in markets with imperfect information: note. Am. Econ. Rev. **10**(3), 442–445 (1983)

Application of Big Data to Precision Marketing in B2C E-commerce

Shi Yin[1,2(✉)] and Hailan Pan[1,2]

[1] Research Center of Resource Recycling Science and Engineering,
Shanghai Polytechnic University, Shanghai, China
yinshi@sspu.edu.cn
[2] School of Economics and Management,
Shanghai Polytechnic University, Shanghai, China

Abstract. With the emergence and the widespread use of big data, more industries gain the strong radiation effect from it, which also brings new opportunities for the development of B2C e-commerce, and the transformation of its marketing mode is more significant. In the new competitive environment, B2C e-commerce market has experienced a change from extensive sales to precise marketing. How enterprises can constantly accurately match the characteristics of its products or services with consumers, and always maintain a relatively stable customer group size and structure, not only depends on the characteristics of its products or services, but also depends on precision marketing. This paper mainly studies precision marketing of B2C e-commerce enterprises using data mining to analyze valuable information from consumers, and make specific analysis on Jingdong Mall. Research on precision marketing strategies of e-commerce enterprises has practical application value.

Keywords: Precision marketing · Big data · Business-to-consumer · E-commerce

1 Introduction

Business-to-consumer (B2C) e-commerce is a business model that enterprises use the Internet to conduct transactions directly with consumers. With the development of information technology and global economic integration speeding up, B2C e-commerce made further development and numbers of typical enterprises have emerged. China's B2C e-commerce websites embrace a host of online shopping categories. In terms of consumer awareness, there are Jingdong Mall and Suning that offer a comprehensive array of electric appliances, Dangdang that deals in books, maternal and child products, cosmetics, clothes, home textiles, Vipshop that covers branded clothes, bags, cosmetics, maternal and child products and household supplies, and AliExpress, an online trading platform of Alibaba which targets the global market and which vendors call "International Taobao". B2C e-commerce share of total online retail transactions in China is growing year by year.

The E-commerce Report China 2018, released by the Ministry of Commerce of China, shows that B2C e-commerce occupied 45.7% of total online retail sales in 2014,

© Springer Nature Singapore Pte Ltd. 2020
M. Atiquzzaman et al. (Eds.): BDCPS 2019, AISC 1117, pp. 731–738, 2020.
https://doi.org/10.1007/978-981-15-2568-1_100

surpassed C2C for the first time in 2015, and growing year by year, accounted for 62.8% of total online retail sales in 2018 (see Fig. 1). The Information Economy Report, published by the United Nations, also shows that among the top ten economies, China had the biggest B2C e-commerce share of e-commerce transactions, which was inseparable from the Chinese government's long-term policy to stimulate domestic demand and drive the domestic economy.

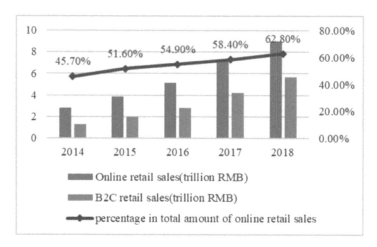

Fig. 1. China's B2C e-commerce transaction volume accounted for the proportion of China's online retail sales and e-commerce transaction scale (2014–2018)

With the deepening of the consumption upgrade, consumers will pay more attention to online shopping brands, quality and services, and the B2C market will gain a greater advantage.

2 B2C E-commerce Precision Marketing on Big Data

The biggest business change brought by big data is marketing change. Businesses now adopt a marketing model that satisfies consumers' individual needs in order to occupy the market. This marketing model is exactly precision marketing.

The concept was first proposed by Lester Wunderman in 1999. It was defined as setting up a personalized customer communication service system to realize a measurable low-cost business expansion path. Later, with the widespread use of big data, precision marketing has a broader definition in application: with scientific management as the basis, with consumer insight as the means, segmenting the market aptly and pertinently and employing intensive marketing to exploit the market, occupy the market, and obtain the expected benefits [1, 2].

Precision marketing, according to its definition, should have the following five characteristics: first, establishing a complete consumer information base by collecting and collating all types of consumer data; second, grasping differentiated needs of consumer segments through data analysis; third, designing according to the segments; fourth, providing differentiated products and services; fifth, further understanding consumers' essential needs and purchase/use habits through data mining, namely increasing consumer stickiness to the product or the company. The five characteristics are in a codependent relationship: establishing a complete consumer information base is the foundation of precision marketing; data analysis is the booster for precision marketing; designing products or services that meet customer needs is the positioning of precision marketing; providing products or services that meet customer needs is the value of precision marketing; mining data and improving customer loyalty is the core of precision marketing [3, 4].

Precision marketing is a growing trend across industries because of its measurable returns to Internet businesses. With the development of the Internet and information technology, the Internet plays a more and more important role, brings great convenience to our life and work, and becomes a part of our daily life. Accompanying all this is an explosion in information—how to get what's useful to consumers and businesses from the overwhelming sea of information has become a pressing issue facing marketers, but Internet businesses have seized this opportunity in time and embarked on a journey to online precision marketing. With the aid of personalized technologies like B2C e-commerce, media information and social networking sites, Internet businesses help users filter out information they don't need from a mass of data for the purpose of precision marketing. It is based on the demand from both consumers and businesses that Internet businesses have stepped in and led to the rise and explosion of precision marketing.

3 Role of Big Data in B2C E-commerce Precision Marketing

Two important factors in precision marketing are precision and marketing. Precision means conveying the right information, products and services to the right people—target customers. B2C businesses rely on big data for precision marketing. There are usually four steps: data management; segmenting and understanding customers based on user data; designing marketing plans based on user data; implementing marketing plans. In this study, the first two steps are defined as the use of big data to pinpoint customers and the last two steps as the use of big data to reconstruct business models. Obviously, the most important thing for businesses to achieve precision marketing is to target customers and the most difficult thing is to make profits on target customers [5]. In view of this, B2C needs to utilize big data to pinpoint customers and reconstruct business models to achieve the value of precision marketing.

3.1 Theoretical Basis for B2C Businesses Using Big Data for Precision Marketing

B2C businesses use big data to achieve precision marketing; they are actually reconstructing business models. Precision marketing is different from marketing in many ways, and all these differences explain why B2C businesses are reconstructing business models when they use big data to achieve precision marketing.

(1) Differences between Precision Marketing and Marketing

First of all, precision marketing is different from traditional marketing in the logic underlying them: traditional marketing puts action over cognition while precision marketing does just the reverse. In traditional marketing, execution comes first and promises success. Yet in the age of big data, blind execution, information surplus and scarcity of attention may lead to overcapacity.

Second, there is a difference in the marketing path: traditional marketing is outside-in while precision marketing is inside-out. In traditional marketing, businesses first build awareness, visibility and loyalty, then promote their products, cover a broad population, and finally narrow the range until they identify the core group [6, 7]. In precision marketing, they first build a reputation and then gain wide visibility; they first identify the core group and then cover the public, which is from personalization to wide coverage and influence.

Third, there is a difference in the marketing tool. The Internet age sees a fundamental change in marketing tools: technology, big data, we-media, topics, content and potential energy can be used as marketing tools, while traditional marketing is based on material, home screen, albums, display, investment brochures, and policies.

Fourth, there is a difference in the way of marketing implementation. Traditional marketing is often implemented by distributors, agents or direct sellers, while precision marketing is usually implemented by customers or businesses. Obviously, precision marketing minimizes the burden of intermediate processes on businesses.

(2) Theoretical Basis for B2C Precision Marketing with Big Data

The fundamental theory for B2C businesses relying on customer data for precision marketing is Gartner Group's customer relationship management (CRM). It emphasizes that a company will significantly increase its profitability, revenue, and customer satisfaction when its organizational structure, product design, various business models including the marketing model, and various processes are customer-centric. CRM is designed to connect customer-oriented front-end and back-end departments of a company, and any department of the company is also responsible for inputting information into and extracting information from the CRM system. The value points of customer relationship include processes, strategies, technologies, resources and functions [6, 8, 9]. It is precisely surrounding these value points that B2C businesses embed customer data in their marketing-sale-service business processes to gain business value and drive processes.

The necessary theory for B2C businesses conducting precision marketing is the customer life cycle. It starts with a customer buying a product or service and ends with the customer giving up the product or service as the product ages or service quality declines. A life cycle can be roughly divided into five stages and during each stage, a customer's loyalty to a product or service is different [10]. At the early stage of purchase, the customer is in the process of understanding the product or service. An impulse purchase may lead the customer to return the product or service. At the advanced stage, the customer gets familiar with the product or service, and with a good first impression, customer satisfaction also goes up. When it comes to the mature stage, performance of the product or service is optimized, and there is chance of cross-selling and up-selling, so that the company can make the most profit. When the product begins to age or service quality declines, there comes the customer decline stage when the customer begins to replace the product or service. And when the customer pays for an alternative, it is time to say goodbye to the product or service.

3.2 B2C Enterprises Targeting Customers Precisely with Big Data

To achieve good results in precision marketing, B2C businesses must first fully understand consumer/customer data in all aspects and carefully research these data to lay the groundwork for precision marketing. This usually takes two steps:

The first step is to build a customer database, classifying the data dispersed in each IT system of a B2C business and the external data of the B2C business (like market research and third-party data), using customer ID as the keyword to extract, convert and load data into another centralized database as the basis for creating customer profiles and for the desired target data design.

The second step is to create the target database. In the 1990s, with the application of big data in marketing, more businesses or people are focused on customer relationship marketing, so there is the customer relationship management system. The goal of CRM is to reduce the operating cost of a company through improvement of business process management and to provide faster, better service to attract more customers or increase customer stickiness.

In the created target database, customers are usually divided into groups based on customer characteristics. To improve the identification of each group, it is best to make groups not similar to one another, so that businesses can take relevant measures to provide products or services that suit each customer group.

3.3 Analysis of Jingdong Mall Using Big Data for Precision Marketing

(1) Overview of Jingdong Mall Marketing

Founded in 2004, Jingdong Mall is one of the most popular and influential B2C e-commerce businesses in China. In 2019, it accounted for 17.3% of China's B2C market, second only to Alibaba (58.2%).

In terms of transaction size, Jingdong Mall occupied 56.9% and retained the first place among China's B2C shopping websites that are based on independent sales. Jingdong Mall mainly targets young people aged 20–35. Arguably, millions of college students or fresh graduates are the important target group or target market of Jingdong Mall. According to the 2017 Jingdong Mall and third-party shop user group analysis, 25% of Jingdong Mall users are the post-90s generation (the new generation), 43% are the post-80s generation (family supporters), and 26% are the post-70s generation (quality consumption representatives). Jingdong Mall reached 305 million active users throughout the year 2018, and the figure exceeded 320 million by August 2019, an increase of 11 million active users.

(2) **Framework of Big Data-Driven Precision Marketing of Jingdong Mall**

Jingdong Mall first collects log data, transaction data, and non-transaction data left by users on the Internet to form the foundation of precision marketing. It then builds models for user behavior, such as user attribute identification, user interest identification, user relationship, user life cycle, and user credit; based on these models, it portrays users and assesses their marketing value and risk, and supplies these underlying data to each marketing system. According to user portraits, it finally employs EDM, SMS, APPUSE, and product advertising to promote products and achieve precision marketing.

(3) **Application of Big Data-Driven Precision Marketing of Jingdong Mall**

In big data-driven precision marketing of Jingdong Mall, the most effective precision marketing strategies and methods are search engines, personalized recommendation engines, EDM, SMS, and coupons.

(1) Search Engines

Jingdong Mall started as a computers/communications/consumer electronics provider, with its business focused on mobile phones and home appliances. As a result, it has invested the most in these two aspects, and its precision marketing strategy has turned out a big success. If we query "smartphone" or "home appliances" on the Baidu search engine, Jingdong Mall will come out among the top search results. Of the brand keywords, Jingdong Mall has paid for Apple, Huawei and other brands, so most of its keywords will appear on the first three pages of the SEO ranking. When it comes to specific product model keywords, some hot product models will also have Jingdong Mall on the first three pages, so customers will naturally think of and go for Jingdong Mall when purchasing a certain product.

Jingdong Mall's SEO web optimization is also clear and user-friendly, which increases the rate of conversion from browsing to purchase.

When entering "headphones" on Jingdong Mall, we can see the categories in the left column, from audio-visual entertainment products to mobile phone accessories and peripheral products, and in the top middle column, there is also relevant search feature, which not only improves user experience but also saves users time and effort to eventually boost the conversion rate.

(2) Personalized Recommendation Engines

Compared with search engines, recommendation engines are more proactive, increasing the conversion rate through internal web applications, when personalized recommendation engines are based on user portraits and product portraits, which are proactive in a two-way fashion.

Jingdong Mall's personalized recommendation is divided into data collection and storage, offline and online analysis, and online recommendation implementation. Data collection and storage rely on real-time clickstream systems as well as historical offline data. At the same time, Jingdong Mall has data from its partners Tencent QQ, WeChat, Paipai. Offline data is mainly analyzed by the JDW platform and partly calculated by Mahout and Spark technologies; online data is analyzed by the JRDW platform, and real-time calculations are made by Storm. Data analysis includes the cleansing and modeling of data.

The cleansing of data is followed by the modeling of product portraits and user portraits. The basic data of product portraits is based on product attributes and related traffic and sales, and the products are mined and modeled by algorithms like LDA, Smash, and FP-Growth. The basic data of user portraits includes online log data, user behavior data, and user transaction data. Technically, mining and modeling are performed using algorithms like GBDT, MAHOUT, and Rank SVM. After the product model is combined with the user model, the online recommendation system is employed to implement personalized recommendation for users.

In addition, Jingdong Mall's recommendation system also adopts swarm intelligence algorithms to analyze user group behavior, similarity between users, and users' personalized needs for niche products, thereby making the recommendation more precise, diverse, and novel.

4 Conclusion

In the age of big data, those who make more effective use of information can seize the opportunity of and take advantage of market resource allocation. E-commerce as an online transaction form has its unique advantages. It can obtain useful information from big data, process such information, and tailor marketing to customer needs to display novel customer care, which can improve customer satisfaction and thus improve corporate business benefits. This paper has investigated the role of big data in precision marketing of B2C e-commerce businesses and, in particular, presented a case study of Jingdong Mall. The successful story of Jingdong Mall provides a positive reference point for B2C businesses using big data for precision marketing.

Acknowledgements. This work was supported by Gaoyuan Discipline of Shanghai–Environmental Science and Engineering (Resource Recycling Science and Engineering), Discipline of Management Science and Engineering of Shanghai Polytechnic University (Grant No. XXKPY1606) and Construction of Ideological Education System for E-commerce Major (Grant No. ZZEGD19021).

References

1. You, Z., Si, Y.-W., Zhang, D., Zeng, X.: A decision-making framework for precision marketing. Expert Syst. Appl. **42**(7), 3357–3367 (2015)
2. Zhu, Z., Zhou, Y., Deng, X., Wang, X.: A graph-oriented model for hierarchical user interest in precision social marketing. Electron. Commer. Res. Appl. **35**, 100845 (2019)
3. Mayank, Y., Zillur, R.: Measuring consumer perception of social media marketing activities in e-commerce industry: scale development & validation. Telematics Inf. **34**(7), 1294–1307 (2017)

4. Erevelles, S., Nobuyuki, F., Linda, S.: Big Data consumer analytics and the transformation of marketing. J. Bus. Res. **69**(2), 897–904 (2016)
5. Calder Bobby, J., Malthouse Edward, C.: Maslowska Ewa: brand marketing, big data and social innovation as future research directions for engagement. J. Market. Manag. **32**(5–6), 579–585 (2016)
6. Xu, Z., Frankwick, G.L., Ramirez, E.: Effects of big data analytics and traditional marketing analytics on new product success: a knowledge fusion perspective. J. Bus. Res. **69**(5), 1562–1566 (2016)
7. Alexandra, A., Paulo, C., Paulo, R.: Moro sérgio:research trends on big data in marketing: a text mining and topic modeling based literature analysis. Eur. Res. Manag. Bus. Econ. **24**(1), 1–7 (2018)
8. Severina, I., Iain, D., Chris, A.-B., Ben, M., Amy, Y.: A comparison of social media marketing between B2B, B2C and mixed business models. Ind. Market. Manag. **81**, 169–179 (2019)
9. Anne, F.H.H.C., Valos, M.J., Michael, C.: E-marketing orientation and social media implementation in B2B marketing. Eur. Bus. Rev. 27(6), 638–655 (2015)
10. Katarina, Ć., Zvonko, M., Tena, S.: Challenges of application of the big data in marketing: case study croatia. WSEAS Trans. Bus. Econ. **15**, 162–170 (2018)

Application of Information Theory and Cybernetics in Resource Configuration

Baifang Liu[1(✉)], Liqiu Sui[2], Meijie Du[1], Jiahui Xia[1], Ziqing Li[1], and Yuchen Liu[1]

[1] Business School, Beijing Language and Culture University,
Beijing 100083, China
liubaifang@blcu.edu.cn
[2] WEI Fang Bank, Weifang 261000, China

Abstract. Information theory and cybernetics pertain to science and technology, but it's theory principle has the role of guiding allocating the enterprise resource at the macro level. From a practical perspective, the assets structure could be regarded as an information system, through the collection, storage and processing of information to realize the control and regulation of the system, thus scheduling the enterprise resources reasonable to make the best use of each asset of the enterprise, avoid wastage and ensure the realization of business objectives.

Keywords: Resource configuration · Information theory · Cybernetics

One of the bottlenecks the modern enterprises are facing is the scarcity of available resources, how to allocate the limited resources of the enterprise reasonable and scientific to make the best use of each asset and avoid idle and waste, thus forming a reasonable capital structure should be an important topic of the science of financial management research. In this respect in the 1940s, the information theory and cybernetics which respectively created by the American mathematician Claude Shannon and Norbert has important guiding significance. Asset structure is the proportion of corporate assets of all types to total assets, and its arrangements should ensure the normal operation of the business activities and to obey and serve the overall goal of business finance [1]. The author believes that the assets structure could be regarded as an information system, through the collection, transmission, reception, storage, processing of information and etc. to study the operation of the asset management system, thereby to achieve the control and regulation of the system. This paper will analyze its mechanism of action, in order to inspire the study of the assets structure of the enterprise.

1 The Basic Principle of the Information Theory, Cybernetics and Their Methodological Significance

Information theory is created by American mathematician Professor Shannon of Bell Labs, Professor Shannon published two papers - "the mathematical theory of communication" and "communication in the noise" respectively in 1948 and 1949 which

M. Atiquzzaman et al. (Eds.): BDCPS 2019, AISC 1117, pp. 739–744, 2020.
https://doi.org/10.1007/978-981-15-2568-1_101

marking the generation of information theory. In these two papers, Professor Shannon discusses the problem about information encoding, transmission, decoding and measurement in detail. The central idea of the discussion is that the value of the information is it can eliminate the uncertainty of the objective world, so as to stabilize people's behavior, in the control implement by the controller to which under control, the important aspect is the grasp of the various behavioral states of which under the control, which requires different behavioral information about which under the control. Due to various disturbances, information obtainment will be very difficult, but whether you can get real and reliable information superiority is the key to the implementation of scientific control [2]. On the difficulty of understanding the objective world, there are brilliant expositions of Bertalanffy: "They are not the object of perception and direct observation, but the conceptual structure. Even the objects in daily life as well, they will never be simply 'given' the sense data or simple perception or...... the difference between real objects and the systems given by observation and 'conceptual' architecture is could not be described with common sense [3]." Therefore, to obtain information about its operational status of objective things is definitely tough task, in the interference of many factors, the way things worked is very complicated.

The principle talked above has profound meaning to the management of corporate assets structure. In the process of arrangement of the corporate assets structure, there needs a profound grasp of the living environment of the enterprise, however, it is often very difficult, because the living environment of the enterprise is not stationary but constantly changing, in addition, the environmental behavior under the stationary state may be very complex. Therefore, managers must deeply understand the importance of information and the difficulty of access to information, only in this way could make the enterprise resource allocation based on a scientific basis, and ultimately enable the enterprises to obtain reasonable and optimal allocation of resources to succeed in the market competition invincible.

Cybernetics is "the infiltrated product of automatic control, electronic technology, radiocommunication, neurophysiology, biology, psychology, medicine, mathematical logic, computer technology, statistical mechanics and other subjects." From the generation base side of control theory, it is the infiltrated product of multidisciplinary, Wiener, founder of cybernetics is a mathematician, but he was extensively involved in each of these disciplines. The direct motivity of the generation of Cybernetics is the military demands of World War II. Wiener realize the basic principles of automatic anti-aircraft artillery from the hunter's hunting action, namely feedback principle. Feedback principle refers to take the output information of the controlled system again as input information to alter the behavior of the controlled system, for automatic control systems often rely on the transmission mechanism designed within the system to achieve the above purpose, for artificial systems, it is the behavior of the controller that make conscious use of the output information by the controlled system to change the operating state of the system [4]. However, it should be noted that the function condition of the feedback principle is the operation process of the controlled system deviates from the initial setup goals of system operation, that is there are various behavior results of system operation, the randomness and controllability of the operating state is the essential prerequisite of the feedback principle, under the different operate state of system, you should choose a behavior that best to achieve the overall goal of system.

Information theory and cybernetics is inseparable, it is decided by the internal logic that both of them works [5]. About the relationship between the two, the founder of cybernetics Wiener has a brilliant summary: "Control engineering issues and information engineering issues is not able to differentiate, and the key of these issues are not surrounded by electrical technology but the more basic message concept, regardless of the news is transformed by the way of electrical, mechanical or dissemination." Information theory and cybernetics as a scientific methodology is an important guiding ideology of enterprise asset allocation.

The author believes that the allocation of resources is mainly to solve the problem of transmission and obtainment of information, on this basis to make rational arrangement and scientific control of the corporate asset structure, Shannon's information theory made people begin to pay attention to the basic role of information in allocating resources, and access to information is a difficult process, which requires not only scientific and efficient means of information transmission, but high demand to the subject and object themself, the real world is very complex, subject only through scientific means can really realize the essential law of object and make conscious use of these laws in order to achieve the desired results. Although cybernetics comes from the natural sciences, but its basic idea is worth learning by social sciences, if social sciences ignored the new methodology brought about by the other scientific developments, it will miss an excellent development opportunity.

2 Reasonable Arrangements for Asset Items

Corporate asset structure itself is a controlled system, it is the business managers use a variety of information to exert purposeful control behavior. Efficient resource allocation behavior need to use a lot of advance information to influence the behavior of the various resource allocation, it is because the corrective measures afterwards often cause some waste, sometimes this corrective action does not exist, because of the huge bias may imposes enterprises a higher burden hard to back that towards bankruptcy [6]. This shows that the information, especially prior information is vital important to the system control and the stable development of system, on this issue, Wei Hongsen pointed out after study: "The size of the amount of information of a system that reflects the degree of organization and complexity of the system, it can make the material system to self-regulate in a very economical way. The more complex the system, the more important the message." Below the author will illustrate this with some of the specific assets items.

2.1 Monetary Funds and Short-Term Investments

Their main role is to meet the repayment requirements of corporate, the allocation of resources on these projects depends on the company's grasp of their own liabilities, including two aspects: the quantity structures of liabilities and term structure of liabilities. Only accurately master the total amount of corporate debt and the repayment request in different periods, companies can arrange monetary funds and the scale of a variety of short-term investments reasonable, so that highly liquid assets of the enterprises can not only meet repayment requirements but also make less money remain in the low income asset items, thus achieve the best combination of liquidity and profitability of asset structure. It is relatively easy to obtain the information about this, in the management of these items, it can be completed only if there is effective communication between the various corporate sector or positions, in the case of the current rapid development of a variety of financial software, these relevant information has been very easy to obtain, companies can monitor their debt situation at any moments and arrange for the scale of repayment funds reasonable, use the cash floating flow intelligent, minimize the cost of corporate to maintain asset structure liquidity.

(1) Account Receivables

The size of the accounts receivable depends on the master of the business to their survival environment and their own capability information. This is because accounts receivable is a double-edged sword for business that use properly can expand the size of the company's sales and market share, thus leading to more profit [7]. However, improper use of account receivable will brought the enterprise a higher burden on opportunity cost of capital and a lot of bad debt, thereby reducing the level of corporate total revenue. The author suggests that arranging the scale of account receivables should have information on the following aspects: Market structure, such as current competitive situation of firms and customer's right and position relative to the enterprise; Customer's credit status, such as the customer's solvency, reputation, etc. As well as the enterprise its own business capacity.

From the customer's credit situation, whether the business-to-customer financing can be take back depends on the customer's repayment ability and their credit rating levels. Financial theorists believe that the measure of a customer solvency standards generally the "5C" system, but these standards generally have static nature which can not reflect the dynamic characteristics of corporate solvency, therefore we should maintain appropriate trust and appropriate vigilance for these solvency standards. In the developed market economies, there is a lot of the credit rating agencies that publish the credit level of each public companies regularly, although they has advantages in terms of credit rating as a professional organization, but the corporate scandals appears in United States and Europe in recent years tells us in the case that information can not be completely symmetrical, the corporate should keep credit scale from any customer to avoid significant impact on their operations due to unexpected bankruptcy of large customers, meanwhile enterprises should seek other sources to know whether the customer's operation is normal and the change of their solvency.

In addition, the company's own operate ability will influence the scale of account receivables, objectively speaking, different companies have their own unique ability that determine their competitive advantage in the marketplace, companies should objectively assess their management capabilities. It must be admitted that due to various constraints, such as the quality of personnel, this ability distributed unevenly among the enterprises, which will exert a certain influence on the allocation of corporate financial resources, companies must based on the fully aware of their operate ability to formulate financial policies and strengthen management of accounts receivable.

(2) Inventory

Inventory is an important current assets of enterprise, but from the perspective of corporate value creation, it is a waste that funds occupied in the inventory [8] Business inventory can be divided into two parts: one is various reserves for the production; the other is new value that created. The optimal goal of enterprise inventory control is: various reserves supplied for the production can be carried out immediately and new value created can be sale immediately. The information that enterprises need for inventory control involves a number of aspects, including market structure, production and operation characteristics of enterprise and so on.

In addition to these information, the effective transfer of information between enterprises and customers as well as suppliers is vital important to inventory management, which involves technical issues of information transmission in information theory. Recently, the development of network technology which greatly improves the efficiency of information transmission between the enterprises which its industrial chain performance vertically, some enterprises take advantage of the latest developments in information technology to control the efficiency of resource allocation in their inventory effectively [9]. The company which achieved optimal inventory control by use of information technology should be the U.S. computer company - DELL company.

(3) Intangible Assets

Intangible assets have become a key factor in gaining access to sustainable competitive advantage and excess profits, such as various brands, patents, non-patented technology and so on. It should be noted that the obtainment of those rights are not without cost, it is also the results of resources input.

According to Shannon information theory, there is active or passive access to information. For enterprises, passive perception is only felt when changes in the external environment will threaten its survival mode; while active perception is the enterprise act as cognitive subject to observing small changes of the living environment of the enterprise initiatively and made a deep understanding of the impact that these changes on the production and operation of enterprises, so that promote the beneficial and abolish the harmful, then avoid misunderstandings and prevent crises before they emerge [10]. Since the establishment of the market economy in China is not long, many companies do not establish the appropriate marketing values and sense of competition. Particularly, the inadequate attention to intangible assets, poor brand awareness and less fully awareness of the fundamental role of knowledge assets to business growth.

From a realistic point of view, multinational companies have become our major competitors, these companies thoroughly tempered in international competition that cultivate a unique competitive capabilities and a strong competitive advantage, if enterprises of China want to compete with them, we must establish the advanced management philosophy, change the resource allocation behavior timely, cultivate the unique resources and core competitiveness of enterprises, build long-term competitive advantage.

In summary, information theory and cybernetics as technology basic science could guide the allocation of resources from the macro level and arrange for asset structure reasonable. As a modern enterprise, smooth financial information transmission system and feedback system must be built based on the need of business operate strategy, then managers could understand the situation beforehand, strengthen control, manage the enterprise resources rationally, improve operational efficiency of assets. And through the information feedback to adjust related instructions timely, reduce business risk and ensure the successful completion of corporate business objectives.

Acknowledgments. 1. This Research findings is Supported by Science Foundation of Beijing Language and Cultural University (supported by the Fundamental Research Funds for the Central Universities. Approval number: 18PT02).

2. This Research findings is supported by the special fund of the basic research of the Central University, project number: 16ZDD01.

References

1. Titman, S.: The effect of capital structure on a firm's liquidation decision. J. Finan. Econ. **13**, 137–151 (1984)
2. Brander, J., Lewis, T.: Oligopoly and financial structure: the limited liability effect. Am. Econ. Rev. **76**, 956–970 (1986)
3. Bolton, P., Scharfstein, D.: A Theory of predation based on agency problems in financial contracting. Am. Econ. Rev. **80**, 93–106 (1990)
4. Maksimovic, V., Titman, S.: Financial reputation and reputation for product quality. Rev. Finan. Stud. **2**, 175–200 (1991)
5. Hicks, J.R.: Marginal productivity and the principle of variation. Economica **35**(35), 79–88 (1932)
6. Holmes, T.J., Levine, D.K., Schmitz, J.A.: Monopoly and the incentive to innovate when adoption involves switchover disruptions. Am. Econ. J. **4**(3), 1–33 (2012)
7. Hombeck, R., Suresh, N.: When the levee breaks: black, migration and economic development in the American South. Soc. Sci. Electron. Publishing **104**(3), 963–990 (2014)
8. Hsieh, C., Klenow, P.J.: Misallocation and manufacturing TFP in China and India. Q. J. Econ. **124**(2), 1403–1448 (2009)
9. Kang, S., Kumar, P., Lee, H.: Agency and corporate investment: the role of executive compensation and corpraterate governance. J. Bus. **79**, 1127–1147 (2006)
10. Kim, Y., Li, H., Li, S.: CEO equity incentives and audit fees. Contemp. Account. Res. **2**, 608–638 (2015)

Structural Optimization of Service Trade Under the Background of Big Data

Guiping Li[✉]

School of Economics, Shanghai University, Shanghai 200072, China
muxueyuan@163.com

Abstract. With the advent of big data, the traditional way of service trade has not meet the needs of the development of international trade. Therefore, we need to use means of big data when you trade and analyze factors affecting the development of service trade in one country to optimize the structure of service trade. Moreover, the efficiency and quality of service trade can also be improved. This article selects the Shandong province service trade structure optimization as the research object and analyze the impact factors that affects the structural optimization of service trade in Shandong province under the era of big data. The grey correlation analysis comprehensively analyze various factors and the connection degree of relative index of the service trade structure optimization. Grey correlation analysis concluded that the level of science and technology, human capital has the highest correlation with the relative index of service trade structure optimization. Therefore, we should promote the optimization and upgrading of service trade structure and continuously improve the service trade competitiveness in Shandong province by increasing the investment of scientific research and education funds to improve the market openness.

Keywords: The era of big data · Structure of trade in services · Factors

1 Introduction

Under the background of big data, the traditional way of service trade cannot meet the needs of the development of service trade, so how to use the advantages of big data in the process of trade is particularly important. From the perspective of scholars' research, there is no definite concept for the definition of big data at present, but when McKinsey consulting company first studied big data, its definition of big data is that when the data exceeds the storage capacity of conventional database, it will get a huge database with the ability of relying on, storing, managing and analyzing. With the development of Internet, data gathering is developing dynamically and growing continuously.

The Internet is characterized by "big data". Scholars believe that big data is a large number of data collection with complex structure. Based on the cloud computing data processing mode, it is formed by data sharing. But at present, based on the background of "big data" of Internet, there are few literatures about influencing factors of service trade in Shandong Province.

M. Atiquzzaman et al. (Eds.): BDCPS 2019, AISC 1117, pp. 745–752, 2020.
https://doi.org/10.1007/978-981-15-2568-1_102

In this paper, the optimization of service trade structure in Shandong Province is selected as the research object. Based on the comprehensive analysis of the current situation of the development of service trade structure in Shandong Province, the influencing factors of the optimization of service trade structure in Shandong Province under the age of big data are analyzed by grey correlation analysis, and the correlation degree between each factor and the relative index of the optimization of service trade structure is comprehensively analyzed [1] (Table 1).

Table 1. Import and export volume of Shandong Province's service trade in 2017 (unit: 100,0000 Yuan)

Industry	Import and export		Export		Import	
	Mount	Growth rate %	Mount	Growth rate%	Mount	Growth rate%
Gross value	306222	18.0	150879	8.2	155343	29.4
Transportation services	43517	−12.3	14255	2.8	29262	−18.2
Tourism services	126949	48.8	21032	12.7	105917	59.0
Building services	86266	15.4	80664	14.6	5601	28.3
Financial services	364	81.9	215	113.1	149	50.3
Insurance services	742	28.9	395	6.1	346	70.9
Telecommunication computer and information services	13979	−12.8	13483	−12.2	496	−26.6
Proprietary royalties	3032	2.2	50	42.5	2981	1.7
Sports culture and entertainment services	116	−10.8	26	83.2	90	−22.7
Other business services	31252	4.4	20755	0.3	10496	13.6
Among them: processing services	14673	−1.1	14645	−0.9	28	−48.8
Maintenance and repair services	2135	22.9	1391	18.1	743	33.0
Other services	14443	8.0	4718	−0.6	9725	12.7

1.1   Import and Export Level Is Lower

In terms of the structure of commodities, transportation construction and tourism, and other traditional service import and export of high latest statistics show that 2017 building tourism and transportation of the three traditional service trade import and export trade volume of 86.27 billion yuan, 126.95 billion yuan respectively, 43.52 billion yuan, up 10.4% 6.7% and a 0.2% drop in, accounts for about 83% of the total trade in services in Shandong province, among them, the transportation construction and tourism services complete export 14.255 billion yuan to 80.664 billion yuan and 21.032 billion yuan respectively, 2.8%, respectively 14.6% and 12.7%, the proportion of the whole province service exports were 8.9% 51.1% and 13.2% respectively, the three combined accounted for 73.2% of 2018 building tourism and transportation of import and export maintained a rapid development momentum, foreign trade volume increased by 15.4% 48.8% and 12.3% respectively, and the insurance financial computer and information consulting movie audio-visual cultural entertainment and other modern service trade of low, although the copyright information Emerging services such as research and development design presents the different levels of growth slow growth, however, is still at the primary stage of development, compared with the developed areas, the industrial structure is unreasonable, provide service, and the low level of overall service trade export of Shandong province level is low, the export of service trade is still focused on labour-intensive services, and capital intensive and technology of knowledge intensive services exports accounted is still low, export structure needs to be improved [2, 3].

1.2 Regional Development Is Uneven

Differences due to the Shandong province regional economic level, service industry and even the development of trade in services is also a regional differences, in 2018, there are 7 municipal services in export volume reached ten billion yuan, total six before the city of Jinan Qingdao Yantai Weihai Weifang and Binzhou has 12 municipal service import and export growth, growth before six Zaozhuang city Weihai is Binzhou Texas respectively Taian and Qingdao in terms of the structure of region, Shandong province regional development gap of service trade, service trade is relatively developed eastern coastal areas, Qingdao, Yantai, 2018 Jinan three cities service exports total accounted for 70% of total exports more than the entire province service regional unbalanced development of service trade is outstanding, inland infrastructure weak economy is not developed and lack of service industry development foundation, Jinan and Qingdao, Yantai Yantai using the geography position advantage and service industry foundation, undertake national area along the strategic priority to the development of service trade policy advantage, at the same time also let the flow of resources such as capital and technology more and more in the region, further widening gap of service trade area, critical [8, 9] (Fig. 1).

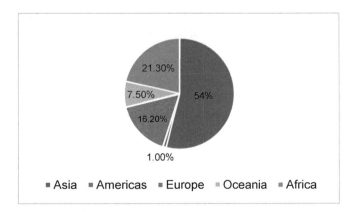

Fig. 1. Market distribution of service trade in Shandong Province 2018 (Unit: percentage)

1.3 The Import and Export of Emerging Services Grew Rapidly

Carry out key policy thinking for the development of modern service industry in our province, the province to enhance the consciousness of service innovation, the service trade has become a new growth point of knowledge technology intensive, rapid growth of import and export of high value-added emerging service, constantly improve the competitive advantage in April 2018, the province's financial telecommunication emerging services such as computer and information continue to maintain rapid growth of imports and exports, insurance services, import and export continued decline in financial services for import and export of 150 million yuan, up 84.6%. Import and export of telecommunications, computers and information services totaled 4.26 billion yuan, up 71.7%; Import and export of proprietary royalties and royalty services amounted to 1.3 billion yuan, up 26.7%; Import and export of sports culture and entertainment services totaled 95.73 million yuan, up 174.6%; The import and export of insurance services totaled 50.74 million yuan, down 93.3%. The import and export of insurance services, telecommunications, computers, information services, financial services and other knowledge and technology-oriented services grew rapidly, higher than the provincial average [10]. However, in general, service trade in Shandong province is just at the initial stage. It needs to deepen the opening of service industry to the outside world in a deeper and broader field, apply information technology more to the traditional service industry, and make the traditional service industry networked and intelligent [4, 5].

2 Empirical Analysis of Influencing Factors of Service Trade Structural Optimization in Shandong Province

2.1 Model Introduction

Grey correlation analysis was first put forward by control science and engineering professor Julong Deng, is a kind of multi-factor statistics analysis method, essence is to reflect a certain we focus on the degree of project is affected by other factors, the size of the measure reflect the degree of the factors associated with the project of grey correlation degree and sorting, analysis results to avoid the multiple regression analysis can only a small number of variables in most cases and the limitation of the linear relationship between the specific steps are as follows:

First, establish the reference sequence and compare the sequence. Set m reference sequences, n comparison sequences, and x0 = {x0(k)|, k = 1, 2 ..., m}, comparison column xi={xi(k)|k = 1, 2 ..., n}is adopted in this paper, reference sequence in the previous chapter CSTS index sequence, sequence, sequence is the need to establish the order of factors, on the basis of reference related literature research, the author choose material capital level of science and technology development, level of foreign direct investment in services trade and human capital as a research of comparative sequence [6, 7].

Second, to do dimensionless processing that is to say normalization. this operation is to reduce the differences of absolute value of the data, they are unified to the approximate range, namely remove various comparc sequence unit, unified into one, and then focuses on the changes and trends, the normalized data processing, the main methods such as initial value, mean and so on. The computation formula is as follows:

Thirdly, calculate the grey correlation coefficient. $\zeta_i(k)$ is the correlation coefficient of sequence x_i to reference sequence x_0 on the k-th index. $\min\limits_{i}\min\limits_{k}|x_0(k) - x_i(k)|$ and $\max\limits_{i}\max\limits_{k}|x_0(k) - x_i(k)|$ are the two-stage minimum difference and the two-stage maximum difference respectively. p is the discrimination coefficient, the value range is 0–1, here p is 0.5, the specific calculation formula is as follows:

$$\zeta_i(k) = \frac{\min\limits_{i}\min\limits_{k}|x_0(k) - x_i(k)| + \rho \cdot \max\limits_{i}\max\limits_{k}|x_0(k) - x_i(k)|}{|x_0(k) - x_i(k)| + \rho \cdot \max\limits_{i}\max\limits_{k}|x_0(k) - x_i(k)|} \tag{1}$$

Then, compute correlation. The correlation coefficient represents the correlation information between the comparison sequence and the reference sequence at k different points. Because the information is too scattered, we need to concentrate the correlation coefficients at different moments into a value for comparison, that is, correlation degree. The correlation degree is the average value of the correlation coefficients at k different moments, which is represented by the letter r_i. The closer r_i is to 1, the greater the correlation io. The specific calculation formula is as follows:

$$r_i = \frac{1}{N}\sum\nolimits_{K=1}^{N}\zeta_i(k) \tag{2}$$

Finally, it is meaningless to sort the correlation degree, that is, to form the correlation coefficient of the correlation order. It mainly compares the correlation degree of the sequence and the reference sequence, and then sorts the correlation degree of the comparison scquence to form the correlation order for comparison.

2.2 Variable Definitions and Data Sources

Among them, the data of import and export volume/GDP of service trade comes from the Department of Commerce of Shandong Province, and the rest of the subsequence data are calculated and sorted out according to the statistical yearbook of Shandong Province from 2013 to 2018. Since the service trade volume after the modification of the service trade statistical caliber in 2014 is in RMB, this paper, for the convenience of statistics, calculates the openness of service trade according to the State Administration of foreign exchange. The published annual average exchange rate will convert the import and export volume of service trade over the years from USD to RMB and then divide it with GDP (Table 2).

Table 2. Definition of each influencing factors

STS (Y)	
Material capital (X_1)	Contribution rate of tertiary industry
Scientific and technological level (X_2)	Granted patent approval
Open trade in services (X_3)	Service trade import and export/GDP
FDI (X_4)	Total amount of foreign capital actually utilized
Volume of trade in goods (X_5)	Total trade in goods
Development level of service industry (X_6)	The service industry in GDP/GDP
Human capital (X_7)	The number of graduates from higher education

2.3   Empirical Analysis and Research Results

Based on DPS data processing system, the data processing steps of grey correlation analysis method are as follows (Table 3):

Table 3. Raw data

Y	X_1	X_2	X_3	X_4	X_5	X_6	X_7
5.7	36.6	76976	0.0679	679451	2671.6	41.7	128297
1.26	37.8	72818	0.0303	542843	2771.2	43.1	147592
1.34	44.5	98101	0.0333	606709	2417.5	44.8	161377
2.11	54.7	98093	0.0450	652874	2342.1	46.8	167440
2.12	56.1	100522	0.0452	689618	2630.6	48	279185

The correlation order of the calculated results is $x_2 > x_5 > x_3 > x_7 > x_6 > x_4 > x_1$. According to the correlation order, the order of the correlation degree of the relative index of service trade structure optimization is, in order, the scientific and technological level of goods trade, import and export, service trade openness, the development level of human capital, foreign direct investment and contribution rate of the tertiary industry (Table 4).

Table 4. Grey correlation coefficient of each factor

X_1	X_2	X_3	X_4	X_5	X_6	X_7
0.3385	0.9304	0.8637	0.4979	0.7024	0.7218	0.8044
0.7108	1.0000	0.9907	0.8494	0.9889	0.7717	0.8741
0.6824	0.9474	0.8192	0.8396	0.9615	0.9790	0.9149
0.8942	0.7727	0.8496	0.9332	0.9183	0.9722	0.9040
0.5217	0.6945	0.6404	0.5895	0.6210	0.5919	0.6124

2.4   Research Conclusion

First of all, from the results of the study, the correlation between science and technology level and CSTS index is the highest. The number of patent authorizations represents the development level of science and technology in a country or region, and the progress of science and technology is an important factor that determines the comparative advantage of service trade and the international pattern. The improvement of science and technology level first improves the efficiency of resource utilization by changing the proportion of labor, capital, technology and other factors in production, and then changes the output of service products. The content of export technology establishes the comparative advantage, optimizes the structure of service trade and enhances the competitiveness of service trade.

Secondly, the import and export volume of goods trade is highly correlated with CSTS index. According to the theory of national competitive advantage put forward by Michael Porter, the development of related industries and pillar industries will affect the export competitiveness of service trade. Service trade and goods trade have strong complementarity. The import and export of goods trade is often accompanied by service trade. For example, international goods transport will inevitably promote the development of insurance, consulting, credit, communication and other service trade, and then improve modern services. The international competitiveness of the trade sector promotes the optimization and upgrading of the service trade structure. Therefore, we should give full play to the leading role of the trade in goods, promote the coordinated development of the two, and complement each other's advantages.

Thirdly, the high degree of association with CSTS index is the openness of service trade, that is, the proportion of import and export volume of service trade in GDP. The openness of service trade reflects the dependence of a country's economic growth on the opening of service trade market. According to the theory of marginal industry expansion, the higher the openness of a country's market is, the more favorable it is for the developed service industry to import into the domestic market. The technology spillover effect is conducive to improving the technical content of services, and then promoting the development of service trade, especially modern service trade. Therefore, the openness of service trade Degree has a high correlation with the relative index of trade structure optimization.

Finally, human capital is related to CSTS index to some extent. In fact, human capital is the same as the transmission way that the level of science and technology affects the structure of service trade. As a high-level factor of production, human capital can be transformed into productivity. Countries or regions with abundant human capital have comparative advantages in the export of knowledge and technology intensive modern service trade, and vice versa.

3   Conclusions

Shandong province as a big province in economy, under the background of reform on the supply side, the service trade show strong momentum of development, service outsourcing become a new economic growth point At the same time there is also a big regional development gap, deficit long-standing and the gap with the developed countries or areas, it shows that Shandong province's labor resources endowment advantage and competitive advantage has not play its important role in promoting the service trade structure optimization. Promoting the service trade structural optimization is beneficial to the development of service industry and economy, in turn, further promote the trade structure optimization the advancement in networking and multi-media technologies enables the distribution and sharing of multimedia content widely. In the meantime, piracy becomes increasingly rampant as the customers can easily duplicate and redistribute the received multimedia content to a large audience.

As for Shandong province, the current main challenges is to consider how to use of big data for the innovation of the structure of service trade, service trade enterprises should make full use of existing big data technology advantage, to broaden the scope of service trade and types, promote the Shandong province service trade structure optimization. This paper analyzes the current situation of the development of service trade in Shandong province, comprehensively research status at home and abroad on the basis of theoretical analysis to select data from 2013 to 2017, by using the grey correlation analysis method and empirically analyzes on the influence factors of service trade structure optimization in Shandong province. The results show that the level of scientific and technology and volume of trade in goods has the highest correlation degree on the basis of our province. And we should strengthen education of science and technology investment funds, pay attention to the development of high-end service personnel, further promote the services market opening, promote the development of trade in goods so as to promote the optimization of the structure of service trade and further improve the international competitiveness.

References

1. Ayoub: Estimating the effect of the internet on international trade in services. J. Bus. Theor. Pract. **6**(3), 65 (2018)
2. Balachi, N., Hoekman, B., Martin, H., Mendez-Parra, M., Papadavid, P., Primack, D., Willemte Vedle, D.: Trade in services and economic transformation. Supporting Econ. Transf. (SET) (2016)
3. Li, M., Gao, S.: Research on the trading development of productive service industry and industrial optimization in China. Appl. Mech. Mater. **20**(3), 123–130 (2014). (In Chinese)
4. Frensch, R.: Trade liberalization and import margins. Emerg. Mark. Financ. Trade **13**(6), 12–14 (2010)
5. Hiziroglu, M., Hiziroglu, A., Kokcam, A.H.: An investigation on competitiveness on services: Turkey Verses European Union. J. Econ. Stud. **109**(3), 102–104 (2013)
6. Crozet, M., Milet, E., Mirza, D.: The impact of domestic regulations on international trade in services: evidence from firm-level data. J. Comp. Econ. **44**(3), 585–607 (2016)
7. Choi, C.: The effect of the internet on service trade. Econ. Lett. **109**(3), 102–104 (2010). ISSN 0165-1765
8. Hoekman, B.: Services trade policies in the East African community and merchandise exports. Case East Afr. Commun. IGC Pap. **24**(3), 02–04 (2016)
9. Ismail, N.W., Mahyideen, J.M.: The impact of infrastructure on trade and economic growth in selected economics in Asia: Asian Development Bank, Working Paper No. 553. (2015)
10. Miroudot, S., Shepherd, B.: The paradox of preferences: regional trade agreements and trade costs in services. World Econ. **37**, 1751–1772 (2014)

Analysis of the Relationship Between Individual Traits and Marketing Effect of E-Commerce Marketers Based on Large Data Analysis

Lei Fu$^{(\boxtimes)}$

Shenyang Polytechnic College, Shenyang, China
1292755032@qq.com

Abstract. In recent years, large data analysis method has been widely used in many fields, such as environmental art design, computer technology application enterprise investment and so on. With the popularity of online consumption and payment, large data technology is more and more widely used in e-commerce. This paper mainly analyzes the relationship between the individual traits of e-commerce marketers and the marketing effect based on large data analysis. First, it simply understands the concept of large data and the traits of large data, and then designs the relationship index and the relationship model according to the relationship between the two. In order to further study the relationship between the individual traits of e-commerce marketers and marketing effect, this paper selects two models: one is the relationship between different individual traits and marketing effect, the other is the relationship between different e-commerce platforms and marketing effect. Finally, it can be concluded that the individual traits of e-commerce marketers have a great impact on the marketing effect, and e-commerce marketers with strong personal charm often get more profits, accounting for 60% of the total marketing results.

Keywords: Large data analysis · Electronic Commerce · Individual traits · Marketing effect

1 Introduction

Large data refers to the data collection that cannot be grabbed, managed and processed by conventional software tools in a certain period of time. Large data has five traits: large amount, high speed, diversity, low value density and authenticity. It has no statistical sampling method, but only observes and tracks what happened. The use of large data tends to be predictive analysis, user behavior analysis or some other advanced data analysis methods.

With the development of economy and the progress of science and technology, large data technology is applied in all aspects, especially in e-commerce marketing mode. In [1], one of the author's purposes is to investigate the extent to which a number of factors that may affect the use of e-commerce have been implemented and used in developing countries. The results show that there is a significant gap in the

© Springer Nature Singapore Pte Ltd. 2020
M. Atiquzzaman et al. (Eds.): BDCPS 2019, AISC 1117, pp. 753–758, 2020.
https://doi.org/10.1007/978-981-15-2568-1_103

implementation of these factors. During the five-year period, there have been no significant changes or developments in technology availability and e-commerce infrastructure. In [2], the author investigated whether the display of products in the form of personalization will affect the amount consumers are willing to pay. The results show that the price of products can be increased by 7% by adding visual anthropomorphic features to the display of products, while the effect of increasing auditory anthropomorphic features is not obvious. In [3], the author reports the results of a randomized experiment on the effects of structural assurance and situational normality on trust transfer in e-commerce between entities and virtual environments. The results show that there is a significant difference between the operation of trust in physical to virtual transmission and virtual to physical transmission. In [4], the author studies and develops a model of consumer intention transfer from traditional (Internet-based) e-commerce to e-commerce. The results show that consumers' intention to change from traditional e-commerce to e-commerce can be explained by factors related to perceived technology and value differences. In [5], the purpose of this study is to explore the possibility of implementing e-commerce to gain competitive advantage. The most significant results show that the existence or relevance of this relationship means that it has an impact from the exclusion of e-commerce to the acquisition of competitive advantage. In [6], this study establishes a research model to explore the impact of key social business burdens on the dimensions of fast relationship and subsequent purchase intention. The results show that interactivity, adhesiveness and public praise have positive effects on mutual understanding, mutual benefit and harmonious relationship. In [7], the author constructs a conceptual model based on refined likelihood model and trust transfer theory, and tests the model with an online survey data on tripadvisor.com. The survey results show that trust is transferred from the review site to the reviewer community, and then to specific reviewers, rather than directly from the review site to specific reviewers. In [8], the author proposes a market management method for anonymous buyers and sellers, which makes honest behavior the most profitable behavior for rational sellers. The market equipment, process and algorithm are described in detail, so that the market operators can easily apply the method in their existing market. In [9], the author discusses the importance of e-commerce in people's daily life. The results show that consumers are still skeptical about the accurate delivery of products purchased online, the applicability of payment methods and the use of personal data obtained by online retailers. In [10], the author conducted two studies on how Trustmark use affects consumer trust, consumer risk perception and thus consumer purchase intention. The results show that the use of Trustmark increases consumers' online trust and purchase intention, and reduces their perceived risks.

This paper mainly analyzes the relationship between the individual traits of e-commerce marketers and marketing effect based on large data analysis. Finally, it can be concluded that the individual traits of e-commerce marketers have a great impact on marketing effect, and the e-commerce marketers with strong personal charm often get more profits, accounting for 60% of the total marketing results.

2   Method

2.1   Overview of Large Data Analysis

Large data analysis is now widely used in a field. The so-called large data refers to a new data developed on the basis of computer, Internet of things, cloud computing and other technologies. At present, large data is widely used in business, medical, education and other fields. Every object has its essence, large data is no exception, and the essence of large data is complete data information. Many scholars use the information of various dimensions and angles to record the behavior track of some things and judge their original traits. Therefore, it can be understood that although the definition of large data has not yet formed a unified conclusion in the current academic circles, the research on large data by scholars provides an important theoretical basis for the further application of large data technology. Large data analysis mainly has four traits: large amount of data, special types of data, special speed of data processing and low value density of data. Many researchers apply large data to various fields of research, and this paper is really to explore the impact of large data analysis on e-commerce personnel and marketing effect.

2.2   Related Index Design

In the context of large data analysis, the consumption mode of consumers has changed completely, from the original offline consumption to online shopping, and gradually consumers start to pay attention to search and sharing. Some e-commerce marketers with personal charm, brand influence and award-winning promotion will attract consumers to consume. This paper is based on the theory of "AISAS model" to study the influence of individual traits of e-commerce marketers on marketing effect. For the influencing factors of the individual traits of e-commerce marketers on consumer attitudes, most consumers are attracted by the personal charm of these marketers, that is, the individual traits. Those marketers with good media image and full of positive energy in life and work are often able to absorb a large number of fans, a kind of buying and selling behavior of fans and these marketers. Naturally, the marketing effect will be very good. Table 1 shows the design of individual traits of e-commerce marketing personnel and marketing effect related indicators.

Table 1. Indicators of influence of individual traits of e-commerce marketers on consumer attitudes

Variable	Index
Prediction variable	(1) personal charm (2) brand influence (3) Prize promotion
Intermediate variable	(1) Cognition (2) emotion
Outcome variable	Purchase intention

2.3   Relationship Model

The process that e-commerce marketers influence consumer attitudes can be divided into the following steps: first, e-commerce marketers attract consumers' attention through e-commerce marketing platforms, such as TAOBAO, JINGDONG, VIPSHOP, etc.; second, consumers get some features of products through browsing these e-commerce platforms, so they are attracted. In addition, due to the influence of the individual traits of e-commerce marketers, consumers' recognition of these products is enhanced, and these consumers who have purchased the products will sell these products to other people, so that the marketing effect of the e-commerce marketers will be greatly improved.

3   Experiment

In order to analyze the relationship between the individual traits of e-commerce marketers and their marketing effects, two models are selected in this paper. The first is to take TAOBAO as an example on the same e-commerce platform and select four e-commerce marketing personnel whose individual traits are strong personal charm, active marketing, occasional marketing and passive marketing, and then analyze their marketing effect. The second is to select the same e-commerce marketer with strong personal charm to analyze the relationship between their individual traits and marketing effect on different e-commerce platforms, such as TAOBAO, JINGDONG, VIPSHOP and GOME.

4   Results

The research results show that in the first model, the individual traits of e-commerce marketers have a great impact on the marketing effect. E-commerce marketers with strong personal charm will attract many fans to consume, so the marketing effect will be better, accounting for 60% of the total marketing results, while the marketing effect of e-commerce marketers with active marketing and occasional marketing will be better. The results are not very good, accounting for 21% and 14% of the total marketing results, respectively. Finally, the marketing effect of e-commerce marketers with negative marketing is the worst, accounting for 5% of the total marketing results. In the second model, different e-commerce platforms may have some influence on the marketing effect, but in general, the marketing effect depends on the individual traits. The marketing effect on TAOBAO accounts for 30% of the total marketing result, the marketing effect on JD accounts for 27% of the total marketing result, the marketing effect on VIPSHOP accounts for 24% of the total marketing result, and the marketing effect on GOME accounts for 24% of the total marketing result. The effect accounts for 19% of the total marketing results. Generally speaking, the marketing effect is not very different. See Figs. 1 and 2 for specific results.

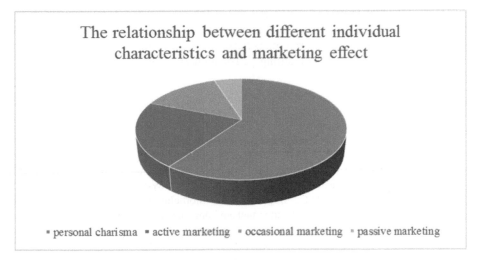

Fig. 1. Relationship between different individual traits and marketing effect

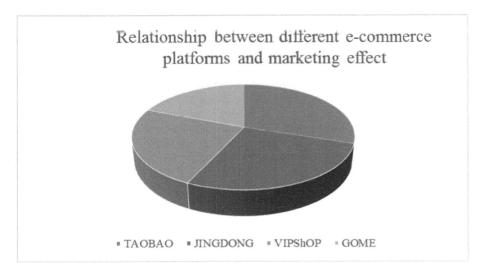

Fig. 2. Relationship between different e-commerce platforms and marketing effect

5 Conclusion

Based on the analysis of large data, this paper analyzes the relationship between individual traits and marketing effect of e-commerce marketers. Firstly, this paper simply describes the concept of large data and the traits of large data. Then, according to the relationship between the two, we design the correlation index and the correlation model. In order to further study the individual traits and marketing effect of e-commerce marketers. This paper selects two models, one is the relationship between different individual traits and marketing effect, the other is the relationship between different e-commerce platforms and marketing effect. Finally, it can be concluded that the individual traits of e-commerce marketers have a great impact on marketing effect, and e-commerce marketers with strong personal charm often get more profits.

References

1. Sila, I.: Antecedents of electronic commerce in developing economies. J. Global Inf. Manag. (JGIM) **27**(1), 66–92 (2019)
2. Yuan, L., Dennis, A.R.: Acting like humans? Anthropomorphism and consumer's willingness to pay in electronic commerce. J. Manag. Inf. Syst. **36**(2), 450–477 (2019)
3. Wingreen, S.C., Mazey, N.C., Baglione, S.L., Storholm, G.R.: Transfer of electronic commerce trust between physical and virtual environments: experimental effects of structural assurance and situational normality. Electron. Commer. Res. **19**(2), 339–371 (2019)
4. Al-Adwan, A.S., Alrousan, M., Al-Soud, A., Al-Yaseen, H.: Revealing the black box of shifting from electronic commerce to mobile commerce: the case of Jordan. J. Theor. Appl. Electron. Commerce Res. **14**(1), 51–67 (2019)
5. Mohammed, W.M., Weli, A.S., Ismael, F.M.: Application of electronic commerce and competitive advantage: a case study of electrical appliances trading companies in Baghdad. J. Eng. Appl. Sci. **14**(9), 3040–3052 (2019)
6. Lin, J., Luo, Z., Cheng, X., Li, L.: Understanding the interplay of social commerce affordances and swift GUANXI: an empirical study. Inf. Manag. **56**(2), 213–224 (2019)
7. Lee, J., Hong, I.B.: Consumer's electronic word-of-mouth adoption: the trust transfer perspective. Int. J. Electron. Commer. **23**(4), 595–627 (2019)
8. Riazati, M., Shajari, M., Khorsandi, S.: An incentive mechanism to promote honesty among seller agents in electronic marketplaces. Electron. Commer. Res. **19**(1), 231–255 (2019)
9. Jimenez, D., Valdes, S., Salinas, M.: Popularity comparison between e-commerce and traditional retail business. Int. J. Technol. Bus. **1**(1), 10–16 (2019)
10. Thompson, F.M., Tuzovic, S., Braun, C.: TRUSTMARKS: Strategies for exploiting their full potential in e-commerce. Bus. Horiz. **62**(2), 237–247 (2019)

Analysis of Urban Street Microclimate Data Based on ENVI-met

Ning Mao$^{(\boxtimes)}$

School of Architecture and Urban Planning, Shenzhen University,
Shenzhen 518060, China
maoning0226@foxmail.com

Abstract. Microclimate is an important factor affecting the behavior and activities of people in cities. How to analyze the influencing factors of urban microclimate through data analysis and put forward the improvement strategy is an urgent problem to be solved under the background of intelligent city construction. In this paper, the microclimate data of four streets in Majialong Industrial area of Shenzhen are analyzed by using ENVI-met data simulation software, and based on several elements of street orientation, sky view faktor and plant, the improvement strategy of microclimate in urban streets is put forward. To achieve the purpose of guiding urban design, so as to promote the construction of smart city.

Keywords: ENVI-met · Microclimate data · Urban street

1 Introduction

With the improvement of people's living standards, urban residents put forward higher requirements for urban microclimate environment. As a public space with high frequency of use, the microclimate environment of urban streets is an important factor to determine the vitality of urban streets and the willingness of residents to activity.

Street microclimate is affected by urban natural environment and artificial environment. In the aspect of natural environment, street microclimate is affected by earth latitude, topography, perennial wind direction and solar radiation. In artificial environment, street microclimate is affected by building height, block plant, street orientation and so on. Therefore, in order to conform to the scientific nature of the study, The study of street microclimate needs to be carried out according to specific urban climate conditions and artificial environment [1].

2 Research on Microclimate

Field measurement is a traditional method of urban microclimate research. A large number of scholars use the method of actual measurement to study the change characteristics of urban microclimate, and to explore the influence of different factors on urban microclimate [2, 3]. However, the limitation of actual measurement is that only a limited number of microclimate data can be obtained, so it is difficult to collect the urban microclimate data comprehensively [4].

© Springer Nature Singapore Pte Ltd. 2020
M. Atiquzzaman et al. (Eds.): BDCPS 2019, AISC 1117, pp. 759–767, 2020.
https://doi.org/10.1007/978-981-15-2568-1_104

In order to solve the limitations of practical measurement, some scholars began to use satellite remote sensing as a new technical platform for urban microclimate research [5]. However, satellite remote sensing technology is a macro observation of urban microclimate, which still can not meet the needs of block scale microclimate research.

The envi-met numerical model has higher refinement characteristics and can better quantitatively analyze the microclimate characteristics of urban blocks [6] just fill the gap of microclimate research in block scale. ENVI-met was first officially released in 1998 [7], over the years, after many scholars have corrected its scientific research, it has been widely used in the study of urban microclimate environment [8–10].

3 Research Technique

Shenzhen is located in the low latitudes of China, and its humid subtropical climate is very suitable for ENVI-met data analysis. Moreover, its climate is hot and warm in summer and winter, and citizens can spend a long time in outdoor spaces such as streets. Street space, as an important carrier of urban life, social activities, interpersonal communication and so on, has a high frequency of daily use. Therefore, it is particularly important to shape a good street microclimate environment.

In order to study the influencing factors of street space microclimate under humid and hot weather in summer, this paper takes four streets in Majialong Industrial area of Shenzhen as the actual research object. The data simulation of street microclimate is carried out by using ENVI-met in order to quantitatively analyze the influence of urban buildings and plant on street microclimate.

3.1 Subject Investigated

Shenzhen is a coastal city in southern China, adjacent to Hong Kong. It is a subtropical oceanic climate. The annual dominant wind direction is easterly to the southeast, the annual average temperature is 22.4 °C, the annual rainfall is 1933.3 mm, and the average annual sunshine hour is 2120.5 h (Table 1).

Table 1. Street information

	Street orientation	Length	Width	Remarks
Hongbu Road	East and west	370 m	8 m	There are mainly multi-storey buildings on both sides of the street
Daxin Road	East and west	450 m	20 m	High-rise commercial buildings are mainly on both sides of the street
Yiyuan Road	North and South	350 m	16 m	There are mainly multi-storey houses on both sides of the street
Baolong Road	North and South	250 m	12 m	There are mainly multi-storey industrial buildings on both sides of the street

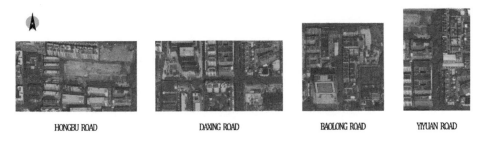

HONGBU ROAD   DAXING ROAD   BAOLONG ROAD   YIYUAN ROAD

Fig. 1. Satellite map of streets

In this paper, four streets in Majialong Industrial area of Shenzhen: Hongbu Road, Daxin Road, Yiyuan Road and Baolong Road are selected as the research objects, and the simulation model of the study area is established by using ENVI-met Fig. 1.

3.2 Digital Simulation

Based on ENVI-met, the microclimate of four streets was simulated on June 23, 2018.

According to the actual meteorological data of Shenzhen collected in Energy Plus, the meteorological data such as wind direction, initial wind speed, initial temperature and initial humidity in the software are calibrated, so that the simulation results are in good agreement with the actual data. The output data include air temperature, relative humidity, wind speed, predicted mean vote (PMV) and so on. The output time period is 0:00 to 24:00, the time is 24 h, and a set of data is output per hour.

4   Interpretation of Result

4.1   Analysis of Air Temperature Data

Through the simulation of four streets, the microclimate data are collected, and the microclimate information inside the street is counted from two angles: the dynamic change of temperature and the spatial distribution of temperature.

Dynamic change of temperature: The temperature changes of the four streets and streets between 10:00 and 16:00 are different. The temperature change of Hongbu road belongs to the upper and lower fluctuating type, and there are two peaks from 10:00 a. m. to 4:00 p.m., which are 29.29 °C and 30.91 °C respectively. The temperature change of Yiyuan Road belongs to the steady growth type, from 10:00 a.m. to 2:00 p.m., the temperature increases from 29.64 °C to 31.88 °C. The temperature change of Daxin Road belongs to mutant type, and the temperature variation range from 10:00 to 12:00 is small, which is kept at 29.60 °C, and the temperature increases rapidly in the afternoon, reaching between 31 °C and 32 °C. The overall temperature of Baolong Road is the smallest, hovering from 29 °C to 30 °C, the highest temperature appears between 12:00 p.m. and 1:00 p.m., and the temperature at 2:00 p.m. was lower than 12:00 p.m. and dropped to 30.62 °C Fig. 2.

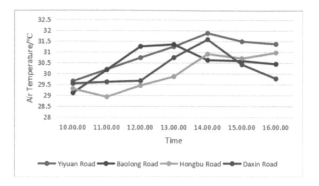

Fig. 2. Air temperature change

Temperature spatial distribution: The temperature distribution data of four streets at 2:00 in the afternoon are extracted. From the point of view of temperature spatial distribution, the temperature difference between the four streets is great, the internal temperature difference of Daxin Road is the largest, and the overall temperature difference of Yiyuan Road is small, but the temperature is the highest, reaching 32. 97 °C. The temperature difference between Hongbu Road and Baolong Road is small, and the temperature field distribution is more uniform Fig. 3.

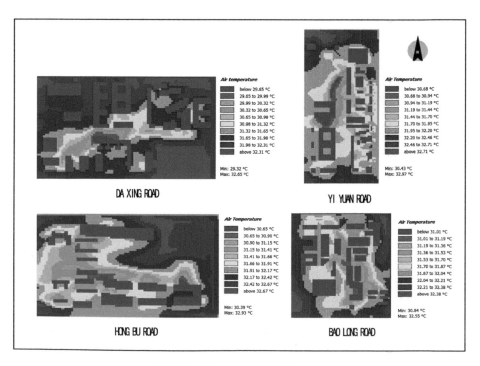

Fig. 3. Air temperature of streets

4.2 Analysis of Wind Speed

According to the results of wind speed distribution, Baolong Road has the lowest wind speed in the valley of the four streets, and 60% of the area inside the main street is below 0.7 m/s. The wind speed of Yiyuan Road varies the most from 0.02 m/s to 9.44 m/s, but the wind speed inside its main street is small, 60% of the range is between 0.96 m/s and 1.90 m/s. The high wind speed areas of Yiyuan Road and Daxin Road are mainly concentrated in the street canyons on the east side of the block, and the local wind speed difference is 4 to 5 times. However, the high wind speed areas of Hongbu Road is mainly concentrated in the open area of the block, and the difference of local wind speed is small Fig. 4.

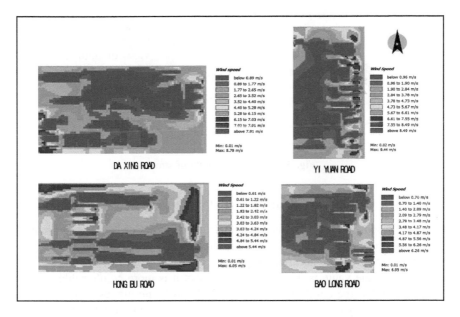

Fig. 4. Wind speed of streets

4.3 Analysis of Relative Humidity

From the data distribution map, it can be concluded that the relative humidity of Yiyuan Road, Hongbu Road and Baolong Road is similar, ranging from 43% to 52%. While the humidity of Daxin Road is relatively high, with 55% in part. It is positively correlated with the distribution of street vegetation Fig. 5.

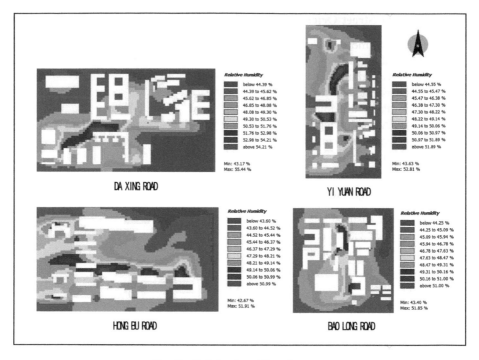

Fig. 5. Relative humidity of streets

4.4   Analysis of Predicted Mean Vote (PMV)

PMV was proposed by the Danish scholar Fanger. ENVI-met can improve the parameters needed to calculate PMV according to the actual situation, so that the obtained data can be more accurate. In general, the value range of PMV is in the range of −4–4, where the value is 0, which indicates that it is more comfortable.

It is found that the comfort value of Daxin Road is the highest, which can reach 2.23, and it is mainly concentrated on the main street. The comfortable area of Yiyuan Road is the most widely distributed, mainly concentrated in the building-intensive area in the east of the block. The comfort area of Hongbu Road is distributed in the neatly arranged area of the west side of the building, while the distribution of the comfortable area of Baolong Road is scattered, and there is no obvious distribution law.

5   Correlation Between Street Space and Microclimate Data

Based on the analysis of microclimate data of four streets, in the street orientation, sky view faktor and plant on street microclimate is analyzed.

5.1 Impact of Street Orientation

The orientation of the street has a certain influence on the wind speed and temperature inside the street. According to the wind speed distribution map and PMV distribution map, the wind speed and PMV of Hongbu Road and Daxin Road facing east and west are obviously better than those of Baolong Road and Yiyuan Road facing north and south. On the PMV distribution map of Hongbu Road, this result is more obvious.

Street orientation has a great impact on the change of street microclimate. Choosing a good street orientation in urban design has a more positive impact on the improvement of street microclimate.

5.2 Impact of Sky View Faktor

In this paper, the distribution map of sky view faktor data derived from ENVI-met is analyzed. The sky view faktor of Daxin Road and Hongbu Road is higher, between 0.20 and 0.49, the local spatial node is more than 0.58, and the sky view faktor of Baolong Road and Yiyuan Road is small, ranging from 0.09 to 0.27. The sky view faktor is influenced by the section of the street and the height of surrounding buildings, and also determines the amount of solar radiation of the street. The aspect ratio of the street affects the spatial distribution of the microclimate inside the street, which has great influence on the microclimate change Fig. 6.

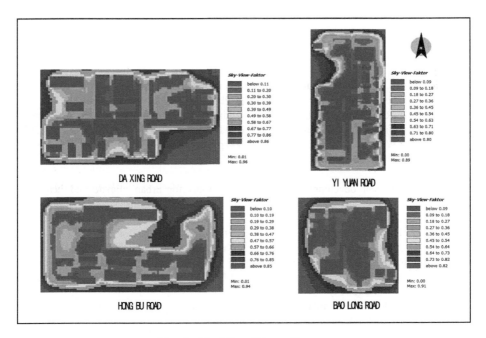

Fig. 6. Sky-View-Faktor of streets

Based on the above research, it is found that the narrow street aspect ratio must promote the street microclimate on wind speed and PMV. In the future urban design, we can consider optimizing the vertical temperature distribution and wind speed distribution of pedestrian street by adjusting the aspect ratio.

5.3 Impact of Plant

It is found that the distribution of vegetation can adjust and optimize the humidity change and temperature distribution in the street to a certain extent [11]. In the four streets involved in this study, the humidity of vegetation cover area is higher and PMV is small, which plays an active role in the regulation of microclimate.

6 Conclusion

The spatial form characteristics of urban streets have a great influence on the micro-climate of streets. This paper uses Envi-met software to simulate street microclimate data. The data show that different street orientation, scale and interface have certain influence on the microclimate distribution inside the street. Through the experimental study in this paper, the effects of urban microclimate change on PMV, air temperature, humidity and wind speed are considered in the design. Through intuitive data to show this impact, to guide the purpose of urban design, so as to promote the construction of smart city.

References

1. Shanglin, W., Yimin, S.: Numerical simulation and improvement strategy for the street micro-climate in the Guangzhou area. Urban Plann. Forum 1, 56–62 (2016)
2. Giridharan, R., Ganesan, S., Lau, S.S.Y.: Daytime urban heat island effect in high-rise and high-density residential developments in Hong Kong. Energ. Build. 36(6), 525–534 (2004)
3. Zhang, K., Wang, R., Shen, C.: Temporal and spatial characteristics of the urban heat island during rapid urbanization in Shanghai China. Environ. Monit. Assess. 169(1–4), 101–112 (2010)
4. Ooka, R.: Recent development of assessment tools for urban climate and heat-island investigation especially based on experiences in Japan. Int. J. Climatol. 27(14), 1919–1930 (2007)
5. Tran, H., Uchihama, D., Ochi, S.: Assessment with satellite data of the urban heat island effects in Asian mega cities. Int. J. Appl. Earth Observ. Geoinf. 8(1), 34–48 (2006)
6. Ambrosini, D., Galli, G., Mancini, B., et al.: Evaluating mitigation effects of urban heat islands in a historical small center with the ENVI-Met (R) Climate Model. Sustainability 6(10), 7013–7029 (2014)
7. Bruse, M., Fleer, H.: Simulating surface-plant-air interactions inside urban environments with a three dimensional numerical model. Environ. Model Softw. 13(3–4), 373–384 (1998)
8. Tsoka, S., Tsikaloudaki, A., Theodosiou, T.: Analyzing the ENVI-met microclimate model's performance and assessing cool materials and urban vegetation applications–a review. Sustain. Cities Soc. 43, 55–76 (2008)

9. Acero, J.A., Arrizabalaga, J.: Evaluating the performance of Envi-met model in diurnal cycles for different meteorological conditions. Theor. Appl. Climatol. **131**(1–2), 455–469 (2016)
10. Ali-Toudert, F., Mayer, H.: Numerical study on the effects of aspect ratio and orientation of an urban street canyon on outdoor thermal comfort in hot and dryclimate. Build. Environ. **41** (2), 94–108 (2006)
11. Wong, N.H., Kwang Tan, A.Y., Chen, Y.: Thermal evaluation of vertical greenery systems for building walls. Build. Environ. **45**(3), 663–672 (2010)

Big Data Security Issues in Smart Cities

Shidie Wu$^{(\boxtimes)}$ and Lirong Sun

School of Economics, Shanghai University, Shanghai, China
wsd5363@163.com

Abstract. Data is the soul of a smart city and the cornerstone of building a smart city. With the rapid rise of emerging technologies dominated by the Internet of Things, cloud computing, and big data, the construction and development of smart cities affect all aspects of people's lives. The issue of big data security has also begun to affect people's livelihood security, economic security, and even national security. With the deepening of the construction and development of smart cities, the issue of big data security risks has become increasingly severe, and it has become a huge obstacle to the sustainable development of smart cities. This paper analyzes the current state of big data security issues from the perspective of intelligent infrastructure, big data collection, transmission and storage, and the management of big data. Several suggestions have been put forward from the perspectives of top-level design, technical protection, information security education, and big data legislation.

Keywords: Smart city · Big data · Top-level design · Privacy protection

1 Introduction

Since 2008, IBM first proposed the concept of "smart city". With the rapid development of new generation information technology and data science, countries around the world have gradually put the construction of smart cities on the agenda. Smart cities have become a new direction for urban development [1]. The foundation of smart city construction is the new generation of information technology based on Internet of Things, cloud computing and big data. Through the dynamic monitoring, collection, integration and analysis of data generated by various departments in the city, the goal of urban intelligent management is realized [2]. Data is the soul of a smart city and the basis for the operation of a smart city system. At present, the architecture of smart city solutions basically follows the "3I" feature proposed by IBM, which is to build the Internet of Things through RFID, barcode and other sensing technologies. In the process of building a smart city, using advanced Internet of Things sensing technology, we will thoroughly and thoroughly perceive urban life, and establish an information interaction process between people, between people and things, and between objects and things. In the process, various information data are collected and stored continuously. Integrate and analyze the massive data obtained through cloud computing and supercomputer technology, and mine the valuable information in the massive data and the possible internal connections to provide support and basis for intelligent management and decision-making in smart cities [3]. Information technology is a double-edged

© Springer Nature Singapore Pte Ltd. 2020
M. Atiquzzaman et al. (Eds.): BDCPS 2019, AISC 1117, pp. 768–773, 2020.
https://doi.org/10.1007/978-981-15-2568-1_105

sword that brings convenience, but it is also accompanied by information security issues [4]. Therefore, in the process of building and developing smart cities, it is crucial to ensure the security of big data information. How to ensure the safety of these data and information is a major issue that must be treated with caution in smart city construction.

2 Smart City Big Data Security Status

Domestic and foreign scholars divide the architecture of smart cities into four levels: the perception layer, the network layer, the platform layer, and the application layer. The perception layer, by embedding smart devices such as sensors and chips into objects, performs object feature recognition, information perception and collection, is one of the sources of information generation, and is the most vulnerable to attack by criminals. At the network layer, high-speed transmission and sharing of information through interconnected networks is the basis for smart cities to efficiently obtain and utilize information, and is also the hardest hit area for big data security issues. At the platform level, through the construction of a unified data center, the various data of the city are integrated, analyzed and predicted to achieve the goal of overall management and application of data. The application layer refers to various application services provided by the public, enterprises and the government. This level involves multiple interest groups such as the government, application providers and users, and contains a large amount of personal privacy information, which is a frequent area for personal information security issues.

2.1 Intelligent Infrastructure

Intelligent infrastructure typically includes, but is not limited to, infrastructure such as communications, networking, electricity, transportation, and government. In the construction of smart cities, a major project is the intelligentization of traditional infrastructure, replacing the original infrastructure with advanced intelligent equipment. Using mature network technology, the original independent infrastructure devices are connected into one whole, and various data are continuously transmitted in this whole. Therefore, the security of information infrastructure is critical to the security of big data.

At present, the core technologies in China's information security field are mainly in the hands of developed countries, and their dependence on foreign countries is very high. In particular, core chips, communication chains, and system software are mostly provided by foreign companies. The hidden backdoors of these products directly threaten urban information security. At the same time, the maintenance of some smart cities and other services are also carried out by foreign companies. In the process of maintaining various software and hardware, the data is always exposed to the risk of leakage. With the continuous deepening of the construction of smart cities, various new technologies and new equipments have been introduced into the smart city system, and their influence has penetrated into all aspects of people's lives, making the safety of basic equipment related to people's livelihood security and economic security, even national security. However, the information acquired and processed by these devices is often the most vulnerable to attack by criminals. In recent years, more and more hackers and offenders have focused on urban infrastructure [5]. Attacks against urban infrastructure are more destructive than traditional information security attacks. Since the intelligent infrastructure devices are connected to each other through the network, they are no longer an independent entity. When an attacker destroys a single device or system, it may cause collective shackles of the entire network device and pose a serious threat to the personal safety of the people.

2.2 Big Data Collection, Transmission and Storage

In the daily operation of smart cities, the accuracy of big data is the key to effective management of the city and making the right decisions. There are a large number of smart devices and users in the smart city system, and the diversity of information sources directly increases the risk of information leakage [6]. A smart city is a complex information system. A large amount of information is transmitted and shared between different systems, and this process often lacks reliable security protection. The data collected at the perception layer is transmitted using different network protocols according to different environmental and security requirements. However, the encryption algorithms and security strengths supported by different protocols are different, and the existing transmission protocols and encryption protection technologies have vulnerabilities, which often lead to great security risks in the transmission process. How to choose a secure data transmission technology in a specific network environment is an issue that needs to be considered. With the continuous development of cloud computing and cloud storage technologies, data storage is decentralized, that is, a single data is stored in multiple different nodes, which greatly increases the risk of data being attacked and leaked. In addition, there are differences in the level of protection of different storage devices, which will further increase data security risks.

2.3 Big Data Management

There are a wide range of big data in smart cities, including sensory data collected through the sensory layer, first-hand data uploaded by users, and various information published by government agencies. These data are updated quickly and the update speed is rising. Due to the complexity of information sources and the existence of information silos, the storage of relevant data is quite scattered. Due to the lack of top-level design of information security system, it is impossible to establish a unified management system and protection standards, and the data faces great security risks. In all data types, personal information has great asset value because it can clearly point to or describe a specific individual [7]. Especially as people's dependence on the Internet increases, online chat, shopping and payment may inadvertently reveal personal privacy information such as personal name, phone number, ID number and address, and the risk of information leakage increases [8]. When personal information falls into the hands of criminals, criminals can analyze personal privacy information, send spam sales messages, and commit financial fraud to specific individuals. In the early days of the construction of smart cities, people generally lacked the awareness of personal information protection and lack of personal information protection means, which led to the unfavorable position of personal information security protection.

The process of building a smart city will result in massive data, including public information that can be obtained from legitimate sources, as well as personal and national secret information that is easily exploited by criminals. This means that different protection measures are needed when facing different types of information. At present, China's legislative activities related to data security mainly focus on the protection of personal privacy information. The laws on protecting other areas of big data are still blank, and the legal governance system is still not perfect. With the advent of the era of big data, the dissemination and collection of information has become convenient, and everyone has become more "transparent" than before, but personal privacy still needs to be protected. In this case, the information boundary becomes more and more blurred, and it is more difficult to distinguish between information sharing and privacy protection. Therefore, it is more necessary to make a clear division by legal form and establish the right and wrong standards.

3   Smart City Big Data Security Protection

Do top-level design and adhere to safety peers. "Top Design" is a term derived from the field of engineering, proposed by Niklaus Wirth in the 1970 s. It is based on a specific design object, using a systematic approach to strategic design from top to bottom. From a global perspective, it is a simplistic, concrete, and stylized design method that manages all aspects, levels, and elements of a design object to achieve structural optimization, functional coordination, and resource integration. Smart city big data security should proceed from system security, use scientific forecasting methods, fully demonstrate the various security issues that may arise in the construction process of smart cities, and formulate corresponding solutions according to their specificity. At the same time, the construction of smart cities is a long-term development process, and big data security issues are emerging one after another. This means that top-level design should also keep pace with the times, and it is a process of continuous improvement. The top-level design should remain flexible as the smart city construction continues to deepen, and it will be constantly updated and adapted according to the changes in the smart city environment and situation. On the one hand, the government should build a standardized safety management system for smart cities, fully absorb the experience of big data security management in the process of smart city construction, continuously improve the system construction and operation process, and do a good overall layout and planning from top to bottom. On the other hand, it is necessary to establish and improve a smart city big data security organization system. Governments at all levels should take the lead in setting up relevant regulatory agencies, clarify the division of responsibilities among various departments, and achieve the cooperation and cooperation of various departments under the supervision of government regulatory departments. Different smart cities specify corresponding management strategies.

Build a security firewall for technical protection. Information infrastructure is one of the sources of information and the most vulnerable to attack by criminals. It can be secured by setting access control, intelligent authentication and verification technology, fake attack detection technology, and data encryption. However, the security of information infrastructure in smart cities differs from the security of traditional information systems. Existing information system security technologies cannot be directly applied to the information infrastructure of smart cities. Therefore, it is necessary to research the information security technology corresponding to the characteristics of the information infrastructure of the smart city to protect the security of the information infrastructure. In the big data sharing and transmission link of smart cities, protocol encryption technology can be used to encrypt network protocols without encryption to prevent data from being tampered or stolen during transmission [9]. In addition, using boundary isolation technology, based on existing network processes and architectures, security measures are set on key nodes of the network to identify and filter boundary information, which can effectively prevent illegal intrusion [10].

Guided by safety education to raise people's awareness of safety protection. In real life, some people's awareness of personal information protection is weak, and the lack of information protection measures and skills makes it easy for people to disclose their own private information in daily network activities, so that criminals engaged in related crimes can take advantage of it. Therefore, the government should pay attention to the comprehensive publicity of big data security, make it the daily content of universal legal education, let information security education enter government agencies, schools, enterprises and communities, and effectively strengthen the confidentiality awareness and common sense of personnel at all levels. Achieving the goal of smart city information security, everyone knows and everyone practices. Government agencies should raise the importance of information security education, raise it to the height of urban development planning strategy. Develop a detailed and scientific information security education plan to comprehensively enhance citizens' information security awareness and common sense. In addition, the government should actively use new media and other means to promote the theme, use big data technology for information mining, and accurately push security education resources.

Strengthen information legislation and provide legal protection. Information security laws and regulations guarantee an important foundation for the construction and operation of smart cities. With the gradual deepening of smart city construction, existing laws and regulations can provide protection to information security to a certain extent, but it is equivalent to developed countries in Europe and America. China still lags behind in big data security legislation, and it is still a blank in some areas. Therefore, in the face of new problems arising in the process of building a smart city, the national information security function department can first formulate and issue relevant departmental regulations. With the continuous construction of smart cities, relevant industries continue to develop, and in the practice of information security law, we continue to accumulate legislative and judicial experience and lessons. When the time is ripe, we will rise to administrative regulations or laws, and raise the level of information security legislation.

4   Conclusion

Information security issues are constantly evolving with the development of information technology. The higher the degree of informationization of urban development, the greater the impact of information security issues. The development and construction of smart cities cannot be separated from the support of big data, and the security of big data directly affects the security of smart city systems. With the rapid development of emerging information technologies, the related applications of smart cities are integrated into all aspects of people's lives. The impact of big data security issues on urban management and people's lives is increasingly obvious. Big data security is related to national security, urban development and Citizens' lives. The rapid development of the new generation of information technology has brought us unprecedented convenience and speed. At the same time, the construction of smart cities will face severe challenges. Therefore, the construction and development of smart cities, network security is always a prerequisite. In the era of digital economy, how to build smart cities, how to protect big data security, how to use the next generation of information technology such as Internet of Things, big data, cloud computing to improve the intelligence level of smart city security protection, these issues are waiting for us to solve.

References

1. Breeden II, J.: The dangers in smart cities (2019). http://nextgov.com/
2. Bibri, S.E.: The anatomy of the data-driven smart sustainable city: instrumentation, datafication, computerization and related applications. J. Big Data **6**(1), 59 (2019)
3. Zhang, K.: Research on data mining security under the background of big data era. In: Proceedings of the 8th International Conference on Management and Computer Science (ICMCS 2018) (2018)
4. Zhang, D.: Big data security and privacy protection. In: Proceedings of the 8th International Conference on Management and Computer Science (ICMCS 2018) (2018)
5. Lee, S., Huh, J.-H.: An effective security measures for nuclear power plant using big data analysis approach. J. Supercomput. **75**(8), 4267–4294 (2019)
6. Yin, M.: Data security and privacy preservation in big data age. In: Proceedings of the 2nd International Conference on Mechatronics Engineering and Information Technology (ICMEIT 2017) (2017)
7. Lin, J., Wu, X., Chen, S., Hu, Y., Liang, C.: Research on the protection of network privacy rights of citizens in the big data era. In: Proceedings of the 2017 5th International Conference on Frontiers of Manufacturing Science and Measuring Technology (FMSMT 2017) (2017)
8. Hou, Y.: Analysis and research of information security based on big data. In: Proceedings of the 2018 International Symposium on Communication Engineering & Computer Science (CECS 2018) (2018)
9. Rasori, M., Perazzo, P., Dini, G.: A lightweight and scalable attribute-based encryption system for smart cities. Comput. Commun. **149**, 78–89 (2019)
10. Liu, D.-L., Li, D., Ma, L., Liu, X., Yu, H., Chang, Y.-X., Chen, J.-F.: Research on electric power information systems network security situation awareness based on big data technology. In: Proceedings of the 3rd Annual International Conference on Electronics, Electrical Engineering and Information Science (EEEIS 2017) (2017)

System Test of Party Affairs Work Management Information System in Higher Vocational Colleges

Xiaoping Zhao[✉]

Chongqing Industry & Trade Polytechnic, Fuling, Chongqing 408000, China
410082087@qq.com

Abstract. As the database design of the party affairs work management information system of higher vocational colleges is completed, and then the design results are tested. This paper will show the results of the system from six aspects: test environment, test objectives, test content, test cases, and test results analysis and test conclusions. It is hoped that it is checked whether the functions of the various modules of the system are stable and can meet the requirements of customers.

Keywords: Higher vocational colleges · Information system · Testing

1 Introduction

With the development of computer technology, computers are widely used to improve personal work efficiency, but in a modern work environment that requires many people to work together, it is necessary to improve overall work efficiency. The rapid development of Internet/Intranet provides technical guarantee for the exchange and sharing of information, the collaborative operation of the team, and also indicates the advent of the networked office era. With the continuous development of higher vocational colleges and their subordinate departments, the task of the party affairs management department is becoming more and more arduous [1]. The database is designed for a specific system environment. In order to meet the requirements of effective data storage and processing, it is necessary to construct an optimal database schema to build the database and its corresponding system. The process of designing and building a database on a specific database management system according to the needs of users during database design is one of the key technologies in the software system development process. In the database field, various types of systems that use databases are generally referred to as database application systems. To this end, the object naming convention should meet the definition criteria when establishing the database, and should conform to the three paradigms of the database; the development database should be designed according to the data dictionary and ER diagram; the user needs should be understood, and the database should be designed based on the requirements [2]. The design of party affairs work management information system in

M. Atiquzzaman et al. (Eds.): BDCPS 2019, AISC 1117, pp. 774–780, 2020.
https://doi.org/10.1007/978-981-15-2568-1_106

higher vocational colleges is based on the analysis of system requirements. The goal of system design is to complete the overall design and planning of the system before the detailed design and implementation of the system. This paper analyzes and designs the database of party affairs work management information system in higher vocational colleges.

2 Test Principles

Before the system test, the software project configuration and test project configuration are first performed. The software project configuration includes three parts: the source code, the design model and the requirements specification; the test project configuration may be part of the software configuration, which includes the test plan and the test case. And the main parts of the test tools.

After the test work is completed, the tester needs to evaluate the test results. If the expected test results are inconsistent or different from the actual test results, the system program has a problem. In this case, the program debugging software tool is needed to locate the defects and correct the errors [3].

After the debugging work is completed, re-package the released program, and then follow the test, evaluation and debugging process again for error verification until the defect or error is completely corrected. Through the collection and evaluation of the test results, some qualitative indicators of the quality and reliability of the system can be determined, and finally the above content forms a complete test report [4].

3 Test Process

In the initial stage of the software project, a system test plan needs to be developed. The test plan describes the system test time, personnel, resources, etc.; when the system requirements specification is completed, the requirements need to be reviewed and tested; After the completion of the statute, it is agreed to review and test the system design; after the system is completed, the code needs to be integrated and tested, and the test is confirmed after the system is formed. If you find a problem during the test, you need to debug the system to find out the cause of the problem and fix it [5–8].

4 Test Environments

The test environment of this system is shown in Table 1.

Table 1. System Test Environment

Equipment and environment	Introductions
Host configuration	CPU model: eight core, Intel Core i5, clocked at 3.5 GHz Memory size: 12 GB Disk capacity: 2 TB
Client configuration	CPU model: dual core, Intel Core i3, clocked at 2.6 GHz Memory size: 4 GB Disk capacity: 500 GB
Network configuration	Huawei 1000 Mbps switch
Software and tools	Introductions
Operating system	Server: Windows Server 2008 Client: Windows 7
Database version	SQL Server 2008
Development Tools	Visual Studio 2010, Dreamweaver, Photoshop
Browser	IE 8.0
Web server	IIS 7.0

5   Test Process and Results

The test of this system covers all functional modules of the party affairs work management information system of higher vocational colleges, including document data management, receipt management, document management, approval flow management, conference management, party work management, instant communication management, and statistical analysis of data, system management, and more. Each functional module has a corresponding functional test case based on the black box test method. Each test case includes the use case number; use case content, test data, target result, and test result. The following uses the common system login and user management functions as an example to introduce the test case design [9, 10].

5.1   The User Logs in to the System Test Case

Mainly to detect the system login function, the test is shown in Table 2.

Table 2. System login test case

Use case number:	FunTest-001_ system login test		
Use case module:	System management	Function point:	User login function

Test Conditions:
The construction of the party affairs work management information system in higher vocational colleges is completed, the system can run normally, and the back-end database runs normally.
Test steps:
1. Go to the system login page system_login.aspx
2. Enter the verification code, username, and password on the system login page
3. Click the "Enter System" button on the login page

Use case scenario	Use case data	Request result	Actual result
Login account pair, login password pair	Account: test_user, password: 87654321	Login successfully entered the system	Pass
Login account pair, login password is wrong	Account: test_user, password: 2wsx3edc	Prompt user name or password is incorrect	Pass
Login account is wrong, login password pair Account: test user	password: 87654321	Prompt user name or password is incorrect	Pass
Login account blank, login password pair	Account: empty, password: 87654321	Prompt user name can not be empty	Pass
Login account pair, login password is empty	Account: test_user, password: blank	Prompt password can not be empty	Pass
The account has been logged in (the same account continues to log in)	Account: test_user, password: 87654321	The user has been logged in	Pass

5.2   Meeting Management Function Test Case

It mainly detects the conference management function of the party affairs work management information system of higher vocational colleges, and its test is shown in Table 3.

Table 3. Conference management function test cases

Test number:	Funtest-001_conference management test		
Use case module:	System management	Function point:	User management function

Test Conditions:

The construction of the party affairs work management information system in higher vocational colleges is completed, and the system can operate normally.

Test steps:

1. After entering the account to log into the party affairs work management information system of the higher vocational college, enter the meeting information management (meeting_manage. aspx) function page.

2. Click the "Add Meeting" button to enter the meeting new management page (meeting_add. aspx.)

3. Enter the corresponding conference information in the user's new interface, and click the "Save" button to complete the business operations of the new conference function.

Use case scenario	Use case description	Required result	Actual result
The conference information entered is correct	Conference Name:The second meeting of the party work, time: 2015-11-11, personnel: all personnel, address: the third conference room	The meeting added new successfully	Pass
The entered meeting time is not standardized	Username: 20153030	The meeting time is incorrectly entered	Pass
The entered conference name is not standardized	Password: 2@@@@11qq	Prompt the conference name is entered incorrectly	Pass
The entered meeting already exists	Conference Name: The second meeting of the party work	indicates that the meeting name already exists	pass
The entered meeting information is empty	None	The meeting information cannot be empty	Pass
The name of the conference entered is blank	Conference Name:	Prompt the conference name cannot be empty	Pass

5.3   User Management Function Test Case

It mainly detects the user management function of the party affairs work management information system of higher vocational colleges, and its test is shown in Table 4.

Table 4. User management function test cases

Test number:	Fun test-001_ system user management test		
use case module:	System management:	Function point:	User management function:

Test conditions:
The construction of the party affairs work management information system in higher vocational colleges is completed, and the system can operate normally.
Test steps:
1. After entering the account to log into the party affairs work management information system of the higher vocational college, enter the user information management (user_manage.aspx) function page.
2. Click the "Add User" button to enter the user's new management page (user_add.aspx).
3. Enter the corresponding user information in the user's new interface, and click the "Save" button to complete the business operation of the user's new function.

Use case scenario	Use case description	Required result	Actual result
The user information entered is correct	Username: test, password: 87654321, work number: 2008120001, address: Chongqing	User added successfully	Pass
The user name entered is not standardized	Username: 000032589	The user name is incorrectly entered	Pass
The password entered is too long	Password: 963852741236974	The password is incorrectly entered	Pass
The entered job number does not exist	ID: PPP9631775	The job number does not exist	Pass
The entered user information is empty	None	Prompt user information is not available	Pass
The user name entered is empty	Username: Password: 87654321	The user name cannot be empty	Pass

6 Conclusions

In short, this paper completes the system testing of the party affairs work management information system of higher vocational colleges, first introduces the software and hardware environment for system testing; then describes the principles and processes of system testing; finally, the process and results of the system test are introduced in detail.

Acknowledgements. This paper is one of the results of the research project of the Party School Work Management Information System of Higher Vocational Colleges (Project No.: ZR201713).

References

1. Li, J., Chai, W.: ASP.NET Application Design. Beijing University of Aeronautics and Astronautics Press, Beijing (2009). (In Chinese)
2. MacDonald, M., Freeman, A., Szpuszta, M.: Bosi Studio Translation. ASP.NET 4 Advanced Programming, 4th Edn. People's Posts and Telecommunications Publishing Society, Beijing (2011). (In Chinese)
3. Zhang, W.: C# Implementation method of SQL server 2000 database deployment. Comput. Programm. Skills Maintenance 71–83 (2010). (In Chinese)
4. Xu, L:. Design and implementation of educational administration system based on SQL server 2000. Inf. Secur. Technol. 41–59 (2012). (In Chinese)
5. Wang, X.: Design and implementation of document processing system for small and medium-sized enterprises, pp. 22–40. Tongji University, Shanghai (2008). (In Chinese)
6. Yin, Z.: Research on the development status and countermeasures of e-government in China, pp. 10–11. Northwest University for Nationalities, Lanzhou (2012). (In Chinese)
7. Li, Y.: Analysis of B/S and C/S architecture, pp. 30–33. Friends of Science, Beijing (2011). (In Chinese)
8. Guo, X.: Development and implementation of graduation design management system based on B/S mode. Comput. Technol. Devel. 45–59 (2010). (In Chinese)
9. Song, Y.: Comparison of C/S and B/S architecture of ERP. Inf. Technol 99–100 (2009). (In Chinese)
10. Huang, S.: Design and implementation of cadre education system based on ASP.NET MVC framework. Comput. Technol. Dev. 59–63 (2010). (In Chinese)

Design Method of Axial Clearance Adjustment Gasket for Mass Production Components Based on Big Data Application

Yaqi Li[✉]

College of Mechanical and Electrical Engineering, Yunnan Land and Resources
Vocational College, Kunming 652501, Yunnan, China
1006115841@qq.com

Abstract. With the in-depth development of Internet information technology, the advent of the big data era is changing the development situation in all fields of our society. In order to ensure that the axial assembly clearance of mass production parts meets the technical requirements and at the same time improve the assembly efficiency, shorten the production cycle and reduce the production cost, this paper, under the guidance of the general principle of magnifying the manufacturing tolerance of parts and reducing the difficulty of machining, takes the axial clearance requirement of a rolling bearing of a reducer as the specific research object, and in accordance with the effectiveness analysis of big data and the application of assembly dimension chain theory and probability theory, analyzes and calculates the error distribution range and distribution law. On this basis, the axial clearance adjustment gasket is designed comprehensively and systematically according to three different design ideas, and the application characteristics of the three design methods are analyzed and compared in detail. The research shows that under the condition of mass production combined with big data, the system design of adjusting gasket can be brought into the process of part design, which can make the gap adjustment method more diversified. At the same time, it can also give more scientific and reasonable guidance to adjust the gasket reserve, so as to change the assembly work from passive to active, and realize the improvement of assembly quality and benefit.

Keywords: Mass production · Big data · Axial clearance · Adjusting gasket · Assembly dimension chain · Probability · Design method

1 Introduction

In assembling mechanical parts, it is one of the commonly used gap adjustment methods to adjust the gasket to make the internal axial clearance meet the technical requirements. In actual production, in order to reduce the machining difficulty and cost of parts, the axial size of parts is usually designed directly as no tolerance in the design stage, and the specific requirements of axial clearance are indicated in the assembly drawing. According to the traditional experience to reserve a small amount of adjustment gasket, leave the gap adjustment problem to the assembly link to solve. For single-room small batch production, this is a simple, flexible, economical and practical

© Springer Nature Singapore Pte Ltd. 2020
M. Atiquzzaman et al. (Eds.): BDCPS 2019, AISC 1117, pp. 781–791, 2020.
https://doi.org/10.1007/978-981-15-2568-1_107

method. However, if the above methods are also used in mass production, there will be problems such as on-site repair quantity or adjusting the temporary production quantity of gaskets, which will lead to low assembly efficiency and prolonged production cycle.

In this paper, taking a reducer under the condition of mass production as the research object, the design method of axial clearance adjustment gasket of rolling bearing is analyzed and discussed using Dimension Chain Theory and Probability Theory, in order to improve the assembly efficiency and shorten the production cycle on the basis of ensuring the assembly accuracy of mass production parts.

2   Analysis and Calculation of Axial Clearance of Rolling Bearing of Reducer

The reducer is a three-stage gear reducer, the axial clearance of each bearing is adjusted by adjusting gasket, and the adjusting gasket is placed between the end cover and the end face of the bearing outer ring. In this paper, the assembly dimension chain is established by taking the second stage driven shaft as an example, and it is analyzed and calculated, as shown in Fig. 1.

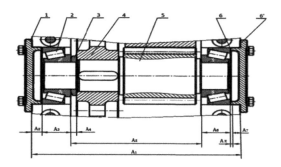

Fig. 1. Structure and dimension chain composition of the second stage driven shaft of the reducer

1. Bearing cover 2. Bearing (32317) 3. Retaining ring 4. Second stage driven gear (helical tooth) 5. Third stage driving gear (straight tooth) 6. Adjust gasket 6'. Adjusting gasket

According to the definition of the closed ring and the constituent ring, the axial clearance (0.07–0.18 mm) of the rolling bearing required after the assembly of the second stage driven shaft of the reducer is the closed ring of the dimension chain, that is, $A_{\Sigma} = 0^{+0.18}_{+0.07}$; while A_1–A_7 is the constituent ring of the assembly dimension chain. Of which:

$A_1 = 460 \pm 0.3$ mm, which is the distance (outer width) between the outer end faces of the bearing seat on the reducer box seat, and it is the increasing ring;

$A_2 = A_7 = 26.5 \pm 0.1$ mm, which is the length of the matching section between the reducer bearing cover and the box bearing housing hole, is the reducing ring;

$A_3 = A_6 = 63.5 \pm 0.2$ mm, which is the width of rolling bearing (32317), is the reducing ring;

$A_4 = 15 \pm 0.1$ mm, which was the thickness of the retaining ring and the decreasing ring;

$A_5 = 280 \pm 0.2$ mm, which is the distance (inner width) between the inner end faces of the bearing seat on the reducer box seat, is the reducing ring.

Among the above dimensions, the tapered roller bearing is the standard part, and the limit deviation of its width (A3 and A6) is determined according to GB/T 307.1-2017 [1]. The limit deviation of the remaining dimensions is usually selected according to the unindicated tolerance, and this example is determined according to the highest precision of the general tolerance of the linear dimension given by GB/T1804-2000 [2] —the precision class (f-level).

The formula for calculating the cumulative error of the closed ring of the assembly dimension chain is as follows [3]:

Upper deviation

$$B_S A'_\Sigma = \sum_{i=1}^{m} B_S \overrightarrow{A}_i - \sum_{i=m+1}^{n-1} B_x \overleftarrow{A}_i \tag{1}$$

Lower deviation

$$B_x A'_\Sigma = \sum_{i=1}^{m} B_x \overrightarrow{A}_i - \sum_{i=m+1}^{n-1} B_s \overleftarrow{A}_i \tag{2}$$

Error range

$$T_{A'_\Sigma} = \sum_{i=1}^{n-1} T_{Ai} = B_s A'_\Sigma - B_x A'_\Sigma \tag{3}$$

The limit deviations (or tolerance values) of each dimension of A_1–A_7 determined by the above method are replaced by formulas (1), (2) and (3) respectively. The results are as follows:

$$B_S A'_\Sigma = 0.3 - (-0.1 - 0.1 - 0.2 - 0.2 - 0.1 - 0.2) = 1.2\,\text{mm}$$

$$B_X A'_\Sigma = -0.3 - (0.1 - 0.1 - 0.2 - 0.2 - 0.1 - 0.2) = -1.2\text{mm}$$

$$\text{Error range}: T'_{A\Sigma} = 1.2 - (-1.2) = 2.4\,\text{mm}$$

From the above calculation, it can be known that under the condition that the limit deviation of each constituent ring (except rolling bearing) is selected according to the precision level (f-level) without injection tolerance, the actual axial assembly clearance (A'_Σ) is formed and changing within the interval [−1.2, 1.2], exceeding the gap requirements of 0.07–0.18 mm.

Because the assembly gap formed by multi-ring (n ≥ 4) dimension chain is a random variable, according to the theory of probability, the distribution is always close to the normal distribution under the condition of batch production (output is more than 300 sets/batch) [4]. Among them, the gap interval with the highest occurrence probability ($2\sigma = 68.27\%$) is [−0.4, 0.4]. Since only the reserve requirement of the adjustment gasket which meets the $A'_\Sigma > A_\Sigma$ (under condition of positive clearance) is indicated in the assembly drawing, and usually the thickness, grouping and number of adjusting gaskets are mostly based on traditional experience (for example, only 8 adjusting gaskets with a thickness of 0.3 mm for the second stage driven shaft are required in the assembly drawing of the reducer). It is bound to cause the reserved adjustment gasket unable to meet the gap adjustment requirements with the highest frequency [7].

3   Study on the Design Method of Axial Clearance Adjusting Gasket Under the Condition of Batch Production

3.1   Adjusting Gasket Design to Ensure that the Assembly Gap Is Equal to or Greater Than the Technical Requirements

In order that the assembly clearance is equal to or greater than the technical requirement (0.07–0.18):

$$B_S A'_\Sigma = 0.07 \tag{4}$$

According to formulas (2) and (4), the lower deviation of the increasing ring or the deviation of the decreasing ring in the dimension chain shall be redistributed:

Rolling bearings are standard parts, and the limit deviations ($B_S A_3$ and $B_S A_6$) of their axial dimensions are fixed values that cannot be changed. Dimension A_5 is a internal size, so it is difficult to process and the limit deviation should not be changed, so that the purpose of $B_X A'_\Sigma = 0.07$ can only be achieved by changing $B_X A_1$ or $B_S A_2$, $B_S A_7$ and $B_S A_4$. The main results are as follows:

(1)   under the condition that the lower deviation ($B_X A_1$) of the increasing ring is constant, the $B_X A'_\Sigma = 0.07$ mm is made by changing the deviation on the decreasing ring ($B_S A_2$, $B_S A_7$ and $B_S A_4$).

With reference to the basic dimension segment of GB/1804-2000, $A_2 = A_7$, which belongs to the same dimension section as A_4, the equal limit deviation can be obtained according to the principle of equal tolerance method, that is, $B_S A_2 = B_S A_7 = B_S A_4$. After bringing the relevant data into the formula (4), the following can be obtained:

$B_S A_2 = B_S A_7 = B_S A_4 = -0.97/3 = -0.324$ mm, according to the precision level (f-level) tolerance, take the tolerance values $T_{2,7,4} = 0.2$, then $B_X A_2 = B_X A_7 = B_X A_4 = -0.324 - 0.2 = -0.524$ mm.

The adjusted limit deviation values are substituted into formulas (1), (2) and (3), respectively, and the following can be obtained:

$$B_SA'_\Sigma = 0.3 - (-0.524 - 0.524 - 0.2 - 0.2 - 0.524 - 0.2) = 2.47\,\text{mm}$$

$$B_XA'_\Sigma = -0.3 - (-0.324 - 0.324 + 0.2 + 0.2 - 0.324 + 0.2) = 0.07\,\text{mm}$$

$$T_{A\Sigma'} = 2.47 - 0.07 = 2.4\,\text{mm}$$

(2) Under the condition that the deviation on the decreasing ring (B_SA_2, B_SA_7 and B_SA_4) is constant, the $B_XA'_\Sigma = 0.07$ mm is made by changing the deviation under the increasing ring (B_XA_1).

The formula is the same as before:

$$B_XA_1 = (0.1 + 0.1 + 0.2 + 0.2 + 0.2 + 0.1) + 0.07 = 0.97\,\text{mm}$$

According to the precision level (f-level) tolerance, if the tolerance value T = 0. 6 is taken, then

$$B_SA_1 = 0.97 + 0.6 = 1.57\,\text{mm}$$

$$B_SA'_\Sigma = 1.57 - (-0.1 - 0.1 - 0.2 - 0.2 - 0.1 - 0.2) = 2.47\,\text{mm}$$

$$B_XA'_\Sigma = 0.97 - (0.1 + 0.1 + 0.2 + 0.2 + 0.1 + 0.2) = 0.07\,\text{mm}$$

$$T_{A\Sigma'} = 2.47 - 0.07 = 2.4\,\text{mm}$$

Through the above calculation, the error band can be changed from [−1.2, 1.2] to [0.07, 2.47] (see Fig. 3), and the assembly gap is equal to or greater than the technical requirement. When the assembly clearance is greater than the specified maximum value (0.18 mm), an adjustment gasket should be installed between the end cover and the rolling bearing (as shown in Fig. 1, part 6).

In mass production, the adjusting gasket shall be grouped and designed in thickness:

The gasket manufacturing tolerance T_{AK} = +0.03 mm, was divided into m = $T_S/T_\Sigma - T_{AK}$ = (2.4 − 0.11)/0.11 − 0.03 = 29 groups, and the grouping grade difference $T_R = T_S/m$ = (2.4 − 0.11)/29 = 0.079 [5].

$$\text{Thinnest gasket } \delta_{min} = T_R = 0.08^{+0.03}\,\text{mm}$$

$$\text{Thickest gasket } \delta_{max} = B_SA'_\Sigma - B_SA_\Sigma = 2.47 - 0.18 = 2.29^{+0.03}\,\text{mm}$$

$$\text{Gasket thickness } \delta_I = \delta_{min} + (i-1)T_R$$

Using the powerful function calculation function of Excel, the relevant data such as the gap grouping interval, the corresponding gasket thickness and the percentage of each thickness gasket can be easily obtained (as shown in Table1).

Table 1. Gasket design analysis of assembly clearance equal to or greater than $B_X A_\Sigma$ (0.07 mm)

Group No.	Range of clearance		Thickness of gasket/mm	Type of gasket	Percentage of gasket	Group No.	Range of clearance		Thickness of gasket/mm	Type of gasket	Percentage of gasket
	Start	End					Start	End			
0	0.07	0.18	0.00	No need for Gasket	0.19	16	1.365	1.444	1.27	Gasket 6	7.44
1	0.18	0.259	0.08	Gasket 6	0.25	...	...	...	...	...	...
...	...	...	...	...	...	19	1.602	1.681	1.50	Gasket 6	5.12
4	0.417	0.496	0.32	Gasket 6	1.00	...	...	...	...	...	...
...	...	...	...	...	...	24	1.997	2.076	1.90	Gasket 6	1.26
9	0.812	0.891	0.71	Gasket 6	4.56	...	...	...	...	...	...
...	...	...	...	...	...	29	2.392	2.47	2.29	Gasket 6	0.12

3.2   Adjusting Gasket Design to Ensure that the Constituent Rings Are Manufactured According to Unindicated Tolerances

When the clearance is $[-1.2, 0.07]$, an adjustment gasket (piece 6′ in Fig. 1) can be installed between the bearing cover and the reducer box seat to make the bearing cover move outward to increase the clearance; when the gap is $[0.18, 1.2]$, the installation position of the adjustment gasket is exactly the same as that of the first method (piece 6 in Fig. 1).

The adjustment of gasket thickness and grouping of mass production in two different positions need to be calculated separately (the calculation formula is the same as before):

The grouping of interval $[-1.2, 0.07]$ adjustment gaskets is calculated as follows:

The number of groups was $1.27/0.11 - 0.03 = 16$ groups, and the grouping grade difference was $T_R = 1.27/16 = 0.079$ mm. Of which:

Thinnest gasket $\delta_{min} = 0.08^{+0.03}$ mm

Thickest gasket $\delta_{max} = 1.27^{+0.03}$ mm

$[0.18, 1.2]$ the grouping of interval adjustment gaskets is calculated as follows:

The group was divided into 13 groups ($m = 1.02/0.11 - 0.03$), and the grouping grade difference was $TR = 1.02/13 = 0.079$ mm. Of which:

$$\text{Thinnest gasket } \delta \min = 0.08^{+0.03} \text{ mm}$$

$$\text{Thickest gasket } \delta \max = 1.02^{+0.03} \text{ mm}$$

The relevant data of the two adjustment gaskets, such as the gap grouping interval, the corresponding gasket thickness and the percentage of each thickness gasket, are shown in Table 2.

Table 2. Gasket design analysis to ensure that the constituent rings are manufactured according to unindicated tolerances (no repair area)

Group No.	Range of clearance		Thickness of gasket/mm	Type of gasket	Percentage of gasket	Group No.	Range of clearance		Thickness of gasket/mm	Type of gasket	Percentage of gasket
	Start	End					Start	End			
1	−0.009	−0.070	Gasket 6'	0.08	7.84	0	0.070	0.180	No need	0.00	10.42
...	...	...	...	...	...	1	0.018	0.259	Gasket 6'	0.08	6.77
6	−0.404	−0.325	Gasket 6'	0.47	5.20	...	...	...	...	...	...
...	...	...	...	...	...	3	0.338	0.417	Gasket 6'	0.24	5.05
11	−0.799	−0.720	Gasket 6'	0.87	1.30	...	...	...	...	...	...
...	...	...	...	...	...	8	0.733	0.812	Gasket 6'	0.63	1.23
16	−1.200	−1.115	Gasket 6'	1.27	0.13	...	...	...	...	...	...
...	...	...	...	...	...	13	1.128	1.200	Gasket 6'	1.02	0.11

On the basis of the above, if the interval of [−0.18, 0.07] is designed as a repair area (the maximum amount of repair is 0.25 mm, the empirical value of the most suitable amount of repair in actual production is generally 0.1–0.3 mm). The gasket group for increasing the gap can be reduced by three groups, and the corresponding thickness of the adjustment gasket 6' of the interference interval [−1.2, −0.18] and the gap interval [0.18, 1.2] are completely equal. The related calculations for the two types of gaskets are exactly the same:

The group was divided into 13 groups (m = 1.02/0.11 − 0.03 = 13groups), and the grouping grade difference was T_R = 1.02/13 = 0.079 mm. Of which:

$$\text{Thinnest gasket } \delta_{min} = 0.08^{+0.03} \text{ mm}$$

$$\text{Thickest gasket } \delta_{max} = 1.02^{+0.03} \text{ mm}$$

The gap grouping interval of the two adjustment gaskets, the corresponding gasket thickness and the percentage of each thickness gasket are shown in Table 3.

Table 3. Gasket design analysis table to ensure that the constituent rings are manufactured according to uninjected tolerances (setting up repair areas)

Group No.	Range of clearance		Thickness of gasket/mm	Type of gasket	Percentage of gasket	Group No.	Range of clearance		Thickness of gasket/mm	Type of gasket	Percentage of gasket
	Start	End					Start	End			
0	−0.18	−0.07	0.00	Repaired area	24.31	0	0.07	0.18	0.00	No need fro Gasket	10.42
1	−0.259	−0.18	0.08	Gasket 6'	6.77	1	0.18	0.259	0.08	Gasket 6'	6.77
...	...	...	...	...	...	...	...	...	...	...	...
3	−0.417	−0.338	0.24	Gasket 6'	5.05	3	0.338	0.417	0.24	Gasket 6'	5.05
...	...	...	...	...	...	...	...	...	...	...	...
Group No.	Range of clearance		Thickness of gasket/mm	Type of gasket	Percentage of gasket	Group No.	Range of clearance		Thickness of gasket/mm	Type of gasket	Percentage of gasket
	start	end					start	end			
8	−0.812	−0.733	0.63	Gasket 6'	1.23	8	0.733	0.812	0.63	Gasket 6'	1.23
...	...	...	...	...	...	...	...	...	...	...	...
13	−1.2	−1.128	1.02	Gasket 6'	0.11	13	1.128	1.2	1.02	Gasket 6'	0.11

3.3   Adjusting Gasket Design to Ensure that the Probability of not Installing the Gasket Is the Maximum

In order to maximize the probability of not installing the adjusting gasket, the distribution center of the error zone of the closed ring must be coincident with the center of its tolerance zone [4] (that is, the error zone $T A \Sigma'$ shifts $\frac{B_x A_\Sigma + B_s A_\Sigma}{2}$ to the right, in this case, $\frac{B_x A_\Sigma - B_s A_\Sigma}{2} = 0.125 mm$), where the length of the negative gap interval and the gap interval are equal. The grouping and corresponding thickness of the two kinds of adjusting gaskets are exactly the same, and the related calculations are exactly the same:

The number of groups m = 1.145/0.11 − 0.03 = 15 groups, and the grade difference was T_R = 1.145/15 = 0.076 mm, of which:

$$\text{Thinnest gasket } \delta_{min} = 0.08^{+0.03} \text{ mm}$$

$$\text{Thickest gasket } \delta_{max} = 1.14^{+0.03} \text{ mm}$$

Because the whole error band moves 0.125 mm to the right, the lower deviation value $B_x A_\Sigma' = -1.075$ mm and the upper deviation value $B_s A_\Sigma' = 1.325$ mm, so the limit deviation of part of the component cycle in the dimension chain must also be adjusted. If

$$B_X A_\Sigma' = -1.075 \tag{5}$$

The main results are as follows:

(1) Under the condition that the lower deviation (B_XA_1) of the increasing ring is constant, the $B_XA'_\Sigma = -1.075$ mm is made by changing the deviation on the decreasing ring (B_SA_2, B_SA_7 and B_SA_4).

Make the $B_SA_2 = B_SA_7 = B_SA_4$. After bringing the relevant data into the formula (5), we can get:

$$B_SA_2 = B_SA_7 = B_SA_4 = 0.175/3 = 0.058 \text{ mm}$$

$$B_XA_2 = B_XA_7 = B_XA_4 = 0.058 - 0.2 = -0.142 \text{ mm}$$

$$B_SA'_\Sigma = 0.3 - (-0.142 - 0.142 - 0.2 - 0.2 - 0.142 - 0.2) = 1.325 \text{ mm}$$

$$B_XA'_\Sigma = -0.3 - (0.058 + 0.058 + 0.2 + 0.2 + 0.058 + 0.2) = -1.075 \text{ mm}$$
$$T'_{A\Sigma} = 1.325 - (-1.075) = 2.4\text{mm}$$

(2) Under the condition that the deviation on the decreasing ring (B_SA_2, B_SA_7 and B_SA_4) is constant, the $B_XA'_\Sigma = -1.075$ mm is made by changing the deviation under the increasing ring (B_XA_1).

The relevant data can be obtained after bringing the relevant data into the formula (4).

$$B_XA_1 = (0.1 + 0.1 + 0.2 + 0.2 + 0.2 + 0.1) - 1.075 = -0.175 \text{ mm}$$

$$B_SA_1 - -0.175 + 0.6 = 0.425 \text{ mm}$$

$$B_SA'_\Sigma = 0.425 - (-0.1 - 0.1 - 0.2 - 0.2 - 0.1 - 0.2) = 1.325 \text{ mm}$$

$$B_XA'_\Sigma = -0.175 - (0.1 + 0.1 + 0.2 + 0.2 + 0.1 + 0.2) = -1.075 \text{ mm}$$

$$T'_{A\Sigma} = 1.325 - (-1.075) = 2.4 \text{ mm}$$

4 Analysis and Comparison of Three Design Methods of Adjusting Gasket

The greatest advantage of the first method is that it can ensure that there is no minimum gap less than that specified in the technical requirements after assembly. When the gap is greater than the specified maximum clearance, only one kind of adjustment gasket (such as gasket 6 in Fig. 1) needs to be installed in the part where the gap can be reduced. However, most of the gaps (99.54% in this case) are larger than the specified gaps, and the amount of compensation is large, and the gasket thickness difference is large, resulting in the difficulty of gasket manufacturing and on-site management, engineering quantity and investment in tooling and inspection tools. In addition, the limit deviation value of part of the constituent cycle needs to be readjusted through dimension chain analysis and calculation, because the adjusted limit deviation value is not equal to the unindicated tolerance specified in the national standard. In the stage of part design, the problem of accuracy excess caused by unindicated tolerance should be fully considered, the relevant design and calculation of gasket should be done well, and the limit deviation value obtained by adjustment and calculation should be marked in the part drawing [8].

The limit deviation of each component ring of the second method can be manufactured according to the unindicated tolerance. compared with the first and third methods, the part design process is not affected by the assembly dimension chain, and the clearance probability of the gasket without adjusting the gasket is much higher than that of the first method. Although the total number of gasket groups is equal to that of the first method, because it is two different types of gaskets installed in different positions, the maximum compensation amount and the maximum thickness difference of gaskets are only about 1 of the first method. If a suitable repair area is designed in the negative gap range near BxA_Σ (the empirical value of the repair quantity is generally 0.1–0.3 mm), the required minimum gap can be obtained by using a general tool in a short period of time. The number of gaskets that increase the gap can be reduced accordingly. However, two different types of adjustment gaskets must be manufactured and stored [9].

The third method can maximize the gap probability without adjusting the gasket. Although two different types of gaskets need to be installed in different positions, the number of groups of the two gaskets is equal to the thickness of the corresponding group. It minimizes the difficulty of gasket manufacturing, storage and management. However, because the position of the error band is moved, similar to the first method, the limit deviation value of part of the constituent cycle must be readjusted through the analysis and calculation of the dimension chain, and the limit deviation value obtained by adjustment and calculation must be marked in the part drawing [10].

5   Conclusion

The main results are as follows:

1. The axes of most mechanical parts usually need to leave a certain axial assembly clearance, and their values are generally small, so even if the axial dimensions of the related parts are not according to the precision grade tolerance specified in the national standard, There is bound to be the problem of poor accuracy. Because the limit deviations of the axial dimensions of the related parts determined according to the latest national standard (GB/T1804-2000) are symmetrical, Therefore, the probability of negative gap is greater than or equal to the probability of positive gap.
2. Under the condition of mass production, the system design of adjusting gasket is brought into the process of part design. On the basis of magnifying the manufacturing tolerance of parts, reducing the difficulty and cost of machining and manufacturing, ensuring assembly accuracy, improving assembly efficiency and shortening the production cycle, the gap adjustment method can be more diversified.

At the same time, it can also give more scientific and reasonable guidance to adjust the gasket reserve, so as to change the assembly work from passive to active, and realize the double harvest of assembly quality and benefit.

3. In the system analysis and design of adjusting gasket, using the powerful function calculation function of Excel, the detailed data such as the interval of gap grouping, the corresponding thickness of gasket and the percentage of each thickness gasket can be easily obtained. At the same time, it also reduces the time and tedious degree of calculation.

4. The probability of gasket with different thickness is different. Because the amount of gasket with high probability is large and the amount of gasket with low probability is small, in order to avoid excessive or insufficient gasket reserve, the reserve quantity of gasket with different thickness can be determined according to the probability of gasket (thickness) [6].

References

1. General Administration of Quality Supervision, Inspection and Quarantine of the People's Republic of China, Standardization Administration of the People's Republic of China. GB/T307.1 2017, Rolling bearings—Radical bearings—Geometrical product specifications (GPS) and tolerance values. China Zhijian Publishing House, Beijing (2017)
2. General Administration of Quality Supervision, Inspection and Quarantine of the People's Republic of China. GB/T1804-2000, General Tolerances Tolerances for Linear and Angular Dimensions Without Individual Tolerance Indications. China Zhijian Publishing House & Chinese Standard Publishing House, Beijing (2005)
3. Yan, C.: Mechanical Equipment Repair Technology, 2nd edn, pp. 24–26. China Machine Press, Beijing (2010)
4. Fang, P.: Calculation method of minimum compensating gasket in assembly dimension chain. Acta Aeronautica et Astronautica Sinica 3, 91–97 (1979)
5. Zhang, R.: Principle of Dimension Chain and Its Application, pp. 200–208. China Machine Press, Beijing (1986)
6. Luo, Y., Wang, J.: Study on the solution of assembly dimension chain by grouping gasket method. J. Hebei Inst. Mechano-Electric Eng. 12(2), 34–2004 (1995)
7. Zeng, F., Chen, X., Li, W.: Clearance adjustment and coaxiality measurement of centrifugal compressor sliding bearing. Equip. Manag. Maintenance 03, 41–42 (2016)
8. Huang, J.: Improved valve clearance adjustment shim assembly process of diesel engine with measuring tool. Guangxi Mach. 02, 25–27 (2001)
9. Xie, Z.: Influence of axial clearance of motor and solutions. Sci. Technol. Innov. 03, 128–129 (2019)
10. Zhou, H.: Adjustment of axial clearance of crankshaft. Friends Agricul. Mach. Tools 05, 15 (1995)

Exploration on the Implementation Path of Drawing Course Under the Information-Based Background

Jilan Yan[✉]

Department of Mechanical and Electrical Engineering,
Shandong Vocational College of Light Industry, Zibo 255300, Shandong, China
yanhan0628@163.com

Abstract. Information-based teaching is a popular teaching means at present, practice has proved that advanced information-based teaching means can improve students' learning interest and enthusiasm. When students read and draw mechanical patterns, they realize the spatial image thinking process of mutual transformation of "objects" and "figures" from object to figure and then from figure to object. Most students have difficulty in adapting to the situation. This paper, from the six aspects of the ideological and political education, blending learning, adopting advanced manufacturing technology, teaching platform, application of teaching resources and science and technology learning associations, will carry out the analysis of teaching implementation path in the context of information-based technology. Based on the characteristics of mechanical drawing course, this paper explores the teaching mode and path under the background of informatization, Writer hopes to provide some references for mechanical drawing teachers in information-based teaching.

Keywords: Information-based · Teaching path · Curriculum ideological and political education · Intelligent evaluation system · Blending learning

1 Introduction

Reading and Drawing of Mechanical Pattern is a required professional basic course for mechanical majors. The course mainly trains students with freehand drawing, tool drawing and computer drawing skills to read mechanical drawing and strictly follow national standards. Mechanical pattern, also known as the technical language of engineering, is a necessary basic skill for mechanical students in higher vocational colleges and the cornerstone of subsequent courses.

Courses are taught to first-year vocational students majoring in mechanical design and manufacturing. This major are mainly composed of students from single enrollment, spring university entrance examination, and summer university entrance examination. Due to the difficulty in recruiting students, most students have poor performance and uneven basic knowledge. Students lack confidence and enthusiasm in learning professional courses and are addicted to mobile network. For freshmen, they are full of longing for their major, curious about new things, and have strong ability to use mobile software. Students can use information teaching platform to assist learning.

© Springer Nature Singapore Pte Ltd. 2020
M. Atiquzzaman et al. (Eds.): BDCPS 2019, AISC 1117, pp. 792–799, 2020.
https://doi.org/10.1007/978-981-15-2568-1_108

From the previous students' learning of mechanical pattern reading and drawing, students have the following deficiencies in the learning process.

(1) Lack of professional quality, the awareness of following national standards has not been firmly established [1, 2], such as random drawing, lack of the spirit of excellence and hard-working.
(2) Lack of awareness of challenging difficulties, lack of independent inquiry ability, lack of ability to analyze and solve problems.

2 The Path Design of Teaching Implementation

Information-based teaching has been popular, but it is only a teaching method. In order to improve students' learning enthusiasm and efficiency and achieve the goal of whole process education, it is necessary to design the implementation path of the course teaching systematically [3]. The implementation path of mechanical pattern reading and drawing mainly includes the following aspects.

2.1 Whole Process of Education, to Introduce Ideological and Political Ideas for Students to Lay a Solid Ideological Foundation

In order to meet the new requirements for the training of technical and skilled personnel in the new era, to conform to the fundamental task of establishing professional teaching standards and implementing cultivation with virtue issued by the Ministry of Education, to cultivate students' independent learning ability, information quality and the professional spirit of following national standards, occupational spirit, and the artisan spirit of striving for perfection and loving labor.

Course design process should carry out the ideological and political education requirements, such as the teaching process will be combined with "Made in China 2025" and the "Strategy of Manufacturing Power", to guide the students to set up the broad ideal and patriotic feelings, then adopt 3D printing technology of manufacturing, and quote the relevant videos, to reflect the new technologies, new processes, new specifications of industry upgrading in the field of mechanical manufacturing. The stories of craftsmen will be illustrated in the teaching, to guide the students to attach great importance to strengthen labor education and carry forward the spirit of labor. To take the drawing error in enterprises will affect the production as an example, to cultivate students for "craftsmen" spirit like devotion to work, striving for perfection, etc.

2.2 Combination of 3D Printing Model with Digital Instructional Model

The traditional teaching model is rigid and lacks some novelty for current students, and the structure is not flexible enough. In order to improve the interest of students' learning, 3D printing intelligent manufacturing technology is introduced into the classroom as the carrier of 3D printing model, and models of different positions are printed according to the learning needs, which increases the intuitiveness and reduces the difficulty of learning [4].

3D printing model is designed and manufactured by students of "3D Maker Society" through the training room of "3D Printing Maker Center" of the department. Students can participate in the whole process and have a preliminary understanding of 3D printing manufacturing technology.

At the same time, the teaching adopts the combination of "real" and "virtual", and introduces the model of 3D software Solidworks into the course teaching. Instead of drawing model, students just watch the model from different angles, and cut them, and then come to the understanding of object to figure.

2.3   Blending Learning, Break the Traditional Boundaries Among Classroom, Practical Training and AutoCAD Learning

Mechanical drawing no longer adopts the simple teaching and practice integration mode, and the course teaching is no longer a simple conversion between textbooks and problem sets. Instead, the traditional mechanical drawing training room and AutoCAD training room are replanned and integrated, the classroom teaching, manual drawing and CAD drawing are organically integrated. The accuracy of CAD drawing is high, which can supplement the inaccuracy of manual drawing by means of CAD drawing [8]. In addition, computer drawing and manual drawing are indispensable in the learning process, and students will complete the learning of the two skills. After the combination, it effectively increases the diversity and interest of learning mechanical drawing.

2.4   Introduce CAD Intelligent Evaluation System into Classroom Teaching

There are many kinds of mechanical drawing methods, and the expression method is flexible and changeable [6]. The complete drawing has its basic rules, drawing methods, dimension annotation methods, surface roughness, limit and fit, geometric tolerance. It is difficult for students to draw complete, correct, reasonable and beautiful drawings according to the standards. The amount of work for teachers to review is also large. It takes 2–3 h to review an assignment, and the time for complex drawings will be longer and there will be omissions [5].

For the first time, the CAD intelligent evaluation system is applied to the evaluation section of CAD drawing [6, 7]. The software can highlight the unqualified part of the inspection results.

Basing on the national mechanical drawing standards, it can check the standardization of CAD graphics drew by students. The students can also test the standardization and accuracy of their drawing, and correct the deviation in the drawing.

At the same time, the application of the Intelligent Scoring increases the objectivity and efficiency of drawing evaluation. The teacher can get to know the mistakes of students, and share the statistical analysis and the mastery status of unit knowledge.

2.5   Rich Technical Resources Are Applied and Flipped Classroom Is Adopted as the Main Teaching Mode

(1) On the Chaoxing learning platform, there have been already systematic teaching resources on the reading and drawing of mechanical pattern [5]. It has been in use since 2018. The teaching platform can realize students' independent learning before class, in-class test, online discussion, classroom implementation process evaluation and testing, also the after-class development, feedback of students' learning results.
(2) Rich teaching resources, such as microlecture, MOOC, intelligent vocational education and icourses.

Through the intelligent vocational education and icourses, relevant teaching resources can be obtained [9] The important knowledge points of the course and the contents that are not easy for students to understand are presented in the form of QR Code. Students can scan the QR Code and watch the explanation of microlecture video repeatedly to expand the learning space and time. According to the 3D model, rotation, cutting and other operations are carried out to realize assistance to teaching and learning.

2.6   Take Surveying and Mapping Modeling and 3D Maker Society as the Platform for Communication and Learning

At the beginning of freshmen's entrance, they can join the surveying and mapping modeling society and 3D maker society. The surveying and mapping modeling society will carry out the basic exercises of mechanical pattern manual drawing and CAD drawing in the first semester, so as to strengthen the learning of professional knowledge of mechanical pattern. 3D maker society mainly carries out 3D basic knowledge, as well as the simple SolidWorks modeling method to print 3D models.

3   The Implementation of Classroom Teaching Process

In the following, the writer will take the half-section view drawing method as an example to illustrate the implementation path of informatization in the teaching process.

The implementation of classroom teaching is divided into three aspects: pre-class, in-class and after-class. Pre-class preview task list is issued through the teaching platform. In class, the class task list will be carried out for consolidation exercise. After class, a training task list will be issued for reinforcement.

3.1   Preparation Before Class

See Table 1.

Table 1. Preparation before class

Preparation before class			
Teaching link	Teaching content	Teacher activities	Student activities
Autonomous learning	1. Comparison of advantages and disadvantages between sectional view and view 2. Print 3D models	Teachers issue pre-class tasks through the hyperstar learning platform	Complete the preview task list before class

3.2   In-Class

See Tables 2, 3, 4 and 5.

Table 2. Classroom implementation-First link

Classroom implementation			
Teaching link	Teaching content	Teacher activities	Student activities
Pre-class learning test	1. Views differ from sectional views 2. Structural Characteristics of the Model 3. Concept of semi-sectional view	1. Teachers Show Students' Pre-class Tasks through Teaching Platform 2. Teachers put forward questions 3. Teachers Propose Better Plans	1. Students Show the difference between View and Section View 2. Students analyze the characteristics of the model and determine the expression scheme

Table 3. Classroom implementation-Second link

Classroom implementation			
Teaching link	Teaching content	Teacher activities	Student activities
Break through the key and difficult points	1. technique of drawingsemi-sectional view 2. Matters needing attention	1. Teachers demonstrate CAD Drawing Semi-sectional View 2. The teachers sum up the main points of drawing	1. Students draw semi-sectional views through CAD 2. Students discuss the application of semi-sectional view

Table 4. Classroom implementation-Third link

Classroom implementation			
Teaching link	Teaching content	Teacher activities	Student activities
Classroom test	1. Select the correct half section view 2. According to the given view, the students draw semi-sectional view through CAD software trought 3.CAD Intelligent Evaluation System evaluates the accuracy of drawing	1. Teachers check students' answers 2. Teachers inspect students' drawing status 3. Teachers give standard answers and use the evaluation system to test them	1. Login platform and Answer questions in class 2. Students Use CAD to Draw Semi-sectional Views of Known Views

Table 5. Classroom implementation-Fourth link

Classroom implementation			
Teaching link	Teaching content	Teacher activities	Student activities
Summary	1. Summary of teaching content: Drawing of semi-scctional view 2. Push expanded content	1. Teachers summarize learning contents 2. Teachers release tasks through teaching platform	1. Students reflect on learning content 2. Students receive after-school assignments

3.3 After-Class

See Table 6.

Table 6. Task after class

Task after class			
Teaching link	Teaching content	Teacher activities	Student activities
Intensive training	Draw B-B semi-sectional view	Teachers release tasks through teaching platform	1. Learning the corresponding resources of "wisdom vocational education" and "love curriculum" 2. Read the view and complete the drawing of semi-sectional view

4 Teaching Ideas and Improvement

4.1 In-Depth Exploration into Resources of Typical Part Cases of Enterprises

Enterprises will be visited, to obtain basic materials of vocational ability through analysis of enterprise experts, and establish the course ability list on reading and drawing of mechanical pattern [10]. Explore the typical part cases produced by enterprises to explain the knowledge, and further combine with the production practice, so as to meet the needs of modern manufacturing enterprises for professional quality.

4.2 Expand the Typical Production Process

The reading and drawing of mechanical pattern serves for the later stage design and processing. The later stage teaching introduces the typical production process of modern intelligent manufacturing into the course resource base as the extended knowledge, so as to improve the professional quality of students.

5 Final Words

High-level information means appeals, if there is lack of effective teaching path, good teaching effect can not be achieved. Only the combination of the two can effectively break through the important and difficult points of teaching and improve the quality of teaching. Blending learning, teaching platform, digital teaching software and application of intelligent evaluation system greatly expand students' learning time and space, break through the limitations of traditional classroom, practical training room and 45-min teaching time, so as to improve the learning efficiency.

References

1. Sokol, R.G., Slawson, D.C., Shaughnessy, A.F.: Teaching evidence-based medicine application: transformative concepts of information mastery that foster evidence-informed decision-making. BMJ Evid.-Based Med. **24**(4), 149–154 (2019)
2. Elske, A., Hackl, W.O., Felderer, M., Hoerbst, A.: Developing and evaluating collaborative online-based instructional designs in health information management. Stud. Health Technol. Inf. **243**, 8–12 (2017)
3. Stefan, D.: Content and instructional design of MOOCs on information literacy. Inf. Learn. Sci. **120**(3/4), 173–189 (2019)
4. Long, J., Li, S., Liang, J., Wang, Z., Liang, B.: Preparation and characterization of graphene oxide and it application as a reinforcement in polypropylene composites. Polymer Compos. **40**(2), 723–729 (2019)
5. MacAulay, M.: Antiviral marketing: the informationalization of HIV prevention. Can. J. Commun. **44**(2), 239–261 (2019)
6. Pollaris, G., Note, S., Desruelles, D., Sabbe, M.: Novel IT application for reverse triage selection: a pilot study. Disaster Med. Public Health Preparedness **12**(5), 599–605 (2017)

7. Xiaolu, H., Huang, H., Pan, Z., Shi, J.: Information asymmetry and credit rating: a quasi-natural experiment from China. J. Banking Financ. **106**, 132–152 (2019)
8. Iskrev, N.: On the sources of information about latent variables in DSGE models. Eur. Econ. Rev. **119**, 318–332 (2019)
9. Macaluso, F.S., Mocci, G., Orlando, A., Scondotto, S., Fantaci, G., Antonelli, A., Leone, S., Previtali, E., Cabras, F., Cottone, M.: Prevalence and incidence of inflammatory bowel disease in two Italian islands, sicily and Sardinia: a report based on health information systems. Dig. Liver Dis. **51**(9), 1270–1274 (2019)
10. Yeom, H., Ko, Y., Seo, J.: Unsupervised-learning-based keyphrase extraction from a single document by the effective combination of the graph-based model and the modified C-value method. Comput. Speech Lang. **58**, 304–318 (2019)

Practical Teaching Mechanism
of Electronic Commerce

Qiong Zhang and Hui-yong Guo[(⊠)]

School of Management, Wuhan Donghu University, Wuhan, China
515275934@qq.com, 345430543@qq.com

Abstract. Practical teaching is very important for cultivating students' practical ability and innovative ability. At present, practice teaching of electronic commerce is insufficient, so it is necessary to construct an effective practice teaching mechanism for the major of electronic commerce. This paper puts forward that the teaching practice process should run through the cultivation of students' independent ability, the teaching process should pay attention to efficiency and effect as well as the combination of basic theory and practical teaching, the teaching practice should pay attention to the mode of cooperation and joint training between schools and enterprises, and practical teaching should pay attention to the various abilities cultivation of the major of electronic commerce, which provides a certain theoretical basis for the establishment of the practical teaching mechanism of electronic commerce.

Keywords: Electronic commerce · Teaching mechanism · Practice

1 Introduction

In recent years, with the popularization of network and the development of economy, electronic commerce is infiltrating into every field of economic life. The demand for professional talents in electronic commerce is expanding and the requirement for talents is also increasing. Due to the late start of electronic commerce education in China, the professional construction and teaching mode are still being explored [1]. There is still a considerable gap of graduates with the major of electronic commerce cultivated in colleges and universities between the structure of knowledge and ability as well as the needs of enterprises. Therefore, how to cultivate applied talents with practical ability, innovative spirit and entrepreneurial consciousness has become the focus of electronic commerce curriculum reform [2, 3].

2 The Logic of Constructing Practical Teaching Mechanism of the Major of Electronic Commerce

2.1 According to the Goal of National Higher Education Reform

The major of electronic commerce should cultivate high-quality and high-skilled talents who are equipped with professional ethics and legal consciousness, use modern

M. Atiquzzaman et al. (Eds.): BDCPS 2019, AISC 1117, pp. 800–805, 2020.
https://doi.org/10.1007/978-981-15-2568-1_109

information technology, are familiar with electronic commerce business processes and skills, can operate and understand the management of electronic commerce front-line. The major of electronic commerce has the characteristic of strong practicality, so colleges and universities should pay more attention to its practical teaching. In the process of making and implementing teaching plans, it must change the educational mode which mainly focuses on theoretical teaching and shift the focus of teaching to practical teaching. In the teaching mode, curriculum setup and teaching method, it should constantly innovate the mode and update the content, and finally construct an all-round and feasible practical teaching system.

2.2 According to the Goal of Educational and Teaching Reform in Applied Undergraduate Colleges

Most talents training modes emphasize to face the needs of the society, strengthen the cultivation of students' ability, strengthen the practice link, cultivate the team spirit, and improve the ability of employment and innovation. In order to ensure its realization, it is necessary to design the teaching objectives, construct the curriculum system, strengthen practical courses, innovate teaching methods and reform assessment methods, among which practical teaching is an important link [4]. For the major of electronic commerce, constructing electronic commerce practice teaching system accords with the development of the major of electronic commerce, the reform of teaching mode in schools, and the goal of cultivating talents in local colleges and universities.

2.3 According to the Talents Cultivation Objectives of the Major of Electronic Commerce in Applied Undergraduate Colleges

In order to meet the needs of the society for electronic commerce professionals, electronic commerce professionals should not only have a certain theoretical basis for business courses, but also have a higher technical ability to operate. Therefore, it is necessary to shift the focus to practice teaching and rebuild the traditional practice teaching based on ensuring the integrity of the theoretical teaching system in order to cultivate qualified electronic commerce talents in colleges and universities.

3 Practical Teaching Problems in E-Commerce Courses

The Proportion of Practical Teaching is Insufficient. Electronic commerce is a practical subject, which requires students to have a strong practical ability. At present, most electronic commerce teaching is still based on classroom theory teaching, and the practical teaching curriculum is insufficient, or does not have actual effect. From the perspective of graduate feedback, their ability is far from meeting the social requirements. Even if some colleges and universities offer practical courses, they are mainly confined to electronic commerce technology courses, which are less interlinked with related disciplines such as marketing, consumer behavior and finance. As a result, it greatly limits the depth and breadth of students' thinking and it is difficult to achieve the ultimate goal of practical teaching.

3.1   The Practical Teaching Contents Are Unreasonable

The unreasonable teaching contents of electronic commerce practice is mainly manifested in the outdated data, the poor operation of practice data as well as the lack of systematicness and pertinence in the curriculum design. First of all, the practice teaching materials are outdated [5]. At present, the electronic commerce technology develops rapidly in China, and the practice teaching in colleges has some lag in technology and resources. The design of training project is usually confined to a pre-designed closed environment. The work requirements and process settings are much simpler than the actual situation, so it is difficult for the trained students to meet the enterprise standards in terms of skills. In addition, the practice of electronic commerce data is not operational. Electronic commerce simulation laboratories always use the data of a certain enterprise directly when selecting experimental data, but lack necessary processing, finishing and scouring extraction, as a result, the material obtained is lack of universality.

3.2   Lack of Initiative and Creativity

After a long period of development, practical teaching has achieved good results, but now it seems that there is still a big gap between the practical teaching of colleges and universities and the expectations of society. In today's college practice, students' own practice and independent practice projects. Very lacking, a lot of practical teaching still stays between teaching and learning, just moving from the classroom to the factory, and can not stimulate students' learning ability and interest in learning [6].

4   Reflections on the Practical Teaching System Construction of the Major of Electronic Commerce

4.1   The Process of Practical Teaching Should Penetrate the Educational Idea of Cultivating Students' Initiative Ability

Students' initiative ability includes the ability of self-regulation in learning, the ability of finding and solving problems, the ability of participating in classroom teaching activities, the ability of innovating and the ability of lifelong learning. It specially emphasize the cultivation and formation of students' learning ability, so that students can grasp the continuing study skill and set up the life-long study consciousness. For the cultivation of electronic commerce professionals, it should actively explore the implementation methods and significance of electronic commerce network entrepreneurship training in practical teaching [7]. It is significant because students can fully experience the operation process of electronic commerce through entrepreneurship and establish the Internet mode of thinking through combining theory and practice. In the process of entrepreneurship, solving a variety of practical problems continuously stimulates students' interest in careers and the desire for success.

4.2 The Teaching Process Should Pay Attention to the Efficiency and Effect

In teaching process, teachers should emphasize the perceptual knowledge of students in the process of learning, such as the demonstration of a certain software operation process and the discussion of a specific problem in the development of electronic commerce, which can be the form of practical teaching. Combining the basic theories and methods of electronic commerce, teachers can analyze the knowledge and skills of electronic commerce targeted to a specific case. Practical training can increase the understanding and application of knowledge through after-school assignments, case studies, project tasks and electronic commerce software. There are many ways and methods used in the teaching of theories for teachers, and their effects are remarkable, such as case teaching method, student-led approach, situational teaching method, and objective teaching method.

4.3 The Teaching Process Should Pay Attention to the Combination of Basic Theory and Practical Teaching

In order to construct a reasonable and effective practical teaching system of electronic commerce, it must firstly integrate the theoretical teaching contents of the major of electronic commerce, and cultivate the thinking mode and ability of students to solve business problems based on some basic theories and operation methods of electronic commerce through using technology. The author thinks that we should pay attention to the following three aspects: Firstly, it should integrate all kinds of existing resources and set up a practical teaching platform covering all aspects of curriculum knowledge and skills training for the major of electronic commerce. The specific measures are to develop excellent online courses, purchase electronic commerce laboratory software, build a training room for college students' entrepreneurship, and improve the resources database of electronic commerce business. Secondly, it should be based on the ability and innovate the practice teaching method [8]. In practical teaching, teachers are the main line and students are the main body. It can improve the quality of practical teaching through interchangeability, competitive and adversarial methods, such as subject investigation, project design and classroom discussion. Thirdly, it can use the related qualification examination tutoring books as teaching materials. The teaching content refers to the knowledge of electronic salesperson qualification examination tutoring textbook, and tries to use the relevant qualification examination tutoring book as the teaching material. For the students majoring in electronic commerce, employers pay more attention to the qualification certificate of electronic salesmen, and college students have formed a "research fever" [9]. At the same time, the teaching materials for electronic commerce qualification examination are adjusted every year, which can quickly reflect the new changes in the field of electronic commerce and the specific practices in reality, and has a strong practicability. Using the electronic salesperson qualification examination teaching material as the teaching material not only serves for the students to participate directly, but also reflects the use of learning teaching ideas. Teachers should pay attention to the integration of teaching theory and knowledge system. According to the relevant courses offered by the major of electronic commerce, the design and selection of the teaching contents can refer to the examination requirements and tutoring materials such as the national professional qualification standards of electronic commerce teachers.

4.4 Practical Teaching Should Pay Attention to the Mode of School-Enterprise Cooperation and Co-cultivation

It is the most direct and effective way for students' practical teaching to construct the practical training base of school-enterprise cooperation and the co-cultivation mode [10], cooperate thoroughly between the school and the enterprise, and construct the internal base in the school. Its disadvantage is that due to the differences between schools and enterprises, such as system, operation mode and so on, there will be some difficulties in connecting with each other. At present, the innovative mode of deep school-enterprise cooperation cultivation that can be used and studied is as following: Establishing a joint cooperative development mechanism. Schools and enterprises work together to complete electronic commerce projects, such as the operation of Taobao stores to achieve the close cooperation between schools and enterprises. In the development process of the project, students can learn and think while practicing through solving practical problems based on a specific background using theory and technology, therefore, they will gradually have a deeper understanding of what they have learned in previous textbooks, and may also produce creative ideas, which can enhance the adaptability of professional posts.

5 Conclusion

The practical teaching of the teaching process, including teaching practice, intro-school practice, off-campus practice and graduation internship, constitutes the current teaching practice system, different practical processes and different learning tasks, thus realizing the realization of education-learning-practice In the coordinated development, new research must be carried out on practice, and practical projects should be arranged according to the characteristics of students and disciplines to help students master theoretical knowledge and reduce barriers to graduation.

Acknowledgments. This research is funded by the school-level teaching research project of Wuhan Donghu University (A talent training research of real estate valuation based on the integration of industry and education) (Item Number: 34).

References

1. Zhang, Y.: Research on the practical teaching and the employment quality improvement of college students in colleges and universities - a case study of the major of electronic commerce in L University. Jiangxi Soc. Sci. **37**(12), 250–256 (2017)
2. Yu, J., Zhu, Z., Liu, M.: Research and practice of project-driven modular teaching model for electronic commerce. Bull. Sci. Technol. **33**(07), 269–274 (2017)

3. Ma, X.: The practical application of network actual warfare method in electronic commerce curriculum. Vocat. Tech. Educ. Forum **23**, 67–70 (2016)
4. Wei, X., Li, Y., Zhao, W.: Exploration of the practice teaching system of electronic commerce triple system. Res. Explor. Lab. **35**(03), 157–160, 172 (2016)
5. Chen, L., Yuan, J., Song, W., Nie, H., Ni, C., Shen, J., Fan, X., Tan, L., Li, X., Xi, J., Jia, S., Wang, Q., Shang, W., Shen, F., Wang, H., Yang, Z.: Comprehensive practice training mode and collaborative innovation mechanism of the major of electronic commerce system. Chin. Vocat. Tech. Educ. (31), 62–68 (2015)
6. Guan, H.: Research on the practical teaching mode of electronic commerce group purchase. Res. Explor. Lab. **34**(08), 255–258 (2015)
7. Chen, Y.: Practical teaching design of project management course for the major of electronic commerce based on the guidance of entrepreneurship education. Exp. Technol. Manag. **32**(01), 205–208 (2015)
8. Wang, L.: The practical teaching system construction of the major of electronic commerce in applied undergraduate colleges. Educ. Vocat. **32**, 161–163 (2014)
9. Ju, C., Yang, J., Tang, X.: Research on the practical teaching mode of mobile learning of the major of electronic commerce and logistics in colleges and universities. Res. High. Educ. Eng. (01), 175–180 (2014)
10. Zhong, T., Zhu, G.: School-enterprise cooperation in excellent engineer education and training program at UESTC. Comput. Educ. **288**(12), 142–144 (2018)

Analysis and Research of Mine Cable Fault Point Finding Method Based on Single Chip Microcomputer and FPGA Technology

Jun Wang and Qiang Liu[✉]

JILIN Engineering Normal University, Changchun, China
719598353@qq.com, 78675985@qq.com

Abstract. Gas exists in most coalmines in China, which leads to complicated production conditions and working environment in mines. If the failure of mine cable is not eliminated in time, it will cause explosion. Based on these factors, it is particularly important to monitor the failure of mine cable and locate the fault quickly and effectively. At present, mine cable online fault location technology is not mature, but underground cable online fault location needs more and more urgent, therefore, this paper analyzes the mine cable fault point search method and field application, hoping to provide useful reference for the network maintenance work.

Keywords: Mine cable · Fault · Detection · Single chip microcomputer · FPGA

1 The Introduction

The cable fault of coalmine will interrupt the power supply system of the mine and affect the normal production of the mine. It is of great practical significance to accurately detect the cable fault type of the mine and timely deal with it.

2 Types of Cable Faults

2.1 Insulation Failure of Mine Cable

The insulation failure of mine cable is due to the low insulation level of mine cable. Core wire phase or insulation resistance to the ground does not meet the requirements, or between the core wire leakage current caused by too much.

2.2 Ground Fault

Complete grounding (also known as dead grounding) is a certain phase of mine cable grounding, If the insulation resistance to the ground is zero measured by a meter; Low resistance grounding the mine cable a phase or a few relatively insulation resistance value is lower than 500 kΩ; High resistance grounding, the mine cable a phase or a few

M. Atiquzzaman et al. (Eds.): BDCPS 2019, AISC 1117, pp. 806–812, 2020.
https://doi.org/10.1007/978-981-15-2568-1_110

relatively insulated resistance above 500 kΩ, but it does not meet insulation requirements.

2.3 Short Trouble

Mine cable short circuit fault has complete short circuit, low resistance or high resistance short circuit; Have two phase simultaneous grounding short circuit or two phase direct short circuit; There is a three-phase short circuit and grounding [1].

2.4 Break Line Fault

One or more phases of mine cable are disconnected or part of the conductor core wire is disconnected.

2.5 Flashover Fault

When the voltage of mine cable reaches a certain value, flashover breakdown occurs between wires or on the ground. When the voltage drops, the breakdown stops. In some cases, the breakdown does not occur even when the voltage value is increased again, and will occur again after a period of time [2].

2.6 Closed Fault

Occurs in the mine cable connector or terminal head, especially occurs in the oil immersed mine cable head. When this kind of fault occurs, sometimes insulation breakdown under a certain voltage, until the insulation recovery, breakdown phenomenon will completely disappear, this kind of fault is called closed fault, because the fault cannot be reproduced, it is difficult to find.

The purpose of mine cable fault classification is to select the detection method.

3 Existing Methods of Cable Fault Location

The existing fault location (also known as fault location) methods for cable systems at home and abroad can be divided into three categories: The first kind of fault location is based on the electric quantity signal (voltage and current) when the fault occurs, which can be divided into traveling wave method and impedance method. And according to the different data source Angle, also includes the single end method and the double end method; The second kind USES the external signal (audio) to carry on the fault location, mainly divides into the sound measurement method and the audio frequency induction method [3]. The third category is mainly divided into capacitance comparison method and electromagnetic induction method.

3.1   Traveling Wave Fault Location Method

The principle of traveling wave fault location method is an algorithm based on the time difference between the traveling wave reflecting to the fault point after arriving at the cable, and then reflecting to the cable from the fault point, or based on the time difference between the initial travel wave arriving at both sides of the cable [4].

Among them, the single-end traveling wave method is a ranging method based on the information of a single end. The key of the single-end traveling wave ranging is to accurately calculate the time difference between the first time travel wave reaches the measuring end and its reflection from the fault point back to the measuring end, including the extraction of fault traveling wave components. Because the travelling wave in characteristic impedance change of catadioptric situation is more complex, such as travelling wave reflection happens to reach the point of failure will also by refraction to the contralateral busbar fault point, the fault line is not "infinite", by measuring point refraction wave component after a certain time in the past, and from a measurement point of refraction back fault line, etc., make the traveling wave analysis and the use of single side have great difficulties in traveling wave fault location accurately. So now the research is more two-end traveling wave method. The key of the two-terminal traveling wave method is to record the time when the current or voltage traveling wave arrives at both ends of the line so as to ensure the fault location error within hundreds of meters [5]. But it requires a dedicated unit of synchronization time. With the rapid development of communication technology, double-terminal fault location has become the focus of research and many valuable algorithms have been proposed. Using the current and voltage information at both ends of the line, the two-terminal measurement fault location can theoretically eliminate the influence of the transition resistance, system impedance and fault type on the location accuracy, and has the ability of accurate location.

In the actual transmission line, the line structure change, different transposition method, the earth resistivity uneven along the transmission line, line parameters along with the change of frequency, the problem such as travelling wave dispersion, leading to failure of travelling wave characteristics cannot be make full use of traveling wave analysis and research, it is difficult and has a high requirement on the device [6].

3.2   Impedance Fault Location Method

Impedance of fault location method is a method to estimate the distance between fault impedance points by solving the impedance value of the line from the fault point to the measurement point. In the establishment of "fault impedance location model" of transmission lines, the fault distance is usually considered as a circuit parameter, and the impedance distance is calculated by solving circuit equation [7].

The principle of the impedance method is based on the assumption that the transmission line is uniform, that the impedance or reactance of the fault circuit is proportional to the distance from the measurement point to the fault point. In case of failure, the measuring device is started by the starting element, and the parameters such as voltage and current at the time of failure are measured, so as to calculate the impedance of the fault circuit. Since the line length is proportional to the impedance, the distance from the installation point to the fault point can be calculated.

The calculation of fault impedance can be carried out by using the information at both ends of the line or by using the measurement information at only one end for approximate processing. At the beginning, people mostly adopt single-end impedance method, and there are many specific algorithms. However, single-end method only USES the voltage and current measurement values on one side of the line, which cannot overcome the influence of transition resistance in theory, and some assumptions need to be made in the ranging algorithm, so its measurement accuracy is difficult to guarantee in many cases [8].

There are two kinds of two-terminal measurement fault location algorithms: one is based on two-end asynchronous data. It does not require synchronous sampling at both ends, but the algorithm is complex and the computation is large. The second is an algorithm based on data synchronization at both ends, which requires synchronous sampling at both ends, but the algorithm is simple and the ranging accuracy is high [9]. At the same time, it is worth noting that the two-terminal distance measurement algorithm proposed so far cannot completely eliminate the influence of the following factors on the distance measurement accuracy: line model, line parameter imbalance, line parameter inaccuracy, load current, synchronous measurement accuracy and base wave component extraction accuracy.

3.3   External Signal Ranging Method

At present, the commonly used fault location methods of external signal cable are acoustic measurement method and audio frequency induction method. The so-called audio method is to apply audio current on the fault cable to make the fault cable emit audio electromagnetic waves, and then use a probe on the ground to receive the signal of electromagnetic field change along the path of the fault cable, and send the signal into the earphone to become sound, and finally judge the location of the fault point according to the intensity of the sound in the earphone.

The so-called acoustic measurement method is to add high voltage on the fault cable, make the fault point discharge to produce the explosion sound, and then use the instrument to directly or indirectly detect the location of the explosion sound, can find the fault point. Because the discharge sound at the fault point is very weak when the fault resistance is very small or the metal property short circuit, the acoustic measurement method is only suitable for the high resistance fault. In practical application, acoustic measurement method is often interfered by the environmental factors of cable fault points, such as excessive vibration and noise, too deep buried cable, etc. In order to solve this contradiction, the acoustic - magnetic synchronous detection method is adopted [6].

3.4   Capacitance Comparison Method and Electromagnetic Induction Method

These two methods are the main principles of "cable breakpoint detection devices" on the market at present, including oscilloscope observation method, dc resistance bridge method, capacitor bridge method, pulse voltage method, pulse current method, etc. However, these methods have many shortcomings. For example, the measurement error is very large, mainly because the detected signal is easily affected by the external environment, and the anti-interference ability is poor. In addition, the detection device itself has a large measurement error, and there are many error reasons that are difficult to break through technically. In addition, these devices can only detect unshielded cables, but not shielded cables. However, the application of these devices will be gradually withdrawn from the market due to the advantages of the transmission performance of shielded cables replacing non-shielded cables.

With capacitive bridge and resistive bridge ranging is the most popular, both have the same principle, it is used for cable open-circuit fault distance measurement, fundamental is open-circuit fault of cable capacitance is proportional to the cable fault distance, as long as the size or use a bridge to measure the capacitance resistance size, and know that the unit length of the cable capacitance values or resistance, can take the cable fault distance. However, the bridge method itself has some disadvantages. Firstly, the bridge method cannot be used for high resistance and flashover faults, because the current flowing through the high resistance and flashover faults is very small, which is difficult to detect by the general sensitive galvanometer. Therefore, high-resistance faults need to be heated with high-voltage dc or ac method for low-resistance faults. However, this is a very difficult job, which not only takes a lot of time, but also easily burns down the fault points, leading to increased resistance reflection at the fault points. If the resistance value of the fault point is too low after heating, it will be a permanent short circuit, which will lead to the failure of accurate location by acoustic measurement. Another disadvantage of bridge method is the use of bridge method ranging, no matter what kind of method and equipment, must know the specific cable line information, otherwise it will lead to the error of ranging results. In addition, the bridge method requires good lead contact, otherwise the existence of ten contact resistance will cause a large error in the ranging results. Equivalent length conversion must be carried out when materials or conductors with different sections are connected together in the same cable line. The bridge method can not be used to locate the fault of ten cable two phase short circuit and two phase break circuit. At present, the ten-bridge method has been gradually replaced by other ranging methods.

4 Application of Single Chip Microcomputer and FPGA Technology in Mine Ranging System

4.1 Single Chip Distance Measurement

Compared with the previous scheme, this scheme has the advantages of on-line detection, high ranging accuracy, good automation and simple operation. Therefore, the measurement error in the first scheme and the problem of whether the shielded cable can be detected are greatly solved technically. However, there will be a new problem, that is, there are strict requirements on the distance of cable breakpoints. Main reason is that single chip microcomputer shang da is less than required its own hardware, known as single chip microcomputer as a computer to perform a basic operation in machine cycle time unit, a machine cycle consists of 12 oscillation period, generally for single chip microcomputer to provide a crystal vibration frequency is about 20 MHz, and electrical signals in the cable in transmission rate and light transmission rate is the same order of magnitude in a vacuum. With crystal oscillator 24 MHz as the oscillation frequency, the electric signal has been transmitted nearly 100 m in the cable in one machine cycle. Therefore, in order to make the measurement precision high in the actual measurement, the required experimental equipment (cable) should be long enough, the longer it is, the smaller the error. Therefore, in a certain error range, the MCU has strict requirements on the minimum distance of the breakpoint of the measured cable.

In addition, the width of pulse generated by single chip microcomputer as pulse signal generator is also at least one machine cycle long. When the minimum distance of the breakpoint of the measured cable is not long enough, it will lead to the phenomenon of overlapping of transmitting pulse and reflecting pulse, so that the blind area of test will appear. Once overlaps occur, it will be difficult for subsequent circuits to distinguish the transmitted signal from the reflected signal. However, due to the influence of experimental conditions, it is impossible to provide a long enough cable.

4.2 Set up the Ranging System Method Based on FPGA

This scheme is a great improvement on the previous two schemes, but there are many similarities in the overall implementation method. Because FPGA has great advantages over single chip microcomputer in processing speed and different ways of executing program language, FPGA can execute multiple parallel statements in a machine cycle. The FPGA can generate a narrow pulse signal with a narrower width, which can greatly reduce the blind area error, thus greatly improving the strict requirements on the minimum distance of the breakpoint of the tested cable. According to the theoretical calculation, there will be no overlap between the transmitting pulse and the reflecting pulse for the transmission signal in the cable of about 10 m, and the measurement precision is also very high, which is within a certain error allowable range.

4.3 Overall Scheme Design of the System

In this paper, using pulse reflection method for cable fault location, the positioning system mainly includes two parts, namely the transmitted pulse and high-speed data acquisition and analysis of signals. First transmitted pulse generated by the pulse transmission circuit into the cable, and then reflected by fault point of the reflection pulse for A/D sampling, the sampling of the waveform to the DSP core processor after wavelet denoising and waveform comparison of attenuation law libraries can detect the fault types, the other transmission pulse cycle and wave velocity by known fault distance can be calculated.

In a word, this scheme is greatly improved in key technical problems, and the system is simple in structure, simple in operation, high in ranging accuracy and good in application.

5 Conclusion

To sum up, scientific and effective search methods can effectively improve the efficiency of mine cable fault point search and fault treatment, and ensure the stability of power system operation. Maintenance personnel need to clarify the cause of cable fault and typical characteristics of all kinds of fault, take appropriate fault point search method combined with the specific performance of the fault, skilled use of all kinds of fault test instruments, in order to accurately locate the cable fault point, improve the fault treatment efficiency and power supply reliability of the grid.

Acknowledgments. The author Jun Wang is a member of the innovation research group of automotive engineering college, Ji Lin engineering normal university, "Automotive parts efficient process development and surface quality control innovation team".

References

1. Lee, H., Mousa, A.M.: GPS traveling fault locator system: investigation into the anomalous measurement related to lightning strikes. IEEE Trans. Power Deliv. **11**(3), 1214–1223 (1996)
2. Navaneethan, S., Soraghan, J.J., Siew, W.H., Mcpherson, F., Gale, P.F.: Automatic fault location for underground low voltage distribution networks. IEEE Trans. Power Deliv. **16**(2), 346–351 (2001)
3. Mashikian, M.S., Bansal, R., Northrop, R.B.: Location and characterization of partial discharge sites in shielded power cables. IEEE Trans. Power Deliv. **5**(2), 833–839 (1990)
4. Inoue, N., Tsunekage, T., Sakai, S.: On-line fault location system for 66 kV underground cables with fast O/E and fast A/D technique. IEEE Trans. Power Deliv. **9**(1), 579–584 (1994)
5. Novosel, D., Hart, D.G., Udren, E., et al.: Unsynchronized two terminal fault location estimation. IEEE Trans. Power Deliv. **11**(1), 130–138 (1996)
6. Chui, C.K., Lian, J.: A study of orthonormal multi-wavelets. J. Appl. Numer. Math. **20**(3), 273–298 (1996)
7. Jiao, L., Pan, J., Fang, Y.: Multiwavelet neural network and its approximation properties. IEEE Trans. Neural Netw. **12**(5), 1060–1066 (2001)
8. Zhang, J.: Wavelet neural networks for function learning. IEEE Trans. Signal Process. **43**(7), 1485–1497 (1995)
9. Miller, J.T., Li, C.C.: Adaptive multiwavelet initialization. IEEE Trans. Signal Process. **46**(12), 3282–3291 (1998)

Emotional Analysis of Sentences Based on Machine Learning

Yan Zhuang[✉]

Telecommunications Engineering with Management,
Beijing University of Posts and Telecommunications, Beijing, China
2017212725@bupt.edu.cn

Abstract. This paper introduces a new model for analysing the feelings sentences contained. At first, more than 100 experts in the field correctly labelled sentences and placed the same category of labels in a folder. Whereas smile was used to extract the acoustic features from the clips, it combines the global features and temporal features of utterances based on IS09_emotion and IS13_ComParE which are the baseline acoustic feature set of the 2009–2013 INTERSPEECH challenges in my Deep Neural Network model. Speech Recognition will be realized by CMUSphinx python and for text processing. It goes without saying that MonkeyLearn python will be used. Reading the text and corresponding label stored in the feature file (corresponding column), and uploading the <text, label> to MonkeyLearn. For all that, TensorFlow is used to do machine learning and the correct intention of the sentence will be output.

Keywords: openSMILE · Deep Neural Network · CMUSphinx · MonkeyLearn · TensorFlow

1 Introduction

Speech intention recognition makes it possible for machines to recognize human speech [1]. It is widely used in the field of artificial intelligence, which connects humans and machines.

Maas, Andrew L reported DNN classifier performance and final speech recognizer word error rates and compared DNNs using several metrics to quantify factors influencing differences in task performance in Building DNN acoustic models for large vocabulary speech recognition. Taddy's Multinomial Inverse Regression for Text Analysis introduced a straightforward framework of sentiment-sufficient dimension reduction for text data. Gao, Jianqing proposed a general framework that established a unified model for diversified speech data with different sampling rates and channels in Mixed-bandwidth Cross-channel Speech Recognition via Joint Optimization of DNN-based Bandwidth Expansion and Acoustic Modelling. The framework was a joint optimization of deep neural network (DNN)-based bandwidth expansion and acoustic modelling to exploit a large amount of diversified training data.

© Springer Nature Singapore Pte Ltd. 2020
M. Atiquzzaman et al. (Eds.): BDCPS 2019, AISC 1117, pp. 813–820, 2020.
https://doi.org/10.1007/978-981-15-2568-1_111

Recent work has shown that speech can be converted to text as accurately as possible by means of audio processing [2]. However, such a process results in a sentence having only one intention which is a simple processing [3]. For this problem, we will use context [4] and an army of machine learning [5] to identify the specific context in which the word is found, and then find the true intention of the sentence.

2   Concept Description

When you heard a word like "football", which circumstances do you think it appears, maybe you can hear it on the court, during daily communication, or even at the scene of the murder. As human, it is hard for us to judge the emotion of a sentence. But with the help of artificial intelligence and deep learning, the voice can convert to text [6] and the audio signal can be collected [7] at the same time.

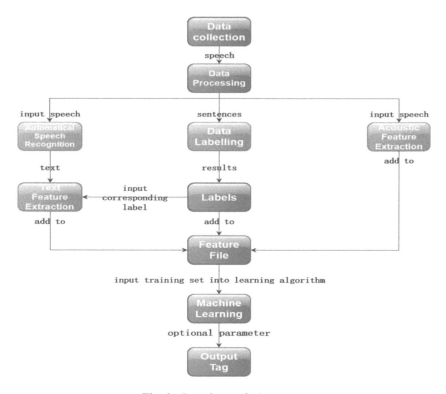

Fig. 1. Intention analysis process

It should be noted that all words should have values to be analysed [8]. For example, prepositions, conjunctions or some words that humans can recognize by themselves are not the focus of this model. The objects of my analysis are words in the sentences which have abstract meaning or referential meaning. In a specific situation, human beings cannot quickly get the intention while these may be simple for machines.

3   Realization Process

Flow chart of this model work is in Fig. 1.

3.1   Data Labelling

The speech will be annotated manually in difference intentions as labels. Labels will be added as a feature to the feature file by using one-hot coding [9].

Intentions recognition is highly subjective [10]. Humans usually disagree to some degree as to what the intention should be expressed in the speech of others. Self-assessment is subjective, so we choose to use multi-assessment. That means more than 50 experts and professors need to be invited to give their own understandings of a specific sentence.

After labelling, each utterance of the same intention will be assigned to corresponding file folder with an utterance ID indicating its intention [11]. For example, if the utterance is the first one be classified into "Encourage", then it will be included in the file folder called "Encourage" with the ID encourage_01.

3.2   Acoustic Feature Extraction

To extract acoustic features from the segments, the openSMILE toolkit [12] with batch process like Fig. 2 will be used to extract a minimalistic expert-knowledge based feature set, i.e., eGeMAPS [13], which contains 88 statistical features calculated by applying various functions over 23 Low-Level Descriptors.

We choose to combine the global features and temporal features of utterances based on IS09_emotion and IS13_ComParE which are the baseline acoustic feature sets of the 2009–2013 INTERSPEECH challenges in our DNN (Deep Neural Network) model in Fig. 3 [14]. The reason why using several sets is that the intention recognition relates to more than emotion features but also the state of speaker and so on [15]. Adding more features is helpful for analysis. Also, to deal with multiple features, we need construct my DNN model carefully. There will be many hidden layers [16].

Output features from openSMILE are stored in .txt file [17]. We will read this .txt file and convert it into a .csv file for better fetch by python code. In .csv file, the row is the utterance ID and the column is the features corresponding [18].

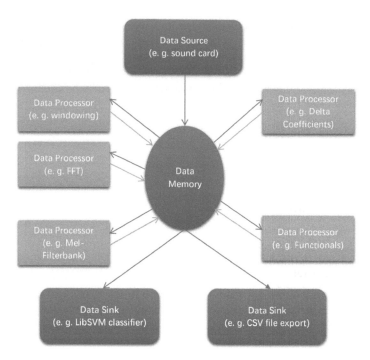

Fig. 2. Overview on openSMILE's component types and openSMILE's basic architecture

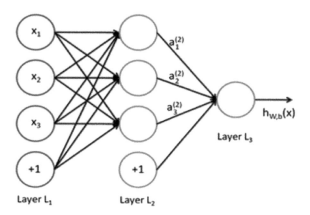

Fig. 3. DNN model

3.3 Speech Recognition

Speech recognition using Jays-PySPEECH, Jays-PySPEECH using SpeechRecognition system and CMU Sphinx to implement speech recognition engine [19]. This step is the prerequisite of step D (Text processing).

The result of this step will be recorded in the feature file but will not be feature we want to input into the DNN model because we don't input a whole text but the results (you can see detail below) of text processing.

3.4 Text Processing

For text processing, we will use MonkeyLearn python [20].

First, since we have already made the intention standards in previous step (Data labelling), we will use the same standard as tags when doing text processing.

Second, we read the text and corresponding label stored in the feature file (corresponding column), and upload the <text, label> tuples to MonkeyLearn. Then MonkeyLearn will generate a text classifier model with optimized parameters for the relation between text and intentions [21]. Then from this model we can extract the value of relevance between each word from the text and labelled intention. And we add them as the features to the feature file [22]. But there are different word numbers for different text, so we will use bubble sort algorithm and filter the ones with least confidence.

Bubble sort is the ability to move small elements forward or large elements backward and its algorithm flow is shown in Fig. 4 [23]. A comparison is a comparison of two adjacent elements, and the exchange also occurs between the two elements. So, if two elements are equal, they don't swap; If two equal elements are not adjacent to each other, then even if the two elements are adjacent to each other through the previous two exchanges, they will not exchange, so the order of the same elements before and after does not change, so bubble sorting is a stable sorting algorithm [24].

For example, there is a text "For example, this is a circle." from intention Explain file folder. From the MonkeyLearn trained, I get the relationship like "For–Explain: confidence = 0.05, example–Explain: confidence = 0.2…" There are 6 features I added, 2 left position will be filled with 0. And for the text "Circle is different from rectangular because they do not have edges." There are 11 features I can add. But it's out of range. So, I will discard 3 tuples with least confidence.

The output of this step will be added to the feature file and be input to my DNN model. That's why we don't use whole text of the utterance.

3.5 Normalization

Features in the feature file should be easily handled by our model, thus normalization is necessary. We may use pre-processing, normalize () to complete this step and then generate a new feature file [25].

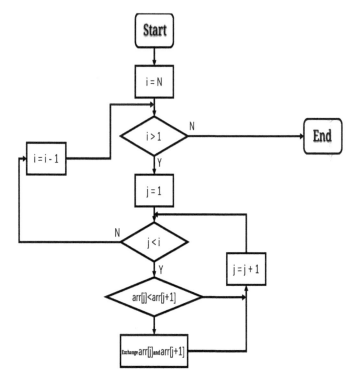

Fig. 4. Bubble sort

4 Expected Outcome

After training, our model can recognize the emotion of one sentence from any people.

For example, we record an utterance from a person and put it to the system which have source code to realize the processing function like get the acoustic features. Then system can give the intention result with optimised parameters in the model which have already been trained.

5 Conclusions

Initially, more than 50 experts in this field correctly labeled sentences. Then we use openSMILE to extract voice features from fragments. Speech recognition will be implemented by CMUSphinx python, and text processing will use Monkey Learn python. Finally, TensorFlow is used for machine learning to output the correct emotion of one sentence.

The output of the whole process is a great success, after testing 1,000 sentences, the accuracy of the output is up to 96%, which proves that the establishment of this model is effective.

Acknowledgment. The author wishes to thank Dr. Marie-Luce for her help on intention analysis technique and automatic speech recognition.

References

1. Gish, H., et al.: Investigation of text-independent speaker identification over telephone channels. In: Proceedings of IEEE ICASSP, pp. 379–382 (1985)
2. Tierney, J.: A study of LPC analysis of speech in additive noise. IEEE Trans. Acoust. Speech Signal Process. **28**(4), 389–397 (1980)
3. Matsui, T., Furui, S.: A text-independent speaker recognition method robust against utterance variations. In: Proceedings of IEEE ICASSP, pp. 377–380 (1991)
4. Tishby, N.Z.: On the application of mixture AR hidden Markov models to text independent speaker recognition. IEEE Trans. Signal Process. **39**, 563–570 (1991)
5. Reynolds, D.A.: A Gaussian mixture modeling approach to text-independent speaker identification. Ph.D. thesis, Georgia Institute of Technology, September 1992
6. McLachlan, G.: Mixture Models. Marcel Dekker, New York (1988)
7. Godfrey, J., Graff, D., Martin, A.: Public databases for speaker recognition and verification. In: Proceedings of ESCA Workshop Automatic Speaker Recognition, Identification, Verification, pp. 39–42 (1994)
8. Holmes, J., Sedgwick, N.: Noise compensation for speech recognition using probabilistic models. In: Proceedings of IEEE ICASSP (1986)
9. Hathaway, R.: A constrained formulation of maximum-likelihood estimation for normal mixture distributions. Ann. Stat. **13**(2), 795–800 (1985)
10. Proakis, J.G.: Digital Communications. Electrical Engineering. McGraw-Hill, New York (1983)
11. Coates, A., Huval, B., Wang, T., Wu, D.J., Ng, A.Y., Catanzaro, B.: Deep learning with COTS HPC. In: Proceedings of the 30th International Conference on Machine Learning, Atlanta, pp. 1337–1345 (2013)
12. Dahl, G.E., Acero, A.: Context-dependent pre-trained deep neural networks for large-vocabulary speech recognition. IEEE Trans. Audio Speech Lang. Process. **20**(1), 30–42 (2012)
13. Gurban, M., Thiran, J.P., Drugman, T., Dutoit, T.: Dynamic modality weighting for multi-stream HMMs in audio-visual speech recognition. In: Proceedings of the 10th International Conference on Multimodal Interfaces, Chania, pp. 237–240 (2008)
14. Hinton, G., Deng, L., Yu, D., Dahl, G., Mohamed, A., Jaitly, N., Senior, A., Vanhoucke, V., Nguyen, P., Sainath, T., Kingsbury, B.: Deep neural networks for acoustic modeling in speech recognition. IEEE Signal Proc. Mag. **29**, 82–97 (2012)
15. Huang, J., Kingsbury, B.: Audio-visual deep learning for noise robust speech recognition. In: Proceedings of the IEEE International Conference on Acoustics, Speech, and Signal Processing, Vancouver, pp. 7596–7599 (2013)
16. Krizhevsky, A., Hinton, G.E.: Using very deep autoencoders for content-based image retrieval. In: Proceedings of the 19th European Symposium on Artificial Neural Networks, Bruges, Belgium (2011)
17. Lan, Y., Theobald, B.J., Harvey, R., Ong, E.J., Bowden, R.: Improving visual features for lip-reading. In: Proceedings of the International Conference on Auditory-Visual Speech Processing, Hakone, Japan (2010)

18. LeCun, Y., Bottou, L.: Learning methods for generic object recognition with invariance to pose and lighting. In: Proceedings of the IEEE Computer Society Conference on Computer Vision and Pattern Recognition, Washington, vol. 2, pp. 97–104 (2004)
19. Lee, H., Grosse, R., Ranganath, R., Ng, A.Y.: Convolutional deep belief networks for scalable unsupervised learning of hierarchical representations. In: Proceedings of the 26th International Conference on Machine Learning, Montreal, pp. 609–616 (2009)
20. Maas, A.L., O'Neil, T.M., Hannun, A.Y., Ng, A.Y.: Recurrent neural network feature enhancement: the 2nd CHiME challenge. In: Proceedings of the 2nd International Workshop on Machine Listening in Multisource Environments, Vancouver, Canada (2013)
21. NVIDIA corporation CUBLAS library version 6.0 user guide. CUDA Toolkit Documentation (2014)
22. Pearlmutter, B.: Fast exact multiplication by the Hessian. Neural Comput. **6**(1), 147–160 (1994)
23. Robert-Ribes, J., Piquemal, M., Schwartz, J.L., Escudier, P.: Exploiting sensor fusion architectures and stimuli complementarity in AV speech recognition. In: Stork, D., Hennecke, M. (eds.) Speechreading by Humans and Machines, pp. 193–210. Springer, Heidelberg (1996)
24. Sutskever, I., Martens, J., Hinton, G.: Generating text with recurrent neural networks. In: Proceedings of the 28th International Conference on Machine Learning, Bellevue, pp. 1017–1024 (2011)
25. Vincent, P., Larochelle, H., Lajoie, I., Bengio, Y., Manzagol, P.A.: Stacked denoising autoencoders: learning useful representations in a deep network with a local denoising criterion. J. Mach. Learn. Res. **11**, 3371–3408 (2010)

Application and Research of Basketball Tactics Teaching Assisted by Computer Multimedia Technology

Lei Yu[✉], Dong Li, Dong Chen, and Wenbin Li

Lanzhou Vocational Technical College, Lanzhou, Gansu, China
Yulei_1979@haoxueshu.com

Abstract. Multimedia computer-aided teaching can comprehensively process sound, image, text, graphics, audio, video, animation and other materials to carry out tactical innovation, so that basketball tactical teaching forms an interactive teaching system. Combining the characteristics of basketball teaching, this paper explores computer multimedia technology to assist basketball tactics teaching, aiming at enriching basketball teaching methods and improving basketball tactics teaching effects. This thesis uses the relevant theories and methods of education, computer science, physical education and sports training to explore computer multimedia technology to assist basketball tactics teaching, and enrich teaching methods with pictures, videos, animations, texts and other multimedia materials to improve teaching quality of basketball tactics.

Keywords: Computer · Multimedia technology · Basketball · Tactics teaching · Research

1 Introduction

The educational infrastructure has taken shape, but the teaching model and teaching methods have not changed much compared to before. In many cases, these facilities are only used for leadership inspections and others [1, 2]. So how can we properly use these tools as a medium to explore their potential and give full play to their value to promote learners' learning? This requires teachers to carefully design and use the tools in a reasonable way. The research on practice and education has deepened and extensive research, and has gradually formed a certain theoretical system. At present, the teaching design ideas adopted by the two major learning theories of behaviorism. However, in the tide of reform and opening up, in the rapid development of modern society, our education must be forward-looking and keep up with the development trend of the world. Teaching design theory must also strengthen the integration with constructivist learning theory. Basketball teaching as a physical education course should be like this. The development of the times requires us to design a teaching method that suits it. The development of computer-aided technology provides a material basis for the integration of constructivist learning theory and basketball tactics [3, 4]. However, computer-aided technology itself will not naturally promote the preliminary design and implementation of computer-aided teaching software for

© Springer Nature Singapore Pte Ltd. 2020
M. Atiquzzaman et al. (Eds.): BDCPS 2019, AISC 1117, pp. 821–828, 2020.
https://doi.org/10.1007/978-981-15-2568-1_112

teaching basketball tactics. Demonstration of the location, it is also necessary to explain the timing of the start. The entire tactical process is abstracted by the teacher's explanation and demonstration, and it is difficult to form a global tactical awareness [4, 5]. Moreover, it takes time and effort, so basketball tactics teaching needs to change the status quo, and computer-aided basketball teaching can use video presentation, tactical animation, timing explanation, interactive operation and other technologies to comprehensively process, not only can virtual construction of basketball venues, basketball, athletes and other materials. It can also simulate the real background of simulated basketball tactics. Therefore, the timing, tactical route and inspiring students' technical innovation of basketball tactics are analyzed, which makes it easy for students to form tactical awareness and sports representation, and improve the efficiency and effectiveness of teachers' basketball tactics teaching.

The connotation and characteristics of computer multimedia technology-assisted instruction Multimedia Computer-Assisted Instruction (MCAI) refers to the powerful hardware system and multimedia material based on multimedia computer. According to the requirements of the syllabus and the needs of teaching, after strict teaching Design, use computer technology to create a new teaching mode for teaching software to assist classroom teaching. Computer-assisted instruction combines the power of text, images, sound, animation, and video to suit human thinking. As a modern teaching method, it can improve the efficiency of classroom teaching, break through the key difficulties, and solve some practical problems that are difficult to solve in traditional teaching [6, 7]. As a teaching medium, it is an auxiliary teaching tool like other teaching media (such as blackboards, projectors, televisions and wall charts). Its application helps to improve the teaching effect, expand the scope of teaching and extend the function of teachers [8]. Basketball tactics are an important part of basketball teaching content. Focusing on personal offense and basic cooperation, the ball-based aggressive and versatile defensive use is more common, and the confrontation is enhanced. It is technical, tactical, physical, intelligent and psychological. A comprehensive counter-mechanism that integrates into one and a four-dimensional space defensive system that is based on the ball, with people, balls, zones, and time. The traditional teaching method is based on the teacher's classroom explanation and demonstration. However, because tactical teaching involves the cultivation of tactical awareness, the multiple choice of tactical forms and approaches, the capture of tactical opportunities, and the implementation of tactical actions, the complexity is far more than that [9, 10]. Technical teaching, because tactical demonstration is a process of simultaneous interaction between multiple people (including offensive players, offensive team members and defensive players), and teachers cannot make standard demonstration actions at each position at the same time, so basketball tactics teaching pairs Teachers put forward higher requirements. In addition to organizing a reasonable teaching process and implementing effective teaching methods, students should also have a strong interest in learning, sufficient practice time and space, in order to achieve the desired teaching results.

2   Concepts and Characteristics of Computer-Assisted Teaching

Using computer-assisted teaching software to create a picture, interaction, dynamic and static, suitable for classroom teaching, it can promote students' eyes, ears, hands, brain and other senses to stimulate, induce and maintain learning. Interest, enhance the initiative to participate in teaching. The computer-assisted simulation function makes the teaching content interesting and interesting, and visualizes the abstract theory and the problems that are difficult to solve with language and teaching aids. It becomes intuitive and attractive. In addition, the unique convenience of multimedia and the superiority of transcending time and space can bring the distance of space closer and shorten the process of time. In the traditional way of teaching, teachers should use chalk to teach on the blackboard, which is time-consuming and labor-intensive. With multimedia-assisted instruction, it is only necessary to scan this information into the computer for proper editing and adjustment. In the classroom, the audio, image, picture and text can be presented in a clear and standardized manner according to the teaching procedure. This saves time in the classroom, avoids the distraction of students' attention, and increases the capacity of the classroom to deliver information. During the class, with just a click of the mouse, the teaching content will be displayed one by one, and the teaching rhythm will be greatly accelerated. Although the teaching content is nearly half that of usual, students can easily learn more knowledge within the pre-scribed time.

3   Teaching Design of Computer Multimedia Technology Assisted Basketball Tactics

Basketball tactical awareness is the process by which athletes take reasonable actions through feeling, observation, thinking, judgment, and dominance. It manifests itself in the specific movements to thinking, and then from thinking to specific actions. The founders of modern cognitive psychology, Simon, Newell and others, describe this psychological process from the perspective of information processing. According to this view, we can regard the formation process of basketball tactical awareness as a cyclical process: accepting information (observation ability) - instant judgment (thinking ability) - tactical action (resilience) - feedback enhancement. In the formation of basketball tactical awareness, thinking ability plays a key role, which is the central link of basketball consciousness. The formation of basketball tactical consciousness is also the specific performance of thinking activities in the execution of tactical actions. Information processing mode as shown in the Fig. 1.

3.1   Apply Picture Material to Mark Key Actions

Picture material can show the details of the action. In terms of basketball tactics teaching, some key actions (such as transfer and matching, sudden distribution, and coordination) are passed through the game. Shoot or intercept NBA star video clips,

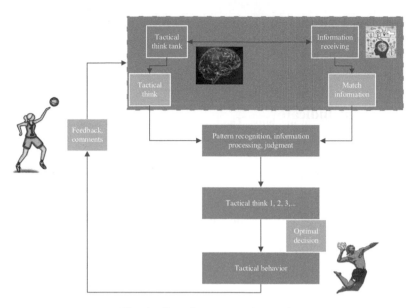

Fig. 1. Conscious processing mode

edit them with Photoshop software, and use the colored arrows, text comments, etc. to wake up the target and note the key actions and notes. Intention. In addition to the physical demonstration, teachers can borrow large screens to display pictures and let students understand the essentials and details. At the same time, there are Chinese and foreign baskets in the picture. Ball stars (such as Yao Ming and Jordan) can stimulate students' interest and enthusiasm for learning.

3.2 Apply Animation Material Decomposing Action

The animation material can deduct the decomposition action. Some actions in the basketball tactics are done in one go, but in the early teaching process, the students need to be clear about the tactical decomposition actions (such as squeezing, mate, bypass, and exchange). It is to create high-quality 3D Flash animations with 3D Flash Animator animation software or create simple animations with PPT, and use interactive on-demand technology to control animation playback (single frame, fast forward, slow release, pause).

3.3 Apply Text Material Annotation Action Essentials

The text material can explain the essentials of the action. In addition to the teacher's oral explanation, the pictures, videos, and animations need to be marked with corresponding texts, so that students can grasp the key points (such as the offensive area pressing defense, attacking mixed defense, and man-to-man defense). Teacher explanations and other media material presentations, appropriate inspiration and guidance can enable students to establish an impression of movement in the brain, quickly grasp the essentials and internal links of basketball tactics, systematically establish a complete concept of basketball tactics, and follow-up learning. It is very important to play in actual combat. The text material should not be complicated, and the color, font size, etc. should be coordinated with the background, so that it is concise and appropriate.

4   Assessment Indicators and Result Discussions

4.1   Indicators

(1)   Master the test of tactical concepts and principles
It is a closed-book theory test on the concept and principle of basketball skills and tactics for students after the completion of teaching and learning.
(2)   The use of offensive and defensive tactics
Basketball games are a highly collective project. Athletes should actively cooperate with their peers, strive to create opportunities for the team to attack and organize tight defensive tactics. This is a specific indicator of tactical awareness. The main statistics are the number of tactical cooperation students use in each game.
(3)   The rationality of the action
The students' rationality in the game is multi-faceted: the match between the players during the match, the offensive and defensive timing, the time of transport, the sudden time, the pass, the time cast, in one go, not dragging the water. Especially in the critical moment, the ability to handle the ball reasonably reflects the strength of the player's basketball tactical awareness. This indicator is mainly subjective assessment by the professional basketball teachers in the experimental class, the tactical awareness of the behavior displayed in the random mixed game of the students.
(4)   Player statistics
This indicator comes from various technical statistics in the student group competition. Technical statistics are the data that best reflects the performance of players on the court. This paper extracts representative assist statistics and error statistics as evaluation indicators, and calculates the average number of assists and turnovers per team.

4.2   Experimental Results and Discussions

As can be seen from the Fig. 2 and Table 1, the theoretical test results of the two groups showed that the average score of the experimental group was better than that of the control group. Technical statistics showed that students in the experimental group were significantly more rational in the use of tactical tactics and tactical actions than the control group. Moreover, the increase in assists and the reduction of mistakes also indicate that students are no longer willing to fight alone in the competition, but actively seek to cooperate with their peers and organize an effective attack and defense system with their peers. It shows that the use of computer multi-media technology to assist teaching, in the basketball tactics teaching can improve the theoretical learning effect of students and cultivate the tactical awareness of basketball.

Table 1. Results of indicator

Indicator	Experimental group	Comparison group
Technical principle concept test	75.43	71.64
Tactical fit usage	62.33	48.71
Rationality of tactical actions (%)	85.24	76.71
Assist statistics	17.38	14.16
Error statistics	23.74	27.82

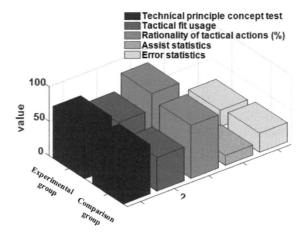

Fig. 2. Results of indicator

Basketball tactical awareness is a kind of special psychological quality and mental function reflex ability that athletes are engaged in basketball practice activities and have accumulated through the active thinking of the brain to correctly reflect the regularity of basketball. According to the theory of modern system theory, information theory and cybernetics: the four chain reaction links of information, control, effect and feedback form the whole process of psychological activities, realize the self-control and regulation of human body self-control, and explain the self-control of human body. And the reflexive activity of regulation also shows that any conscious process of human beings is the highest form of mental activity completed by thinking and refining based on the perception of surrounding objective things, and is expressed through action, in basketball tactical awareness. In the formation process, thinking ability plays a key role. Therefore, basketball tactical awareness is essentially an individual's psychological activity. The famous psychologist Tehla has come to the conclusion that 83% of the knowledge gained by human learning is through vision, 11% through hearing, and only 6% through olfactory touch and taste. That is to say, 94% of human information is obtained through sight and hearing. In the basketball tactics teaching, the application of computer multimedia assisted teaching in the theoretical class, the use of human-computer interaction technology, the combination of audio-visual, broadcast, problem-oriented, hypothetical discovery and imagination, etc., enable students to combine their brains and hands, and mobilize students' senses in all directions. The ability to participate in learning through intellectual and non-intellectual factors is more in line with the physiological and psychological mechanisms of students' learning knowledge, and promotes students' understanding of knowledge and flexible use, promoting the development of thinking and opening up creativity. At the same time, the application of multimedia technology to assist teaching can also use the original rigid and monotonous basketball tactics theory to appear vivid, greatly improve students' interest in learning, and lay a solid and flexible theoretical thinking foundation for cultivating basketball tactical awareness. In addition, multimedia technology can shorten the time process, the distance of the space is close, so that the performance of the information has a large time span and spatial range. The static image can also be used to display some continuous and complicated tactical movements. Decompose or combine on the screen. Students sitting on the court in front of the computer, as well as access to a variety of tactical information, have greatly helped the development of tactical awareness.

In basketball teaching, demonstration is a teaching method that is often used. Due to the influence of time and space factors and the teaching environment and conditions, the use of demonstration has great limitations. The demonstration of basketball tactics can only be accomplished by means of chalk and blackboard in traditional teaching, and the teaching effect is poor. It is difficult for students to link the tactical methods on the blackboard with the movements and movements on the court. The teachers spend a lot of time talking about the tactical intentions, and the students can only comprehend some relatively simple and rigid running routes, which can not display the tactics at all. Multimedia technology can show both standard tactics and tactical changes. Applying multimedia techniques to practice tactics can maximize the realistic effects of tactical exercises. In addition, the teaching and training of basketball often requires the creation of new tactics, and multimedia technology is the most ideal tool for creation. By changing the limitations of using the students to do demonstration exercises in person, it is more convenient to create them by combination, splicing, creativity, transformation, etc. When students use multimedia technology to practice various tactics, they are consciously placed in the game. They can operate according to fixed tactics, and can also consider which tactics to use according to changes in the field. Therefore, multimedia technology assisted teaching is the second venue for the training of basketball tactics.

5 Conclusions

In any case, the cultivation of basketball tactical awareness is always inseparable from the actual training and competition. The application of multimedia technology to basketball tactics has improved some shortcomings in basketball tactics teaching after teaching reform. However, we must clearly understand that multimedia technology is only a supplementary teaching technology. Teachers are always the most active and active factors in teaching. They should guide students to explore and encourage students to participate actively, and fully develop the subjective initiative of teachers in multimedia teaching. With the enrichment, perfection and maturity of computer multimedia technology, the correct application of this technology to teaching will surely receive good teaching results.

References

1. Hennessy, S., Hassler, B., Hofmann, R.: Pedagogic change by Zambian primary school teachers participating in the OER4Schools professional development programme for one year. Res. Papers Educ. **31**(4), 399–427 (2016)
2. Bradbury, A., Roberts-Holmes, G.: Creating an Ofsted story: the role of early years assessment data in schools' narratives of progress. Br. J. Sociol. Educ. **38**(7), 943–955 (2017)
3. Koparan, T.: Teaching game and simulation based probability. Int. J. Assess. Tools Educ. **6**(2), 235–258 (2019)
4. Rekik, G., Khacharem, A., Belkhir, Y., Bali, N., Jarraya, M.: The instructional benefits of dynamic visualizations in the acquisition of basketball tactical actions. J. Comput. Assist. Learn. **35**(1), 74–81 (2019)
5. Ayres, P., Marcus, N., Chan, C., Qian, N.: Learning hand manipulative tasks: when instructional animations are superior to equivalent static representations. Comput. Hum. Behav. **25**(2), 348–353 (2009)
6. Shin, D., Park, S.: 3D learning spaces and activities fostering users' learning, acceptance, and creativity. J. Comput. High. Educ. **31**(1), 210–228 (2019)
7. Pedra, A., Mayer, R.E., Albertin, A.L.: Role of interactivity in learning from engineering animations. Appl. Cogn. Psychol. **29**(4), 614–620 (2015)
8. Marcus, N., Cleary, B., Wong, A., Ayres, P.: Should hand actions be observed when learning hand motor skills from instructional animations? Comput. Hum. Behav. **29**(6), 2172–2178 (2013)
9. Khacharem, A., Spanjers, I.A., Zoudji, B., Kalyuga, S., Ripoll, H.: Using segmentation to support the learning from animated soccer scenes: an effect of prior knowledge. Psychol. Sport Exerc. **14**(2), 154–160 (2013)
10. Wong, A., Marcus, N., Ayres, P., Smith, L., Cooper, G.A., Paas, F., Sweller, J.: Instructional animations can be superior to statics when learning human motor skills. Comput. Hum. Behav. **25**(2), 339–347 (2009)

Accurate Calculation of Rotor Loss of Synchronous Condenser Based on the Big Data Analysis

Aijun Zhang[1], Huadong Xing[1], and Jingdi Zhou[2(✉)]

[1] Inner Mongolia Power Research Institute, Hohhot 010020,
Inner Mongolia, China
[2] School of Electrical and Electronic Engineering,
North China Electric Power University, Beijing 102206, China
Zhoujingdi400@163.com

Abstract. The synchronous condenser is used to supply reactive power for HVDC converter station through the power transformer. The DC power that invades the power transformer biases the operating point of the magnetic field and produces harmonic components on the side of the synchronous condenser. In this paper, the analysis method of big data is adopted to analyze rotor loss of synchronous condenser under the DC bias of the transformer for the goal of accurate calculation of rotor loss. Lots of high-frequency harmonics in terminal voltage of synchronous condenser are generated by DC-bias of transformer. The rotor loss is calculated and compared under lagging phase and leading phase of synchronous condenser. The big data method is used to analyze the harmonic which produce the largest rotor loss. Some conclusions about the rotor loss caused by stator harmonic current are obtained.

Keywords: DC bias · Rotor loss · Synchronous condenser · The big data

1 Introduction

As the rapid development of power grid, the malconformation of power supply and load becomes more and more severe. Therefore, the technology of the direct current (DC) transmission is applied to long distance and high capacity power transmission [1]. When DC transmission systems operate in the monopole mode using ground return, a certain amount of direct current are infused into the earth through the ground electrodes [2].

The DC bias of the power transformer can make the parts of the core saturate in half cycle [3]. This can have some adverse effects in the power transformer, such as grew noise, more losses, and harmonic voltages on the secondary side [4]. When the secondary side of transformer is connected to a generator, the harmonic voltage will be applied to the stator winding of generator. The waveform distortion of voltage results in the increase of rotor loss and heating of generator [5]. The second and fifth harmonic voltages develop primarily negative-sequence currents [6–9]. The fourth and seventh harmonics develop primarily positive sequence currents [10]. The positive-sequence

© Springer Nature Singapore Pte Ltd. 2020
M. Atiquzzaman et al. (Eds.): BDCPS 2019, AISC 1117, pp. 829–837, 2020.
https://doi.org/10.1007/978-981-15-2568-1_113

currents produce the magnetic field rotating forward at a frequency of the harmonic in the air gap. The negative-sequence currents produce magnetic field rotating backward to the rotor rotation at a frequency of the harmonic in the air gap. The induced currents in rotor slot wedge and iron core increase the loss of the generator.

Due to reliability constraints of power electronic device, the large capacity synchronous condenser is used to supply reactive power for HVDC converter station. In this paper, the influence of DC bias on the loss of synchronous condenser is studied. Firstly, the electromagnetic field of the synchronous condenser is calculated according to the time step finite element method. Then, the rotor loss due to the late phase of the synchronous condenser and the high frequency component of the stator current at the leading phase is calculated. Finally, some conclusions about the rotor loss caused by stator harmonic current are obtained.

2   The Influence of DC Bias on SEMF of the Transformer

2.1   Transformer Model Considering the Effect of DC Bias

Synchronous condenser can work as a transformer load or power supply, the terminal voltage of the synchronous condenser is closely related to the Secondary Electromotive Force (EMF) of the transformer. In order to obtain the influence of DC bias on the operating characteristics of the synchronous condenser, a large-scale transformer model considering the influence of DC bias is established and the influence of DC bias on the secondary electromotive force of the transformer is studied. Figure 1 shows the structure and flux density distribution of the transformer.

2.2   Influence of DC Bias on Secondary Electromotive Force of Transformer

After the synchronous condenser is connected to the secondary side of the transformer, the voltage on the secondary side of the transformer is closely related to the terminal voltage of the synchronous condenser. By calculating the induced electromotive force of the secondary side under different DC bias, the results are shown in Fig. 2. The comparison shows that due to the DC bias of the transformer, the induced electromotive force waveform is distorted and the harmonic content is increased. The result of Fourier analysis shows that in addition to the original fundamental and odd harmonics, the induced electromotive force waveforms also contain even-order harmonic components such as 2, 4 and 6. With the increase of DC current, the fundamental induction electromotive force decreases, while the content of odd harmonics such as 3, 5, 7 and so on increased, the even order harmonic content such as 2, 4, 6 increased first and then decreased. The analysis shows that when the DC bias magnetic flux is near the inflection point of no-load characteristic, the corresponding magnetic permeability of the magnetic flux produced by the positive and negative half-wave greatly varies, resulting in serious asymmetry of the positive and negative half-wave of the magnetic flux and a larger even harmonic component. Ensuring proper use of copyrighted multimedia content has become increasingly important.

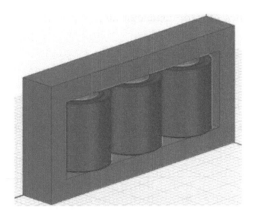

Fig. 1. Three-dimensional Model of Transformer

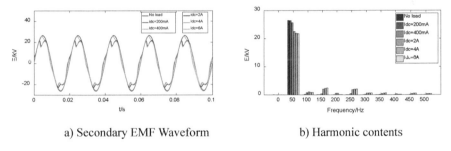

a) Secondary EMF Waveform   b) Harmonic contents

Fig. 2. Secondary EMF Waveform and Harmonic Contents

3 Time-Stepping Finite Element Model Used to Calculate the Rotor Loss Caused by High-Frequency Current

3.1 Field-Circuit-Network Coupled Equations

According to the actual rotor structure of the 300-MVar synchronous condenser shown in Fig. 3, the T-S FEM of synchronous condenser is as follow:

$$
\begin{bmatrix} K & -C & -C_f \\ 0 & -R_s & 0 \\ 0 & 0 & r_f \end{bmatrix} \begin{bmatrix} A \\ I_s \\ i_f \end{bmatrix} + \begin{bmatrix} D_s + D_r & 0 & 0 \\ -l_{ef}C_s^{\mathrm{T}} & -L_s & 0 \\ -l_{ef}C_f^{\mathrm{T}} & 0 & l_f \end{bmatrix} \frac{d}{dt} \begin{bmatrix} A \\ I_s \\ i_f \end{bmatrix} = \begin{bmatrix} 0 \\ U_l \\ u_f \end{bmatrix} \quad (1)
$$

where, D_s, D_r are the eddy currents matrixes; C_s and C_f are the incidence matrices of stator and field currents; U_l is the stator voltage; K is the stiffness matrix; I_s and i_f are the stator and field currents; R_s is the matrix of stator resistance, L_s is the matrix of the end-leakage inductance of the stator winding.

The current densities of rotor conductor are:

$$
J_e \begin{cases} J_s = -\sigma_s \dfrac{\partial A}{\partial t} \cdots \cdots \cdots area \textcircled{1} \\ J_r = -\sigma_r \dfrac{\partial A}{\partial t} \cdots \cdots \cdots area \textcircled{2} \end{cases} \tag{2}
$$

where σ_s is the conductivity of the rotor slot wedge, σ_r is the conductivity of the rotor iron core.

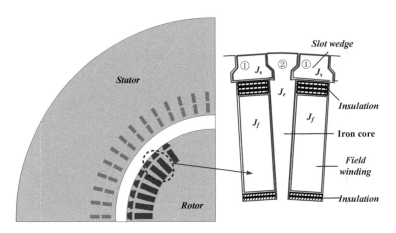

Fig. 3. Typical cross section of the synchronous condenser

3.2   Rotor Slot Wedge and Iron Core Loss Considered High-Frequency Current

The air-gap flux density of the synchronous condenser contains basic components and harmonics. The basic magnetic field and the harmonics produced by the excitation current are motionless refer to the rotor, while the basic magnetic field produced by the stator current is motionless refer to the rotor. Therefore, these fields do not generate losses. When the harmonic magnetic field produced by the stator basic current and the stator harmonic current are rotated refer to the rotor, these fields generate losses in the rotor. For example, the pole number and speed of the 5th harmonic Magneto-motive Force (MMF) generated by the fundamental current of stator current is 5 and 1/5 times that of the fundamental MMF generated by fundamental current. This magnetic field produces 6th harmonic eddy currents of rotor conductor; while 5th harmonic current generated by the 5th harmonic terminal voltage produces MMF with 1 time pole pair number and 5 times speed of MMF generated by the fundamental current. This magnetic field also produces 6th harmonic eddy currents of rotor conductor. These losses in the rotor slot wedge (P_{SW}) and iron core (P_{IC}) are calculated by the finite-element method as follow [11]:

$$
\begin{cases}
P_{SW} = \frac{1}{T}\int_0^T \sum_{i=1}^k \frac{1}{\sigma_s} l_{ef} S_{s\_e} J_{s\_e}^2 \\
P_{IC} = \frac{1}{T}\int_0^T \sum_{i=1}^k \frac{1}{\sigma_r} l_{ef} S_{r\_e} J_{r\_e}^2
\end{cases}
\tag{3}
$$

where, S_{s_e}, S_{s_r} are the element area of rotor slot wedge and iron core; J_{s_e}, J_{r_e} are the element eddy current density of rotor slot wedge and iron core.

3.3 Harmonic Current of Synchronous Condenser Considering Transformer DC Bias

The magnetic fields generated by terminal harmonic voltage caused by transformer DC bias are different from that generated by the fundamental voltage. The phase sequence relations of harmonic current generated by harmonic voltage are written in (4)–(10). Table 1 shows the rotation direction of the harmonic magnetic field and the frequency of the eddy current in rotor slot wedge and iron core. It can be seen from Table 1 that 3rd and 6th harmonic currents cannot produce the rotation magnetic field and rotor eddy current. The higher the harmonic voltage is, the smaller the amplitude is. Therefore, the rotor eddy current corresponding to the higher harmonic voltage is negligible. Since the three phases of the transformer are usually star-connected and the neutral point is not directly grounded and the multiples of 3rd harmonic harmonics have the same phase, they do not generate the corresponding currents and magnetic fields in the stator windings, nor will they induce eddy currents in the rotor. Both 2nd and 4th harmonic voltages induce eddy currents in multiples of 3 in the rotor; The 5th and 7th harmonic voltages all induce vortices in multiples of 6 in the rotor.

$$
\begin{cases}
I_a^1 = I_m^1 \cos(\omega t) \\
I_b^1 = I_m^1 \cos(\omega t - 120) \\
I_c^1 = I_m^1 \cos(\omega t + 120)
\end{cases}
\tag{4}
$$

$$
\begin{cases}
I_a^2 = I_m^2 \cos(2\omega t) \\
I_b^2 = I_m^2 \cos(2\omega t - 240) = I_m^2 \cos(2\omega t + 120) \\
I_c^2 = I_m^2 \cos(2\omega t + 240) = I_m^2 \cos(2\omega t - 120)
\end{cases}
\tag{5}
$$

$$
\begin{cases}
I_a^3 = I_m^3 \cos(3\omega t) \\
I_b^3 = I_m^3 \cos(3\omega t - 360) = I_m^3 \cos(3\omega t) \\
I_c^3 = I_m^3 \cos(3\omega t + 360) = I_m^3 \cos(3\omega t)
\end{cases}
\tag{6}
$$

$$
\begin{cases}
I_a^4 = I_m^4 \cos(4\omega t) \\
I_b^4 = I_m^4 \cos(4\omega t - 480) = I_m^4 \cos(4\omega t - 120) \\
I_c^4 = I_m^4 \cos(4\omega t + 480) = I_m^4 \cos(4\omega t + 120)
\end{cases}
\tag{7}
$$

$$
\begin{cases}
I_a^6 = I_m^6 \cos(6\omega t) \\
I_b^6 = I_m^6 \cos(6\omega t - 720) = I_m^6 \cos(6\omega t) \\
I_c^6 = I_m^6 \cos(6\omega t + 720) = I_m^6 \cos(6\omega t)
\end{cases}
\tag{8}
$$

$$
\begin{cases}
I_a^5 = I_m^5 \cos(5\omega t) \\
I_b^5 = I_m^5 \cos(5\omega t - 600) = I_m^5 \cos(5\omega t + 120) \\
I_c^5 = I_m^5 \cos(5\omega t + 600) = I_m^5 \cos(5\omega t - 120)
\end{cases}
\tag{9}
$$

$$
\begin{cases}
I_a^7 = I_m^7 \cos(7\omega t) \\
I_b^7 = I_m^7 \cos(7\omega t - 840) = I_m^7 \cos(6\omega t - 120) \\
I_c^7 = I_m^7 \cos(7\omega t + 840) = I_m^7 \cos(6\omega t + 120)
\end{cases}
\tag{10}
$$

Table 1. The harmonic voltage corresponds to the frequency of the rotor eddy current

Harmonic order	Rotation direction of the field	Frequency of the rotor eddy current
2	Reverse rotation	3f
3	–	–
4	Forward rotation	3f
5	Reverse rotation	6f
6	–	–
7	Forward rotation	6f

4   The Rotor Loss of Synchronous Condenser Caused by DC Bias

4.1   The Effect of Different Harmonics Voltage on the Rotor Loss Under the Lagging Phase Operation of the Synchronous Condenser

The rotor loss are calculated in the lagging phase operation of synchronous condenser under the terminal voltage considered as the ideal voltage source, the voltage source with the 2th harmonic component, the voltage source with the 4th harmonic component, the voltage source with 5th harmonic component. In four cases, the influence of different harmonic voltage on the rotor slot wedge and iron core loss are compared and analyzed. From Fig. 4 and Table 2, some conclusion can be obtained as follow:

(1)   No matter whether the harmonic components is considered or not, the loss of rotor slot wedge is less than that of iron core obviously; the loss of iron core is about 4.5 times than that of rotor slot wedge under the ideal voltage source.
(2)   The losses of rotor slot wedge and iron core under harmonic voltage source are larger than that under the ideal voltage source. The increment of the iron core loss is larger than that of the rotor slot wedge.
(3)   As the increase of harmonic order of the voltage, the rotor losses decline gradually. The rotor losses under the voltage source with the 2th harmonic component are the largest; which are about 2 times than that under the ideal voltage source.

Table 2. The average loss in rotor slot wedge and iron core in four cases under lagging phase operation

	P_{SW}/kW	ξ_{SW}	P_{IC}/kW	ξ_{IC}
Ideal terminal voltage	8.77	–	37.89	–
With 2th harmonic voltage	18.09	2.1	75.40	2.0
With 4th harmonic voltage	13.12	1.5	67.12	1.8
With 5th harmonic voltage	11.97	1.4	60.37	1.6

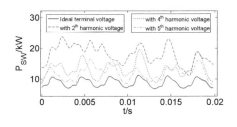 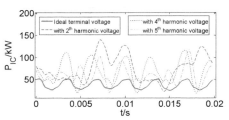

a) Loss of rotor slot wedge   b) Loss of rotor iron core

Fig. 4. Rotor loss under lagging phase operation of synchronous condenser

4.2 The Effect of Different Harmonics Voltage on the Rotor Loss Under the Leading Phase Operation of Synchronous Condenser

The rotor slot wedge and the iron core losses under the leading phase operation are compared and analyzed in four cases. Figure 5(a) is the loss of rotor slot wedge in four cases, and Fig. 5(b) is the loss in the rotor iron core in four cases. Table 3 shows the average loss of the rotor slot wedge and the iron core in four cases. From these results, some conclusion can be obtained as follow:

(1) The loss of the rotor wedge and core is significantly lower in the leading phase than in the lagging phase. Under the ideal voltage source, the loss of the rotor wedge and the rotor core during the hysteresis operation is 28% and 33% respectively in the leading phase operation.

(2) As the increase of the harmonic order of the voltage, the rotor losses decline gradually. The rotor losses under the voltage source with the 2th harmonic component are the largest; which are about 5.5 times and 4.4 times than that under the ideal voltage source.

Table 3. The average loss in rotor slot wedge and iron core in four cases under lagging phase operation

	P_{SW}/kW	ξ_{SW}	P_{IC}/kW	ξ_{IC}
Ideal terminal voltage	2.49	–	12.48	–
With 2th harmonic voltage	12.36	5.0	55.09	4.4
With 4th harmonic voltage	7.22	2.9	46.14	3.7
With 5th harmonic voltage	5.91	2.4	37.84	3.0

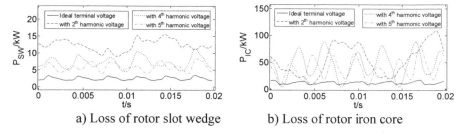

a) Loss of rotor slot wedge   b) Loss of rotor iron core

Fig. 5. Rotor loss under lagging phase operation of synchronous condenser

5   Conclusions

(1) The loss of the rotor wedge and core is significantly lower in the leading phase than in the lagging phase. Under the ideal voltage source, the loss of the rotor wedge and the rotor core during the hysteresis operation is 28% and 33% respectively in the leading phase operation.

(2) The losses of rotor slot wedge and iron core under harmonic voltage source are larger than that under the ideal voltage source. The increment of the iron core loss is larger than that of the rotor slot wedge.

(3) With the increase of the voltage harmonic order, the losses of rotor slot wedge and iron core decline gradually. The losses of the rotor slot wedge and iron core under the voltage source with the 2th harmonic component are the largest.

Acknowledgments. This research was supported by Technology project of Inner Mongolia Power (Group) CO., LTD.

References

1. Zhang, W., Yu, Y., Li, G.: Researches on UHVDC technology. Proc. CSEE **27**(22), 1–7 (2007). (in Chinese)
2. Zhang, B., Zhao, J., Zeng, R., He, J.: Numerical analysis of DC current distribution in AC power system near HVDC system. IEEE Trans. Power Deliv. **23**(2), 960–965 (2008)

3. Pan, Z., Wang, X., Mei, G., Liu, Y., Yao, W., Liu, H., Wen, X.: A transformer neutral current balancing device to restrain half-cycle saturation induced by HVDC monopolar operation. Electr. Power Syst. Res. **4**(132), 104–114 (2016). (in Chinese)

4. Bernabeu, E.: Single-phase transformer harmonics produced during geomagnetic disturbances: theory, modeling, and monitoring. IEEE Trans. Power Deliv. **30**(3), 1323–1330 (2015)

5. Gish, W.B., Feero, W.E., Rockefeller, G.D.: Rotor heating effects from geomagnetic induced currents. IEEE Trans. Power Deliv. **30**(3), 712–719 (1994)

6. Bernabeu, E.: Modeling geomagnetically induced currents in Dominion Virginia Power using extreme 100-year geoelectric field scenarios—part 1. IEEE Trans. Power Deliv. **28**(1), 516–523 (2013)

7. Pirjola, R.: Effects of space weather on high-latitude ground systems. Adv. Space Res. **36** (15), 2231–2240 (2005)

8. Lesher, R.L., Porter, J.W., Byerly, R.T.: SUNBURST- a network of GIC monitoring systems. IEEE Trans. Power Deliv. **9**(1), 128–137 (1994)

9. Girgis, R., Vedante, K.: Effects of GIC on power transformers and power systems. In: IEEE Power and Energy Society Transmission and Distribution Conference and Exposition, vol. 8, no. 8, pp. 1–8 (2012)

10. Xu, G., Liu, X., Luo, Y., Zhao, H.: Influence of different practical models on the first swing stability of turbine generators. Electr. Power Compon. Syst. **43**(2), 212–223 (2015). (in Chinese)

11. Wang, L., Li, W., Huo, F., Zhang, S., Guan, C.: Influence of underexcitation operation on electromagnetic loss in the end metal parts and stator step packets of a turbogenerator. IEEE Trans. Energy Convers. **29**(3), 748–757 (2014)

A Parallel Compressed Data Cube Based on Hadoop

Jingang Shi[1(✉)] and Yan Zheng[2]

[1] School of Information and Control Engineering, Shenyang Jianzhu University,
Shenyang, Liaoning, China
shijingang@163.com
[2] Shenyang DONFON Titanium Industry Co., Ltd., Shenyang, Liaoning, China

Abstract. Aiming at the on-line analytical processing technology, this paper proposes a parallel compressed data cube algorithm based on Hadoop architecture. The algorithm divides a single data cube into several independent sub-compressed data cubes, and then uses Hadoop architecture to realize the parallel construction and query of the entire data cube. Experiments show that the parallel compressed data cube algorithm combines the parallelism and high scalability of the Hadoop architecture on the one hand, and on the other hand, it can realize faster query operation on data cube by means of a self-indexing of the compressed data cube. So it has good research value and practical application significance.

Keywords: Data cube · Hadoop · Parallel

1 Introduction

On-line analysis and processing applications need aggregation of a large number of data, which requires high query performance. If a data cube is computed and instantiated beforehand, the query response time can be shortened [1]. However, as the number of dimensions increases, the size of data cube will increase dramatically. The size of storage space of data cube becomes an important problem [2]. In order to reduce the storage size of data cube, many data cube compression methods have been studied. Among them, Condensed Cube [3], Quotient Cube [4] and Dwarf [5, 6] algorithms are more effective.

Dwarf data cube achieves the purpose of cube compression by eliminating the prefix redundancy and suffix redundancy of data. It has a high compression ratio for cube and is an effective data cube compression algorithm. However, with the rapid increase of data volume, the traditional Dwarf cube is constructed by a single node to a single data file, so the construction speed of the traditional Dwarf cube becomes very slow, and the structure of the data cube becomes very complex, which is far from meeting the needs of practical application.

MapReduce [7–9] is a distributed computing framework first proposed by Google that can process massive data concurrently in large computer clusters. It is a simplified distributed programming model. Users only need to consider how to implement Map and Reduce processes to meet business needs. Data cutting, task scheduling, and node

M. Atiquzzaman et al. (Eds.): BDCPS 2019, AISC 1117, pp. 838–845, 2020.
https://doi.org/10.1007/978-981-15-2568-1_114

communication are automatically completed by MapReduce architecture. Hadoop framework [10] is an implementation of MapReduce architecture using Java.

Aiming at the problem of the traditional Dwarf cubes do not support parallel construction and query operations, this paper implements parallel construction, storage and query algorithms of Dwarf cubes based on Hadoop architecture.

2 Traditional Dwarf Data Cube

A Dwarf data cube compresses and stores data elements with the same prefix and suffix, eliminates redundant information of these two types, greatly reduces the storage space of data cube, and reduces a fully instantiated data cube to a very dense data structure.

Table 1. An example of fact table

D_A	D_B	D_C	D_D	M
A1	B1	C2	D2	70
A1	B1	C3	D1	40
A1	B2	C1	D1	90
A1	B2	C1	D2	50
A2	B1	C3	D1	60
...	...	...	...	...

In the example of the fact table shown in Table 1, there are four dimensions and one numeric metric attribute. The data of fact table are sorted according to the values in each dimension. The aggregation function of data cube is introduced with SUM as an example.

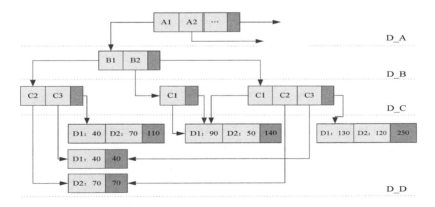

Fig. 1. An example of Dwarf data cube

In Fig. 1, it illustrates a Dwarf data cube structure derived from the fact table in Table 1. Here, the aggregation function SUM is introduced as an example. The height of the Dwarf tree equals the number of the dimensions, and its root node contains two fields: [key, pointer]. And each cell represents a different value in the current dimension. Each Cell's pointer points to the next node connected by different keys, which is pointed by a Cell, and all the cells inside are dominated by the Cell.

3   Segmentation of Data Cube

For the Dwarf data cube in Fig. 1, we divide it to several sub-cubes according to the first dimension. For the values of the first dimension $A1$, $A2$, ..., An, we split their sub-trees T_{A1}, T_{A2}, ..., T_{An} to generate several independent new Dwarf cube trees D_{A1}, D_{A2}, ..., D_{An}. For the data node of all cell in the dimension D_A, it is discarded directly, and its value is dynamically computed and generated by the Dwarf sub-trees D_{A1}, D_{A2}, ..., D_{An}.

In order to improve the security and reliability of Dwarf cube files, and to be easily processed and accessed in parallel, we use the distributed file system DFS to store the physical files of Dwarf data cubes. In order to organize and query sub-Dwarf cubes effectively, we construct a physical file index table of Dwarf data cubes. Each row of data in the index table consists of a binary *<dim, path>*. And the *dim* records a value of the top-level dimension, the *path* records a full storage path of the distributed file corresponding to the sub-Dwarf cube. The physical file index table of Dwarf cube is stored in the distributed file system as a text file.

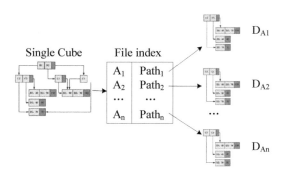

Fig. 2. Storage of parallel Dwarf cube

As shown in Fig. 2, logically, a whole Dwarf data cube is represented by a file index table, which records each value of the top-level dimension and all file paths of the sub-Dwarf files corresponding to each dimension value. Dwarf cube is divided according to the dimension D_A. Each value of the dimension D_A is $A1$, $A2$, ..., An. And each attribute value corresponds to a sub-tree. In the file index table, each attribute value of the dimension D_A corresponds to a tuple of data, which records the attribute value A_i and the distributed file full path $path_i$ of the corresponding sub-Dwarf cube respectively.

4   Parallel Dwarf Data Cube

Combining with the method of Dwarf cube segmentation in the previous section, we describe in detail the construction and query process of parallel Dwarf data cube based on MapReduce architecture in this section.

4.1   Construction of Parallel Dwarf

The construction of parallel Dwarf data cube is divided into three stages, namely Map, Shuffle and Reduce. The whole construction process is shown in Fig. 3.

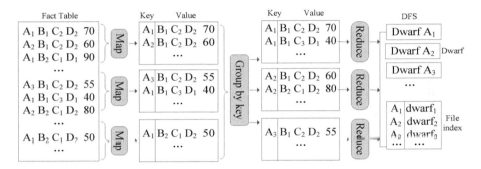

Fig. 3.   Construction of parallel Dwarf

In Map stage, the system first reads the tuples of data from the disordered source record files in block by block, and then generates <*key, value*> pairs according to the segmentation condition of the sub-Dwarf data cube. The *key* is composed of a compound key <*group, offset*>, and the *group* is a grouping identity of the sub-Dwarf to which the record belongs, the *offset* is used to sort the records in the group, the *value* is the current whole record tuple. For example, in the example given in Fig. 3, for a tuple $(A_1, B_1, C_2, D_2, 70)$, the Map task divides the tuple into <*key, value*> pairs, where the *group* in *key* is the attribute value A_1 of dimension D_A, the *offset* consists of the remaining dimension values (B_1, C_2, D_2) after removing dimension D_A, and the *value* consists of all remaining values after the tuple is removed dimension D_A. Here the value consists of the remaining values $(B_1, C_2, D_2, 70)$.

In Shuffle stage, the system groups <*key, value*> pairs generated in Map stage, assigns the same *group* value in *key* to the same group, sorts all values in the same group according to the *offset* in *key*, and finally transmits the same group of data to a single Reduce node. For example, the Shuffle stage divides all the tuples with D_A dimension A_1 in the source record into the same group, and then sorts all records belonging to group A_1 according to the *offset* values, namely dimension D_B, D_C and D_D.

Each Reduce task processes the list with the same group of values. At this time, all records have been segmented according to sub-Dwarf segmentation conditions, and have been sorted according to dimension values, so the construction conditions of a single-machine Dwarf data cube have been satisfied. The system calls the algorithm of building Dwarf data cube in a single node to build the sub-Dwarf tree, stores the sub-Dwarf data cube files in the distributed file system DFS, and generates *<dim, path>* pairs as the output of the Reduce task. Finally, the system collects all output *<dim, path>* pairs of all Reduce tasks and stores them in the physical file index table of the parallel Dwarf data cube.

4.2   Query of Parallel Dwarf

The query of parallel Dwarf data cube is also divided into three stages: Map, Shuffle and Reduce. The whole parallel query process is shown in Fig. 4.

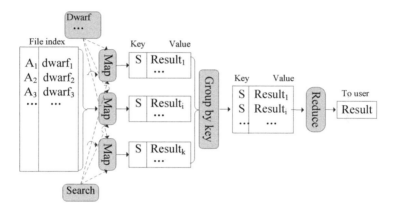

Fig. 4. Query of parallel Dwarf

In Map stage, the system reads data item by item from the distributed file index table of the Dwarf data cube. Each Map task filters the segmentation value of each data in the file index table according to the retrieval conditions of the query, and finds out all sub-Dwarf cubes that meet the retrieval conditions. Then, for the sub-Dwarf data cubes that meet the retrieval conditions, the system retrieves the corresponding physical files in distributed file system and generates the retrieved results into *<key, value>* pairs, in which the *key* is composed of query statements and the *value* stores the retrieval results of the current sub-Dwarf data cube.

In Shuffle phase, the system merges all *<key, value>* pairs generated by each Map task into a set. And then it is transmitted to the only Reduce task.

In Reduce phase, the system combines the search results of several sub-Dwarf data cubes to generate the final query results, and forms the key value pair *<search, result>* of the query results, and returns the final query results to the front-end users.

5 Experiments

The running environment of the experiment is that the cluster contains 10 server nodes, each node has 4 servers with 2.00 GHz Intel Xeon (R) CPU and 4G main memory, Linux Red Hat 5.1 operating system and Hadoop version 1.1.0. The experimental data set is generated by TPC-DS [11], which contains 10 dimension attributes and one metric attribute, totaling 1 million tuples. The potential of each dimension is 10, 100, 100, 1000, 1000, 2000, 5000, 5000 and 10000 respectively. The aggregation function is SUM.

5.1 Construction Performance of Parallel Dwarf

In Fig. 5, we compare the construction time of the traditional single Dwarf and parallel Dwarf. The experiment runs on a single node and a cluster of five nodes. And the experiment compares the different running situations when the data contain 6 to 10 dimensions. As can be seen from the figure, the construction time of the parallel Dwarf is much shorter than that of single Dwarf. When the number of dimensions of data sets is small, the effect of the parallel Dwarf is not obvious. As the number of dimensions of data sets increases, the advantage of the parallel Dwarf becomes more and more obvious. When it reaches 10 dimensions, the construction time is less than one sixth of that of single Dwarf. And with the increase of dimension, the construction time of single Dwarf cube increases rapidly, while the construction time of the parallel Dwarf cube increases slowly.

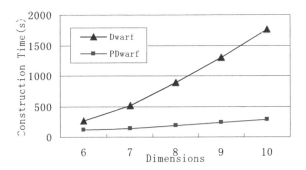

Fig. 5. Comparison of construction time

5.2 Query Performance of Parallel Dwarf

In Fig. 6, we compare the query performance of the traditional single Dwarf cube, the parallel Dwarf cube and general parallel data cube based on MapReduce architecture. As can be seen from the figure, the common parallel cube implemented by MapReduce needs to scan the data completely because it does not use index mechanism, and its query time is the largest and the query efficiency is the lowest. And the query time of the parallel Dwarf cube is slightly longer than that of single Dwarf cube.

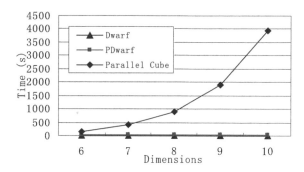

Fig. 6. Comparison of query

6   Conclusions

This paper presents a parallel compression data cube construction algorithm based on Hadoop framework. The algorithm divides the traditional single data cube into several independent sub-cubes, and then uses Hadoop architecture to construct and query the data cube in parallel. Experiments show that the parallel compression data cube algorithm can provide efficient data cube construction performance, and can provide efficient queries with the help of self-indexing mechanism of Dwarf cube. The parallel compression data cube has good scalability, and has good practical application value.

Acknowledgements. This work was supported by the National Natural Science Foundation of China (No. 61702345).

References

1. Chaudhuri, S., Dayal, U.: An overview of data warehousing and OLAP technology. SIGMOD Rec. **26**(1), 65–74 (1997)
2. Golfarelli, M., Rizzi, S.: Designing the data warehouse: key steps and crucial issues. J. Comput. Sci. Inf. Manag. **2**(3), 13–22 (1999)
3. Wang, W., Lu, H.J., Feng, J.L., et al.: Condensed cube: an effective approach to reducing data cube size. In: Proceedings of the 18th International Conference on Data Engineering, pp. 155–165 (2002)
4. Lakshmanan, L.V.S., Pei, J., Han, J.W.: Quotient cube: how to summarize the semantics of a data cube. In: Proceedings of the 28th International Conference on Very Large Data Bases, pp. 778–789 (2002)
5. Sismanis, Y., Deligiannakis, A., Roussopoulos, N., et al.: Dwarf: shrinking the PetaCube. In: Proceedings of the 2002 ACM SIGMOD International Conference on Management of Data, pp. 464–475 (2002)
6. Xiang, L.G.: Construction and compression of dwarf. J. Zhejiang Univ. SCI. **6**(1), 519–527 (2005)
7. Lammel, R.: Google's MapReduce programming model-revisited. Sci. Comput. Program. **70** (1), 1–30 (2008)

8. Dean, J., Ghemawat, S.: MapReduce: simplified data processing on large clusters. Commun. ACM **51**(1), 107–113 (2008)
9. Dean, J., Ghemawat, S.: MapReduce: a flexible data processing tool. Commun. ACM **53**(1), 72–77 (2010)
10. Landset, S., Khoshgoftaar, T.M., Richter, A.N., et al.: A survey of open source tools for machine learning with big data in the hadoop ecosystem. J. Big Data **2**(1), 1–36 (2015)
11. Othayoth, R., Poess, M.: The making of TPC-DS. In: Proceedings of the 32nd International Conference on Very Large Data Bases, pp. 1049–1058 (2006)

Development Trends of Customer Satisfaction in China's Airline Industry from 2016 to 2018 Based on Data Analysis

Yawei Jiang and Huali Cai[✉]

Quality Management Branch, China National Institute of Standardization,
Beijing, China
583451349@qq.com

Abstract. Airline industry plays a significant role in national economic and social development, and it is a common need of the people to enjoy a high-quality airline services while traveling. This paper conducted a nationwide survey of six airlines in China during 2016–2018. The questionnaire has 29 questions, covering six structural variables that related to brand, expectations, quality, price, satisfaction and loyalty, and there were nearly 5000 samples collected after the survey. By analyzing the three-year data, the results revealed that the service quality of China's airline industry was experienced a continuous improving progress, it was remained in the "relatively satisfied" range. In terms of structural variables, all of them showed a gradual upward tendency. In addition, the customer satisfaction score of the six airline companies in China shows that the service quality in Air China, Sichuan Airlines and Eastern Airlines has been increasing year by year.

Keywords: Data analysis · Service quality · Consumer satisfaction · Airline industry

1 Introduction

As an important strategic industry and an advanced transportation mode, airline industry plays a significant role in promoting the development of the overall national economy. Studying the quality of airline industry could help to improve transportation quality and increase international competitiveness, as well as better promote the implementation of national development strategy and better meet the needs of people.

Many scholars have carried out research on customer satisfaction of civil aviation service. Koklic examined the antecedents and consequences in the fields of customer satisfaction by conducted a questionnaire survey to passengers in airline enterprises [1]. Zhibin introduced the importance-performance-impact analysis model to identify the influenced index of customer satisfaction on different airline resources [2]. Stelios used the multicriteria satisfaction analysis method to calculate customer s' satisfaction and found weak aspects of airlines [3]. Bellizzi proposed an ordered logit model which based on passengers' satisfaction to analysis the influence of service factors on service quality [4]. Gupta ranked airlines through VIKOR method and found that tangibility, reliability, security and safety are the most important factors of service quality [5].

© Springer Nature Singapore Pte Ltd. 2020
M. Atiquzzaman et al. (Eds.): BDCPS 2019, AISC 1117, pp. 846–851, 2020.
https://doi.org/10.1007/978-981-15-2568-1_115

Stamolampros used several analytical subjects that were related to interest rates, fuel prices and market concentration to analyze airline service quality. The analytical subject includes on-time performance, cancelled flights, mishandled baggage, and passenger's complaints [6]. Park applied the SEM method to estimate the influenced coefficient of customers' sentiment on airline services [7]. Lu found that airline service quality would play a positive effect to repurchase intention [8]. Catherine analyzed the impacts of brand experience, brand love and service quality on customer engagement in civil aviation [9] as well as studied the passenger's airport selection and destination choice within different airline service quality [10].

This paper applied the structural equation modeling method to build a customer satisfaction model which could measure the customer satisfaction indexes from airline industry. By using the calculated data in 2016–2018 and analyzing the comparations between the six major airlines in China, we can discover the service quality developing progress in China's airline industry, and find its advantages and weaknesses. Those findings would be useful to airline policy makers.

2   Customer Satisfaction Model

We build the customer satisfaction model based on the structural equation modeling method. The model is shown in Fig. 1:

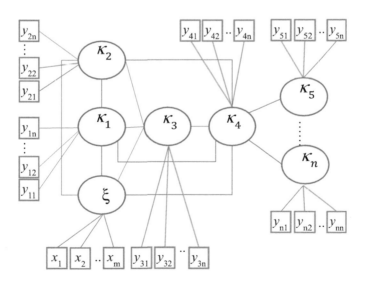

Fig. 1. The structural equation model

The mathematical model is:

$$\eta = B\eta + \Gamma\xi + \zeta \tag{1}$$

Where, η is the endogenous latent variables that would be affected by other variables. ξ is the exogenous latent variables which only affects other variables rather than be affected by others. B is the endogenous latent variable coefficient matrix which depicts the relationships between different endogenous latent variables. Γ is the exogenous latent variable coefficient matrix which denotes the effects of exogenous latent variables on endogenous latent variables. ζ is a residual term that reflects the part of the equation that cannot be explained.

For the relationship between indicators and latent variables, it is usually written as the following measurement equation:

$$X = \Lambda_x \xi + \delta \tag{2}$$

$$Y = \Lambda_y \eta + \varepsilon \tag{3}$$

Where, X is the observed variable of exogenous latent variable; Λ_x is the matrix relationship between the observed variables and exogenous latent variables; Y is the observed variable of endogenous latent variable; Λ_y is the matrix relationship between the observed variables and endogenous latent variables.

Thus, we can build the customer satisfaction model of in airline industry. The model is shown in Fig. 2:

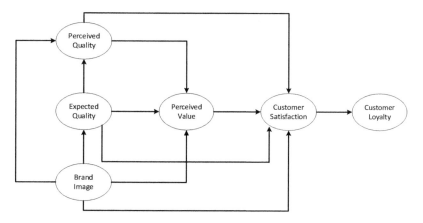

Fig. 2. Customer satisfaction model

In the model, the brand image is passengers' evaluation of brand impression and characteristics. The expected quality is passengers' expectations for airline services before selecting airlines. The perceived quality is passengers' evaluation of airline service after receiving services from airlines and it includes airline punctuality, catering quality, cabin comfort, cabin crew service, etc. The perceived value is passengers' evaluation of the matching degree on price and quality. The customer satisfaction includes passengers' overall satisfaction and passengers' satisfaction in terms of expected service, ideal service and other airlines' service. The customer loyalty includes passengers' reselect willingness and recommend willingness of an airline.

We conducted a questionnaire survey in 2016–2018, and collected 9000 effective samples from 6 airline enterprises: Hainan Airlines, Xiamen Airlines, Sichuan Airlines, Southern Airline, Eastern Airlines, Air China. The overall customer satisfaction of airline industry in 2016–2018 is shown in Fig. 3.

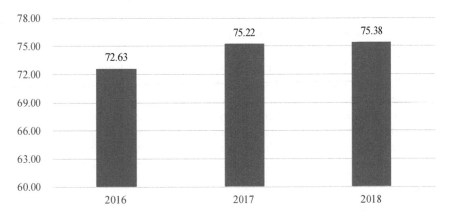

Fig. 3. Customer satisfaction of airline industry in 2016–2018

As can be seen from Fig. 1, in 2016–2018, the customer satisfaction of airline industry showed a gradual upward tendency. In general, it has been in the "satisfactory" range for the past three years that reflects the continuous improvement of the quality of airline industry services and the airline service have already met the main needs of people.

The structural variable scores are shown in Fig. 4.

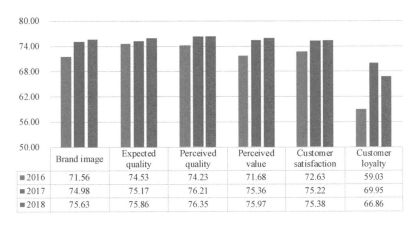

	Brand image	Expected quality	Perceived quality	Perceived value	Customer satisfaction	Customer loyalty
2016	71.56	74.53	74.23	71.68	72.63	59.03
2017	74.98	75.17	76.21	75.36	75.22	69.95
2018	75.63	75.86	76.35	75.97	75.38	66.86

Fig. 4. Structural variable scores in 2016–2018

In 2016–2018, the brand image, expected quality, perceived quality and perceived value of airline industry were all showed a gradual upward trend. It reflects that the customers' brand awareness and brand impressions of airline industry have been improved continuously, as well as the customers' expectations, actual feelings and cost performance in airline industry. In addition, the customer loyalty of airline industry showed a "rising-decreasing" trend. In 2017, the loyalty score was the highest, and in 2018, the loyalty decreased slightly. It revealed that the customers' recommendation for airline transportation and the willingness to re-select have a slight fluctuation.

The scores of customer satisfaction of the six airlines in 2016–2018 are as follows.

As can be seen from Fig. 5, customer satisfaction of Air China, Sichuan Airlines and Eastern Airlines have been increasing year by year. In 2018, Air China's satisfaction index was the highest among all the companies and shown a significantly higher score when compared with other companies. In addition, the customer satisfaction of China Southern Airlines and Hainan Airlines showed a "rising-decreasing" tendency, especially when it comes to the Hainan Airline, its customer satisfaction was significantly reduced in 2018. Moreover, Xiamen Airlines' customer satisfaction showed a "decline-up" trend, but its fluctuations was small and its overall customer satisfaction remained stable in general.

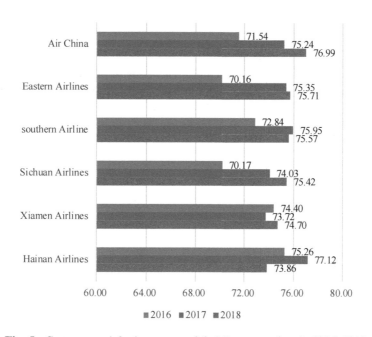

Fig. 5. Customer satisfaction scores of 6 airline enterprises in 2016–2018

3   Conclusions

This paper investigates the customer satisfaction of China's airline industry in 2016–2018. By calculations, we find that the customer satisfaction of China's airline industry is in the "satisfactory" range for three consecutive years. And it can be seen from the evaluation results that all the structure variables of airline industry have gradually improved, reflecting the continuous improvement of China's airline service quality and the tendency of high-quality development in airline industry. In addition, customer satisfaction of Air China, Sichuan Airlines and Eastern Airlines has been increasing year by year, that means the service quality construction of those enterprises has been achieved remarkable progress.

Acknowledgments. This work is supported by the National Social Science Fund Major Project —The research of quality management system and policy of high-quality development promotion. NO. 18ZDA079.

References

1. Kos Koklic, M., Kukar-Kinney, M., Vegelj, S.: An investigation of customer satisfaction with low-cost and full-service airline companies. J. Bus. Res. **80**(10), 188–196 (2017)
2. Zhibin, L., Ilias, V.: An advanced analytical framework for improving customer satisfaction: a case of air passengers. Transp. Res. Part E: Logist. Transp. Rev. **114**(6), 185–195 (2018)
3. Stelios, T., Theodosios, K., Angelos, P.: A multiple criteria approach for airline passenger satisfaction measurement and service quality improvement. J. Air Transp. Manag. **68**(5), 61–75 (2018)
4. Bellizzi, M.G., Eboli, L., Forciniti, C., Mazzulla, G.: Air transport passengers' satisfaction: an ordered logit model. Transp. Res. Procedia **33**, 147–154 (2018)
5. Gupta, H.: Evaluating service quality of airline industry using hybrid best worst method and VIKOR. J. Air Transp. Manag. **68**(5), 35–47 (2018)
6. Stamolampros, P., Korfiatis, N.: Airline service quality and economic factors: an ARDL approach on US airlines. J. Air Transp. Manag. **77**(6), 24–31 (2019)
7. Park, E., Jang, Y., Kim, J., Jeong, N., Kunwoo, B., Angel, P.: Determinants of customer satisfaction with airline services: an analysis of customer feedback big data. J. Retail. Consum. Serv. **51**(10), 186–190 (2019)
8. Lu, C., Yongquan, L., ChihHsing, L.: How airline service quality determines the quantity of repurchase intention - mediate and moderate effects of brand quality and perceived value. J. Air Transp. Manag. **75**(3), 185–197 (2019)
9. Catherine, P., Xuequn, W., Sandra Maria, L.C.: The influence of brand experience and service quality on customer engagement. J. Retail. Consum. Serv. **50**(9), 50–59 (2019)
10. Catherine, P., Mariam, K.: The role of airport service quality in airport and destination choice. J. Retail. Consum. Serv. **47**(3), 40–48 (2019)

Intelligent Outlet Based on ZMCT103C and GSM

Hongyu Zhao, Zhongfu Liu[✉], Ziyang Liu, Kaixuan Zhao, and Yuqi Shi

College of Information and Communication Engineering, Dalian Minzu
University, Dalian 116600, Liaoning, China
103086817@qq.com

Abstract. With the increasing number of electrical equipment in the home, the safety of household appliances, electricity statistics and remote transmission of information are becoming more and more important. In order to solve this problem, this paper uses STM32F103C8T6 single chip microcomputer as the main control chip to design an intelligent socket system which can measure the power consumption, and display the power consumption information locally, and control the wireless communication with the mobile phone through GSM. The intelligent socket system adopts low power supply, and the collection of power consumption is completed by current transformer ZMCT103C. The communication mode of the system is GSM communication, and the user can plug in the intelligent plug through the mobile phone APP at any place. The system sends text messages to control on-off and timing, so as to achieve the need to save electricity. In this paper, the system software is designed, and the mainstream diagram of the system and the flow chart of the main control modules are given. Finally, the measurement and actual power consumption data of the system are given, and the physical diagram of the system is given. After various tests, the system has the advantages of high reliability and strong measurement and control function, and can be widely used in smart home control system.

Keywords: Intelligent outlet · AC transformer · Wireless communication

1 Introduction

Current outlet in the market have relatively simple function, and sometimes cannot meet the needs of users. For example, users may forget to pull the plug sometimes when they plan a few hours to charge the battery, so, long charging time the battery will cause some damage. Or users forget to switch the power off while they go out for a long time. When users remember that, they could not cut down the power immediately, thus resulting the waste of electricity. Sometimes the electricity outlet position may be outside the house, such as a garage electric vehicle charging. We need to remember the charging time and then go to the garage when the charging time is enough, which a lot of people may not be willing to go out in the evening. All sorts of situations have

© Springer Nature Singapore Pte Ltd. 2020
M. Atiquzzaman et al. (Eds.): BDCPS 2019, AISC 1117, pp. 852–862, 2020.
https://doi.org/10.1007/978-981-15-2568-1_116

exposed some shortcomings of the traditional outlet: no timer service and cannot control remotely.

In recent years, researchers at home and abroad have conducted in-depth researches on Intelligent outlet. Lin has studied an expandable IOT-Based smart outlet system [1]. AI-Hassan also designed an smart power outlet for monitoring and controlling electrical home appliances [2].

Aiming at the above situation, this essay designs an outlet that can set the power-on time, can be remotely controlled by the mobile phone, and can display the electricity information on the screen and send the information to the mobile terminal at any time. You can remotely control the on-off and timing of the outlet by simply sending a specific command to the GSM module on the outlet through the handpiece [3]. This essay will describe the composition and design of the outlet in detail.

2 System Scheme Design

Using STM32F103C8T6 MCU as the core processing unit of this intelligent outlet, the collection of electric quantity is completed by the AC transformer. After the transformer collects the current value, the data is sampled by the ADC and sent to the MCU for processing. The processed data can not only be displayed on the screen, but also sent to the user via GSM. At the same time, code special functions to certain characters for the information GSM receives, such as the receiving of "JDOF" is to power off the outlet. The on-off of outlet is controlled by the relay. The information received by GSM is processed by single chip microcomputer to distinguish what function is performed. Using the selected OLED screen, single-chip computer is able to read and write through the I2C bus.

3 System Hardware Circuit Design

The hardware system includes the following parts: smallest SCM system, power collection module, GSM module and a relay circuit. The power collection module is mainly composed of ZMCT103C module and its surrounding circuit. The function of GSM module is mainly to send power information and receive control commands. The overall system scheme design is shown in Fig. 1:

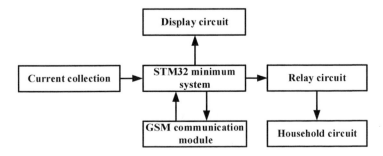

Fig. 1. The overall system scheme design

3.1   Power Circuit

The power supply circuit includes two parts. One part is to convert 220 V AC to 5 V DC for AC transformer module and GSM. The other part converts 5 V to 3.3 V for STM32.

3.1.1   220 V to 5 V Transfer Circuit

The circuit uses LM7805 as the voltage regulator. After the 220 V AC voltage is passed through the transformer, the output voltage is about 9 V to 10 V, and after rectification and filtering [4], it is sent to the input end of the LM7805. A 470uf electrolytic capacitor and a 105 capacitor are connected at the output terminal to further filter the ripple and obtain a 5 V regulated power supply [5]. The light emitting diode serves as an indicator. A schematic of this circuit is shown in Fig. 2:

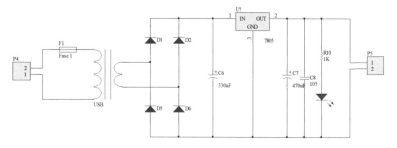

Fig. 2. 220 V to 5 V transfer circuit

3.1.2   5 V to 3.3 V Transfer Circuit

STM32 microcontroller stability needs stable voltage, thus converting 5 V to a stable voltage, 3.3 V for MCU. This design uses the AMS1117-3.3 V linear regulator chip, to convert input 5 V voltage to an output of 3.3 V. This part of the circuit diagram is shown in Fig. 3. "VCC5" is 5 V voltage; "IN" is the input pin of the regulator chip; output pin "OUT" is 3.3 V voltage; the left side are capacitor 104, capacitor 220uF used for filtering effect; and the light emitting diode is used for indication [6].

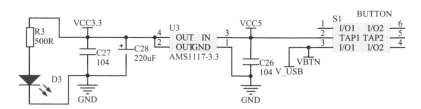

Fig. 3. 5 V to 3.3 V transfer circuit

3.2 Smallest System of Microcontroller Circuit

The smallest STM32 SCM system mainly consists of the following components: STM32 chip, reset circuit, clock circuit, power supply circuit and a downloaded circuit.

The system selects STM32F103C8T6 as the central control chip. STM32F103C8T6 is based on ARM Cortex-M core and has the advantage of high-performance, low-cost and low-power. On-chip resources include 48 KB SRAM, 64 KB Flash, four timers, three serial ports, a USB interface, three 12-bit ADC and so on, which are qualified to meet system requirements. The smallest system diagram of STM32F103C8T6 is shown in Fig. 4:

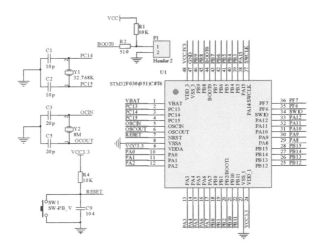

Fig. 4. The smallest system circuit of STM32F103C8T6

3.3 Current Collecting Circuit

The core device of the current acquisition circuit is the AC transformer ZMCT103C Type A, which is a ring structure, and the current value can be measured by the principle of electromagnetic induction by passing the wire to be tested through the middle hole. The main technical parameters of this module are shown in Table 1. ZMCT103C Type A has two pins, one for signal ground and the other for MCU sampling. This module can measure the AC current within 5 A, and the corresponding analog output is 5 A/5 mA, namely 1000:1. The schematic diagram of the current acquisition circuit part is shown in Fig. 5. The analog current output by the AC transformer passes through the half-wave rectifying circuit [7, 8], and the voltage drop on the resistance is collected by PA0. PA0 is the ADC pin of the single-chip microcomputer. "C1" is the filter capacitor, making the output voltage more stable. "C2" can filter high-frequency clutter signal.

Table 1. Main parameters of AC transformer

Parameters	Numerical value
Rated input current	5 A
Rated output current	5 mA
Ratio	1000:1
Phase difference (at rated input)	Less than or equal to 20′(100 Ω)
Linear range	0–10 A (100 Ω)
Linearity	0.2%
Precision level	Level 0.2
Isolation withstand voltage	3000 V measurement
Sealing material	Epoxy resin
Operative temperature	−40 °C–+70 °C

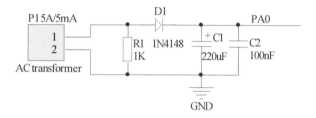

Fig. 5. Schematic of current collecting module

3.4 Relay Circuit

A relay is a device using "small current" to gain control of "high current" devices. When the change amount of input to meet the requirements, the output circuit in the electrical manipulation is controlled to a predetermined amount of step change occurs. The relay has an interaction between the control system and the controlled system. Briefly, in this design, the small current of the microcontroller is used to control the large current in the household circuit. There are springs, moving contacts, normally open contacts, normally closed contacts, coils, etc. [9], inside the relay. The circuit shown in Fig. 6. "4" indicates a normally open contact, "5" a normally closed contact and "3" a movable contact. The transistor is turned on when the JD connected to the MCU pin at low voltage. After the current flows from "1" to "2" through the triode, the coil between "1" and "2" is powering. Due to electromagnetic induction, a magnetic field is created, and the movable contact "3" is attracted to "4", and the right circuit (outlet) is turned on. The transistor is turned off when JD set at high voltage. There is no current is between "1" and "2". By the action of spring, the movable contact "3" returns to "5", then the right circuit is disconnected. Therefore, if you want to control the on/off of the current in the outlet through the electromagnetic relay, you can control the continuity of the outlet by simply connecting one pin of the microcontroller to the

"JD" to control the high and low levels of the pin. According to the Fig. 6, "D2" is a light-emitting diode. When the relay picked up, the diode glows, which indicates the powering in the outlet.

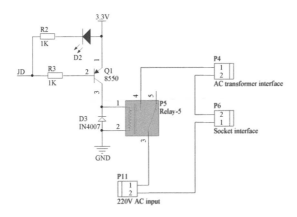

Fig. 6. Relay circuit diagram

3.5 Human-Computer Interaction Module

The human-computer interaction module includes the following two parts: the display part and the operation part.

The display part uses OLED. Compared with LCD, OLED has the following advantages: low power consumption, high contrast, thin thickness, wide viewing angle, fast response, and a variety of interfaces. It is read and written using the I2C bus. For a monolithic integrated circuit, only two pins are needed to control the OLED, one to control the clock and the other to transmit data. The contents to be displayed include current value, power value and timing time. Because the OLED used does not have its own word library, it is necessary to use the software to obtain the hexadecimal value of the written Chinese characters.

The operative part is the key, which can be used to manually control the relay and GSM or debug some other functions.

3.6 Communication Module

The outlet communicates with the handset via GSM. The module used is SIM868. This module provides two types of TTL interfaces, one of which is compatible with TTL levels of various voltages, can be directly connected with 5 V or 3.3 V single-chip microcomputers, and provides SMS, voice and GPRS data transmission functions. The schematic diagram of the circuit is as shown in Fig. 7. According to the Fig. 7, "TXD" and "RXD" are used to communicate with the serial port of MCU. "SIM_RST" is the reset pin, and the MCU can reset it at a high level. "SIM_VDD" is the power pin of the chip, the GSM module requires an operative voltage not less than 5 V, so this pin should be connected to the 5 V power supply.

GSM has an AT instruction to set interface, which is controlled by sending AT instructions. The communication mode between GSM and MCU is serial port, so connect RX and TX of serial port of MCU to "TXD" and "RXD" of SIM868 respectively, connect GND of them together again, after power-on, the MCU can send AT instruction to GSM, set its working mode, data communication, etc.

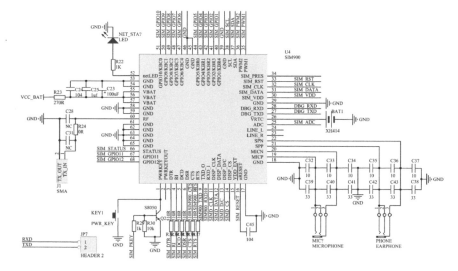

Fig. 7. Circuit diagram of GSM

4 System Software Design

The main part of the software is data processing, which can be roughly divided into the following sections: The ADC collects the analog voltage value of the AC transformer output, converts the collected analog voltage value into actual current value, GSM transmission and reception data, controls the relay according to the data received by GSM, and displays relevant information on the OLED. Using Modular programming mode, encapsulate the above parts for later calls in the main program.

4.1 Main Program Design

The main function of the software is designed to build the equivalent of a "framework". The process can be described as follows: First, apply the initialization function to the microcontroller pins, clocks, and each module. Only after initialization the system can work. Then use serial interface to send AT commands to the GSM, so that GSM can register network and configure the text mode. The program then enters a loop comprising the following components: obtaining the ADC sample value, detecting whether there is a button press, detecting whether the GSM receives the short message, and refreshing the display. After obtaining the ADC sample value, calculate according to the conversion relationship between the actual current and the sampling value; the specific operation can be performed by acquiring the state of the button, such as sending the power consumption information to the handset. Instruction is detected after the GSM received, further control relay on and off and timing. The main implementation process is shown in Fig. 8:

4.2 Current Acquisition Program Design

This program module includes ADC collection, processing and calculating a power value of a GSM transceiver information. ADC is first initialized. Comprising the following steps: turn on the ADC clock; configure the ADC operating mode, trigger mode, channel, data alignment, conversion time, etc., then enable the ADC. Following enters a loop. In the Fig. 2, the microcontroller collects the analog voltage value through the PA0 pin, the analog voltage value is the voltage on the 1K resistor in Fig. 2. Divided it by 1000 to get the output current value of the current transformer module. According to 5 A/5 mA output characteristics of the module, calculate the size of the household current. ADC acquired current program flowchart is shown in Fig. 9:

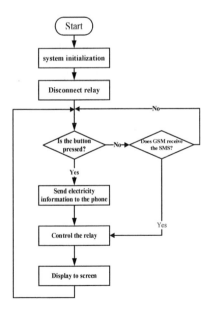

Fig. 8. Main program flow

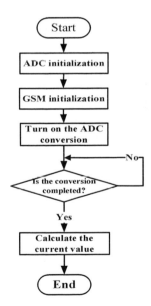

Fig. 9. Current acquisition software design flow chart

4.3   GSM Communication Module Programming

The control of GSM generally includes the following steps: First, detect whether the module can normally communicate with the MCU, send an "AT" with the serial port. If the GSM can communicate with the MCU, it will return "OK" through the serial port, otherwise, it needs to detect whether there is a problem with the circuit connection or whether the SIM card insertion direction is correct. Secondly, set the module network registration prompt, SMS mode, select TE character set and SMS not transcoding, etc. The above various AT instructions can be found in the corresponding AT instruction set, here do not elaborate. For information received by GSM, determine whether to open or close the relay by resolving specific characters preset in the program. At the same time, detect whether the key controlling the GSM is pressed. If it is pressed, send the electricity message to the preset mobile phone number.

5   System Debugging

Debug the outlet with a fan with adjustable power: After the system powers on, wait for the GSM module to register the network to be successful, and then turn on the fan, 1–3 gears correspond respectively to 20 W, 30 W, 40 W. Turn off the relay and observe the current and power information displayed on the OLED, which is shown in the Table 2. And the errors are within the acceptable range. Press the button that sends the text message, and you can receive the text message with the electricity message on the mobile phone. Use a mobile phone to send a text message with specific instructions to the mobile phone card number on the GSM, which can control the relay. For example, if you send "JDON", you can see that the indicator light glows, the relay is off, and the outlet is conductive; when "JDOF" is sent, the indicator light is off, the relay is turned on, and the outlet is powered off. Send "time00:10", you can see that the OLED displays a 10 s countdown, and the relay cuts off the current after 10 s. Figure 10 shows a photograph of the entire system, and Fig. 11 shows a photograph of the sending interface on the mobile phone.

Table 2. Debugging measured data

Index	Measurements		
Measured current value	0.104 A	0.148 A	0.192 A
Measured power value	0.023 KW	0.033 KW	0.042 KW
Actual fan power	20 W	30 W	40 W

Fig. 10. System photo   **Fig. 11.** Mobile phone console

6 Conclusion

The intelligent outlet designed in this essay not only has the function of the traditional outlet, but also incorporates the more humanized design such as remote control and screen display, which can basically meet the needs of most users. However, the design still exists some problems, such as the accuracy, because the circuit capacitance of the power acquisition module does not release the loop [10], which needs to be corrected in the program. At the same time, the integration of the system is not high and further optimization is needed. To summarize, the design has a good practical value.

Acknowledgments. Fund project: National Innovation Training Program for College Students (201912026046).

References

1. Lin, Y.-B., Huang, C.-M., Chen, L.-K., Sung, G.-N., Yang, C.-C.: MorSocket: an expandable IoT-based smart socket system. IEEE Access **6**, 53123–53132 (2018)
2. Al-Hassan, E., Shareef, H., Islam, Md.M., Wahyudie, A., Abdrabou, A.A.: Improved smart power socket for monitoring and controlling electrical home appliances. IEEE Access **6**, 49292–49305 (2018)
3. Bharadwaj Manda, V.L.K., Kushal, V., Ramasubramanian, N.: An elegant home automation system using GSM and ARM-based architecture. IEEE Potentials **37**(5), 43–48 (2018)
4. Yin, S., Liu, Y., Liu, Y., Tseng, K.J., Pou, J., Simanjorang, R.: Comparison of SiC voltage source inverters using synchronous rectification and freewheeling diode. IEEE Trans. Ind. Electron. **65**(2), 1051–1061 (2018)
5. Mannen, T., Fujita, H.: A DC capacitor voltage control method for active power filters using modified reference including the theoretically derived voltage ripple. IEEE Trans. Ind. Appl. **52**(5), 4179–4187 (2016)

6. Michal, V.: Switched-mode active decoupling capacitor allowing volume reduction of the high-voltage DC filters. IEEE Trans. Power Electron. **31**(9), 6104–6111 (2016)
7. Esmail, E.M., Elkalashy, N.I., Kawady, T.A., Taalab, A.-M.I., Lehtonen, M.: Detection of partial saturation and waveform compensation of current transformers. IEEE Trans. Power Deliv. **30**(3), 1620–1622 (2015)
8. Ghosh, N., Kang, W.P., Davidson, J.L.: Half-wave rectification and envelope detection utilising monolithic nanodiamond lateral field emission diode. Electron. Lett. **47**(21), 1187–1189 (2011)
9. Tekdemir, I.G., Alboyaci, B.: A novel approach for improvement of power swing blocking and deblocking functions in distance relays. IEEE Trans. Power Deliv. **32**(4), 1986–1994 (2017)
10. Lee, T.-J., Tsai, T.-Y., Lin, W., Chio, U.-F., Wang, C.-C.: A dynamic leakage and slew rate compensation circuit for 40-nm CMOS mixed-voltage output buffer. IEEE Trans. Very Large Scale Integr. (VLSI) Syst. **25**(11), 3166–3174 (2017)

Internal Quality Control and Evaluation Method of Food Microorganism Detection

Mingxiao Jin[✉] and Yanhua Zhou

Jilin Engineering Normal University, Changchun 130052, Jilin, China
tougao_004@163.com

Abstract. With the continuous improvement of living standards, people attach great importance to the quality of food safety. Food microbial detection can better reduce food safety risks. Food microbial detection is a very complex process, and its detection results are affected by many aspects. In this paper, the internal quality control methods of food microbial detection are proposed from the perspectives of personnel, equipment and environment, sample collection and processing, culture medium and reagents. In the aspect of quality evaluation, three-level indicators were selected to calculate qualitatively and quantitatively by analytic hierarchy process (AHP) and coefficient of variation (CV) method, and a fuzzy comprehensive evaluation system was established to evaluate the quality of food microorganism detection. Taking the detection process of Staphylococcus as an example, the results showed that the quality of detection was improved after the fuzzy comprehensive evaluation according to the internal quality control method of microbial detection proposed in this paper.

Keywords: Food microbial detection · Internal quality control · Quality evaluation · Fuzzy comprehensive evaluation

1 Introduction

Food quality problems are related to people's production and life. If people eat unhealthy food, it will affect people's mental state for a whole day or even a long time, which will lead to worse quality of work and life, and even people will lose their lives because of poor quality food. Therefore, food safety inspection is particularly important. Among them, food microbial detection is a very important part. As we all know, microorganisms are numerous and complex, which are difficult to detect in the detection process, but microorganisms are the biggest threat to food safety. Therefore, the internal quality control and evaluation of food microbial detection is very important.

In food detection, food quality has always been the focus of attention. In [1, 2], the author points out that the legal framework does not ensure that food will never pose a risk to any consumer. Develop risk management procedures to control potential risks associated with food consumption. Quality assurance is a key factor in producing reliable and repeatable analytical results. In [3], the authors observed people's hygienic habits. The study assessed and compared the food safety concepts of 100 food processors and 295 consumers. In restaurants, microbial E. coli is found in salad appliances, food counters and consumer mobile phones [4]. Food safety in food services

© Springer Nature Singapore Pte Ltd. 2020
M. Atiquzzaman et al. (Eds.): BDCPS 2019, AISC 1117, pp. 863–869, 2020.
https://doi.org/10.1007/978-981-15-2568-1_117

needs to be improved and the risk of foodborne diseases reduced. Here it is necessary to introduce the food microbiological test. Food microbiological testing is an important part of food inspection, which mainly refers to bacteriological testing, including the total number of bacteria, coliform bacteria and pathogenic bacteria testing [5, 6]. The main methods of food microbial inspection are microbiological microscopy, physiological and biochemical reactions, isolation and cultivation methods to test the number and species of food microorganisms, so as to provide a more reliable evaluation of food quality and hygiene in the process of processing [7–9]. One of the purposes of microbiological testing is to prove that food is qualified or not. Another goal of microbiological testing is to detect whether there is a non-negligible deviation in the production process of food, that is to say, to detect whether the semi-finished products of goods are contaminated [10].

In order to provide accurate experimental data and strictly control the quality of food safety and hygiene, this paper studies the internal quality control and evaluation methods of food microorganisms in order to make people have no worries when enjoying delicious food.

2   Internal Quality Control of Food Microorganism Detection

In the process of food microbial detection, personnel, equipment and environment, sample collection and processing, culture medium and reagents will have an impact on quality control. Next, the internal quality control methods are put forward from these four aspects.

(1) The quality of inspectors is directly related to the level of laboratory inspection. The responsibility and technical ability of inspectors are the core of the quality of inspectors. Relevant inspectors should have the basic abilities of skilled use of testing instruments and mastery of test methods, indicators, processes, etc. They should also have the ability to analyze the final test results, so as to ensure the smooth progress of testing.

(2) The environment and equipment of food microorganism testing laboratory will influence the results of testing. Temperature has a great influence on the survival of microorganisms. Reasonable control of ambient temperature and reduction of the influence of ambient temperature on data can improve the reliability of test results more accurately. In addition, laboratory humidity and sanitary environment are also influencing factors. During the inspection period of the instrument and equipment in use, the necessary period verification should be carried out according to the frequency of use, the performance of the instrument and so on, so as to ensure that the instrument and equipment are in normal working state.

(3) Sample collection and processing. Microbial testing samples are sensitive to transportation and storage temperature, time and other factors. External contamination should be strictly prevented during sampling, sterile operation must be followed, storage and transportation should meet the corresponding conditions, and corresponding records should be made. As far as possible, the samples should be detected immediately in accordance with standard methods after sampling. The

microbiological laboratory should record the status of samples in detail and be reflected in the test report. In addition, the sampling date, sampling conditions, sample acceptance date and acceptance conditions should be recorded in detail. If there are insufficient samples, temperature discomfort, package breakage, incomplete labeling and so on, the microbiology laboratory should communicate with customers, and then decide to test or refuse to test. For sample labels and packaging may be contaminated, contamination and diffusion should be avoided during storage and transportation.

(4) Medium and reagent. When purchasing and using reagents and media in laboratories, special personnel should be arranged to manage reagents and media in strict accordance with national standards, pay attention to the storage environment, avoid dampness, and inventory frequently. The first opening date should be recorded after opening, and the bottle should be sealed after opening. Commercial dehydration media should be sterilized in strict accordance with the instructions and refuse to use contaminated media. Usually make good use plan, buy reagent culture medium to ensure that it is used up in the trial period.

3 Evaluation Method of Food Microbial Detection Quality

A fuzzy comprehensive evaluation system was established to evaluate the quality of food microbial detection. The steps of fuzzy comprehensive evaluation are as follows.

3.1 Determining the Weight of Indicators

There are two main methods to determine the weight of indicators, namely, coefficient of variation method and analytic hierarchy process. Analytic Hierarchy Process (AHP) is used to deal with the primary and secondary indicators here. The variation coefficient method was used to deal with the quantitative three-level indicators.

Analytic Hierarchy Process
In the analytic hierarchy process, the scoring criteria are divided into 10 levels, each level is scored by experts in the field, and then the total weight is determined by indicators, of which 1 indicates that one element corresponds to the lowest of other elements, while 10 represents the highest. The importance matrix is obtained by comparing the indicators, and then normalized. The eigenvalues and eigenvectors of the normalized matrix are calculated respectively. Then, the weight vector is determined. The vector corresponding to the largest eigenvalue is selected as the weight vector. The weight vector of the first index is $W = (w_1, w_2, \cdots, w_n)$, n is the number of the first index, the second index of the first index is $W_i = (w_{i1}, w_{i2}, \cdots, w_{ij})$, and the W_{ij} is the weight of the second index of the j under the first index.

Coefficient of variation method
If the indicators differ greatly from each other, the more information can be expressed by the selected indicators, which also shows that the selected indicators are broadly

representative and more convincing. The coefficient of variation among different indicators is calculated by:

$$V_i = \frac{\delta_i}{X_i} \quad i = 1, 2, \cdots, n \tag{1}$$

In the above formula, δ_i is the mean square deviation of the first index, V_i is the coefficient of variation of the second index, and n is the number of indexes.

3.2 Constructing Qualitative Scoring Set

After determining the weight by the way mentioned in the previous section, the next step is to create a set of qualitative scores, that is, to classify the evaluation objects. For the qualitative score set, this can be set according to their actual situation, such as the construction of such a qualitative score set {excellent, good, general, poor}. Of course, in the case of specific description, the score set can be constructed by numerical value. Correspond different grades to different values, and arrange them in order from large to small. The higher the evaluation, the greater the value, and vice versa.

3.3 Evaluation Methods Among Different Indicators

Quantitative indicators and qualitative indicators constitute the indicators of the evaluation system proposed in this paper. Considering the differences between the two indicators, different methods are used to evaluate them. The specific methods are as follows:

Quantitative indicators
Qualitative score set has four levels, corresponding to the description of {excellent, good, general, poor}, in the form of characters for W1 (Xi), W2 (Xi), W3 (Xi), W4 (Xi), here, (Xi) refers to the value of the first evaluation object. The next step is to find the most suitable membership function for each rating level. This process requires constant attempts to modify the fine-tuning. There are many methods available, such as piecewise representation and undetermined coefficient method. Usually, if there are two subjects to be evaluated, one has a higher and excellent degree of membership, while the remaining one is relatively low. It can be concluded that the second rank also has a higher level of membership.

Qualitative indicators
Delphi method was used to evaluate the qualitative indicators. The specific implementation method is as follows: let many people rating the selected indicators through the given scores, and the total score of each person's score is kept at 1. Scoring results will continue to optimize until there is no significant difference between the scores of each person. Another method is to divide the scores of many people on the same index into grades, and calculate the average scores of each grade, and then get the final results.

3.4   Fuzzy Computing

This process begins with the lowest level of indicators, and calculates the evaluation results of the upper level based on the weights mapped by each level of indicators. The operation of fuzzy transformation is applied here. Then, according to the process described above, repeat this step and stop the whole process after calculating the evaluation results of the top level indicators. Then, the assignment vector is transposed, and the final evaluation set is multiplied by the changed vector. Then the final quality evaluation index comes out. Then according to the results of the quality evaluation index to find out the corresponding quality level.

4   Results

Result 1: Testing process
The process of food microbial detection is relatively complex. Here, the process of Staphylococcus detection is taken as an example. The specific process of detection is shown in Fig. 1. The whole test environment is between 35 and 37 °C.

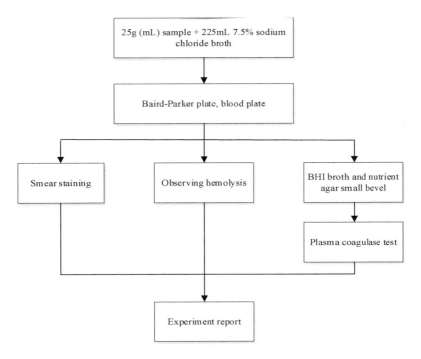

Fig. 1. Examination of Staphylococcus

Result 2: Comparison of basic quality indicators

Starting from the secondary indicators, we compare the importance of indicators at different levels, and finally get the score. The comparison scores of basic quality indicators experts are shown in Table 1.

Table 1. Comparison of importance of basic quality indicators

	A_{11}	A_{12}
A_{11} main physical and chemical indicators	1	4
A_{12} other nutrients	1/4	1

As can be seen from Table 1, the main physical and chemical indicators are as important as their own, so they are all 1. Experts believe that the main physical and chemical indicators are more important than other nutrients, and the relative importance is 4. The reciprocal of the importance of other nutrient elements relative to the main physical and chemical indicators is 1/4. The importance of I relative to j is reciprocal to that of J relative to I.

5 Conclusion

Food safety is a social problem that must be solved. Food safety is one of the basic material needs of the nation. Food microbial detection experiment is a very complex process, but the related technology of food microbial detection is not advanced enough. Nowadays, there are many links in food microbial detection and inspection, and the process of inspection is complex, which often results in inspection errors, thus the accuracy of the results of food microbial inspection can't reach the desired level. In order to do a good job in food microbial detection, we should strictly grasp the quality control from various aspects, and improve the relevant evaluation system.

References

1. De Boer, A., Bast, A.: Demanding safe foods–safety testing under the novel food regulation. Trends Food Sci. Technol. **72**, 125–133 (2018)
2. Leder, R., Ščitnik, V., Vukoja, M., Boras, A., Petric, I.V., Antunac, N., Banović, M.: Quality assurance of wine testing results in analytical laboratory. Hrvatski Časopis za Prehrambenu Tehnologiju Biotehnologiju i Nutricionizam-Croatian J. Food Technol. Biotechnol. Nutr. **12**(3/4), 146–154 (2017)
3. Her, E., Seo, S., Choi, J., Pool, V., Ilic, S.: Assessment of food safety at university food courts using surveys, observations, and microbial testing. Food Control **103**, 167–174 (2019)
4. Zhang, F., Xu, W., Wang, H., Gao, F., Zhi, J., Cui, S.: ATP bioluminescence method for the evaluation of microbial contamination of tableware. J. Food Saf. Qual. Test. **7**(3), 911–916 (2016)
5. Wang, Q., Yang, R., Yan, L., Ji, A.: Detection and analysis of the pathogen and hygienic microbial indicators of Jingmaishan. J. Food Saf. Qual. Test. **9**(2), 342–348 (2018)

6. Yu, Y.: Exploring the characteristics of food microbial testing. Chem. Manag. (2), 60 (2018)
7. Chen, Y.: Common techniques in food microbial testing. Chin. J. Pharm. Econ. **13**(10), 124–127 (2018)
8. Lin, X., Chen, W., Jiang, J., Zhang, B., Yin, W., Xiao, J.: Analysis and discussion of microbial factors in vinegar turbidity. J. Food Saf. Qual. Test. **8**(4), 1293–1297 (2017)
9. Kianpour, S., Ebrahiminezhad, A., Mohkam, M., Tamaddon, A.M., Dehshahri, A., Heidari, R., Ghasemi, Y.: Physicochemical and biological characteristics of the nanostructured polysaccharide-iron hydrogel produced by microorganism Klebsiella oxytoca. J. Basic Microbiol. **57**(2), 132–140 (2017)
10. Cheng, J., Xie, Y., Liu, J., Chen, L., Jiang, L.: Safety hazards and improvement measures in the production of traditional fermented soybean products. J. Food Saf. Qual. Test. **8**(8), 3092–3098 (2017)

Application of Big Data in B2C E-commerce Market Analysis

Yanqi Wang[✉], Hongke Shen, and Guangning Pu

Chengdu Neusoft University, Chengdu 611844, China
wangyanqi@nsu.edu.cn

Abstract. Big data provides new tools for market operations. In market analysis, people fully exploit their hidden business value and help companies make more accurate decisions by collecting, organizing and using data information. This paper aims to find a breakthrough in market analysis methods in the field of e-commerce through the collation and research of B2C e-commerce market analysis methods in the context of big data, and provide a certain reference for the development of e-commerce.

Keywords: Big data · B2C E-commerce · Market analysis

1 Introduction

In the era of big data, e-commerce is becoming more and more intense, prompting enterprises to continuously explore and develop information technology. Ma Yun said: "We have never regarded ourselves as an e-commerce company. We are actually a data company, a data platform!" Big data itself has many types, high application value, fast access, and large capacity. And it is a new technology for mining, creating and using information for a series of analysis such as collection-storage-association of data with scattered distribution, large quantity and variable format. At present, big data is developing rapidly. The first-tier cities such as Beishangguang and the big data industries in some western cities have developed rapidly [1]. In addition, big data will be widely used in transportation, tourism, and medical fields. According to the research report, by 2020, China's big data related products and services business revenue will exceed 1 trillion yuan [1, 4] (Fig. 1).

On the one hand, the arrival of the era of big data has impacted and changed people's way of thinking, work and living habits, and people have become data sources of information in daily life. On the other hand, people's lives are rapidly integrated with e-commerce [5]. However, the quality has emerged in an endless stream, and problems have frequently occurred in recent years. Major e-commerce companies have sought new breakthroughs in technology and management. Accurate market analysis is an important prerequisite for companies to implement a series of activities such as SPT and 4P. E-commerce companies rely on the advantages of innate platforms, and based on the support of big data, can deeply explore and utilize existing data to achieve full and accurate collection of market information in a "customer-centric" marketing concept. Excavate the commercial value of market information, thereby improving the

© Springer Nature Singapore Pte Ltd. 2020
M. Atiquzzaman et al. (Eds.): BDCPS 2019, AISC 1117, pp. 870–876, 2020.
https://doi.org/10.1007/978-981-15-2568-1_118

China's Big Data Industry Develops into a Growth Trend.

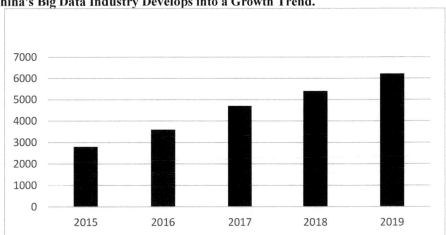

Fig. 1. China's big data industry market scale and forecast in 2015–2019 (1 trillion yuan) Source: iiMedia Research

precision of market analysis and guiding the healthy and rational development of business activities. Market research in the context of big data is of great significance in the field of e-commerce. However, there is not much research on this aspect today, and research on this issue.

2 Related Concepts

Big Data
Victor Meyer-Schoenberg believes that big data is the analysis of all the data, the amount of information is often large, and the data is more accurate and accurate. Big data has the characteristics of high speed, mass, authenticity, diversification and value [2, 6]. Data can truly feedback consumers' information and help companies predict and evaluate huge market value. Big data can also be easily understood as collecting data from various customers in various ways.

B2C E-commerce
As a business-to-customer retail model, B2C e-commerce has the characteristics of complete variety and one-stop shopping [7]. In addition, the company provides a series of multi-functional platforms such as payment and security to achieve resource sharing, cost reduction and efficient operation [8]. In recent years, the development trend of the B2C e-commerce platform market has steadily increased, showing the situation of several giants.

China's B2C Network Retail Platform Market Share Is Clearly Divided
Nowadays, China's large-scale e-commerce companies have launched online and offline business transformations, expanding their business and making consumers more

intuitive. Even so, e-commerce merchants are mixed with dragons and snakes, and there are often problems such as credit risk, wrong goods, and imperfect payment systems. In addition, many businesses currently have inaccurate positioning problems, which will directly affect the management of the enterprise marketing system. Therefore, market analysis as the most critical link in the production and operation of enterprises must use effective tools to understand the market, analyze customer needs, form precise positioning, and promote the healthy and stable development of the industry value chain (Table 1 and Fig. 2).

Table 1. 2018 China's B2C network retail platform market share in the first half of the year. Source: iiMedia Research

Ranking	B2C Online retail platform	Share
1	Tmall	52.5%
2	JD	31.3%
3	Vipshop	5.7%
4	Suning Online Market	5.6%
5	Gome	3.7%
6	Other	1.2%

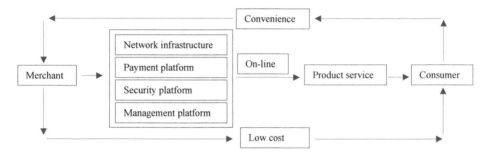

Fig. 2. E-commerce operation process

Market Analysis

Market analysis is an analysis of factors affecting market supply and demand changes, market supply and demand changes and development trends. Market analysis is based on content including market demand forecast analysis, market demand level and market demand analysis of various regions, estimating product life cycle and saleable time [9]. Its purpose is to better understand the relationship between supply and demand of goods. Market analysis can help companies identify market opportunities, create development conditions, strengthen control of sales, identify and solve problems in business, provide a basis for marketing activities, and ensure the smooth implementation of corporate marketing strategies. It provides an important basis for enterprises to make production and operation decisions, enhance management capabilities and competitiveness, and promote healthy and stable development (Fig. 3).

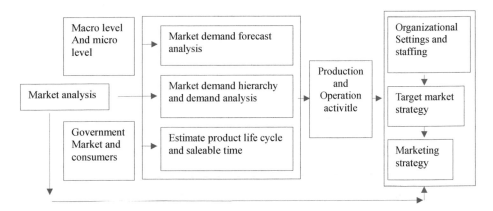

Fig. 3. The process and role of market analysis

3 Big Data Application in B2C E-commerce Market Analysis

3.1 Characteristics of Consumers and B2C E-commerce Companies in the Context of Big Data

On the one hand, consumers as an important part of the market, its new features are related to the overall market trends. Different from traditional sales and online trading activities, consumers are pursuing convenient, fast and convenient shopping forms. Consumption evaluation as one of the forms of interaction between consumers and merchants after purchase, it has the characteristics of wide spread and long retention time. Usually, consumer evaluation is an important reference standard for consumers who cannot know the purchase price information of other consumers. In addition, individual large-scale e-commerce giants have formed information on guiding products such as "guess you like" and "recommended list" based on the data information of consumers' browsing and consumption habits, and will also affect consumers' purchases.

On the other hand, B2C e-commerce has the characteristics of being able to achieve sales patterns across time and across regions, attracting more and more fine customer groups. At the same time, market analysis based on big data can rely on the advantages of e-commerce platform to achieve rapid and accurate collection of large consumer information, and fully tap and refine potential business value. For example, according to the frequency of browsing or purchasing, the system judges to give a classification of great concern and comparative preference; according to the consumer's delivery address or location, determine the overall consumption level, personal consumption level of the city or region in which it is located, and whether the area has the physical store of the item or other related information. However, the degree of information asymmetry between enterprises and consumers is getting deeper and deeper. Enterprises with a large amount of consumer information can use this information to organize and analyze, and price products. In addition, some platform opaque promotional prices and cumbersome rules are known to exist. Consumers have been passive in this process.

Under the background of big data, the new situation of consumers and B2C e-commerce enterprises forces enterprises to seek technological innovation to meet the changes and development of the market. Market analysis, as an important early stage of the company's production and operation activities, must attract the attention of enterprises.

3.2   Application of Big Data in B2C E-commerce Market Analysis

3.2.1   Analysis of B2C E-commerce Market Based on Big Data

Nowadays, the core position of business management is to achieve precise business activities. How to be truly accurate is related to the process of enterprise systematization. Therefore, enterprises must correctly position market demand, analyze market demand, product life cycle, sales time, etc., thereby guiding production and operation activities, and fully understand market analysis as an important content. Internet-based B2C e-commerce enterprises must constantly improve the tool technology of big data, and take the accuracy of market analysis as the focus of survival and development.

Big data technology usually manifests as search engine, information flow, etc. Through statistical analysis of raw data, it evaluates market conditions, predicts future market development, and achieves a reasonable distribution between market demand and production operations. It is an important tool technology to realize the forecast analysis of enterprise market demand (Fig. 4).

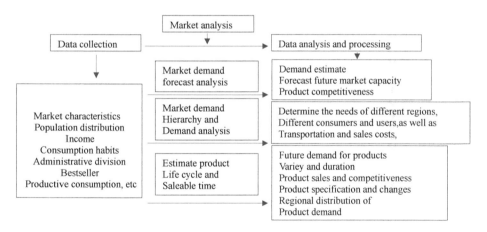

Fig. 4. Market analysis based on big data

3.2.2   The Role of Big Data in the Analysis of B2C E-commerce Market

Accurate market positioning. Big data can provide accurate market demand information, enabling companies to easily and accurately access market information and make strategic decisions in a timely and accurate manner. The main operating mode of B2C e-commerce in China has also expanded from the initial platform mode to platform mode, social mode, special sale mode and preferred mode.

Cost reduction can be achieved by professional processing such as digitization of market information collected in the early stage (Fig. 5).

Some market demand information obtained through surveys

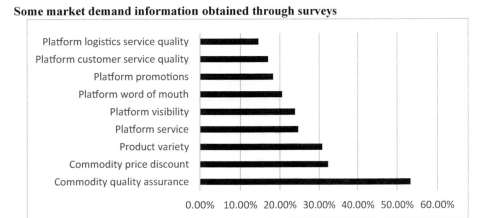

Fig. 5. 2018Q3 China B2C e-commerce user selection platform operation mode platform attention factor survey Source: iiMedia Research

Breaking through the limitations of time and space, the business activities can be carried out anytime and anywhere, which can improve efficiency.

Open. Through technological innovation, enterprises can capture global market information and bring breakthroughs in production and operation activities [10].

Resource sharing. Through the innovation of big data technology, small and medium-sized enterprises realize the sharing of information resources and further improve their competitiveness.

Data sink. Providing accurate data information is the key to using big data to discover and solve problems [3]. After a large amount of data is collected, data can be calculated, analyzed, and made into a decision-making direction. With the platform of big data, different information of the market can be classified and integrated. The need to develop a corresponding strategy is conducive to optimizing operations and marketing.

3.3   Problems and Improvements

With the increasing connection between the Internet and big data, B2C e-commerce enterprises have exposed certain problems while using big data. The security of information is not guaranteed, the management of data information is regulated, and data information and electronic evidence are shared. Identify and other issues.

In the future, B2C e-commerce companies should use Big Data for market analysis while doing:

Data ecosystem compounding. Big data runs through demand-production-sales. Achieving government policy analysis - enterprise market analysis - consumer purchase behavior analysis and other stakeholder associations, is conducive to improving the ability of enterprises to judge the development trend of the industry.

Strengthen the data alliance. Breaking down the barriers of enterprises and industries, realizing the sharing of data with schools and industries, relying on theoretical and practical breakthroughs to ensure the security of information, at the same time, realize the understanding of the macro development of the industry, understand the satisfaction degree of different industries, and estimate the market of different industries. Capacity, assessment of the matching of corporate resource levels with the competitive environment in the industry, and potential assessment of different industries. Thereby achieving enterprises to capture the overall development of the industry.

Strengthen data processing technology and management. Big data can help companies achieve accurate market analysis and improve the accuracy of business management and decision making. Therefore, combining big data with cloud computing to ensure the quantity of data while improving the quality management of data makes data become the core competitiveness of enterprise market analysis, thus providing more decision-making consultation for the industry.

References

1. Niu, D.: Analysis of agglomeration pattern and spatial reconstruction of urban agglomerations in Western China. Financ. Econ. Theory Res. (06), 34–48 (2018)
2. Qi, W.: Research on brand management of small and medium enterprises in the era of big data. China Mark. (32), 126–140 (2018)
3. Yang, M.: Innovative research on financial management mode of construction enterprises in the age of big data. Trade Qual. (16), 169 (2018)
4. Industry data. Printed today (11), 4 (2019)
5. Wu, L.: Research on the rise and development of e-commerce economies. Mod. Econ. Inf. (09), 332 (2017)
6. Zhang, Y.: On the simple application of big data in library. Cont. Bridge Vis. (08), 239 (2017)
7. Lin, R.: Overview of the development prospects of commodity customization platform under C2B e-commerce model. Financ. News (11), 137 (2017)
8. Zhang, M.: Research on e-commerce model of clothing professional market. Sci. Technol. Innov. Rev. (03), 11–122 (2016)
9. Han, L.: Discussion on the application and influence of electronic commerce in international trade. China Informatiz. (06), 185 (2013)
10. Wu, Z.: Research on market demand forecast based on enterprise competition. Business (06), 12–15 (2013)

Judicial Statistics Based on Big Data Context

Xiao Wan and Pengnian Jin[✉]

Zhejiang University, Hangzhou 310008, Zhejiang, China
wanxiaozju@163.com, pnj2018@163.com

Abstract. Today's big data is changing the social relationship between people. Its wide application has brought us changes in various fields. This is an unprecedented opportunity and challenge in the field of judicial statistics, which will inevitably have a broad and far-reaching impact on the traditional legal practice model. Data analysis is very important in the prevention of illegal and criminal activities, the effective conduct of reconnaissance activities, the collection of criminal evidence, the acquisition of judicial intelligence, the criminal ruling of similar cases and the reflection of popular legislative demands, judicial injustice, and judicial corruption effect. In addition, the era of big data has a positive effect on the construction of a country's complete judicial system, including institutional legal institutions and the selection of professionals. At the same time, it also has a positive application and development of judicial disclosure and judicial statistics profound impact.

Keywords: Big data context · Judicial statistics · Judicial disclosure

1 Introduction

In the research of judicial statistics, big data technology contains incalculable social value. Its wide variety, huge order of magnitude and unimaginable update speed have great potential for development. Big data provides a life chain for decision-making and makes decisions for decision makers. It is extremely important and valuable to make correct and reasonable decisions.

Its value is mainly reflected in three levels, the first is the predictive alarm function. That is, we pass the previous data inventory, constantly analyze the direction of the data, predict a general rule, to find the user industry, the abnormalities in the case, can have a buffer time, can write a good plan for planning and handling. The second is the value that can be reflected in real time. A large amount of data can form a real-time monitoring mechanism through the computer system to analyze and understand the effectiveness of our decisions and plans, and let us prepare in advance. The third is a summary of the value. The accumulation of data allows us to summarize the shortcomings and shortcomings of the previous work, and provide reference for the improvement of work or the improvement of statistics [1]. Its value in the field of judicial statistics:

(1) Forecast value. Through the superior data receiving and processing capabilities of the computer to analyze the massive data in the early stage, determine the direction of the data, estimate the possibility of occurrence of related things, find

© Springer Nature Singapore Pte Ltd. 2020
M. Atiquzzaman et al. (Eds.): BDCPS 2019, AISC 1117, pp. 877–884, 2020.
https://doi.org/10.1007/978-981-15-2568-1_119

the abnormalities in the parties, the case, and come up with solutions in a timely manner. Have a brain with powerful analytical skills, pay attention to details, and even find clues that are difficult to mine. Observing and its small evidence are the common strengths of excellent criminal investigation experts, and linking various clue evidence through their own analysis and judgment. Get up and infer the facts of the crime and what happened on the spot. At the same time, in addition to the traditional collection of information such as file search, visit transcripts, interview recordings, and telephone records, the police can also use a large amount of network information [2]. The predictive value brought by these big data to our public security, courts, and procuratorates is simply incalculable.

(2) Helping the collection and investigation of case evidence. The application of information technology and networks in the 21st century big data era has penetrated into all aspects of our lives. It is very difficult for criminals or suspects to leave no personal information or electronic information. Relevant data and information can be an important source of evidence and clues for collecting cases [3]. In addition to the information entry and storage of criminals, it is possible to conduct investigations and comparisons in the investigation of similar cases in the future to determine the suspects and speed up the detection of the cases. Realize the sharing of relevant crime information. The criminals or suspects' passports, air tickets, online login accounts and ID cards will all show his situation, which is of great help to the investigation work.

(3) Promoting the sunshine and efficiency of judicial statistics. China's judicial statistics work mainly records, collects, analyzes, archives and other data activities related to various activities related to court trials, and of course includes the processing of data information linked to other social phenomena related to trials. Traditionally, we mainly complete the judicial statistics work through manual reports. It is difficult to avoid errors and omissions in human work and record analysis. On the other hand, the statistics of paper files are difficult to correct, and they have to go through complicated procedures and search. Previously, our case analysis method relied mainly on sample analysis. This analysis method has large defects and cannot accurately reflect the correlation between the characteristics of the case and other information. The judicial statistics work supported by big data technology will be greatly improved. The computer input, analysis and storage will greatly improve the efficiency of work and avoid artificial leakage. At the same time, the analysis of good data and related information through the platform of the network to inform the masses will greatly promote the sunshine of the court's judicial statistics work [4]. The analysis and summarization of these data is conducive to the judge's reference to the trial of the case and the legitimacy of the guarantee procedure, and promotes the reform and efficient implementation of the judicial work.

(4) The criminal ruling reference of similar cases and the reflection of the public's legal interests. Establish a national case database, classify and summarize each case, compare them, and have a relative reference basis for trials and rulings in similar cases, reducing the incidence of wrong cases. People's speeches on the Internet and on information platforms such as WeChat, Renren, and Weibo can conduct effective data analysis, find valuable information related to the judiciary,

and seek public dissatisfaction and revision opinions on the existing laws. Whether the guarantee is in place and whether relevant laws and regulations have been effectively implemented. The network platform is a large collection of data. We can even find information on judicial injustice and corruption, so as to strengthen the supervision of relevant institutions and individuals to promote justice and integrity.

2 Method

Literature research: Using the literature survey method, through the collection, collation, analysis and research of the subject of judicial statistics research based on big data context, the research on the development and changes of the research subjects during this period is carried out. Based on the real situation of the original literature, with reference to the research results and experience that the predecessors and others have obtained, through the reasoning and analysis, to further understand the hotspots of foreign research and the frontiers of academic dynamics, and to deeply understand the connotation of judicial statistics research based on the context of big data. Complete literature review and theoretical summary.

Comparative analysis method: Comparative analysis, also called comparative analysis, compares objective things to achieve the understanding of the nature and laws of things and to make a correct evaluation. This paper mainly compares the results of the current situation of judicial statistics work in the grassroots courts of Y City, H province, compares the service themes in the judicial statistics research based on big data context, summarizes the similarities and differences between them, and analyzes them. The shortcoming between. Further research on the content, characteristics and methods of network information services; through the current situation of network information services based on the judicial research of big data context, the existence of network information services based on the context of big data context The problems and causes are compared and analyzed, and the other countries in foreign countries based on the big data context of judicial statistical research network information service work worthy of reference [5].

Case study method: In the process of investigating the status quo of network information services in university libraries, a wide range of judicial research cases based on big data contexts at various levels are widely investigated.

3 Experiment

This paper experiments the current situation of the judicial statistics work of the grassroots courts in Y City of H Province. The court where the author is located belongs to the H province. At present, the judicial statistics work is mainly completed through the Tongdahai trial management system and the Ziguang Huayu Judicial Statistics Software (version 1.8.9.9): after the basic information such as the case settlement case is entered, the judicial statistical report automatically generates data in the system synchronization; check the table and table. After the relationship is correct, the data will be docked to the judicial statistics software, and will be reported to the higher court after being aggregated. The basic court of Y City is one of the representative courts of the H province. An empirical investigation of the judicial statistics work of the grass-roots courts in Y City can reflect the current situation of the judicial statistics work of the courts of the H province.

(1) Judicial statisticians are mostly part-time and have limited professional competence. There are 16 persons engaged in judicial statistics in 16 grassroots courts in Y City. From the educational background and full-time of judicial statisticians, 15 people are law majors, 1 is a computer major, and there are no statistical related professionals. At the same time, only 4 people are full-time judicial statisticians, and the remaining 12 are part-time judicial statisticians (also concurrently two or three tasks such as the audit office, research office, office, and trial court).

Judging from the working years and work experience of judicial statisticians, there are 12 people who have been engaged in judicial statistics for five years or less, two people who have been working for five years or more, and two or more. Among them, the majority of the staff with a working life of less than five years are undergraduate law graduates who have entered the court and have no experience in trial business. Only 4 people have experience in trial business.

Judicial statistics is a highly technical and professional job, and requires a high level of comprehensive ability for judicial statisticians. It requires a certain amount of statistical knowledge, but also requires a certain level of computer operation and trial business experience. Judicial statisticians can only be able to bypass the judicial work with a combination of three skills. The professional and academic structure of judicial statisticians in Y City Court is single, and the lack of practical personnel for long-term full-time judicial statistics has bound the possibility of diversified development of judicial statistics to a certain extent, and it is difficult to meet the "large pattern" of judicial statistics under big data. "The realization."

(3) Data mining methods are limited and data is rare. Massive trial data information is stored in a huge computer system. The key to realizing the value of data lies in the way data is mined and "purified". The more comprehensive the valuable data is obtained within the set caliber, the more practical the judicial statistics and analysis are. Data mining is a prerequisite for judicial statistics. The author used a sample of 50 staff members in the Y-layer Ethylene Court to conduct a sample survey of the data mining methods of grass-roots court staff. The sample covers the number of judges, judge assistants, clerks and judicial administrators [6]. The data shows that the excavation data of the hospital is mainly based on the summary of judicial statistical software and manual inquiry by the trial management system, and the following departments are responsible for submitting data and other methods (see Fig. 1).

Semi-supervised clustering algorithm is introduced into the character image recognition algorithm of non-linear kernel residual network to preprocess the judicial statistical data based on large data context. Before the data enter the parameter training network, the number of clusters n (n is the number of categories of known data sets) is given according to the demand, and the clustering operation is carried out to generate N known clustering centers. In the process of operation, the cluster is optimized around the known center in the class (not the whole data set). Finally, the pre-processing

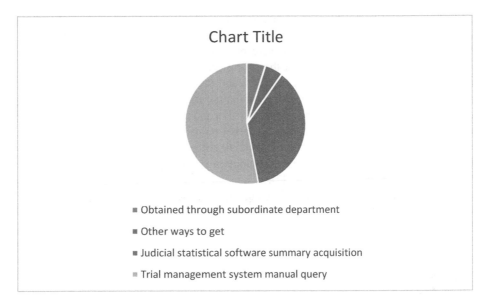

Fig. 1. Mining data scale diagram

operation is completed by assistant stroke number detection. Not only can good results be obtained, but also the intra-class iteration method reduces the operation of searching clustering centers by large iterations in the original method, thus reducing the training time. The method consists of five steps: sample allocation, intra-class updating and clustering center.

Step 1: Initialize the clustering center, According to the characteristics of different data sets, representative samples of n (known) categories are selected as the initial clustering number and clustering center.

Step 2: Assign samples XJ to adjacent clustering sets. Sample allocation is based on:
   (Formula 1)

Step 3: Update the clustering center within the category (not the whole data set) according to the result of step 2:
   (Formula 2)

Step 4:
   (Formula 3)
   If the iteration reaches the maximum iteration step or the difference between the two iterations is less than the set threshold, that is to say, the formula (1.2) is satisfied, then the algorithm is completed; otherwise, step 2 is repeated.

Step 5: Put the trained data into the stroke number detection system for auxiliary detection to increase the clustering reliability.

The method of obtaining data through judicial statistics software can quickly count the basic data such as the number of cases accepted, the number of cases closed, the rate of settlement, the rate of withdrawal, the method of closing the case, and the number of litigation targets. However, since the data of the judicial statistics software is automatically synthesized by the computer, the formula set by the program is fixed and the type is limited, and it cannot completely cover all the demand information items. For example, the statistical administrative non-litigation review and enforcement case settlement method only shows that the ruling grant execution and the ruling are not allowed to execute two settlement methods, and the data of the withdrawal item is not set, so it is still necessary to manually combine the trial management system. Auxiliary queries can know the number of withdrawals. If you can't get the data in two ways, you can only get it through various business departments and subordinate departments, or other forms, such as the number of judicial aid cases and the amount of grants, community corrections, and judicial assistance in Hong Kong, Macao and Taiwan, the crimes of women and minors, etc. [7].

With the advent of the era of big data, the demand for judicial data is larger and involves a wider range of issues, which may involve the operation of trial cases, the application of laws, and the evaluation of effects. However, under the existing conditions, it is difficult for the people's courts to obtain such precious data. Even if it is barely obtained, the data is incomplete and has no reference value, so it cannot accurately and completely reflect the overall picture of judicial activities.

4   Results and Analysis

Regarding the points mentioned above, we have given the following suggestions and countermeasures. (1) Optimizing the statistical model to collect information data comprehensively and objectively; First, change the original reporting method based on single information data reports, and establish a sound statistical model with case information as the core and statistical data as supplementary. It can reduce the phenomenon of underreporting, and can reflect the judicial trial work in more dimensions and perspectives. Second, improve the survey and summary model, establish and improve the statistical survey summary model with comprehensive investigation, key investigation and sample survey, and actively carry out special investigations on judicial hotspots, social conditions and public opinion, and comprehensively collect judicial information data. Third, the case information synchronization online mode is implemented. In the judicial trial process, one case and one card, one card and one recorded information reporting mode are required to ensure the timeliness of judicial statistical information data. At the same time, increase the number of case information in the statistical process and the automatic verification of the referee documents to ensure the objective authenticity of the information data.

(2) Establishing a sense of big data and comprehensively expanding the function of judicial statistics; the biggest impact of big data on judicial statistics is the transformation of the way of thinking. In the era of big data, judicial statistics should not be regarded as a simple report, but should actively promote the role of judicial statistics in information consultation and service decision-making. First, strengthen the awareness of big data and enhance the understanding of responsibility. The leaders of the courts at all levels should put the work of judicial statistics into the key work of the court, form the overall responsibility of the president, be in charge of the leadership, the work pattern of the coordinated development of various departments, and get rid of the old

ideas of "re-trial, light statistics" [8]. Second, under the premise of effectively protecting the legitimate rights and interests of litigants, the judicial statistics will be gradually disclosed in accordance with the law, and the social credibility of judicial information data will be improved. Third, focus on the integration and utilization of information resources. Conduct special investigations on the statistical analysis of judicial information in a timely manner, and research and improve the special system and the introduction of research and statistical analysis to enable the research and statistical analysis to be backed up in judicial practice. Fourth, strengthen judicial forecasting and analysis. Through the summarization and research of judicial statistical information and data, timely release the empirical knowledge of crime prevention and resolution and settlement of litigation disputes to the society, through the full exploitation of the social laws and characteristics behind judicial statistics information and data, through work reports, judicial advice, etc. The party committee government and relevant departments provide resolutions to play a functional role in the decision-making of judicial statistics services for social and economic development.

(3) Strengthening the supervision of data to ensure that the information is accurate, effective, objective and true. First, all judicial statistics will be automatically generated by the case information management system. In this context, the grass-roots courts should strengthen the departments of the hospital. Regulatory efforts to ensure timely and accurate entry of case information, making the judicial statistics to be more precise and accurate [9]. The intermediate courts should adopt a combination of regular inspections and occasional random inspections, increase the supervision of grass-roots courts, prevent grass-roots courts from falsifying data in order to improve performance rankings, and strengthen the development of trial information disclosure. Strengthen external supervision. Second, increase the punishment for judicial statistics fraud. If it is found that there is fraudulent behavior on judicial statistics, the court should cancel its pre-qualification and reduce its performance ranking, and also respond to the relevant leaders of the hospital. Responsibility is checked with staff to ensure that the data obtained from judicial statistics is objective, accurate, and reliable.

5 Conclusion

Promote the combination of judicial statistical software and judicial practice based on the context of big data; try to avoid the separation or backwardness of software and systems and practices, increase the investment in the informationization of judicial statistics, and enable our judicial department to have sufficient funds for System transformation and upgrade [10]. Constantly optimize the work of judicial statistics. In the process of computerization and networking of judicial statistics, it is imperative to find out a special investigation and gradually explore a judicial statistics that is in line with the actual situation of China's trial work.

References

1. Zhu, W.: Research on the change of judicial statistics in the age of big data. Rule Law Forum (4), 243–249 (2017)
2. Chen, F., Chen, F., Chen, W., et al.: Statistical analysis based on big data platform. Value Eng. **37**(22), 259–262 (2018)
3. Research on log statistics and analysis system based on big data technology. Comput. Knowl. Technol. **12**(34), 9–11 (2016)
4. Research on government statistical workflow optimization based on big data. Northwest University (2018)
5. Ren, P., Jiang, S., Wang, L.: Research on national PM2.5 pollution characteristics based on big data statistical analysis. Comput. Netw. **44**(14), 53 (2018)
6. Zhang, X.: On the introduction and construction of the "big data" model of judicial statistics in courts. Shandong University (2016)
7. Anonymous. Research on data cleaning based on big data. Jiangxi Sci. **36**(168(4)), 120–123
8. Wang, Z., Yu, G., Zhao, L.: Research on the demand index of talent demand in China's new economic service industry based on big data. In: Proceedings of 2016 National Statistical Modeling Competition (2016)
9. Anonymous. DT law firm referee data analysis system research. Northwest University (2018)
10. Meng, S.: Discussion on the development of engineering cost in the context of big data. Arch. Build. Decor. (21), 143 (2017)

Empirical Study on Intercultural Adaptability of Undergraduates Under the Background of Big Data

Yejuan Wu[1] and Lei Zhang[2(✉)]

[1] Yunnan Chinese Language and Culture College,
Yunnan Normal University, Kunming, China
[2] School of Tourism and Geographical Science,
Yunnan Normal University, Kunming, China
zlei2308@163.com

Abstract. This study adopts the reference of the revised psychological capital scale, social cultural adaptation scale, sensitivity to cross-cultural communication and cross-cultural communication self-efficacy scale of Yunnan Normal University, 75 undergraduate students to study abroad in psychological, academic and cultural three aspects of cross-cultural adaptability to investigation and study, and comparing to the students to go to the Thailand and Vietnam analysis, found that respondents psychological ability to adapt is relatively lack, cultural adaptation and academic ability to adapt is relatively good, to the students in Thailand and Vietnam significant difference on psychological adaptation and change trend are similar, differences in cultural adaptation and academic adaptation.

Keywords: Undergraduates · Cross-cultural adaptation · Empirical study

1 Introduction

The concept of cross-cultural adaptation was first proposed by anthropologist Redfield et al. From the perspective of anthropology, they believed that acculturation refers to the change of cultural pattern caused by the constant contact between two different cultural groups [1]. Gordon believes that acculturation is when individuals finally adapt and assimilate into the mainstream culture [2]. Black believes that cross-cultural social adaptation consists of three dimensions: general adaptation, work adaptation and communicative adaptation, and that individual personality factors will influence cross-cultural adaptation [3, 4]. Ward and his colleagues believe that cross-cultural adaptation can be divided into psychological adaptation and social acculturation, and the classification proposed by them is accepted by most researchers [5, 6]. According to psychologist Berry, acculturation refers to the learning process of individuals in continuous life and contact with two different cultures. Acculturation can be divided into two processes: maintenance of the original culture and connection with the new culture [7]. Yang, a domestic scholar, believes that cross-cultural adaptation refers to the process in which participants transform in a new environment and form a good communicative ability to cope with the new environment, but has not classified it [8]. Zhu defined

© Springer Nature Singapore Pte Ltd. 2020
M. Atiquzzaman et al. (Eds.): BDCPS 2019, AISC 1117, pp. 885–893, 2020.
https://doi.org/10.1007/978-981-15-2568-1_120

cross-cultural adaptation as the adjustment process of individual psychology and behavior in coping with new cultural situation, and added one dimension of academic adaptation on the basis of psychological adaptation and social acculturation to investigate the integration process of academic system between foreign students and host countries [9, 10]. This study adopts the viewpoint of Zhu guohui to analyze from three dimensions of psychological adaptation, academic adaptation and acculturation.

2 Study Design

2.1 Research Methods

The questionnaire of this study consists of two parts: the first part investigates the basic information of undergraduates, including 6 items: gender, age, country of study, whether they have been to Thailand/Vietnam before study, Thai/Vietnamese level, and English level. The second part survey undergraduates cross-cultural adaptability, researchers on psychological capital Scale, social cultural adaptation Scale, Sensitivity to cross-cultural communication Scale ISS (Intercultural Sensitivity Scale) and cross-cultural communication self-efficacy Scale IES (Intercultural Effective Scale) were modified, the revised questionnaire for five Likert Scale, with 30 items are divided into three modules: 1 to 10 tests of psychological adaptation, 11 to 20 tests of cultural adaptation, 21 to 30 tests of academic adaptation. Each test question was marked with "strongly disagree", "disagree", "uncertain", "agree" and "strongly agree" to evaluate the direct feelings of the subjects, with the values of "1", "2", "3", "4" and "5" respectively.

2.2 Reliability and Validity of the Questionnaire

After calculation, Cronbach's Alpha coefficient of the internal reliability index of the two parts of the questionnaire was higher than 0.60, indicating the overall reliability of the questionnaire was good. The Kaiser-Meyer-Olkin coefficients of the questionnaire are all greater than 0.50, and the significance probability of Bartlett chi-square statistics is all 0.000, less than 0.05, indicating that the scale's structural validity is reasonable.

2.3 Survey Implementation

This study selected undergraduate students majoring in Chinese education from Yunnan Normal University as the survey objects. According to the Southeast Asian languages they have learned, the students of this major go to universities in Thailand and Vietnam for a one-year overseas study in their third year of college, which is a representative research object of cross-cultural adaptability. SPSS19.0 software was used for statistical analysis of the survey results.

2.4 Sample Structure

In this study, 87 questionnaires were sent out and 75 were collected, with a recovery rate of 86.2%. Among the respondents, there are 14 males and 61 females, with a lower proportion of males. The age of the subjects was about the same, and they were all between 20 and 25 years old. There were 39 students studying in Thailand and 36 students studying in Vietnam. Most of the students have not been to the country where they studied before, and they think they have a relatively high proportion of students who think their Thai/Vietnamese and English level is average or not very good.

3 Results and Analysis

3.1 Mental Adaptation

According to the analysis, the average of the psychological adaptability of the subjects is between 2–3, which is in the low level, it's standard deviation $S_{distance\ and\ strangeness}1.091 > S_{pressure}1.090 > S_{opinions\ from\ others}1.057 > S_{conmmunicating\ obstacle}1.027 > S_{emotion\ control}1.017 > S_{aloneness}0.955 > S_{processing\ power}0.889 > S_{conmmunicate\ avoiding}0.836 > S_{feeling\ of\ getting\ alone}0.780 > S_{willingness\ of\ conmmunication}0.772$, the degree of data dispersion is large, which indicates that the psychological adaptability of the subjects varies in different items.

The KMO value of the psychological adaptation questionnaire was 0.702, and the associated probability given by the Bartley spherical test was 0.000, less than the significance level of 0.05, which was suitable for factor analysis. Using principal component analysis (pca), the results of factor extraction after factor analysis were rotated with maximum variance. After calculation, there are 4 factors with feature root greater than 1, and the cumulative variance interpretation rate is 68.568%. The characteristic values of the 4 common factors change significantly, and then the change tends to be stable, indicating that the extraction of the 4 common factors can play a significant role in the information description of the original variables (Table 1).

Table 1. Psychological adaptation rotation component matrix[a]

	Ingredients			
	1	2	3	4
Willingness to communicate	−.241	.049	−.763	.181
Loneliness	.282	.759	−.179	−.011
Communication disorders	.748	.232	.033	.292
Get along with feeling	.723	−.041	.141	−.208
Difficult to deal with	−.090	−.116	−.043	.938
Distance and strangeness	.386	.559	.228	−.184
Avoid communication	.769	.132	.217	−.064
Emotional control	−.030	.472	.635	−.012
Pressure	−.066	.829	.283	.000
Others view	.265	.112	.768	.148

Extraction method: principal component analysis.
Rotation: the Kaiser standardized four-point rotation.
[a]Rotation converges after 5 iterations.

According to the standard that the absolute value is not less than 0.5, the first factor integrates the three original variables that the load is not less than 0.723, namely "communication obstacle", "getting along with feeling" and "avoiding communication". The second factor of load is not less than 0.559 3 original variables, namely the "left", "distance and strangeness" "pressure", the third factor of load is not less than 0.635 2 original variables, namely, "emotion control", "others", the fourth factor of load is equal to 0.938 a "difficult to deal with" the original variables.

The researchers used the variance analysis method to examine the differences in the psychological adaptation of the subjects, and the results showed that there were no significant differences in the psychological adaptation of the subjects in terms of gender, the experience of going to Thailand/Vietnam before studying abroad, and the level of Thai/Vietnamese.

3.2 Cultural Adaptation

By the analysis of the subjects psychological adaptability lie somewhere in between 3–4 more than average, in the average level, the standard deviation $S_{culture\ contrast}$ $1.184 > S_{Eating\ habits}$ $1.110 > S_{The\ concept\ of\ time}$ $1.049 > S_{Behavior}$ $1.011 > S_{Values}$ $0.972 > S_{Customs\ and\ habits}$ $0.903 > S_{History\ and\ culture}$ $0.881 > S_{Values}$ $0.878 > S_{Common\ customs}$ 0.744, traditional clothing data dispersion degree is bigger, show the culture of the subjects to adapt to the differences on different projects.

The KMO value of acculturation questionnaire was 0.654, and the associated probability given by Bartley spherical test was 0.000, less than the significance level of 0.05, which was suitable for factor analysis. Using principal component analysis (pca), the results of factor extraction after factor analysis were rotated with maximum

Table 2. Culture adaptation rotation component matrix[a]

	Ingredients		
	1	2	3
History and culture	.378	−.449	.415
Customs and habits	.295	−.042	−.560
Behavior	−.338	.673	.275
Eating habits	.728	−.185	.000
Values	.728	.042	.005
Traditional festivals	.807	−.204	−.009
Cultural comparison	.018	.670	−.242
Common customs	.401	.501	.360
The traditional dress	.210	−.075	.745
The concept of time	−.133	.443	−.009

Extraction method: principal component analysis.
Rotation: the Kaiser standardized four-point rotation.

[a]Rotation converges after 5 iterations.

variance. After calculation, there are three factors whose feature root is greater than 1, and the cumulative variance interpretation rate is 52.163%. The characteristic values of the three common factors change significantly, and then the change tends to be stable, indicating that the extraction of the three common factors can play a significant role in the information description of the original variable (Table 2).

According to the standard that the absolute value is not less than 0.5, the first factor synthesizes the three original variables that the load is not less than 0.728, namely "eating habits", "values" and "traditional festivals". The second factor synthesizes three original variables whose load is not less than 0.501, namely "behavior habit", "cultural contrast" and "cultural common", and the third factor synthesizes one original variable "traditional clothing" whose load is equal to 0.745.

The researchers used a nova to examine the differences in acculturation of the subjects, and the results showed that there was no significant difference in gender, experience of going to Thailand/Vietnam before going abroad, and level of Thai/Vietnamese.

3.3   Academic Adaptation

According to the analysis, the average value of the academic adaptability of the subjects mostly lies between 3–4, which is in the upper middle level. It's standard deviation $S_{practical\ activity}$ $1.237 > S_{courses}$ $1.089 > S_{relationship\ between\ knowledge\ and\ employment}1.055 > S_{teaching\ form}1.047 > S_{collaborative\ learning}1.044 > S_{Southeast\ Asian\ language\ level}0.966 > S_{compliance}0.935 > S_{attitude\ with\ teacher}$ $0.916 > S_{ability\ of\ understanding\ classes}0.891 > S_{study\ pressure}0.869$, the degree of data dispersion is large, which indicates that the academic adaptability of the subjects varies in different items.

Academic part questionnaire (KMO value of 0.588, Bentley's spherical inspection by the given probability is 0.000, less than the significance level of 0.05, suitable for researchers do factor analysis using principal component analysis (pca), the results of extracted by factor analysis factor orthogonal varimax rotation through calculation, the characteristic root is greater than 1 factor has four, accumulative total variance explained at a rate of 63.737%, four common factor characteristic value change is very obvious, leveled off after the change, extracted four public factor can be information description of the original variables have a significant effect (Table 3).

According to the standard that the absolute value is not less than 0.5, the first factor synthesizes the three original variables that the load is not less than 0.638, that is, getting along with the teacher, understanding the classroom requirements and complying with the school system. The second factor of load is not less than 0.599 3 original variables, namely the curriculum teaching practice, the third factor of load is not less than 0.649 2 original variable knowledge and employment of mutual learning two indicators, the fourth factor of load is not less than 0.540 2 original variables, namely the southeast Asian language learning pressure level.

The results show that there are no significant differences in gender, experience of going abroad to Thailand/Vietnam and Thai/Vietnamese level.

Table 3. Academic adaptation rotation component matrix[a]

	Ingredients			
	1	2	3	4
Attitude towards teachers	.638	−.087	.475	−.115
Course offered	−.129	.794	.003	−.012
Teaching Method	.475	.599	−.111	−.018
Understand classroom requirements	.827	.058	.082	−.030
The relation between knowledge and employment	.145	.353	.649	−.191
Academic stress	−.203	−.502	−.213	.540
Abide by the rules	.700	.052	−.029	.192
Southeast Asian language proficiency	.223	.244	.184	.774
Practical activity	.062	.658	.234	.293
Collaborative learning	.053	.006	.790	.267

Extraction method: principal component analysis
Rotation: the Kaiser standardized four-point rotation

[a]Rotation converges after 5 iterations.

3.4   A Comparative Study on the Intercultural Adaptability of Students from Thailand and Vietnam

3.4.1   Comparison of Psychological Adaptation

Figure 1 shows that the differences of psychological adaptation between the students going to Thailand and the students going to Vietnam are small and the trend of change is similar. Compared with the students going to Thailand, the psychological adaptation of the students going to Vietnam is relatively stable.

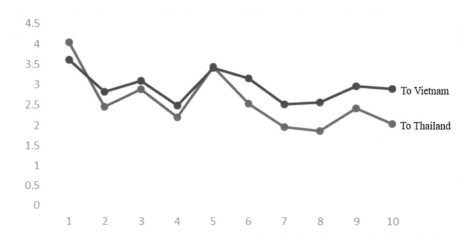

Fig. 1. Comparison of psychological adaptation between to Thailand and Vietnam

3.4.2 Cultural Adaptation Comparison

Figure 2 shows that the cultural adaptation of the students who go to Thailand and Vietnam fluctuates greatly, and their average values are relatively close in item 1 "history and culture", item 8 "common customs" and item 9 "traditional clothing", and the average values of the other seven items are quite different.

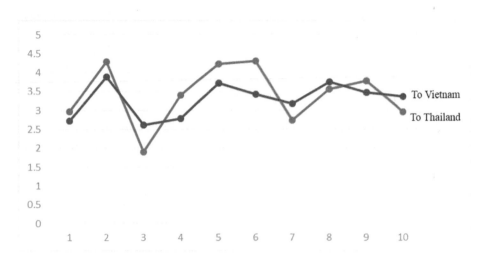

Fig. 2. Comparison of cultural adaptation between to Thailand and Vietnam

3.4.3 Comparison of Academic Adaptation

Figure 3 shows that the academic adaptation of the students who go to Thailand and Vietnam is not stable, and the mean value of the two is quite different in project 1 "attitude towards teachers", project 3 "teaching methods", project 6 "learning pressure", project 8 "Southeast Asian language level" and project 9 "practical activities".

Fig. 3. Comparison of academic adaptation between to Thailand and Vietnam

4 Conclusions and Suggestions

It is found that the psychological adaptability of Chinese majors is generally low, and the cultural and academic adaptability are relatively good. Factors such as communication barriers, feelings of getting along, eating habits, values, getting along with teachers, understanding classroom requirements and school system have significant effects on students' psychological adaptation, cultural adaptation and academic adaptation. There is no significant difference in psychological, academic and cultural adaptation between undergraduates with different gender, overseas experience and Southeast Asian language level. The psychological adaptation of the students to Vietnam is better, and the cultural adaptation and academic adaptation of the students to Thailand and Vietnam fluctuate greatly and have obvious differences.

It is suggested to offer intercultural communication courses and psychological education courses for students before going abroad, organize students to participate in intercultural communication activities, and enhance students' psychological adaptability and communication ability; before going abroad, offer cultural overview courses to help students understand the history, culture, customs and habits of countries studying abroad, and require students to respect and understand other cultures, establish a multicultural outlook, and reduce Less cultural conflicts and misunderstandings; while studying abroad, we should strive to improve our language level, understand and adapt to the teaching methods and teaching styles of teachers in our country, actively participate in all kinds of cultural exchange activities organized by the school, and complete overseas studies as required; at the same time, according to the different countries studying abroad, we should focus on training and education of students to reduce their cross-border education as much as possible. Collision and conflict in cultural communication can better adapt to overseas study and life.

References

1. Robert, R., Herskovits, M.: Memorandum on the study of acculturation. Am. Anthropol. **38**(38), 149–152 (1936)
2. Gordon, M.M.: Assimilation in American life. Oxford (1964)
3. Black, S.: The relationship of personal characteristics with the adjustment of Japanese expatriate managers. MIR. Manag. Int. Rev. **30**(2), 119–134 (1990)
4. Black, J.S.: Locus of control, social support, stress, and adjustment in international transfers. Asia Pac. J. Manag. **7**(1), 1–29 (1990)
5. Ward, C., Kennedy, A.: Locus of control, mood disturbance, and social difficulty during cross-cultural transitions. Int. J. Intercult. Relat. **16**(2), 175–194 (1992)
6. Furnham, A., Bochner, S.: Culture shock. Psychological reactions to unfamiliar environments (1986)
7. Berry, J.W.: Acculturation: living successfully in two cultures. Int. J. Intercult. Relat. **29**(6), 697–712 (2005)

8. Yang, J.: Study on intercultural adaptation of international students in China. East China Normal University (2005)
9. Zhu, G.: Study on cross-cultural adaptation of international students from universities to China. East China Normal University (2011)
10. Zhu, G.: A review of the study of intercultural psychological adaptation of western international students. Innov. Entrep. Educ. **02**(2), 51–55 (2011)

Big Data and Evaluation of Tourism Competitiveness of Counties in Guang-Dong Province

Heqing Zhang, Jing Ma, and Xiaobo Su$^{(\boxtimes)}$

Tourism College of Guangzhou University, Guangzhou 510006, China
lyzhq8007@gzhu.edu.cn, 911192460@qq.com,
suxiaobo_2000@yahoo.com

Abstract. The county area of Guang-Dong province is vast. Supported by county tourism, Guang-Dong province is a big and strong tourism province in China. The tourism industry has developed rapidly in recent years. It is of great significance to understand the current situation and problems of the tourism competitiveness of Guang-Dong county and to improve the tourism competitiveness of Guang-Dong county. Taking 21 counties in Guang-Dong province as an example, this study selected representative indicators to build an index system of tourism competitiveness of the county, and evaluated its tourism competitiveness by combining big data and factor analysis. On the basis of comprehensive analysis, this paper makes a scientific evaluation on the tourism competitiveness of Guang-Dong county. It also provides scientific reference for the further development of Guang-Dong county tourism. The results of this paper show that the tourism competitiveness of Guang-Dong counties is large in guangshen area and relatively backward in other cities. In tourism competitiveness, tourism market competitiveness and tourism support competitiveness contribute more; there is a positive relationship between tourism competitiveness and regional economic development.

Keywords: Big data · County tourism · Tourism competitiveness · Factor analysis

1 Introduction

With the continuous improvement of the level of economic and social development, people's basic material needs are constantly met, and travel demand also increases. Therefore, tourism has maintained a sound momentum of rapid development throughout the country, and gradually received more and more attention [1]. Guang-Dong province is a big tourism province and a strong tourism province in China. In 2018, the tourism income accounted for 11.54% of the provincial GDP, far exceeding the national average level of 4.96%. The total number of overnight tourists reached 362,252 million, with a year-on-year growth of 10.57%, ranking among the top in China [2]. The position of county tourism areas in Guang-Dong province as an important tourist destination, source of tourists and distribution center of outbound tourism in China is further strengthened [3]. With the rapid development of information technology and the constant change of consumer demand, online tourism services have become the general trend of tourism development, so it is of great significance to study the improvement of tourism experience based on big data [4]. It is helpful for online tourism enterprises to further build and improve their business marketing model, so that they can seize the new opportunities brought by big data, increase the number of users and improve their status [5]. It is believed that this will play an important role in promoting the future development strategy of the enterprise and the enterprise entering the online travel service market. In order to further understand and evaluate the tourism competitiveness of Guang-Dong county and its possible problems, this paper, relying on big data technology and using factor analysis method, is of great significance to enhance the tourism competitiveness of Guang-Dong county and promote the sound development of tourism in Guang-Dong county [6].

In the 1960s, foreign scholars conducted in-depth studies on foreign county tourism experience from different perspectives, covering the connotation of tourism experience and measurement of tourism experience quality [7]. However, the study on county tourism in China began in the late 1990s [8]. Now opens a branch to leaf has big data technology, as a result, deep in the big data technology development momentum of Guang-Dong province to play their own advantages in the fierce market competition, a new competition pattern, should make full use of big data to carry out the marketing, dedicated to provide users with unique, personalized travel experience, including online and offline experience, improve the core competitiveness of tourism enterprises, so as to seize market share, occupy the industry of higher status, further improve the image of the tourism enterprise [9, 10].

In order to more accurate and objective evaluation of the tourism competitiveness of the county of Guang-Dong province, this study adopts the quantitative analysis method, factor analysis, to reflect the current state of the 21 county of Guang-Dong province tourism development index as the original data, through the SPSS statistical software indicates that the main factors affecting the tourism competitiveness of the county and the ranking of competitiveness, assessment analysis of tourism competitiveness of each county, and on this basis for the development of the county of Guang-Dong province tourism provides corresponding Suggestions.

2 Factor Analysis Method

2.1 Basic Idea and Principle of Factor Analysis Method

Factor analysis originated in the early 20th century and was proposed by k. Pearson, c.pearman and other scientists for the measurement of intelligence. With the emergence of computer, it has been further developed as a statistical method, widely used in medicine, psychology, economics and other scientific fields as well as socialized production.

Factor analysis is to find out a few random variables that can integrate all variables by studying the internal dependence of correlation coefficient matrix among multiple variables, which are called factors. Then, variables are grouped according to the size of correlation, which means that variables in the same group have a higher correlation, while variables in different groups have a lower correlation, and all variables can be expressed as linear combinations of common factors. The goal is to reduce the number of variables and replace all variables with a few factors to analyze the whole economic problem. The starting point of factor analysis is to replace most of the information of the original variables with fewer independent factor variables, which can be expressed by the following mathematical model:

$$
\begin{aligned}
X_1 &= a_{11}F_1 + a_{12}F_2 + \ldots + a_{1m}F_m \\
X_2 &= a_{21}F_1 + a_{22}F_2 + \ldots + a_{2m}F_m \\
&\ldots \\
X_p &= a_{p1}F_1 + a_{p2}F_2 + \ldots + a_{pm}F_m
\end{aligned}
\tag{1}
$$

The purpose of factor analysis is to use several potential random variables that cannot be directly observed to describe the covariance relationship of multiple variables that reflect things. Since there is a certain correlation between variables, measured variables can be converted into a few irrelevant comprehensive indicators to reflect the comprehensive information in variables. By using factor analysis method to evaluate the tourism competitiveness of Guang-Dong counties, it is to find several comprehensive variables as factors from several variables reflecting the tourism competitiveness of these counties through calculation, and calculate the factor scores, so as to further rank and cluster the tourism competitiveness of Guang-Dong counties.

The factor analysis method is specifically described as follows: suppose N samples are obtained through data collection, and there are p indicators reflecting the characteristics of each sample, then the collected data can be represented by matrix A, where A can be expressed as:

$$
A = \frac{a_y - \frac{1}{p}\sum_{i=1}^{n} a_y}{\sqrt{\frac{1}{p}\sum_{i=1}^{n}\left(a_y - \frac{1}{p}\sum_{i=1}^{n} a_y\right)^2}}
\tag{2}
$$

The score of principal factors can be calculated through A, and sorted and analyzed according to the score.

2.2 Establishment of Indicator System

Tourism is a comprehensive activity, the tourism competitiveness measurement index should have the corresponding integrity, and can fully reveal the city's tourism development status. The principles of comprehensiveness, operability, objectivity and scientificity are the principles to be followed in selecting and establishing the evaluation index system of urban tourism competitiveness. Therefore, eight statistical indexes that can comprehensively reflect the tourism competitiveness of each county in Guang-Dong can be selected and the corresponding statistical index system can be established. Specific indicators are shown as follows: X1: total tourist times; X2: total tourism revenue; X3: tourism revenue/GDP; Total number of X4 hotels; X5: number of travel agents; X6: number of tourism practitioners; X7: number of tourist attractions; X8: bus passenger volume.

These eight indicators respectively reflect the tourism competitiveness of 21 cities in Guang-Dong province from the aspects of the strength of urban tourism industry, the potential of urban tourism development and the capacity of urban tourism organization and reception.

3   Experimental

The first step is to determine whether the variable to be analyzed is suitable for factor analysis. SPSS18.0 was used to conduct KMO and Bartlett tests on the original data. The closer the KMO statistic is to 1, the better the effect of factor analysis. In the actual analysis, the KMO statistic is more than 0.7 with good effect, and less than 0.5 is not suitable for factor analysis. Bartlett test is used to test whether the correlation matrix is an identity matrix. If it is, all variables are independent, and factor analysis method is invalid.

The second step is to conduct standardized processing on the data to eliminate the impact of dimensional differences in the data.

The third step is to calculate the eigenvalue, variance contribution rate and cumulative variance contribution rate of the correlation matrix, and determine the number of factors according to the eigenvalue size. Generally, if the cumulative contribution rate of the first m eigenvalues is >0.85, the factor load matrix can be obtained.

The fourth step is to calculate the factor load matrix, use the maximum variance method, and rotate the extracted factors to obtain more obvious practical meaning. The m column vectors of the factor load matrix obtained by the orthogonal rotation transformation with the maximum variance correspond to m principal factors.

Fifth, the score of each factor was obtained by linear regression analysis, and then the comprehensive score of tourism competitiveness was obtained by weighting the variance contribution rate of each factor.

4   Discuss

4.1   Analysis of Experimental Results

According to the obtained original data, the SPSS software was applied to conduct KMO and Bartlett spherical tests, and the KMO statistical value was = 0.732 > 0.5, indicating that the information overlap between the variables was high, which could be used for factor analysis. In Bartlett's spherical test, chi-square statistical value = 185.574, freedom of freedom line = 37, and associated probability = 0, which rejected the assumption that all variables were independent of each other and showed that common factors existed among all variables, which was suitable for factor analysis of dimensionality reduction of original data.

Factor load matrix is shown in Table 1. As can be seen from the table, X1 total tourist person-times, X2 total tourist income and X3 total tourist income/GDP have a large load on the first factor F1, while other indicators have a small load on F1. Therefore, F1 can be named as the tourism industry strength factor. The total number of X4 hotels, X5 travel agencies and X6 tourism practitioners are the second factor, and F2 has a large load. Other indicators have a small load. Therefore, F2 can be named as the factor of urban tourism development potential. Similarly, in the third factor F3, the number of X7 tourist attractions and the passenger volume of X8 bus account for a large load. F3 can be named as the factor of reception capacity of urban tourism organization.

Table 1. Factor load matrix after rotation

Evaluation index	F1	F1	F1	Evaluation index	F1	F2	F3
X1	0.721	0.104	0.265	X2	0.864	0.156	0.175
X3	0.623	0.150	0.093	X4	0.032	0.813	0.831
X5	0.234	0.917	−0.014	X6	−0.053	0.786	0.786
X7	−0.073	0.175	0.895	X8	0.303	0.330	0.186

Lambda 1 = 9.24 is the first three eigenvalues of the correlation coefficient matrix obtained by SPSS. Lambda 2 = 3.05; Lambda 3 = 1.61, the corresponding contribution rate is 56.19%, 28.93%, 6.22% respectively, the cumulative contribution rate is 91.34%, indicating that the first three factors reflect 91.34% of the original index information, the original competitiveness contribution rate index information is shown in Fig. 1 below. Therefore, these three factors can be used as comprehensive indicators to reflect the tourism competitiveness of 21 counties in Guang-Dong province.

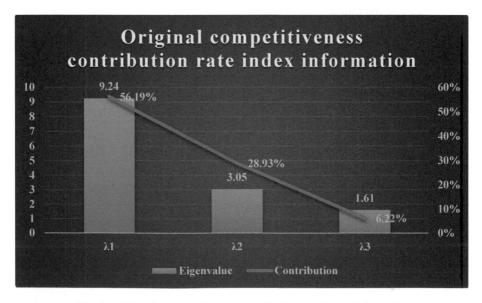

Fig. 1. Original competitiveness contribution rate index information

4.2 Analysis and Evaluation

The evaluation of comprehensive competitiveness needs to determine the weight of each part so as to obtain the score formula of comprehensive competitiveness. And variance is used to measure the deviation degree between the random variables and their average index, so the greater the variance of competitiveness score of each part, the competitive power, the greater the difference between regions, and in a suitable environment, as long as the system, the indicators of competitiveness can be greatly improved, so should pay more attention to, which should be given a higher weighting. Therefore, with the variance of each item's competitiveness score as the weight, the comprehensive tourism competitiveness score can be obtained by weighting.

The weight of sub-competitiveness represents the contribution rate of sub-competitiveness to tourism competitiveness. It can be seen that the tourism market competitiveness and tourism support competitiveness have a greater weight, which indicates that the original tourism market share, market growth, public environment and public transportation have a greater role in tourism support for tourism competitiveness.

In the tourism competitiveness, the tourism market competitiveness and tourism support competitiveness have a greater impact. Among them, market size factor, tourism development factor and infrastructure support factor account for a higher proportion. It shows that the existing tourism market scale, tourism development scale and transportation infrastructure support play an important role in improving the tourism competitiveness of Guang-Dong province.

5 Conclusion

The structure of the tourism competitiveness of each county of Guang-Dong province constitute different counties should be according to the characteristics of the regional tourism competitiveness, focus on develop strengths, supplement the short board, while improve the comprehensive competitiveness, strive to build characteristic tourism reputation, expand the essence resort the visibility and vigorously construction of public services, to strengthen the management of scenic spots and service, improve the tourists experience, enhance its appeal to tourists. Therefore, on the one hand, understand the needs of domestic and foreign tourists, vigorously improve the tourist intention. Each region needs to make efforts to build a differentiated tourism model according to its own characteristics. To build different types of carrier elements, improve the attraction to tourists, so that tourists come, satisfaction and return; On the other hand, it is necessary to improve the tourism traffic, do a good job of supporting the traffic around the scenic area, further improve the bus line of the scenic area, according to the travel needs of tourists, design the bus volume and operating time flexibly, and accelerate the construction of the bus network in the densely populated areas of the scenic area. In this way, the attraction of the city to tourists can be enhanced, and the tourist experience in the area can be improved to obtain the best evaluation.

References

1. Ruíz, M.A.C., Bohorquez, S.T., Molano, J.I.R.: Colombian tourism: proposal app to foster smart tourism in the country. Adv. Sci. Lett. **23**(11), 10533–10537 (2017)
2. Huang, W.J., Hung, K., Chen, C.C.: Attachment to the home country or hometown? Examining diaspora tourism across migrant generations. Tour. Manag. **68**, 52–65 (2018)
3. Joshi, O., Poudyal, N.C., Larson, L.R.: The influence of sociopolitical, natural, and cultural factors on international tourism growth: a cross-country panel analysis. Environ. Dev. Sustain. **19**(3), 825–838 (2017)
4. Tsokota, T., Von Solms, R., Van Greunen, D.: An ICT strategy for the sustainable development of the tourism sector in a developing country: a case study of Zimbabwe. Electron. J. Inf. Syst. Dev. Ctries. **78**(5), 1–20 (2017)
5. Wijaya, N., Furqan, A.: Coastal tourism and climate-related disasters in an archipelago country of Indonesia: tourists' perspective. Procedia Eng. **212**, 535–542 (2018)
6. Nisco, A.D., Papadopoulos, N., Elliot, S.: From international travelling consumer to place ambassador. Int. Mark. Rev. **34**(3), 425–443 (2017)
7. Nene, G., Taivan, A.: Causality between tourism and economic growth: evidence from sub Saharan Africa (SSA). J. Dev. Areas **51**(2), 155–169 (2017)
8. Roult, R., Domergue, N., Auger, D.: Cross-country skiing by Quebecers and where they practice: towards 'dwelling' outdoor recreational tourism. Leisure/Loisir **41**(13), 1–21 (2017)
9. Fatima, J.K., Ghandforoush, P., Khan, M., et al.: Mobile learning adoption for tourism education in a developing country. Curr. Issues Tour. **22**(2), 1–8 (2018)
10. Kim, J.J., Kim, I.: Chinese international students' psychological adaptation process in Korea: the role of tourism experience in the host country. Asia Pac. J. Tour. Res. **24**(3), 1–18 (2018)

Fused Image Quality Assessment Based on Human Vision

Lei Zhang[1], Zhao Yao[1], Ou Qi[1(✉)], and Jian Han[2]

[1] Army Academy of Amored Forces, Changchun, China
haikuotiankongru@163.com
[2] Logistic University of PAP, Tianjin, China

Abstract. Favorable human eye recognition is the only criterion for false color image fusion. In order to overcome this problem, a new method for detecting the fusion quality of pseudo-color images is proposed. This method makes full use of the structure-sensitive characteristics of human beings and approximates human vision through algorithm. The results show that this method can achieve better results.

Keywords: Human eye · Fusion quality · Structure-sensitive

1 Introduction

With the development of computer technology and sensor technology, pseudo-color image fusion algorithms have emerged in large numbers. How to evaluate the quality of pseudo-color image fusion is an urgent problem to be solved [1]. Fused image quality evaluation used to be done in a subjective way. The process of subjective evaluation is very tedious and cannot be embedded in image fusion algorithm, which can not meet the real-time and automation requirements of image fusion system [2–4]. Therefore, the objective quality evaluation which is consistent with human subjective feeling is the main direction of current research. In 2000, Petrovic and Xydeas proposed a pixel-level fusion image quality evaluation index based on image edge information [5–9]. In 2003, Piella and Heijmans proposed a fusion image quality evaluation index based on structural similarity. The index of Petrovic and Piella effectively solves the problem of gray image evaluation. At present, the mechanism of color evaluation is not perfect. Many image fusion systems use the quality of gray component of pseudo-color fusion image instead of its overall quality [10–13]. This method was once used to evaluate pseudo-color fusion image, but this method of abandoning color component is not perfect. It can evaluate the effect of color on human eyes and bring great damage to the accuracy of quality evaluation of pseudo-color fusion [4, 14].

Bad Recently, Zhang Xiuqiong proposed an index for evaluating pseudo-color fusion image, which uses global gray structure similarity and pixel-level hue and saturation similarity to obtain the quality of fusion image. When the input image is a color image, it is consistent with the subjective feeling of human beings; but when the input image is a gray image, it will give errors. The result of mistake has great limitation in application. In this paper, a general index is proposed to solve the

M. Atiquzzaman et al. (Eds.): BDCPS 2019, AISC 1117, pp. 901–907, 2020.
https://doi.org/10.1007/978-981-15-2568-1_122

problems of narrow application scope and difficult application of the current evaluation index.

2 Construction of Pseudo-Color Image Fusion Framework

Wang Zhou and Bovik put forward the concept of structural information on the basis of many years' research on image processing and image quality. Wang Zhou and Bovik believe that the human eye can extract the structured information from the image highly adaptively, and complete the function of human visual observation by processing the information. Therefore, effective measurement of image structural distortion is an effective method to evaluate image quality. Based on this idea, an image quality evaluation index, structural similarity evaluation index, is developed, which combines the characteristics of HVS. For an image, the average brightness, contrast and structure information are independent of each other, so the overall quality of the image can be measured by comparing the size of the three components between the evaluated image and the input image.

If X and Y are respectively the source input image and the image to be evaluated, then the structural similarity index of the two images is defined by the following sub-definitions:

$$SSIM(X, Y) = l(X, Y) * c(X, Y) * s(X, Y) \tag{1}$$

In style

$$l(X, Y) = \frac{2\mu_X \mu_Y}{\mu_X^2 + \mu_Y^2} \tag{2}$$

$$c(X, Y) = \frac{2\sigma_X \sigma_Y}{\sigma_X^2 + \sigma_Y^2} \tag{3}$$

$$s(X, Y) = \frac{\sigma_{XY}}{\sigma_X \sigma_Y} \tag{4}$$

The values of these quantities are calculated by the following formulas:

$$\mu_X = \bar{X} = \frac{1}{M \times N} \sum_{i=1}^{M} \sum_{j=1}^{N} X(i, j) \tag{5}$$

$$\mu_Y = \bar{Y} = \frac{1}{M \times N} \sum_{i=1}^{M} \sum_{j=1}^{N} Y(i, j) \tag{6}$$

$$\sigma_X^2 = \frac{1}{M \times N - 1} \sum_{i=1}^{M} \sum_{j=1}^{N} [X(i, j) - \bar{X}]^2 \tag{7}$$

$$\sigma_Y^2 = \frac{1}{M \times N - 1} \sum_{i=1}^{M} \sum_{j=1}^{N} [Y(i,j) - \bar{Y}]^2 \tag{8}$$

$$\sigma_{XY} = \frac{1}{M \times N - 1} \sum_{i=1}^{M} \sum_{j=1}^{N} [X(i,j) - \bar{X}][Y(i,j) - \bar{Y}] \tag{9}$$

Mean and variance of the image reflect the corresponding changes of brightness and contrast of the image, while covariance is used to evaluate the degree of change of relative source image nonlinearity. In summary, L (X, Y), C (X, Y), s (X, Y) can be used as the contrast of brightness, contrast and image structure of two images. For images, brightness, contrast and structure are independent of each other. Changing one item alone will not cause changes in the other two items. S(X, Y) is mathematically the angle cosine of vectors X-X and Y-Y. S (X, Y) cannot describe the construction of an image, but it can describe the structural similarity of two images.

In case $(\mu_X^2 + \mu_Y^2)/(\sigma_X^2 + \sigma_Y^2) \neq 0$, for images X and Y, the formulas of structural similarity are expressed as follows:

$$SSIM(X, Y) = \frac{4\mu_X\mu_Y\sigma_{XY}}{(\mu_X^2 + \mu_Y^2)(\sigma_X^2 + \sigma_Y^2)} \tag{10}$$

When $(\mu_X^2 + \mu_Y^2)$ or $(\sigma_X^2 + \sigma_Y^2)$ is near 0, the small noise in source input images X and Y will strongly affect the evaluation results of SSIM. This is not expected in the process of quality evaluation. In order to improve the stability of evaluation, the above formula is revised as follows:

$$SSIM(X, Y) = \frac{(2\mu_X\mu_Y + C_1)(2\sigma_{XY} + C_2)}{(\mu_X^2 + \mu_Y^2 + C_1)(\sigma_X^2 + \sigma_Y^2 + C_2)} \tag{11}$$

3 The Methods of Evaluating Quality of Fused Images

The methods of evaluating the quality of fused images based on structural similarity theory are mainly as follows:

1. Quality Evaluation Method of Pseudo-color Image Fusion Based on Global Average

   In this method, reference image R and quality image F are divided into non-overlapping, equal-sized blocks, and the number of blocks is counted as T. The structural similarity value of each block is calculated by formula, and then the overall quality of all blocks is obtained by summing and dividing T.

2. Quality Evaluation Method of Pseudo-color Image Fusion Based on Block Weighting

The sensitivity of the human eye to all parts of the image is different. For example, the human eye is sensitive to regions with sharp gray changes, but insensitive to regions with gray equalization []. In view of the different sensitivities of the human eye to each part of the image, the global average method is used to evaluate the fused image, and the evaluation results are inconsistent with the human eye perception. Considering these shortcomings, image quality evaluation methods have been developed to block weighted evaluation methods.

(1) Weighting method with variance as feature

Similar to the global average evaluation method, the source input image R and the quality of the image F to be evaluated are subdivided into blocks without overlap, and the structural similarity values of each subblock are calculated one by one. Let the coefficient of each subblock be wj. Then the overall structure similarity value of the fused image is:

$$WMS(F,R) = \frac{1}{T}\sum_{j=1}^{T}[w_j \cdot SSIM(F_j, R_j)] \qquad (12)$$

In the formula, WJ is the variance feature, and if the variance of the jth sub-fasts of the input image F is Dj, then. The overall structural similarity of the fused image is within the range of [0, 1]. The closer the value is to 1, the better the fusion effect will be. The farther the value is from 1, the worse the fusion effect will be.

(2) Weighting Method on the premise of Flat/Edge Region

There are various regions in the image. In the region, some parts have uniform pixel brightness and single line. Because these regions change little, we call them flat areas. Others have strong variations in pixel brightness, obvious contrast and strong visual impact. Most of these parts are located in the edge zone between regions. We call these pixels edge regions. The edge area changes dramatically and the lines are obvious, which is the main body of the image structure, and the human eye is more sensitive to it; while the flat area occupies a larger area, but its change is monotonous, contains less information, and the human eye is less sensitive to it. In view of the characteristics of human eyes, different weighted values are used to calculate the quality of fused images for edge and flat regions, which is called the weighted method based on flat/edge regions.

Like the global averaging method, the original input image R and the quality of the image F to be evaluated are first subdivided into non-overlapping, equal-size blocks. The structural similarity values of each sub-block are calculated separately. Then the image segmentation operation is performed on the source input image R and the quality image F to be evaluated, and the edge part BWF of the flat part BWR is obtained. The number of pixels of BWR and BWF is recorded as SR and SF. The global average structural similarity formula is used to solve the structural similarity of edges and flat areas, which are expressed as

BMS and PMS respectively. The weighted summation of edges and flat regions gives the overall quality as follows:

$$WBMS(F,R) = PMS \cdot \min(p,q) + BMS \cdot \max(p,q) \tag{13}$$

The closer the WBMS is to 1, the better the quality of the fused image is.

3. Quality Evaluation Method of Pseudo-color Image Fusion Based on Gradient Structure Similarity

Based on the above definition, the gradient on coordinates (x, y) is defined by using f (x, y) to represent the gray level of the image.

$$grad[f(x,y)] = \begin{bmatrix} \frac{\partial f}{\partial x} \\ \frac{\partial f}{\partial y} \end{bmatrix} \tag{14}$$

Set $G[f(x, y)] = |grad(f(x, y))|$

$$G[f(x,y)] = \sqrt{\frac{(\partial f(i,j)/\partial i)^2 + (\partial f(i,j)/\partial j)^2}{2}} \tag{15}$$

Where

$$\frac{\partial f(i,j)}{\partial i} = f(i,j) - f(i-1,j) \tag{16}$$

$$\frac{\partial f(i,j)}{\partial j} = f(i,j) - f(i,j-1) \tag{17}$$

Gradient similarity between source input image R and pseudo-color fusion image F can be defined as g(F, R):

$$g(F,R) = \min(sim(G_R, G_F), sim(G_F, G_R)) \tag{18}$$

$$sim(G_R, G_F) = \frac{\sum_i \sum_j G_R(i,j) G_F(i,j) \mid C}{\sum_i \sum_j [G_R(i,j)]^2 + C} \tag{19}$$

C is a small constant, mainly to prevent the denominator from being zero. The gradient structure similarity of pseudo-color image fusion is as follows:

$$GF(F,R) = l(F,R) \cdot c(F,R) \cdot g(F,R) \tag{20}$$

$$l(F,R) = \frac{2\mu_R \mu_F + C_1}{\mu_R^2 + \mu_F^2 + C_1} \tag{21}$$

$$c(F, R) = \frac{2\sigma_R \sigma_F + C_2}{\sigma_R^2 + \sigma_F^2 + C_2} \tag{22}$$

Where $C_1 = (K_1 L)2$, $C_2 = (K_2 L)2$, $L = 255$, $K2 = 0.03$, $K1 = 0.01$.

4 Conclusion

The study of human visual characteristics shows that the human eye can perceive objects is an important function. The human eye uses the tracking of the edge information of unknown objects to recognize unknown objects. Image quality is greatly affected by edge information, and human eyes are sensitive to edge and strip structure. The image with good image quality has clear edges. The gradient magnitude change of gray level of image pixels can represent the edge information of image. Gradient similarity can be used to measure the quality of pseudo-color fusion image better.

References

1. Ji, H., Luo, X.: 3D scene reconstruction of landslide topography based on data fusion between laser point cloud and UAV image. Environ. Earth Sci. **78**(17) (2019)
2. El-Hoseny, H.M., Kareh, Z.Z., Mohamed, W.A., Banby, G.M., Mahmoud, K.R., Faragallah, O.S., El-Rabaie, S., El-Madbouly, E., El-Samie, F.E.A.: An optimal wavelet-based multi-modality medical image fusion approach based on modified central force optimization and histogram matching. Multimed. Tools Appl. **78**(18), 26373–26397 (2019)
3. Yao, Y.-C., Lin, H.-H., Chou, P.-H., Wang, S.-T., Chang, M.-C.: Differences in the interbody bone graft area and fusion rate between minimally invasive and traditional open transforaminal lumbar interbody fusion: a retrospective short-term image analysis. Eur. Spine J. **28**(9), 2095–2102 (2019)
4. Leger, T., Tacher, V., Majewski, M., Touma, J., Desgranges, P., Kobeiter, H.: Image fusion guidance for in situ laser fenestration of aortic stent graft for endovascular repair of complex aortic aneurysm: feasibility, efficacy and overall functional success. Cardiovasc. Interv. Radiol. **42**(10), 1371–1379 (2019)
5. Ch, M.M.I., Riaz, M.M., Iltaf, N., Ghafoor, A., Ahmad, A.: Weighted image fusion using cross bilateral filter and non-subsampled Contourlet transform. Multidimens. Syst. Signal Process. **30**(4), 2199–2210 (2019)
6. Kanmani, M., Narasimhan, V.: An optimal weighted averaging fusion strategy for remotely sensed images. Multidimens. Syst. Signal Process. **30**(4), 1911–1935 (2019)
7. Wu, S., Wu, W., Yang, X., Lu, L., Liu, K., Jeon, G.: Multifocus image fusion using random forest and hidden Markov model. Soft. Comput. **23**(19), 9385–9396 (2019)
8. Hou, R., Nie, R., Zhou, D., Cao, J., Liu, D.: Infrared and visible images fusion using visual saliency and optimized spiking cortical model in non-subsampled Shearlet transform domain. Multimed. Tools Appl. **78**(20), 28609–28632 (2019)
9. Sun, J., Wang, W., Zhang, K., Zhang, L., Lin, Y., Han, Q., Da, Q., Kou, L.: A multi-focus image fusion algorithm in 5G communications. Multimed. Tools Appl. **78**(20), 28537–28556 (2019)

10. Shuang, L., Deyun, C., Zhifeng, C., Ming, P.: Multi-feature fusion method for medical image retrieval using wavelet and bag-of-features. Comput. Assist. Surg. **24**(1), 72–80 (2019)
11. Dao, P.D., Mong, N.T., Chan, H.P.: Landsat-MODIS image fusion and object-based image analysis for observing flood inundation in a heterogeneous vegetated scene. GIScience Remote Sens. **56**(8), 1148–1169 (2019)
12. Yamada, R., Bassaco, B., Wise, C., Barnes, L., Golchin, N., Guimaraes, M.: Radiofrequency wire technique and image fusion in the creation of an endovascular bypass to treat chronic central venous occlusion. J. Vasc. Surg. Cases Innov. Tech. **5**(3), 356–359 (2019)
13. Jiang, M.-S., Yang, T., Xu, X.-L., Zhang, X.-D., Li, F.: Rapid microscope auto-focus method for uneven surfaces based on image fusion. Microsc. Res. Tech. **82**(9), 1621–1627 (2019)
14. Shopovska, I., Jovanov, L., Philips, W.: Deep visible and thermal image fusion for enhanced pedestrian visibility. Sensors **19**(17), 3727 (2019)

Intelligent Monitoring and Simulation of the Whole Life Cycle of Coal Mining Process and Water Quantity and Quality in Coal and Water Coordinated Development Under Large Data

Yu-zhe Zhang, Xiong Wu$^{(\boxtimes)}$, Ge Zhu, Chu Wu, Wen-ping Mu,
and Ao-shuang Mei

School of Water Resource and Environment, China University of Geosciences,
Beijing 100083, China
wuxiong@cugb.edu.cn

Abstract. Coal is China's main energy, and this situation in the future in a long period of time is unlikely to have fundamental changes. As an underground activity, coal mining inevitably causes local damage and pollution to underground water cut system due to improper application of coal mining technology. And a large number of mine drainage, not only need to spend a lot of drainage costs, increase the cost of coal production, but also waste very valuable water resources. In this paper, based on the water - coal, on the basis of coordination development, through big data analysis methods, proposes the establishment process of coal mining and mine water quality, and the whole life cycle of the coal - water coordination development precision control technology system, through a series of research on the laws of the evolution of time and space and nonlinear multiobjective model building and simulation development plan, research and development of the evaluation method of mine water resource utilization efficiency and benefit, help to realize the coordinated development of the coal - water control in whole life cycle of technology. So as to realize the exploitation and utilization of mine drainage and turn waste into treasure, it can not only relieve the tight situation of water supply in coal mines, but also reduce the environmental pollution caused by coal mining to the mining areas and surrounding cities, and realize the unity of economic and environmental benefits.

Keywords: Large data analysis · Coordinated development · Full life cycle · Evaluation method · DEA model

1 Introduction

Mine drainage is the underground water near the coal seam and the opening roadway in the mining area. When the mine area is short of water, a large amount of mine water is directly discharged, which is not only a waste of water resources, but also a great harm to the environment of the mine area [1]. A coal - water coordinated development and

© Springer Nature Singapore Pte Ltd. 2020
M. Atiquzzaman et al. (Eds.): BDCPS 2019, AISC 1117, pp. 908–914, 2020.
https://doi.org/10.1007/978-981-15-2568-1_123

carries on the whole life cycle of monitoring and simulation should be in the maintenance and improvement of regional coordinated development of society, ecology and environment at the same time, in the coal resources development and protection of water resources, the mine water quality, improving the quality of mine water resource utilization and ecological, drive the sustainable development of economy and ecology [2, 3].

The concept of Life Cycle is widely used, especially in many fields such as politics, economy, environment, technology and society. Its basic meaning can be popularly understood as the whole process of Cradle to Grave. For a product, is from the natural to the whole process of return to nature, which is both needed to manufacture products of raw materials acquisition, processing, such as production process, also includes the circulation of products storage and transportation process, and also includes the use of the product process and discarded or disposal of waste to return to the natural process, this process constitute a complete product life cycle [4, 5]. The life cycle of mine water is the whole process from its generation to recycling and finally to waste water resources. With the development of artificial intelligence and big data, about the whole life cycle of intelligent monitoring and simulation, is also one of the hot spot of academic research in recent years, domestic scholars are analyzed from the points of view of sustainable development of the whole life cycle of the intelligent monitoring necessity and important significance, but the research has not yet reached full maturity stage [6]. With the increasingly serious environmental problems, the research focus of the whole life cycle gradually turns to energy issues and solid waste. Due to the increase of coal resource mining year by year, the mine water pollution is increasingly serious in the mining process, which promotes the research on the whole life cycle to be gradually widely applied in the field of coordinated coal development [7, 8].

Coal - water in this paper, by analyzing the different region coordinated development status, the algorithm using the DEA model to different regions within the area of coal - water coordination development evaluation and clustering, for intelligent monitoring of water quality, and the whole life cycle and simulation of the formation based on life cycle evolution of coal - water coordination development strategy system [9, 10]. Through the classification and sequencing of different regions in the region, the status of coal-water coordinated development in various regions in the region is graded and sorted, which is of more targeted and guiding significance in the later mining construction and the implementation of the plan of optimizing coal-water coordinated development in the life cycle.

2 Method

2.1 Algorithm of DEA Model

Data envelopment analysis (DEA) is a system analysis method developed on the basis of the concept of "relative efficiency evaluation" by famous operations research experts A. Harnes and W. W. Cooper. In this study, DEA's ccr-i model is selected as the basic platform for the evaluation of the utilization efficiency of water resources in mines. The CCR model USES the ratio of outputs to synthesis to evaluate efficiency, and

USES the envelope concept to project the outputs and synthesis of all decision units (DUM) into the hyperplane to find the "effective front surface" with the highest or lowest synthesis. DUM located on the effective front surface is called DEA effective, and DUM located within the effective front surface is called DEA ineffective. Meanwhile, projection method is used to point out the reasons for non-dea effective and weak DE effective, as well as the direction and degree of improvement. Equation (1) gives the value of ecological efficiency of mine water utilization for the calculation of HP, and Eq. (2) can calculate the use variables and residual variables in areas where the ecological efficiency of mine water resources is not effective, and the obtained results serve to realize DEA optimization.

Assuming that the utilization efficiency of mine water resources in n regions will be evaluated and that each DMUP produces Yjp through input Xip, the efficiency value of DUM p is as follows:

$$\max h_p = \frac{\sum_{j=1}^{s} U_j Y_{jp}}{\sum_{i=1}^{m} V_i X_{ip}} \tag{1}$$

Where: Xik – output of item I of the KTH DUM; Yjk – the KTH DUM j item is used comprehensively; Vi – output weight of item I; Uj – weight of comprehensive use right of item j; HP – efficiency value of the p-th DUM (relative efficiency); Epsilon – non - Archimedes infinitely small.

$$S.T \frac{\sum_{j=1}^{s} U_j Y_{jk}}{\sum_{i=1}^{m} U_i X_{ik}} (\leqslant 1 \vee k, k = 1, 2, 3, \cdots\cdots, n) \tag{2}$$

2.2   Evaluation Index and Model Determination Method

Output and comprehensive use of mine water are the basis of ecological efficiency evaluation of mine water utilization. After the DUM that needs evaluation is determined, indicators of output and comprehensive use that can be based on the same type of DUM need to be provided. In this paper, combined with the regional characteristics of different coal mining areas in China, and drawing on various evaluation index systems of ecological efficiency of mine water resources, the regional average energy output, regional average mine fishery output and regional average benefit output were taken as DUM's coal mining output. The gross coal production (GDP) of the region represents the expected output of the mining area, and the COD emission, sulfur dioxide emission, ammonia nitrogen emission, wastewater emission and soot emission are determined as the non-expected output of mining area. Since the ccr-i model requires that the comprehensive use index must be positively correlated with the output index, the non-expected output of mine water is regarded as the concomitant amount of expected output, and the output index of mine water use efficiency evaluation is finally determined to be god-gdp load, so2-gdp load, ammonia nitrogen-gdp load and wastewater-gdp load. Finally, a nonlinear multi-objective coal-water dual resource regulation model with dynamic replacement of indicators was established to form a whole life cycle regulation technology of coal-water coordinated development.

3   Experiment

The monitoring and simulation of the whole life cycle of mine water based on DEA model is based on the DEA model algorithm. The main idea is to make statistics on the coordinated development of coal and water in n regions and the utilization of mine water, find out the effective and invalid DEA, and calculate the utilization evaluation index of mine water in each region. In this way, not only the regional classification is made according to the correlation of mine water utilization in all regions, but also the sustainable development of coal mining and mine water utilization can be realized.

The following steps of the algorithm based on DEA model are given:

Step 1: determine the number of classes and set the initial cluster center. Firstly, the final classification number is determined according to the coal mining situation and the utilization level of mine water in the region, generally no less than two, and no more than 7. The sustainable development status of mine water in n regions is finally determined to be divided into n categories, and the standardized evaluation vector of n regions is randomly selected to form the initial clustering center.

Step 2: statistics and calculation of data in different regions. Classify and sort out the collected data and put them into publicity, so as to obtain the validity and invalidity of DEA.

Step 3: classification. According to the calculation of Step 2, the effective and ineffective DEA can be classified, and the DUM output and DUM comprehensive use can be classified again, so as to obtain the utilization evaluation indexes of mine drainage in different areas and make relevant classification according to the evaluation indexes.

Step 4: sort all regions by the above steps. After the realization of in-class and inter-class ranking, the ranking of evaluation results of mineral mining and mine water sustainable development in all areas in the whole region can be realized.

4   Discuss

4.1   Calculation and Analysis of Ecological Efficiency of Mine Water

With data envelopment analysis software DEAP2.1 as the computing platform, ccr-i model was selected to calculate the collected data, and the ecological efficiency of mine water in 7 provinces in China from 2010 to 2018 and the average value of each year were obtained. The final arrival statistics are shown in Table 1 below.

Table 1. Statistics of mine water ecological efficiency in 7 provinces

Region	2010	2011	2012	2013	2014	2015	2016	2017	2018	Average
XinJiang	0.805	0.791	0.641	0.745	0.695	0.735	0.647	0.714	0.725	0.710
YunNan	0.815	0.760	0.801	0.785	0.724	0.702	0.697	0.699	0.679	0.712
ShanXi	0.732	0.394	0.463	0.573	0.579	0.601	0.647	0.638	0.635	0.597
HeNan	0.726	0.820	0.735	0.643	0.645	0.703	0.721	0.716	0.708	0.704
AnHui	0.860	0.830	0.685	0.729	0.689	0.658	0.675	0.698	0.654	0.647
HeBei	0.725	0.625	0.721	0.712	0.699	0.711	0.720	0.700	0.688	0.659
NeiMeng	0.673	0.685	0.643	0.720	0.637	0.663	0.658	0.685	0.694	0.648

4.2 Analysis on the Evolution Rule of the Whole Life Cycle of Mine Water Ecological Efficiency

According to Table 1, you can see that the seven provinces in 9 years mine water ecological efficiency of DEA is invalid, that in these areas are unrealized five index under the existing level of output minimizes the comprehensive use of several indicators, and other ecological efficiency requires effective area by changing the output and the comprehensive use, in order to achieve the ecological use of mine water DEA effective. In a word, there are differences in the ecological efficiency space of mine drainage utilization in all provinces, and none of them can reach DEA efficiency. Even in some areas, the utilization rate of mine drainage is lower than that of other areas. This is mainly caused by the unrealized development of coal-water downflow. In the mining process, a large amount of mine water is wasted or even caused serious pollution due to the unreasonable underground technology or scheme. The inefficient utilization of mine water caused by high yield and low use should be changed to the intensive economic growth mode of "low consumption, low pollution and high benefit" in the future, so that the coordinated development of coal and water can be realized.

The optimization of mine water ecological efficiency will be carried out according to the following three principles: (1) DUM landing on the efficiency front surface is a sufficient and necessary condition for DEM to be effective; (2) the optimization of land use ecological efficiency gives priority to the emission control and reduction of pollutants, and improves the fixed part of secondary index factors in the optimization; (3) in the realization of DEA effectiveness, resource constraint conditions are not taken into consideration, that is, the actual objective conditions do not restrict the output and comprehensive utilization of coal mining process and realize the change of DEA effective output and comprehensive utilization. In this paper, the indicators in Table 1 are involved. Therefore, on the basis of DEA data, the whole life cycle regulation technology of coordinated development of coal and water is formed. The evaluation method of water resource utilization efficiency and benefit of mine was developed, and the technical system of efficient utilization of dual resources was formed. The specific optimization design scheme is shown in Table 2, and people's attitude towards DEM is shown in Fig. 1.

Table 2. Optimization design of mine water utilization in 5 cities in 2019

Reg	Out optimization				Using the optimization				
	S_1^-	S_2^-	S_3^-	S_4^-	S_1^+	S_2^+	S_3^+	S_4^+	S_5^+
YunNan	0.28	3.1	0	1.49	0.04	0	211.96	0	12.57
ShanXi	254.54	0.65	53.45	0	0.4	1905.32	1535.78	7318.51	0
HeBei	0	3.22	55.13	1.32	0.08	556.6	159.59	0	0
AnHui	0.41	2.35	0	0	0.06	114.77	359.93	664.86	0
HeNan	0	2.42	17.94	1.15	0	0	122.38	830.89	44.85

Note: represents the reduction of coal mining output; Represents reductions in other energy output; Represents the reduction of mine fishery output; Represents the decrease in capital output; In comprehensive use, represents the increase of wastewater - GDP load; Represents the load increase of so2 - GDP; Represents the load increase of ammonia nitrogen - GDP; Represents the load increment of COD-GDP; Represents soot - the increase in GDP load.

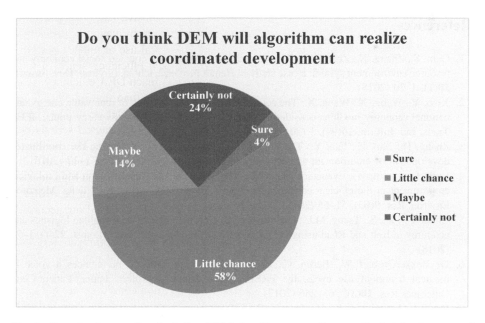

Fig. 1. Investigation results of whether DEM algorithm can realize coordinated development of coal and water

5   Conclusion

It can be seen from this paper that China has not fully realized the coordinated development of coal-water resources in coal mining, which to a large extent has a negative impact on further coal mining and sustainable development of economic ecology. Therefore, it is necessary to realize the coordinated development of coal and water resources and realize the efficient utilization of mine water resources. Increase the strength of scientific research experiment of mine water, strengthen technical communication, introduce advanced technology, increase propaganda, change people's concept of mine water recycling, in the mine planning, scientific research, preliminary design and construction drawing design stages, underground drainage must be taken as water resources to develop and use. Therefore, it is necessary to establish a nonlinear multi-objective coal-water dual resource regulation model with dynamic replacement of indicators, and form a whole life cycle regulation technology of coal-water coordinated development. The evaluation method of water resource utilization efficiency and benefit of mine was developed, and the technical system of efficient utilization of dual resources was formed. Only the linear target model obtained by DEA algorithm and the optimization scheme designed accordingly can realize the efficient utilization of mine water resources.

Acknowledgements. This work was supported by National Key R&D Program of China (No. 2018YFC0406400).

References

1. Qian, S., Zhang, X., Zhang, H.: Coordinated development of a coupled social economy and resource environment system: a case study in Henan Province, China. Environ. Dev. Sustain. **20**(1), 1–20 (2018)
2. Xiao, Y., Wang, X., Wang, X.: The coordinated development path of renewable energy and national economy in China considering risks of electricity market and energy policy. IEEE Trans. Ind. Inform. **16**(99), 1 (2017)
3. Zhang, H., Zhu, Z., Fan, Y.: The impact of environmental regulation on the coordinated development of environment and economy in China. Nat. Hazards **91**(3), 1–17 (2018)
4. Ruili, G., Linlin, W.: Evaluation of coordinated development of urbanization and ecological environment in the efficient ecological economic zone of the yellow river delta. Meteorol. Environ. Res. **9**(03), 51–55 (2018)
5. Chen, Y., Lan, S., Tseng, M.L.: Coordinated development path of metropolitan logistics and economy in Belt and Road using DEMATEL–Bayesian analysis. Int. J. Logist. **22**(1), 1–24 (2018)
6. De BekkerGrob, E.W., Berlin, C., Levitan, B.: Giving patients' preferences a voice in medical treatment life cycle: the PREFER public-private project. Patient-Patient-Cent. Outcomes Res. **10**(3), 263–266 (2017)
7. De, S.J., Antonissen, K., Hoefnagels, R.: Life-cycle analysis of greenhouse gas emissions from renewable jet fuel production. Biotechnol. Biofuels **10**(1), 64 (2017)
8. Lan, S., Tseng, M.L.: Coordinated development of metropolitan logistics and economy toward sustainability. Comput. Econ. **52**(4), 1–26 (2017)
9. Grier, A., Mcdavid, A., Wang, B.: Neonatal gut and respiratory microbiota: coordinated development through time and space. Microbiome **6**(1), 193 (2018)
10. Tenenbaum, C.M., Misra, M., Alizzi, R.A.: Enclosure of dendrites by epidermal cells restricts branching and permits coordinated development of spatially overlapping sensory neurons. Cell Rep. **20**(13), 3043–3056 (2017)

On Blended Learning Mode of Comprehensive English Based on MOOC

Linhui Wu[✉]

Department of Applied Foreign Languages, Chengdu Neusoft University,
Chengdu, Sichuan, China
15413460@qq.com

Abstract. As an important open educational resource, MOOC provides a kind of new teaching mode and learning method for teachers and students, and integrates online education with traditional education. The blended learning mode of comprehensive English based on MOOC has realized the construction of teaching mode of pre-class learning, face-to-face teaching and after-class learning. This model can fully mobilize students' learning enthusiasm, cultivate students' autonomous learning ability, strengthen the role of teachers in teaching guidance, and effectively improve the creativity and effectiveness of comprehensive English teaching activities for college English majors.

Keywords: MOOC · Comprehensive English · Blended Learning mode

1 Introduction

With the rapid development of information technology in higher education, information technology has constantly changed the way students learn. Social media and mobile communication technology are more and more widely used. Educational resources based on the concept of sharing and opening have become the main development trend in the future education industry. MOOC is a new kind of teaching. The rapid rise of learning mode in the global scope has greatly broken the limitations of time and space, and helped students improve their learning autonomy. As the core basic course for English majors, Comprehensive English course plays a very important role in the teaching of English majors. With the advancement of teaching information technology, the traditional single classroom teaching mode cannot meet the teaching needs of the course. Under this background, the blended learning mode has been introduced into the teaching of Comprehensive English course. How to create an information-based teaching model, set up the guiding role of teachers, improve students' learning initiative, create an interactive learning environment, and promote the diversified development of basic teaching for English majors is a problem that must be considered in the reform of classroom teaching for English majors and the improvement of teaching quality for English majors.

M. Atiquzzaman et al. (Eds.): BDCPS 2019, AISC 1117, pp. 915–919, 2020.
https://doi.org/10.1007/978-981-15-2568-1_124

2 Summary of MOOC and Blended Learning Model

2.1 The Concept and Application of MOOC

MOOC is a large-scale open online course, Massive Open Online Course, which is the product of "Internet + education" [1]. MOOC is originated in Canada. It was first proposed by George Siemens and Stephen Downsky in Canada and the world's first CMOOC-type curriculum was created. The curriculum model emphasizes the human-computer interaction learning model, which integrates curriculum designers, lecturers, learners and teaching resources [2]. A learning community is set up to promote learners of different thinking types and learning modes to learn in the interactive mode of human-computer and human-computer interaction, thus triggering knowledge transfer and knowledge creation. MOOC teaching has a wide range of openness, which can provide more students with English curriculum resources, so that learners can learn foreign language knowledge at anytime and anywhere and choose their favorite courses according to their personal needs [3].

Due to the unique teaching value concept and learning-based teaching concept, many universities in China have launched online course collaboration, introduced the network courses of internationally renowned universities, and joined the trend of Mu-curricula teaching. The teaching mode of MOOC can enrich the comprehensive English teaching system, enhance students' self-study ability and improve teaching effect [4].

2.2 Overview of Blended Learning Model

The concept of Blended Learning was first put forward by foreign scholars. Bonk and Graham define "Blended Learning" in the Manual of Blended Learning [5]. They point out that blended learning is a new mode that combines face-to-face teaching and multi-media teaching. Professor He Kekang first officially advocated this concept in China. He believed that Blended Learning combines the advantages of traditional learning methods with those of E-learning. It not only plays the leading role of teachers in guiding and monitoring teaching, but also fully reflects students' initiative, enthusiasm and creativity in the learning process [6].

The advantages of Blended Learning mode are mainly embodied in: (1) diversification of teaching technology. This teaching mode is based on Internet technology and combines video, audio, text, graphics and other multimedia technologies; (2) The combination of students' active learning and teacher-guided teaching. This teaching mode embodies the idea that students are the main body of learning and teachers are the supporters of learning [7]. In this teaching mode, teachers are no longer a single knowledge transmitter, but the designers and supporters of students' learning. With their own teaching experience, combined with online learning and students' self-study, teaching interaction; (3) diversification of teaching forms. Blended Learning mode combines traditional face-to-face teaching and network diverse learning environment [8]. It effectively mixes learning environment, learning resources and learning methods to better meet students' individual differences in learning habits and learning levels and help students achieve better learning results.

3   Construction of Blended Learning Mode of Comprehensive English Based on MOOC

3.1   Pre-class Preparation Stage

Teachers decompose the teaching content of the teaching unit according to the teaching objectives of the unit before class. It can be roughly divided into: preview task, text comprehension and appreciation, language knowledge, unit tasks. In the course of curriculum design, teachers need to make full use of the advantages of network resources, reasonably design teaching content and control the length of video, and cover as much knowledge and information as possible. Meanwhile, in the course of video design, teachers should grasp the classroom structure comprehensively so as to make the video content interesting and interactive.

3.2   Course Implementation Stage

The implementation of curriculum teaching is mainly divided into three parts: pre-class learning, classroom teaching and online learning after class. After making the video, teachers can combine the idea of curriculum construction and carry out curriculum teaching activities with the help of the curriculum teaching system. Before the implementation of classroom teaching, teacher should upload the video of lessons in the teaching system. By means of Internet and mobile learning, students can make personalized learning. They can study and analyze the important and difficult problems in video independently with reference to translation materials, and can also publish their own views. The discussion groups can use Apps like WeChat or QQ to share learning materials, exchange learning results and fully communicate and interact with teachers online. After the communication is completed, students can acquire relevant information before face to face teaching process. In the system, teachers can carry out exercises on it while students' exercises are evaluated and corrected.

Secondly, in the process of classroom teaching, the first is to answer the students' doubts in the process of learning on the MOOC system, collate the knowledge information in the video, and supplement the learning materials such as electronic documents or blackboard books. Teachers should give full play to the role of organizing, guiding and monitoring the teaching process, and give immediate answers to the problems encountered in learning.

After each unit is completed, students complete unit tasks and unit tests. The process is closely linked. Teachers can observe learning progress on the learning software give feedback.

3.3   Course Evaluation

Blended learning mode determines the diversity of evaluation mechanism. The subject, objective, content and method of evaluation have changed accordingly [9]. The main body of evaluation has changed from teacher evaluation to group leaders, student representatives and students themselves participating in the evaluation. Such an evaluation method can diversify the channels of evaluation acquisition and make the results more objective. Through self-evaluation and mutual evaluation, students can turn themselves into active evaluators. In the process of evaluating themselves and others, students learn to revise their learning goals, examine their learning effects, learn to reflect, and change passive learning into active learning [10]. The diversification of evaluation objectives mainly includes the use of language knowledge, learning attitude and learning strategies. The evaluation content mainly includes the evaluation of students' learning process, learning skills and learning effect. The evaluation method mainly consists of formative evaluation and final evaluation. Specific ways include learning effect record, teacher record, cooperative mutual evaluation record, learning effect evaluation, personal self-evaluation. The proportion of each way is set according to the specific tasks of classroom teaching. Various evaluation methods can be combined alternately.

4   Conclusion

As a new type of teaching mode, MOOC is of great significance in promoting the construction of open educational resources and students' individualized learning [11]. It is also an inevitable trend of the development of English teaching in colleges and universities. Blended Learning mode based on MOOC can play a very positive role in English teaching. It has changed the traditional classroom to teaching-oriented and student-centered classroom, maximizing the students' learning initiative and teachers' leading role. At the same time, this teaching mode also gives higher requirements for teachers. Teachers should not only pay attention to how to teach, but also to how to promote learning. Blended Learning has changed the traditional mode of classroom communication. Teachers should pour their emotions into the process of communicating with students and combine classroom communication with online communication. The introduction of Blended Learning mode requires teachers to use the network platform more skillfully, track and feedback students' learning records, and lay a foundation for students' learning and growth.

References

1. Song, M.Y.: English teaching reform in applied undergraduate colleges and universities in the age of MOOC. Teach. Educ. (High. Educ. Forum) (2015)
2. Jonassen, D.H.: Thinking technology: context is everything. Educ. Technol. (6) (1993)
3. Zhang, Y.: The influence of the development of MOOC on China's higher education and its countermeasures. J. Hebei Univ. (Educ. Sci. Ed.) (2014)
4. Dong, J.: The development of MOOC and its impact on higher education. Master's Degree Thesis of Shandong Normal University (2015)
5. Jared, M.C.: Blended learning design: five key ingredients (2005)
6. Hu, J., Wu, Z.: Research on MOOC-based college English flip classroom teaching model. Foreign Lang. Audiov. Teach. (6), 22 (2014)
7. Wen, Q.: Output-driven-input-driven hypothesis: an attempt to construct the theory of college foreign language classroom teaching. Foreign Lang. Educ. China (2) (2014)

8. Huo, H.: Preliminary exploration of English teaching reform in higher vocational colleges under the background of MOOC. Vocat. Tech. Educ. **8**, 43–45 (2014)
9. Wang, Q., Huang, G.: Construction of multimedia learning environment. China Distance Educ. (Compr. Ed.) (4) (2001)
10. Pu, Z.: College English teaching reform in the age of MOOC. J. Kiamusze Vocat. Coll. (8) (2015)
11. He, G., Ren, Y.: A preliminary study on the mechanism of college English teaching reform. Heilongjiang High. Educ. Res. **15**(3) (2010)

Optimized Event Identification Fused with Entity Topic

Xutong Chen[✉]

International Academy, Beijing University of Posts and Telecommunications,
Beijing 100000, China
chenxutong@bupt.edu.cn

Abstract. Event recognition model is very important for machine learning. In the existing event learning model, there are many kinds of models. For example, in the study of event recognition oriented to emergencies, it is mainly based on semantics and syntax, involving feature vectors and factors. For Internet events, there has also been progress from Boolean model, vector space model to probability model and language model. Aiming at the learning of news events, a more novel recognition model is expected to be developed and processed on the basis of the existing model, so that machine learning can have new ideas and more feature points to consider, the model will be more complete and the accuracy rate will be improved. Based on the existing topic model, we update the relationship between entity, entity topic, word topic and word. In the previous stage, we add the consideration of feature words, which we will mention later, and use concise graph theory to express. By comparing the predicted content with the original content and calculating the relationship with the threshold, the event recognition result and classification can be determined. The accuracy of the new topic recognition model is obviously improved compared with the original model.

Keywords: Topic prediction · Event identification · Optimized model · Multiple dimensions

1 Introduction

With the continuous development of the internet, there are too many kinds of information swarm into our life, and people are gradually entering into the era of We-Media. Communication mode by applications like Wechat, QQ, instagram etc. bring lots of complex structure of news also include news stream. Due to the complexity of the structure and content, the amount of machine learning calculation using the original basic method will become huge, and it is unable to understand and classify information quickly and accurately by itself. However, the current market needs to quickly and accurately find the information needed for readers to judge the type, main content and value of information. At present, there are quite a number of relevant researches, including the design of news event identification system and the design of Internet event identification system. Efficient models, such as the construction of trigger word bank, the selection of topic sentences and further analysis, provide new ideas for event identification. Our research is based on the knowledge of the event topic. According to the study of the original model, we combined the original topic discovery model with the new prediction model, and improved the event identification system by setting the threshold.

1.1 Advantages and Disadvantages of Different Topic Identification Models

Boolean model is a simple retrieval model, which is used to find the information returned in the query with the value of 'true'. It can accurately give the correlation through simple binary judgment and easily return intuitive results from individuals. For example, if the feature item in the text is set as a unit vector, each text is a vector composed of many feature items. The value of this vector is 0 or 1. If there are corresponding feature items, it can be set as 1; otherwise, it can be set as 0. Boolean model provides a framework of user easy to learn, easy to understand, but also has many shortcomings, first of all, only a binary decision will reduce the administrative levels, only 0 and 1 options are sometimes unable to accurately determine the relevant level which leads to the topic judge's mistake. Secondly, the lack of the concept of document classification. Compared with Boolean model, vector space model is improved to some extent. As a computing tool for document representation and similarity, it has strong computability and operability, and can be used for text retrieval. In terms of dimension value, it is no longer a binary judgment, but a corresponding weight can be used. The different importance and reference degree of feature items to the text can be considered, but the correlation between each feature item cannot be considered in the process of building the model [1–3].

Based on the above two models, the probabilistic model and the language model add the uncertainty consideration, which can connect different feature terms and take into account the relationship between different feature terms. However, the probability model and the language model have introduced the prediction function to some extent. In our research stage, we decided to use the simple vector space model to judge the topic first, and then use the event topic model to predict and calculate the similarity and compare with the threshold.

1.2 Event Topic Learning Model Introduce

The main related work of events is usually through event extraction, event discovery and tracking and event knowledge base construction. But most of the knowledge is static entity and entity relationship, but lack of dynamic event knowledge. Further study of the basic event topic model can lead to the event topic model integrated into the entity. Event topic model that integrates entities combines hierarchical analysis of entities. Existing mature entity topic models are as follows: CorrLDA, ETM and CorrLDA2. CorrLDA - the model of CorrLDA - assumes that the process of document generation generates words, generates entities, and does not fully reflect the importance of entities. Nor can you model relationships between entities [4, 5]. Entity topic model ETM - given entities, topics and generated words. Reflect the importance of entities, but cannot model the relationship between entities. Related topic model CorrLDA2 - first generate words, then entity, not fully embodies the importance of the entity. Entity topics can be used to model the relationship between entities. In this study, we further optimized CorrLDA and CorrLDA2, changing the relationship between the original topic and sub-topic into a further model. There are two alternative models, namely linear prediction model and multi-element joint prediction.

2   Innovation Research

In the event recognition, vector space model is used for text recognition. For each document, the words in the sentence of the document were first divided and the set W was generated. Then, referring to the eigenvectors of each class, the existing document W was represented as the corresponding document vector V. If there are related words, multiply them by 1, and if not, multiply them by 0 to get the vector projected by the corresponding weight. For example, the original key vector for {work, department director, files, team members, director of}, the corresponding weight vector is {0.3, 0.2, 0.1, 0.15, 0.1, 0.15}, the original news after word segmentation results is {work, graduation, director, file, universities, youth corps committee, organize}, and the corresponding vector is {0.3, 0,0.15, 0.15, 0, 0} [6, 7].

The original event topic model is shown in the Fig. 1. It can intuitively see the relationship between entity, entity topic, word topic and word.

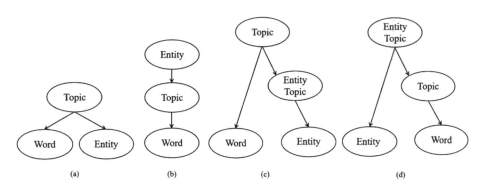

Fig. 1. Original models

The pictures above are CorrLDA (a), ETM (b) and CorrLDA2 (c), Figure (d) is the further optimization of ETM model. In order to solve the disadvantages of the existing model, entity as the core is selected to model entity topic as the distribution of word topic for the study and prediction of documents. In the current study, we used the following two models to optimize CorrLDA and CorrLDA2, we call them ECorrLDA (e) and ECorrLDA2 (f) (Fig. 2):

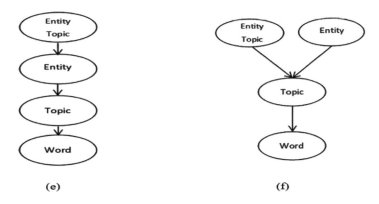

Fig. 2. Innovative models

2.1   ECorrLDA

ECorrLDA fully reflects the importance of entity by solving the original problem of related topic model. After the model is proposed, experiments are carried out on each document in the collection of event documents and related topics to calculate the confusion, entropy and so on. And we can check the new algorithm as follows: First for each document in the collection of documents, from the prior distribution Dirichlet distribution, sampling an entity in the alpha topic for each entity subject sample entities subject to the physical distribution, and distribution entities to words. For each word, sampling word distribution subject to the words, and then use Collapsed Gibbs from to solve the model, calculation and comparison under different model of confusion degree, entropy and sKL. PS. confusion refers to the standard indicator in the field of information theory, which measures how well the model fits the data. The smaller the value, the better the model's effect [8–10].

2.2   ECorrLDA2

ECorrLDA2 predicts the topic through the interpretation of the entity and the entity topic, and then generates words and phrases from the topic to form the required vectors and sentences, thus generating a complete article. In this model, both the entity topic and entity are studied and understood, and the topic is positioned from two dimensions to obtain a more accurate prediction of the topic. The algorithm: First of all, for each document in the document collection, one entity topic distribution is sampled from the prior distribution Dirichlet alpha, entity topic to topic distribution is sampled for each entity topic, and word topic distribution is sampled for each entity topic. For each word topic, sample word topic to word distribution, word topic to entity topic distribution, and word topic to entity distribution. The model is solved using Collapsed after Gibbs from, calculation and comparison under different model of confusion degree, entropy and sKL.

The main formulas used are as follows:

$$p\left(\tilde{z}_i = \tilde{k} | e_i = e, \widetilde{z_{\neg i}}, e_{\neg i}, \alpha, \tilde{\beta}\right) \propto \frac{N_{m_{\neg i}}^{\tilde{k}} + \alpha}{\sum_{k=1}^{\tilde{k}} N_{m_{\neg i}}^{\tilde{k}} + \tilde{K}\alpha} \cdot \frac{N_{k_{\neg i}}^e + \tilde{\beta}}{\sum_{e=1}^{\tilde{V}} N_{k_{\neg i}}^{\tilde{e}} + \tilde{V}\tilde{\beta}} \tag{1}$$

$$p\left(Z_i = k, x = \tilde{k} | w_i = w, z_{\neg i}, \tilde{z}, w_{\neg i}, \tilde{\beta}\right) \propto \frac{N_m^{\tilde{k}} + 1}{\tilde{N} + \tilde{K}} \cdot \frac{N_{m_{\neg i}}^{\tilde{k}} + \gamma}{\sum_{k=1}^{\tilde{K}} N_{k_{\neg i}}^{\tilde{k}} + K\gamma} \cdot \frac{N_{k_{\neg i}}^w + \beta}{\sum_{w'=1}^{V} N_{k_{\neg i}}^{w'} + V\beta}$$

$$\tag{2}$$

3   Experimental Verification and Result Evaluation

For the number of topics and the number of entity topics, we try different values to find the value that makes the least degree of confusion. The comparison methods include LDA, CorrLDA2, and ETM. Evaluation indexes include confusion degree, which is used to measure the degree of data fitting of the model, entropy which is used to measure the purity of entity topic to entity clustering, sKL which is used to measure the distance between entity topic and the effect of entity clustering. Experiments were conducted to evaluate the effects of the model, and the results were as follows (Tables 1, 2 and 3):

Table 1. Different kind of models' confuse degree

Data sample	Confuse degree–different kinds of models				
	LDA	ETM	CorrLDA	ECorrLDA	ECorrLDA2
SET1	2478.3	1563.1	1449.0	1420.5	1334.2
SET2	29372.3	23981,3	17983.1	17272.6	13744.4
SET3	38142.2	31876.9	24511.1	22809.0	19030.4

Table 2. Different kinds of models' entropy

Data sample	Entropy–different kinds of models		
	CorrLDA	ECorrLDA	ECorrLDA2
SET1	0.7883	0.7745	0.5390
SET2	1.4874	1.4023	0.9436
SET3	3.8991	3.4500	2.8531

Table 3. Different kinds of models' sKL

Data sample	sKL–different kinds of models		
	CorrLDA	ECorrLDA	ECorrLDA2
SET1	2.3392	2.4617	3.0844
SET2	4.8753	4.9239	5.2624
SET3	0.3543	0.2353	0.7328

The results obtained by vector space model and predicted by topic learning have been described above. Conclusion can be obtained – the new topic learning model can reduce the result confusion degree, that is, the data fitting degree becomes better, and the effect of the second model is more obvious. The entropy decreases, the physical topics become more pure, and the sKL distance increases, which means the distance between the physical topics increases.

4 Conclusion

Through the optimization of the event recognition model, we find a new correction direction. Firstly, the traditional vector space model is used for event identification, and then the event identification and prediction methods integrated into the entity topic are used for calculation and comparison after the classification of large categories, so as to obtain more accurate event classification and make the event identification more accurate. In the event topic knowledge learning module, we choose to further optimize the original model and obtain two new learning models. The overall effect of the two models is better than that of the original model, but there is no further comparison with the EETM model, and the overall effect is good. Compared with the two models, ECoorLDA2 model is more efficient and has better performance after optimization.

References

1. Granroth-Wilding, M., Clark, S.: What happens next? Event prediction using a compositional neural network model. In: Proceedings of the 30th Conference on Artificial Intelligence, pp. 2727–2733. AAAI (2016)
2. Chen, Y., Zhou, C.: Named entity recognition from Chinese adverse drug event reports with lexical feature based BiLSTM-CRF and tri-training. J. Biomed. Inform., 103252 (2019). https://doi.org/10.1016/j.jbi.2019.103252
3. Shi, Y., Wang, Y.: An event recognition method for Φ-OTDR sensing system based on deep learning. Sensors (2019). https://doi.org/10.3390/s19153421
4. Ghosh, S., Vinyals, O., Strope, B., Roy, S., Dean, T., Heck, L.: Contextual LSTM (CLSTM) models for large scale NLP tasks. In: KDD Workshop on Large-Scale Deep Learning for Data Mining (DL-KDD). ACM (2016)
5. Zou, L., Packard, J.L.: Morphological and whole-word semantic processing are distinct: event related potentials evidence from spoken word recognition in Chinese. Front. Hum. Neurosci., 133 (2019). https://doi.org/10.3389/fnhum.2019.00133

6. Katzouris, N., Artikis, A., Paliouras, G.: Parallel online event calculus learning for complex event recognition. Future Gener. Comput. Syst. (2018). https://doi.org/10.1016/j.future.2018.11.033

7. Hasan, M., Paul, S., Mourikis, A.I., Roy-Chowdhury, A.K.: Context-aware query selection for active learning in event recognition. IEEE Trans. Pattern Anal. Mach. Intell. (2018). https://doi.org/10.1109/tpami.2018.2878696

8. García-Bajos, E., Migueles, M., Aizpurua, A.: Different bias mechanisms in recall and recognition of conceptual and perceptual information of an event. Psicológica J. (2018). https://doi.org/10.2478/psicolj-2018-0011

9. Zhang, X., Zhang, H., Zhang, Y.: Deep fusion of multiple semantic cues for complex event recognition. IEEE Trans. Image Process.: Publ. IEEE Signal Process. Soc., 1033–1046 (2016). https://doi.org/10.1109/tip.2015.2511585

10. Pichotta, K., Mooney, R.J.: Learning statistical scripts with LSTM recurrent neural networks. In: Proceedings of the 30th Conference on Artificial Intelligence, pp. 2800–2806. AAAI (2016)

Innovation of "Internet + English" Teaching Mode Driven by Big Data

Yongxin Li[✉]

Department of Textile and Clothing Engineering, Shandong Vocational College
of Light Industry, Zibo 255300, Shandong, China
louiseyt2006@163.com

Abstract. With the continuous promotion of the age of big data, "Internet + education" has further promoted the innovation and development of the English education model in China. Using big data to teach has become a learning trend, and also meets the needs of the times, ensuring the same pace of learning with the domestic and foreign society, and cultivating compound talents in line with the needs of the times. Taking English education in higher vocational colleges in China as the research object, combined with the status of English teaching in China, the reform of English teaching from the angle of big data information technology is analyzed, with a view to finding out effective and feasible English teaching methods and strategies, so as to further improve the English teaching effect and strengthen the communication ability and the overall quality of students in higher vocational colleges.

Keywords: Big data · Internet + education · English teaching

1 Introduction

In the information age, the continuous development of big data technology has brought great changes to people's work and life, as well as to English teaching [1–3]. Compared with data information, English textbook knowledge is obviously more limited, and information content is often lagging behind [4–6]. Therefore, business English needs to continue to innovate to better meet the demand of the information society. From the traditional teaching mode in China, business English teaching tends to focus on the imparting of theoretical knowledge, and there is little teaching of English practical ability [7–10]. Based on the relevant information, this paper analyzes the reform of English teaching from the perspective of big data information technology.

In recent years, with the wide application of big data information technology in all walks of life in China, the related resources and information have been greatly integrated and optimized. In particular, the rapid development of modern education technology has promoted the popularization of big data in teaching to some extent. Compared with the traditional teaching mode, the widespread popularity and application of big data technology in the education industry not only greatly reduces the cost of teaching, but also greatly expands the teaching method and further improves the vitality and effectiveness of teaching. For English teachers, they should fully absorb and use big data information technology, constantly optimize their own teaching mode,

M. Atiquzzaman et al. (Eds.): BDCPS 2019, AISC 1117, pp. 927–930, 2020.
https://doi.org/10.1007/978-981-15-2568-1_126

create a pleasant teaching environment for students, and fully tap students' inner emotions about English learning while carrying out English teaching to achieve the purpose of enhancing English teaching effect.

2 Current Situation of English Teaching

Nowadays, there are some main problems in English teaching, such as boring classroom teaching content, mechanical evaluation model, low student participation, etc. The existence of these problems greatly limits students' enthusiasm for English learning and seriously affects the effectiveness of teachers' teaching. In addition, due to various reasons, some teachers fail to fully consider the actual situation of students and the overall goal of teaching in the process of designing English teaching content, resulting in a large deviation between teaching content and practice.

2.1 Outdated Teaching Ideas

In fact, for students, English learning is not only for examination, but more importantly for application and practice, which can be effectively used in work. However, in current English teaching, teachers attach too much importance to the education of theoretical knowledge and seldom consider the actual needs of enterprises, which results in a big deviation between students' English learning and market demand. Moreover, teachers seldom enhance the cultivation of students' communicative competence in English. Therefore, it is obviously urgent to innovate teachers' teaching concepts.

2.2 Old-Fashioned Teaching Methods

As the main way to spread information technology, the Internet and mobile phone have been widely used in people's daily life. Nowadays, teachers and students in colleges and universities rely heavily on electronic equipment. However, it is not difficult to see that teachers still tend to use the oral mode in their English classroom instruction. They seldom resort to advanced electronic information technology and media. Compared with the current advanced social development and market development, this lagging teaching mode obviously fails to stimulate students' interest in English learning, enrich English class content and improve students' English literacy. In addition, teachers are also accustomed to using test scores to evaluate students' English learning, and seldom combine the students' learning process to make a comprehensive assessment.

3 The Significance of Constructing "Internet + English" Teaching Mode

The main purpose of English courses is to enhance students' English communication and use ability, and to cultivate compound talents for transnational business, trade and management. However, the teaching content of traditional English in China is too dull, the evaluation model is relatively mechanical, and the students' participation is relatively low. These problems greatly limit the students' enthusiasm for English learning and seriously affect the teaching effect of teachers. To better meet the demand of the current social development, English teaching combined with advanced information technology can not only greatly enrich the content and methods of English teaching, but also increase the positive interaction between teachers and students, and make up for the shortcomings of traditional teaching. Moreover, with the help of the Internet model, students' English learning resources can be further expanded and their English practice can be enhanced.

4 Teaching Practice of "Internet + English"

The main purpose of English majors is to cultivate high-quality talents with strong English communication skills and a multicultural international perspective. It is this particularity that makes the development of English teaching not only pay attention to the basic English theoretical knowledge, but also attach importance to the cultivation of students' English practical skills. In the current social environment of "Internet+ ", the openness and sharing of knowledge has greatly changed the language service industry, bringing great vitality and changes to it, and creating a good environment for the reform English teaching mode. However, the relationship between the Internet and English should be effectively integrated in the actual teaching process. Teachers should actively innovate their own teaching concepts and methods, fully absorb the advantages of the Internet, and effectively integrate and optimize high-quality teaching resources.

Making full use of the big data analysis in the process of evaluating the teaching effect is conducive to objectivity, scientization and standardization of evaluation and analysis. The so-called formative assessment usually includes two levels: One is the evaluation of online participation, mainly including attendance rate, homework completion rate, and discussion group participation. The other is the evaluation of offline classroom participation, which mainly covers classroom homework testing, classroom discussion, and presentation. In the analysis and evaluation of students' learning effect, teachers can fully use the statistical analysis of relevant information to record and analyze students' daily learning in detail, and constantly improve and optimize their teaching through the evaluation of students' learning process.

5 English Teaching Reform Measures from the Perspective of Big Data

Big data technology not only provides new ideas and methods for the development of teaching work, but also provides a large number of advanced teaching technology and tools. However, teachers still need to constantly improve their abilities in the development of practical teaching, carefully analyze the specific functions and characteristics of different technologies, and select appropriate teaching tools. As an educator, teachers are not only responsible for the transmission of knowledge, but also important organizers and participants in teaching activities. With the continuous deepening and development of "Internet + education", the application of the Internet in English teaching is inevitable. Only by setting up correct concepts and paying attention to the dominant position of students can teachers better realize the effective integration of the Internet and English teaching.

6   Conclusion

Under the social environment of big data, teachers should actively improve the deficiencies faced by current business English teaching, absorb advanced teaching ideas and techniques, select appropriate teaching methods, and constantly improve and optimize business English teaching. Meanwhile, teachers should pay attention to students' dominant position, enhance the cultivation of students' English practical ability, and adopt scientific and effective evaluation methods to help students improve their English literacy and practical ability to the greatest extent.

Acknowledgments. This research was supported by 2018 Zibo City School Integration Development Plan: Public Training Base for Textile and Garment Specialty.

References

1. Wang, H.: English teaching in the big data era. Overseas Engl. (13), 117–119 (2019). (in Chinese)
2. Wang, Q.Y.: The way to achieve the training target of English professionals in the big data era. Think Tank Era (27), 279–280 (2019). (in Chinese)
3. Chen, M.X., Yu, W.T.: On the educational research paradigm shift in the informatization process. J. High. Educ. (12), 47–55 (2016). (in Chinese)
4. Zhou, R.: Application of blue Moyun class in basic course of computer application. PC Fan (8), 133–134 (2018). (in Chinese)
5. Lu, Y.: Research on artificial intelligence promoting college English teaching reform. Educ. Mod. **6**(58), 46–47 (2019). (in Chinese)
6. Yan, Z.S., Xu, J.H.: Research on online and offline integration of "Internet+" personalized teaching mode. Chin. Vocat. Tech. Educ. (5), 74–78 (2016). (in Chinese)
7. Xu, X.L.: Analysis of English teaching in higher vocational colleges under big data environment. Course Educ. Res. (36), 104 (2019). (in Chinese)
8. Hu, E.J.: Research on multimodal English teaching in higher vocational colleges in the age of big data. Nei Jiang Ke Ji **40**(7), 131–132 (2019). (in Chinese)
9. Niu, Z.: An analysis of the mixed English teaching model in higher vocational education based on the big data era. Overseas Engl. (16), 247–248 (2019). (in Chinese)
10. Zhao, Y.: The innovation and informatization development of college English teaching mode in the big data era. China J. Multimed. Netw. Teach. (08), 50–51 (2019). (in Chinese)

PPI Inference Algorithms Using MS Data

Ming Zheng and Mugui Zhuo[✉]

Guangxi Colleges and Universities Key Laboratory of Professional
Software Technology, Wuzhou University, Wuzhou, China
370505375@qq.com

Abstract. With the development of proteomics, the focus of research has begun to focus on the establishment of all human protein interaction (PPI) networks. Mass spectrometry has become a representative method for predicting protein interaction. Mass spectrometry is one of the main experimental means to construct protein interaction network. Based on mass spectrometry, a large number of protein purification data, such as AP-MS data and PCP-MS data, have been generated. These data provide important data support for PPI network construction, but it is not only inefficient but also unrealistic to construct PPI network by manual means. Therefore, the network inference algorithm for PCP-MS data is one of the hot topics in bioinformatics. In this paper, the algorithm of constructing a kind of mainstream mass spectrometry data (PCP-MS data) PPI network is studied. Considering the existing bottlenecks, the goal of constructing a high-quality PPI network is achieved. The existing PPI network inference algorithm for PCP-MS data is still in its infancy, and there are few related methods. At the same time, there are some problems in the quality of the results of the algorithm, such as: (1) many wrong interactions are included in the results of different inference algorithms, while some correct interactions are omitted; (2) different inference algorithms perform differently on the same data set; (3) for different data sets, the volatility variance of the performance of the same algorithm is large.

Keywords: MS data · PPI network · Direct protein interaction · Correlation analysis · Sequencing integration

1 Introduction

Protein is the ultimate executor of cell activity and function. Each protein does not perform its function independently. It usually interacts with other proteins to form temporary or stable complexes to perform specific functions. Therefore, large-scale and high-throughput protein-protein interaction (PPI) [1] research emerges as the times require. Its purpose is to map the network of protein-protein interaction in the whole proteome under specific physiological conditions of cells. Based on these relationships, we can really elucidate the function of a protein, and further study the occurrence and regulation mechanism of various physiological reactions in cells, and ultimately reveal the essence of life [2].

Mass Spectrometry (MS) [3] analysis technology is one of the most important biochemical experimental means used to construct PPI network. Several representative

© Springer Nature Singapore Pte Ltd. 2020
M. Atiquzzaman et al. (Eds.): BDCPS 2019, AISC 1117, pp. 931–936, 2020.
https://doi.org/10.1007/978-981-15-2568-1_127

methods of PPI network construction based on mass spectrometry technology, Affinity Purification-Mass Spectrometry (AP-MS) [4], Protein Correlation Profiling Mass Spectrometry (PCPMS) [5] and Cross-Linking Mass Spectrometry (XLMS) [6]. Among them, AP-MS technology continuously selects bait protein (bait), captures many prey proteins which interact with the protein, and then identifies these proteins by mass spectrometry technology, so as to obtain the interaction between proteins, and finally establish PPI network. PCP-MS technology separates protein complexes by sucrose density gradient centrifugation (SGF) [7], isoelectric focusing (IEF) [8] and ion exclusion chromatography (IEX) [9] based on the characteristics of density, isoelectric point and hydrophobicity, so as to obtain the relationship between proteins. In recent years, researchers have begun to study PPI network construction for PCP-MS data [10].

In recent years, PPI network inference algorithm for MS data has become the mainstream method of PPI network inference. The research on PPI network inference for PCP-MS experimental data has just begun. The quality of existing algorithms is far from enough. Existing network inference methods for PCP-MS data usually convert protein interaction inference into classification problem, that is, using classifiers to infer whether there is interaction between proteins. In the process of classifier model training, some existing methods use part of the reference set data to train the model, which may have some problems such as over-fitting and unreasonable performance evaluation. Therefore, the main problem to be solved in this paper is the construction of PPI network for PCP-MS data.

This paper presents a PPI network construction algorithm for PCP-MS data. The method takes the initial PCP-MS experimental data as input, including two steps: first, calculating the protein interaction fraction and measuring the protein correlation. It should be noted that there are many groups of experimental results because different separation techniques are usually used to separate protein complexes during the experimental process. Then, a series of experimental results were integrated by sequencing and integration to obtain a more comprehensive and comprehensive protein interaction relationship. Based on this result, a PPI network was constructed. To sum up, the main innovations of this paper are as follows: Aiming at the problem of PPI network construction for PCP-MS data, a method of PPI network construction based on correlation analysis and sorting integration is proposed. Compared with existing supervised learning methods, this method does not introduce reference set information to train models, so the results obtained are more convincing. In addition, this method is essentially a PPI scoring strategy based on unsupervised learning, which avoids the problems of parameter selection, optimization and over-fitting. Through the integration of multi-group results to obtain more comprehensive and reliable PPI, build a high-quality PPI network.

2 Method

The first step is correlation analysis. PCP-MS data contains quantitative information of each protein in different fragments during protein separation experiments. If two proteins interact with each other, they are likely to appear in one fragment at the same time. Therefore, the correlation coefficient between proteins is calculated to measure the degree of correlation between two proteins. This paper selects three correlation coefficients, Pearson correlation coefficient, Pearson correlation coefficient Noised-Pearson based on noise model and weighted cross-correlation coefficient Wcc. For each group of initial PCP-MS data, multiple groups of results can be obtained by calculating the correlation coefficient between proteins. Each experimental group corresponds to a PPI ranking list. The larger the coefficient, the more likely there is interaction between proteins.

The second step is sorting and integrating. For the PPI sorting list obtained in the previous step, the sorting and integration method is used to integrate it into a comprehensive PPI score, which can comprehensively and comprehensively represent the protein interaction relationship. In this paper, Reinforcement method is used to integrate multiple sorting lists. In this paper, an unsupervised learning method based on correlation analysis and ranking integration is proposed, which can effectively predict protein-protein interaction through the above two steps. Compared with the supervised method, a stable and reliable PPI network can also be built without using the standard answer.

In PCP-MS experiment, the protein of the same complex was co-eluted by biochemical fractionation technology, and the protein-fragment data were obtained by mass spectrometry analysis. There are M fragments, each of which records the quantitative information of N proteins. Specifically, each protein and Faction form a "Protein-Faction" vector. Because the proteins identified in the same Fraction may originate from the same protein complex, that is, there may be interactions between proteins. Therefore, the correlation coefficient is calculated to measure the correlation between proteins.

Firstly, for each protein, the frequency of its occurrence in the fragment is calculated. The formula is as follows.

$$B_{ik} = \frac{A_{ik}}{\sum_{i=1}^{N} A_{ik}} \tag{1}$$

Among them, B is the normalized matrix of matrix A, and the normalized vector of each protein in each behavior in the matrix. If two proteins appear in the same Fraction several times at the same time, and the number of both proteins is higher when they appear at the same time, Pearson correlation coefficient can measure the interaction between them more effectively. However, if the number of two proteins is low when they appear at the same time, the measurement effect of Pearson correlation coefficient is not ideal. Therefore, in order to solve the above problems, noise can be introduced into the original MS data and Pearson correlation coefficient can be calculated.

Since each pair of PPIs corresponds to multiple correlation coefficients, how to integrate them into a score as a reliable basis for ranking is the key to the problem. The fractional integration problem is essentially a Rank Aggregation problem, that is, given a set of sorted lists, it can be integrated into a better list in some way. Sorting integration methods are mostly based on the following strategies: whether relying on ranking information and whether relying on score information. The corresponding method is called ranking-based method and score-based method. In ranking-based methods, elements are ranked according to their ranking or ranking. In the fraction-based method, the elements in the list are sorted according to the size of the fraction, or the converted fractions are sorted. Obviously, the multiple correlation coefficients of each PPI correspond to multiple fractions, so this paper adopts the integration method based on fractions.

At present, there are many different integration methods of fractional ranking, ranging from simple weighted average to various complex statistical methods. Aiming at the noise of mass spectrometry data, a Reinforce method based on hypothesis test and error rate control is proposed. This method can be used not only for PPI network inference of PCP-MS data, but also for PPI network construction of AP-MS data.

3   Result

The experimental data in this paper are from the Human Soluble Protein Complex (http://human.med.utoronto.ca). In order to verify the validity of this method, the initial LC-MS/MS experimental data is selected as the data set, and CORUM data set is selected as the reference set (http://human.med.utoronto.ca). If we think that there is an interaction relationship between proteins in the same complex, but not between proteins in the same complex, we can construct a PPI reference set in this way and evaluate the experimental results.

In order to test the effect of the PPI scoring method based on similarity analysis and sorting integration proposed in this paper, two groups of experiments are conducted for unsupervised learning method and supervised learning method respectively, and the results of two methods are compared.

Before the experiment, two points need to be considered. First, this method uses RF classifier, which has the problem of parameter selection. Whether the parameters will have an important impact on the experimental results remains to be verified. If there is an impact, how to select the appropriate parameters and get the optimal model is a problem to be concerned. Secondly, there is randomness in the construction of positive and negative examples of the model. Whether this randomness will affect the experimental results is also worth considering.

In view of the above two problems, experiments are carried out to verify them. Firstly, the influence of parameters is explored. A set of positive and negative examples is generated randomly. Based on the data set, the results of RF under different parameters are tested. Let $n = 100$, depth values are 2, 4, 6, 8 and 10, respectively, as shown in Fig. 1. It can be seen that when the RF classifier chooses different parameters, the trend of the curve is basically the same, and the AUC value fluctuates between 0.67 and 0.75, which shows that the parameters have some influence on the model, but not the most important factor. When the maximum depth of the tree increases gradually from 2 to 8, the corresponding AUC value increases gradually from 0.67 to 0.75; when depth continues to increase from 8 to 10, the AUC value decreases gradually from 0.75 to 0.71. When depth is 8, the AUC value is the largest and the effect is better. The results network in this paper can be seen as below:

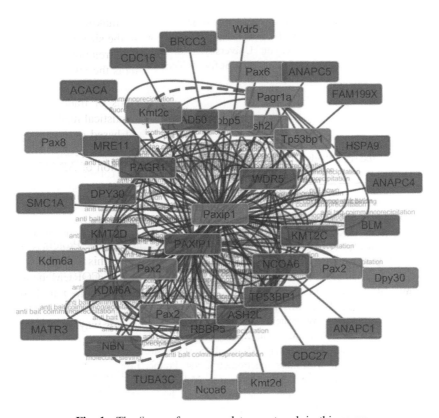

Fig. 1. The figure of gene regulatory network in this paper

4 Conclusion

Aiming at the problem of PPI network construction for PCP-MS data, this paper proposes a method of PPI network construction based on correlation analysis and ranking integration, which includes the following two steps: the first step is correlation analysis. With the initial PCP-MS experimental data as input, the correlation coefficients between proteins were calculated, and the corresponding PPI results of several groups of experiments were obtained. The second step is sorting and integrating. Reinforce method was used to integrate multiple PPI results into a single PPI score, indicating the degree of protein-related interaction. Experimental results show that the proposed method is comparable with the existing supervised learning methods. Since the method presented in this paper is an unsupervised learning method and does not draw on standard reference data, it is better than supervised learning in terms of interpretable method. In addition, it is possible to avoid overfitting problems in supervised learning methods. In the future work, how to extend the application scope of the model is worth studying.

Acknowledgement. This work was supported by grants from The National Natural Science Foundation of China (No. 61862056), the Guangxi Natural Science Foundation (No. 2017GXNS FAA198148) foundation of Wuzhou University (No. 2017B001), Guangxi Colleges and Universities Key Laboratory of Professional Software Technology, Wuzhou University.

References

1. Wang, E.C., Weng, G.Q., Sun, H.Y., et al.: Assessing the performance of the MM/PBSA and MM/GBSA methods. 10. Impacts of enhanced sampling and variable dielectric model on protein-protein Interactions. Phys. Chem. Chem. Phys. **21**(35), 18958–18969 (2019)
2. Jumppanen, M., Kinnunen, S.M., Valimaki, M.J., et al.: Synthesis, identification, and structure-activity relationship analysis of GATA4 and NKX2-5 protein-protein interaction modulators. J. Med. Chem. **62**(17), 8284–8310 (2019)
3. Li, K., Zhang, Z.X., Ma, S.W., et al.: Effects of NH4H2PO4-loading and temperature on the two-stage pyrolysis of biomass: analytical pyrolysis-gas chromatography/mass spectrometry study. J. Biobased Mater. Bioenergy **14**(1), 76–82 (2020)
4. Wu, D., Li, J.W., Struwe, W.B., et al.: Probing N-glycoprotein microheterogeneity by lectin affinity purification-mass spectrometry analysis. Chem. Sci. **10**(19), 5146–5155 (2019)
5. Crozier, T.W.M., Tinti, M., Larance, M., et al.: Prediction of protein complexes in Trypanosoma brucei by protein correlation profiling mass spectrometry and machine learning. Mol. Cell. Proteomics **16**(12), 2254–2267 (2017)
6. Chavez, J.D., Mohr, J.P., Mathay, M., et al.: Systems structural biology measurements by in vivo cross-linking with mass spectrometry. Nat. Protoc. **14**(8), 2318–2343 (2019)
7. Johnson, M.E., Bustos, A.R.M., Winchester, M.R.: Practical utilization of spICP-MS to study sucrose density gradient centrifugation for the separation of nanoparticles. Anal. Bioanal. Chem. **408**(27), 7629–7640 (2016)
8. Kahle, J., Stein, M., Watzig, H.: Design of experiments as a valuable tool for biopharmaceutical analysis with (imaged) capillary isoelectric focusing. Electrophoresis **40** (18–19), 2382–2389 (2019)
9. Shaddeau, A.W., Schneck, N.A., Li, Y., et al.: Development of a new tandem ion exchange and size exclusion chromatography method to monitor vaccine particle titer in cell culture media. Anal. Chem. **91**(10), 6430–6434 (2019)
10. Zhang, B., Wu, Q., Xu, R., et al.: The promising novel biomarkers and candidate small molecule drugs in lower-grade glioma: evidence from bioinformatics analysis of high-throughput data[J]. J. Cell. Biochem. **120**(9), 15106–15118 (2019)

Foreign Language Education in the Era of Artificial Intelligence

Yanxia Hou[(⊠)]

Nanchang Institute of Technology, Nanchang 330099, China
hou_yanxia621@hotmail.com

Abstract. With the development of big data, cloud computing and super-computers, artificial intelligence is widely used in various fields. The deep integration of artificial intelligence technology and education has changed the mode of modern teaching and learning, which also brought opportunities and challenges to foreign language education. Artificial intelligence has a profound impact on foreign language teachers, learners, teaching mode and methods, the ways of evaluation, etc. However, the application of artificial intelligence technology in foreign language education is far from satisfaction. The improvement of current situation needs the efforts from whom it may concern. The college should take some measures to motivate the application of artificial intelligence and provide financial and technical support; foreign language teachers must change their concepts first, improve their abilities of applying new information technology into teaching process, adapt to the goal of future foreign language education, and keep up with the pace of education reform in the Era of artificial intelligence so as to promote the development of foreign language education.

Keywords: Artificial Intelligence · Foreign language education · Impact and countermeasures

1 Introduction

The rapid development of science and technology has pushed the wave of intelligent revolution to spread over the world. Under the influence of artificial intelligence, profound changes are taking place in the field of education. In August 2017, the "Development Planning for a New Generation of Artificial Intelligence" issued by the State Council officially upgraded the application of artificial intelligence to the national strategy. The plan clearly pointed out that in the future, intelligent technology should be used to speed up promotion in talent cultivation pattern and reform of teaching methods and construct the new education system including intelligent learning and interactive learning, promote application of artificial intelligence in teaching, management, resource construction and other processes, develop three-dimensional and comprehensive teaching field as well as online teaching and education platforms based on big data intelligence [1]… At present, the field of higher education has become one of the important applications of artificial intelligence.

© Springer Nature Singapore Pte Ltd. 2020
M. Atiquzzaman et al. (Eds.): BDCPS 2019, AISC 1117, pp. 937–944, 2020.
https://doi.org/10.1007/978-981-15-2568-1_128

Now the post-00's generation becomes the main force of universities. They have distinct characteristics in learning, cognitive models and application habits of information technology. They are considered to be "indigenous people" in the digital age [2]. Traditional teaching methods and channels for getting information cannot meet their needs. They are more inclined to use intelligent means to obtain digital learning resources. However, in the face of such a large number of foreign language learning resources, how to realize personalized learning and improve learning efficiency and effectiveness becomes more urgent than ever before. It is necessary for foreign language teachers to explore and study a new foreign language teaching mode based on artificial intelligence technology [3].

2 The Impact of Artificial Intelligence on Foreign Language Education

Artificial Intelligence (AI) is a new technical science involving computer science, statistics, information theory, cybernetics, neuroscience and neurocognition, linguistics, psychology, and learning science. AI is widely used in manufacturing, transportation, health, education, agriculture, management and other fields [4]. With the continuous maturity of AI technology, the penetration of AI technology into the field of foreign language education continues to increase. It not only reshapes the ecological environment of foreign language teaching, but also revolutionizes the ways of foreign language learning. It is an important carrier of learning resources as well as a tool for foreign language learning. Language learning, being similar to AI, is closely related to neuroscience, resulting in a series of AI techniques for language teaching, such as text recognition, speech recognition, speech synthesis, image recognition, machine translation, human-machine dialogue, and natural language comprehension. These technologies have begun to be put into practical use; what's more, some achievements have greatly changed the traditional language education, giving birth to a profound impact on foreign language teachers, learners, teaching mode, methods and assessments.

2.1 Impact on Foreign Language Teachers

With the increase of intelligent learning tools and the continuous upgrading of big data analysis technology, the application of big data and AI in foreign language teaching has become more and more extensive. As modern teaching technology, the integration of big data, AI and foreign language education has greatly improved the teaching efficiency and quality. However, the popularization and application of intelligent, digital and networked information technology also puts higher demands on teachers: in addition to the professional knowledge, they also need to have good capabilities of applying modern information technology, processing and analyzing data, which are both opportunities and challenges for teachers [5].

In the era of AI, the role of teachers must be changed from the traditional one to modern one. AI technology can replace or assist teachers to complete a series of teaching work, including the collection of information resources, the transmission of knowledge, the correction of homework, answering questions for students, and it is even possible for students to interact directly with an intelligent robot with native British or American accents. Then instructors have more time and energy to participate in the design of foreign language teaching. In addition, the foreign language learning software supported by AI can comprehensively test the learners' levels; the speech recognition technology can accurately correct their pronunciation in real time; and the listening and speaking test system with intelligent speech technology can perform automated oral examination and objective scoring and so on [6]. These repetitive and data-based jobs done by AI will undoubtedly help foreign language teachers, but at the same time it also makes foreign language teachers reflect on themselves so as to adjust their roles to adapt to this new education environment.

But we must clearly realize that in the intelligent era of human-computer symbiosis, the tasks can be done by AI are the things that need to be repeated, data accumulated and analyzed, and the things that need precise positioning. In fact, teachers' responsibility cannot be replaced. Teachers need to help and guide students to deal with and choose learning resources, and teach students how to learn independently. In the information age, without the guidance and edification of teachers, the mixed information will have a negative impact on them. Fragmented learning does harm to their concentration and deep learning. Improper use of mobile phones, computers and other electronic products for a long time not only weakens students' self-discipline, but also ruins their vision, thus affecting their physical and mental health. Avoiding these problems, teachers should take the role of guidance, telling the students the directions, which illustrate the irreplaceability of teachers.

2.2   Impact on Learners

In the era of AI, thanks to the wide application of the Internet and big data, learners have more approaches to information. Learning resources are no longer ones from traditional textbook or electronic materials originating from books, but massive sharing resources from big data platforms. In addition, various intelligent platforms, software, and online education can greatly promote students' independent learning by means of video teaching, online question and answer, self-assessment, and learning progress tracking. Students no longer sit bored in the classroom listening to the teacher, but develop an effective learning plan according to their own needs [7]. With more resources and means, they can conduct personalized language learning at any time and any place. In the era of AI, students are required to have a high level of self-management, abilities of identifying and screening resources, and skills of using intelligent tools.

2.3   Impact on the Mode and Methods of Foreign Language Education

In recent years, foreign language teaching has tried to change from a "teacher-centered" model to a "student-centered" one. The implementation of "Flipped Classroom" is an exploration of this new model, but this transformation is not successful thoroughly. At present, foreign language teachers generally use a mixed teaching method that combines classroom teaching with extracurricular intelligent APP learning. However, the teachers give much priority to the classroom teaching, but neglect giving supervision to the students who adopt intelligent tools to study outside the classroom. So the situation that students are poor in foreign language communication has not been significantly improved.

The emergence of AI technology liberates teachers from cumbersome homework corrections and counseling. Mechanized parts can be performed by AI, such as grammar, phrases, sentence patterns, vocabulary memories, questions practice, format writing, etc., so teachers have more time and energy to supervise online learning process and effect, which is the mode of a combination of online learning and offline counseling. Teachers make timely analysis and evaluation according to the students' learning data, clearly grasp the students' learning dynamics, record their strengths and weaknesses, and take targeted measures to help them improve [8].

2.4   Impact on Evaluation

The traditional evaluation of foreign language teaching is usually based on the final evaluation of language proficiency test, lacking the process evaluation and personalized evaluation. Under the influence of AI technology, foreign language evaluation is more purposeful and targeted. The students' attendance, task completion, and classroom performance are recorded and analyzed through big data technology, and learning effects were comprehensively analyzed, then the process evaluation and targeted evaluation of students are realized.

Manual process evaluation is a long-term and cumbersome process. The big data-based intelligent analysis system can accurately determine the problems that students have after each practice and form a tracking solution. Due to AI, it is possible to simulate human standard pronunciation and obtain feedback to enhance accuracy of pronunciation; intelligently reviewing essays, producing diagnostic reports, tracking development, and giving recommendations have become a reality [9]. In foreign language classroom, the use of AI to create scenarios, human-machine dialogue, remote course sharing, and intelligent assessment system are also gradually popularized.

The evaluation of foreign language learning by AI technology is mainly embodied in the records of each student's performance in the form of knowledge list and task section. It is reasonable to combine the final evaluation with the process evaluation under the influence of AI technology, which helps students to have a overall understanding of their foreign language proficiency, wipe out their problems, improve their learning methods and ability. At the same time, AI also enables teachers to better grasp the language mastery of students and take countermeasures accordingly in the next step.

3   Application Status of AI in Foreign Language Education

Nowadays, although most foreign language teachers have realized the importance of AI, only a few of them have applied it in practical teaching, and the result of AI application is far from satisfaction. Though many intelligent learning tools have entered the field of language education, the rate of application is really low, and there is still much room for improvement. Many foreign language teachers don't make investigations of the existing intelligent products in the market, and they are not familiar with the operation of some AI technology either due to the solidification of outdated foreign language teaching mode, the lag of education concepts, the weak technical and financial support, the difficulties of implementing incentive mechanisms, and the lack of evaluation mechanisms, etc. [5]. Currently, the factors affecting the effective application of AI are more complicated than we can imagine. To achieve long-term effectiveness in foreign language education, the key is to break through these constraints.

4 Countermeasures to Promote the Effective Application of AI in Foreign Language Education

4.1 Countermeasures for College

At present, colleges and universities have recognized the importance of AI technology in the field of education, but further measures are needed to promote the effective application of AI, such as providing skills-developing programs for teachers, increasing technical and financial support, strengthening the construction and maintenance of AI software and hardware, establishing incentive evaluation mechanism for AI application, developing domestic and foreign cooperation projects, introducing education resources and technology with high quality, organizing teachers' open class activities to promote sharing and exchanges, drawing on excellent experience and achievements [5] … College authorities also need to change their concepts and have foresight in the application of AI in the future.

4.2 Countermeasures for Teachers

4.2.1 Update Teaching Concepts and Reshaping the Role of Teachers

Foreign language teachers must first change the concept of teaching: from the traditional "teacher-centered" to "student-centered", taking the students' internal and external autonomous and cooperative learning as the center, from single "offline learning" to "combination of online and offline learning", from "focus on knowledge and skills" to "focus on literacy and ability development". Teachers are not only the instructors of knowledge and skills, the organizers of classroom teaching, but also the planners and guide of personalized learning. Teachers can intelligently prepare lessons, give lectures and answer questions, which are beneficial to solve the problems of inefficiency and low participation in traditional classroom, and promote the development of teaching in the direction of intelligence, precision and individuality. At the same time, teachers should be aware of educational AI tools that are more suitable for language learning in addition to recognizing that AI does not replace teachers but help teachers to teach.

4.2.2 Follow the Latest Technology to Improve Information Literacy and Intelligent Education

Due to professional limitations, most of the foreign language teachers are not good at the application of technology. Facing the changes in education and the continuous advancement of information technology, teachers need to keep pace with the times, actively learn AI technology, participate in distance education training, and strengthen exchange and cooperation between peers, master the intelligent learning tools commonly used in foreign language teaching, know how to use advanced information technology to guide students to carry out independent learning, personalized learning and inquiry learning, and explore the teaching mode in the new era.

4.2.3 Reform the Teaching Mode and Optimize the Language Teaching Process

Teachers are encouraged to build a new intelligent teaching mode based on big data and AI. In this smart teaching mode, teachers take on the following duties and tasks:

First, through the big data learning and analysis technology, teachers can accurately grasp the students' learning attitude, learning style, and learning needs, provide students with matching learning resources, select suitable teaching methods for students, and promote the precision of teacher guidance and the individuality of student learning.

Secondly, by using human-computer interactive speech recognition system, visualization technology and VR technology, teachers can create a simulated immersive learning environment for students to listen, speak, read, and write in a near-real and vivid language environment to improve communication skills.

Third, using intelligent assessment techniques to automate the assessment of students' learning (pronunciation, vocabulary, grammar, listening, reading, writing, etc.) to form a visual diagnostic report, teachers can modify the teaching plan according to the diagnosis in time, so as to build a more efficient class for students and maximize the quality of teaching.

Fourth, teachers can make use of intelligent, personalized, and information-based online learning platforms, mobile language learning APPs, and various smart education products to track, monitor, and record the progress, duration, and effectiveness of students' after-school language learning, and promptly remind and urge them. Then students' extracurricular language learning can be normalized, long-lasting and precise.

4.3 Countermeasures for Students

AI learning is a very complicated process. The elements involved here include: interaction between teachers and students, interaction between students and students, interaction between teachers, students and knowledge, interaction between people and machines, etc. The intelligent teaching environment is a completely new style. Students should get rid of laziness and dependence on teachers first, and then learn to use intelligent learning tools rationally, formulate reasonable learning plans according to their needs and characteristics. In particular, they should learn to filter effective information from massive resources and make up for the shortcomings based on intelligent feedback, taking knowledge they learned as a "part of the body" [10].

5 Conclusion

AI brings challenges and opportunities to foreign language education. The emergence of AI promotes the innovation of teaching mode, changes the teaching mode of teachers, updates the learning methods of students, and provides a new idea for curriculum reform. Foreign language teaching reform based on big data and artificial intelligence is imperative. However, the application of AI to teaching is still in the initial stage. Foreign language teachers' belief in teaching informatization is inconsistent with teaching practice. Advanced teaching concepts and techniques have not been implemented in foreign language education. The information technology level of teachers is far from the requirements of informatization teaching. The future is an era of collaboration between human and AI, which will call higher requirements for the educational technology capabilities of foreign language teachers in colleges and universities. In view of this, colleges and universities should increase the support of AI application; teachers should also change their concepts and improve the capabilities of big data and AI application. The majority of front-line teachers should take the duty to promote the application and popularization of information technology in foreign language education, create favorable conditions for foreign language teaching reform, inject vitality into foreign language classrooms, to make the foreign language education reach a higher stage.

Acknowledgments. This research was supported by Jiangxi Educational Science Project during the 13th Five-Year Plan Period (project number: 19YB238), and Instructional Reform Project of Jiangxi Educational Commission (project number: JXJG-15-18-7).

References

1. Notice of the State Council: Development Planning for a New Generation of Artificial Intelligence [EB/OL] (2017). http://www.gov.cn/zhengce/content/2017-07/20/content_5211996.htm. (in Chinese)
2. Chen, J., Jia, Z.: A tentative study on IT-based FL learning modes in the big data era. Teach. Foreign Lang. Electrification (4), 3–8+16 (2017). (in Chinese)
3. Li, C.: Application and research hotspot of artificial intelligence in foreign language teaching. Chin. J. ICT Educ. (6), 29–32 (2019). (in Chinese)
4. Gan, R., He, G.: Exploring learning analytics in foreign language teaching in big data era. Technol. Enhanced Foreign Lang. Educ. (3), 40–46(2016). (in Chinese)
5. Zhang, H.: Research on the status and countermeasures of English teaching informationization under the background of artificial intelligence + big data. Engl. Teach. (5), 77–83 (2019). (in Chinese)
6. Li, Y.: Research on the application of artificial intelligence in vocational college English teaching under the information environment. J. Liaoning Provincial Coll. Commun. (3), 88–91 (2019). (in Chinese)
7. Yan, Y.: Study on the way of deep learning in English teaching in the age of artificial intelligence. J. Teach. Manag. (27), 106–108 (2019). (in Chinese)

8. Yang, Y.: Challenges and opportunities for foreign language teachers in the age of artificial intelligence. Think Tank Era (27), 241+248 (2018). (in Chinese)
9. Ma, R.: Application of artificial intelligence in blended foreign language learning and the construction of its ecological chain. Theory Pract. Contemp. Educ. (1), 115–121 (2019). (in Chinese)
10. Hong, C.: A study on the construction of college English ecological teaching. Technol. Enhanced Foreign Lang. Educ. (12), 29–34 (2018). (in Chinese)

Effective Information Extraction Algorithm of Dynamic Logistics Big Data Based on SVM

Bixia Fan and Xin Rao[✉]

Hubei Business College, Wuhan 430079, China
raoxin@hbc.edu.cn

Abstract. With the advent of the era of big data, data plays a more and more important role. At the same time, the massive data also contains a variety of invalid information. The existing data information extraction methods do not perform well in dealing with this kind of data, which makes the results deviate greatly. To solve these problems, this paper proposes an efficient information extraction algorithm based on SVM for dynamic logistics big data. First of all, considering the unstructured characteristics of logistics dynamic big data, preprocessing data, including cleaning data, data specification and transformation, as well as data dimensionality reduction, to facilitate subsequent operations. Then, the approximate support vector regression function is used to mine the information features effectively. Through the simulation experiment, the results show that the accuracy of the proposed algorithm is about 82.3%, which is significantly higher than the other two algorithms. This shows that the method proposed in this paper has obvious effect on the effective information extraction of logistics big data.

Keywords: Dynamic logistics big data · Information extraction algorithm · Support vector machine (SVM) · F-measure

1 Introduction

With the development of social economy, e-commerce industry and other hot global, led to the vigorous development of the field of logistics. In the long run, a large amount of data has been generated in the field of logistics. Massive data contains rich potential valuable information, and the mining of logistics big data will contribute to the healthy development of logistics field [1]. Dynamic logistics big data is to integrate and extract the existing logistics data with the help of information feedback technology and data perception technology, extract effective information and filter redundant information with the help of data mining methods, so as to achieve the goal of obtaining valuable information [2].

In the field of logistics data and other fields of data analysis and processing, many experts and scholars have conducted research and achieved fruitful results. In [3], the author points out that cloud manufacturing entangled with the Internet of things has been awakened to realize the final intelligent manufacturing. The author uses production logic and time stamp to link RFID data to help different end users simplify their daily operations. In [4], the author investigates the research and application of big data analysis in logistics and supply chain management, and emphasizes that big data

© Springer Nature Singapore Pte Ltd. 2020
M. Atiquzzaman et al. (Eds.): BDCPS 2019, AISC 1117, pp. 945–950, 2020.
https://doi.org/10.1007/978-981-15-2568-1_129

business analysis and supply chain analysis should be understood as strategic assets, which should be integrated across business activities to achieve integrated enterprise business analysis. In [5], the author proposes big data analysis of intelligent manufacturing workshop based on physical Internet, and uses Internet of things and wireless technology to convert typical logistics resources into intelligent manufacturing objects, so as to create an intelligent workshop environment supporting RFID. In [6], the author proposes the performance evaluation of equipment logistics based on data-driven social network analysis, and finds that this method can evaluate the logistics related to equipment scheduling and planning, so as to enhance equipment management by improving decision-making. Support vector machine (SVM) has strong generalization ability, and its goal is to minimize structural risk. SVM has been applied in many places. In [7], the author uses SVM to classify medical data, and conducts experiments on breast cancer data set and Parkinson's data set. The experimental results show that the proposed method can not only obtain more suitable model parameters, but also greatly reduce the calculation time, thus improving the classification accuracy. In [8], SVM is applied to predict global solar radiation to study the future feasibility of solar energy potential. Through the experiments of monthly, seasonal and annual meteorological data, the importance of SVM prediction model as a qualified strategy for short-term and long-term prediction of solar energy prospect in this research area is determined. In [9], the author applies SVM to the spatial prediction of landslide disaster. Using aerial photos, satellite images and field surveys, a landslide inventory map with 282 landslide locations was determined. 70% of landslides are used as training data sets and the rest are used as verification data sets. The results show that the prediction model has high accuracy. In [10], the author applies SVM to early fault diagnosis of bearing, and trains feature fault recognition through SVM. The results show that this method is more effective than the traditional method in the feature extraction and early fault diagnosis of weak impulse signals. Considering the advantages of SVM, we consider the application of SVM in the extraction of effective information of dynamic logistics big data.

In this paper, aiming at the problems of high difficulty in information extraction of dynamic logistics big data and low accuracy of traditional methods, an effective information extraction algorithm based on SVM is proposed. SVM is used to extract effective features to improve the quality of data mining process, in order to provide a reference for the related big data information extraction.

2 Method

2.1 Big Data Overview

In the era of big data, data is huge. Mining the data and extracting the effective information from it will help the decision-making of enterprise managers and promote the healthy development of the company. One of the goals of big data analysis is to find the hidden pattern hidden in a large amount of data. Big data has the characteristics of low value density because of its different characteristics and huge quantity. In addition, there are too many invalid data in big data, so eliminating invalid data is also a checkpoint that big data analysis must go through. Logistics big data shows a dynamic trend of change, so in the process of information extraction, we should consider the speed requirements of data processing.

2.2   Information Feature Preprocessing

Considering the unstructured characteristics of logistics dynamic big data, the existence of this unstructured characteristic makes a lot of trouble for the subsequent data mining process. The preprocessing of information features is considered to ensure the effectiveness of later information. There will be a lot of noise and outliers in the logistics data. Here we deal with many outliers by feature dimension reduction. In data preprocessing, the main operations are cleaning data, data specification and transformation. Set the variable starting point of logistics data to t_0, and the valid information is distributed in i-th layer. Then the return status $x_0(t_k)(k = 1, 2, \cdots, i)$ has the following expression:

$$Y = AY + B[f(Y) + u] \tag{1}$$

The set S_K is defined as a bit string B_K with m bits. The initial values are all set to 0. At this time, after map reduce processing steps, the effective data and invalid data will be separated, which can help to extract useful information more conveniently. In data dimensionality reduction, principal component analysis is mainly used, and then the data variables are measured by variance. If the value of var (F_i) is larger, the more information F_i carries.

2.3   Effective Information Extraction Algorithm of Dynamic Logistics Big Data Based on SVM

After data preprocessing, data extraction begins. Firstly, the nonlinear function of the distance between the sample point and the center of the class is used to reduce the influence on the noise point and the isolated point. Then select the super parameters in SVM, and use SVM method to get an approximate support vector regression function for the preprocessed correlation dimension features. Considering the unilateral weighted fuzzy support vector machine model used to solve the two classification problem, the data set will contain the regression function. Finally, we need to define the data information. At this time, we must pay attention to the principle of information specification and design a variety of filters to filter the information. In this way, we can realize the effective mining of information characteristics based on SVM algorithm, and then improve the efficiency of information mining.

Suppose the sensitive parameter of SVM algorithm is $\{S_j, j = 1, 2, \cdots, N - 1\}$, and the shortest path node of this parameter is N_j. The distance formula between them can be expressed as follows:

$$d_{\min} = \min\{d_j\} \ (0 < j < N - 1) \tag{2}$$

3 Experiment

In order to verify the effectiveness of the information extraction algorithm of logistics big data based on support vector machine, this paper designs experiments to verify. In this study, the PC system used is win 10, and the running memory is 8 GB. The simulation experiment is completed on MATLAB 2014b. In the process of experiment, considering that the model and algorithm are adjusted, the dynamic logistics model is constructed. The model is built by the sensor feed-forward control and the reverse control device. When the sensor feed-forward control device collects data, it needs to design a continuous power supply device. According to the real-time discrete point data collected by the monitoring instrument, in the filter design of dynamic logistics big data, support vector machine data row number is 50 * 1000 = 20000, column number is 50. There are three kinds of data collected in this paper: standard data set, feature set and redundant noise data set.

4 Results

Result 1: data extraction process based on SVM
In the effective information extraction method of dynamic logistics big data based on support vector machine proposed in this paper, the first is to select data samples. The next step is information filtering and information filtering. After filtering out invalid information, useful information can be obtained. Finally, the information extraction is carried out. Through the above steps, the data extraction process based on SVM is completed. The flow chart of this process is shown in Fig. 1.

Result 2: Comparison of data information extraction effects of different algorithms
In order to verify the effectiveness of the data information extraction method proposed in this paper, the SVM based dynamic logistics big data information extraction algo-rithm proposed in this paper is compared with ANN and PSO algorithms. The com-parison index is F-measure. It is necessary to explain that F-measure is a weighted harmonic average of precision and recall, and it is a common evaluation standard in the field of information retrieval. It is often used to evaluate the quality of classification models. Through the measurement of F-measure, the comparative experiment is carried out, and the experimental results are shown in Fig. 2.

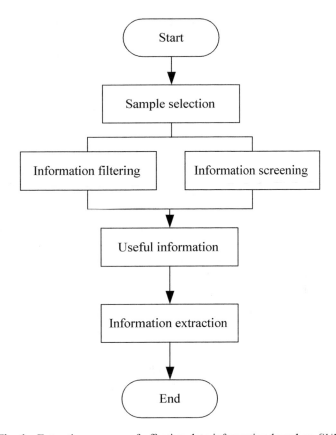

Fig. 1. Extraction process of effective data information based on SVM

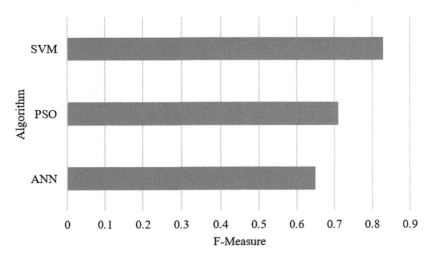

Fig. 2. F-measure evaluation of different algorithms

As can be seen from Fig. 2, the SVM based information extraction method proposed in this paper achieves the best results in the measurement of F-measure indicators, with an accuracy of about 82.3%. This shows that the method proposed in this paper has obvious effect on the effective information extraction of logistics big data.

5   Conclusion

Dynamic logistics big data contains a lot of valuable information, which is of great practical significance to the mining of logistics data. In reality, the data is complex and diverse. When extracting big data information, we should not only be able to mine the data, but also be able to accurately identify the data. In this paper, aiming at the problems of high difficulty in information extraction of dynamic logistics big data and low accuracy of traditional methods, an effective information extraction algorithm based on SVM is proposed. The experimental results show that the method presented in this paper has high accuracy. It is hoped that the research in this paper can provide a reference for the extraction of big data information.

References

1. Wei, Y.: Analysis of the mining and application of logistics information in the era of big data. Comput. Knowl. Technol. **13**(18), 1–2 (2017)
2. Yu, X.: Analysis of the application of big data technology in cold chain logistics. Logist. Eng. Manag. **40**(288(6)), 104–1057 (2018)
3. Zhong, R.Y., Lan, S., Xu, C., Dai, Q., Huang, G.Q.: Visualization of RFID-enabled shopfloor logistics Big Data in Cloud Manufacturing. Int. J. Adv. Manuf. Technol. **84**(1–4), 5–16 (2016)
4. Wang, G., Gunasekaran, A., Ngai, E.W., Papadopoulos, T.: Big data analytics in logistics and supply chain management: certain investigations for research and applications. Int. J. Prod. Econ. **176**, 98–110 (2016)
5. Zhong, R.Y., Xu, C., Chen, C., Huang, G.Q.: Big data analytics for physical internet-based intelligent manufacturing shop floors. Int. J. Prod. Res. **55**(9), 2610–2621 (2017)
6. Liu, C., Ji, W., AbouRizk, S.M., Siu, M.F.F.: Equipment logistics performance measurement using data-driven social network analysis. J. Constr. Eng. Manag. **145**(5), 04019033 (2019)
7. Shen, L., Chen, H., Yu, Z., Kang, W., Zhang, B., Li, H., Yang, B., Liu, D.: Evolving support vector machines using fruit fly optimization for medical data classification. Knowl.-Based Syst. **96**, 61–75 (2016)
8. Deo, R.C., Wen, X., Qi, F.: A wavelet-coupled support vector machine model for forecasting global incident solar radiation using limited meteorological dataset. Appl. Energy **168**, 568–593 (2016)
9. Hong, H., Pradhan, B., Jebur, M.N., Bui, D.T., Xu, C., Akgun, A.: Spatial prediction of landslide hazard at the Luxi area (China) using support vector machines. Environ. Earth Sci. **75**(1), 40 (2016)
10. Liu, R., Yang, B., Zhang, X., Wang, S., Chen, X.: Time-frequency atoms-driven support vector machine method for bearings incipient fault diagnosis. Mech. Syst. Signal Process. **75**, 345–370 (2016)

Brand Shaping Strategy of Characteristic Towns Under the Background of "Internet +" Based on Rooted Culture

Yan Li[✉], Yangyang Deng, Chunya Li, and Yang Yang

Business School of Nantong Institute of Technology, Nantong, Jiangsu, China
muziliyan@163.com

Abstract. The construction of small towns with Chinese characteristics is in full swing and has achieved initial results. How to highlight the characteristics of characteristic towns has become the key feature and difficulty of brand building of characteristic towns. We should take advantage of the "Internet +" as the sharp weapon to subvert the traditional industries. From the perspective of embeddedness, this paper tries to explain the three forms of rooted culture such as natural genetic genes, social capital and market demand preferences, recognizing the significance of rooted culture in shaping the characteristics of distinctive town brands, and exploring the characteristic town brand strategy based on rooted culture in the context of "Internet +". In addition, rooted in the local area, do a good job of featured articles, implement differentiated development strategy, improve the image recognition system (CIS), use the "Internet +" micro economic means to carry out the dissemination of the characteristic town brand; Build and improve the infrastructure of the characteristic town; innovative the brand shaping subject of characteristic Town; take full advantage of the "Internet +" era to enhance the brand elements.

Keywords: Internet + · Rooted culture · Characteristic town · Brand building

1 Introduction

Since 2002, Urbanization in China has developed at an average rate of 1.39%. Besides the contribution of big cities, the role of small towns cannot be ignored. It is worth pondering that some small towns have begun to find a unique way to achieve a win-win situation of uranization, culture and industrial modernization which is worthy of our study and discussion. At present, Zhejiang Province, with its profound historical accumulation, abundant natural resources and the innovative spirit of its predecessors, has built a number of small towns with distinctive industries, beautiful landscapes, cultural heritage, community functions and integration, such as Yunxi Town of West Lake, Zhuji Socks Town, Longquan Celadon Town, Ninghai Smart Car Town. All of these are recognized as provincial demonstration towns.

However, it is worth pondering that the urbanization of small towns is developing at an alarming speed at this stage, but the shaping of characteristic brands have been neglected in the process of development. Some small towns neglect their own characteristics, blindly follow the trend to imitate big cities, and demolish large-scale

© Springer Nature Singapore Pte Ltd. 2020
M. Atiquzzaman et al. (Eds.): BDCPS 2019, AISC 1117, pp. 951–958, 2020.
https://doi.org/10.1007/978-981-15-2568-1_130

construction to build complex, central business district (CBD), and these almost identical towns have been use the model prevails, abandon the local natural endowment, ignore its own history, culture, historical buildings… The town lost its vitality and charm, and its immense intangible value was put to the fire. How to excavate the characteristics of Characteristic Towns in the context of "Internet +" and build competitive brands around characteristics is a problem that needs to be explored urgently. From the perspective of embeddedness, this paper explores the shaping strategies of characteristic town brand to promote the sustainable and healthy development of characteristic towns.

2   Overview of Grounding Theory

2.1   The Concept of Grounding

Karl Polanyi, an economist from Hungary, first put forward the concept of grounding in the book 《The Great Transformation》 in the mid-20th century, which is also called embedding type [1, 2]. Rooted in the local resources or industrial base, through the inherent, basic, long-term power to affect the growth or decline of local industries [3]. Because of the differences in natural endowment, geographical location, social capital and so on, there will be great differences in local industries and forms of expression. When the town is rooted in the local resources endowment, pays attention to the regional characteristics, melts into the local soil and water everyone, and takes over the land gas, it can be said that it has endogenous forces and basic support. With the traction of market and policy and other sources, the vigorous development of the town is imminent.

2.2   Expressions of Rooting

According to the division of natural productivity and social productivity based on classical economic principles, it can be understood that grounding performance is as follows: natural endowment gene, social capital base and market demand preference.

2.2.1   Natural Endowment Gene

In general, the development of industry is related to the difference of natural resources. Introducing congenital factors into the industrial field as the primary factor of production can clearly understand the foundation of the industrial system, as well as the various links faced, such as favorable and unfavorable conditions, support and restriction conditions, rich and shortage conditions. This first natural factor mainly includes two aspects: one is location; the other is factor endowment or unique natural conditions (quantity and quality). The former provides spatial basis for industry, while the latter provides material conditions for industry.

"Factor endowment or unique natural conditions" refers to a series of factors related to the region, such as the supply level of production factors and the abundance of natural resources, which can be regarded as the factors with comparative advantages in location. They decide whether to accept or not, or whether they can support the development of a leading industry (system) and the exertion of its ability for a long time. For elements, elements of different nature will have different influence over time. Some exhausting factors play a very important role at the beginning of the industrial development. With the continuous development of the industry, the comparative advantage of these resources will gradually weaken. Most resource-based cities are relying on a certain type of energy and mineral resources to construct the industrial system and the characteristics of this resource economy are obvious [4].

2.2.2   Social Capital Base

Social capital refers to the state and characteristics of the close relationship between social subjects (individuals, groups, organizations, regions, countries, societies, etc.), which is mainly manifested in social networks, trust, norms, consensus, authority and social morality (Granovett 1985). From the connotation of this concept, we can see the rooting of the region. In social capital, labor resources and government orientation are two important contents. As for labor resources, it includes the quality of workers, the long-term accumulation of labor skills, the organization and scale of labor force, the flow and renewal of labor force, etc. These contents are formed in the long-term history of specific areas, and are the basic factors affecting whether an industry can "implant", "take root" and "sustain development" in one place. The government's actions and policies, including those of support, restriction and even prohibition, play a vital role in the development, growth and growth of the industry. Generally speaking, industries that are compatible with the government's Macro-Objectives and policies can win a more relaxed and favorable development environment, thereby gainning more support and operating at a lower cost; and industries that are contrary to government policies and principles are less supportive, more expensive and more restricted or banned [5]. Therefore, in the process of choosing the leading industry, the root of the relationship with social capital is also the factor that must be considered.

2.2.3   Market Demand Preference

Market demand is directly related to the formation and development of a local productivity. The development of an industry must depend on its demand and preference in the market on the basis of the first essential factor. This requires that the choice of leading industry should be adapted to the market demand, and the production capacity should be allocated according to consumers' preferences in a certain segment or level of the market, so as to form specialized production, large-scale production and characteristic production, so as to play the role of leading industry and promote the development of regional economy. The market includes consumer goods market, means of production market, labor market, financial market, technology market, real estate market, property rights trading market and information market. The system also involves market mechanism, market rules and market service intermediaries, which constitute the grounding of local market preferences.

In addition to the above three strong roots, grounding also has some important characteristics, such as network characteristics, dynamic expansion and two-way interaction characteristics, which should not be ignored in the process of leading industry selection. In practical work, it is the key to the success of leading industry construction whether we can sort out the industrial development foundation and context of a place clearly.

3 Understanding the Significance of Embedded Culture to the Brand-Building of Characteristic Towns

In July 2016, the three ministries and commissions of the state jointly issued the Notice on Developing the Cultivation of Characteristic Towns, which proposed that the cultivation of Characteristic Towns should adhere to the prominent characteristics and prevent a rush to the side of thousands of towns. "Characteristic" is rooted, so we can say that it is very important to understand the rooted culture for shaping the brand of characteristic towns.

3.1 Differential Orientation of Characteristic Towns

At present, some characteristic towns ignore their own conditions and see other big cities create "CBD" successfully. They follow the trend of those big cities to develop constructions. Finally, the local economy is dragged down by such industries, resulting in industrial convergence and waste of resources. Rooted in the local natural endowment, social capital and market demand, the study of the rooting of characteristic towns can differentiate positioning, avoid similarities, and cultivate the unique personality of the brand of characteristic towns.

3.2 Avoid Acclimatization of Characteristic Towns

Understand the rooted culture, sort out the context of local resources, economic development and industry, avoid investing resources, manpower and material resources in industries that are not satisfied with the water and soil or are not grounded in gas, stop losses in time, and let those industries that are suitable for rooting in local areas take root, germinate and thrive.

3.3 Sustainable Development of Characteristic Towns

The study of grounding provides a new way of thinking and direction for the development and growth of characteristic towns. Understanding the rooted culture can provide basic and deep support for sustainable development for the development of characteristic towns, avoid short-term behavior, and make the style show lose its soil for growth, combining with local natural endowments and historical culture. Let the characteristic towns go further and develop continuously.

4 Brand Shaping Strategy of Characteristic Towns Under the Background of "Internet +" Based on Rooted Culture

4.1 Grounded in the Local Area, Do a Good Job of Characteristic Articles, and Implement the Strategy of Differentiated Development

The brand positioning of characteristic towns needs to be rooted in the town's natural endowment, cultural connotation and characteristic industries [6]. It should be positioned from the aspects of industry positioning, development direction, development theme, development goals, core functions, and tourist market and development strategy. From the list of the first and second batches of national characteristic towns published, we can see that the construction of characteristic towns focuses on the word "special" [7]. Highlighting the existing characteristics and tapping the potential characteristics are the right way for the development of characteristic towns [8]. For example, the carving industry of Yang Ping Town in the city of Yixian, the special industry with the core of Hong Mao medicinal liquor as the carving industry, and the characteristics of Hakka town in the town of Yalong, and the special natural resources of the ceramic culture and the market of the Yao town in the city. Hekou Town is characterized by commercial culture, ancient street style and historical relics. Local governments should play the card of "history and culture" and dig the historical position of Hekou Town as a political, economic and cultural town. The difference between the characteristic town and the traditional town lies in its unique culture, ecology and industrial resources. If we want to develop, the first thing we have to do is to change the pattern of imitation and cottage in the past, grasp the lifeblood of our own characteristics and take the road of differentiation. Humanistic differences, environmental resources differences and specialized industrial chains are the development direction of characteristic towns. As the first batch of Characteristic Towns in China, Gubei Water Town in Miyun District, Beijing, has achieved great success by attaching great importance to the two characteristic elements of "Great Wall" and "Water Township", supplemented by local folk culture.

4.2 Improving the Image Recognition System (CIS)

Emblem logo is a symbol of a place or an organization, which can best represent local characteristics. The essence of characteristic towns is described intuitively by the way of brand visual image design, which is an important content of shaping brand towns from the visual angle. The visual identity design of the brand featured town can give the town a deeper cultural connotation and enhance the attraction of the town to the potential tourists in the surrounding cities. The visual image design of characteristic towns, in fact, is to give new implications in the continuous improvement of the image of small towns. It is a supplement and packaging of the content of the visual image of small towns. It can refine the visual elements of small towns and strengthen the visual symbols of small towns, which can enhance the cognitive memory of the brand of small towns. In addition, it is also in the utmost care for the emotional and psychological needs of the audience in the town, to meet the individual pursuit of the audience in the daily life of the town and to visit the rest and vacation. Setting up the visual image of the brand town is the key content of the image design that accords with the town's image. The brand visual image design of the characteristic town is the core of constructing the brand name and differentiated town image card, and is an important thrust in the construction and development of the characteristic town [9].

4.3 Using the "Internet +" Micro Economic Means to Spread the Characteristic Town Brand

With the development of information technology, the propaganda work has become simple, but at the same time, new difficulties have arisen. On the one hand, with the emergence and rise of new media, the development of communication technology has expanded the ways and channels of propaganda; on the other hand, the development of communication technology has expanded the ways and channels of propaganda. Moreover, the increasing channels of information dissemination also means that people receive more information. People receive a lot of spam information, which leads to the related propaganda can not attract people's attention. At this time, we need to innovate the propaganda work. Special towns should attach importance to such platforms as Weibo, Weixin official platform, Weixin applet and so on, so as to open up a way for the public to understand the town. For example, the establishment of the town's own website, the opening of official Weibo public number and Weixin public number, etc., provide the public with opportunities and ways to understand the town. In the era of "Internet+", characteristic towns should also attach importance to the use of mobile marketing. After establishing their own official platform of Weibo and Weixin, it is better to authenticate and enhance the credibility of the account. At the same time, it conforms to the habit of fragmented reading of the public, and tries to use typesetting tools to beautify the pushed articles, improve the interest, practicability and readability of the pushed messages. Towns should make good use of live broadcasting, self-media, community marketing and other means to invite industry experts to recommend small towns. Characteristic towns strive to provide good service experience for talents and tourists and promote their active sharing. By organizing entrepreneurship activities, introducing creative space, improving the town's exposure, conveying the characteristics of the town, enhancing the brand and influence of the town. Nowadays, society has entered an era of attention-oriented economy and social communication. A microfilm, an advertisement and a propaganda slogan can be disseminated through microblog, micro-mail and other platforms to enhance the popularity of small towns.

4.4 Constructing and Perfecting the Infrastructure of Characteristic Towns

At the same time, we should strengthen the construction and improvement of infrastructure, strengthen the protection of historic buildings, promote the renovation of town appearance, improve the reception service level, such as transportation, and form the word-of-mouth effect. Where is the "special" of a characteristic town and how to locate its brand play an important role in the competition between regions under the upsurge of construction of a characteristic town. At the beginning of brand building of characteristic towns, brand positioning and shaping can be based on the principle of rooted culture and taking the lead. Only in this way we can ensure the continuity of the brand of small towns, shape the sense of community that drives the residents to participate spontaneously, complete the industrial transformation and upgrading and create industrial characteristics. Only by fully studying the five characteristic resources of local people, culture, land, property and scenery, extracting the key spiritual theme and representative cultural symbols of small towns in order to establish an effective brand image of small towns with characteristics and convey the local cultural connotation and brand influence to the society and the public.

4.5   Innovative the Brand Shaping Subject of Characteristic Town

Statistical characteristics of small town brand path need the participation of stake-holders and benign interaction. At present, the shaping of characteristic small town brand is mainly based on three main bodies: first, mainly by enterprises or investors, second, collective organizations or management agencies of township government, third, other operators or contractors, as detailed in Table 1 [10].

Table 1. Brand shaping subject of characteristic town

	Investment	Brand shaping subject	Brand ownership	Brand maintenance subject
Path one	Enterprises or investors	To make clear	To make clear	To make clear
Path two	Village-level collective organizations of township governments	Vague	It depends on the circumstances	It depends on the circumstances.
Path three	Contracting operation contract or franchisers	Vague	Vague	Vague

In the era of brand competition, it is a general trend for consumers to participate in brand building of characteristic towns, and the rapid development of science and technology makes it possible for consumers to participate. Fournier put forward the theory of brand relationship, constructed a model, and divided the relationship between consumers and brands into four levels: consumer-product relationship, consumer-brand relationship, consumer-consumer relationship, consumer-enterprise relationship [11]. This theory is helpful to understand and guide consumers and characteristics. It provides a solid theoretical foundation for the brand building relationship of the characteristic towns. Consumers are indispensable in the process of brand building of characteristic towns. They are brand identification, participation, shaping of holder, experiencer and shaper. It can be said that there is no characteristic town brand without consumers.

4.6   Make Full Use of the Wisdom and Technology in the Internet + Era to Enhance the Brand Elements

The characteristic town enterprises should make full use of the advanced technologies such as Internet, big data, cloud computing, smart city and other "Internet +" era, combine with the town industry, enhance the independent innovation ability of the town, solve the core technical problems, create their own brand, and promote the industrial upgrading. For example, Haining leather town and Tongxiang sweater town can take the road of "Internet plus design" to build a famous brand belonging to the town itself, or take the road of "Internet + advanced customization + personalized gift platform", so as to enhance the profit margins of small town products. Small towns should also take advantage of the "Internet +" era of intelligence and technology, pay attention to docking with universities and research institutes, research and develop core technologies, and enhance the R & D capability of small towns [12].

Acknowledgements. This work was supported by the 13th Five-Year Plan of Jiangsu Province "Key Construction Discipline Project of Business Administration Level 1" Project number: SJY201609.

This work was supported by the Philosophy and Social Science Fund of Education Department of Jiangsu Province (Project number: 2018SJA1282).

References

1. Fei, H.: Transformation and development of characteristic towns in China from the perspective of embeddedness. J. Chin. Med. (4), 89–93 (2018)
2. Polanyi, K.: The economy as instituted process. In: Le Clair, E., Schneider, H. (eds.) Economic Anthropology. Holt, Rinehart and Winston, New York (1968)
3. Fan, B.: Research on the construction mechanism of small towns with Chinese characteristics from the perspective of embeddedness theory. Sports Sci. (1), 84–89 (2018)
4. Shao, S., Yang, L.: Abundant natural resources and resource industries depend on China's regional economic development. Managing World (9), 26–44 (2010)
5. Fu, X., Jiang, Y.: Based on the perspective of embeddedness, the development mode of Characteristic Towns in China is discussed. China Soft Sci. (08), 102–111 (2017)
6. Chen, D.: Research on brand image orientation of characteristic towns based on contextualism—take Guluba international research town as an example. Art Technol. **30** (11), 15–16 (2017)
7. Sheng, S.H., Zhang, W.: Characteristic town: an industrial spatial organization form. Zhejiang Soc. Sci. (3), 36–38 (2016)
8. Li, Y., Yang, Y.: Research on brand formation mechanism of characteristic towns based on SEM. Market Weekly (12), 84–87+96 (2018)
9. Liu, Z.: Brand-building strategies for tourist towns with characteristic: a case study of Hekou town. Tourist Surv. (Second Half Month) (8), 75–76 (2016)
10. Zhang, Y.: Research on brand building and dissemination of sports featured towns from the perspective of brand relations. J. Jilin Inst. Phys. Educ. (8), 7–12 (2019)
11. Fournier, S.: Consumer and their brands: developing relationship theory in consumer research. J. Consum. Res. **2**(4), 3 (1998)
12. Zhang, X.X.: Research on the construction of "sports town" in the Internet + perspective. J. Soc. Sci. **31**(04), 18–22 (2017)

Application Research of Big Data Technology in Cross-Border E-Commerce Store Operation

Zhitan Feng[✉]

School of Commercial, Nantong Institute of Technology,
Nantong 226002, Jiangsu, China
Fengzhtan@126.com

Abstract. The development of the Internet has brought rapid development of e-commerce in China, while traditional foreign trade transactions have encountered new opportunities and challenges. Cross-border e-commerce has flourished in China. This model changes the product sales model and sales targets. More extensive, involving more technology and people. However, in the process of cross-border e-commerce development, enterprise stores are faced with problems such as product selection and construction, attracting traffic, store maintenance and cross-border logistics and distribution. Based on the analysis of the problem, this paper deals with various aspects from the perspective of big data technology. In the near future, big data technology will be more deeply applied in the cross-border field. This kind of application will face greater pressure on enterprises. As long as the data analysis is continuously improved and the essence of data analysis is achieved, accurate customer sales can be achieved. In order to ensure the long-term development of the store.

Keywords: Big data · Big data technology application · Cross-border e-commerce · Store operation

1 Introduction

Cross-border e-commerce refers to a transnational commercial trade activity in which merchants and consumers use the e-commerce platform to transport goods across different borders to consumers for commodity trading and payment settlement. Cross-border e-commerce mainly includes cross-border business-to-business (B2B) and business-to-consumer (B2C). Cross-border e-commerce business processes include cross-border goods transactions, that is, cross-border transactions, cross-border cargo transportation, that is, cross-border logistics, cross-border payment of goods price transportation, that is, cross-border payment, and finally the after-sales service of goods. These four parts after the sale. Cross-border electricity business calculation is an extension of foreign trade business. It is characterized by fast delivery, small amount and high frequency. In a narrow sense, cross-border e-commerce only includes business-to-consumer (B2C), that is, only cross-border retail trade, or cross-border e-commerce platform to trade goods with consumers in different countries, usually using international express postal parcels. Or third-party logistics or other cross-border logistics models to transport goods to consumers. In the context of

M. Atiquzzaman et al. (Eds.): BDCPS 2019, AISC 1117, pp. 959–964, 2020.
https://doi.org/10.1007/978-981-15-2568-1_131

"Internet + international trade", cross-border e-commerce has risen with the development of e-commerce, which has had a violent impact on China's traditional foreign trade. Traditional foreign trade is limited by time and space, while exporting cross-border electronics is more flexible. Big data technology needs to be based on certain goals in the cross-border e-commerce field. By extracting relevant data from the cross-border e-commerce model, and collecting, sorting, analyzing, summarizing and researching its problems and development trends based on the data, Provide a foundation for better product applications in the future [1, 2]. Big data technologies cover technologies and tools in the areas of cloud computing, data sharing, and analytics.

2   The Main Problems in the Operation of Cross-Border E-Commerce Enterprises in China

2.1   Selection Issues

The development of cross-border e-commerce provides a good platform and opportunity for the export of products of Chinese enterprises. Some enterprise products are very good, but the analysis of good competitive categories before they are put on shelves often leads to the unpopular category of hard-to-order orders, which affects sales and performance [4, 6, 7]. As a company, how to choose which products to sell according to different platforms and different product characteristics, and even if the productive enterprises sell their own platforms, which should be placed in the "Blue Ocean" or "Red Sea" areas. These problems are the primary problems that plague cross-border e-commerce companies.

2.2   Product Shelves, Traffic Problems

After the product is put on the shelf, the flow and exposure become more important and an important factor in verifying the effectiveness of the product. As long as there is sufficient traffic, a certain conversion rate can be formed. In addition to natural traffic, these traffic needs to be obtained through other channels, all of which require analysis of factors that can affect traffic. FBA, keywords, customer characteristics, prices, advertising, etc. all have an impact on cross-border e-commerce traffic.

2.3   Store Maintenance Issues

When a series of products are put on the shelves, it is necessary to face how to carry out store maintenance. The quality of store maintenance determines the problem of traffic introduction and re-transformation of sales. Which products are exploding and rising? What are the small sums, how to establish a link between them, and achieve a balance between sales and revenue. This question involves optimization and analysis of pages, content, information, and videos and images.

2.4   Logistics Issues

After several years of development, cross-border e-commerce has formed a large scale, but there is no effective coordination of logistics supporting development, especially in the central and western regions of China. Cross-border logistics cost is an important factor restricting the development of e-commerce. Choosing an efficient logistics path directly affects the cost and customer experience of cross-border logistics [10].

3   Research on the Application of Big Data Technology in Cross-Border E-Commerce

In the application of cross-border e-commerce, big data technology analyzes data on cross-border platforms and data mining to find out various types of data that affect store operations, thus providing a basis for store operation decision-making and analysis.

Big data technology aims to extract weight through data analysis, processing and mining [8].

3.1   Application of Big Data Technology in Selected Products

Only through a large number of platform and store data analysis can we find "inspi ration" and "innovation", as operators can find better channels and verify the results through further data analysis. The selection is such a process, foreign consumption habits, aesthetic concepts and domestic differences are very different, if you still choose according to domestic standards, this result may not achieve your expected results; even flow cannot be produced. Therefore, before the selection, it is necessary to analyze and evaluate from the perspective of competitors and customers. From the perspective of competitors, you can dig out which products are sold in the data of each platform, and what characteristics of this product are used as a reference to choose or improve your product. From the customer's point of view, we must observe the needs and ideas from the customer's point of view, find out the rules through the data, and provide the basis for product production. For example, we can view the customer's evaluation of the product and even obtain the customer evaluation data through the channels such as FACEBOOK and WeChat. Data analysis and mining, deleting fake data and extracting useful information. In the selection process, you must first determine the target of the selection, and then select the products through the platform BEST sellers, or you can use a certain category to dig deep, copy excellent stores, analyze product prices and social media and other aspects of data. Selection. Through the category search, the data such as the product racking quantity is obtained, and the market state in which the product is located is evaluated, whether it is a completely competitive market or a monopoly market.

3.2   Application of Big Data Technology in Product Launching Traffic

The quality of the shelves is mainly analyzed by the traffic indicator. In the process of drainage, the store first needs to identify and analyze the customers to find the target customer group and potential customers. Then, according to the characteristics of the target customers, such as age, nationality, cultural habits, etc., the data that has an impact on the flow rate is mined. By obtaining data from similar stores, analyzing related keywords, prices and other factors, a data analysis model is established, which involves keywords, prices, advertisements, and the like. A bad keyword, even if the advertising is invested more, it is difficult to form an effective order, and the shelf product needs to calculate the weight of each keyword. For example, as a good keyword, you should evaluate and brush from the number of query pages, whether there are high reviews, the number of high-quality listings, and the product property settings with a lot of exposure. Selected. By excavating the behavior of consumers, we construct customer portraits, record and analyze the purchase purpose and habits of consumption, as well as the browsing, price comparison and purchase, and establish a customer consumption model. Through the model, the corresponding increase can be Attract customers' key points to increase traffic and increase conversion rates. The main goal of store introduction of traffic is to capture customers, achieve purchase conversion, achieve accurate marketing, and finally reach a deal, and provide suitable products or services to foreign consumers through cross-border platforms.

3.3   Big Data Technology Application in Store Maintenance

Through data analysis, we analyze a certain type of products on a cross-border platform for explosions, ascending funds and small amounts. There have been a large number of orders for the explosions, and there are considerable benefits; the sales of the increased products may become the explosion of the future stores. Once the flow of these products is well transformed, the benefits will be significantly improved; the number of products is huge, but the amount is small, and the effort on conversion rate may have little effect. No matter what kind of product traffic conversion, you need to set the corresponding conversion target first, and then make more articles on store maintenance. The image data of the product is analyzed, and the customer is more willing to look at the picture than the text, which requires that the product related data be displayed as much as possible on the picture. At the same time, this store maintenance also introduces the introduction of product design, user Q&A, page, price and video optimization.

Taking the Amazon platform as an example, Amazon uses the A9 algorithm to ensure that consumers search for "the products they really want to buy" the fastest and most accurately, and the ultimate goal is to maximize the buyer's benefit under the premise of customer satisfaction. Based on this, the most relevant index correlation, conversion rate and customer retention rate correlation in the A9 algorithm are derived from the consistency of the search results and the customer's true purchase intention. The client comes from the words used by the customer to search, that is, the product keywords we usually call; and the seller's content, which can correspond to the product keywords searched by the customer, is mainly reflected in the product title, the five elements, the product description and the search. In the Tern keyword. Of course, for some products, product attributes, brand names, technical parameters, etc., to a certain extent, are also the elements of the A9 algorithm to identify product relevance. If correlation is the basis of the match, the conversion rate is a test of the match.

Therefore, the conversion rate of the A9 algorithm occupies a considerable proportion, and the conversion rate directly affects the display result of a listing evaluation by the A9 algorithm. According to experience, on the Amazon platform, the factors of shadow conversion rate mainly include sales, ranking, buyer reviews (review quantity and star rating) product images (especially the main image) and price users determine whether a product meets its needs through the above factors. And whether the quality can reach its good value. If it is, the purchase is easy to form, the purchase rate is high, and the A9 algorithm will increase the display weight for it later. The basis of the A9 algorithm is data. Through the collection of big data to customer information and product information, mutual configuration can be realized. The store can optimize the store conversion rate by using keyword optimization, keyword optimization, content optimization and advertisement optimization [11].

3.4 Application of Big Data in Cross-Border Logistics

Big data involves cross-border logistics distribution paths, delivery times, and distribution models for cross-border logistics [3, 5]. In China, cross-border logistics distribution postal parcels, third-party logistics, overseas warehouses, etc., but logistics and distribution models such as postal parcels have long problems in logistics and delivery, and poor customer experience [9]. These problems need to be optimized in domestic and foreign distribution routes, especially in inland cities and regions. Big data analysis is needed to achieve intra-regional or even nationwide logistics integration, thereby saving logistics costs, improving distribution efficiency, and improving customer satisfaction. For example, the Amazon platform, through the customer's address and inventory location, select the appropriate warehouse for distribution, and route optimization, which greatly saves time and cost, and the customer's experience is also high, which also brings about an increase in traffic. At the same time, the quality of distribution can be analyzed and evaluated through four aspects: logistics cost, operation efficiency, technical level and logistics quality. These four aspects need to collect data and establish corresponding data model.

4 Summary

The development of the Internet has brought rapid development of e-commerce in China, while traditional foreign trade transactions have encountered new opportunities and challenges. Cross-border e-commerce has flourished in China, and many companies have shifted from domestic platforms to cross-border platforms for product sales. The transformation of sales patterns and sales targets involves a wider range of technologies and people. However, in the process of cross-border e-commerce development, enterprise stores are faced with problems such as product selection and construction, attracting traffic, store maintenance and cross-border logistics and distribution. Based on the analysis of the problem, this paper deals with various aspects from the perspective of big data technology. In the near future, big data technology will be more deeply applied in the cross-border field. This kind of application will face greater pressure on enterprises. As long as the data analysis is continuously improved and the essence of data analysis is achieved, accurate customer sales can be achieved. In order to ensure the long-term development of the store.

Acknowledgements. Jiangsu University Philosophy and Social Science Research Fund Project "Research on Innovation Mechanism of Rural E-Commerce Development Model Based on Synergistic Effect". Project number: 2019SJA1475; First Level Key Built Discipline Projects of the Business Administration under Jiangsu Provincial "the 13th Five-Year Plan". Project number: SJY201609; Nantong Institute of Research professor and doctoral research project (201823).

References

1. Hu, L., Li, S.: The impact of cross-border e-commerce development on China's foreign trade model transformation. Bus. Econ. Res. **19**, 142–145 (2019)
2. Huang, Y.: Application of big data technology in cross-border e-commerce industry chain. Foreign Trade **07**, 75–77 (2019)
3. Gao, T.: Analysis of the construction of cross-border e-commerce supply chain platform under the background of big data. Shopp. Mall Mod. **12**, 70–71 (2019)
4. Deng, Z.: Analysis strategy of cross-border e-commerce platform selection based on big data. Spec. Econ. Zone **06**, 135–137 (2019)
5. Li, W., Zhao, R.: Research on cross-border e-commerce logistics chain optimization based on supply chain perspective. Bus. Econ. Res. **12**, 76–79 (2019)
6. Zhang, W.: Exploring the application of big data technology in cross-border e-commerce. Mod. Mark. (Late Issue) **06**, 204–205 (2019)
7. Deng, Z.: Research on the overall operation mode of cross-border e-commerce under the background of big data. Econ. Res. Guide **13**, 162–164+177 (2019)
8. Sun, Y.: Application of big data technology in cross-border e-commerce. Coop. Econ. Technol. **07**, 90–91 (2019)
9. Huang, Y., Ma, G.: The status quo, problems and countermeasures of cross-border e-commerce development in Zhejiang Province under the "New Normal". E-commerce **03**, 28–29 (2019)
10. Li, S.: Analysis of new retail import cross-border e-commerce logistics model. Coop. Econ. Technol. **03**, 78–79 (2019)
11. Ye, P.: Amazon Cross-Border E-Commerce Operations Combat. China Railway Publishing Co., Ltd (2019)

Path of the Reform of the Psychological Health Education Model in Higher Vocational Colleges in the Big Data Era

Tiantian Zhang[✉]

Jiangxi Vocational College of Mechanical and Electrical Technology,
Nangchang, China
1012924152@qq.com

Abstract. With the continuous development of science and technology, the data in the field of mental health education is also explosive growth. The traditional mode of mental health education has been unable to meet the growing practical needs, which urgently needs to change the mode of mental health education. In the era of big data, we should consider introducing big data technology into the mental health education model of Higher Vocational colleges. By analyzing the characteristics of large data, and using data mining method based on spectral clustering to mine mental health data. The results show that the application of big data technology can effectively improve the effectiveness of mental health education. Through data integration, information resources can be shared, and personalized mental health education can be provided for students, so as to explore the way to change the mode of mental health education in Higher Vocational Colleges to meet the requirements of the era of big data.

Keywords: Big data · Mental health education in Higher Vocational Colleges · Model change · Data mining

1 Introduction

With the development of economy and society, the pressure of competition in society keeps rising, which undoubtedly adds pressure to the people who struggle for life. In the era of big data, information shows explosive growth [1]. The complicated information also brings severe test to the mental health of students in Higher Vocational colleges [2]. In the era of big data, it is of great significance to overcome the shortcomings of the existing mental health education model, explore and reform the mental health education model, and improve the effectiveness of the current health education. Especially under the impetus of big data means, it provides new ideas for mental health education in Higher Vocational colleges [3].

There are many studies on mental health education. In [4], the authors point out that people with severe mental illness are increasingly turning to popular social media, including Facebook, Twitter or YouTube, to share their illness experiences or seek advice from others in similar health situations. Mental health problems are closely related to the environment. In [5], with the help of virtual reality (VR) and computer-

© Springer Nature Singapore Pte Ltd. 2020
M. Atiquzzaman et al. (Eds.): BDCPS 2019, AISC 1117, pp. 965–970, 2020.
https://doi.org/10.1007/978-981-15-2568-1_132

generated interactive environment, individuals can experience the problems they encounter repeatedly and learn how to overcome them through evidence-based psychotherapy. According to a report commissioned by the United Nations High Commissioner for Refugees, taking the mental health of the Syrian population as an example, it is pointed out that Syrians affected by the conflict may encounter a wide range of mental health problems [6]. Studies have shown that high stress, especially chronic stress, can greatly promote the development of anxiety and depression symptoms. However, researchers have also found that people with high resilience can be free from stress and thus exhibit lower levels of anxiety and depression symptoms [7, 8]. No matter where you are, people from all walks of life will more or less suffer from mental stress and lead to mental health problems. In the era of big data, big data technology may be used as a means to promote mental health. Large data technology can bring valuable information to the research field. In [9], the author uses big data analysis technology to realize strategic business value, which gives the company competitive advantage, which makes data information an important basis for company decision-making. In [10], the author applies big data analysis technology to the medical and health care industry, identifies the important value chain through the application model, and provides practical insights for managers. Big data technology extracts information from massive data to discover potential rules. In the era of big data, the reform of mental health education in higher vocational colleges can be implemented with the help of big data technology, thus promoting the improvement of mental health education.

In view of the challenges faced by traditional mental health education in the era of big data, this paper proposes to apply big data technology to mental health education in Higher Vocational Colleges in order to make up for the limitations of traditional mental health education, make complex work simple and bring practical results to mental health education.

2 Method

2.1 Overview of Big Data

The reason why the era of big data is called the era of big data is that there are huge amounts of data in this era. Big data has the following characteristics: large storage space, large amount of calculation work; abundant data sources and various formats; data generation speed is very fast, so matching data processing speed is required; low value density of big data, its value is inversely proportional to the size of the total amount of data, the more data, the lower value density. In the era of big data, mental health data are also huge.

2.2 Web Crawler Technology

In the era of big data, one of the important sources of data acquisition is through crawlers. The process of reptiles is as follows:

(1) The seed URL is used as the initial URL to join the URL queue to be crawled.
(2) Web crawling. Connect the server through the relevant protocols of the network, grab the data on the web page and save it locally.
(3) Web page storage. After crawling the web page data, the collected web page URLs, page contents and files are stored.
(4) Web page parsing. The URLs of out-of-chain pages are parsed from the collected pages, and the formats of out-of-chain URLs are transformed to get complete and standardized URLs. The URL format contained in the page is sometimes a relative path, which needs to be transformed into a uniform absolute path format and then passed to the URL de-duplication module for processing.
(5) According to the selected crawler crawling strategy and the configuration of the network crawler, it is decided whether to add the de-duplicated URL to the queue of the URL to be crawled or to end the crawling task.

2.3   Data Mining Method Based on Spectral Clustering

Spectral clustering algorithm is a data and graph correspondence algorithm, which realizes clustering by graph partition. The results of spectral clustering can obtain the global optimal partition, and have better clustering effect for non-convex distribution data. Firstly, a partition cost function is defined, and the approximate optimal solution of the minimum cost function is calculated by relaxation.

For data set $\{x_1, x_2, \cdots, x_n\}$, these points are divided into k clusters. The purpose is to make the data in the same cluster have higher similarity and the data in different clusters have higher similarity. Graphs consist of vertices and edges. In this paper, the vertex v_i of the data x_i analog graph is taken as the weight w_{ij} between vertices. The formula for calculating the weight is as follows:

$$w_{ij} = \sqrt{\sum_{i=1}^{n} (x_i - y_i)^2} \tag{1}$$

In this way, an undirected graph based on similarity is constructed. The original data clustering problem is transformed into a partition problem based on undirected graph G, which divides graph G into k subgraphs, so that the internal weight of the partitioned subgraphs is as large as possible and the weight between subgraphs is as small as possible.

Laplace matrix plays a key role in spectral clustering algorithm and contains important information. The Laplacian matrix is computed by:

$$L = D - W \tag{2}$$

In the above formula, D is the degree matrix of graph G, which is a diagonal matrix with the sum of similarity degree from vertex to vertex as the diagonal element, and W represents the similarity matrix.

3   Data

In order to obtain the relevant data of mental health education, crawler technology was used to crawl the relevant information on the web page, including students' truancy information, learning attitude, peacetime behavior and so on. The truancy information includes the beginning of truancy time, the period of truancy most, the subjects of truancy, and the classmates who truancy together. The related data of learning attitude include academic achievement, classroom performance, homework completion after class, etc. Usual behavior-related data are mainly obtained by describing the surrounding classmates, which can ensure the authenticity of data information. Through the massive data of mental health education, the valuable information is mined by using the data mining method based on spectral clustering, and the path of the transformation of mental health education mode in Higher Vocational Colleges in the era of big data is explored.

4   Results

Result 1: Integration of students' psychology and behavior under big data
At this stage, the integration of students' information in schools is not enough. Many students have many cards issued by schools. There are many students' data in different systems, but the data between different systems are not interlinked. In the exploration of mental health education model in Higher Vocational colleges, this paper proposes to build an integrated system, which integrates students' psychological test information, campus card consumption information and academic performance, and discovers valuable information through data mining. The integration system of students' psychology and behavior under big data is shown in Fig. 1.

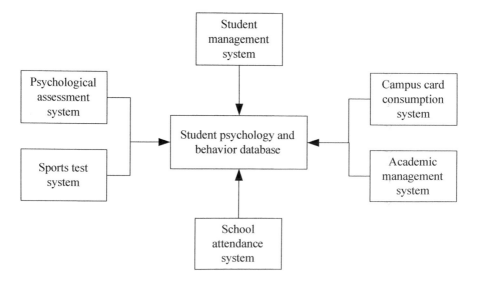

Fig. 1. Integration system of students' psychology and behavior based on big data

In the era of big data, students' psychological and behavioral data should include not only scale data, but also big data everywhere in school, life and network, including students' personal data, academic achievement data, college students' health physical test data, college students' mental health assessment data, student management data, school attendance data and network information data. The application of big data technology is to strengthen the concept of mental health work, to make use of each other's resources, to carry out curriculum reform combined with students' ideological dynamics, and to improve the management level of mental health work.

Result 2: Personalized psychological service

In view of the traditional mode of mental health education, this paper conducts a survey on the satisfaction of teachers and students in Higher Vocational colleges. At the same time, the evaluation of mental health education model combined with big data technology in teachers and students is counted. The satisfaction of teachers and students under these two modes is shown in Fig. 2.

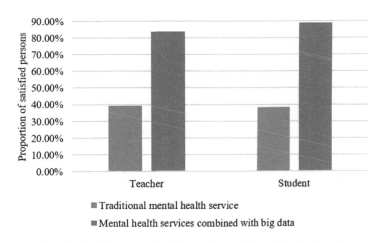

Fig. 2. Satisfaction with different forms of psychological work

As can be seen from Fig. 2, compared with the traditional mode of mental health education, the mode of mental health education combined with big data technology is more welcomed by teachers and students in Vocational colleges. This shows that in the era of big data, the reform of mental health education mode in higher vocational colleges is of practical significance. Under the influence of big data technology, teachers' workload is reduced, and students are more inclined to this mode.

In the era of big data, it is necessary to introduce big data technology into the path change of mental health education. First of all, we should integrate the data of students to make the related work more scientific and reasonable. Data related to students can better describe the situation of students themselves, making mental health work more targeted. Secondly, it is necessary to share mental health information resources with each other. Let mental health educators not because of time and space constraints make it difficult to carry out work, facilitate access to information, and lay a solid foundation for better service to students. In addition, personalized mental health services can also be provided through large data analysis.

5 Conclusion

In the era of big data, data is the source of power. At this stage, some universities have not fully realized the role of data in the era of big data. The application of big data can broaden the network information channels and provide practical health education programs. With the passage of time, driven by the era of big data, I believe that relevant staff will attach great importance to big data and give high expectations. We need to be brave in exploring changes, giving full play to the role of big data, and introducing big data technology into the work of mental health education in peacetime.

References

1. Sun, H.: Research on intelligent retailing of physical retailing in big data environment. J. Xi'an Univ. Finan. Econ. **29**(2), 41–46 (2016)
2. He, Y.: Discussion on psychological problems and educational strategies of new students in higher vocational colleges. Educ. Obs. **5**(8), 45 (2016)
3. Li, X.: The path of mental health education reform in higher vocational colleges in the era of big data. J. Zhenjiang Coll. **32**(01), 88–91 (2019)
4. Naslund, J.A., Aschbrenner, K.A., Marsch, L.A., Bartels, S.J.: The future of mental health care: peer-to-peer support and social media. Epidemiol. Psychiatr. Sci. **25**(2), 113–122 (2016)
5. Freeman, D., Reeve, S., Robinson, A., Ehlers, A., Clark, D., Spanlang, B., Slater, M.: Virtual reality in the assessment, understanding, and treatment of mental health disorders. Psychol. Med. **47**(14), 2393–2400 (2017)
6. Hassan, G., Ventevogel, P., Jefee-Bahloul, H., Barkil-Oteo, A., Kirmayer, L.J.: Mental health and psychosocial wellbeing of Syrians affected by armed conflict. Epidemiol. Psychiatr. Sci. **25**(2), 129–141 (2016)
7. Gloria, C.T., Steinhardt, M.A.: Relationships among positive emotions, coping, resilience and mental health. Stress Health **32**(2), 145–156 (2016)
8. Fang, J.: Relationship between anxiety and depression and work stress in psychiatric nursing staff. Mod. Health (2), 120 (2017)
9. Grover, V., Chiang, R.H., Liang, T.P., Zhang, D.: Creating strategic business value from big data analytics: a research framework. J. Manag. Inf. Syst. **35**(2), 388–423 (2018)
10. Wang, Y., Kung, L., Wang, W.Y.C., Cegielski, C.G.: An integrated big data analytics-enabled transformation model: application to health care. Inf. Manag. **55**(1), 64–79 (2018)

RMB Exchange Rate Prediction Based on Bayesian

Wenyuan Hu[(✉)]

Shanghai University, Shanghai 201800, China
1844270208@qq.com

Abstract. The traditional time series analysis and prediction methods do not take into account the prior information of samples and parameters, resulting in a large deviation between the prediction results and the actual data, and the Bayesian parameter estimation method can make full use of the prior information of the parameters. The variance of the estimated parameters is smaller, the estimated results are more accurate, and the predicted results are more real. In order to correctly analyze and predict the changing trend of RMB exchange rate, this paper selects exchange rate data of 995 working days from August 1, 2015 to August 30, 2019 to model the exchange rate of RMB against US dollar in time series autoregressive model. By using the MCMC method, I carry out the model parameters with Gibbs sampling estimation, which makes the prediction of the model more accurate.

Keywords: Autoregressive model · Bayesian parameter estimation · MCMC method · Gibbs sampling

1 Introduction

Since the Reform and Opening-up, the fluctuation of the exchange rate has increased, cross-border capital flows have become increasingly complex, and an orderly exchange rate system has been used to maintain macroeconomic stability. It is very important to build an orderly financial system to achieve the goal of financial sustainable development. Therefore, the correct analysis and prediction of exchange rate fluctuations is of great significance in the formulation of macro-policy by the government. However, the traditional time series analysis method for capturing the complex characteristics of time series data is not accurate, and the prediction deviation is also relatively large. In contrast, the Bayesian method based on Bayesian principle not only makes use of model information and sample data information [1], but also integrates the prior information of the unknown parameters of the model.

In 1986, Litterman applied the Bayesian method to the analysis and prediction of the time series model for the first time in his paper "Forecasting with Bayesian Vector Autoregressions-Five Years of Experience". Compared with the traditional parameter estimation method, the Bayesian time series analysis method takes into account people's empirical knowledge [2]. As a result, the prediction accuracy of the model also increases [3].

© Springer Nature Singapore Pte Ltd. 2020
M. Atiquzzaman et al. (Eds.): BDCPS 2019, AISC 1117, pp. 971–979, 2020.
https://doi.org/10.1007/978-981-15-2568-1_133

This paper selects 995 daily exchange rate data of RMB to US dollar from August 1, 2015 to August 30, 2019. First of all, the exchange rate data is used to establish the time series autoregressive model. Second, by using the MCMC method, this paper integrates the prior information of the unknown parameters of the AR model, and then carries out the model posterior parameters with Gibbs sampling estimation and establish the Bayesian time series model. Next, I predict the values in the next 10 working days based on the time series autoregressive model and the Bayesian time series model. Finally, by using the R software, I compare the prediction effects between the Bayesian time series model and the traditional time series model on the RMB exchange rate series. In conclusion, the Bayesian time series model is more precise than the AR model.

2 Bayesian Parameter Estimation of AR(p) Model

2.1 AR (P) Model

If the highest order of a system is p and the random perturbation term not related to the previous sequence, that is, $E(x_s \varepsilon_t) = 0, s < t$, and random perturbation term sequence is $\{\varepsilon_t\}$, which is a white noise sequence, then the constructed model is called the p-order autoregressive model and is denoted as the AR (P) Model:

$$y_t = \beta_0 + \beta_1 y_{t-1} + \beta_2 y_{t-2} + \cdots + \beta_p y_{t-p} + \varepsilon_t \tag{1}$$

$\{\varepsilon_t\}, (t = 1, 2, 3, \cdots)$ is the random error terms of the Eq. (1), which are independent of each other and are subject to the normal distribution, $\beta = (\beta_1, \beta_2, \cdots \beta_p)^T$ is the parameter of the AR (P) model, and p is the order of the model.

2.2 Likelihood Function

Denoted the formula (1) in the form of a matrix $Y_t = X\beta + \varepsilon_t$. Thus, the maximum likelihood function of β and σ^2 is

$$F(Y_t|\beta, \sigma^2) = (2\pi\sigma^2)^{-\frac{T}{2}} \exp\{-\frac{(Y_t - \beta X_t)^T (Y_t - \beta X_t)}{2\sigma^2}\} \tag{2}$$

2.3 Prior Distribution

Through the analysis of the first step, it is found that the prior distribution of β is normal distribution, and the prior distribution of the σ^2 is gamma distribution. The joint prior distribution of β and σ^2 is $P(\beta, \sigma^2) = P(\beta|\sigma^2)P(\sigma^2)$.

2.4   Posterior Distribution

On the basis of the bayesian formula, the joint posterior distribution of β and σ^2 is

$$H(\beta, \sigma^2 | Y_t) = \frac{F(Y_t | \beta, \sigma^2) \times P(\beta, \sigma^2)}{F(Y)} \tag{3}$$

We can also denote the Formula (3) in the following form: $H(\beta, \sigma^2 | Y_t) \propto F(Y_t | \beta, \sigma^2) \times P(\beta, \sigma^2)$.

3   Gibbs Sampling

Gibbs sampling is an algorithm for Markov Monte Carlo that is used to estimate the posterior distribution values of complex parameters [4]. $\pi(x_1, x_2, \ldots, x_n)$ is a n-dimensional probability distribution, a new sample can be obtained by rotating samples on n axes. For the transfer to any coordinate axis x_i, $P(x_1 | x_1, x_2, \ldots, x_{i-1}, x_{i+1}, \ldots, x_n)$ is the state transition probability of Markov chain, the n-1 coordinate axis is fixed and moves in a certain coordinate axis.

The specific algorithm process is as follows:

Step 1: Input the stationary distribution $\pi(x_1, x_2, \ldots, x_n)$ or the conditional probability distribution of all the corresponding features, set the threshold of the number of state transitions n_1, and the number of samples n_2;

Step 2: Random initialization of initial state values $\left(x_1^{(0)}, x_2^{(0)}, \ldots, x_n^{(0)}\right)$.

Step 3: t from 0 to $n_1 + n_2 - 1$:

(a) Samples x_1^{t+1} are obtained from conditional probability distribution $P(x_1 | x_2^{(t)}, x_3^{(t)}, \ldots, x_n^{(t)})$;

(b) Samples x_2^{t+1} are obtained from conditional probability distribution $P(x_2 | x_1^{(t+1)}, x_3^{(t)}, x_4^{(t)} \ldots, x_n^{(t)})$;

(c) Samples x_n^{t+1} are obtained from conditional probability distribution $P(x_n | x_1^{(t+1)}, x_2^{(t+1)} \ldots, x_{n-1}^{(t+1)})$;

Finally, the sample set $\left\{\left(x_1^{(n_1)}, x_2^{(n_1)}, \ldots, x_n^{(n_1)}\right), \ldots, (x_1^{(n_1+n_2-1)}, x_2^{(n_1+n_2-1)}, \ldots, x_n^{(n_1+n_2-1)}\right\}$ is the sample set corresponding to the required stationary distribution.

4   Empirical Strategy

4.1   Data

The AR model established in this paper involves a variable, that is, the exchange rate of RMB to the US dollar. The sample range is 995 working days from August 1, 2015 to August 30, 2019. The data are from the State Administration of Foreign Exchange.

4.2 Stationarity Test

Before establishing the AR model, it is necessary to judge whether the variable is stationary or not. Generally speaking, if the time series is unstable, it will lead to the phenomenon of pseudo regression, that is, there is no correlation between variables, but the goodness of fit is high and the variable coefficient is also significant, which makes the statistical test meaningless. Any non-stationary sequence is stable after the difference through appropriate order difference, then the AR model of the difference sequence can be fitted.

4.2.1 Sequence Diagram

According to the known data, drawing the time series diagram of the RMB exchange rate series from January 1, 2015 to August 30, 2019, it can be found that the data do not fluctuate around a certain average, and it is preliminarily determined that the exchange rate sequence has a certain trend [5, 6] (Fig. 1).

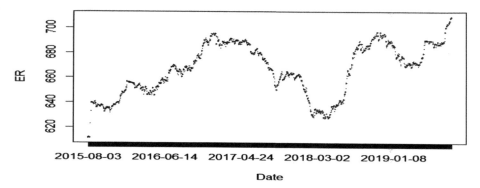

Fig. 1. Sequence diagram

4.2.2 ADF Test

Table 1. ADF test

Variable	ADF statistic	1% level	5% level	10% level	P-value
ER	−1.792	−3.436709	−2.864236	−2.568258	0.3847

Further through the ADF test, as shown in Table 1, at the significant levels of 1%, 5% and 10%, the test values of ADF were −3.436709, −2.864236 and −2.568258 were less than the significant level of −1.792, respectively. And P value 0.3847 is greater than 0.05, so the test cannot pass, the time series is non-stationary. Therefore, the first-order difference of the sequence is carried out, and the sequence diagram after the difference is as follows (Fig. 2):

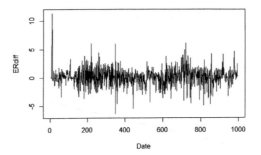

Fig. 2. Sequence diagram after the first-order difference

After the difference, the ADF unit root test is carried out, and the results are shown in Table 2. The test results show that the P value of the ADF test of RMB exchange rate is less than 0.05, at the significant level of 1%, 5%, 10%. The test values of ADF are greater than the values of ADF statistics, and the series becomes a stationary time series.

Table 2. ADF test after the first-order difference

Variable	ADF statistic	1% level	5% level	10% level	P-value
dER	−28.18587	−3.436716	−2.864239	−2.568259	0.0000

4.3   Model

The autocorrelation and partial autocorrelation graphs of the first order difference sequence are drawn as Figs. 3 and 4.

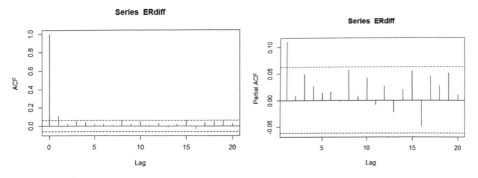

Fig. 3. ACF after the first order difference   **Fig. 4.** PACF after the first order difference

By observing the ACF and PACF diagrams of the sequence, it can be found that ACF has trailing property, while PACF is truncated after order 1. According to the minimum information criterion of model order, the AR (2) model is established as follows:

$$Y_t = 0.0978 + 0.1092Y_{t-1} + 0.0008Y_{t-2} + \varepsilon_t, t = 2, 3, \ldots, 995 \qquad (4)$$

4.4   Gibbs Sampling

Next, the MCMC estimation of the exchange rate series is analyzed. Gibbs sampling was carried out according to the established AR (2) model. Because, if, the prior distribution setting of the model accuracy parameters may be unreasonable, which will lead to the problem of the posterior distribution of the parameters. In practical application, the initial value of the parameter is accompanied by a prior distribution without information [7, 8]. In this paper, the parameter's prior distribution of the original value is in the form of normal-Gamma conjugate prior distribution, and the distribution of the parameters are as follows:

(i)   The prior distribution of the coefficient β obeys the standard normal distribution, the mean value is 0 and the variance is 1;

(ii) The prior distribution of the variance parameter σ^2 obeys the inverse Gamma distribution: $P(\sigma^2) \sim \Gamma^{-1}\left(\frac{T_0}{2}, \frac{\theta_0}{2}\right)$.

In order to make the parameters of the model reach a steady state, 15000 iterations are simulated. And then the first 4000 Gibbs pre-iterations of the orbit are discarded to reduce the turbulence of the initial value. The posterior estimates of the parameters after the iteration are shown in Table 3:

Table 3. The posterior estimates of AR (2) model

Parameters	Mean	Variance	5%	95%	P-value
Alpha	3.069761	0.8027444	3.053016	3.086507	<2.2e−16
Beta1	1.106322	0.0005054	1.105902	1.106742	<2.2e−16
Beta2	−0.110796	0.0005034	−0.1111216	−0.110377	<2.2e−16
Sigma	1.182471	0.002122	1.181610	1.183332	<2.2e−16

Table 3 shows that the P value of the t statistic of the model parameters is far less than the significant level 0.05, and the posterior mean values of the parameters Beta1, Beta2 and Sigma are 1.106322, −0.110796 and 1.182471, respectively. The 95% confidence intervals of Beta1, Beta2 and Sigma were (1.105902, 1.106742), (−0.1111216, −0.110377) and (1.1816101, 183332), respectively. The figures of the posterior parameters are as follows (Figs. 5, 6, 7 and 8):

To sum up, the MCMC estimation results of, AR (2) model under Gibbs sampling are as follows:

$$Y_t = 3.069761 + 1.106322Y_{t-1} - 0.110796Y_{t-2} + \varepsilon_t, t = 2, 3, \ldots, 995 \qquad (5)$$

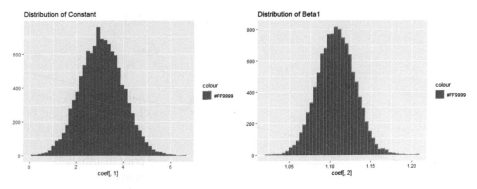

Fig. 5. Distribution of constant   **Fig. 6.** Distribution of Beta

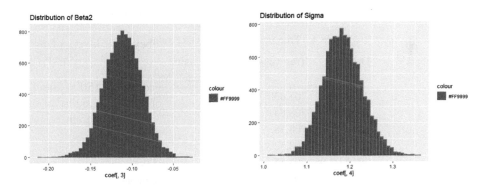

Fig. 7. Distribution of Beta2   **Fig. 8.** Distribution of Sigma

4.5 Prediction

Figure 9 shows the unified time series prediction method, the black line represents the actual values, the green line represents the predicted values, we find that the trend between the predicted values and the real values is quite different.

Figure 10 shows the prediction results of the AR (2) model after Gibbs sampling, with the red line represents the predicted exchange rate values and the black line represents the original values. We find that the two trends are roughly the same, indicating that the Bayesian parameter estimation method makes full use of sample information and the model information [9, 10]. It makes the prediction of the model more accurate.

Table 4 lists the real value of RMB exchange rate, AR (2) model forecast and the MCMC forecast value in the next 10 working days.

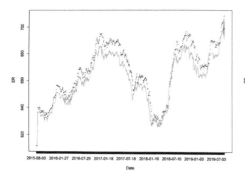

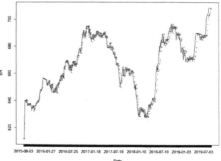

Fig. 9. Prediction of AR (2) model   **Fig. 10.** Prediction of MCMC

Table 4. True value and predictive value

Date	True value	Predictive value	
		AR (2) model	MCMC
2019-09-02	708.83	709.0065	708.4790
2019-09-03	708.84	709.2095	708.3732
2019-09-04	708.78	709.3999	708.2622
2019-09-05	708.52	709.5784	708.1447
2919-09-06	708.55	709.7457	708.0303
2019-09-09	708.51	709.9026	707.9291
2019-09-10	708.46	710.0497	707.8196
2019-09-11	708.43	710.1877	707.7307
2019-09-12	708.46	710.3170	707.6188
2019-09-16	706.57	710.4383	707.5332

5   Conclusion

In this paper, a prior distribution function of the model parameters is constructed according to the conditional likelihood function of the time series AR model, and the Bayesian estimation method is introduced into the parameters, because the Bayesian method is difficult to estimate. The MCMC method and the R software are used to estimate the RMB exchange rate series model, which makes it easier to apply the Bayesian method to the parameter estimation of the time series model, and can make up for the shortcomings of the traditional estimation methods. The conclusion is that prediction effects of the Bayesian time series model is more accurate than the traditional time series model on short-term time series.

References

1. Di Persio, L., Frigo, M.: Gibbs sampling approach to regime switching analysis of financial time series. J. Comput. Appl. Math. **300**, 43–55 (2016)
2. Dey, T., Kim, K.H., Lim, C.Y.: Bayesian time series regression with nonparametric modeling of autocorrelation. Comput. Stat. **33**(4), 1717–1731 (2018)
3. Nariswari, R., Pudjihastuti, H.: Bayesian forecasting for time series of count data. Procedia Comput. Sci. **157**, 427–435 (2019)
4. Vosseler, A., Weber, E.: Forecasting seasonal time series data: a Bayesian model averaging approach. Spring J. **33**(4), 1733–1765 (2018)
5. Das, M., Ghosh, S.K.: semBnet: a semantic Bayesian network for multivariate prediction of meteorological time series data. Pattern Recogn. Lett. **187**, 256–281 (2017)
6. Ohyver, M., Pudjihastuti, H.: ARIMA model for forecasting the price of medium quality rice to anticipate price fluctuations. Procedia Comput. Sci. **135**, 707–711 (2018)
7. Krafty, R.T., Rosen, O., Stoffer, D.S., Buysse, D.J., Hall, M.H.: Conditional spectral analysis of replicated multiple time series with application to nocturnal physiology. J. Am. Stat. Assoc. **112**(520), 1405–1416 (2017)
8. Zorzi, M.: Empirical Bayesian learning in AR graphical models. Automatica **109**, 677–692 (2019)
9. Huang, Z., Shen, Q., Wu, S.: A Gibbs sampling method to determine biomarkers for asthma. Comput. Biol. Chem. **67**, 255–259 (2017)
10. Kolm, P., Ritter, G.: On the Bayesian interpretation of Black Litterman. Eur. J. Oper. Res. **258**(2), 564–572 (2017)

Shared Smart Strollers

Rui Peng, Yanwen Xin, Yongjie Lei, Shilin Li$^{(\boxtimes)}$, and Jiayue Wang

Information School, Hunan Institute of Humanities, Science and Technology,
Loudi 41700, China
609354425@qq.com

Abstract. Designing a shared smart stroller eliminates the hassle of carrying a stupid stroller out of the way to solve the shortcomings of traditional stroller safety facilities; smart strollers can provide some equipment to protect the baby and facilitate the parents. It is very convenient to unlock the smart baby car through the mobile APP. It adopts electric forward and labor-saving, real-time monitoring and review in the car, backtracking of the movement track, setting the electronic fence access alarm, real-time positioning and other functions to facilitate parents to take care of the baby, using air quality testing. And purification, light UV detection, rain detection, obstacle avoidance, automatic roof, to protect the baby from unnecessary external damage. The emergence of shared smart strollers can greatly facilitate parents to carry children with children, and provide people with a high-quality intelligent experience, giving the baby a comfortable and safe travel tool while promoting people's understanding and use of smart strollers.

Keywords: Shared smart baby stroller · Mobile APP · Real-time positioning · Automatic roof

1 Introduction

The smart baby carriage is mainly composed of a processor, a networked terminal, a power part, a smart protection part and a smart auxiliary part. This work uses STM32F103 as the central processor to process the collected data. The networked terminal is composed of a GPRS module and a GPS module, wherein GPRS is used for data transmission between the baby carriage and the database, GPS provides real-time positioning for the smart baby carriage, and the data is transmitted through the GPRS module, and the user can view the location through the mobile phone APP [9, 10].

The power section uses a high-density lithium-ion battery as the power source for the entire system. It uses two DC motors with a reduction ratio of 30 to provide forward power. The intelligent protection part uses ultrasonic waves for distance measurement and monitoring of the road surface to detect obstacles or large pits and steps. When buzzers or pits and steps are detected, buzzers and vibration motors are activated to drive the handlebars [1]. The vibration reminds the user to avoid the function of collision avoidance and road condition detection; the contact capacitance is used to detect whether the user's hands are loose or not, and if it is released, the baby carriage automatically decelerates to stop; the air quality detection module is installed in the vehicle for real-time detection. Ambient air quality, raindrop sensor detects water drop,

© Springer Nature Singapore Pte Ltd. 2020
M. Atiquzzaman et al. (Eds.): BDCPS 2019, AISC 1117, pp. 980–984, 2020.
https://doi.org/10.1007/978-981-15-2568-1_134

UV light intensity sensor detects ambient light intensity or outdoor UV intensity [3]. When the air quality is poor or the light intensity is too high or water drops, the STM32F103 processor automatically controls the rudder [2]. The machine rotates to lower the roof of the roof to avoid damage to the baby caused by the surrounding environment [4, 5].

In addition, when the air quality is poor, the system will automatically start the air purifier to draw the surrounding air through the HPAE filter to filter the negative ion generator and then pass through the fan and The catheter is introduced into the baby carriage; the baby carriage carries the smart night When the ambient light is dark, the LED in front of the car will automatically turn on the front road, and the red LED behind the car will alert the rear pedestrians and the incoming car to ensure the safety of the parents at night [6]. The intelligent assistant part is mainly to help parents better bring their children. The HD camera installed in the car can store the captured baby pictures to the memory card and look back when they are accidentally sent [8].

At the same time, the pictures will be captured in real time. It is transmitted to the LCD screen above the handle of the car for the user to check the status of the baby car in real time; the temperature and humidity detecting device detects whether the baby is urinating and defecating; the temperature of the baby car is detected by the temperature sensor, and the fan is turned on to dissipate heat when the temperature is high [7].

2   Hardware System Structure Diagram

The structure of the smart stroller system is shown in Fig. 1:

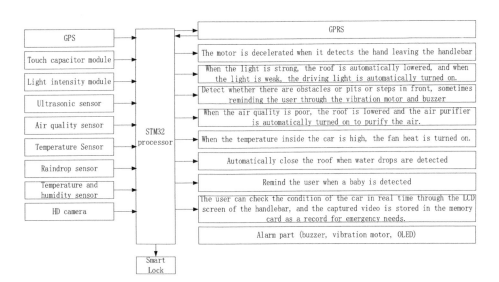

Fig. 1. Smart baby car system structure

The mobile phone APP and the single-chip communication structure diagram shown in Fig. 2:

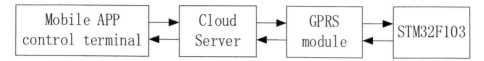

Fig. 2. Schematic diagram of communication structure between mobile APP and single-chip microcomputer

3   Program Flow Chart

(See Fig. 3).

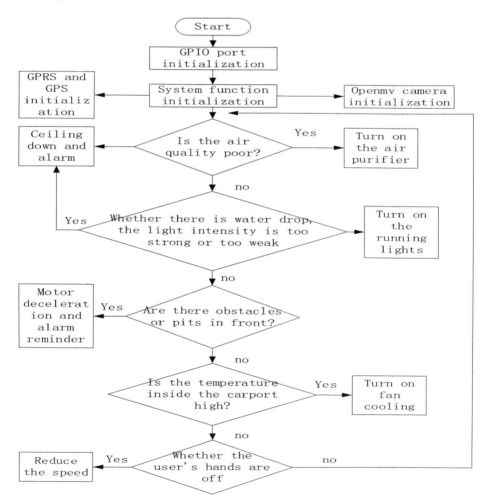

Fig. 3. Program flow chart

4 Testing and Analysis

According to the survey results in the above table, we have carried out many measurements. The above table only tests a small part of us, which basically meets our ideal goal. It can realize the intelligent and sharing function of sharing the baby carriage, which greatly solves the parents. The problem of carrying a baby to carry a baby stroller is not convenient. It not only promotes the smart baby stroller, but also gives the parents a deeper understanding and better experience of the smart baby stroller. It also provides a comfortable and safe travel tool for the baby (Table 1).

Table 1. Data test

Measure serial number	1	2	3	4	5	6	7
Test project	Light intensity project test	Air quality project test	Anti-watering project test	Anti-collision and pit ladder project test	Temperature monitoring	Motor driven	Camera test
Actual Happening	The roof shed is normally open and closed.	Air purifier works normally	The roof is lowered normally	The motor decelerates normally, the buzzer and the vibration motor work normally	The fan is normally turned on.	Stable driving	The camera can shoot and store video normally
Completion level compared to expected goals	Good	Good	Good	Good	Good	Good	Good
Test project completion is not ideal	No	No	No	No	No	No	No

5 Conclusion

This product function is well-found, the rich through sensor structures, to implement the intelligent of the pram, the environment around the smart baby stroller can real-time monitoring the pram road conditions, air quality, light intensity and rainfall, etc., and through the rise and fall of air purification and automatic ceiling and other automated equipment to reduce environment impact on the baby, in addition, the car camera can be real-time or back to look at the situation of the stroller. The combination of sharing and intelligence not only makes it easier for parents to take their babies out, but also provides them with a safe means of travel.

Acknowledgements. This work is supported by the Research Foundation of Education Committee of Hunan Province, China (18C0896), the Foundation of Hunan University of Humanities, Science and Technology (2018).

References

1. Cao, Y., Lu, C., Chen, J., Xia, Z., Gao, Q., Ren, Z., Chen, X., Cheng, J.: Design of multifunctional safety smart baby stroller. South. Agric. Mach. **50**(03), 179–180+184 (2019)
2. Guo, J.: Exploration on the development trend of smart stroller. Pop. Lit. Art **20**, 228–229 (2018)
3. Liu, B., Liu, H.: Design of data communication intelligent safety baby carriage based on STM32. Technol. Innov. (16), 62–63+67 (2018)
4. Li, S., Tian, H., Tan, M., Gao, F., Zhao, X.: Design of intelligent following baby carriage based on STM32. China Sci. Technol. Inf. (09), 81–82+12 (2018)
5. Ren, M., Wang, Z.: Study on the design of baby intelligent products. Design (19), 16–17 (2017)
6. Xu, F., Guo, H., Xiao, X.: Remote control smart baby car. Sci. Technol. Outlook **26**(25), 120–122 (2016)
7. Wee, H.-L., Li, S.-C., Xie, F., et al.: Are Asians comfortable with discussing death in health valuation studies? A study in multi-ethnic Singapore. Health Qual. Life Outcomes **4**(1), 93 (2006)
8. Shen, Y.: Design of real-time location system of school bus based on Android. Comput. Prod. Circ. (05), 129 (2019)
9. Fang, Z., Wang, S., Xie, Z.: Design of intelligent vehicle locator based on BDS and GPRS. China New Technol. New prod. (22), 22–24 (2018)
10. Burden, M.: Mobile app connects with customers. Automot. News (2019)

Port Competitiveness Analysis Based on Factor Analysis

Wendi Lu[(✉)]

Department of Economics and Management, Dalian University,
Dalian, Liaoning, China
Lwd1187401676@163.com

Abstract. Qingdao Port was built in 1892 and has a history of 125 years. And it is China's key state-owned enterprise, China's second foreign trade billion tons of large ports. The waters in the port are wide and deep, the four seasons are open to navigation, and the harbor and mouth are large. It is a famous and excellent port in China. In view of the fact that Qingdao Port is an important part of China's port group and has a positive significance for promoting the "Belt and Road". This paper takes Qingdao Port as the main research object, and then uses factor analysis and SPSS software to access the top ten ports of Ningbo Port, Shanghai Port, Guangzhou Port, Tianjin Port, Dalian Port, Xiamen Port, Yingkou Port, Shenzhen Port and Fuzhou Port. Conduct an empirical analysis of competitiveness.

Keywords: Port competitiveness · Qingdao Port · Factor analysis

1 Introduction

With the development of global economy, as an important transportation hub, ports play an important role in promoting the development of urban economy. Its functions are increasingly rich, and its status and role are becoming more and more important. Qingdao Port is one of the few deep-water ports in the world. It has excellent navigation conditions and can operate for 350 days throughout the year. As one of the important ports of China's port group, Qingdao Port urgently needs to establish a method to evaluate port competitiveness. Then, we will put forward relevant strategies for port development to achieve the improvement of our port competitiveness. The importance and superiority of the port is mainly reflected in its competitiveness. Port competitiveness refers to the ability of two or more participants to form a comprehensive index that evolves from each other. This is not a single indicator, but a mutual indicator. Only through mutual comparison and competition can we generate ability. At the same time, port competitiveness is the ability to show size or strength relative to the comparison, which is an indicator that is difficult to determine. Many internal conditions or external conditions of the port will affect the competitiveness of the port. Port competitiveness plays an important role in regional economic development. It is necessary to evaluate and analyze the main factors affecting the competitiveness of the port, so as to find out their own deficiencies and disadvantages, so that they can be rectified, improved and optimized in a targeted manner.

© Springer Nature Singapore Pte Ltd. 2020
M. Atiquzzaman et al. (Eds.): BDCPS 2019, AISC 1117, pp. 985–991, 2020.
https://doi.org/10.1007/978-981-15-2568-1_135

Hoare [1] pointed out the new concept of the hinterland division of the port. It is believed that the hinterland between the ports can overlap each other. Winkelmans [2] proposed that ports need to have their own irreplaceable advantages in order to effectively improve the competitiveness of the port. Robinson [3] considers the port as part of the value chain system from the perspective of value the system. The port should provide continuous value to customers. It should not only consider the competition factors between ports. Ha [4] selected nine container ports in Northeast Asia to evaluate the quality of service, indicating the availability of information on port activities, port location, port transit time, facility availability, port management, port rates and customers. Van de Voorde and Winkelmans [5] consider the route to be an important factor that the ship often considers when selecting a port. Wu [6] used a principal component analysis to propose eight evaluation indicators for port competitiveness, and demonstrated that port throughput is one of the important output indicators of port competitiveness. Previous studies on traditional port performance evaluation focused on two aspects of input and output: input factors for facility investment, or expansion of port services, and output factors for higher quality and value-added services. However, Song and Panayides [7] pointed out that the impact of supply chain management on port competitiveness should not be limited to terminal performance. Wang [8] used the Fuzzy-AHP method to propose the determinants of the cruise port facilities, the natural environment of the hinterland, the tourist charm and the convenience of connectivity. Yang [9] studied the selection of sea and land routes, port facilities, and cruise terminal facilities. At the same time, a comparative analysis of the advantages and disadvantages of the Wusongkou International Cruise Terminal in Shanghai, Yokohama Port and Jeju Island in Japan was carried out. Lee [10] emphasized the convenience of cruise terminals and passengers. The port depth and current, as well as the importance of expanding the special berths for cruise ships.

It can be seen from the above research review that the depth and breadth of the port competitiveness of the current research needs to be further improved. The research on port competitiveness has developed into a multi-factor and multi-field stage, and various evaluation methods have been applied to the competitiveness analysis in order to have a reasonable evaluation of port competitiveness. This paper uses factor analysis method to study Qingdao Port as the main body and analyze the competitiveness of China's top ten ports.

2   Data Processing and Establishment of Evaluation Indicators

There are many factors affecting the competitiveness of the port, mainly including internal factors and external factors. According to the previous literature, internal factors mainly refer to the internal management system, port facilities and equipment, resources of the port itself and natural factors. External factors mainly refer to the social and cultural environment, the government's policies and regulations, and the economic situation of the city. These two factors determine the competitiveness of the port in different degrees. There is a certain internal logical relationship between each index, which can not only reflect the interrelationship between the indexes of port competitiveness scientifically and objectively, but also reflect the mutual restriction of the indexes to form an organic whole. In some cases, except for the size of index data, other situations are relatively stable, so these index data are comparable.

According to the development of the port, nine indicators were selected. The selected nine indicators are the total number of berths (X1), berths above 10,000 tons (X2), port cargo throughput (X3), foreign trade throughput (X4), container throughput (X5), total GDP of the city where the port is located (X6), the total import and export volume of the city where the port is located (X7), the GDP growth rate of the city where the port is located (X8), and the growth rate of total import and export in the city where the port is located (X9). The data mainly comes from the *China Port Yearbook, Liaoning Statistical Yearbook* and various journal articles and Internet portals. Table 1 summarizes the port data of Qingdao Port, Shanghai Port, Guangzhou Port, Ningbo Zhoushan Port, Tianjin Port, Dalian Port, Xiamen Port, Yingkou Port, Shenzhen Port and Fuzhou Port.

Table 1. Raw data for each indicator

Index	Qingdao Port	Shanghai Port	Guangzhou Port	Ningbo Zhoushan Port	Tianjin port
X1	624	1195	694	121	174
X2	157	172	41	66	47
X3	92291.11	70176.55	54356.12	51463	55055.75
X4	43147.94	38012.28	12643.52	34301	29692.8
X5	2156.71	3713.33	1884.97	1805	1451.92
X6	8686.49	28178.65	19547.44	10011.29	17885.39
X7	94923.22	433768.2	129309	65581.15	102655.95
X8	8.53	12.16	7.99	7.64	8.15
X9	−5.42	−3.87	−3.41	−6.6	−10.17
Index	Dalian Port	Xiamen port	Yingkou Port	Shenzhen port	Fuzhou port
X1	196	164	78	156	179
X2	78	75	59	46	57
X3	43660.01	20910.78	35217.02	21400	14515.65
X4	13909.51	9866	7955.13	18000	5883.08
X5	958.3	916.37	608.73	2397.93	268.12
X6	6730.33	3784.27	1300	19492.6	6197.64
X7	51443.88	77176.81	54.07	398438.92	31978.54
X8	−12.95	9.18	−14.12	11.37	10.32
X9	−8.19	−7.34	1.1	−9.95	−4.09

It must be standardized before processing the above raw data. The processed results are shown in Tables 2, 3, and 4. The common values of variables represent the degree of interpretation of the extracted common factors by the original data. Generally speaking, the degree of sharing of variables More than 75% of the common factors have strong explanatory power for each variable. As shown in Table 2 below, the common factor is basically over 75%, and only the common factor X8 is not exceeded.

Table 2. Common factor variance

Index	Initial	Extract
X1	1.000	.919
X2	1.000	.791
X3	1.000	.914
X4	1.000	.945
X5	1.000	.940
X6	1.000	.882
X7	1.000	.887
X8	1.000	.588
X9	1.000	.901

As can be seen from Table 3 below, the SPSS software self-selects the first three initial eigenvalues greater than one. The initial value characteristics in the table are 4.908, the second is 1.179, the third is 1.066, and the remaining 6 are less than 1, and the variance of these three terms is 86.313% of the variance of the main components. Therefore, it is very reasonable to use these three components to represent the principal component.

The composition matrix after rotation using the maximum variance rotation method is shown in Table 4 below. Including the total number of berths, container throughput, the GDP of the city where the port is located, the total volume of foreign trade imports and exports of the city where the port is located. The first principal component has a higher load on the first principal component and is referred to as the port city logistics and the economic development level factor. The number of berths above 10,000 tons, cargo throughput, and foreign trade cargo throughput have higher loads in the second principal component, and the second master becomes the port cargo throughput factor. The GDP growth rate of the city where the port is located has a high load on the third principal component, which is the factor of port development potential.

Table 3. Explain the total variance

Ingredients	Initial eigenvalue			Extract square sum loading		
	Total	Variance%	Accumulation%	Total	Variance%	Accumulation %
1	4.908	54.528	54.528	4.908	54.528	54.528
2	1.179	19.938	74.466	1.794	19.938	74.466
3	1.066	11.846	86.313	1.066	11.846	86.313
4	.554	6.160	92.472			
5	.404	4.486	96.958			
6	.211	2.344	99.302			
7	.044	.486	99.788			
8	.019	.210	99.998			
9	.000	.002	100.000			

Table 4. Component matrix after rotation

	Ingredients		
	1	2	3
X1	.851	.383	.166
X2	.462	.862	−.099
X3	.227	.867	.269
X4	.012	.860	.413
X5	.840	.409	.291
X6	.679	.093	.653
X7	.820	.423	.150
X8	.007	.249	.880
X9	.811	−.161	−.351

Then calculate the principal component weights and the port synthesis. Construct the following function based on the factor analysis model and factor score coefficients:

$$F1 = 0.243X1 + 0.007X2 - 0.094X3 - 0.179X4 + 0.231X5 \\ + 0.222X6 + 0.224X7 - 0.072X8 + 0.351X9 \tag{1}$$

$$F2 = 0.003X1 + 0.429X2 + 0.375X3 + 0.373X4 - 0.015X5 \\ - 0.266X6 + 0.037X7 - 0.107X8 - 0.155X9 \tag{2}$$

$$F3 = -0.003X1 - 0.333X2 - 0.043X3 + 0.077X4 + 0.086X5 \\ + 0.461X6 - 0.027X7 + 0.611X8 - 0.249X9 \tag{3}$$

$$F = 0.545 * F1 + 0.199 * F2 + 0.118 * F3 \tag{4}$$

From the above results, the comprehensive scores of each port can be calculated. The calculation results are shown in Table 5.

Table 5. Common factor scores and rankings

Port	F1	Ranking	F2	Ranking	F3	Ranking	F	Comprehensive ranking
Ningbo	0.178	6	0.885	1	0.327	8	0.313	3
Shanghai	0.830	1	0.584	2	0.691	4	0.651	1
Guangzhou	0.529	2	−0.092	9	0.700	3	0.352	2
Qingdao	0.090	9	0.320	3	0.585	5	0.182	5
Tianjin	0.066	10	0.202	5	0.837	2	0.176	6
Dalian	0.130	7	0.258	4	−0.003	9	0.122	10
Xiamen	0.122	8	0.028	6	0.451	7	0.125	8
Yingkou	0.340	4	0.023	7	−0.293	10	0.155	7
ShenZhen	0.387	3	−0.091	8	0.936	1	0.304	4
FuZhou	0.202	5	−0.176	10	0.475	6	0.131	9

Table 5 shows the cargo throughput of Qingdao Port, and the second principal component factor ranks third in terms of competitiveness. However, the port logistics and economic development level and the port logistics potential development factor rank relatively backward and the competitiveness is not strong. The comprehensive factor ranks ninth, and there is always a great potential for development.

3 Conclusion

This paper analyzes the factors of the competitiveness data of Qingdao Port and several other large ports in China. With SPSS software, the data is processed, modeled, and the port competitiveness is studied more scientifically and rationally. The following three strategic suggestions for improving the competitiveness of Qingdao Port are proposed. First, Qingdao Port has natural conditions to become the world's international hub port. Compared with other ports, Qingdao Port is competitive in terms of cargo throughput, and can strengthen the berth advantage to develop more berths. At the same time, the construction of the port infrastructure should be strengthened to make full use of Qingdao Port resources. Second, while developing berths and strengthening infrastructure, we should strengthen the cultivation of talents, better serve customers and strengthen management to increase port efficiency. Third, Qingdao Port is relatively backward in terms of logistics and economic development level and hinterland economy. This requires the vigorous development of the economy and the improvement of Shandong's economic strength to further enhance the competitiveness of Qingdao Port.

Acknowledgements. The author is thankful to the reviewer and the editor for their constructive suggestions and comments to improve the initial version of our manuscript.

References

1. Hoare, A.G.: British ports and their export hinterlands: a rapidly changing geography. Geogr. Ann. **68**(1), 29–40 (1986)
2. Notteboom, T., Winkelmasn, W.: Structural change in logistics: how will port authorities face the challenge. Marit. Policy Manag. **28**, 71–89 (2001)
3. Robinson, R.: Ports as elements in value-driven chain system: the new paradigm. Marit. Policy Manag. **29**, 241–255 (2002)
4. Ha, M.S.: A comparison of service quality at major container ports: implications for Korean ports. J. Transp. Geogr. **11**(2), 131–137 (2003)
5. Van de Voorde, E., Van Hooydonk, E., Verbeke, A.: Port Competitiveness: An Economic and Legal Analysis of the Factors Determining the Competitiveness of Seaports, pp. 67–87. De Boeck Ltd., Antwerp (2002)
6. Tongzon, J., Wu, H.: Port privatization, efficiency and competitiveness: some empirical evidence from container ports (terminals). Transp. Res. Part A **39**(5), 405–424 (2005)
7. Song, D.W., Panayides, P.M.: Global supply chain and port/terminal: integration and competitivmess. Marit. Policy Manag. **35**(1), 73–87 (2008)

8. Wang, Y., Jung, K.-A., Yeo, G.-T., Chou, C.-C.: Selecting a cruise port of call location using the fuzzy-AHP method: a case study in East Asia. Tour. Manag. **42**, 262–270 (2014)
9. Yang, J.: A Study on Cruise Homeport Selection Factor: Focusing on Comparative Study of Major Ports of Korea, China and Japan. Jeju National University, Jeju(Korea) (2015)
10. Lee, C.B., Lee, J., Noh, J.: A study on key successful factors of cruise port. J. Korea Port Econ. Assoc. **29**(2), 12–15 (2013)

Application of English Virtual Community in Teaching Based on Campus Network Construction

Yang Song(✉)

Shandong Women's University, Jinan 250002, Shandong, China
19508960@qq.com

Abstract. With the rapid development of network technology and information technology, various kinds of network applications emerge in an endless stream, which brings many opportunities for English teaching. The English virtual community is a new way of teaching relying on network technology. Campus network is a kind of campus local area network, which can provide convenient network services for teachers and students. It enables teachers and students to share resources, transfer information, network connection and other contents, and realizes the practical needs of all teachers and students for network services. As an educational virtual community, English virtual community based on campus network can provide high-quality teaching services and an open teaching environment. It is constructed by campus network. Its main users are teachers and students in school. It not only has the function of communication in teaching, but also has the function of promoting self-learning and ensuring teaching. It provides great convenience for English teaching. According to the survey results of teachers and students in schools, 83.9% of the respondents hold a positive attitude towards the construction of English virtual community, which shows that the application research has very important practical significance.

Keywords: Campus network · Virtual community · English teaching · Autonomous learning

1 Introduction

The rapid development of network technology has brought new opportunities and ways to modern teaching, which also provides a new teaching mode and means for Contemporary College English teaching. Nowadays, network has become an indispensable part of people's daily life. Participating in various virtual communities has become a normal way of College Students' network activities. According to the survey, nowadays college students will join many different types of virtual communities. They are active in virtual communities such as forums, post bars, social groups, but these virtual communities are not targeted and targeted. Using campus network to construct English virtual community, combining virtual community with teaching, and creating a virtual community conducive to college students' English teaching has great practical significance for English teaching and the growth of College students.

© Springer Nature Singapore Pte Ltd. 2020
M. Atiquzzaman et al. (Eds.): BDCPS 2019, AISC 1117, pp. 992–998, 2020.
https://doi.org/10.1007/978-981-15-2568-1_136

Campus network is an indispensable basic network of Contemporary University campus. There are many applications based on campus network, and good practical results have been achieved. In [1], the author relies on the campus network and uses PHP to realize the graphical management interface of the campus network. Finally, a complete campus network management system is displayed through the web page, which brings convenience to the administrator in network management and maintenance. In [2], the author analyses the new changes of information-based intelligent education service under the campus network environment, combines the value and significance of modern education, analyses the construction of intelligent campus system under the education cloud, and puts forward corresponding countermeasures. In [3], a campus distributed bicycle sharing system is proposed to enhance the convenience of bicycle sharing on campus, which greatly facilitates the teachers and students on campus, and the response time and the average raw data of the system are lower. In [4], aiming at the real-time problem of student attendance management in large lecture halls, the author proposes an effective solution, using high-speed technology, which can quickly respond and accurately identify students.

Based on the advantages of campus network, this paper applies virtual community to teaching activities and proposes to build an English virtual community based on campus network. English virtual community is a kind of virtual teaching community based on campus network. It is more convenient, interactive, sharing and autonomous than the real English community. It can provide more effective support and help for students' English teaching. The results of the survey show that most of the teachers and students have a positive attitude towards the English virtual community based on the campus network. Therefore, the English virtual community based on campus network proposed in this paper is of great significance in teaching.

2 Method

2.1 Virtual Community

A key reason why virtual communities are different from real communities is that they have no real geographical constraints [5–7]. The key characteristic of the network is that it has no time and space constraints. In daily life, people in different areas use the network connection to form a "society" of daily communication, mutual assistance, sharing and even material exchange. This is a "virtual existence", which is what we call virtual community. Because of the network characteristics of virtual community, participants in virtual community can communicate freely, which is not the characteristic of real network, so the development of virtual community is very rapid.

In modern colleges and universities, the network has become an indispensable part of College Students' daily life. With the help of the network, students exchange and share information, and virtual communities formed by the network become platforms for college students to exchange and learn. Virtual community is of great significance to daily teaching activities depending on its virtual Characteristics, sharing and freedom.

2.2 Campus Network

Campus network is a kind of local area network. Its users and service objects are mainly teachers and students in school. It has certain user and geographical restrictions. Generally, it can only be accessed and used within the campus area. Campus network is a kind of LAN technology, which can provide convenient network services for teachers and students. It enables teachers and students to share resources, transfer information and connect network, and realizes the practical needs of all teachers and students for network services [8, 9]. In addition, the campus network can connect the communication products in the campus network and the external network through other network technologies, which is an efficient, safe and convenient means of network sharing. When teachers carry out teaching activities, campus network can make use of its unique network advantages to provide reliable and efficient support and help for teachers' teaching activities, facilitate learning exchanges between teachers and students, and share and query learning resources, and provide more ways for teachers' daily teaching activities.

2.3 Building English Virtual Community

With the continuous development and progress of information technology means, the teaching work of teachers in our country is constantly changing. Integrating information technology means with teaching work is one of the important innovative directions of education and teaching work. Relying on the advantages of campus network to build English virtual community, the traditional teaching mode of "blackboard + textbook" will undergo tremendous innovation. English virtual community is a convenient way to show the teaching video, audio, courseware, doubts and other aspects of teachers' daily learning, which greatly facilitates teachers' daily English teaching work. Due to the expansion and freedom of English virtual community, the English virtual community constructed by campus network can not only carry out English teaching work, but also mobilize students' autonomy and enthusiasm in learning English. At the same time, it can also popularize and transmit foreign cultural knowledge, cultural ideas and other extracurricular knowledge in English teaching work. English virtual community provides an open platform for English teaching, facilitates the learning and communication between teachers and students, removes the space and time constraints of the fixed classroom, stimulates students' enthusiasm for learning English, and improves students' understanding of foreign cultural ideas, so that students can use the English virtual community platform to understand English, understand English and love English.

3 Design Analysis and Data

The construction of English virtual community is not only a supplement to college English teaching, but also a platform for students to learn English independently. English virtual community has four functional divisions, namely "Famous Teachers Online", "Community Service", "Special Learning", "Question-answering and Question-solving". Its functions include not only English course learning, but also extra-curricular English learning. It also includes IELTS's special English learning, daily data downloading and question-answering divisions, which can meet the teaching needs of College English learning. Ask. Its functions are practical, easy to operate and use, beautiful interface and generous, to meet the needs of teachers and students in English teaching.

After the development of English virtual community, it has been tested and modified many times to make its use more convenient. In order to understand the effect of community construction, 186 teachers and students were randomly surveyed by questionnaires, and their satisfaction with English virtual community was counted, so as to find out the teaching effect of English virtual community.

4   Results

4.1   Architecture Design and Implementation Effect

English virtual community based on campus network is a platform for English teaching and communication on campus network. After the completion of the system development, it is necessary to ensure that it can run smoothly on the campus network. Any communication equipment connected with the campus network can be used by browser [10]. The server side of the system requires environmental support for the operation of the whole system.

Based on the above requirements, the English virtual community system uses a three-tier architecture based on B/S. In this three-tier architecture, the user's access interface can be accessed by browser. Except a few simple transaction logic is completed on the browser side, all other transaction logic is completed on the server side. This not only makes the development and maintenance of the system simpler, but also makes the use of teachers and students more convenient. As shown in Fig. 1, the

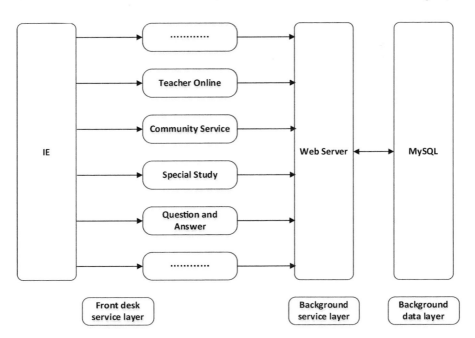

Fig. 1. English virtual community system architecture

system architecture of English virtual community consists of front-end service layer, back-end data layer and back-end processing layer.

According to the above development needs and the network characteristics of campus network, and according to the simple and convenient design principles, we design the English virtual community, which basically achieves the above requirements. The English virtual community based on campus network is called "English Community", and the home page interface is shown in Fig. 2.

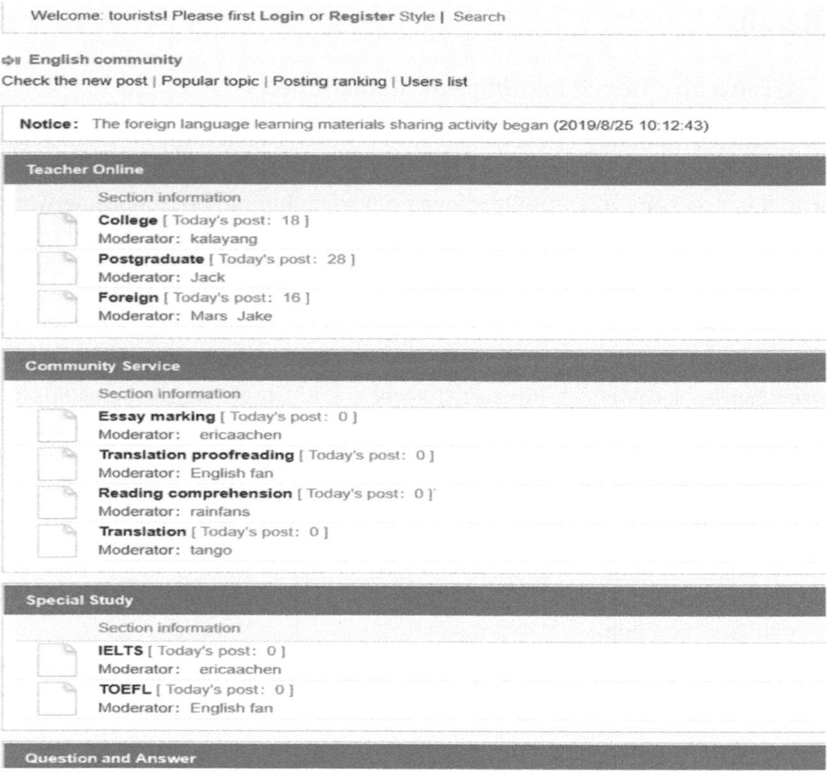

Welcome: tourists! Please first **Login** or **Register** Style | Search

🖒 **English community**
Check the new post | Popular topic | Posting ranking | Users list

Notice: The foreign language learning materials sharing activity began (2019/8/25 10:12:43)

Teacher Online

Section information

College [Today's post: 18]
Moderator: kalayang
Postgraduate [Today's post: 28]
Moderator: Jack
Foreign [Today's post: 16]
Moderator: Mars Jake

Community Service

Section information

Essay marking [Today's post: 0]
Moderator: ericaachen
Translation proofreading [Today's post: 0]
Moderator: English fan
Reading comprehension [Today's post: 0]
Moderator: rainfans
Translation [Today's post: 0]
Moderator: tango

Special Study

Section information

IELTS [Today's post: 0]
Moderator: ericaachen
TOEFL [Today's post: 0]
Moderator: English fan

Question and Answer

Fig. 2. English virtual community interface diagram based on campus network

4.2 Community Effectiveness Survey Statistics

In order to further improve the virtual English community, 186 teachers and students were randomly investigated anonymously, as shown in the Fig. 3.

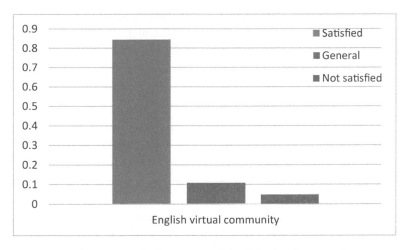

Fig. 3. The satisfaction survey of English virtual community

Through anonymous survey results, it can be clearly seen that 83.9% of the respondents have a positive attitude towards the construction of English virtual community, which shows that the application research has very important practical significance. Only 11.3% of the respondents said that they are general, and only 4.8% of the respondents expressed their dissatisfaction. Through the analysis of the results of the survey and the communication between teachers and students, we can draw the following conclusions about the effectiveness of the virtual English community based on campus network:

(1) Generally speaking, the teachers and students surveyed gave a positive evaluation of the effectiveness of the use of English virtual communities, indicating that English virtual communities play an active role in teaching and have given satisfactory attitudes. The development of English virtual communities has achieved our desired teaching results.

(2) By communicating with teachers and students with general and unsatisfactory attitudes, we find that the functions of the English virtual community are not perfect enough, and there is still much room for improvement in the form of communication between teachers and students, which requires us to further improve.

5 Conclusion

The virtual community of English based on campus network proposed in this paper can bring great convenience to college English teaching. The construction of English virtual community is an important way to innovate college English teaching mode. It extends the traditional classroom teaching mode and integrates the teaching means of information network. It provides an important way for teachers and students to communicate and interact in English learning, and greatly promotes teachers' and students' cooperative learning and autonomous learning ability in English. English virtual community has important practical significance for English teaching.

References

1. Ren, Y.: Application of network certification management system in campus network of secondary vocational schools. Electron. Technol. Softw. Eng. (3), 12 (2017)
2. Xu, W.: Analysis on the construction of intelligent campus system under the education cloud. Comput. Inf. Technol. **26**(06), 57–59 (2018)
3. Pan, C., Zhang, W., Lv, H., et al.: APP design of campus bike-sharing service system. Shanxi Archit. (3), 257–258 (2018)
4. Mohammed, K., Tolba, A.S., Elmogy, M.: Multimodal student attendance management system (MSAMS). Ain Shams Eng. J. **9**(4), 2917–2929 (2018)
5. Aroles, J.: Performance and becoming: rethinking nativeness in virtual communities. Games Cult. **13**(5), 423–439 (2018)
6. Wu, J., Xie, K., Xiao, J., et al.: Effects of customer heterogeneity on participation performance in virtual brand community: a two-stage semiparametric approach. Int. J. Electron. Commer. **22**(2), 289–321 (2018)
7. Thoma, B., Brazil, V., Spurr, J., et al.: Establishing a virtual community of practice in simulation: the value of social media. Simul. Healthc. **13**(2), 124–130 (2018)
8. LaBorie, T.: Computer literacy over the campus network. Coll. Res. Libr. News **54**(2), 70–74 (2019)
9. Ammu, A.M., Nair, B., Lolitha, P., et al.: An SDN approach to ARP broadcast reduction in an unstructured campus network. Int. J. Pure Appl. Math. **114**(12), 109–118 (2017)
10. Kim, S., Song, K., Coppersmith, S.: Creating an interactive virtual community of linguistically and culturally responsive content teacher-learners to serve english learners. Contemp. Issues Technol. Teach. Educ. **18**(2), 442–466 (2018)

Analysis and Research on the Innovative Teaching Mode of Collegiate Athletic Education Under the Background of Big Data

Wensheng Huang[✉]

Dalian Polytechnic University Sports Teaching Department,
Dalian 116034, China
hws02@163.com

Abstract. Under the background of Big Data, collegiate athletic education courses use the collection and analysis ability of Big Data on the Internet to establish various sports skills and health education learning platforms, reform the traditional single sports teaching mode, innovative diversified and intelligent sports teaching fields, increase the ways for college students to learn sports and health common sense, and improve the self-consciousness of college students to acquire sports knowledge. Interest points, strengthen the ability of self-exercise and physical cultivation, complete the teaching objectives of collegiate athletic education, and use Big Data technology to achieve the reform and innovation of collegiate athletic education teaching mode.

Keywords: Big Data · Internet · Collegiate athletic education teaching mode · Reform and innovation

The application and research of Big Data technology in the field of education has attracted great attention of the Ministry of education of the people's Republic of China. In the action plan for promoting the development of Big Data and the 13th five-year plan for educational information, it is clearly proposed to explore and play the supporting role of Big Data in the reform of education mode, the promotion of education equity and the improvement of education quality, and it is emphasized to actively play the role of Big Data in education and learning. Learning platform plays an important role in network construction [1]. According to the requirements of the Ministry of education to change the education mode in the context of Big Data, it is necessary to carry out the analysis and research of the reform and innovation according to the actual situation of the current university physical education teaching mode.

1 Characteristics and Application of Big Data

The development of science and technology and the Internet is promoting the advent of the era of Big Data, and all walks of life are generating a huge amount of data fragments every day, including traditional structural data and non structural data, such as text, pictures, audio and video, etc., with huge data volume, many data types, strong real-time, and great value contained in data [2]. It has brought great changes to modern

© Springer Nature Singapore Pte Ltd. 2020
M. Atiquzzaman et al. (Eds.): BDCPS 2019, AISC 1117, pp. 999–1003, 2020.
https://doi.org/10.1007/978-981-15-2568-1_137

people's life and learning by using such technical means as rapid analysis, sorting and timely push of Big Data [3]. In the field of education, Big Data is conducive to students' personalized and differentiated learning. The teaching mode of collegiate athletic education should make use of this advantage of Big Data for reform and innovation. Teachers and students should change the traditional single way of thinking and teaching mode of teaching and learning, teaching in a diversified way and evaluating students in a diversified system [4].

2   The Current Situation of the Teaching Mode of Collegiate Athletic Education

2.1   There Are Fewer Physical Education Teachers, Less Physical Education Teaching Facilities and Fewer Courses

With the expansion of university enrollment, the situation of fewer physical education teachers is increasingly prominent, which is common in universities. In addition to extracurricular activities, PE teacher should also help students train and compete. The work pressure is high and the intensity is high, which affects the teaching effect of physical education in class [5]. The lack and obsolescence of sports venues and equipment lead to fewer types of physical education courses and narrow the scope of course selection. College students can't choose physical education courses according to their own preferences, thus losing their enthusiasm for physical education courses, even generating boredom, and it's difficult to achieve the teaching objectives of collegiate athletic education courses.

2.2   It Pays Too Much Attention to the Training of Sports Skills, Neglects the Health Education of Physical Quality, and Has a Single Evaluation System

There is a common phenomenon in the teaching of collegiate athletic education, which attaches great importance to the training of sports skills and neglects the health education of physical quality, and the evaluation of College Students' physical education is single [6]. The thinking inertia over the years has led to the misunderstanding that the main teaching task of collegiate athletic education course is to cultivate the special skills of college students. In the classroom, teachers are still the main part, and students are the auxiliary part. The subjective initiative of college students has been hindered. The scoring standard of special skills has been used to evaluate the learning situation of College Students' physical education course, and the difference of students' physical quality has been ignored. In fact, the singleness of the evaluation standard and the low evaluation of the students with poor physical quality make these students lack of confidence and enthusiasm in the study of physical education class and fail to realize the teaching concept of health first [7].

2.3 There Are Few Class Hours of Sports Theory, and the Knowledge of Sports Theory and Sports Health Care of College Students Is Too Lacking

The theory of physical education is an important part of the content of physical education. Through the study of the theory of physical education, it is not only helpful to understand and master the special technical actions, but also helpful to enrich the knowledge of physical health and health care. Due to various reasons, the phenomenon of less physical education theoretical class hours is more common, less than 4 class hours per semester accounts for more than 60%, and even some universities do not arrange physical education theoretical class, physical education theoretical class is not paid attention to, affecting the enthusiasm of students to acquire physical health knowledge.

3 Innovating the Teaching Mode of Collegiate Athletic Education Under the Background of Big Data

3.1 Innovate Teaching Thinking Mode and Establish Online Interactive Physical Education Teaching Mode

Using Big Data technology to establish online interactive physical education teaching mode is one of the best ways to achieve the goal of Collegiate athletic Education and cultivate qualified college students with the concept of health first. Under the background of Big Data, to reform and innovate the teaching mode of collegiate athletic education, first of all, we need to change the traditional teaching thinking mode, introduce the open online course education platform like MOOC into the traditional collegiate athletic education classroom, and according to the Big Data statistics and screening of College Students' browsing amount and browsing time, we can set up the types and safety of physical education courses. Schedule class time. College students can watch the sports open courses of our school or other universities for free, preview and review the sports skills they have learned. Even if they are unable to attend classes due to illness and other reasons, they can learn and master the contents of the course through online interaction with teachers. This new online interactive sports teaching mode makes college students more flexible in acquiring sports knowledge and improves self-learning. Ability of sports health knowledge [8]. Under the background of Big Data, the online interactive innovative physical education teaching mode has been applied and developed.

3.2 Build and Introduce Intelligent Sports Venues and Sports Sensing Equipment, and Innovate the Learning Mode and Sports Health Experience of College Students

By using Big Data information analysis and push technology, we can combine the learning process of College Students' physical education with intelligent physical education sensor materials, and transfer their own various sports physiological indicators, sports data and sports skill proficiency to personal mobile phones for easy viewing, so that each college student can fully grasp their own sports state, and then make their own physical education learning plan. So as to improve the level of sports and enhance the fun of sports health [9]. Physical education teachers can refer to the data generated by intelligent physical education sensing equipment, assign physical education tasks according to the sports load index of students, make physical education prescription and accuracy, give full play to the consciousness and initiative of students' physical education learning, and provide more sports experience for students.

3.3 Creating a New Teaching Mode of Diversified Physical Education for College Students

Diversification of College Students' physical education is the inevitable trend of the reform and development of physical education in the era of Big Data. The significance of physical education is to enable students to master physical skills, cultivate learning interests, meet sports needs, and realize the concept of health first. Using Big Data information technology, through multi-form and multi-level physical education courses, this concept can be fully realized [10]. College students use the online internet physical education platform to easily have more learning space, and can make an objective evaluation of their own physical education, so as to realize the diversified new mode of physical education teaching with the participation of physical education teachers and college students themselves. Compared with the traditional mode of physical education teaching evaluation, it improves the participation of college students and improves the physical education teaching list dominated by teachers. Lack of oneness.

3.4 Innovative and Diversified Evaluation System of College Students' Physical Education

1. Using Big Data technology, in the aspect of physical education theory learning, college students can use all kinds of open online courses physical education platform for self-study and examination to evaluate their physical education theory level. In terms of sports skills, college students can interact with PE teachers online, and teachers can explain key points, skills and easy to make mistakes online. They can also learn repeatedly through micro class, combining practice with theory, which helps students master new sports skills faster.
2. Using Big Data technology, the free self-evaluation system of physical education is established. Under the guidance of PE teacher, university students exercise after class by using fitness software such as APP to give full play to their initiative and realize self-evaluation.
3. Using Big Data technology, according to the fitness situation of college students and the practical sports situation in and out of class, PE teachers objectively evaluate the sports situation of college students.

4 Conclusion

With the development of Big Data technology, the traditional physical education teaching mode of college students has been impacted. The reform and innovation of collegiate athletic education teaching thinking, concept and diversified teaching mode is the inevitable result of adapting to the times. The traditional single course type, old teaching method and evaluation mode can't meet the students' sports needs. The diversified and interactive PE teaching method will become the new mode of College PE teaching. Under the background of Big Data, the teaching mode of collegiate athletic education should make full use of the advantages brought by Big Data, play its good role in the exploration of the frontier technology of collegiate athletic education skills, the collection of new physical education knowledge, the discussion of new teaching methods and the analysis of new teaching evaluation system, and use more methods and ways to better innovate the teaching mode of collegiate athletic education. Provide data support and improvement measures.

References

1. Fan, J.: Research on the application of "Big Data" in physical education. Phys. Educ. **141**, 81–83 (2016)
2. Liu, G., Liu, H.: On the influence of big data era on physical education and teaching research. J. Changjiang Norm. Univ. **4**, 121–122 (2016)
3. Shi, Q., Yu, N.: Research on the influence of big data technology on physical education in primary and secondary schools. Contemp. Sports Technol. **26**, 245–246 (2015)
4. Xie, C.: Research and thinking on collegiate athletic education classroom teaching in the era of big data. Curric. Teach. **35**, 106–107 (2016)
5. Sun, H., Zheng, Q.: Application status and development trend of core technology of education big data. Distance Educ. J. (2016)
6. Zhao, G.: Historical Opportunities in the Era of Big Data. Tsinghua University Press, Beijing (2013)
7. Zhao, Y.: Big Data Revolution - Theoretical Model and Technological Innovation. Electronic Industry Press, Beijing (2014)
8. Liu, X.: Discussion on the teaching mode of collegiate athletic education theory course in the era of big data. Autom. Instrum. **01**, 208–209 (2017)
9. Wang, M.: The construction of the innovation system of collegiate athletic education teaching mode for sunshine sports. J. South. Norm. Univ. Nat. Sci. Ed. **9**, 185–188 (2014)
10. Hou, X., Yang, Y.: The development and application of multimedia curriculum resources in collegiate athletic education. J. Inner Mongolia Univ. Finance Econ. **12**(4), 136–138 (2014)

Application of Big Data and Smart City Technology in New Urbanization Planning

Minghui Long[(⊠)]

Wuhan Donghu University, Wuhan, Hubei, China
642604051@qq.com

Abstract. In recent years with the further development of new urbanization, the Internet of Things represented by sensing technology, the Internet represented by smart mobile devices, communication network and cloud computing technologies have been widely used in the construction of smart cities. Smart city is the use of information and communication technology to sense, analyze and integrate the key information of the core system of urban operation, so as to make an intelligent response to various needs, including people's livelihood, environmental protection, public safety, urban services, business activities. Smart city not only changes the social and economic life style, but also promotes people's new views on physical and cognitive space and cognitive ability. However, smart cities bring convenience, but also bring many challenges and problems. This paper expounds the characteristics of smart city and the current situation of its management and operation, and explores the road of smart city development by drawing lessons from the experience of foreign smart city construction.

Keywords: New urbanization · Smart city · Big data · Information security

In March 2014, the Central Committee of the Communist Party of China and the State Council issued *the National New Urbanization Planning (2014–2020)*. It points out that the new urbanization should focus on the connotation and quality of the city, and puts forward four strategic tasks of "urbanization of agricultural transfer population, coordinated development of large and medium-sized cities, sustainable urban development and integration of urban and rural development", and introduces the construction of smart cities into the national strategic plan for the first time. By 2020, a number of smart cities with distinctive characteristics will be built. Following the concept of "Smart Earth" put forward in 2008, governments at all levels from the central to the local attach great importance to the construction of smart cities. Many cities take it as the focus of development and actively explore and make pilot projects, providing a good reference for the future development of cities in China, and also contributing to the scientific and rational planning of cities.

1 The Application Prospect of Big Data

China is currently in the process of rapid urbanization. Rapid population growth poses enormous challenges to the ecological environment, social life, urban management and public services. Moreover, with the application and popularization of the Internet and

M. Atiquzzaman et al. (Eds.): BDCPS 2019, AISC 1117, pp. 1004–1012, 2020.
https://doi.org/10.1007/978-981-15-2568-1_138

mobile Internet, the public's requirements for knowing, participating and supervising the work of the government are constantly improving, and the requirements for the fairness, equality, service quality and timeliness of public services are constantly improving, which puts forward a test for the governance ability and level of the government. In August 2014, in response to the problems exposed in the process of pilot construction of smart cities, with the consent of the State Council, the State Development and Reform Commission and other eight ministries and commissions jointly issued *Guidance on Promoting Healthy Development of Smart Cities*, emphasizing that "to strengthen top-level design, local governments of cities should study and formulate smart city construction plans from the overall strategic perspective of urban development" [1].

1.1   Realize High Efficiency Operation of Cities

In order to ensure the safety and efficiency of urban operation, the construction of smart city needs to collect, integrate, store and analyze massive data resources, and use intelligent perception, distributed storage, data mining, real-time dynamic visualization and other data technology to achieve the rational allocation of resources. Therefore, urban big data is the key support to realize urban wisdom, and it is also an important engine to "promote political exchanges, benefit the people and revitalize industries". With the continuous progress of data processing technology, people's awareness of data application is constantly improving. The data generated by people's life and the operation of various industries show explosive growth, forming large urban data. At present, most of urban big data are government big data and industry big data, so the main promoter of urban big data should be a city's government and related enterprises with a certain data scale.

1.2   Promote a Large Number of Emerging Industries

The construction of smart city is to dig out the value information needed for the operation of a city from such a huge amount of data ocean with the help of cloud computing, big data, chips, algorithms and other basic capabilities. Full data aggregation and full data calculation have become the consensus of the industry. It can be said that the construction of smart city is undergoing a major transformation. The emergence of various large data processing and application technologies has accelerated the evolution of smart city construction to a higher level. At the same time, a variety of business scenarios in the city also give rise to different user needs and applications. Smart city construction cannot be separated from the support of Internet of Things, cloud computing and other technologies, and the technology involved in the Internet of Things is a large integration, which will lead to the formation of large-scale industrial chain, including Internet of Things equipment and terminal manufacturing, Internet of Things network services, Internet of Things infrastructure services, Internet of Things support industry, Internet of Things software development.

1.3 Satisfy the Happy Life of Residents

Under the new urbanization planning, in order to conform to the development trend of smart city and information age, it is necessary to build a big data analysis center. Systematic data can be obtained by using big data technology. Comprehensive analysis of urban conditions can be carried out through the model, which will help urban planning to be more rational and scientific. Smart home creates a safe, intelligent and comfortable home environment for residents, and smart transportation makes residents travel more smoothly and efficiently. Smart medical system enables residents to receive timely and effective care for their physical and mental health, and to effectively solve the current problems of limited medical resources and unbalanced distribution at the same time. Smart city safety emergency system can effectively monitor the urban public security situation, effectively deal with urban crime and emergencies, and create a safe urban environment [2].

2 Challenges of Smart City Development

There are many problems in the development of data-driven smart cities. Although local governments at all levels and enterprises are actively exploring the construction of smart cities, there are still some problems, such as unclear characteristics, poor experience and insufficient sharing. The root of the problem lies in the failure to achieve a good integration of urban big data resources and urban business.

2.1 The Uneven Development of the Industry and a Shortage of Professionals

Under the tide of big data, many enterprises have devoted themselves to the study of how big data can be applied to smart cities. However, based on many industry standards and management mechanisms, the concept of big data has not been clearly planned. There are some enterprises doing things by irregular ways, and the development of the industry is uneven. In addition, the lack of data analysis personnel and the lack of professional algorithms to support analysis make the application of big data still have many limitations.

2.2 Barriers to Big Data Pull-Connect Sharing

The construction of a new type of smart city with big data as its core is also facing many new difficulties in the process of practice. One of the key problems is that there are serious barriers to data pull-connect sharing. As the infrastructure of a new smart city, big data naturally requires to stand at the height of the city to connect all kinds of data generated by the government, enterprises and other urban entities in the production and operation. Although the Chinese government has 80% of the information resources of the whole society, and most provinces and cities in China have established government data open platforms, these information resources can not be effectively utilized because of the restrictions on the interests of various departments or regions. In addition, the inconsistency of departmental data standards hinders the flow of information, which leads to the ineffective use of large data hoarding. The management system of each government department is linear, and each department is responsible to its superior department and lacks the power to connect with other departments horizontally, which makes it difficult to form a comprehensive data pulling-connecting and sharing, and ultimately leads to the difficulty of consolidating the foundation of smart city construction.

2.3   Lack of Innovation in Technology Application

Technological innovation is also a bottleneck in China. At present, the application of big data in China is mostly concentrated in government affairs, urban management, video traffic, etc. In many aspects of people's livelihood, such as medical treatment, education, community, tourism, etc., big data value has not been fully exploited; data collation and analysis are too single, and technology application is lack of innovation, resulting in the inefficient use of a lot of data. Government enterprises need to strengthen technology exchanges at home and abroad, and consider how to make full use of big data, clarify the new direction of technology application and development, so as to make smart cities more intelligent [3].

2.4   Lack of System Specification for Big Data Management and Application

Due to the lack of top-level design and overall planning at the data level in the historical construction, there are some problems in the availability of data in various sectors of the city, which are manifested in the poor quality of data. After data fusion and association, the availability of data is directly affected. Many scenarios need to update the data in real time, and even realize the whole process from data generation to data collection and storage. However, the frequency of data acquisition in many government departments is low and data governance is lacking [4].

In the daily operation of smart cities, it needs to constantly process and analyze data, and produce classified data. The management and application of data need to be treated systematically. And it needs to activate the resources from the top level design, and establish the system and specification of large data management and application. Only in this way can it avoids the phenomenon of data unattended collection, updating and maintenance, and unattended use.

3   Strategies of Smart City Construction Based on Big Data

3.1   Establish Urban Data Management System

The urban big data platform has established a unified standard for data governance to improve the efficiency of data management. Through unified standards, it can avoid data confusion and conflict, one data from multiple sources and other issues. By centralized processing, the "validity period" of data is extended, and multi-angle data attributes are quickly mined for analysis and application. Through quality management, problems such as uneven data quality, data redundancy and data missing can be found and solved in time. The urban big data platform standardizes the sharing and circulation of data among various business systems, and promotes the full release of data value. Through overall management, we can eliminate the "privatization" of information resources in various departments and the mutual restrictions between departments, enhance the awareness of data sharing, and improve the motivation of data opening. Through effective integration, the utilization level of data resources can be improved [5].

3.2   Spatial Planning of Smart City

The most critical link in the process of building a smart city is the need for rational and scientific planning. Now we can strengthen the application of big data, which can effectively integrate the urban space, further promote the urban development, and promote the city to have better development prospects in the future. In addition, in the process of urban spatial planning, it is necessary to establish scientific and reasonable development goals according to the actual situation and characteristics of the city. Especially for the current smart city, we can use big data to integrate smart city land resources, strengthen land planning, and integrate data and network to ensure that smart city has a higher rationality and scientificity. In addition, in the process of urban and rural planning, it not only needs to strengthen the top-level design, but also to achieve a large number of innovations in terms of technical content. Especially in the planning process, it is necessary to adhere to the principle of overall planning and consideration, and fundamentally guarantee the scientificity and rationality of top-level design. In addition, the construction and planning process of smart city can also effectively allocate urban resources to ensure the realization of the goal of smart city spatial planning.

3.3   Raise the Level of Government Public Service Through Accurate Analysis

In the field of transportation, through real-time traffic monitoring such as satellite analysis and open cloud platform, we can perceive traffic conditions and help citizens optimize travel plans; in the area of safe city, through centralized monitoring and analysis of behavior track, social relations and public opinion, it provides strong support for command decision-making and intelligence research and judgment of public security departments [6].

The government owns the government data resources. Internet enterprises often have advanced data technology and professional team with Internet thought. Local enterprises have a clearer and more accurate understanding of local human resources, market environment, industrial development and other factors. They need to activately use the resources of the government, Internet enterprises, local enterprises and other parties to participate in the construction and operation of the platform. In the field of government service, relying on the unified Internet e-government data service platform, we can achieve "data does more, people do less"; in the field of health care, through the exchange of health records, electronic medical records and other data, it can not only improve the quality of medical services, but also timely monitor the epidemic situation and reduce the medical risk of citizens [7].

3.4   Establish Big Data Analysis Center and Promote the Development of Urban Digital Economy

Use large data mining, analysis, prediction algorithms, models, and innovate intelligence analysis ideas to expand intelligence analysis means, build large data analysis and prediction tools, fully analyze information, provide multi-dimension information association function, realize intelligent collision and comparison of data, intelligence prediction and automatic pushing of results. In the current process of urban planning development, in order to effectively ensure that smart urban planning conforms to the development trend of the information age, we should build a large data analysis center. In other words, we need to use big data to obtain systematic information, and we can also analyze the model effectively, so as to ensure the scientific and rational urban planning. In addition, the government and enterprises should form a good cooperative relationship with each other. At the same time, the local and central cooperation mode should be constructed to promote the integration of enterprise power, government power and folk power, so as to meet the realistic needs of urban and rural planning. In addition, a large amount of information and data can be collected from the city for further mining the value of big data, which can provide better urban services for the people, and also help urban and rural planning work more scientific and reasonable. Strengthen the management of urban big data and realize the standardized management of the whole process from data acquisition to data capitalization. Clear data ownership and benefit distribution, as well as personal information protection, data life cycle management responsibility issues. And clarify the classification and hierarchical management of data resources and improve the standards of data resources management [8].

Open the sharing large data platform will promote data docking between government and enterprise, and stimulate social forces to participate in urban construction. On the one hand, enterprises can obtain more urban data, excavate business value and improve their business level. On the other hand, the data of enterprises and organizations contributed to a unified big data platform can also "feed back" to government data, support the fine management of cities, and further promote modern urban governance.

3.5   Promote Platform Construction

The data management and operation system of urban big data platform is quite complex. There is no fixed model for the platform construction mode and path. It is necessary to give full play to the subjective initiative of all parties, adapt to local conditions, find local advantages, highlight local characteristics, and provide strong support for urban big data decision-making, and strengthen the top-level design of the platform. Scientific and reasonable top-level design is the key to the construction of urban big data platform. It needs to start from the implementation of national macro-policy, combine with local actual needs, take into account the platform objectives, data sovereignty, key technologies, legal environment and functions. Carry out the top-level design of the platform with "high starting point, high positioning and stable implementation" to ensure that the construction of the big data platform in cities is continuously promoted with goals, directions, paths and rhythms. According to the progress of the project, keep iterating to update and bring forth new ideas. The construction and application of urban big data platform should be combined to avoid the phenomenon of emphasizing platform construction over platform use. The data resources of government, industry and city are extremely complex, so it is necessary to clarify the rights and attributes of the platform data resources and ensure the ownership of the data [9].

Perfect the supporting mechanism of the platform. The construction and operation of urban big data platform must have corresponding supporting guarantee mechanism, and give full play to the guiding and supporting role of the guarantee mechanism to ensure the coordination of platform planning and the realization of the overall effectiveness of the platform. For example, we should establish the management mechanism of urban big data resources, clarify the centralized management department of data content, data acquisition unit and sharing open mode, establish the operation management mechanism of urban big data platform, clarify the data, process, security and other content and management standards in the use of the platform, so as to ensure the sustainable and stable operation of the platform.

3.6   Comprehensive Evaluation of Urban Big Data

Provincial and municipal big data authorities should formulate long-term operation mechanism and evaluation methods of the platform, establish perfect reporting, inspection and evaluation mechanism, design quantitative assessment content and standards, strengthen data quality control of the platform, and make good use of the urban big data platform. It needs to strengthen post-evaluation and project audit of large urban data platform projects, and strengthen audit supervision over data resources construction, open data sharing, data quality and security. To scientifically construct the comprehensive evaluation index system of urban big data platform, carry out the comprehensive evaluation of the effectiveness of the construction of urban big data platform, guide the construction of urban big data platform in all regions, and constantly improve the application effect of the construction of urban big data platform.

3.7   Strengthen Platform Data Security and Build Wisdom Prevention and Control

Urban Big Data Platform contains a large number of government and industrial data, involving national interests, public security, business secrets, personal privacy. It has high sensitivity. Therefore, it is necessary to strengthen the capacity building of platform data security. It needs to implement basic systems such as hierarchical protection, security assessment, electronic certification and emergency management, establish security assessment mechanisms for data acquisition, transmission, storage, use and opening, and clarify the scope, subject, responsibility and measures of data security protection. It also needs to study and formulate data rights guidelines, data benefit distribution mechanism and data flow trading rules, clarify the subject of data responsibility, and strengthen the protection of technology patents, digital copyright, digital content products, personal privacy, etc. [10].

Use Internet of Things technology to build a diversified intelligent perception network, aiming at key areas and important entrances and exits through the construction of front-end equipment such as face recognition, car face recognition, video surveillance, mobile card, etc., to achieve three-dimensional intelligent information acquisition. Use large data stream computing, real-time computing and other technologies to perceive and collect massive information for security situational awareness and assessment. Combined with PGIS, it provides intuitive security situation monitoring capability, discovers security hot spots in time, and controls the overall security situation macroscopically.

4 Conclusions

To sum up, the concept and methodology of big data have gradually become one of the standardized thinking of people. In order to speed up the construction of smart cities, it is necessary to use modern information technology in the process of urban planning and construction, especially in the fields of data collection, data governance and structural analysis. It can be predicted that big data, as an information infrastructure in the new era, will surely form a more comprehensive coverage in the construction of smart cities. But nowadays, the application of big data is not thorough and comprehensive enough. It is necessary for relevant people to strengthen the in-depth study of this aspect, so that big data technology and smart city technology can play a better role in the process of urban and rural planning.

Acknowledgements. This work was supported by the grants from Wuhan Donghu University School level educational research project (2019) No. 25 Document, 190008.

References

1. Wu, P.: Research on the evaluation of the development potential of information service in smart city. Xiang Tan Univ. (12), 1–3 (2017). (in Chinese)
2. Yu, H.: Application research of intelligent city information service based on big data theory: a case Study of Qiqihar City in Heilongjiang Province. Electron. Commer. (05) (2019). (in Chinese)
3. Guo, Y.: Application of big data and smart city technology in urban and rural planning. Build. Mater. decor. (05) (2019). (in Chinese)
4. Lv, W.: The practice and application of big data in smart city research and planning. Smart City (06) (2019). (in Chinese)
5. Lin, W.: Applying big data to promote urban intelligence. J. Wuyi Univ. (04) (2019). (in Chinese)
6. Xu, L.: Research on urban and rural planning and smart city construction in the era of big data. Technol. Inf. (03) (2019). (in Chinese)

7. Cao, Z.: Research on the application of big data and Internet of Things technology in smart city. Mod. Econ. Inf. (02) (2019). (in Chinese)
8. Wang, W.: Application of big data and Internet of Things technology in smart city. Commun. World (03) (2018). (in Chinese)
9. Xue, Z.: Application of big data and cloud computing technology in smart city. Electron. Test. (06) 2018. (in Chinese)
10. Liu, Z.: Application of big data and Internet of Things technology in smart city. Inf. Comput. (08) 2018. (in Chinese)